Contemporary Clinical Neuroscience

Series editor

Mario Manto
Division of Neurosciences, Department of Neurology, CHU-Charleroi,
Charleroi, Belgium, University of Mons, Mons, Charleroi, Belgium

Contemporary Clinical Neurosciences bridges the gap between bench research in the neurosciences and clinical neurology work by offering translational research on all aspects of the human brain and behavior with a special emphasis on the understanding, treatment, and eradication of diseases of the human nervous system. These novel, state-of-the-art research volumes present a wide array of preclinical and clinical research programs to a wide spectrum of readers representing the diversity of neuroscience as a discipline. The book series considers proposals from leading scientists and clinicians. The main audiences are basic neuroscientists (neurobiologists, neurochemists, geneticians, experts in behavioral studies, neurophysiologists, neuroanatomists), clinicians (including neurologists, psychiatrists and specialists in neuroimaging) and trainees, graduate students, and PhD students.

Volumes in the series provide in-depth books that focus on neuroimaging, ADHD (attention deficit hyperactivity disorder and other neuropsychiatric disorders, neurodegenerative diseases, G protein receptors, sleep disorders, addiction issues, cerebellar disorders, and neuroimmune diseases. The series aims to expand the topics at the frontiers between basic research and clinical applications. Each volume is available in both print and electronic form.

More information about this series at http://www.springer.com/series/7678

Ioan Opris • Mikhail A. Lebedev
Manuel F. Casanova
Editors

Modern Approaches to Augmentation of Brain Function

Springer

Editors
Ioan Opris
Department of Neurosurgery
University of Miami
Miami, FL, USA

Mikhail A. Lebedev
Center for Bioelectrical Interfaces
National Research University Higher
School of Economics
Moscow, Russia

Manuel F. Casanova
Greenville Health System
University of South Carolina
Greenville, SC, USA

ISSN 2627-535X ISSN 2627-5341 (electronic)
Contemporary Clinical Neuroscience
ISBN 978-3-030-54563-5 ISBN 978-3-030-54564-2 (eBook)
https://doi.org/10.1007/978-3-030-54564-2

This Springer imprint is published by the registered company Springer Nature Switzerland AG
The registered company address is: Gewerbestrasse 11, 6330 Cham, Switzerland

Ioan "John" Opris (1957–2020)

Dr. Ioan Opris, a well-known international scholar and an enthusiast of brain augmentation research, passed away in Miami, FL, on October 16, 2020. He is survived by his children Ioan Opris Jr. and Iris Mihaela Opris as well as by his loving wife Anca Liliana Opris.

Ioan Opris was born in Barsana, Maramures County, Romania, and received both his undergraduate and PhD degrees in Physics/Biophysics from the University of Bucharest. In 1990, Ioan began his academic career as an assistant professor within the Faculty of Physics at the University of Bucharest. Years later he would introduce to his university a Master of Neuroscience course and establish together with Professor Ioana Moisil the Romanian Society of Neural Networks. In 1995, Ioan started his neuroscience research in the USA by working with Professor Randall Nelson (University of Tennessee, Memphis) on the role of the neostriatum in coding movement kinematics and motor control. Later work with Professor Vincent Ferrera from Columbia University would lead him to a McDonnell Pew Award (2000) on the neural correlates of decision mechanisms by the prefrontal cortex. He further pursued the work of his McDonnell Pew Award in the laboratories of Professors Charles Bruce and Patricia Goldman-Rakic at Yale University. Ioan then moved to Wake Forest University as a scientific researcher in the laboratory of Dr. Sam Deadwyler. In this laboratory he expanded his expertise of memory prosthetics. Ioan's academic career began to soar as he participated in several articles demonstrating the functional role of the prefrontal cortical minicolumns in executive control. His studies demonstrated the restoration of cognitive function through a neuroprosthesis that used neural activation specific to the minicolumn in the prefrontal cortex of nonhuman primates. His interest in brain augmentation led to a collection of research publications which won the 2017 Frontiers Spotlight Award.

Ioan's focus on cortical modularity established him as an heir to Vernon Mountcastle. He used to think of the stereotyped translaminar connections of the cell minicolumn in striking analogy to the quantum jumps of electrons across different energy levels. This conceptualization propelled Ioan into the exploration of the physics of the mind and brain disorders. This effort culminated in a book for which the Romanian Academy of Sciences bestowed the distinguished Nicolae Simionescu award.

At the time of his death, Ioan was an associate professor at the University of Miami working for the Department of Biomedical Engineering and the Miami Project to Cure Paralysis. Along with Brian Noga, Jim Guest, and Vance Lemon, he studied locomotor behavior using a wide range of tools, including the multichannel recording of brainstem neuronal activity combined with optogenetics and deep brain stimulation.

In his unquenchable curiosity, Ioan Opris was the archetype of the Renaissance man. His mind was always active. He drew ideas from a significant number of complex subjects to solve specific problems.

By keeping an open mind, he was always excited about what the world of neuroscience would bring and how he could explore the same. Indeed, according to Ioan, "It must be a fascinating concert, that of the mind paralleled by the brain's physiology. I dreamed of articulating for the field this concert for forty years." We have to believe that in the end his biggest academic regret was that there was so much more left to be explored.

Goodbye good friend, until we meet again,
Manny and Misha

Dedication

This book is dedicated to Dr. Jon Howard Kaas who helped to unravel the organization of the mammalian brain, articulated the workings of many domains of the cerebral cortex, and promoted the idea of neuroplasticity. It took someone with the skills of a polymath of the neurosciences to mold our views of the sensory and motor brain systems, while revealing how, from an evolutionary standpoint, their organization is altered during brain development. As heir to Socrates in asking all of the pertinent question, Jon Howard Kaas opened the door that many others would follow. Indeed, the field of brain augmentation now solidly stands on the shoulders of a giant.

Many neuroscientists support the reductionist approach; "You, your joys and your sorrows, your memories and ambitions, your sense of personal identity and free will are in fact no more than the behavior of a vast assembly of nerve cells and their associated molecules" (Crick 1995). Jon Kaas and others have proposed a utopian dream where technological advancements will allow us to surpass some of the limitations of our biological brain. External interventions are finally upholding the tenet of the self-help community in that we can do better than we are born with. The static properties of the elemental constituents of the brain are, in this regard, an end to manipulations engendering the emergence of unforeseen properties.

Foreword

I have been invited to provide a brief introduction to this important volume on modern approaches to augmentation of brain function. Although I am not an expert on this topic and I feel fortunate having been invited to contribute a chapter to this volume, I do have the distinction of having been in the field of neuroscience research for a long time. At the start of my career, there were very few methods and the early generations of computers were very limited. The field was growing, but progress was slow. As technical advances made more approaches viable, rapid progress in understanding of brain organization and function became possible. Yet, there were early efforts to find ways to augment brain functions, especially for those that had sensory impairments. As a prime example, in the days of my postdoctoral training in the laboratory of Clinton Woolsey at the University of Wisconsin in the late 1960s, I was sent to represent the laboratory at a meeting at NIH to discuss the feasibility of developing a visual prosthesis as an aid to the blind. The goal was to create visual images in the blind by directly stimulating primary visual cortex with an array of implanted electrodes. We all know that it did not go well, such that a recent review concluded that the development of a prosthetic device for the blind is still in its infancy. There were obviously many technical issues in the 1960s, and many, but not all, have gone away. One such issue was presented by a speaker who was representing the blind. This I did not anticipate. The speaker said that the blind would not accept a device that required bulky stimulating and processing equipment that would occupy, at the least, a huge backpack. This surprised some researchers and university administrators looking for funds, but it made sense. We are highly social primates, and we all want to fit in. Toward the end of practicality, there has been much progress as the miniaturization of computers, electrode arrays, and other equipment has been one of the signs of great progress. The well-known auditory prosthetic devices have been a huge success as they depend on few stimulating electrodes and only access to a sensory nerve rather than to the brain. But even this success took years to develop, and many animal studies. For some of us, including me, the ability to help the hearing impaired so much with only a few electrode sites to stimulate was unexpected, but it worked. The needed number of stimulation sites continues to be a problem for augmenting the visual system, where high acuity images are needed.

More recently, as I got to know neurosurgeons at Vanderbilt, I became more aware of the effectiveness of using an implanted electrode to disrupt subcortical circuits that became dysfunctional in Parkinson's disease. In many adults with this problem, a tremor that interferes with normal hand function is of practical and social concern. In Nashville, I saw a retired professional guitar player suddenly recover his ability to play by turning on his stimulation device. This device greatly improved his quality of life. As a result of this experience, while visiting in northern Wisconsin, I recommended to the best friend of my sister that her brother with a Parkinson tremor inquire about such a subcortical stimulating device. In a subsequent visit, I ran into her by chance, and she told me about the wonderful difference this stimulation device made in her brother's life.

There is also great hope and expectations for the development of ways to alter, redirect, or focus the functions of the normal brain by electrically stimulating relatively few neurons or by stimulating afferent or motor peripheral nerve pathways. Much of this topic is covered in the wonderful chapters of this book, as are other approaches. Some transformations are so simple that we do not even think of them as augmentation. For example, for researchers and others interested in studying the high sound frequency calls of bats, it is a simple matter to record and replay the calls in real time as sounds we can hear. If need be, we can experience sensory stimuli beyond the capabilities of our sensory receptors. And in a plastic brain, experience is always altering brain functions. Thus, understanding how such plasticity is mediated, and applying suitable protocols, we can induce and improve desired functions. As I age, I have become more aware of available programs to help me retain mental abilities, but my ability for denial has put off such training. Please enjoy these informative chapters.

Jon H. Kaas
Vanderbilt University

Jon H Kaas
Distinguished Professor of Psychology
Vanderbilt University

Preface

Human Intellectual Capacity and Its Growth

William James Sidis was born in the United States on April Fool's Day, 1898, to Jewish emigrants from Ukraine. His father, Boris Sidis, was a prominent psychiatrist who founded the New York State Psychopathic Institute and the *Journal of Abnormal Psychology*. Boris undoubtedly felt fortunate to escape political prosecution and be accepted by the Boston Society upon his arrival in the United States. Consequently, for Boris, anything seemed possible, and along with his wife, Sarah, they were the prototypical power couple. Sarah, herself, went to Boston University and then graduated from its school of medicine. As they were expanding their social circle, Boris established numerous friendships with the intelligentsia of his time. Indeed, William Sidis' name came from Boris' good friend, teacher, and colleague, the famous American philosopher and psychologist William James. More "apropos" with our theme of brain augmentation, Boris became a controversial figure in psychology for his philosophical perspectives on education, believing that an appropriate nurturing environment promoted one's intellectual capacity. Heredity indeed plays a role, but genius can be created! As its gel-like consistency might betray, the brain is quite a malleable organ and, given proper information, experience, guidance, and discipline, there was a tacit assumption that its full potential could be unleashed. Thus, William Sidis became a well-publicized experiment for the views of his father. As a child prodigy with exceptional mathematical and linguistic skills, William entered Harvard University at age 11 and soon became conversant in 25 different languages. As an observer to the rearing experiment of this precocious child, William James told his audiences that we only use a fraction of our full potential. In this regard William Sidis thus became the rallying cry of a popular culture aimed at changing our minds and therefore our destinies. As John Campbell, a science fiction writer and editor, once said, "no man in all history ever used even half of the thinking part of his brain" (Campbell 1961). From then on, the self-help community believed we could do far better than what we are born with.

Humans tend to flatter themselves in asserting their uniqueness in the world. Some propose that we are fallen angels rather than risen apes. Indeed, we can all agree that humans do possess many cognitive abilities not seen in any other species. As to an answer for the age old "question of all questions" concerning man's place in nature, humans are different from other primates by their larger brain in proportion to body size, and a frontal cortex that endows them with the capacity for complex thought. This wonderful organ accounts for only 3% of the total weight of the human body but consumes 20% of the body's energy. Given this fact, the reports of early electrophysiologists were puzzling for many scientists. These reports revealed areas of the brain that appeared to be "silent," that is, not having a recognized function. Wilder Penfield (1891–1976) who expanded neurosurgical techniques and identified the homunculus could not assign a function for a large expanse of the cerebral cortex. This led to a misinterpretation that stemmed from the shortcomings of early electrophysiological experiments; namely, that we only use 10% of our brains.

To clarify, the 10% myth aside, humans use the full extent of their brains all of the time. Contrary to a muscle that is active only during contraction, the brain is continuously active day and night, even during sleep. Our frontal cortex, which is heavily involved in thought and judgment while we are awake, mediates normal sleep physiology and is intimately involved in the sleep-deprivation phenomena. However, akin to muscles, the brain can be trained to improve its capabilities, encompassing memory, attention, and learning. Accepting the fact that brain function can be improved, the more crucial question remains as to whether they can also be enhanced.

Modern Approaches to Brain Augmentation

The brain is a biophysical system that consists of hundreds of billions of elements, called neurons, interconnected through the connectome into neuronal circuits that perform complex information processing and generate network states dynamically distributed across the entire brain. The operation of brain circuitry is governed by the fundamental laws of physics, such as minimization of entropy under the principles of hierarchical dynamics. Brain dysfunctions—exorbitantly costly, both economically and sociologically—damage the fine and sophisticated brain architecture and lead to debilitating disorders. We are witnessing the explosive development of highly technological methods for brain diagnostics and therapeutics, such as neuroimaging techniques and nanotechnology. These new technologies hold promise to revolutionize medicine and lead to remedies to severe neurological conditions previously deemed uncurable. In addition to clinically relevant applications, these technologies facilitate the advancement of fundamental Neuroscience and bring us closer to resolving the greatest challenge in science—elucidation of brain functionality.

Modern neural technologies are also applicable to augmenting brain functions in healthy humans—the idea that belonged exclusively to the realm of science fiction not so long ago but currently is becoming a reality owing to the advances in Neuroscience and Neuroengineering. Augmentation technologies have emerged in sensory, motor, and cognitive domains, and if they become both safe and efficient it is likely that they will become widespread. The development of augmentation technologies takes place on different scales—micro- and macrocircuits of the brain—and with different recording/stimulation methods, some invasive and some noninvasive. This book overviews the existing and emerging methods for brain augmentation; a multiplicity of approaches are discussed.

If Ray Kurzweil is correct and we are approaching singularity (Kurzweil 2005), brain augmentation technologies can be viewed as an advancement toward a merger between the brain and machines, the development that will cause irreversible changes in society. This book describes the recent trends in this direction. Different chapters in this book cover advances in the neural sciences by which we can enhance sensory, motor, and cognitive functions, as well as mood and emotions. Additionally, insights are made into the application of brain augmentation approaches to the treatment of devastating neural disorders such as spinal cord injury, Parkinson's disease, depression, dementia, and autism. Given the potential impact of brain augmentation approaches on the society, it was also important to discuss the philosophical and ethical issues arising from the use of technologies to enhance neural processing of information, memory, attention, and emotions.

The book is organized in seven parts containing 30 chapters. The Introduction is written by Professor Jon Howard Kaas, member of the US Academy of Sciences. The seven parts are summarized below.

Part I: Stimulating the Brain

A multitude of neurostimulation methods have emerged during the last several decades for treating neural disorders, rehabilitation, and brain augmentation, including electromagnetic and optogenetic stimulation and pharmacological approaches. Jon H. Kaas and Iwona Stepniewska open this narrative by laying down the idea that brain functions can be directly augmented by electrically stimulating ensembles of neurons to potentiate their actions. Their chapter (Chap. 1) is entitled "Using Electrical Stimulation to Explore and Augment the Functions of Parietal-Frontal Cortical Networks in Primates."

Jim Guest and his colleagues in the chapter entitled "Multi-system Benefits of Epidural Stimulation Following Spinal Cord Injury" (Chap. 2) describe recent epidural stimulation discoveries that have changed our understanding of the capabilities of the spinal cord, leading to a shift away from detailed hierarchical cortical and brainstem control of spinal function to one in which the spinal cord itself executes complex locomotor programs.

Elyahoodayan and his colleagues address the possibility of building prosthetic memory using neurostimulation. In Chap. 3, "Neurostimulator for Hippocampal Memory Prosthesis," they describe a neurostimulator design that is desirable in a variety of neural interface applications, particularly hippocampal memory prosthesis aiming to restore cognitive functions by reinstating neural code transmissions in the brain.

In Chap. 4 entitled "Modern Approaches to Augmenting the Brain Functions," Opris and his colleagues review a range of augmentation approaches based on stimulation, including neuromodulation, pharmacological methods, brain-computer and brain-to-brain interfaces. The chapter covers recent advances in neural technologies, such as microtechnology, spintronics, nanotechnology, optogenetics, and minimally invasive electrode arrays.

Part II: Brain-Computer Interfaces

Brain-computer interfaces are artificial systems that connect the brain with external devices, such as limb prostheses and means of communication. Brain-computer interfaces are applicable to both treatment of patients and brain augmentation in healthy people. In the opening chapter for this part, Chap. 5 entitled "Brain Machine Interfaces Within a Critical Perspective," Zippo and Biella describe the brain as a complex mechanistic machinery that executes functions and processes information. Brain-computer interfaces either enable invasive physical communication between the nervous system and the artificial devices or noninvasive communication using recording methods such as electroencephalography and magnetoencephalography, and stimulation methods such as transcranial magnetic stimulations.

In Chap. 6 entitled "An Implantable Wireless Device for ECoG and Cortical Stimulation," Romanelli describes the first prototype of a wireless ECoG system that provides an innovative approach to detect seizure foci in medically refractory epilepsy and to perform BCI procedures.

Kaya, Bohorquez, and Özdamar continue the theme of brain-computer interfaces with the chapter entitled "BCI Performance Improvement by Special Low Jitter Quasi-Steady State VEP Paradigm" (Chap. 7). Visual Evoked Potential (VEP) based brain-computer interfaces are common because of their ease of implementation and a number of advantages. Recently, transient VEP retrieval from linear regression and random code estimation has gained attention as a brain interface method. When applied to electroencephalographic recordings, this method improves the performance of brain-computer interfaces.

In Chap. 8 entitled "Communication with Brain-Computer Interfaces in Medical Decision-Making," Glannon argues that brain-computer interfaces can restore or augment motor functions that have been impaired or lost from traumatic brain injury and neurodegenerative disease. The author discussed how "locked-in" patients and some patients in the minimally conscious state could use a brain-computer interface to express consequential decisions about life-sustaining therapy.

Part III: Augmenting Cognitive Function

The possibility of augmenting cognitive functions with neural technologies has received considerable attention. Gonzalez-Lima contributes Chap. 9, "Neuroprotection and Neurocognitive Augmentation by Photobiomodulation," where animal and human studies are described using red to near-infrared lasers and LEDs, a noninvasive and relatively inexpensive intervention, which are applicable to neuroprotection and for the augmentation of cognitive brain functions.

In Chap. 10 entitled "Avoiding Partial Sleep: The Way for Augmentation of Brain Function," Pigarev and Pigareva describe a study that demonstrates that cortical areas that process signals from extero- and proprioreceptors during wakefulness switch to the processing of interoceptive information during sleep. They suggest that during sleep the computational power of the brain is directed to the restoration of the vital functions of internal organs.

In Chap. 11, "Augmentation of Brain Functions by Nanotechnology," Opris and his colleagues describe an unprecedented increase that occurred during the last decade in the successful application of nanotechnology methods to basic neuroscience and to clinical practice. The authors review how nanotechnology (nanoparticles, nanowires, carbon nanotubes, devices, sensors, interfaces) is employed to augment, record, stimulate, repair, and regenerate brain circuits.

Chapter 12, "The Impact of Aging and Age-Related Comorbidities on Stroke Outcome in Animal Models and Humans," written by Popa-Wagner and his colleagues, makes parallels between animal models of stroke and clinical data and summarizes the impact of aging and age-related comorbidities on the efficacy of various therapies and stroke outcome. The authors conclude that the unsuccessful bench-to-bedside translation of therapies that showed efficacy in young animal models, to aged stroke comorbid patients, is, most likely, due to the negative impact of comorbidities and advanced age on the efficacy of restorative therapies.

In Chap. 13 entitled "Diagnostic Markers of Subclinical Depression Based on Functional Connectivity," Zhu, Bohorquez, and Opris review the subclinical depression prediction model that detects abnormal functional connections through machine learning. Overall, it is concluded that functional connection analysis based on the thalamus and habenula nucleus can provide a high-accuracy biomarker of subclinical depression.

Casanova and his colleagues continue with Chap. 14, "Transcranial Magnetic Stimulation in Autism Spectrum Disorders: Modulating Brainwave Abnormalities and Behaviors," where they describe how transcranial magnetic stimulation (TMS), especially at low frequencies, has proven to be a safe intervention for children with ASD. This is the first treatment targeting a core pathological abnormality of autism. Outcome measures reveal improvements in executive functions as shown by a normalization of error monitoring (i.e., detection, evaluation, and correction of errors) and attendant ERP components.

In Chap. 15, "Neurofeedback Training with Concurrent Psychophysiological Monitoring in Children with Autism Spectrum Disorder with Comorbid Attention Deficit/Hyperactivity Disorder," Sokhadze and his colleagues suggest that neurofeedback training can be used as a treatment modality of potential use for improving self-regulation skills in autism spectrum disorder (ASD). They report significant reduction in irritability and hyperactivity subscales of the ABC, decrease of T-score on SRS-2, and decrease in attention deficits scores.

Part IV: Futuristic Approaches to Augmentation

Some approaches to brain augmentation are futuristic, but still feasible. In Chap. 16, entitled "Augmentation Through Interconnection: Brain-Nets and Telemedicine," Lebedev and his colleagues discuss the futuristic approach to brain augmentation based on brain-nets comprising several brains connected into a network that enacts information exchange between several subjects. Brain-nets have extended the bidirectional brain-computer interface approach, where information is extracted from brain activity simultaneously with the delivery of information to the brain with neurostimulation. Furthermore, Lebedev and his colleagues discuss how brain-nets could be used in telemedicine. The authors foresee that brain-nets and telemedicine will merge to advance new, highly effective neurotechnologies.

In Chap. 17, "Cognitive Augmentation via a Brain/Cloud Interface," Angelica, Opris, and Boehm review the future discipline of neural nanorobotics that sets the goal of developing nanorobotic species designated as endoneurorobots, gliabots, and synaptobots that hold promise to facilitate a direct and seamless interface between the human neocortex and the cloud/edge, referred to as a Brain/Cloud Interface (B/CI). This chapter explores these paradigm-shifting possibilities for cognitive augmentation toward envisioning what may be possible within the next few decades.

Ma et al. contribute the chapter "Augmentation of Neuromarketing by Neural Technology" (Chap. 18) where they suggest that marketing and management are fundamental economic activities rooted in the cognitive abilities of the human brain. In this regard, decision-making is a cognitive process of selecting the best option between two or more options. The main research areas discussed in the chapter include augmentation of neuromarketing, neuromanagement, neuro-information systems, decision neuroscience, and neuroeconomics.

In Chap. 19, "Augmentation of Nutrition by Nanotechnology," Sonea, Lupusoru, and Opris provide insights into the augmentation of nutrition process by means of nanotechnology. This technology has evolved more toward the food industry in the domains of food production, processing, conservation, packaging, safety, sensing, functional food, and nutraceutical delivery.

In Chap. 20, "Neural Spintronics: Noninvasive Augmentation of Brain Functions," Barnes et al. review the emerging field of Spintronics that is attracting a lot of attention with its noninvasive abilities to sense the magnetic field of neurons and to modulate their firing with spintronics devices. Two such tools are critical: transcranial magnetic stimulation (TMS), and magnetic encephalography (MEG). Nano-TMS/MEG devices can be used in a broad range of research fields and technological medical applications where programmable focusing magnetic field is required.

Part V: Augmenting Behavior

The part on augmenting behavior starts with the chapter by Mirabella, entitled "Does the Power to Suppress an Action Make Us Free?" (Chap. 21). Mirabella notes that there are still no clear answers to the problem of free will. Yet, the bottom-up approach of neuroscience seems to be more promising than that of philosophy, which is deductive or top-down by nature. Overall, the experimental evidence gathered so far suggests that, except specific medical conditions, we are free of choosing how to act as much as we are responsible for what we do.

In Chap. 22, "Deep Brain Stimulation for Parkinson's Disease: Clinical Efficacy and Future Directions for Enhancing Motor Function," Luca and his colleagues discuss the method of deep brain stimulation (DBS) that produces a significant improvement in motor symptoms in patients with advanced Parkinson's disease (PD). Recent advances in technology provide the ability to deliver stimulation adaptively based on cortical and subcortical brain signals, and subsequently a more physiological and precise modulation of the impaired motor network.

Chapter 23, "Neuromodulation for Gait Disorders," written by Chang and his colleagues, provides an overview of the neurophysiology and pathophysiology of common gait disorders. The authors review the most promising invasive and noninvasive strategies being investigated to augment one of our most fundamental daily activities—walking.

Onose and his colleagues contribute the chapter "Augmentation and Rehabilitation with Active Orthotic Devices" (Chap. 24), where they address the treatment methods for severe disabilities caused by lesions to the nervous systems. Such conditions are a major health and socio-economic problem and a challenge to scientific research. The authors argue that active orthotic devices are an emerging domain for augmenting function in disabled patients.

Part VI: Augmenting Cognition and Emotion

Cognition and emotion can be augmented with neural technologies. In Chap. 25, entitled "Effects of rTMS on Behavioral and Electrocortical Measures of Error Monitoring and Correction Function in Children with Autism Spectrum Disorder," Sokhadze and his colleagues suggest that error monitoring and correction is one of the executive functions that is important for effective goal-directed behavior. The chapter provides rationale to use ERN and Pe, along with behavioral performance measures as functional outcome measures to assess the effectiveness of neuromodulation (e.g., rTMS and rTMS combined with neurofeedback) in children with autism and thus may have important practical implications.

Chapter 26, "Affective Virtual Reality Gaming for Autism," written by Li and his colleagues, reviews emotional impairment, which is one of the common symptoms of many mental diseases. Being able to learn the emotional reactions from subjects using nonintrusive human-computer interactions (HCI) would provide a novel and efficient approach to assist existing intervention and therapy. Psychologists conducted research using virtual reality (VR) as a tool for exposure starting from decades ago. This chapter reviews the methodologies commonly used in affective computing and related research projects using VR exposure as an intervention for people with special needs.

Bălan, Moldoveanu, and Leordeanu provide Chap. 27, "A Machine Learning Approach to Automatic Phobia Therapy with Virtual Reality." The authors present a new automated approach to phobias therapy, based on the integration of virtual reality technology, artificial intelligence, and affective computing—the ability of computational machines to recognize, adapt, and respond intelligently to human emotions.

Part VII: Pharmacological Augmentation

Pharmacological agents have been long explored as an approach to treat neurological disorders and augment the brain of healthy people. In Chap. 28, "Vision Augmentation by Pharmacological Enhancement of the Visual Experience," Vaucher reviews the research examining whether potentiation of the central cholinergic and/or monoaminergic systems aids visual perception and restoration. Vaucher argues that irreversible vision can be prevented by combining visual training with commercially available pharmacological agents.

In Chap. 29, "Cognitive Enhancing Substances and the Developing Brain: Risks and Benefits," Urban and Gao discuss the current trends in cognitive enhancement among adolescents and adults. The authors examine several prescription drugs commonly used or being examined for their potential cognitive enhancing effects and discuss the potential risks of each substance. They also discuss the state of over-the-counter nootropics and the lack of reliable research into their efficacy and safety.

In the last chapter of the book, Chap. 30, entitled "Pharmacological Approaches in the Augmentation and Recovery of Brain Function," Mureșanu and his colleagues describe several pharmacological brain enhancers and assess their capacity to support the brain's endogenous defense mechanisms.

Miami, FL, USA — Ioan Opris
Moscow, Russia — Mikhail A. Lebedev
Greenville, SC, USA — Manuel F. Casanova

Contents

Part III Augmenting Cognitive Function

Part IV Futuristic Approaches to Augmentation

Part I
Stimulating the Brain

Using Electrical Stimulation to Explore and Augment the Functions of Parietal-Frontal Cortical Networks in Primates

Jon H. Kaas and Iwona Stepniewska

1 Introduction

For over 10 years, members of our research group have been using electrical stimulation to explore parietal-frontal networks in the cortex of primates to determine how they are organized, how they interact, and how they connect to other parts of the brain to mediate their functions (see Kaas et al. 2013; Kaas and Stepniewska 2016, for review). In brief, our research group and others (Graziano et al. 2002; Cooke et al. 2003) have provided evidence for at least eight parallel networks involving a series of action-specific modules or domains in a portion of posterior parietal cortex that project to functionally matched domains in premotor cortex and primary motor cortex. We propose that these domains interact largely via the suppression of nonmatching domains as they compete for selecting a specific action, and via the facilitation of functionally matched domains to promote a specific behavior, such as reaching to a location. Visual and other sensory inputs to posterior parietal cortex domains provoke actions based on ongoing sensory information, while cognitive and motivational inputs to premotor domains largely come from prefrontal cortex and cingulate motor areas. Domains in primary motor cortex (M1) receive inputs from premotor (PMC) and posterior parietal (PPC) domains, as well as other cortical areas and the motor thalamus, and they dominate the action selection process (Stepniewska et al. 2014; Cooke et al. 2015). Subcortical targets of cortical domains include thalamic, midbrain, brainstem, and spinal cord structures, but especially the putamen of the basal ganglia (unpublished data). Electrical stimulation of cortical domains reveals their involvement in specific, ethologically relevant behaviors, and important aspects of how they interact (Stepniewska et al. 2020). We have characterized these interactions as revealing important features of a

J. H. Kaas (✉) · I. Stepniewska
Department of Psychology, Vanderbilt University, Nashville, TN, USA
e-mail: jon.h.kaas@vanderbilt.edu

I. Opris et al. (eds.), *Modern Approaches to Augmentation of Brain Function*, Contemporary Clinical Neuroscience,
https://doi.org/10.1007/978-3-030-54564-2_1

decision-making process. Alternatively, it may be more accurate to refer to the process as one of response selection. Here, we propose that the parietal-frontal networks include a sequence of three domains of different levels of response selection that successively involve PPC, PMC, and M1. Domains directly activate subcortical centers that are involved in further mediating the selection process and motor components of the selected behaviors. Of course, electrical stimulation is not the same as the natural stimulation of neurons in these networks by other neurons, including those relating information about the environment, the state of the body, and past history. Some have called such stimulation as "highjacking" the nervous system, but electrical stimulation, in our view, provides the most direct information about the functions of motor-related structures in the brain. Here we consider how electrical stimulation informs us about the working of cortical networks, but we also consider how electrical stimulation can be used to alter and augment brain function.

2 Electrical Stimulation: A Selective and Brief History

Electrical stimulation of brain tissue is one of the oldest methods of exploring the functions of the brain (Taylor and Gross 2003; Gross 2007). The use of electrical stimulation was possible well before modern equipment for recording from neurons, such as amplifiers and oscilloscopes emerged in the 1920–30s. Stimulating the surface of the brain with electrodes produced movements and allowed the overall organization of motor-related cortex in primates and other mammals to be revealed. Thus, Ferrier (1874) used electrical stimulation to determine the basic organization of what is now known as primary motor cortex in the frontal cortex of monkeys nearly 150 years ago, and described more complex movements that emerged as stimulation continued. The simple movements that were evoked by a brief series of electrical pulses were recognized as first movements or threshold movements (e.g., Grunbaum and Sherrington 1903). The body part that moved depended on where the stimulating electrode was placed, thereby revealing the well-known sequence of representation of foot to mouth in a medial-lateral progression in frontal cortex. The more complex movements that were produced by longer trains of pulses did not fit the overall somatotopic sequence very well, and they were largely ignored.

Stimulation of other regions, including premotor cortex, the frontal eye field, somatosensory cortex, and even posterior parietal cortex, also produced patterns of movements depending on the location of the stimulating electrode, but usually at higher levels of current (Grunbaum and Sherrington 1903; Woolsey et al. 1953). When the cortex of humans was stimulated electrically, movement patterns evoked in motor cortex were similar to those evoked from motor cortex in monkeys (Penfield and Boldrey 1937). In addition, awake humans could state what they felt when their sensory cortex was stimulated during neurosurgery. Thus, the orderly toe-to-tongue representation of touch in a medial-lateral sequence could be demonstrated in somatosensory cortex, just behind the central fissure. Stimulating subcortical brain structures in awake mammals sometimes produced even more dramatic results.

Most notably, in 1963, Jose Delgado dramatically demonstrated that a charging bull in a bull ring could be stopped in its tracks by remotely activating an electrode placed in the hypothalamus deep in the brain (see Mashour et al. 2005).

As microelectrodes came into common use, first for recording the response properties of single neurons, and later for microstimulating small groups of neurons in sensory cortex (Clark et al. 2011), researchers demonstrated that activating even a few neurons electrically in the sensory cortex could influence and alter perception. One early example came from microstimulating neurons in the middle temporal visual area, MT, as primate visual area (Allman and Kaas 1971) that became known for having neurons responsive to visual motion (Maunsell and Newsome 1987). MT provided this information to other visual areas, especially in the dorsal stream of visual processing for actions (Kaas and Baldwin 2020). Salzman et al. (1990, 1992) and Salzman and Newsome (1994) combined electrical stimulation of clusters of neurons at sites in MT that responded best to visual stimuli moving in a specific direction with the presentation of a field of random moving dots. They found that such stimulation influenced monkey perceptual judgments of the overall direction of the global motion. Other cortical areas have also been microstimulated to influence behavior, including prefrontal cortex in monkeys (Opris et al. 2005a; Opris and Ferrera 2014), the frontal eye field (Opris et al. 2005b), and primary visual cortex in both monkeys and humans (see Cicmil and Krug 2015 for review). In humans, microstimulation, or even surface stimulation, of primary visual cortex (V1) can produce the sensation of a flash of a small spot of light (a phosphene) located in the places in the contralateral visual hemifield corresponding to the electrodes' location in the retinotopic map of V1 (Bosking et al. 2017). The phosphenes produced by electrical stimulation of visual cortex have long suggested that such stimulation over an array of electrodes could be used as a prosthetic device to provide crude vision to the blind (Lewis and Rosenfeld 2016). Similar evidence that microstimulation of locations in sensory cortex can influence perception has been obtained from primary somatosensory cortex (area 3b) of monkeys. Neurons representing the skin of the hand in area 3b in monkeys are segregated within small clusters or columns of "rapidly adapting" neurons that are activated by stimulation of the hand with periodic vibrations and adjacent columns of "slowly adapting neurons" that respond to sustained pressure on the hand. Thus, the clusters represent two different aspects of touch (Sur et al. 1981). Microstimulation of rapidly adapting clusters of neurons in area 3b at frequencies matching that of a real vibration that the monkeys were trained to detect, could substitute for the real vibration on the hand in frequency discrimination experiments (Romo et al. 1998, 2000). Thus, the monkeys appear to feel vibration when electrical pulses are delivered at vibration frequencies to the rapidly adapting columns of neurons in primary somatosensory cortex. While less is known about auditory sensations evoked by stimulating clusters of neurons in primary auditory cortex, this cortex is tonotopically organized in monkeys and other mammals (see Kaas 2011, for review). At one time, we planned elaborately for experiments to determine if complex sounds could be evoked by electrical stimulation of multiple sites in auditory cortex of monkeys. Thus, we planned to record the response patterns of clusters of neurons with arrays of electrodes implanted in primary auditory cortex, A1, while monkeys were exposed to recordings of species-specific communication calls, and were thereby provoked to respond behaviorally, without training. Then, we proposed to replicate the spatial and temporal patterns of neuronal responses to these vocalizations by electrically stimulating the previously recorded neurons with electrical pulses. Unfortunately, only preliminary efforts were possible, and long-term funding did not emerge. Yet, positive results seemed possible, given the tonotopic organization of auditory cortex and the innate behavioral responses to monkey calls. More importantly, the overall evidence that electrical stimulation of neurons in sensory cortex produces sensation and perception is strong, and that such stimulation could augment brain function. Of course, electrical stimulation of nerve fibers of the inner ear with a cochlear implant in the profoundly deaf "is the single most successful neuroprosthetic device available today" (Wallace 2017).

As electrical stimulation of sensory cortex produces sensations, electrical stimulation can be also used to provide new information to the brain that is otherwise unavailable. Such stimulation of the auditory nerve by a series of electrodes has now been used for years to evoke patterns of sounds in the brains of hearing impaired or deaf humans that resemble normal sounds in language. However, recovery of useful hearing also depends on considerable practice and the plasticity of language areas in the brain (Giraud et al. 2001). As another tactic, Hartmann et al. (2016) electrically activated neurons in somatosensory cortex of rats with signals from infrared light, something that rats do not see. As a result, the rats were endowed with a new sense that they could learn and use to locate food. Thus, information about sensory stimuli that is unavailable because of a lack of appropriate receptors can be made available to sensory cortex and the brain via artificial receptors and electrical stimulation of sensory systems.

3 The Action-Specific Parietal-Frontal Network of Primates

Our interests in studying parietal-frontal networks in primates started when Jon Kaas visited the laboratory of Michael Graziano at Yale University and observed a monkey performing during an experiment. The monkey had an electrode implanted in the motor cortex, and whenever the experimenter delivered a half-second train of electrical pulses, the fully awake monkey brought its hand to its mouth, which is a complex movement. For many years, various investigators had mapped motor cortex in monkeys and other mammals, using a short 60 miliseconds burst of electrical pulses at near-threshold levels to reveal a simple movement, such as a twitch of a finger. Earlier researchers such as Grunbaum and Sherrington (1903) had noted that longer trains of electrical pulses produced more complex movements, but they used only the first movements to reveal the somatotopic maps in primary motor cortex. Later, when such "somatotopic" maps were considered in detail with fine grain microelectrode mapping, the maps turned out to have a more complex, and somewhat confusing organization. In an early study of motor cortex of monkeys, we found that the hand-arm region of primary motor cortex, M1, did not have a simple somatotopy, but rather had repeats of simple movements scattered in a seemingly random pattern across cortex (Gould et al. 1986). We called this a "mosaic" pattern of representation. Taking a term from the reported disorder of somatosensory maps in the cerebellum, the maps could also be called "fractured" (Shambes et al. 1978). These maps of fractured somatotopy made more sense when considered in the light of the maps of M1 that were obtained with long trains of electrical pulses by Graziano and coworkers (Graziano et al. 2002, 2005). These investigators reported that long-train stimulation of M1 in macaque monkeys revealed seven small functionally distinct subdivisions that represented seven classes of complex movements, such as reaching, hand-to-mouth, body or head protection, grasping or manipulation, and climbing (Fig. 1). As in short-train maps of M1, complex movements that included the hindlimb were evoked from the more medial cortex, while arm and hand movements were evoked from the more lateral parts of M1, and face movements were evoked from again the more lateral parts of M1. The fractured local somatotopy of short-train maps only starts to make sense in terms of them being the various components of the long-train maps of the action-specific modules that we call domains (Kaas 2012).

Our studies focused on the organization of posterior parietal cortex as revealed by long-train stimulation, but soon expanded to the primary motor and premotor cortex. Studies in macaque monkeys had already revealed several domain-like regions in the posterior parietal cortex, and these regions were known to have connections with frontal premotor and motor areas (Fig. 1). Thus, there was an anterior intraparietal region (AIP) where recorded neurons were activated during grasping (Taira et al. 1990; Sakata et al. 1995). More recently, longer trains of microstimulation

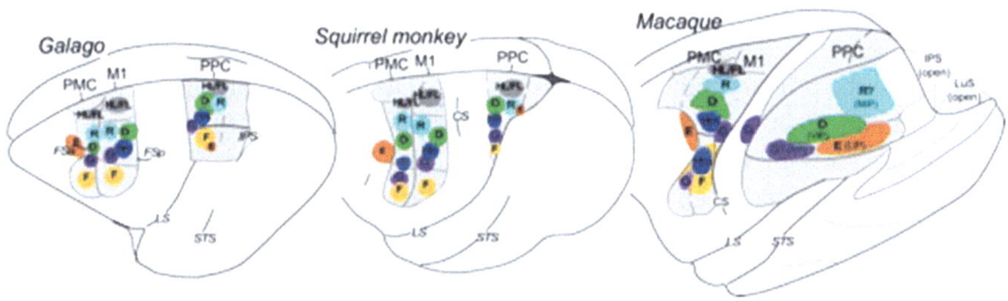

Fig. 1 Locations of action-specific domains in frontal and posterior parietal regions (shaded) in three different primates, galagos, squirrel monkeys, and macaques. In galagos and squirrel monkeys, domains for grasping (G), hand-to-mouth or body (Hm), reaching (R), defense of body (D) and face (F), eye movements or looking (E), and combined forelimb and hindlimb movements as if running or climbing (HL/FL), were identified in M1, PMC, and rostral PPC with long-train microstimulation. There is also microstimulation evidence for existence of such domains in macaque monkeys, mostly in M1-PMC region, but also in ventral intraparietal area (VIP) of PPC, and some homologs have been also suggested for PPC domains in other areas within the intraparietal sulcus (IPS). *AIP* anterior intraparietal area, *LIP* lateral intraparietal area, *MIP* medial intraparietal region, *VIP* ventral intraparietal area. Matching domains in M1, PMC, and PPCr are marked with the same color. *CS* central sulcus, *FSa and FSp* anterior and posterior frontal fissure, *LS* lateral sulcus, *LuS* lunate sulcus, *STS* superior temporal sulcus

have been used to evoke finger movements from the AIP region of macaques (Rathelot et al. 2017). With long-train stimulation, we later identified grasp domains in M1 and premotor cortex, and found that they receive inputs from the AIP region in macaques (Gharbawie et al. 2011a). More caudally, in a part of the ventral intraparietal cortex (VIP) that was first recognized by having inputs from visual area MT (Maunsell and Van Essen 1983), neurons were found to respond to visual and tactile stimuli (Colby et al. 1993). More recently, long-train microstimulation of neurons in VIP produced defensive or protective movements of the face and body (Cooke et al. 2003). An adjacent region, the lateral intraparietal area (LIP), was first distinguished by connections with the frontal eye field (FEF) (Andersen et al. 1990) and later by eye movements evoked by electrical microstimulation (Thier and Andersen 1998; Murphey and Maunsell 2008). While short trains of pulses produced eye movements in LIP, longer trains were more effective (Kurylo and Skavenski 1991). Thus, LIP can be considered to be an eye movement domain, corresponding to the eye movement domains revealed by long-train electrical stimulation in PPC of galagos (Stepniewska et al. 2009) and New World monkeys (Kaas et al. 2018). The corresponding domain in frontal cortex in all these primates is the frontal eye field (Stepniewska et al. 2018). In addition, recorded neurons in a posterior parietal reach region (PRR) or medial intraparietal area (MIP) responded as macaque monkeys prepared to reach (Snyder et al. 2000), suggesting the location of a reach domain. These findings, as well as others, suggest that posterior parietal cortex of macaques has functional divisions that closely resemble the domains described in other primates (Fig. 1). Clearly, more research is needed to further identify and describe domains in posterior parietal cortex of macaque monkeys.

Less is known about the functional organization of posterior parietal cortex in humans, but regions of cortex have been homologized with areas AIP, LIP, and PRR of macaques (see Kaas and Stepniewska 2016 for review). Nevertheless, hand-to-mouth movements have been evoked by electrical stimulation with surface electrodes from the primary motor cortex in humans (Desmurget et al. 2014). The much bigger human brain presents a challenge, as humans are likely to have even more domains as parts of networks than monkeys. Thus, a dorsal stream network for speech production has been proposed (Hickok and Poeppel 2004; Rauschecker and Scott 2009; Hartwigsen et al. 2016). In humans, there is also evidence for the existence of a parietal-frontal network for tool use (Johnson-Frey 2004).

Our research has been with prosimian galagos and with small New World owl and squirrel monkeys because most of the relevant motor, premotor, and posterior parietal cortex is exposed on the brain surface where we can accurately locate the positions of the stimulating electrodes. In galagos, we used long-train microstimulation to identify eight action-specific domains in rostral posterior parietal cortex (PPCr) (Stepniewska et al. 2005). Domains for eye movements, face aggressive, and face defensive were located most laterally, followed more medially by forelimb domains for grasping, hand-to-mouth, forelimb defensive and reaching movements. Combined movements of hindlimb were evoked from the most medial sites in PPCr.

Thus, the organization of this cortex included a fixed relationship between eight functionally distinct domains that roughly corresponded to a head-to-arm-to-hindlimb body sequence, but not a fine-grain somatotopic sequence (Stepniewska et al. 2005). Quite possibly, a crude somatotopic pattern in prenatal development predates the development of this functional pattern (Arcaro et al. 2019).

Similar patterns of domain placements exist in squirrel monkeys (Fig. 2) and owl monkeys (Gharbawie et al. 2011b). If we consider the relative placements of these domains in galagos and owl monkeys, the grasping domain is always most rostral, even merging into area 2. Eye and reaching domains are always more caudal. If the proposed grasping, eye movements, and reaching domains in macaque monkeys have been correctly identified, the grasping domain is most rostral, followed more caudally by the eye movement domain (LIP) and the reach domain most caudally within the parietal reach region (PRR). Thus, in macaques, the pattern is rotated so that the grasping region is most rostral, the defensive region using the arm (VIP) and the eye movement region (LIP) are more caudal, and the reaching region (PRR) is most caudal. A running or climbing region may be located more caudally and medially, but it has not been found yet. The rotation of the sequence of domains likely reflects an expansion of dorsomedial parts of posterior parietal cortex in macaques. In humans, too little is known to speculate on the positions or classes of domains, but a medial-to-caudal rotation of the body part representations is expected.

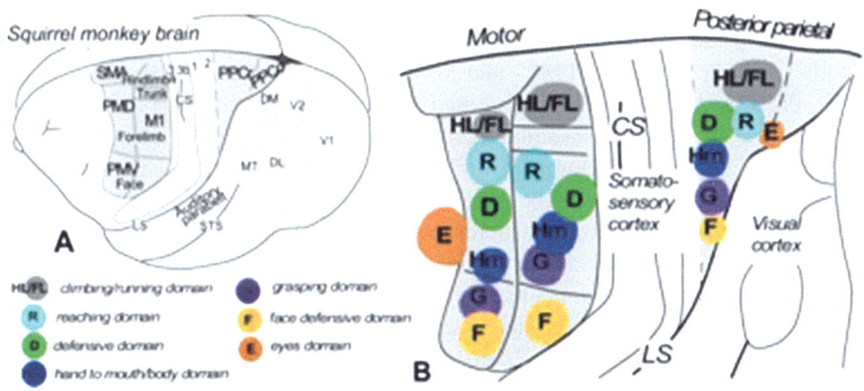

Fig. 2 Frontal and posterior parietal regions (shaded in gray) with action specific domains in a squirrel monkey brain. (**a**) Lateral view of the brain with motor cortical areas indicated. *PMD* and *PMV* dorsal and ventral premotor areas, *M1* primary motor area, *SMA* supplementary motor area, *PPCr* and *PPCc* rostral and caudal posterior parietal areas (**b**). Enlarged view of the M1, PMC, and PPC areas with action-specific domains marked with different colors. Matching domains are the same color. Some other areas and sulci are shown for reference. *DL* and *DM* dorsolateral and dorsomedial visual areas, *MT* medial temporal area, *V1* and *V2* primary and secondary visual areas, *3a, 1, 2* somatosensory areas. Sulci: *CS* central sulcus, *LS* lateral sulcus, *STS* superior temporal sulcus

4 The Functions of the Action-Specific Domains in Primates

Electrical stimulation of neurons in the domains provides rather direct information about their functions. In each of these regions of cortex PPCr, PMC, and M1, electrical stimulation reveals a specific behavior, such as reaching. The reaching is to a position relative to the body, and when the hand gets there from any previous position, the movement stops even if the stimulation continues. Stimulation of other domains results in other behaviors. But why are there three locations where stimulation evokes the same or very similar action? Further observations start to suggest answers. First, higher levels of current are necessary to evoke the actions from PPCr than PMC or M1. This likely indicates that M1 has more direct and more effective connections with the subcortical motor centers responsible for the action. Further evidence comes from the use of tracers to reveal patterns of cortical and subcortical connections (Fig. 3). PPCr domains send dense feedforward projections to behaviorally matching domains in PMC and M1. During electrical stimulation of PPCr domains, only matching domains in PMC and M1 are highly active as revealed by optical imaging of local blood flow increases (Stepniewska et al. 2011). Deactivation of M1 domains by cooling, blocking agent injections, or ablation renders the stimulation of functionally matched PPCr domains ineffective (Stepniewska et al. 2014; Cooke et al. 2015). In addition, electrically stimulating sets of two functionally matched domains in different regions, such as PPCr and M1, at the same time, results in a faster and sometimes more extensive movements (Stepniewska et al. 2020). Altogether, these results support the contention that the PPCr domains mainly function to activate PMC- and M1-matching domains, and that matching PPCr, PMC, and M1 domains form a hierarchy (Fig. 3). We suggest that this is a decision-making or action selection hierarchy to promote the behavior revealed by the electrical stimulation. Domains in PPC, PMC, and M1 form a chain of excitation that extends from M1 to subcortical motor centers. Domains in PPC and PMC project to the same subcortical motor centers (Galea and Darian-Smith 1994; Rathelot et al. 2017), but probably much less effectively. At the same time, we propose that the most excited domains function to suppress nonmatching domains via excitatory connections that terminate mainly on local inhibitory neurons in other domains (Kaas and Stepniewska 2016). This suppressive effect is mediated within regions by intrinsic connections, and by feedback connections from M1 to PMC, and from M1 and PMC to PPCr. Thus, in optical imaging of PPCr domains, only the stimulated domain and a few clusters of adjacent neurons are highly active, while other PPCr domains are suppressed (unpublished data; also see Brock et al. 2013). In M1, a similar overall result occurs, but local activation is more widespread, allowing excitement of functionally related domains. As expected from connection patterns, stimulating pairs of mismatched domains in PPCr at once, results in mutual and sometimes alternating suppression so that neither movement is completed, although sometimes alternating partial movements may occur, or even one movement but not the other will be seen (Stepniewska et al. 2020). Similar results were obtained from stimulating mismatched domains in M1, with one difference. Some

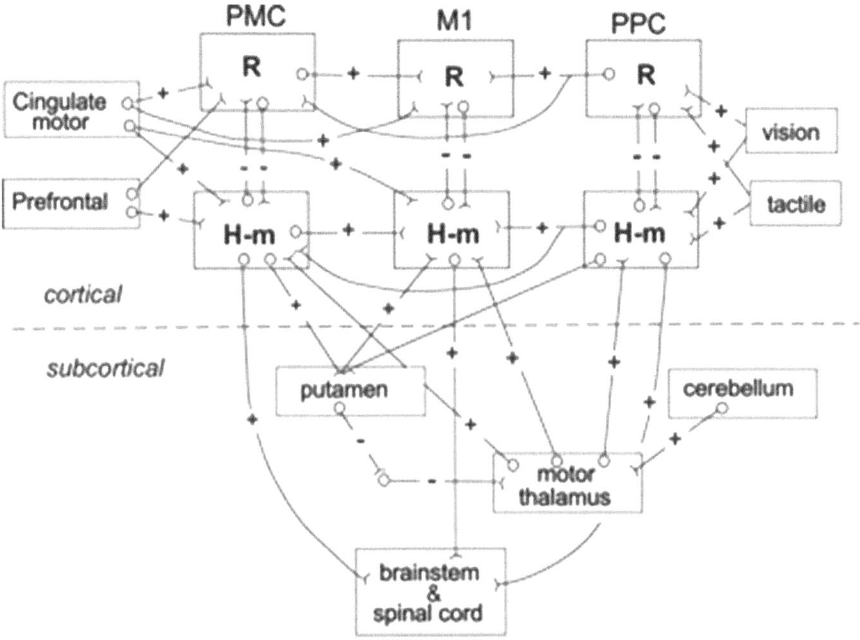

Fig. 3 Interactions of two chosen movement domains: reach (R) and hand-to-mouth (H-m) in posterior parietal (PPC), premotor (PMC), and primary motor cortex (M1) at the cortical and subcortical levels. Each of these three regions contains the sets of matching domains (see text) that are involved in the selection of complex behaviors. Nonmatching domains in each region are in competition with each other (via mutual inhibition) to mediate one of several possible behavioral outcomes. The outcome of the PPC competition is largely based on sensory information (e.g., visual and somatosensory). The most activated PPC domain further activates functionally matching domains in PMC and M1, where further competition takes place. PMC and M1 domains have access to other sources of information (e.g., from prefrontal and cingulate motor areas), and thus the final outcome may change. Interactions between domains may also take place at subcortical levels. Overlapping projections from matching M1, PMC, and PPC domains to the ipsilateral putamen may allow some interaction between cortical domains. Inputs from basal ganglia and cerebellar nuclei to the motor thalamus may modify the behavioral selection by projections from the motor thalamus to cortical domains. Here we indicate some of the subcortical structures that interact with the hand-to-mouth domains, but this information is relevant for other domains as well. (+) excitatory connections, (−) inhibitory connections

domains or parts of domains in M1 appear to potentiate each other so that an emerged movement or a sequence of two movements may occur. Thus, interactions between domains are mainly suppressive in PPCr and both suppressive and sometimes facilitative in M1.

But why are there three cortical levels of action selection? We suggest that there are many levels of action selection that proceed from the most general to the details of the specific movement. The overall role of domains in PPCr is to use sensory information to promote behaviors that are guided by this information. Visual information from mainly the dorsal stream of visual processing (Stepniewska et al. 2016;

Kaas and Baldwin 2020) is the dominant source of sensory information in PPC. A nearby food object will activate the reach domain, while an object aimed at the head will activate a defensive or protective domain. Recent evidence from inactivating LIP in the posterior parietal cortex and recording from the frontal eye field suggests that posterior parietal domains are selectively activated by the most salient visual stimuli and that information is forwarded to the frontal eye field or other domains in PMC, depending on the activated PPC domain (Chen et al. 2020). Domains in PMC evaluate the inputs from matching PPCr domains in the context of other information received, for example, from prefrontal cortex. Perhaps the food item should not be retrieved in the view of a more dominant monkey, so that past experience plays a role. Microcircuits in prefrontal cortex with outputs to premotor cortex provide "executive control" over behavior (Opris et al. 2012). M1 domains consider matching inputs from both PPCr and PMC domains, but also information from the motor thalamus, cingulate motor areas, and the supplementary motor areas (Picard and Strick 1996). Thus, the outcome from the M1 domains may differ from those promoted by PPCr and PMC domains, and some combination between different M1 domains may occur to produce sequences or combinations of actions.

More research is needed to determine the relevance of the different inputs to parietal-frontal domains and the three levels of response selection they represent. However, the selection process is not over with M1, and other outputs, especially to the basal ganglia, need to be considered.

5 The Basal Ganglia: Further Steps in the Action Selection Hierarchy

Most and perhaps all areas of cortex project to the basal ganglia, that include the putamen and the caudate nucleus of the striatum, which then projects to other parts of the basal ganglia (Alexander and Crutcher 1990). A major part of the striatum receives inputs from sensorimotor areas of parietal and frontal cortex, including PPCr, PMC, and M1, where the action-specific domains are located. Thus, it is not surprising that these areas all project to the putamen and to the adjoining parts of the caudate. What is surprising to us at least is that injections of tracers in two or three of the functionally matched and mismatched domains in PPCr, PMC, and M1 reveal an orderly representation of the different classes of domains in a serial order in the basal ganglia that matches the order in cortex, while functionally matched domains project in an overlapping fashion in the basal ganglia (our unpublished data; see also Fig. 3). While there are various theories on what are major functions of the basal ganglia, a prominent view is they uniquely use inhibitory neurons to permit the selection of specific actions or movements, and suppress others (Redgrave et al. 1999; Mink 1996). Thus, direct projections from M1, as well as PPCr and PMC, to motor centers in the brainstem and spinal cord could prime these centers for certain actions, and then be over-ruled by the basal ganglia selection systems in a "last chance to change your mind" sort of way. This would allow motivational inputs to the basal ganglia to cancel and change ongoing behaviors in view of danger and injury. While this role is far from certain, the basal ganglia add another level in the selection process. In addition, there are more basic levels of selection related to the details on how to meet behavioral goals in brainstem and spinal cord motor structures (Barriere et al. 2008). Thus, there are likely multiple levels of the selection process for specific behaviors and ways of mediating them, while the basal ganglia are in a position to stop ongoing behaviors and resolve conflict (Mink 1996; Redgrave et al. 1999). Defensive and aggressive behavior domains in parietal-frontal cortex project to the basal ganglia. The amygdala is another structure in the decision process and it is highly involved in promoting defensive behaviors (Roelofs 2017).

6 Using Electrical Stimulation to Augment Action Selection

Our experiments on the parietal-frontal action specific networks suggest they can be used to augment brain functions in several ways. One demonstrated way is to use recordings from parts of the brain to extract information about intended movements that cannot be performed because spinal cord outputs that would mediate the movement are not available due to the spinal cord injury or other damage to the nervous system. Thus, there has been much research on versions of the brain and machine interface, which involves recording from neurons in the brain, which carry information about intended action, extracting this information in real time with microprocessors, and using this information to control some device that could perform the needed action or a reasonable substitution for that action (Nicolelis and Chapin 2002; Nicolelis and Lebedev 2009; Lebedev and Nicolelis 2017). The usual approach in both monkeys and humans is to put arrays of electrodes in some part of cortex, record from large number of neurons across the electrode arrays, and use this information to selectively trigger a programed movement of, for example, a mechanical arm, that would retrieve an object, or do something else that is useful. Other such tasks could be to instruct a mechanical device that would allow a person in a wheelchair to stand up or sit down, or even stand and walk. The question is, "Where to place the electrode arrays?"

To some extent, it may not matter. If all that is needed is a start or stop signal, then a sudden change in neural activity anywhere in the cortex, or elsewhere in the brain, could trigger the response and thereby reward the person or monkey, and the rewarded individual could gradually or even rapidly, learn how to evoke the necessary neural activity, even if it is totally unrelated to the task. Thus, to stand, one could think of a sad time, for example. But most investigators logically conclude that one can do better by selecting a brain region normally involved in the behavior. Thus, the longstanding evidence that posterior parietal and frontal motor areas are highly involved in motor behavior has led to the placement of electrode arrays in these regions of cortex. In addition, one would suppose that the crude, but overall, somatotopy of these regions would guide the electrode placement. Thus, electrodes have been placed in the arm-hand region of primary motor cortex of monkeys and humans in an effort to use recordings to control a mechanical arm (e.g., Downey et al. 2016, 2018), and comparable regions of posterior parietal cortex have been explored (Aflalo et al. 2015). In such cases, there is still an element of learning by the subject to create the necessary cortical activity, and learning by the microprocessor on how to interpret the neural code. Quite possibly, the need for such learning could be reduced by placing the electrodes more precisely after identifying action-specific domains. However, more research will be necessary before this can be done effectively in macaque monkeys, the most common nonhuman primate model, and in humans.

For macaques, the action-specific domains should be more fully identified and delimited. Fortunately, a number of functionally specific action domains have been identified by Graziano and coworkers (see Graziano 2006). Much of M1 on the anterior bank of the central sulcus has been unexplored, and only two action-specific domains have been identified by electrical stimulation in posterior parietal cortex, LIP for eye movements (Thier and Andersen 1998) and VIP for defensive movements (Cooke et al. 2003). Although several grasping domains have been identified in macaque PPC, the key region of AIP has not been fully explored (Gharbawie et al. 2011a; Rathelot et al. 2017). For humans, we know very little about such action-specific domains, but useful information is starting to emerge (Desmurget and Sirigu 2015; Desmurget et al. 2009, 2014).

As there are three sets of functionally matched domains, one in each of M1, PMC, and PPCr, which set of domains would provide the most useful information? Taken together, the collective results of our experiments suggest that domains in M1 would be most useful. This is because M1 domains appear to be the dominant source of cortical projections to the relevant subcortical motor centers. Stimulating PPCr or PMC domains becomes ineffective in evoking action-specific movement when the functionally matched M1 domain has been deactivated, and current levels necessary for evoking movements are lower for M1 domains, suggesting a more powerful effect on subcortical motor centers. In addition, anatomical results suggest that M1 domains are influenced by more sources of information. But practical concerns are also important, as domains on the cortical surface are more accessible for electrode placement. Finally, electrode arrays should be placed in the domains most closely related to the required behavior, thus controlling a mechanical arm for reaching would be achieved most effectively with recording from cortical reach domains.

There are also other ways of augmenting brain functions. As any great basketball player knows, to keep at the top of your game, you need to practice and then practice some more. Our sensorimotor systems are being shaped by experience, especially in development, but also throughout life. Training on specific tasks not only improves performance, but also long-term practice is needed to maintain performance. Thus, it is tempting to suppose that electrical stimulation of selected domains would increase the effectiveness of their projections to other parts of cortex and subcortical targets, and reduce brain stimulation thresholds for effectiveness as a result of activity-dependent plasticity. In our experiments, we do not stimulate domains many times over long periods of time, so this proposal has not been specifically tested. However, others have noticed that even with an electrical stimulation period of hours, evoked movements may change slightly over repeated stimulations (Nudo et al. 2009). As domains not only evoke movements when stimulated, and suppress other movements from other domains, stimulations of domains at current levels too low to evoke movements, might nevertheless suppress competing movements, allowing another way of altering, and perhaps augmenting brain function. Finally, electrical stimulation of various cortical sites could simply be used to make some actions more likely or some stimuli more noticeable (Mirpour et al. 2010; Chen et al. 2020).

References

Aflalo T, Kellis S, Klaes C, Lee B, Shi Y, Pejsa K, Shanfield K, Hayes-Jackson S, Aisen M, Heck C, Liu C, Andersen RA (2015) Decoding motor imagery from the posterior parietal cortex of a tetraplegic human. Science 348:906–910

Alexander GE, Crutcher MD (1990) Functional architecture of basal ganglia circuits: neural substrates of parallel processing. Trends Neurosci 13:266–271

Allman JM, Kaas JH (1971) A representation of the visual field in the caudal third of the middle temporal gyrus of the owl monkey (Aotus Trivirgatus). Brain Res 31:85–105

Andersen RA, Asanuma C, Essick G, Siegel RM (1990) Corticocortical connections of anatomically and physiologically defined subdivisions within the inferior parietal lobule. J Comp Neurol 296:661–113

Arcaro MJ, Schade PF, Livingstone MS (2019) Body map proto-organization in newborn macaques. Proc Natl Acad Sci U S A 116:24861–24871

Barriere G, Leblond H, Provencher J, Rossignol S (2008) Prominent role of the spinal central pattern generator in the recovery of locomotion after partial spinal cord injuries. J Neurosci 28:3976–3987

Bosking WH, Sun P, Ozker M, Pei X, Foster BL, Beauchamp MS, Yoshor D (2017) Saturation in phosphene size with increasing current levels delivered to human visual cortex. J Neurosci 37:7188–7197

Brock AA, Friedman RM, Fan RH, Roe AW (2013) Optical imaging of cortical networks via intracortical microstimulation. J Neurophysiol 110:2670–2678

Chen X, Zimsak M, Vega GM, Govil E, Lomber SG, Moore T (2020) Parietal cortex regulates visual salience and salience-driven behavior. Neuron 106:177–187.e4

Cicmil N, Krug K (2015) Playing the electric light orchestra—how electrical stimulation of visual cortex elucidates the neural basis of perception. Philos Trans R Soc Lond Biol Sci 370:20140206

Clark KL, Armstrong KM, Moore T (2011) Probing neural circuitry and function with electrical microstimulation. Proc Biol Sci 278:1121–1130

Colby CL, Duhamel JR, Goldberg ME (1993) Ventral intraparietal area of the macaque: anatomic location and visual response properties. J Neurophysiol 69:902–914

Cooke DF, Taylor CSR, Moore T, Graziano MSA (2003) Complex movements evoked by microstimulation of the ventral intraparietal area. Proc Natl Acad Sci U S A 100:6163–6168

Cooke DF, Stepniewska I, Miller D, Kaas JH, Krubitzer L (2015) Reversible deactivation of motor cortex reveals functional connectivity with posterior parietal cortex in the prosimian Galago (Otolemur Garnettii). J Neurosci 35:14406–14422

Desmurget M, Sirigu A (2015) Revealing humans' sensorimotor functions with electrical cortical stimulation. Philos Trans R Soc Lond B Biol Sci 370:20140207

Desmurget M, Reilly KT, Richard N, Szathmari A, Mottolese C, Sirigu A (2009) Movement intention after parietal cortex stimulation in humans. Science 324:811–813
Desmurget M, Richard N, Harquel S, Baraduc P, Szathmari A, Mottolese C, Sirigu A (2014) Neural representations of ethologically relevant hand/mouth synergies in the human precentral gyrus. Proc Natl Acad Sci U S A 111:5718–5722
Downey JE, Weiss JM, Muelling K, Venkatraman A, Valois J-S, Hebert M, Bagnell JA, Schwartz AB, Collinger JL (2016) Blanding of brain-machine interface and vision-guided autonomous robotics improves neuroprosthetic arm performers during grasping. J Neuro Eng Rehab 13:28
Downey JE, Schwed N, Chase SM, Schwartz AB, Collinger JL (2018) Intracortical recording stability in human brain-computer interface users. J Neural Eng 15:046016
Ferrier D (1874) Experiments on the brain of monkeys. Proc R Soc Lond 23:409–430
Galea MP, Darian-Smith I (1994) Multiple corticospinal neuron populations in the macaque monkey are specified by their unique cortical origin. Cereb Cortex 4:518–540
Gharbawie OA, Stepniewska I, Qi H, Kaas JH (2011a) Multiple parietal-frontal pathways mediate grasping in macaque monkeys. J Neurosci 31:11660–11677
Gharbawie OA, Stepniewska I, Kaas JH (2011b) Cortical connections of functional zones in posterior parietal cortex and frontal cortex motor regions in New World monkeys. Cereb Cortex 21:1981–2002
Giraud AL, Price CJ, Graham JM, Frackowiak SJ (2001) Functional plasticity of language-related brain areas after cochlear implantation. Brain 124:1307–1316
Gould HJ III, Cusick CG, Kaas JH (1986) The relationship of corpus callosum connections to electrical stimulation maps of motor, supplementary motor, and the frontal eye fields in owl monkeys. J Comp Neurol 247:297–325
Graziano MSA (2006) The organization of behavioral repertoire in motor cortex. Ann Rev Neurosci 29:10–134
Graziano MSA, Taylor CSR, Moore T (2002) Complex movements evoked by microstimulation of precentral cortex. Neuron 34:841–851
Graziano MSA, Aflalo TNS, Cooke DF (2005) Arm movements evoked by electrical stimulation in the motor cortex of monkeys. J Neurophysiol 94:4209–4223
Gross CG (2007) The discovery of motor cortex and its background. J Hist Neurosci 16:320–331
Grunbaum A, Sherrington C (1903) Observations on the physiology of the cerebral cortex of the anthropoid primates. Proc R Soc Lond 72:152–155
Hartmann K, Thomson EE, Zea I, Yun R, Mullen P, Canarick J, Huh A, Nicolelis MA (2016) Embedding a panoramic representation of infrared light in the adult rat somatosensory cortex through a sensory neuroprosthesis. J Neurosci 36:2406–2424
Hartwigsen G, Weigel A, Schuschan P, Siebner HR, Weise D, Classen J, Saur D (2016) Dissociating parieto-frontal networks for phonological and semantic word decisions: a condition-and-perturb TMS study. Cereb Cortex 26:2590–2601
Hickok G, Poeppel D (2004) Dorsal and ventral streams: a framework for understanding aspects of the functional anatomy of language. Cognition 92:67–99
Johnson-Frey SH (2004) The neural bases of complex tool use in humans. Trends Cogn Sci 8:71–78
Kaas JH (2011) The evolution of auditory cortex: the core areas. In: Winer JA, Schreiner CE (eds) The auditory cortex. Springer, New York, pp 407–427
Kaas JH (2012) Evolution of columns, modules, and domains in the neocortex of primates. Proc Natl Acad Sci U S A 109:10655–10660
Kaas JH, Baldwin MKL (2020) The evolution of the pulvinar complex in primates and its role in the dorsal and ventral streams of cortical processing. Vision:1–19
Kaas JH, Stepniewska I (2016) Evolution of posterior parietal cortex and parietal-frontal networks for specific actions in primates. J Comp Neurol 524:595–608
Kaas JH, Ghrabawie OA, Stepniewska I (2013) Cortical networks or ethologically relevant behaviors in primates. Am J Primatol 75:407–414
Kaas JH, H-x Q, Stepniewska I (2018) Evolution of parietal cortex in primates. In: Vallar G, Coslett HB (eds) Handbook of clinical neurology. The parietal lobe, vol 151. Elsevier, Oxford, pp 31–52

Kurylo DD, Skavenski AA (1991) Eye movements elicited by electrical stimulation of area PG in the monkey. J Neurophysiol 65:1243–1253
Lebedev MA, Nicolelis MA (2017) Brain-machine interfaces: from basic science to neuroprstheses and neurorehabilitation. Physiol Rev 97:767–837
Lewis PM, Rosenfeld JV (2016) Electrical stimulation of the brain and the development of cortical visual prostheses: an historical perspective. Brain Res 1630:208–224
Mashour GA, Walker EE, Martuza RL (2005) Psychosurgery: past, present, and future. Brain Res Rev 48:409–419
Maunsell JH, Newsome WT (1987) Visual processing in monkey extrastriate cortex. Annu Rev Neurosci 10:363–401
Maunsell JH, Van Essen DC (1983) The connections of the middle temporal visual area (MT) and their relationship to a cortical hierarchy in the macaque monkey. J Neurosci 3:2563–2586
Mink JW (1996) The basal ganglia: focused selection and inhibition of competing motor programs. Prog Neurobiol 50:381–425
Mirpour K, Ong WS, Bisley JW (2010) Microstimulation of posterior parietal cortex biases the selection of eye movement goals during search. J Neurophysiol 104:3021–3028
Murphey DK, Maunsell JHR (2008) Electrical microstimulation thresholds for behavioral detection and saccades in monkey frontal eye fields. Proc Natl Acad Sci U S A 105:7315–7320
Nicolelis MA, Chapin JK (2002) Controlling robots with the mind. Sci Am 287:46–53
Nicolelis MA, Lebedev MA (2009) Principles of neural ensemble physiology underlying the operation of brain-machine interfaces. Nat Rev Neurosci 10:530–540
Nudo RJ, Jenkins WM, Merzenich MM (2009) Repetitive microstimulation alters the cortical representation of movements in adult rats. Somatosens Mot Res 7:463–483
Opris I, Ferrera VP (2014) Modifying cognition and behavior with electrical microstimulation: implications for cognitive prostheses. Neurosci Biobehav Rev 47:321–335
Opris I, Barborica A, Ferrera VP (2005a) Microstimulation of the dorsolateral prefrontal cortex biases saccade target selection. J Cogn Neurosci 17:893–904
Opris I, Barborica A, Ferrera VP (2005b) Effects of electrical microstimulation in monkey frontal eye field on saccades to remembered targets. Vision Res 45:3414–3429
Opris I, Hampson RE, Gerhard GA, Berger TW, Deadwyler SA (2012) Columnar processing in primate pFC: evidence for executive control microcircuits. J Cogn Neurosci 24:2334–2347
Penfield W, Boldrey E (1937) Somatic motor and sensory representation in the cerebral cortex of man as studied by electrical stimulation. Brain 60:389–443
Picard N, Strick PL (1996) Motor areas of the medial wall: a review of their location and functional activation. Cereb Cortex 6:342–353
Rathelot JA, Dum RP, Strock PL (2017) Posterior parietal cortex contains a command apparatus for hand movements. Proc Natl Acad Sci U S A 114:4255–4260
Rauschecker JP, Scott SK (2009) Maps and streams in the auditory cortex: nonhuman primates illuminate human speech processing. Nat Neurosci 12:718–729
Redgrave P, Prescott TJ, Gurney K (1999) The basal ganglia: a vertebrate solution to the selection problem? Neuroscience 89:1009–1023
Roelofs K (2017) Freeze for action: neurobiological mechanisms in animal and human freezing. Philos Trans R Soc Lond B Biol Sci 372(1718):20160206
Romo R, Hernandez A, Zainos A, Salinas E (1998) Somatosensory discrimination based on cortical microstimulation. Nature 392:387–390
Romo R, Hernandez A, Zainos A, Brody CD, Lemus L (2000) Sensing without touching: psychophysical performance based on cortical microstimulation. Neuron 26:273–278
Sakata H, Taira M, Murata A, Mine S (1995) Neural mechanisms of visual guidance of hand action in the parietal cortex of the monkey. Cereb Cortex 5:429–438
Salzman CD, Newsome WT (1994) Neural mechanisms for forming a perceptual decision. Science 264:231–237
Salzman CD, Britten KH, Newsome WT (1990) Cortical microstimulation influences perceptual judgments of motion direction. Nature 346:174–177

Salzman CD, Murasugi CM, Britten KH, Newsome WT (1992) Microstimulation in visual area MT: effects on direction discrimination performance. J Neurosci 12:2331–2355
Shambes GM, Gibson JM, Welker W (1978) Fractured somatotopy in granule cell tactile areas of rat cerebellar hemispheres revealed by micromapping. Brain Behav Evol 15:94–140
Snyder LH, Batista AP, Andersen RA (2000) Intention-related activity in the posterior parietal cortex: a review. Vision Res 40:1433–1441
Stepniewska I, Fang PC, Kaas JH (2005) Microstimulation reveals specialized subregions for different complex movements in posterior parietal cortex of prosimian galagos. Proc Natl Acad Sci U S A 102:4878–4883
Stepniewska I, Fang PC, Kaas JH (2009) Organization of the posterior parietal cortex in prosimian Galago: I. Functional zones identified by microstimulation. J Comp Neurol 517:765–782
Stepniewska I, Friedman RM, Gharbawie OA, Cerkevich CM, Roe AW, Kaas JH (2011) Optical imaging in galagos reveals parietal-frontal circuits underlying motor behavior. Proc Natl Acad Sci U S A 108:725–732
Stepniewska I, Gharbawie OA, Burish MJ, Kaas JH (2014) Effects of muscimol inactivations of functional domains in motor, premotor, and posterior parietal cortex on complex movements evoked by electrical stimulation. J Neurophysiol 111:1100–1119
Stepniewska I, Cerkevich CM, Kaas JH (2016) Cortical connections of the caudal portion of posterior parietal cortex in prosimian galagos. Cereb Cortex 26:2753–2777
Stepniewska I, Pouget P, Kaas JH (2018) Frontal eye field in prosimian galagos: intracortical microstimulation and tracing study. J Comp Neurol 526:626–652
Stepniewska I, Friedman RM, Miller DJ, Kaas JH (2020) Interactions within and between parallel parietal-frontal networks involved in complex motor behaviors in prosimian galagos and squirrel money. J Neurophysiol 123:34–56
Sur M, Wall JT, Kaas JH (1981) Modular segregation of functional cell classes within the postcentral somatosensory cortex of monkeys. Science 212:1059–1061
Taira M, Mine S, Georgopoulos AP, Murata A, Sakata H (1990) Parietal cortex neurons of the monkey related to the visual guidance of hand movement. Exp Brain Res 83:29–36
Taylor CS, Gross CG (2003) Twitches versus movements: a story of motor cortex. Neuroscientist 9:332–342
Thier P, Andersen RA (1998) Electrical microstimulation distinguishes distinct saccade-related areas in the posterior parietal cortex. J Neurophysiol 80:1713–1735
Wallace MT (2017) Cooperation between hearing and vision in people with cochlear implant. Proc Natl Acad Sci U S A 114:10003–10005
Woolsey CN, Travis AM, Barnard JM, Ostenso RS (1953) Motor representation in the postcentral gyrus after chronic ablation of precentral and supplementary motor areas. Fed Proc 12:160

Spinal Cord Injury and Epidural Spinal Cord Stimulation

Andrea J. Santamaria, Pedro M. Saraiva, Stephano J. Chang, Ioan Opris, Brian R. Noga, and James D. Guest

1 Introduction: Revisiting Low Expectations for Neurological Recovery After "Complete" SCI

After SCI, deficits of motor, sensory, and autonomic functions are commensurate with injury severity and level (Jaja et al. 2019; Failli et al. 2012). Advances in clinical care have reduced morbidities, increased survival, and improved neurological recovery after SCI (Freed et al. 1966; Stauffer 1975; Closson et al. 1991; Badhiwala et al. 2021). These advances have shifted the focus from complication management to improved life quality and independence with increasingly greater emphasis on recovery. For this review, by recovery, we specifically mean *a measurable enduring improvement linked to improved spinal cord circuit function.* This review does not focus on technological substitution of function using bypass technology such as a brain-machine interface, as covered in other chapters. Rather, we wish to discuss methods by which additional function can be obtained from limited residual connections.

Registries that systematically track spontaneous recovery after SCI report limited neurological recovery after severe ASIA Impairment Scale (AIS) "complete" A injuries (Aimetti et al. 2019; Ottersen and Helm 2002; Wolpaw 2018; Brown and McCouch 1947; Feigin et al. 1951). The prognosis for neurological recovery after SCI has been defined by this apparent clinical "completeness" observed as a complete absence of voluntary motor function, sensory perception, and disturbed autonomic regulation on systematic clinical exams (Kirshblum et al. 2021). These observations overall have affected the design, scope, and investment in rehabilitation practices. For decades,

A. J. Santamaria · P. M. Saraiva · S. J. Chang · I. Opris · B. R. Noga · J. D. Guest (✉)
Department of Biomedical Engineering, University of Miami and Center for Neural Engineering and Miami Project to Cure Paralysis, Miller School of Medicine, Miami, FL, USA
e-mail: jguest@med.miami.edu

I. Opris et al. (eds.), *Modern Approaches to Augmentation of Brain Function*, Contemporary Clinical Neuroscience,
https://doi.org/10.1007/978-3-030-54564-2_2

complete SCI was conceptualized as *effectively being transected,* reducing expectations for recovery potential. This view was further supported by a lack of measurable success in numerous attempted treatments (Geisler et al. 2001; Jones et al. 2010), and failed replication of smaller encouraging studies (Hansebout et al. 1993; Cardenas et al. 2007) when tested in pivotal clinical trials (Cardenas et al. 2014) with only small effect sizes, if any, observed (Cohen 1992). Furthermore, SCI classification, reduced to five categories, submerges the considerable differences in residual connectivity and recovery potential between individuals with complete SCI (AIS A) (Kwon et al. 2010; Tigchelaar et al. 2019; Tanadini et al. 2014).

We examine how new knowledge from spinal cord neuromodulation studies is changing our understanding of clinically complete spinal cord injuries. We introduce the prospect for synergy when suitable neuromodulation is combined with other therapeutics such as biologics and drugs to amplify recovery. It was previously thought that only axonal regeneration could produce neurological recovery (Vidalsanz et al. 1987; Chen et al. 2000). Despite early experience in animal models showing the emergence of new reflex patterns after SCI (Sherrington and Laslett 1903) and describing neuroplasticity (Eccles 1976), the idea that the spinal cord was "hard-wired" and showed only abortive regeneration and limited structural plasticity (Peschanski et al. 1993) solidified over many decades (Ottersen and Helm 2002; Wolpaw 2018; Brown and McCouch 1947; Feigin et al. 1951). Spasticity has long been regarded as a complication but is now being reinterpreted as a window into preserved but imbalanced circuits (Sangari et al. 2019, 2021; Sangari and Perez 2019).

2 Reinterpreting the Lack of Apparent Success of Drugs, Cell Transplants, and Biologics

Tissue repair within the injury region is an essential theme in restorative neurotherapeutics. Despite extensive progress at the molecular and cellular levels, evidence of successful spinal cord axonal regeneration linked to function that improves the quality of life of people living with SCI has not been achieved. To date, testing of transplanted tissues and cells in SCI clinical trials has produced limited evidence of efficacy in human SCI (Fehlings et al. 2018; Curtis et al. 2018; Kucher et al. 2018; Wirth et al. 2001; Santamaria et al. 2020). Relatively recently, a convergence of evidence has challenged the concept of limited recovery potential from complete injury. Scientific evidence shows that whereas regeneration over multi-segmental distances is minimal, there is substantial neuroplasticity (Bareyre et al. 2004) potential (Weidner et al. 2001) after injury. This has resulted in a shift in focus to strengthening residual circuits and fostering new local connections (Jacobson et al. 2021).

Studies of spinal cord neuromodulation have revealed unexpected capabilities in people with complete chronic SCI in the context of intensive locomotor training (Harkema et al. 2011). Implanted or transcutaneous electrical stimulators influence the residual spinal cord circuitry (Dimitrijevic et al. 1998a), with one effect being to enable voluntarily activated movement (Courtine et al. 2009). Notably, this can be observed without a need for multiple sessions, in some cases, immediately (Darrow et al. 2019) upon stimulation establishing the presence of latent connections. This raises the intriguing possibility that one cannot determine the efficacy of some reparative therapeutics due to an adverse excitation state of the spinal cord below the injury, even if there has been some limited repair. Another theme has been that corticospinal tract (CST) repair is essential to achieve neurological recovery. While the CST is the critical motor tract, growing evidence indicates that circuit reorganization involves other systems, notably the propriospinal, reticulospinal, and rubrospinal (Baker and Perez 2017). This is important because many efforts to achieve effective CST regeneration in preclinical research have not been successful, and complete CST regeneration over multiple segments is not feasible. In contrast, CST plasticity is often observed (Peschanski et al. 1993) and likely contributes to natural recovery (Jacobson et al. 2021; Wang et al. 2020; Rosenzweig et al. 2019).

Thus, converging knowledge is overturning the long-held view that recovery from apparently complete SCI is not possible. These themes include improved understanding of post-SCI spinal cord inhibitory interneuron function, clinical rehabilitation paradigms emphasizing locomotor training, and the established safety profile of therapeutic epidural stimulation (ES) in other indications such as intractable pain. Specifically reviewed here are spinal cord neuroplasticity, the concept of "discomplete" SCI, latent circuits, and the importance of newly formed propriospinal and interneuron relay circuits. Finally, from human studies, we will briefly review the discovery of intrinsic locomotor circuits, the effects of locomotor training, and the critical role of sensory inputs to regulate and shape motor output. It is not intended to argue that ES can provide major recovery in complete injuries but rather to illustrate that considerable benefits can occur with limited connectivity.

2.1 Early Neuromodulation: Electrical Stimulation of Muscle to Restore Function

The first systematic application of electrical stimulation for recovery after SCI in people was direct nerve-muscle activation, functional electrical stimulation (FES). This neuromodulation approach was first applied to SCI paralysis in 1962 (Moe and Post 1962) based on motor neuron survival, allowing muscle excitability after SCI (Peckham et al. 1976; Peckham 1987). Several notable applications of FES have been explored in persons with SCI, including restoration of limb function and bladder, bowel, and respiratory system improvements (Peckham and Knutson 2005). Some subjects undergoing intensive FES experienced spontaneous reciprocating locomotor activity in their legs (Calancie et al. 1994; Calancie 2006), revealing central pattern generator-like function in people.

A limited number of muscles can support posture, initiate gait, and enable specific arm and hand functions. This makes it possible to restore some functions with only a few muscle output channels using FES, such as the Freehand system (Mulcahey et al. 2004) for people with tetraplegia (Hoshimiya et al. 1989) and Parastep for walking (Gallien et al. 1995; Klose et al. 1997). Both systems were provided to numerous subjects with SCI, but commercialization proved difficult.

Some technical problems were observed as implanted Teflon-coated wires suffered breakage and lead to inflammation or infection. Externally applied systems have been successful for exercise benefits such as FES cycling and for hand function restoration (Prochazka et al. 1997). Reduced muscle atrophy and bone preservation have been observed (Belanger et al. 2000). However, muscle fatigue is a limitation of the FES approach (Merletti et al. 1992). Overall, these factors have reduced the acceptance to place stimulating and recording devices into muscles for long-term applications (Knutson et al. 2002; Kern et al. 2002; Stein and Mushahwar 2005). A concept emerged that it might be more "physiological" to stimulate upstream of muscle within the central nervous system (CNS).

2.2 Spinal Cord Stimulation

Devices that interface with the nervous system have become increasingly prevalent and validated clinically; these include cardiac pacemakers, deep brain stimulation for Parkinson's disease, vagal stimulators for epilepsy, and implanted ES electrodes for intractable pain. The idea to test if electrical epidural spinal cord stimulation could reduce pain was based on the Gate Control theory hypothesizing that dorsal column stimulation could block nociceptive transmission (Melzack and Wall 1965). ES was also tested to reduce spasticity in the 1960s and 1970s (Mortimer et al. 1970; Shealy et al. 1967). Thus, the application of epidural stimulation (ES) to people with SCI is not a new idea, but only recently, the combination with intensive rehabilitation principles learned from body weight support training has provided critical insights into the latent intrinsic capabilities of the spinal cord after SCI (Harkema et al. 2011; Rejc et al. 2017a; Calvert et al. 2019; Wernig et al. 1995). As early as 2004, ES had been shown to activate intraspinal stepping circuits (Minassian et al. 2004). Thus, convergent multidisciplinary experience was pivotal to generate a new direction for the use of ES.

3 Concepts that Supported the Testing of Epidural Stimulation for Recovery in SCI

3.1 Discomplete Spinal Cord Injury

The finding in people with clinically complete SCI of evidence of some voluntary influence over their legs is called "discomplete injury." This was an essential step to challenge the concept that complete SCI is equivalent to anatomical transection (Dimitrijevic et al. 1983). In detailed studies by Dimitrijevic, Sherwood, and colleagues, subjects with complete motor and sensory chronic SCI were shown to have subtle residual brain influence on spinal cord function (Dimitrijević 1988; Sherwood et al. 1992), such as voluntary suppression of reflexes (Dimitrijevic et al. 1984; Cioni et al. 1986) even when voluntary EMG could not be evoked.

3.2 Early ES Observations in SCI Subjects to Treat Spasticity

An early target of ES in people with chronic SCI was to diminish spasticity symptoms (Adams and Hicks 2005; Richardson and McLone 1978).

Spasticity is "*a motor disorder characterized by a velocity-dependent increase in tonic stretch reflexes (muscle tone) with exaggerated tendon jerks, resulting from hyperexcitability of the stretch reflex, as one component of the upper motoneuron syndrome*" (Lance 1980). Lack of supraspinal inhibition after SCI leads to a state of disinhibition of stretch reflexes which can cause intrusive spasms. After SCI, other contributing factors are protracted limb immobilization that progressively evolves into muscular changes, including decreased sarcomeres and an increase in connective tissue (Tabary et al. 1972; Järvinen et al. 2002; Olsson et al. 2006), making the muscles stiffer (Mirbagheri et al. 2002; Schleip et al. 2006). The muscle additionally experiences a fiber type change with a decrease of the fatigue-resistant type I and less-fatigable type IIa fibers and an increase in the most-fatigable type IIb or IId/x fibers (Malisoux et al. 2007).

Dimitrijevic and Sherwood studied ES for spasticity management in a series of subjects in the 1980s (Gottlieb et al. 1985; Hugenholtz et al. 1988). In six individuals, ES at 30 Hz, 200 μs pulse width, with amplitudes ranging from 2 to 8 mA, reduced spasticity with post-stimulation carry-over effects (Dimitrijevic et al. 1987). During ES, there was a suppression of spasticity and the ability to increase or decrease segmental excitability by adjusting the stimulation intensity (Campos et al. 1987). However, interest in the application of ES for spasticity declined in the 1990s with the introduction of Baclofen pump delivery systems (Penn et al. 1989). Later studies by Dimitrijevic and colleagues included a series of eight chronically injured patients implanted with quadripolar electrodes spanning the L1 to L3 vertebral segments. Frequencies between 50 and 100 Hz, amplitudes of 2–7 V, and a pulse width of 210 μs suppressed severe muscle hypertonia (Pinter et al. 2000). In the course of these studies, Dimitrijevic and colleagues found that stimulation at several frequencies elicited sustained tonic and rhythmic activity, with lower limb extension or stepping-like movements representing different levels of muscle synergies (Minassian et al. 2004).

3.3 Neuroplasticity

Neuroplasticity normally occurs in some structures throughout life (Sagi et al. 2012). In rehabilitation, targeted neuroplasticity is a strategy to improve function in which the Hebbian concept that "*cells that fire together will wire together*" (Hebb 2005) is applied by using task repetition. Conceptually, there are two forms of neuroplasticity important in SCI. The first is the synaptic remodeling of existing neural circuits as a result of denervation, and the second is an axonal structural change in which an intact neuron creates de novo connections through neurite elaboration (Rosenzweig et al. 2010). Some connections are revealed using drugs and are considered to be "latent" (Chen et al. 2018).

Early evidence of neuroplasticity after injury derived from cortical reorganization (Jones 2000; Green et al. 1999). In primates with cervical hemisections, the cortex previously associated with the hand was found to be activated by input from cortical areas representing the face (Jain et al. 1997, 2000). These findings indicate that the topographic representations in the motor cortex are dynamic. Across several species, cortical motor territories controlling certain body parts tend to enlarge and invade those that have lost their peripheral targets (Raineteau and Schwab 2001). In humans, cortical and corticospinal reorganization has been assessed using fMRI and transcranial magnetic stimulation (TMS) after SCI (Levy Jr et al. 1990; Topka et al. 1991). Extensive studies in cats and other models also revealed circuit plasticity in the spinal cord after injury with a capability for walking activity and postural regulation in the absence of supraspinal input (Martinez and Rossignol 2011). In addition, in the cat model, spinal learning resulted from repetitive practice (de Leon et al. 1998; Lovely et al. 1986). In these studies, functionally recovered animals did not have regeneration of descending pathways (Lovely et al. 1986; Roy et al. 1998, 1999; Roy and Acosta Jr. 1986). Spinal "learning" has also been shown in both animal models and people (Harkema et al. 1997, 2012; Edgerton et al. 2001a). Studies of activity-dependent plasticity have shown that both synaptic and structural *neuroplasticity* may improve motor functions (Dubner and Ruda 1992; Behrman et al. 2017).

3.4 The Emerging Understanding of Intraspinal Connectivity

In older concepts of spinal cord function, the emphasis was on white matter tracts that terminated at a level and conveyed "command signals." Modern conceptualizations recognize the spinal cord's integrated intrinsic circuitry as capable of executing and regulating complex functions. These circuits consist of motoneurons, ascending projection neurons, several classes of spinal interneurons, and preganglionic neurons (Zholudeva et al. 2018). Local interneurons with short projections are involved in reflex circuits whereas propriospinal interneurons project to spinal segments different from their origin (Flynn et al. 2011) and are critical for intersegmental coordination (Laliberte et al. 2019). Calancie and colleagues described interlimb connections in SCI subjects from the feet to the hands that became unmasked after SCI (Calancie et al. 1996; Calancie 1991). These "*reflexes*" could only be evoked in those with a motor-complete injury. A critical discovery was that after SCI, supraspinal neurons could establish new connections through propriospinal neurons in rodents and nonhuman primates (Bareyre et al. 2004). After mid-thoracic dorsal transection, rats can exhibit CST sprouting into the cervical gray matter, in addition to an increased number of direct contacts between long propriospinal axon terminals and lumbosacral motoneurons (Bareyre et al. 2004). This plasticity is even more notable in primates, and reticulo-propriospinal contacts from damaged reticulospinal axons have also been reported (Filli et al. 2014).

The increase in connections from descending inputs to propriospinal interneurons provides alternative pathways to spared lumbar circuits. This has also been convincingly tested through staggered hemisection paradigms (Cowley et al. 2008). Spontaneous recovery of stepping shows that interneurons can form detours around lesions to allow transmission of descending input to hindlimb lumbar neuronal pools (Courtine et al. 2008; May et al. 2017). Propriospinal interneurons in lumbar segments are critical for locomotion. As an example, removing dI3 interneurons by eliminating their synaptic transmission abolished functional recovery, indicating that dI3 interneurons play a role in plasticity-induced recovery of locomotion (Bui et al. 2016). Further evidence shows that mice lacking V2a interneurons lose the capacity for left-right alternation at high speeds (Crone et al. 2009).

3.5 *Intraspinal Circuits for Reciprocating Activity*

Numerous activities critical to life do not require conscious input. The question of whether spinal cord functions are dominated by supraspinal command circuits, mainly reflexive to external stimuli, or based on autonomous intrinsic circuits has long been of interest to experimental physiologists (Weiss 1941). In the uninjured state, integrated spinal cord circuits (Orlovskii and Fel'dman 1972; Prentice 1999) allow people to carry out complex tasks like walking in the absence of conscious thought with constant adaptation to the changing environment (Edgerton et al. 2001b). After SCI, this regulation is impaired, peripheral sensory inputs remain, but adequate supraspinal regulation is lost in severe injuries.

The term "central pattern generators" (CPGs) describes interconnected circuits comprised of excitatory and inhibitory neurons that can autonomously generate fictive locomotion in the absence of supraspinal input (Grillner 2006). These circuits are evolutionarily conserved across species (Guertin 2009) to coordinate motion. In mammals, CPG circuits derive from four major classes of ventral interneurons, V0, V1, V2, and V3 (Goulding 2009; Grillner and Jessell 2009), and the system has both rhythm and pattern generating pools. Rhythm pools impose locomotor timing and pace comprised of excitatory ipsilaterally projecting interneurons (Grillner 2006; Hägglund et al. 2013; Kiehn 2016). Downstream of these, V2a glutamatergic interneurons either drive ipsilateral motoneurons or project to commissural interneurons crossing the midline (Dougherty et al. 2013; Kiehn 2011). The pattern-generating neuronal pools induce the sequential activation of motoneuron pools. Flexor-extensor circuits composed of ipsilaterally projecting inhibitory V1 and V2b neurons are involved in alternation (Zhang et al. 2014), and commissural interneurons projecting to motoneurons control the left and right pattern. The pathways change activation patterns during different types of gaits that require alternation or synchronous activity. An important discovery is that excitatory neurons can be activated in isolation to produce locomotor-like rhythmic bursting in flexor or extensor motoneuron pools (Hägglund et al. 2013) and that reciprocal inhibition is not a prerequisite for rhythmicity. Thus, changes in the activation of burst generators allow for variations in the coordination between agonists and antagonist muscles during complex locomotion patterns (Yokoyama et al. 2016). Normally, CPGs integrate visual, vestibular, and proprioceptive inputs (Rossignol et al. 2006).

3.6 Spontaneous Locomotor Activity Observed After Chronic Human SCI

Studies over many decades, especially experiments in cats, demonstrated that fully transected animals could show context-dependent weight support, stepping, and adaptation to different treadmill speeds (Martinez et al. 2012; Forssberg et al. 1980). Spontaneous rhythmic involuntary leg movements have now been confirmed in several studies of people with SCI (Calancie 2006; Nadeau et al. 2010). Early observations include the report of a cervically injured subject who developed rhythmic myoclonic movements of extensor muscles below the injury level; peripheral stimulation of flexor reflex afferents could induce, slow, or interrupt the rhythmic activity (Bussel et al. 1988). Later, Calancie and colleagues found that a chronic incomplete cervical SCI individual could elicit involuntary rhythmic alternating movements of the legs when lying supine and extending his hips. The rate of the step cycle could be altered by specific sensory stimuli and halted by anesthetizing the hip joint.

4 Epidural Electrical Stimulation, Spinal Cord Injury, and Locomotion

When ES is applied to treat neuropathic pain, the concept has been to interfere with dorsal column sensory transmission, and stimulators are placed at the T9/10 level, and tonic stimulation patterns are utilized. However, for standing and walking, the stimulator lead is placed over the conus medullaris and more complex patterns of stimulation are used to optimize standing, walking, and quality of locomotion. These effects are believed to be mediated by discharging the dorsal roots because they show local segmental specificity, but more complex effects on the spinal cord are likely (Minassian et al. 2004; Edgerton et al. 2008).

4.1 Locomotor Activity Resulting from Electrical Epidural Stimulation

Driven by the stepping findings in previous experiments (Minassian et al. 2004), Dimitrijevic and colleagues further explored these observations by implanting quadripolar epidural electrodes spanning vertebral levels T11 to L1 in six chronically injured AIS A individuals (Dimitrijevic et al. 1998b). They observed that stimulus of 5–9 V, pulse width 0.2–0.5 ms, and frequencies between 25 and 50 Hz applied over the L2 spinal cord segment elicited rhythmic step-like EMG discharges with flexion/extension movements in the lower limbs in all individuals while lying supine (Dimitrijevic et al. 1998b), providing the first major evidence of stimulation-evoked complex stepping circuits in people with SCI. The rhythmic motor patterns were later reproduced in ten ASIA A and B individuals implanted with a similar quadripolar linear array. Stimulation frequencies between 2 and 130 Hz, and intensities up to 10.5 V, elicited rhythmic EMG activity in quadriceps, hamstrings, tibialis anterior, and triceps surae with various synchronous and reciprocal relationships between the muscles in seven individuals. The majority of the rhythmic activity was generated with the cathode over the L2-L4 spinal cord segment.

In recent studies of ES to improve standing and assisted stepping, Harkema and colleagues initially studied a subject with a motor complete C7-T1 injury level. Of note, only limited improvements were found after 170 locomotor training sessions using bodyweight support (BWS) and treadmill facilitation. With this training as a baseline, the AIS B subject was implanted with a 16-electrode array (5-6-5 Specify, Medtronic) over spinal cord segments L1-S1. Configurations of 15 Hz and 8 V of ES facilitated sustained standing during progressive BWS reduction and increased EMG amplitude with increasing stimulation intensities. This study confirmed that improved stepping quality was dependent on task-specific sensory cues, including manually facilitated stepping with load alternation. Surprisingly, it also demonstrated that voluntary supraspinal control of toe extension and ankle and leg flexion emerged with ES (Harkema et al. 2011). A subsequent study by this group was conducted in four chronic AIS A and B subjects implanted with a similar 16-lead array (Medtronic, RestoreAdvanced). At study entry, none had motor evoked responses during TMS nor volitional control when measured by EMG. However, all individuals were able to execute volitional ankle dorsiflexion during ES. Furthermore, subjects could synchronize their voluntary movement to the peak or trough levels of a wave displayed on a computer screen exhibiting control (Angeli et al. 2014). Thus, despite the clinical diagnosis of complete sensorimotor function loss, adequately modulating the remaining spared connections' excitability has generated reproducible clinical functions. Controversy exists as to the optimal stimulation methods with some groups favoring tonic stimulation and others spatially and temporally selective stimulation (Wagner et al. 2018).

4.2 *Locomotor Training and Sensory Input*

The emerging popularity of body weight supported (BWS) locomotor rehabilitation on treadmills, robotic devices, or overground in the 1990s was a "paradigm shift from compensation for deficits to rehabilitation as an agent for walking recovery" (Behrman et al. 2006). Dietz and colleagues explored the effects in sub-acute SCI (4–5 weeks post-SCI), including BWS, treadmill walking, and manually assisted stepping in complete and incomplete individuals. Even though coordinated stepping movements were induced in both cohorts, only incomplete patients showed recovery during the locomotor training progressing to produce unsupported stepping movements (Dietz et al. 1994). As the popularity of this therapy expanded, more rigorous clinical trials further defined the intensity and duration required and the durability of benefits. In a cohort of 20 chronic AIS C and D participants undergoing 8 weeks of BWS + treadmill training with a driven-gait orthosis, significant improvements in the subject's gait velocity, endurance, and performance of functional tasks

(Wirz et al. 2005) were reported. A decrease in the need for BWS with increased locomotor endurance was also reported in a single AIS D individual undergoing 100 training sessions (Morrison et al. 2012).

This training has been shown to have health benefits and reduce medical costs, although a large number of sessions can be required (Morrison et al. 2018). Dynamic task-specific training in the presence of ES for 43 weeks while using a front-wheeled walker and trainer assistance enabled independent stepping on a treadmill and overground. A dynamic training paradigm appears to re-educate spinal networks providing extrinsic facilitation of afferent feedback and compensatory spinal cord adaptation (Gill et al. 2018). Bodyweight load and hip joint afferent inputs are needed to achieve suitable training locomotor patterns. Certain drugs such as clonidine, an alpha 2 adrenergic agonist used clinically for spasticity and pain, have been shown to potentiate locomotion in spinal cats and increase improvements during step training in some people with SCI when combined with a serotonergic agonist (Chau et al. 1998; Fung et al. 1990). Intrathecal clonidine delivery was effective in reducing spasticity and improving gait (Remy-Neris et al. 1999).

4.3 Reported Effects of ES on Autonomic Functions

While ES translation for recovery after SCI first focused on motor systems, data emerging from these studies has indicated beneficial impacts on other consequences on SCI. Autonomic dysregulation after SCI leads to serious clinical problems that include lack of blood pressure regulation, autonomic dysreflexia, impaired bladder voiding, slow bowel transit, and loss of thermoregulation. Even from the first reports of ES for spasticity, now 42 years ago, secondary benefits have been reported (Richardson and McLone 1978; Richardson et al. 1979). The first report of objectively documented improved bladder function was by Meglio, Rossi, et al., investigating seven chronic SCI patients with neurogenic bladder. The study found improvements in micturition pressure, detrusor coordination, and bladder capacity with a course of 20 min of ES every 3 h, using frequencies between 60 and 120 Hz and pulse widths between 0.1 and 1 ms (Meglio et al. 1980). A recent study showed that specifically directed stimulation improved reflexive bladder emptying resulting in lower post-void residuals (Herrity et al. 2018). Severe autonomic dysregulation pain states such as reflex sympathetic dystrophy have responded to spinal cord stimulation (Kemler et al. 2000).

Low blood pressure and orthostatic hypotension result in associated cerebral hypoperfusion, and light-headedness, dizziness, blurred vision, and dyspnea symptoms (Krassioukov and Claydon 2006). Studies by Harkema and colleagues in motor complete SCI highlighted the importance of specific task parameter optimization (Rejc et al. 2017b), and the latest studies by the group focus on electrode stimulation parameters optimized to cardiovascular responses. In a group of four individuals implanted with ES (5-6-5 Specify, Medtronic), stimulation configuration was optimized in each subject to maintain a target systolic blood pressure of 110–120 mmHg. After sufficient mapping sessions, a protocol consisting of five 2 h sessions with continuous blood pressure monitoring was applied in an upright sitting position. The results showed significant stabilization of blood pressure, and in another study, BP was stabilized during standing without increasing EMG leg activity, indicating specific targeting of BP (Harkema et al. 2018; Aslan et al. 2018).

In a study by different investigators, two ES-stimulated participants submitted to the tilt-table challenge resulted in higher systolic blood pressure while tilted during stimulation as compared to tilting with no stimulation. Additionally, cognitive function testing also improved during the tilting with stimulation. The study also reported partial restoration of volitional urination in one participant (Darrow et al. 2019).

5 Transcutaneous Stimulation

While ES studies involve implanting electrodes in the epidural space, transcutaneous stimulation (TCS) has been explored to noninvasively improve spinal circuit activity after SCI. Lumbar TCS was shown to improve voluntary leg muscle activation in motor complete subjects, and this was notably potentiated by the serotonergic agonist drug Buspirone 7.5 mg twice daily (Gerasimenko et al. 2015). TCS applied for cervical injury has also been explored. Improved grip and pinch force has been reported when combined with training (Gad et al. 2018; Inanici et al. 2018) with carry-over benefits reported after the training period (Inanici et al. 2021). Based on the improvements and the ability to observe changes during a single session, it is possible that the use of TES may also facilitate the screening of potential ES candidates.

6 Potential ES Synergies

A core idea is that the enablement of voluntary activation of movement by neuromodulation will reinforce synaptic neuroplasticity to amplify recovery of function. That is, by making some voluntary movement possible using stimulation, these connections can be more effectively reinforced. The stimulation techniques may also potentiate the effects of therapeutic drugs synergistically. For example, serotonergic agonists and KCC2 enhancers to normalize neuronal function (REFs) and 4-aminopyridine, a drug that improves function in partially demyelinated axons. 4-aminopyridine was extensively studied preclinically and then in a series of clinical trials as a slow-release compound (REF). The trials in SCI did not meet the primary end-point, likely due to the heterogeneity of study subjects. Several new clinical trials use new stratification methods and advanced biomarkers to detect responders (REFs).

7 The Essential Role of Spinal Cord Repair Research

Restoration of voluntary function requires sufficient circuitry, ultimately settings a limit on recovery. Thus far, there are no reports of combining spinal cord stimulation with biologics in people. This appears to be an essential next step to provide more potent substrates for the effects of neurorehabilitation and stimulation enhanced plasticity. Biologics include cells, neurotrophic molecules, therapeutic enzymes, and some biomaterials. Recently using optogenetic tools, it was reported that transplanted neural stem cells could form integrated circuits after implantation in rodents (Ceto et al. 2020). We reported recovery of subclinical circuits after Schwann cells transplantation. If there is a significant loss of motor neurons after SCI, neuromodulation efficacy will be limited, and cell implantation (Erb et al. 1993; Grumbles et al. 2013) or de novo neurogenesis will be essential. Several biologics including antibodies to reduce neural growth inhibition have advanced to clinical trials (Kucher et al. 2018) and might have greater effects if combined with ES or TCS. Conceptually, these agents should amplify neuroplasticity and allow targeted activity-dependent plasticity to be more effective. The effects of electrical stimulation on how newly created circuits form and are stabilized is an important goal that is now possible to study.

8 Conclusions

It is remarkable that trains of stimuli, input to the spinal cord by fairly simple electrode arrays, can favorably affect several discrete functions after chronic SCI. Spinal cord stimulation has not yet met the standards to become an accepted clinical therapy to improve recovery after SCI. Currently, there are limited expert centers, no published SCI-specific randomized trials, and no regulatory approvals for the SCI indication other than research study IDEs. While the experimental results have been impressive, there are still gaps for the functional restoration to meet commonly accepted clinical endpoints such as the Spinal Cord Independence Measure. For example, safe, functional walking in the community requires sufficient postural control with sensorimotor integration to provide balance unless walking aids are used. Postural stability may be addressed through closed-loop control systems, but this adds complexity to the stimulation paradigms that are yet to be technically resolved. With the technology that is available and in development, neuromodulation, including epidural stimulation, will likely become a significant part of the rehabilitation process for those living with SCI. There will remain a need to repair lost connections through axonal regeneration and substantially amplified neuroplasticity (Fig. 1).

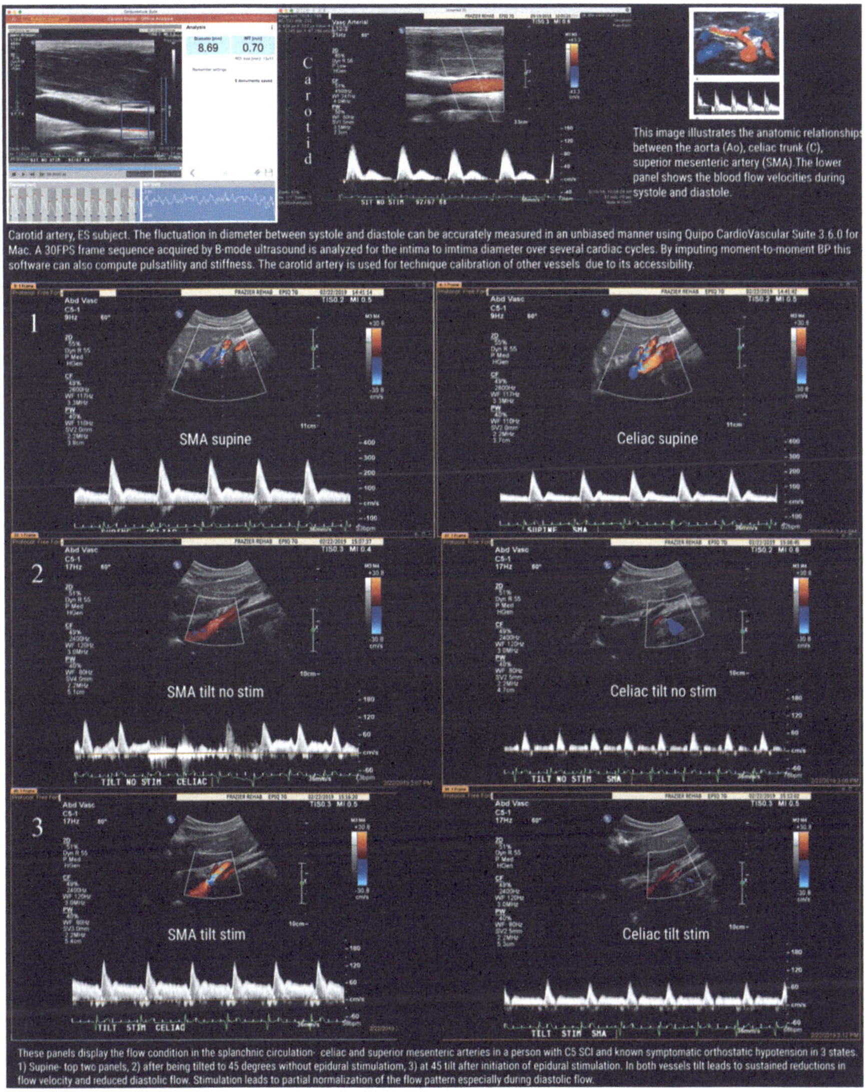

Fig. 1 Epidural stimulation is shown to improve blood flow in a tetraplegic with severe postural hypotension during a 70° tilt. Shown are color flow and velocity ultrasound measurements with peak waveforms during systole followed by the diastolic phase. The upper panel shows the carotid artery, images on the left indicate the change in intimal diameter during several cardiac cycles quantified in Carotid Studio, Quipo Software, Pisa, Italy. Lower panels show the Doppler signal recorded from the superior mesenteric (SMA) and celiac arteries in (1) Supine, (2) Tilt without ES stimulation and (3) Tilt with ES stimulation with breathing held during insonnation. In the tilt, no-stim state, the diastolic Phase is collapsed. When stimulation is initiated, the diastolic flow is recovered

Acknowledgments The ultrasound data were collected at the University of Louisville within the SCI neuromodulation program. Data was analyzed by J Guest at the University of Miami. Specific contributors are Drs. Susan Harkema, Jill Wecht, Alex Ovechkin, and Bonnie Legg, Jessie Fisher, and Shelly Wade.

References

Adams MM, Hicks AL (2005) Spasticity after spinal cord injury. Spinal Cord 43(10):577–586

Aimetti AA et al (2019) Natural history of neurological improvement following complete (AIS A) thoracic spinal cord injury across three registries to guide acute clinical trial design and interpretation. Spinal Cord 57(9):753–762

Angeli CA et al (2014) Altering spinal cord excitability enables voluntary movements after chronic complete paralysis in humans. Brain 137(5):1394–1409

Aslan SC et al (2018) Epidural spinal cord stimulation of lumbosacral networks modulates arterial blood pressure in individuals with spinal cord injury-induced cardiovascular deficits. Front Physiol 9:565

Badhiwala JH et al (2021) The influence of timing of surgical decompression for acute spinal cord injury: a pooled analysis of individual patient data. Lancet Neurol 20(2):117–126

Baker SN, Perez MA (2017) Reticulospinal contributions to gross hand function after human spinal cord injury. J Neurosci 37(40):9778–9784

Bareyre FM et al (2004) The injured spinal cord spontaneously forms a new intraspinal circuit in adult rats. Nat Neurosci 7(3):269–277

Behrman AL, Bowden MG, Nair PM (2006) Neuroplasticity after spinal cord injury and training: an emerging paradigm shift in rehabilitation and walking recovery. Phys Ther 86(10):1406–1425

Behrman AL, Ardolino EM, Harkema SJ (2017) Activity-based therapy: from basic science to clinical application for recovery after spinal cord injury. J Neurol Phys Ther 41(Suppl 3):S39–S45

Belanger M et al (2000) Electrical stimulation: can it increase muscle strength and reverse osteopenia in spinal cord injured individuals? Arch Phys Med Rehabil 81(8):1090–1098

Brown JO, McCouch GP (1947) Abortive regeneration of the transected spinal cord. J Comp Neurol 87(2):131–137

Bui TV et al (2016) Spinal microcircuits comprising dI3 interneurons are necessary for motor functional recovery following spinal cord transection. eLife 5:e21715

Bussel B et al (1988) Myoclonus in a patient with spinal cord transection: possible involvement of the spinal stepping generator. Brain 111(5):1235–1245

Calancie B (1991) Interlimb reflexes following cervical spinal cord injury in man. Exp Brain Res 85(2):458–469

Calancie B (2006) Spinal myoclonus after spinal cord injury. J Spinal Cord Med 29(4):413–424

Calancie B et al (1994) Involuntary stepping after chronic spinal cord injury. Evidence for a central rhythm generator for locomotion in man. Brain 117(Pt 5):1143–1159

Calancie B, Lutton S, Broton JG (1996) Central nervous system plasticity after spinal cord injury in man: interlimb reflexes and the influence of cutaneous stimulation. Electroencephalogr Clin Neurophysiol 101(4):304–315

Calvert JS et al (2019) Emergence of epidural electrical stimulation to facilitate sensorimotor network functionality after spinal cord injury. Neuromodulation 22(3):244–252

Campos RJ et al (1987) Epidural spinal cord stimulation in spastic spinal cord injury patients. Appl Neurophysiol 50(1–6):453–454

Cardenas DD et al (2007) Phase 2 trial of sustained-release fampridine in chronic spinal cord injury. Spinal Cord 45(2):158–168

Cardenas DD et al (2014) Two phase 3, multicenter, randomized, placebo-controlled clinical trials of fampridine-SR for treatment of spasticity in chronic spinal cord injury. Spinal Cord 52(1):70–76
Ceto S et al (2020) Neural stem cell grafts form extensive synaptic networks that integrate with host circuits after spinal cord injury. Cell Stem Cell 27(3):430–440.e5
Chau C, Barbeau H, Rossignol S (1998) Effects of intrathecal alpha(1)- and alpha(2)-noradrenergic agonists and norepinephrine on locomotion in chronic spinal cats. J Neurophysiol 79(6):2941–2963
Chen MS et al (2000) Nogo-A is a myelin-associated neurite outgrowth inhibitor and an antigen for monoclonal antibody IN-1. Nature 403(6768):434–439
Chen B et al (2018) Reactivation of dormant relay pathways in injured spinal cord by KCC2 manipulations. Cell 174(6):1599
Cioni B et al (1986) Voluntary supraspinal suppression of spinal reflex activity in paralyzed muscles of spinal cord injury patients. Exp Neurol 93(3):574–583
Closson JB et al (1991) Rehabilitation in spinal cord disorders. 3. Comprehensive management of spinal cord injury. Arch Phys Med Rehabil 72(4-S):S298–S308
Cohen J (1992) A power primer. Psychol Bull 112(1):155–159
Courtine G et al (2008) Recovery of supraspinal control of stepping via indirect propriospinal relay connections after spinal cord injury. Nat Med 14(1):69–74
Courtine G et al (2009) Transformation of nonfunctional spinal circuits into functional states after the loss of brain input. Nat Neurosci 12(10):1333–U167
Cowley KC, Zaporozhets E, Schmidt BJ (2008) Propriospinal neurons are sufficient for bulbospinal transmission of the locomotor command signal in the neonatal rat spinal cord. J Physiol 586(6):1623–1635
Crone SA et al (2009) In mice lacking V2a interneurons, gait depends on speed of locomotion. J Neurosci 29(21):7098–7109
Curtis E et al (2018) A first-in-human, phase I study of neural stem cell transplantation for chronic spinal cord injury. Cell Stem Cell 22(6):941–950.e6
Darrow D et al (2019) Epidural spinal cord stimulation facilitates immediate restoration of dormant motor and autonomic supraspinal pathways after chronic neurologically complete spinal cord injury. J Neurotrauma 36(15):2325–2336
de Leon RD et al (1998) Locomotor capacity attributable to step training versus spontaneous recovery after spinalization in adult cats. J Neurophysiol 79(3):1329–1340
Dietz V, Colombo G, Jensen L (1994) Locomotor activity in spinal man. Lancet 344(8932):1260–1263
Dimitrijević MR (1988) Residual motor functions in spinal cord injury. Adv Neurol 47:138–155
Dimitrijevic MR et al (1983) Motor control in man after partial or complete spinal cord injury. Adv Neurol 39:915–926
Dimitrijevic MR et al (1984) Suprasegmentally induced motor unit activity in paralyzed muscles of patients with established spinal cord injury. Ann Neurol 16(2):216–221
Dimitrijevic MR et al (1987) Epidural spinal cord stimulation and carry-over effect in chronic spinal cord injury patients. Stereotact Funct Neurosurg 50(1–6):449–450
Dimitrijevic MR, Gerasimenko Y, Pinter MM (1998a) Evidence for a spinal central pattern generator in humans. In: Kiehn O et al (eds) Neuronal mechanisms for generating locomotor activity. New York Acad Sciences, New York, pp 360–376
Dimitrijevic MR, Gerasimenko Y, Pinter MM (1998b) Evidence for a spinal central pattern generator in humans. Ann N Y Acad Sci 860(1):360–376
Dougherty KJ et al (2013) Locomotor rhythm generation linked to the output of spinal shox2 excitatory interneurons. Neuron 80(4):920–933
Dubner R, Ruda MA (1992) Activity-dependent neuronal plasticity following tissue injury and inflammation. Trends Neurosci 15(3):96–103
Eccles JC (1976) The plasticity of the mammalian central nervous system with special reference to new growths in response to lesions. Naturwissenschaften 63(1):8–15
Edgerton VR et al (2001a) Retraining the injured spinal cord. J Physiol 533(1):15–22

Edgerton VR, Roy RR, de Leon RD (2001b) Neural Darwinism in the mammalian spinal cord. In: Patterson MM, Grau JW (eds) Spinal cord plasticity: alterations in reflex function. Springer, Boston, MA, pp 185–206

Edgerton VR et al (2008) Training locomotor networks. Brain Res Rev 57(1):241–254

Erb DE, Mora RJ, Bunge RP (1993) Reinnervation of adult rat gastrocnemius muscle by embryonic motoneurons transplanted into the axotomized tibial nerve. Exp Neurol 124(2):372–376

Failli V et al (2012) Functional neurological recovery after spinal cord injury is impaired in patients with infections. Brain 135(11):3238–3250

Fehlings MG et al (2018) Rho inhibitor VX-210 in acute traumatic subaxial cervical spinal cord injury: design of the SPinal Cord Injury Rho INhibition InvestiGation (SPRING) clinical trial. J Neurotrauma 35(9):1049–1056

Feigin I, Geller EH, Wolf A (1951) Absence of regeneration in the spinal cord of the young rat. J Neuropathol Exp Neurol 10(4):420–425

Filli L et al (2014) Bridging the gap: a reticulo-propriospinal detour bypassing an incomplete spinal cord injury. J Neurosci 34(40):13399–13410

Flynn JR et al (2011) The role of propriospinal interneurons in recovery from spinal cord injury. Neuropharmacology 60(5):809–822

Forssberg H et al (1980) The locomotion of the low spinal cat. II. Interlimb coordination. Acta Physiol Scand 108(3):283–295

Freed MM, Bakst HJ, Barrie DL (1966) Life expectancy, survival rates, and causes of death in civilian patients with spinal cord trauma. Arch Phys Med Rehabil 47(7):457–463

Fung J, Stewart JE, Barbeau H (1990) The combined effects of clonidine and cyproheptadine with interactive training on the modulation of locomotion in spinal-cord injured subjects. J Neurol Sci 100(1–2):85–93

Gad P et al (2018) Non-invasive activation of cervical spinal networks after severe paralysis. J Neurotrauma 35(18):2145–2158

Gallien P et al (1995) Restoration of gait by functional electrical stimulation for spinal cord injured patients. Paraplegia 33(11):660–664

Geisler FH et al (2001) The Sygen multicenter acute spinal cord injury study. Spine (Phila Pa 1976) 26(24 Suppl):S87–S98

Gerasimenko YP et al (2007) Epidural spinal cord stimulation plus quipazine administration enable stepping in complete spinal adult rats. J Neurophysiol 98(5):2525–2536

Gerasimenko YP et al (2015) Noninvasive reactivation of motor descending control after paralysis. J Neurotrauma 32(24):1968–1980

Gill ML et al (2018) Neuromodulation of lumbosacral spinal networks enables independent stepping after complete paraplegia. Nat Med 24(11):1677–1682

Gottlieb GL et al (1985) Evaluation of cervical stimulation for chronic treatment of spasticity. Neurology 35(5):699–704

Goulding MJNRN (2009) Circuits controlling vertebrate locomotion: moving in a new direction. Nat Rev Neurosci 10(7):507–518

Green JB et al (1999) Cortical motor reorganization after paraplegia: an EEG study. Neurology 53(4):736–743

Grillner S (2006) Biological pattern generation: the cellular and computational logic of networks in motion. Neuron 52(5):751–766

Grillner S, Jessell TM (2009) Measured motion: searching for simplicity in spinal locomotor networks. Curr Opin Neurobiol 19(6):572–586

Grumbles RM et al (2013) Acute stimulation of transplanted neurons improves motoneuron survival, axon growth, and muscle reinnervation. J Neurotrauma 30(12):1062–1069

Guertin PA (2009) The mammalian central pattern generator for locomotion. Brain Res Rev 62(1):45–56

Hägglund M et al (2013) Optogenetic dissection reveals multiple rhythmogenic modules underlying locomotion. Proc Natl Acad Sci U S A 110(28):11589–11594

Hansebout RR et al (1993) 4-Aminopyridine in chronic spinal cord injury: a controlled, double-blind, crossover study in eight patients. J Neurotrauma 10(1):1–18
Harkema SJ et al (1997) Human lumbosacral spinal cord interprets loading during stepping. J Neurophysiol 77(2):797–811
Harkema S et al (2011) Effect of epidural stimulation of the lumbosacral spinal cord on voluntary movement, standing, and assisted stepping after motor complete paraplegia: a case study. Lancet 377(9781):1938–1947
Harkema SJ et al (2012) Balance and ambulation improvements in individuals with chronic incomplete spinal cord injury using locomotor training-based rehabilitation. Arch Phys Med Rehabil 93(9):1508–1517
Harkema SJ et al (2018) Normalization of blood pressure with spinal cord epidural stimulation after severe spinal cord injury. Front Hum Neurosci 12:83
Hebb DO (2005) The organization of behavior: a neuropsychological theory. Psychology Press, New York
Herrity AN et al (2018) Lumbosacral spinal cord epidural stimulation improves voiding function after human spinal cord injury. Sci Rep 8(1):8688
Hoshimiya N et al (1989) A multichannel FES system for the restoration of motor functions in high spinal cord injury patients: a respiration-controlled system for multijoint upper extremity. IEEE Trans Biomed Eng 36(7):754–760
Hugenholtz H et al (1988) Cervical spinal cord stimulation for spasticity in cerebral palsy. Neurosurgery 22(4):707–714
Inanici F et al (2018) Transcutaneous electrical spinal stimulation promotes long-term recovery of upper extremity function in chronic tetraplegia. IEEE Trans Neural Syst Rehabil Eng 26(6):1272–1278
Inanici F et al (2021) Transcutaneous spinal cord stimulation restores hand and arm function after spinal cord injury. IEEE Trans Neural Syst Rehabil Eng 29:310–319
Jacobson PB et al (2021) Elezanumab, a human anti-RGMa monoclonal antibody, promotes neuroprotection, neuroplasticity, and neurorecovery following a thoracic hemicompression spinal cord injury in non-human primates. Neurobiol Dis 155:105385
Jain N, Catania KC, Kaas JH (1997) Deactivation and reactivation of somatosensory cortex after dorsal spinal cord injury. Nature 386(6624):495–498
Jain N et al (2000) Growth of new brainstem connections in adult monkeys with massive sensory loss. Proc Natl Acad Sci U S A 97(10):5546–5550
Jaja BNR et al (2019) Association of pneumonia, wound infection, and sepsis with clinical outcomes after acute traumatic spinal cord injury. J Neurotrauma 36(21):3044–3050
Järvinen TAH et al (2002) Organization and distribution of intramuscular connective tissue in normal and immobilized skeletal muscles. J Muscle Res Cell Motil 23(3):245–254
Jones EG (2000) Cortical and subcortical contributions to activity-dependent plasticity in primate somatosensory cortex. Annu Rev Neurosci 23:1–37
Jones LA et al (2010) A phase 2 autologous cellular therapy trial in patients with acute, complete spinal cord injury: pragmatics, recruitment, and demographics. Spinal Cord 48(11):798–807
Kemler MA et al (2000) Spinal cord stimulation in patients with chronic reflex sympathetic dystrophy. N Engl J Med 343(9):618–624
Kern H et al (2002) Denervated muscles in humans: limitations and problems of currently used functional electrical stimulation training protocols. Artif Organs 26(3):216–218
Kiehn O (2011) Development and functional organization of spinal locomotor circuits. Curr Opin Neurobiol 21(1):100–109
Kiehn O (2016) Decoding the organization of spinal circuits that control locomotion. Nat Rev Neurosci 17(4):224
Kirshblum S et al (2021) Characterizing natural recovery after traumatic spinal cord injury. J Neurotrauma

Klose KJ et al (1997) Evaluation of a training program for persons with SCI paraplegia using the Parastep 1 ambulation system: part 1. Ambulation performance and anthropometric measures. Arch Phys Med Rehabil 78(8):789–793
Knutson JS et al (2002) Electrode fracture rates and occurrences of infection and granuloma associated with percutaneous intramuscular electrodes in upper-limb functional electrical stimulation applications. J Rehabil Res Dev 39(6):671–683
Krassioukov A, Claydon VE (2006) The clinical problems in cardiovascular control following spinal cord injury: an overview. Prog Brain Res 152:223–229
Kucher K et al (2018) First-in-man intrathecal application of neurite growth-promoting anti-Nogo-A antibodies in acute spinal cord injury. Neurorehabil Neural Repair 32(6–7):578–589
Kwon BK et al (2010) Cerebrospinal fluid inflammatory cytokines and biomarkers of injury severity in acute human spinal cord injury. J Neurotrauma 27(4):669–682
Laliberte AM et al (2019) Propriospinal neurons: essential elements of locomotor control in the intact and possibly the injured spinal cord. Front Cell Neurosci 13:512
Lance JW (1980) The control of muscle tone, reflexes, and movement: Robert Wartenberg Lecture. Neurology 30:1303–1313
Levy WJ Jr et al (1990) Focal magnetic coil stimulation reveals motor cortical system reorganized in humans after traumatic quadriplegia. Brain Res 510(1):130–134
Lovely RG et al (1986) Effects of training on the recovery of full-weight-bearing stepping in the adult spinal cat. Exp Neurol 92(2):421–435
Malisoux L et al (2007) Effect of long-term muscle paralysis on human single fiber mechanics. J Appl Physiol 102(1):340–349
Martinez M, Rossignol S (2011) Changes in CNS structures after spinal cord lesions: implications for BMI. In: Schouenborg J, Garwicz M, Danielsen N (eds) Brain machine interfaces: implications for science, clinical practice and society. Elsevier Science Bv, Amsterdam, pp 191–202
Martinez M et al (2012) Effect of locomotor training in completely spinalized cats previously submitted to a spinal hemisection. J Neurosci 32(32):10961
May Z et al (2017) Following spinal cord injury transected reticulospinal tract axons develop new collateral inputs to spinal interneurons in parallel with locomotor recovery. Neural Plast 2017:1932875
Meglio M et al (1980) Epidural spinal cord stimulation for the treatment of neurogenic bladder. Acta Neurochir 54(3–4):191–199
Melzack R, Wall PDJS (1965) Pain mechanisms: a new theory. Science 150(3699):971–979
Merletti R, Knaflitz M, De Luca CJ (1992) Electrically evoked myoelectric signals. Crit Rev Biomed Eng 19(4):293–340
Minassian K et al (2004) Stepping-like movements in humans with complete spinal cord injury induced by epidural stimulation of the lumbar cord: electromyographic study of compound muscle action potentials. Spinal Cord 42(7):401–416
Mirbagheri MM et al (2002) The effects of long-term FES-assisted walking on intrinsic and reflex dynamic stiffness in spastic spinal-cord-injured subjects. IEEE Trans Neural Syst Rehabil Eng 10(4):280–289
Moe JH, Post HW (1962) Functional electrical stimulation for ambulation in hemiplegia. J Lancet 82:285–288
Morrison SA et al (2012) NeuroRecovery Network provides standardization of locomotor training for persons with incomplete spinal cord injury. Arch Phys Med Rehabil 93(9):1574–1577
Morrison SA et al (2018) Longitudinal recovery and reduced costs after 120 sessions of locomotor training for motor incomplete spinal cord injury. Arch Phys Med Rehabil 99(3):555–562
Mortimer JT, Shealy CN, Wheeler C (1970) Experimental nondestructive electrical stimulation of the brain and spinal cord. J Neurosurg 32(5):553–559
Mulcahey MJ et al (2004) Implantation of the freehand system during initial rehabilitation using minimally invasive techniques. Spinal Cord 42(3):146–155
Nadeau S et al (2010) Spontaneous motor rhythms of the back and legs in a patient with a complete spinal cord transection. Neurorehabil Neural Repair 24(4):377–383

Olsson MC et al (2006) Fibre type-specific increase in passive muscle tension in spinal cord-injured subjects with spasticity. J Physiol 577(Pt 1):339–352
Orlovskii GN, Fel'dman AG (1972) Classification of lumbosacral neurons by their discharge pattern during evoked locomotion. Neurophysiology 4(4):311–317
Ottersen OP, Helm PJ (2002) How hardwired is the brain? Nature 420(6917):751–752
Peckham PH (1987) Functional electrical stimulation: current status and future prospects of applications to the neuromuscular system in spinal cord injury. Paraplegia 25(3):279–288
Peckham PH, Knutson JS (2005) Functional electrical stimulation for neuromuscular applications. Annu Rev Biomed Eng 7:327–360
Peckham PH, Mortimer JT, Marsolais EB (1976) Upper and lower motor neuron lesions in the upper extremity muscles of tetraplegics. Paraplegia 14(2):115–121
Penn RD et al (1989) Intrathecal baclofen for severe spinal spasticity. N Engl J Med 320(23):1517–1521
Peschanski M, Leforestier N, Rapisardi S (1993) Neuroplasticity as a basis for therapeutics in spinal-cord injuries and diseases. Restor Neurol Neurosci 5(1):87–97
Pinter MM, Gerstenbrand F, Dimitrijevic MR (2000) Epidural electrical stimulation of posterior structures of the human lumbosacral cord: 3. Control of spasticity. Spinal Cord 38(9):524–531
Prentice SD (1999) Biological neural networks: hierarchical concept of brain function. Konstantin V Baev 74(2):247–247
Prochazka A et al (1997) The bionic glove: an electrical stimulator garment that provides controlled grasp and hand opening in quadriplegia. Arch Phys Med Rehabil 78(6):608–614
Raineteau O, Schwab ME (2001) Plasticity of motor systems after incomplete spinal cord injury. Nat Rev Neurosci 2(4):263–273
Rejc E et al (2017a) Motor recovery after activity-based training with spinal cord epidural stimulation in a chronic motor complete paraplegic. Sci Rep 7(1):13476
Rejc E et al (2017b) Effects of stand and step training with epidural stimulation on motor function for standing in chronic complete paraplegics. J Neurotrauma 34(9):1787–1802
Remy-Neris O et al (1999) Effects of intrathecal clonidine injection on spinal reflexes and human locomotion in incomplete paraplegic subjects. Exp Brain Res 129(3):433–440
Richardson RR, McLone DG (1978) Percutaneous epidural neuro-stimulation for paraplegic spasticity. Surg Neurol 9(3):153–155
Richardson RR et al (1979) Percutaneous epidural neurostimulation in modulation of paraplegic spasticity. Six case reports. Acta Neurochir (Wien) 49(3–4):235–243
Rosenzweig ES et al (2010) Extensive spontaneous plasticity of corticospinal projections after primate spinal cord injury. Nat Neurosci 13(12):1505–1510
Rosenzweig ES et al (2019) Chondroitinase improves anatomical and functional outcomes after primate spinal cord injury. Nat Neurosci 22(8):1269–1275
Rossignol S, Dubuc R, Gossard JP (2006) Dynamic sensorimotor interactions in locomotion. Physiol Rev 86(1):89–154
Roy RR, Acosta L Jr (1986) Fiber type and fiber size changes in selected thigh muscles six months after low thoracic spinal cord transection in adult cats: exercise effects. Exp Neurol 92(3):675–685
Roy RR et al (1998) Training effects on soleus of cats spinal cord transected (T12–13) as adults. Muscle Nerve 21(1):63–71
Roy RR et al (1999) Differential response of fast hindlimb extensor and flexor muscles to exercise in adult spinalized cats. Muscle Nerve 22(2):230–241
Sagi Y et al (2012) Learning in the fast lane: new insights into neuroplasticity. Neuron 73(6):1195–1203
Sangari S, Perez MA (2019) Imbalanced corticospinal and reticulospinal contributions to spasticity in humans with spinal cord injury. J Neurosci 39(40):7872–7881
Sangari S et al (2019) Residual descending motor pathways influence spasticity after spinal cord injury. Ann Neurol 86(1):28–41

Sangari S et al (2021) Distinct patterns of spasticity and corticospinal connectivity following complete spinal cord injury. J Physiol

Santamaria AJ et al (2020) Neurophysiological changes in the first year after cell transplantation in sub-acute complete paraplegia. Front Neurol 11:514181

Schleip R et al (2006) Passive muscle stiffness may be influenced by active contractility of intramuscular connective tissue. Med Hypotheses 66(1):66–71

Shealy CN, Mortimer JT, Reswick JB (1967) Electrical inhibition of pain by stimulation of the dorsal columns: preliminary clinical report. Anesth Analg 46(4):489–491

Sherrington CS, Laslett EE (1903) Observations on some spinal reflexes and the interconnection of spinal segments. J Physiol 29(1):58–96

Sherwood AM, Dimitrijevic MR, McKay WB (1992) Evidence of subclinical brain influence in clinically complete spinal-cord injury—discomplete SCI. J Neurol Sci 110(1–2):90–98

Stauffer ES (1975) Spinal cord injuries. Clin Orthop Relat Res 112:2–3

Stein RB, Mushahwar V (2005) Reanimating limbs after injury or disease. Trends Neurosci 28(10):518–524

Tabary JC et al (1972) Physiological and structural changes in the cat's soleus muscle due to immobilization at different lengths by plaster casts. J Physiol 224(1):231–244

Tanadini LG et al (2014) Identifying homogeneous subgroups in neurological disorders: unbiased recursive partitioning in cervical complete spinal cord injury. Neurorehabil Neural Repair 28(6):507–515

Tigchelaar S et al (2019) MicroRNA biomarkers in cerebrospinal fluid and serum reflect injury severity in human acute traumatic spinal cord injury. J Neurotrauma 36(15):2358–2371

Topka H et al (1991) Reorganization of corticospinal pathways following spinal cord injury. Neurology 41(8):1276–1283

Vidalsanz M et al (1987) Axonal regeneration and synapse formation in the superior colliculus by retinal ganglion-cells in the adult-rat. J Neurosci 7(9):2894–2909

Wagner FB et al (2018) Targeted neurotechnology restores walking in humans with spinal cord injury. Nature 563(7729):65–71

Wang X et al (2020) Nogo receptor decoy promotes recovery and corticospinal growth in non-human primate spinal cord injury. Brain 143(6):1697–1713

Weidner N et al (2001) Spontaneous corticospinal axonal plasticity and functional recovery after adult central nervous system injury. Proc Natl Acad Sci U S A 98(6):3513–3518

Weiss P (1941) Autonomous versus reflexogenous activity of the central nervous system. Proc Am Phil Soc 84(1):53–64

Wernig A et al (1995) Laufband therapy based on rules of spinal locomotion is effective in spinal-cord injured persons. Eur J Neurosci 7(4):823–829

Wirth ED 3rd et al (2001) Feasibility and safety of neural tissue transplantation in patients with syringomyelia. J Neurotrauma 18(9):911–929

Wirz M et al (2005) Effectiveness of automated locomotor training in patients with chronic incomplete spinal cord injury: a multicenter trial. Arch Phys Med Rehabil 86(4):672–680

Wolpaw JR (2018) The negotiated equilibrium model of spinal cord function. J Physiol 596(16):3469–3491

Yokoyama H et al (2016) Distinct sets of locomotor modules control the speed and modes of human locomotion. Sci Rep 6:14

Zhang J et al (2014) V1 and v2b interneurons secure the alternating flexor-extensor motor activity mice require for limbed locomotion. Neuron 82(1):138–150

Zholudeva LV et al (2018) The neuroplastic and therapeutic potential of spinal interneurons in the injured spinal cord. Trends Neurosci 41(9):625–639

Neurostimulator for Hippocampal Memory Prosthesis

Sahar Elyahoodayan, Wenxuan Jiang, Huijing Xu, and Dong Song

1 Introduction

The pursuit of brain augmentation requires advanced neural interface technology. In recent years, neural interface technology has been deployed for basic neuroscience research and for providing therapies to patients with neurological damages or diseases. Researchers have been using neural interface as a tool to record and manipulate neural circuits to study neural correlates of sensory, motor, and cognitive functions (Nimmagada and Weiland 2018; Brecht et al. 2004; Salas et al. 2018). In the clinic, deep brain stimulation (DBS) has provided treatments to various neurological disorders such as epilepsy (Lee et al. 2015), depression (Schläpfer and Kayser 2014), Parkinson's disease (Voges et al. 2007), memory loss (Hamani et al. 2008), and Tourette's syndrome (Ackermans et al. 2006). In neural prosthesis applications, neural interface has been used to convert sensory input signals to neural stimulations as in cochlea prostheses (Holden et al. 2013) and retinal prostheses (Weiland et al. 2011), or decode motor cortical output signals into movements as in motor prostheses (Velliste et al. 2008).

Hippocampal memory prosthesis, an example of brain augmentation, is a novel form of neural prosthesis that aims to restore cognitive functions lost in injures or diseases due to destruction of neurons and their connections in a specific region of

S. Elyahoodayan (✉) · W. Jiang · H. Xu
Department of Biomedical Engineering, University of Southern California,
Los Angeles, CA, USA
e-mail: elyahood@usc.edu

D. Song (✉)
Department of Biomedical Engineering, University of Southern California,
Los Angeles, CA, USA

Neuroscience Graduate Program, University of Southern California, Los Angeles, CA, USA
e-mail: dsong@usc.edu

I. Opris et al. (eds.), *Modern Approaches to Augmentation of Brain Function*,
Contemporary Clinical Neuroscience,
https://doi.org/10.1007/978-3-030-54564-2_3

the brain (Berger et al. 2011). It relies on a computational model that mimics the nonlinear dynamical multi-input, multi-output (MIMO) properties of the neural circuit to be replaced (Song et al. 2007, 2018). The MIMO model enables the prosthesis to stimulate a downstream brain region with appropriate output spatiotemporal patterns of neural codes predicted from input spatiotemporal patterns of neural activities recorded from an upstream brain region (Berger et al. 2010; Song et al. 2009). By reinstating the neural code processing and transmission, the damaged brain region is bypassed, and the cognitive function is thus restored (Hampson et al. 2012, 2013; Berger et al. 2011; Deadwyler et al. 2018). To implement a hippocampal memory prosthesis, it is essential to be able to deliver temporally and spatially distributed neural code–based stimulation patterns to the brain tissue with multiple electrodes.

Existing neural interface technologies for brain implantation, as in DBS, utilize fixed interval trains of pulses, with a single or small number of stimulation electrodes. However, to enable neural code-based stimulation, experimentalists need a system for real-time precise delivery of large-scale spatiotemporal patterns of electrical pulses. In addition, the ability to stimulate and record from the same electrode is essential for mimicking the neural code (Fig. 1). This capability would maximize the number of electrodes for recording or stimulation, provide feedback from stimulated tissue for validating stimulation effects and optimizing stimulation parameters, and enable building single neuron-level MIMO model.

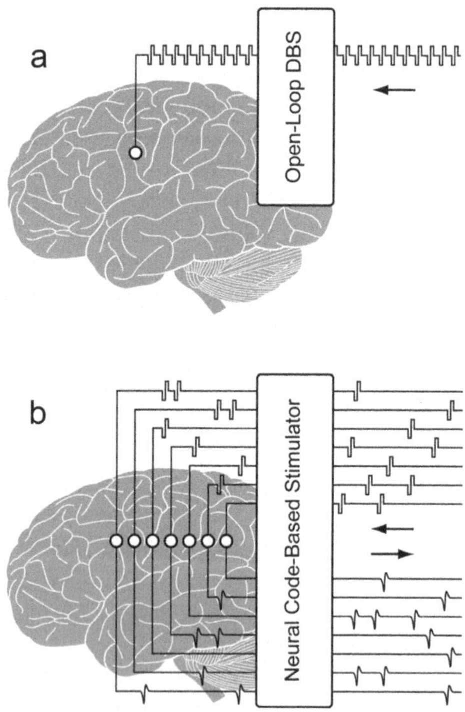

Fig. 1 Illustration of (**a**) a conventional open-loop DBS and (**b**) a configurable multichannel neurostimulator with stimulus artifact suppression for implementing closed-loop processing from the region of stimulation

The main challenge of generating large-scale stimulation pulses is hardware efficiency. Power and area are crucial parameters of an implantable system and apparent solution of separate pulse generators for each electrode is inefficient for a hippocampal memory prosthesis. This is because the natural rate of neural activity is slow, so each pulse generator would need to spend a major amount of time in quiescent mode, which stills consumes power. Whereas time multiplexing a single pulse generator into multiple channels enables combining several current sources into a single high rate pulse generator. Furthermore, neural spiking activity is sparse in nature and simultaneous pulse generation from multiple electrodes would not result in sparse neural code. On the other hand, asynchronous low-magnitude stimulation pulsing has the potential of generating sparse neural code.

The main challenge of recording in conjunction with stimulation is due to the prolonged saturation of the recording amplifier caused by stimulus artifacts, which masks neural activities of interest. Stimulus artifact may last for tens of milliseconds or even hundreds of milliseconds depending on the amplifier, electrode property, tissue property, and stimulus parameters (Rolston et al. 2009). Minimizing such recording contaminants is vital for better recording evoked neural activities and controlling neural interface devices.

In this chapter, the reader will learn how to design a highly configurable asynchronous multichannel neurostimulator that can be driven by a MIMO model–based computational unit to continuously generate neural code–like spatiotemporal patterns of stimulation pulses with adjustable pulse parameters in real time. In addition, generating stimulation pulses in real-time driven by an external source (i.e., the output of the MIMO model) and feedback from neural response to stimulation from the region of stimulation is outlined. For closed-loop applications, each stimulation channel can be equipped with a switching mechanism designed to reduce recovery period from the artifact. The reader will learn about the design, fabrication, and characterization of such system in phantom preparations and then evaluation in the hippocampi of rats. Electrode design and characterization is also briefly described for same electrode stimulation and recording to allow recording single-unit and multi-unit activity and also flexibility of charge storage capacitance to enable proper adjustment of pulse parameters.

2 Materials and Design

The principal elements of the design include a stimulation pattern generator, a multiplexer, a microprocessor-based controller, and a set of serially controlled CMOS switches for stimulus artifact suppression (SAS).

2.1 Stimulation Pattern Generator Circuit

The stimulator consists of a configurable constant current biphasic and monopolar waveform generator and a pattern generator independently specifying spatiotemporal timings and magnitudes of pulses across 32 stimulating electrodes. First, you need to design and test a single-channel configurable current source capable of providing charge-balanced biphasic pulses as follows. An adjustable output voltage is required to drive an adjustable constant current source. One way to do this is to use a low-power microcontroller (MSP430G2553, Texas Instrument) and program it to generate a 40-kHz pulse-width modulator (PWM). Next, a simple op-amp integrator with a cut-off frequency of 4 kHz should be designed to average the PWM to output a negative DC voltage. By adjusting the duty cycle of the PWM, the output DC voltage would change accordingly. For example, if the duty cycle is 50%, the output DV voltage would be Vdd (supply voltage)/2.

Next, the output is converted to a positive DC voltage using an inverting amplifier with a gain of −1. After generating the desired positive and negative DV voltage, you need to dictate the polarity and duration of each to generate biphasic pulses. To do this, you need to generate three other signals from the microcontroller to drive analog switches such as (TS5A22362) to pass positive, negative, or ground. In this way, the system can be programmed to generate cathodic or anodic first stimulation pulses across individual channels. This choice is dependent on the region of the brain being stimulated as one waveform would manifest a lower threshold than the other.

Next, to generate constant current biphasic pulses, an op-amp-based current source you want to design an op-amp-based current source (Howland circuit is commonly used) to convert the output voltage biphasic pulses to constant current biphasic pulses. To expand a single channel to 32 channels, you can then feed the output of the current pulse into a multiplexer (Fig. 2). It is important to place a DC blocking capacitor at the input of the multiplexer to block input offset. Furthermore, since the application of the design requires sparse and low pulse rate, charge build-up is not expected as the electrode should be shorted to ground after each stimulation pulse.

To power the circuit, you need to look at power requirements by each chip provided in its respective datasheet and does not exceed the maximum supply voltage. For the parts provided here, you should use two 3.5-V coin batteries connected in series to obtain ±3.5 V. Here, the absolute maximum supply rating is determined by the microcontroller, which is +4.1 V. Other chips have a maximum voltage rating of ±5 V. As this system is battery operated, voltage regulators are required. To minimize hardware design for possible future miniaturization of the PCB for chronic applications, voltage regulators can be eliminated since all chips can operate at a minimum voltage of ±3 V. To ensure the output DC voltage from the PWM stays constant even with voltage supply drop, the microcontroller should be programmed to sample the supply voltage and adjust the PWM duty cycle according to Eq. (1):

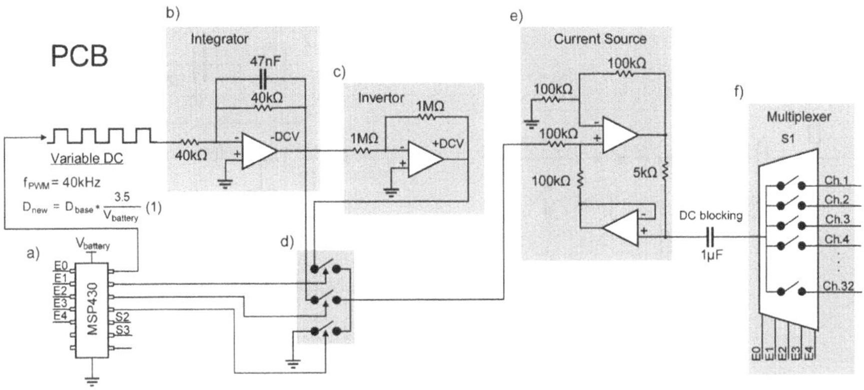

Fig. 2 Design of a 32-channel configurable constant current biphasic neurostimulator. It includes (**a**) a microcontroller (MSP430), which generates a PWM at a 40-kHz frequency and an adjustable duty cycle to configure the stimulus amplitude. The duty cycle is also automatically adjusted as the battery supply voltage drops, governed by Eq. (1), avoiding the need for a regulator. PWM then goes through (**b**) an integrator to generate a constant −DCV, (**c**) an inverter to generate a +DCV, (**d**) a set of analog switches dictating polarity and duration of the pulse controlled by the microcontroller, (**e**) a voltage-to-current converter and a DC-blocking capacitor to generate safe single-channel constant current biphasic pulses. (**f**) Lastly, a multiplexer is used to expand the design to 32 channels

$$D_{\text{new}} = D_{\text{base}} * \left(\frac{3.5}{V_{\text{battery}}} \right), \quad (1)$$

where D_{new} is the adjusted duty cycle; D_{base} is the target duty cycle for when the battery voltage is 3.5 V; and V_{battery} is the voltage of the battery at a time point.

2.2 *Stimulus Artifact Suppression Technique*

It is assumed that you will have a recording system suitable for recording in vivo neural signals. Typical neural recording systems consist of small-signal (typically 10 μV–10 mV) voltage amplifiers with adjustable gain and bandpass filters. Beyond power supply voltage after amplification, the amplifier cannot produce amplification of the input signal. Not only does a large signal cause signal distortion and loss of neural signal, the excess power transfer may damage the amplifier over time. A stimulation pulse applied to an electrode is a large signal and when used in conjunction with a neural recording system, it produces high-amplitude artifacts with long recovery period caused by amplifier saturation and filter ringing.

One way to prevent blockade of neural data by stimulus artifact is to use CMOS switches (ADG714) to block the stimulation current from transmitting to the recording system. You need to connect each switch between the electrode and the recording

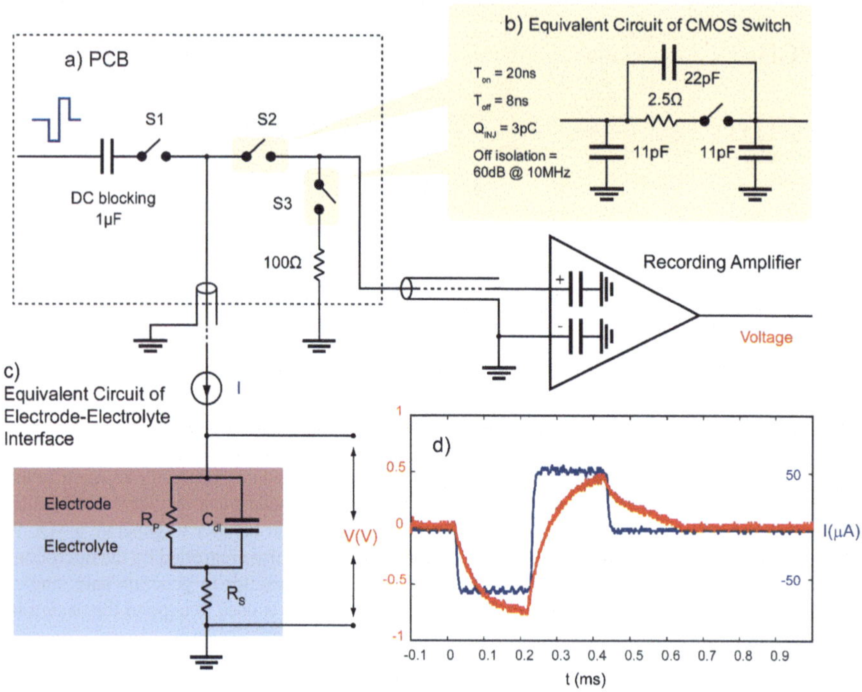

Fig. 3 Stimulation artifact suppression setup. (**a**) Stimulation is synchronized with a set of serially controlled CMOS switches (S2) to block out the stimulus from the recording amplifier during stimulation while connecting the amplifier input to a 100-Ω resistor via S3 to prevent ringing caused in the recording system due to any extra charge coupled across S2. Charge coupling across S2 is due to (**b**) parasitic components of analog switches. The PCB is connected to the recording amplifier and (**c**) the electrode using a coaxial cable. The timing of the switches in (**a**) is determined by (**c**) the electrochemical properties of the electrode, and (**d**) the voltage transient across a microelectrode (red) in response to a biphasic cathodic first current pulse (blue). (**c**) R_S represents the electrolyte resistance, R_P represents faradaic reaction, and C_{dl} represents capacitive reactions at the interface

system and synchronize it with the stimulator to be triggered a short time before and after the stimulation pulse. During this time, the switches connecting the electrodes to the recording module (S2) should be kept open and the switches connecting the stimulator to the same electrodes (S1) should be closed (Fig. 3a).

One challenge to using CMOS analog switches to block the stimulus from the recording system is they contain parasitic components that affect the AC performance of the device (Fig. 3b). This means that part of the stimulus may still couple from the source to the drain of S2 during stimulation. S2 acts to minimize charge coupling into the amplifier, but since the amplifier is typically set to a gain of at least 60 dB, even small voltages may generate long contaminated signals.

To dissipate the charge coupled across the switch during stimulation, a 100-Ω resistor may be used at the input of the amplifier during stimulation to suppress ringing. This resistor is also used to absorb any instantaneous charge coupled from the electrode to the amplifier when the electrode is reconnected to the amplifier. The resistor should be disconnected from the amplifier when recording is resumed using another CMOS switch (S3) (Fig. 3a). The timing of the switches depends on the shape of the voltage transient waveform across the electrode in response to a given stimulation pulse and is determined by the electrochemical properties of the interface as described in the next section.

2.3 Electrochemical Properties

The shape of the voltage transient across a microelectrode is a factor of the electrochemical processes at the interface, which can be estimated with an electrical equivalent circuit model. The model consists of an electrolyte resistance (R_s) in series with the parallel combination of a double-layer capacitance (C_{dl}) and an impedance of faradaic reactions (R_P) (Fig. 3c) (Conway 1991). Equation (2) describes the relationship between an applied constant current pulse (I) and the resulting voltage (V) across the interface:

$$V(t) = IR_s + IR_p\left(1 - e^{-\frac{t}{R_P C_{dl}}}\right) \tag{2}$$

An example of the voltage transient in response to a stimulation current pulse of 60 μA, 200 μs is shown in Fig. 3d.

In this model, capacitive charge injection represented by C_{dl} involves physical absorption and desorption of ions in an electrolyte. Faradaic reaction represented by R_P involves local donation of electrons through oxidation and reduction reactions, which is less desirable than a capacitive process since it involves formation of new species (Cogan 2008). Both charge injection mechanisms may be involved during stimulation (irreversible faradaic reactions are to be avoided) (Merrill et al. 2005). Thus, the time constant of the electrode is governed by:

$$\tau = R_p * C_{dl} \tag{3}$$

After the termination of the stimulation pulse, the electrode is left with an initial polarization voltage (V_{p-i}) and takes several τ's to reach to a value close to its initial bias level (V_b) (Fig. 4). V_{p-i} is defined as:

$$V_{p\_i} = V_b + (\Delta V - V_a), \tag{4}$$

where V_a is the instantaneous voltage drop after the termination of the current pulse and ΔV is the maximum voltage the electrode reaches at the end of the pulse (Bard and Faulkner 2001). V_b is typically a few millivolts with respect to a large return electrode of the same material, as is the case of electrophysiology experiments. It is

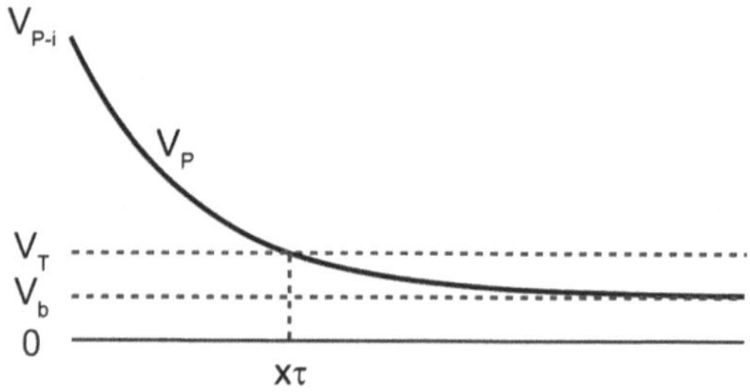

Fig. 4 Polarization voltage (V_P) across the electrode with respect to a reference of the same material after termination of the stimulation pulse with an initial value of V_{P-i}. The voltage must drop below V_T before recording is resumed, to avoid amplifier saturation, which takes a time duration of multiple factors of τ represented as $x\tau$

often difficult to identify V_a in a voltage transient plot; thus, V_{p_i} cannot be accurately determined. Alternatively, V_{P_max}, the maximum polarization voltage across an electrode to avoid potential exertion beyond the water window, may be used as V_{p-i}. This is a worst-case scenario. This value is experimentally determined and is electrode material dependent (Cogan 2008).

It is important that the polarization voltage on the working electrode (V_p) with respect to a reference of the same material, drops to below a threshold before recording is resumed to avoid saturation of the amplifier. This threshold is governed by the settings on the recording amplifier such that:

$$V_p < \frac{V_{\text{max-amp}}}{\text{Gain}} = V_T, \tag{5}$$

where $V_{\text{max-amp}}$ is the maximum output voltage from the recording amplifier, and Gain is the gain of the amplifier.

After the termination of the stimulation pulse, the electrode will take several time constants ($x\tau$) to approach to below V_T, which dictates the time duration the electrode must stay disconnected from the amplifier. The timing and state of the switches are shown in Fig. 5. (1) The stimulator and the resistor at the input of the amplifier are disconnected while the electrode is connected to the amplifier. This is the recording phase. (2) 200 μs before the initiation of the stimulation pulse, S1 and S3 close. (3) S2 then opens to disconnect the recording system and the stimulation pulse is applied to the electrode. (4) The switches stay in that state for the duration of the pulse plus a predefined $x\tau$. (5) S1 then disconnects the electrode from the stimulator and S2 reconnects the electrode to the amplifier while the input of the amplifier is still connected to ground through the 100-Ω resistor to absorb any instantaneous charge injected from the electrode. (6) 200 μs later, the resistor is then disconnected by S3 and recording is resumed.

It is important to note that the electrochemical properties of an electrode must be characterized for different media to determine different time constants of the electrode-electrolyte. Factors that affect the value of the time constant include

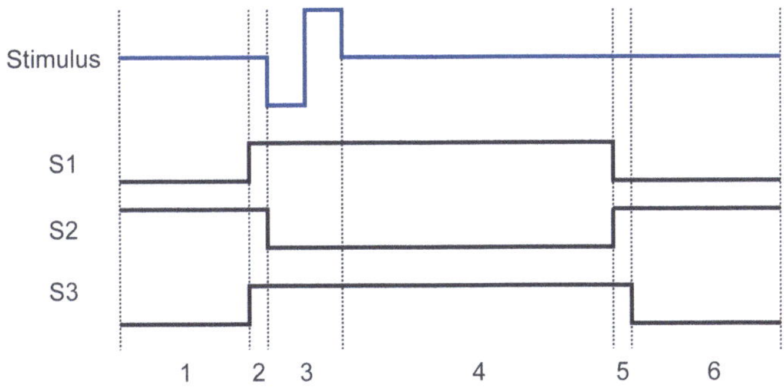

Fig. 5 The state of the switches (1 = closed; 0 = open) shown in Fig. 3a with respect to the stimulation pulse. In all cases when the stimulator is not pulsing, it is connected to ground. States 1 to 6 are as follows: (1) Recording: The electrode is connected to the recording amplifier while the stimulator and the resistor are disconnected. (2) Grounding: 200 μs before stimulation, the stimulator is connected to the electrode and the resistor is connected to the input of the amplifier. (3) Stimulation: The amplifier is disconnected from the electrode. The stimulation pulse is then applied to the electrode. (4) Discharge period: The electrode stays disconnected from the amplifier until the polarization voltage on the electrode falls below V_T. (5) Discharge of residual charge (200 μs): The electrode is disconnected from the stimulator and connected to the recording system. The resistor remains connected to the input of the amplifier to absorb any charge injection due to residual offset on the electrode. (6) Recording resumed: The termination resistor is disconnected, and recording is resumed

electrode material, geometric and effective area of exposed region, and the electrolyte impedance. Thus, an electrode should be separately characterized for in vivo or in vitro experiments.

2.4 System Architecture

Figure 6 illustrates the block diagram and data stream of the neurostimulator with the stimulus artifact suppression technique described earlier. You should run the microcontroller clock frequency at 10 MHz or above. The block diagram shows 4 blocks of control units: MIMO model simulator, pulse pattern generation control unit, multiplexer control unit, and switch control unit. To test the functionality of the system, the MIMO model simulator simulates inputs from an external source such as the output of the MIMO model. You may simulate the output of the MIMO model using a random number generator (RNG), generated by the microcontroller through recording noise from a floating general-purpose pin. Furthermore, you may use another set of random numbers to vary the magnitude of the stimulus simulating feedback from neural response to stimulation. Once this information is generated, you should collapse the timing and magnitude information across 32 channels into a single array in embedded systems software. In case two channels need to be stimulated simultaneously (which is a rare event), one may be delayed by the duration of

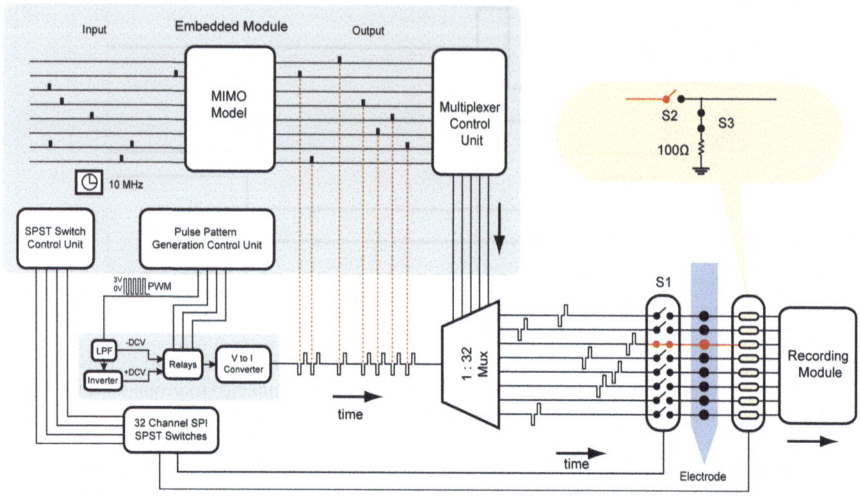

Fig. 6 Block diagram and data stream of the multichannel neurostimulator with stimulus artifact suppression (SAS). The hardware is programmed by the embedded system to generate highly configurable constant current monopolar biphasic pulses. Each block in the embedded system represents an algorithm to control the hardware. It consists of a MIMO nonlinear dynamical model simulator, using a random number generator. The data are inputted to a multiplexer control unit to set the appropriate select lines of the multiplexer. The multiplexer control unit is synchronized with the SPI switch control unit to control CMOS switches

the biphasic pulse. The role of the pulse pattern generation unit is to translate the information into commands controlling the single-channel neurostimulator to generate biphasic pulses with varying timing and magnitude. Next, you should program a multiplexer control unit to activate proper select lines of the multiplexer to send each stimulus to the target channel.

Lastly, the switch control unit, synched with the MIMO model simulator, should be programmed to control timing and state of each switch used to suppress the stimulus artifact. The switches for each channel may be controlled individually, but because the stimulus artifact will contaminate all channels within the same media, the stimulation should be blocked from all recording channels.

2.5 *Power Calculation*

The total power dissipation is dependent on the load current and the quiescent current. The output load current is dependent on the driving load (the electrode) and the stimulus waveform (pulse amplitude, pulse duration, and pulse rate), which is a variable defined by the user. Quiescent power consumption is the product of current drawn by the supply (Icc) and supply voltage (Vcc). All active parts used here except for the microcontroller and the multiplexer have a combined quiescent power dissipation of 815 μW. The multiplexer consumes 60 μW. Thus, without multiplexing the system power consumption would be 815 μW*32 = 26 mW, whereas multiplexing reduces this number to 875 μW. The microcontroller power consumption is dependent on usage of general-purpose input output pins. To generate arbitrary pulse patterns in real time through a single channel, 4 pins are required by the pulse pattern generation control unit (Fig. 6). Without multiplexing, the number of required pins would be 32*4 = 128 for 32 channels. Whereas, with multiplexing, this number would be reduced to 9. Thus, multiplexing greatly reduces power consumption and real-estate usage.

2.6 System Cost

The cost to fabricate and assemble the PCB is approximately $100. The minimum requirements are a personal computer, a TI MSP430 launchpad, and the Code Composer Studio to upload the code.

3 Experimental Methods

3.1 Design Characterization

By now, you have programmed the microcontroller to generate random numbers dictating both the timing and amplitude of pulses across each of the 32 channels independently. The range of amplitudes (I_{max}) should be selected based on the user need and is limited to the supply voltage (V_{supply}) and the total impedance of the electrode-tissue interface ($|z|_{electrode}$)

$$I_{max} = \frac{V_{supply}}{|z|_{electrode}} \quad (6)$$

$|z|_{electrode}$ includes the electrolyte resistance plus the polarization impedance across the electrode-electrolyte interface, which is frequency dependent. An electrochemical impedance spectroscopy (EIS) of the electrode should be used to determine $|z|_{electrode}$ at a frequency equal to the inverse of the pulse duration. This frequency is a reasonable approximation for nonsinusoidal pulses.

A multielectrode array of penetrating electrodes suitable for brain implant of rats (or an animal of interest) should be used to evaluate the system's capability to minimize stimulus artifact. The microelectrode should first be characterized by its electrochemical properties, namely, the magnitude of each element in the equivalent circuit model shown in Fig. 3c, its charge storage capacity and its τ using Cyclic voltammetry (CV), and EIS performed by Gamry Reference 600 potentiostat (Gamry Instruments, Warminster, PA). The return electrode made of the same material as the working electrode (WE) but many times larger in area (WE) should be used.

You should first perform stimulation and recording experiments in a phantom to compare the artifact with and without the SAS component. Brain tissue impedance is approximately 0.25S/m (Kandadai et al. 2012). To mimic this impedance 1/6 diluted PBS can be used. To mimic neural activity, a known input of 1-kHz sinusoidal signal can be applied to another microelectrode in the same solution. 1 kHz is chosen because it is within the spectral range of single-unit activity. Also, this frequency over the frequency of evoked potentials provides better visualization for earliest point at which biological signals may be recovered from the stimulus artifact (Fig. 7). The electric field from this electrode to the recording electrode would be attenuated due to distance and electrolyte impedance, so choose a peak-to-peak amplitude that would result in 100 μV to 1 mV. Attenuation is also the case for in vivo recording as the source of an action potential is in millivolts and distance and tissue impedance from the neuron to the recording electrode results in recordings of few hundred microvolts. Next, you need to measure the time duration when no sinusoidal signal can be recorded due to the artifact. To record LFP, single-unit and multi-unit activities, you need to set the recording system to at least 60-dB gain and a 10 Hz–10 kHz band pass filter.

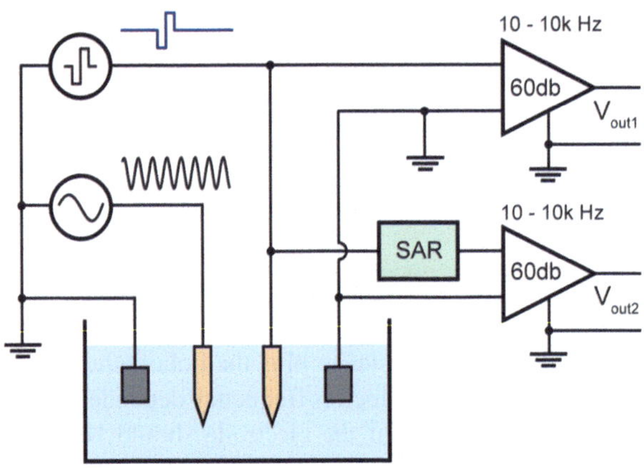

Fig. 7 Test bench setup for evaluating the stimulus artifact with (V_{out2}) and without (V_{out1}) the SAS technique. The setup consists of diluted PBS mimicking neural tissue and a sinusoidal signal (1 kHz, 10 mV peak to peak) applied to a neighboring electrode mimicking neural activity

3.2 *In Vivo Evaluation*

To test this system in vivo, the reader may implant the electrode of their choice in anesthetized rat hippocampus and stimulate through one electrode and record from the same and neighboring electrodes with increasing stimulation amplitudes. You may record neural responses with and without the SAS component for comparison. Keep in mind that a recovery period of more than 1-min should be allowed for neural tissue to return to baseline.

Example of short latency extracellular evoked responses obtained from the CA1 cell body layers following stimulation is shown in Fig. 8. A total of 80 response curves are generated to monitor changes 2.5 ms following the initiation of 10 separate stimulation pulses across a coated electrode and recording across all 8 electrodes. The results in Fig. 8 can be separated into two categories depending on whether only population spikes were potentiated or population spikes plus EPSP were potentiated. At low amplitudes (1–5 μA), population spikes are potentiated in the absence of EPSP. At amplitudes above 10 μA, potentiation of population spikes was always accompanied by potentiation of EPSP. When only population spike was potentiated, there is almost no change in the recorded amplitude with increasing stimulation amplitude. In contrast, when EPSP plus population spikes were potentiated, the calculated amplitude increased with the stimulus magnitude and saturated at 40 μA.

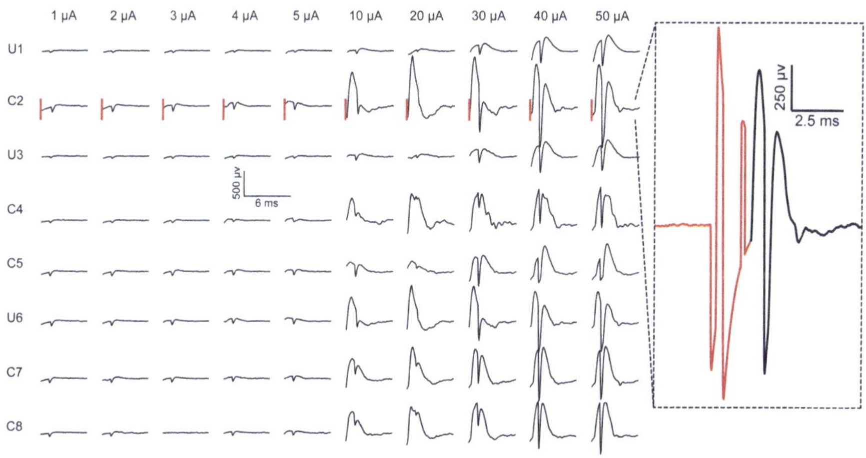

Fig. 8 Electrically evoked fEPSP and PS recorded from all eight electrodes in response to biphasic current pulsing through electrode # 2 in anesthetized rat CA1 region of hippocampus. "U" and "C" written on the left side of the plot represent the uncoated and coated electrodes respectively followed by the electrode number (electrode configuration as sketched in Figure 1a). The vertical lines to the left of each response for C2 represent a time 2.5 ms after the initiation of the stimulation pulse (pulse amplitude labelled above each line). Stimulus pulse amplitudes of 1 μA to 5 μA induced potentiation of PS without fEPSP. Pulse amplitudes of 10 μA to 50 μA cause potentiation of PS accompanied with fEPSP. An example of stimulus artifact (*red*) followed by neural response (EPSP and PS) is presented on the right

4 Notes

In this chapter, the reader learned how to design and test a versatile and cost-efficient neurostimulator that can deliver precise spatiotemporal patterns of stimulation pulses with arbitrary magnitudes and intervals through 32 channels continuously and in real time. This stimulator can be controlled by an external source such as a MIMO nonlinear dynamical model to achieve localized and patterned microstimulation to the brain. The design also consists of a stimulus artifact suppression component that allows stimulation and recording from the same electrodes with a very short delay. The reader also learned how to systematically test this stimulator in a phantom preparation and then in anesthetized animals. Neural responses evoked by microstimulations are characterized and compared with and without the stimulus artifact suppression technique to show its efficacy.

Highly configurable neurostimulators, and the ability to recording neural responses after stimulation is critical for implementing closed-loop neuromodulation or hippocampal memory prosthesis where consistent neural responses are desirable but often difficult to maintain. For example, variations in neural response may occur weeks or months after implantation due to inflammation. Inflammation causes glial cell encapsulation around the electrodes and thus weakens the neuron-electrode interaction, e.g., reduction of the stimulation effect and recorded signals (Polikov et al. 2005). Furthermore, neural plasticity may also contribute to variations of neural responses as the underlying neural circuits are constantly altered by behaviors (Månsson et al. 2016; Kerr et al. 2011). These variations can be compensated by adjusting stimulation parameters based on the feedback signals provided by the recording electrodes. The design here enables future studies to explore this possibility.

This neurostimulator is particularly suitable for building hippocampal memory prosthesis. In a hippocampal memory prosthesis system, spatiotemporal patterns of stimulation to a downstream brain region are calculated based on the ongoing spatiotemporal patterns of neural activities in an upstream brain region using a predictive MIMO nonlinear dynamical model. The stimulation patterns mimic the endogenous neural signals, which intrinsically are sparse, asynchronous and involve multiple channels. The neurostimulator provides away for delivering such patterns when connected to the output of a computational unit that contains the MIMO model.

A key feature of this design is the use of a multiplexer to save power and real estate to handle large numbers of electrodes. Higher channel counts are achievable with simple hardware and software modifications, with a complexity that scales sublinearly with the channel counts. Future work includes a study to determine an optimum input-output ratio of the multiplexer for a hippocampal memory prosthesis. Another feature of the design is that it uses off-the-shelf components making it a reproducible and inexpensive design for use as a neuroscience tool, which may be easily miniaturized for mounting on rat's skull.

Stimulation and recording from the same electrode is another key feature required by a hippocampal memory prosthesis. The challenge in realizing this feature is imposed by direct connection of the stimulator and amplifier through a recording electrode. If a stimulation pulse causes amplifier saturation, the input signal would be clipped at the amplifier's maximum input range. Consequently, neural response would be completely masked with this artifact and cannot be recovered. On the other hand, if recording is from a neighboring electrode to the stimulation electrode, or if the recording is from the stimulation electrode but the applied stimulus magnitude is small enough, the amplifier may not get saturated. In this case, often back-end signal processing may be used to separate artifact from neural response (Zhou et al. 2018).

Thus, a fundamental limitation of back-end signal processing approach is that it relies on unsaturated recordings of neural signals and stimulus artifacts, which are often unavailable due to the commonly encountered saturation of recording amplifiers. Even back-end signal processing of unsaturated recordings such as the ones proposed by Wagenaar et al. (Wagenaar and Potter 2002) and by Wichmann et al. (Wichmann 2000) faces difficulties in separating neural activity from stimulus artifact due to their overlap in both time and frequency domains. More recently, Limnuson et al. (2015) developed a more sophisticated real-time artifact suppression technique based on template subtraction, which requires a VLSI chip.

To avoid amplifier saturation, front-end artifact reduction has been implemented, which typically involves increasing the dynamic range of the amplifier to withstand larger voltages. This approach sacrifices power efficiency by using a higher voltage supply (Rolston et al. 2009) and still requires back-end signal processing to reduce the stimulus artifact. Another front-end approach is to subtract the artifact at the negative input of the amplifier based on a model that replicate the electrode-tissue properties (Nag et al. 2015). This technique holds promise but involves relatively complex computation and is to be validated in biological preparations.

Blanking techniques similar to our system have previously been used to reduce the artifact recorded from nonstimulation electrodes, where residual charge left on the electrode after termination of the stimulation pulse does not need to be accounted for (Venkatraman et al. 2009; Cheng et al. 2017). The system that accounts for the discharge period of the stimulation electrode for same electrode stimulating and recording is by DeMichele et al. (DeMichele and Troyk 2003). This system does not demonstrate data from bench-top system or neural tissue. Hottowy et al. (2012) designed a system with capability to generate spatiotemporal pattern of stimulation and stimulus artifact reduction. The range of stimuli used in this system to test the artifact rejection technique in vitro are low (0.43 μA, 100 μs), which did not cause amplifier saturation.

The system is capable of recording and stimulating from the same electrode when a large stimulus of 60 μA, 200 μs is applied to the electrode. A large stimulus saturates the amplifier and causes ringing for tens to hundreds of milliseconds depending on the time constant of the electrode and the input impedance of the amplifier. To minimize charge coupling between the stimulator and the amplifier, prevent ringing, and allow discharge of the stimulation electrode, we describe the electrochemical characterization of the electrode to determine the timing of the 3 switches illustrated in Fig. 5. It is our hope that this system to be a valuable tool for studying neurobiological basis of cognitive functions and a critical component for building cortical prostheses for restoring and enhancing cognitive functions.

Acknowledgments This work was supported by the National Science Foundation (INSPIRE, CBET-1343193).

References

Ackermans L, Temel Y, Cath D, van der Linden C, Bruggeman R, Kleijer M, Nederveen P et al (2006) Deep brain stimulation in Tourette's syndrome: two targets? Mov Disord 21(5):709–713. https://doi.org/10.1002/mds.20816

Bard AJ, Faulkner LR (2001) Electrochemical methods: fundamentals and applications, 2nd edn. Wiley, New York

Berger TW, Song D, Chan RHM, Marmarelis VZ (2010) The neurobiological basis of cognition: identification by multi-input, multioutput nonlinear dynamic modeling. Proc IEEE 98(3):356–374. https://doi.org/10.1109/JPROC.2009.2038804

Berger TW, Hampson RE, Dong S, Goonawardena A, Marmarelis VZ, Deadwyler SA (2011) A cortical neural prosthesis for restoring and enhancing memory. J Neural Eng 8(4):046017. https://doi.org/10.1088/1741-2560/8/4/046017

Brecht M, Schneider M, Sakmann B, Margie TW (2004) Whisker movements evoked by stimulation of single pyramidal cells in rat motor cortex. Nature 427(6976):704–710. https://doi.org/10.1038/nature02266

Cheng C, Tsai P, Yang T, Cheng W, Yen T, Luo Z, Qian X et al (2017) A fully integrated closed-loop neuromodulation SoC with wireless power and bi-directional data telemetry for real-time human epileptic seizure control. In: 2017 Symposium on VLSI Circuits, pp C44–C45. https://doi.org/10.23919/VLSIC.2017.8008541

Cogan SF (2008) Neural stimulation and recording electrodes. Annu Rev Biomed Eng 10(1):275–309. https://doi.org/10.1146/annurev.bioeng.10.061807.160518

Conway BE (1991) 'Technical Papers' electrochemical science and technology transition from 'Supercapacitor' to 'Battery' behavior in electrochemical energy storage. 138(6)

Deadwyler SA, Hampson RE, Song D, Robinson BS, Fetterhoff D, Dakos AS, Roeder BM, She X, Wicks RT, Witcher MR, Couture DE, Laxton AW, Munger-Clary H, Popli G, Sollman MJ, Whitlow CT, Marmarelis VZ, Berger TW (2018) Developing a hippocampal neural prosthetic to facilitate human memory encoding and recall. J Neural Eng 15(3):36014. http://stacks.iop.org/1741-2552/15/i=3/a=036014

DeMichele GA, Troyk PR (2003) Stimulus-Resistant Neural Recording Amplifier. In: Proceedings of the 25th Annual International Conference of the IEEE Engineering in Medicine and Biology Society (IEEE Cat. No.03CH37439), vol 4, pp 3329–3332. https://doi.org/10.1109/IEMBS.2003.1280857

Hamani C, McAndrews MP, Cohn M, Michael O, Zumsteg D, Shapiro CM, Wennberg RA, Lozano AM (2008) Memory enhancement induced by hypothalamic/fornix deep brain stimulation. Ann Neurol 63(1):119–123. https://doi.org/10.1002/ana.21295

Hampson RE, Dong S, Chan RHM, Sweatt AJ, Riley MR, Gerhardt GA, Shin DC, Marmarelis VZ, Berger TW, Deadwyler SA (2012) A nonlinear model for hippocampal cognitive prosthesis: memory facilitation by hippocampal ensemble stimulation. IEEE Trans Neural Syst Rehabil Eng 20(2):184–197. https://doi.org/10.1109/TNSRE.2012.2189163

Hampson RE, Dong S, Opris I, Santos LM, Shin DC, Gerhardt GA, Marmarelis VZ, Berger TW, Deadwyler SA (2013) Facilitation of memory encoding in primate hippocampus by a neuroprosthesis that promotes task-specific neural firing. J Neural Eng 10(6):66013. https://doi.org/10.1088/1741-2560/10/6/066013

Holden LK, Finley CC, Firszt JB, Holden TA, Brenner C, Potts LG, Gotter BD et al (2013) Factors affecting open-set word recognition in adults with Cochlear implants. Ear Hear 34(3):342–360. https://doi.org/10.1097/AUD.0b013e3182741aa7

Hottowy P, Skoczeń A, Gunning DE, Kachiguine S, Mathieson K, Sher A, Wiącek P, Litke AM, Dąbrowski W (2012) Properties and application of a multichannel integrated circuit for low-artifact, patterned electrical stimulation of neural tissue. J Neural Eng 9(6):66005. https://doi.org/10.1088/1741-2560/9/6/066005

Kandadai MA, Raymond JL, Shaw GJ (2012) Comparison of electrical conductivities of various brain phantom gels: developing a 'brain gel model. Mater Sci Eng C 32(8):2664–2667. https://doi.org/10.1016/j.msec.2012.07.024

Kerr AL, Cheng S-Y, Jones TA (2011) Experience-dependent neural plasticity in the adult damaged brain. J Commun Disord 44(5):538–548. https://doi.org/10.1016/j.jcomdis.2011.04.011

Lee B, Zubair MN, Marquez YD, Lee DM, Kalayjian LA, Heck CN, Liu CY (2015) A single-center experience with the NeuroPace RNS system: a review of techniques and potential problems. World Neurosurg 84(3):719–726. https://doi.org/10.1016/j.wneu.2015.04.050

Limnuson K, Lu H, Chiel HJ, Mohseni P (2015) A bidirectional neural Interface SoC with an integrated spike recorder, microstimulator, and low-power processor for real-time stimulus artifact rejection. Analog Integr Circuits Signal Process 82(2):457–470. https://doi.org/10.1007/s10470-015-0489-z

Månsson KNT, Salami A, Frick A, Carlbring P, Andersson G, Furmark T, Boraxbekk C-J (2016) Neuroplasticity in response to cognitive behavior therapy for social anxiety disorder. Transl Psychiatry 6(February):e727

Merrill DR, Bikson M, Jefferys JGR (2005) Electrical stimulation of excitable tissue: design of efficacious and safe protocols. J Neurosci Methods 141(2):171–198. https://doi.org/10.1016/j.jneumeth.2004.10.020

Nag S, Sikdar SK, Thakor NV, Rao VR, Sharma D (2015) Sensing of stimulus artifact suppressed signals from electrode interfaces. IEEE Sensors J 15(7):3734–3742. https://doi.org/10.1109/JSEN.2015.2399248

Nimmagada K, Weiland JD (2018) Retinotopic responses in visual cortex elicited by epiretinal electrical stimulation in normal and retinal degenerate rats. Transl Vis Sci Technol 7(5):33

Polikov VS, Tresco PA, Reichert WM (2005) Response of brain tissue to chronically implanted neural electrodes. J Neurosci Methods 148(1):1–18. https://doi.org/10.1016/j.jneumeth.2005.08.015

Rolston JD, Gross RE, Potter SM (2009) A low-cost multielectrode system for data acquisition enabling real-time closed-loop processing with rapid recovery from stimulation artifacts. Front Neuroeng 2(July):12. https://doi.org/10.3389/neuro.16.012.2009

Salas MA, Bashford L, Kellis S, Jafari M, Jo H, Kramer D, Shanfield K et al (2018) Proprioceptive and cutaneous sensations in humans elicited by intracortical microstimulation. Elife 7:1–11. https://doi.org/10.7554/eLife.32904

Schläpfer TE, Kayser S (2014) Deep brain stimulation for treatment-resistant depression. Klinische Neurophysiologie 45(2):113–117. https://doi.org/10.1055/s-0034-1375605

Song D, Chan RHM, Marmarelis VZ, Hampson RE, Deadwyler SA, Berger TW (2007) Nonlinear dynamic modeling of spike train transformations for hippocampal-cortical prostheses. IEEE Trans Biomed Eng 54(6):1053–1066. https://doi.org/10.1109/TBME.2007.891948

Song D, Chan RHM, Marmarelis VZ, Hampson RE, Deadwyler SA, Berger TW (2009) Nonlinear modeling of neural population dynamics for hippocampal prostheses. Neural Netw 22(9):1340–1351. https://doi.org/10.1016/j.neunet.2009.05.004

Song D, Robinson BS, Hampson RE, Marmarelis VZ, Deadwyler SA, Berger TW (2018) Sparse large-scale nonlinear dynamical modeling of human hippocampus for memory prostheses. IEEE Trans Neural Syst Rehabil Eng 26(2):272–280. https://doi.org/10.1109/TNSRE.2016.2604423

Velliste M, Sagi P, Chance Spalding M, Whitford AS, Schwartz AB (2008) Cortical control of a prosthetic arm for self-feeding. Nature 453(7198):1098–1101. https://doi.org/10.1038/nature06996

Venkatraman S, Elkabany K, Long JD, Yao Y, Carmena JM (2009) A system for neural recording and closed-loop Intracortical microstimulation in awake rodents. IEEE Trans Biomed Eng 56(1):15–22. https://doi.org/10.1109/TBME.2008.2005944

Voges J, Koulousakis A, Sturm V (2007) Deep brain stimulation for Parkinson's disease. Acta Neurochir Suppl 97(Part 2):171–184. https://doi.org/10.1007/978-3-211-33081-4_19

Wagenaar DA, Potter SM (2002) Real-time multi-channel stimulus artifact suppression by local curve fitting. J Neurosci Methods 120(2):113–120. https://doi.org/10.1016/S0165-0270(02)00149-8

Weiland JD, Cho AK, Humayun MS (2011) Retinal prostheses: current clinical results and future needs. Ophthalmology 118(11):2227–2237. https://doi.org/10.1016/j.ophtha.2011.08.042

Wichmann T (2000) A digital averaging method for removal of stimulus artifacts in neurophysiologic experiments. J Neurosci Methods 98(1):57–62. https://doi.org/10.1016/S0165-0270(00)00190-4

Zhou A, Johnson BC, Muller R (2018) Toward true closed-loop neuromodulation: artifact-free recording during stimulation. Curr Opin Neurobiol 50:119–127. https://doi.org/10.1016/j.conb.2018.01.012

Modern Approaches to Augmenting the Brain Functions

Ioan Opris, Brian R. Noga, Mikhail A. Lebedev, and Manuel F. Casanova

1 Introduction

Augmentation as a concept denotes the process of making something greater or better. Brain augmentation aims to enhance brain performance in both health and disease (Lebedev et al. 2018). Targeted functions for brain augmentation fall under the purview of the sensory, motor, emotive, and cognitive domains. Few examples of brain augmentation include cognitive training (e.g., computer games, virtual reality), nootropic agents, interventions during sleep, and brain–computer interfaces (BCIs).

The approaches to brain augmentation can be classified into three categories: (1) neuromodulation, where activity of neural circuits is targeted with stimulation, such as electrical stimulation and drug infusion, (2) BCIs that extract useful information

I. Opris
Department of Neurosurgery, University of Miami, Miami, FL, USA

B. R. Noga
Miami Project to Cure Paralysis, Department of Neurological Surgery,
Miller School of Medicine, University of Miami, Miami, FL, USA

M. A. Lebedev
Department of Neurobiology, Duke University, Durham, NC, USA

Center for Bioelectric Interfaces of the Institute for Cognitive Neuroscience, National Research University Higher School of Economics, Moscow, Russia

Department of Information and Internet Technologies of Digital Health Institute,
I.M. Sechenov First Moscow State Medical University, Moscow, Russia

M. F. Casanova (✉)
University of South Carolina School of Medicine, Greenville Health System,
Greenville, SC, USA
e-mail: manuel.casanova@louisville.edu

I. Opris et al. (eds.), *Modern Approaches to Augmentation of Brain Function*,
Contemporary Clinical Neuroscience,
https://doi.org/10.1007/978-3-030-54564-2_4

from brain activity and utilize this information to augment various functions, and (3) futuristic ideas, such as brain-to-brain interfaces.

While the term "augmentation" usually describes the improvement of brain functions in healthy people, the term "repair" refers to functional improvements in patients suffering from various neurological disabilities. Opris et al. (2013), for example, suggested that interlaminar microstimulation of cortical microcircuits could be used to repair cognitive dysfunction. Building artificial minicolumns as basic modules to repair the damaged cortical tissue will become a valuable approach in cognitive neuroprosthetics. Such a microcircuit-based neuroprosthesis will be designed specifically to restore and/or repair disrupted cognitive functions. For example, multi-input/multi-output (MIMO) stimulation (Hampson et al. 2012) and transcranial magnetic stimulation (TMS; Sokhadze et al. 2012) are both valuable potential options to repair/treat neuropsychiatric disorders. The multitude of deficits observed in cortical microcircuits include microanatomic disconnections between layers or within minicolumns (autism, schizophrenia, Alzheimer), intra- and interlaminar neuromodulation (drug addiction, aging), and the lack or excess of inhibition (ADHD, depression) (Opris and Casanova 2014). In the future, the use of microcircuit-based prostheses is expected to provide useful therapies for patients with neuropsychiatric disorders.

2 Augmentation Approaches

A recent research topic hold by the *Frontiers in Neuroscience* and entitled "Augmentation of brain function: facts, fiction and controversies" (https://www.frontiersin.org/research-topics/1563/augmentation-of-brain-function-facts-fiction-and-controversy) by Lebedev et al. (2018) will be briefly summarized here.

2.1 Neuromodulation

Neuromodulation is commonly used as a treatment for movement disorders and chronic pain. Neuromodulation can be applied to different regions. Thus, spinal cord stimulation (SCS) is the approach, where a specific portion of the spinal cord is electrically stimulated to block pain signals from reaching the conscious brain.

Vagus nerve stimulation (VGS) affects the brain by electrically stimulating the vagus nerve (VGN) that runs on each side of the body, from the brainstem through the neck into the chest and abdomen. The stimulator is implanted under the skin of the chest with a wire being "wound around" the nerve (Usami et al. 2013). VGS is most often used to treat patients with either epilepsy or depression when other therapies have not worked.

Peripheral nerve stimulation (PNS) is the approach, where a specific nerve is stimulated to relieve chronic pain locally (Herschkowitz and Kubias 2018). PNS is a safe, effective, and durable treatment for movement disorders and/or chronic pain. Indeed, over 60% of patients with pain had a significant improvement following PNS (Mobbs et al. 2007). The technique has been shown to improve a patient's quality of life dramatically. PNS improves sleep, eliminates the need for surgery, and reduces the use of drugs that may expose the patient to tolerance, abuse and toxicity (Kumar et al. 1998; Burchiel et al. 1996; Barolat et al. 2001; Kemler et al. 2004).

Deep brain stimulation (DBS) is an electrical stimulation through an electrode implanted either in the subthalamic nucleus, the internal globus pallidus, or the ventral intermediate nucleus of the thalamus. DBS influences movement control and is used in the treatment of neurological conditions such as essential tremor, Parkinson's disease, dystonia, and epilepsy (Seier et al. 2018; Toleikis et al. 2012; Noga et al. 2017; Kern et al. 2019; Popa et al. 2016). DBS patients have experienced long-term improvement of motor function in Parkinson's disease, essential tremor and dystonia (Krack et al. 2003; Sydow et al. 2003; Vidailhet et al. 2007; Oyama et al. 2014; Huss et al. 2015; Jahanshahi et al. 2014).

Neuromodulation can be based on pharmacological therapy and drugs. Urban and Gao (2014) reviewed the main classes of drugs with potential to enhance cognitive functions in healthy individuals. However, these authors expressed concern about the development of addictive behaviors and detrimental effects of drug overdose and other unwanted consequences of these pharmacological approaches. These drugs could induce brain plasticity in such a way as to interfere with normal brain functions and their development, particularly in young individuals.

Neuromodulation can be applied during sleep. A couple of articles by Diekelmann (2014) and Pigarev and Pigareva (2014) addressed the potential of sleep for augmentation of cognitive functions like attention, language, reasoning, learning, and memory. Diekelmann described how "memory processing during sleep can be augmented by cueing memory reactivation" with olfactory and auditory stimuli. They also discussed electrically inducing brain oscillations characteristic to sleep, and how specific neurotransmitter systems are modulated pharmacologically. Furthermore, Pigarev and Pigareva highlighted "partial sleep" and "visceral processing during sleep" that are relevant to brain-augmenting approaches.

Neuromodulation can be exerted with transcranial lasers. A novel idea by Gonzalez-Lima and Barrett (2014) proposed that transcranial stimulation with infrared lasers may affect the bioenergetics of the brain in a positive way to augment executive frontal cortex functions, like attention, working memory, and affective state (Gonzalez-Lima and Barrett 2014).

Neuromodulation can employ a MIMO model. To perform cognitive augmentation, interlaminar recordings are analyzed via a nonlinear MIMO model, whose output is then converted into patterns of microstimulation (Berger et al. 2011; Opris 2013). In these studies, MIMO models use a precise topographically matched stimulation by extracting the patterns of firing that relate to the successful behavioral performance. This allows us to substitute the firing patterns (task-related) of layer L5 neurons with patterns of electrical stimuli/pulses in the same recording location during columnar transmission from layer L2/3 at the time of object/location target selection. Such stimulation improves normal task performance, but more importantly, recovers performance after being impaired by a pharmacological disruption of decision making (Hampson et al. 2012). Moreover, because the stimulation-evoked spatial tuning (measured in % correct performance) for spatial task trials is quasi-similar to neural tuning, it indicates that cross-laminar microcircuits in prefrontal cortex play a "causal role" to executive control functions (Opris et al. 2005a, b, 2012a, b, 2013, 2014). These findings have the merit of providing the first successful proof for a "microcircuit-based neuroprosthesis" to restore and/or repair a disrupted cognitive function.

The use of both invasive MIMO stimulation (Hampson et al. 2012) and noninvasive TMS (Sokhadze et al. 2012) are valuable potential options to repair or treat such microcircuit dysfunctions. Microcircuit-based neuroprostheses, including the MIMO-based memory implants and/or the decision chips, are promising to provide treatment for various neurological disorders that emerge from compromised microcircuits (Berger et al. 2011; Hampson et al. 2012). Targeting cortical microcircuitry might be a key starting point for the development of next-generation medical treatments.

2.2 *Brain–Computer Interfaces*

The new developments in the BCI field include BCIs for controlling bipedal walking (Wall et al. 2015; Lisi et al. 2014; Zarka et al. 2014; Bouyarmane et al. 2014; Li 2014; Solopova et al. 2015; Song et al. 2015; Onose et al. 2016), and technologies that enhance the human ability to predict future events (Pezzulo et al. 2016). Some other BCIs include those that modulate attention (Astrand et al. 2014; Ordikhani-Seyedlar et al. 2016) and cognition or action (Mandonnet and Duffau 2014). While Mandonnet and Duffau (2014) proposed that the cortical circuitry engaged in cognition and action should be thoroughly investigated in terms of effective BCI methods, Taya et al. (2015) argue that a connectome approach should be combined with cognitive enhancement methods.

BCIs are crucial for the next generation of human neuroprostheses (Schicktanz et al. 2015). Because few pilot projects have tested patients' abilities to control BCIs with the offered technologies, more clinical BCI studies need to be explored in a disease-related environment. Based on their findings, a multifold approach was suggested to the development of clinical BCIs, rooted in the participatory technology development. A whole-body neuroprosthetics can be built using the BCI approach. The integration of BMI with exoskeletons will incorporate recordings of large-scale brain activity, real-time advanced decoding algorithms, and the rapid integration of artificial sensory feedback (Lebedev et al. 2011).

2.3 Futuristic Transfer of Brain Ability

One futuristic idea is the transfer of learning from one brain to another. Deadwyler et al. (2013) reports an augmentation approach, where memory content of one animal (rat) is transferred to the brain of another animal via electrical stimulation. The information has been read out from the "donor" rat's hippocampus performing a long-delay memory task. This neural signal was then processed by means of a MIMO model and delivered to the CA3 hippocampal subfield of the "recipient" animal that used this "memory trace" to reproduce the donor's task performance.

Carr (2002) demonstrated that using a lateral hypothalamic self-stimulation (LHSS) augments the rewarding (i.e., threshold lowering) effect of diverse drugs of abuse (under food restriction). The rewarding effect is attributed to increased sensitivity of a neural substrate, rather than a change in drug pharmacokinetics, because it is preserved when drugs are injected directly into the lateral cerebral ventricle (Carr 2002). The possibility of increased DA receptor function in limbic forebrain dopamine (DA) terminal areas is suggested by findings that rewarding and motor-activating effects of direct DA receptor agonists are augmented by food restriction, and the augmented behavioral effects of amphetamine are reversed by an otherwise subthreshold dose of D-1 antagonist (Carr 2002).

Brain-to-brain interface can be employed for real-time sharing of sensorimotor information (see Fig. 1). Pais-Vieira et al. (2013) developed a brain-to-brain interface (BTBI) for real-time transfer of sensorimotor information between the cortices of two rats. In their BTBI, an "encoder" animal performed a behavioral task where it selected from two choices of visual or tactile stimuli. While the encoder rat executed the task, its cortical activity was sampled and transmitted to the cortex of a "decoder" rat in the form of intracortical microstimulation. The decoder rat learned to respond to microstimulation by making the behavioral selections as the encoder rat. These findings show that BTBIs can enable exchange of information between different brains and, hence, serve as the basis for the development of novel types of neuroprosthetic devices (Pais-Vieira et al. 2013).

An organic computing device can be based on multiple interconnected brains. Pais-Vieira et al. (2015) developed Brainets incorporating several brains. Brainets enable cooperation and information exchange in real time. In one of Pais-Vieira et al.'s experiments, a Brainet enabled an "organic computer" that consisted of four adult rat brains. This Brainet solved several tasks such as classification and storage of information.

Noninvasive technology based on brain-to-brain communication in humans. BTBIs have been demonstrated in humans as well (Grau et al. 2014). These technologies utilized the noninvasive BCI approach to enable communication between human subjects. Grau et al. (2014) utilized binary streams for encoding words. The words were sent from the emitter to receiver. The emitter operated a motor imagery-controlled EEG-based BCI, and the receiver received the messages as TMS of the visual cortex. The TMS induced perception of phosphenes, which served as bits of information transferred. These findings provide a proof-of-principle demonstration, which is important for future development of BTBIs.

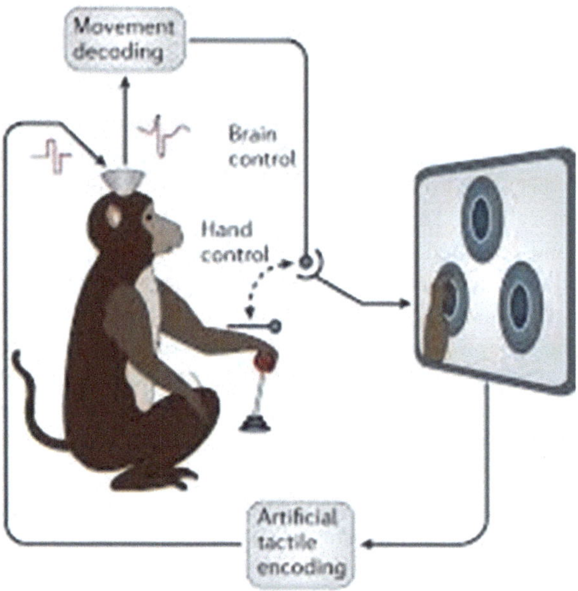

Fig. 1 Bidirectional brain–machine interface. The monkey explored the space with a virtual hand (or avatar) to find which of the three visually identical targets had the right "texture." The avatar was controlled by signals recorded in the primary motor cortex (M1); tactile sensations of texture were elicited by stimulating the primary somatosensory cortex (S1) with patterned pulse trains. When the avatar was over two of the three visual targets, the monkey received one of two pulse trains of intracortical microstimulation (ICMS). The third target did not trigger ICMS. The animal was successful if it correctly identified the target associated with the rewarded ICMS train. Over time, the animal learned to identify the rewarded target. (The figure is modified, with permission, from O'Doherty et al. (2011), Nature)

3 Augmenting Brain Functions

As already defined, the augmentation of brain functions may be regarded as the enhancement in brain abilities to process sensory, motor/occulomotor, cognitive, or emotion/reward signals. Executive abilities emerge from cortico-cortical interactions between interlaminar prefrontal cortical microcircuits and before being conveyed to lower structures. The perception-to-action hierarchy allows an upsurge of information that makes possible the enhancements of sensory, motor, and cognitive brain functions.

3.1 Sensory Augmentation

Visual experience is the outcome of the sense of sight, while visual perception is the ability to interpret the surrounding stimuli using light in the visible spectrum reflected by the objects in the environment. ExtVision is a novel system used to augment visual experiences by generating and projecting an image in peripheral vision (Kimura and Rekimoto 2018). A peripheral projection is a technique to enhance visual experiences.

It has been suggested that color vision can be augmented. Color vision deficiency, also called "color blindness," is the inability to distinguish certain shades of color, with very few sufferers completely blind to all colors (Mitchell 2017). This kind of color blindness affects the ability to see shades of red and green, or the ability to see blue and yellow. To combat this problem, researchers from the University of California in San Diego have developed an "augmented reality" solution for patients like these. This is called CHROMA: "a wearable, real-time augmented reality (AR) app that utilizes the Google Glass to address the real-life issues that people with color blindness face." CHROMA is a digital aid for patients, and can provide information about shades of colors that the user cannot determine. The results were striking! Patients in some cases could even achieve the same results on the Ishihara test as people with perfect color vision.

Retinal prostheses strive to restore vision in people with retinal degeneration (Nimmagadda and Weiland 2018). The electrophysiological response that occurs in the visual cortex after electrical stimulation of the retina shows a dose–response characteristic with respect to stimulus amplitude in both normal rats and rats with retinal degeneration. Cortical responses exhibited retinotopy only in normal rats, not in the ones with retinal degeneration. If a similar loss of retinotopy takes place in humans, retinal prostheses should be designed to provide for flexibility that account for variability in individual patients.

Cortical visual prostheses strive to restore vision by applying electrical stimulation to the visual cortex (Najarpour Foroushani et al. 2018). During such stimulation, subjects perceive phosphenes in different parts of their visual field, which supports the method of creating percepts made of several phosphenes. Additionally, electrical microstimulation of visual cortex increases the thresholds for detection of light (Cone et al. 2018). Niketeghad and Pouratian (2018) recently reviewed the current state of research on visual prostheses.

Hearing loss in aging people is a "multifactorial condition" that affects more than 33% of the aging population (Vaisbuch and Santa Maria 2018). The most common type of hearing loss is that related to high frequency as a result of the aging process (presbycusis). This type of hearing loss is a slow process that affects audibility and perception of speech consonants, but not vowels. Conventional hearing aids provide increased sound pressure in the ear canal for detection of sounds that might otherwise be soft or inaudible (Atcherson et al. 2015). Hearing aids can be used by patients with a wide range of hearing loss severity. As the hearing loss severity increases, the hearing aids for speech perception decrease. Implantable devices such as cochlear devices, middle ear implants, and bone-anchored implants can perform the function of the damaged part of the ear and provide sound signals to the brain. Hearing assistive technology devices augment hearing by providing information of auditory, visual, or tactile modality. These assistive devices include wireless assistive listening device systems, closed captioning, hearing aid-compatible telephones, and other devices.

Tactile sensations can be augmented (see Fig. 1). Brain–machine–brain interfaces (BMBIs) enable bidirectional communication between the brain and external devices. O'Doherty et al. (2011) described a BMBI that controlled exploratory movements of a virtual-reality arm and enacted artificial tactile feedback using ICMS of the primary somatosensory cortex. In this experiment, monkeys executed an active exploration task where a virtual-reality arm was controlled directly by motor cortical activity. Artificial tactile sensations were delivered each time the virtual-reality arm scanned virtual objects. Two monkeys learned to control this BMBI to locate one of three visually identical objects. These results suggest that neural prostheses can benefit from artificial tactile feedback.

3.2 Motor Augmentation

Augmentation of brain function relies on the ability to modify the structure and function of a number of brain regions. A specific example is the shaping of receptive fields in order to become sensitive to novel inputs in primary sensory cortices (Di Pino et al. 2014). When the motor cortical areas are targeted, we can refine the limb movement kinematics by adapting the neurons' firing rates to the new neuroprosthesis representation. For such purpose, the learning process recruits more motor and premotor cortical outputs and decreases attentional recruitment, focusing on the activity on sensorimotor areas and increases in the basal ganglia drive on the cortex (Di Pino et al. 2014). The augmentation approach relies heavily on the executive control mechanism in the frontoparietal network (Di Pino et al. 2014). Microstimulation of the motor cortex can evoke complex, multijoint movements including movements of the arm and hand (Graziano et al. 2005; Graziano 2016). Specifically, the premotor cortex, which owns the motor representation tool, participates in learning to control an external effector, while in the intraparietal sulcus are extracted the visual spatial features of movement (Di Pino et al. 2014). Thus, sensorized prostheses may "provide" the critical sensory afferences to evolve the exploitation of tools through their embodiment, reshaping the body representation and the sense of the self (Di Pino et al. 2014).

When transcranial direct current stimulation (tDCS) is applied to the primary motor cortex (M1), it affects motor function, enhancing or decreasing performance of both healthy participants and brain-damaged patients (Convento et al. 2014). Beyond M1, controlling and guiding movement crucially involves the posterior parietal cortex (PPC). Therefore, the modulation of cortical excitability within PPC can also affect hand motor function in healthy right-handed participants. Experiments using anodal tDCS in the left PPC show selective facilitation of action planning, while the same stimulation of the right M1 shows modulated action execution only (Convento et al. 2014). Based on this finding, it may be possible that movement improvements induced by left PPC and right M1 stimulations rely on a substantially different mechanism. This is opening up new perspectives in the neurorehabilitation for stroke patients with motor and apraxic disorders.

Many natural tasks involve planning sequences of movements, and it would be of interest to enable such behaviors with a BMI (Shanechi et al. 2012). Such a BMI could decode the sequence before it is executed. Using population algorithms, Shanechi et al. (2012) investigated two subpopulations of neurons in monkey premotor cortex. These populations encoded two planned targets of a movement sequence. Based on these results, a BMI was developed that decoded the full motor sequence (Shanechi et al. 2012).

Goal-directed locomotion is initiated by signals arising from either the cerebral cortex processing intentional control or from the limbic system in case of emotional processing (Takakusaki 2013). While the initiation of locomotion takes place in the top nodes of the hierarchy, the integration of sensory and motor signals occurs in the brainstem and the spinal cord (Takakusaki 2013). The basic pattern for locomotion is generated by the spinal interneuronal networks comprising the central pattern generators (CPGs). Intentional modification of gait requires motor programming in the premotor cortices. These motor programs are transmitted to the brainstem by the cortico-reticulo-spinal system that enable the cortico-spinal system to generate limb trajectory and achieve accurate foot placement. To accomplish this, loops from the motor cortical areas to and from the basal ganglia and the cerebellum are engaged to regulate the timing and intensity of descending signals for posture-gait control (Noga and Opris 2017; Takakusaki 2017).

Exoskeleton robots (wearable mobile machines) can assist humans to perform daily activities by mimicking or augmenting the body's own movements (Deng et al. 2018). Deng and colleagues proposed a hierarchical control scheme consisting of two layers that enabled an exoskeleton robot to cooperate with humans. In the low-level control of the upper limb the exoskeleton robot employed an adaptive neural network controller to enable the robot to be back drivable. During the "learning phase," Deng's robot observed and learned how a demonstrator performed a specific impedance-based task successfully, while during the "reproduction phase," the robot provided the subjects with "enough assistance" by extracting human skills from demonstrations to prevent the motion of the robot's end effector to deviate from desired states, e.g., when faced with environmental disturbances.

In terms of augmentation, there are several distinctive places based on the level in the brain hierarchy (see Fig. 2) that we can intervene: (1) at the cortical level (top of hierarchy) intentions are deciphered after detecting electrical signals from the human scalp, being translated into control commands used to operate in real-time external devices, computer displays, and virtual objects, i.e. BCIs (Tariq et al. 2018); (2) at the subcortical level (Marchesotti et al. 2017); (3) deep in the brainstem (Noga and Opris 2017; Takakusaki 2013, 2017); and (4) in the spinal cord (Krucoff et al. 2016; Noga and Opris 2017; van den Brand et al. 2015) and muscles (Irastorza-Landa et al. 2017; Sarasola-Sanz et al. 2017).

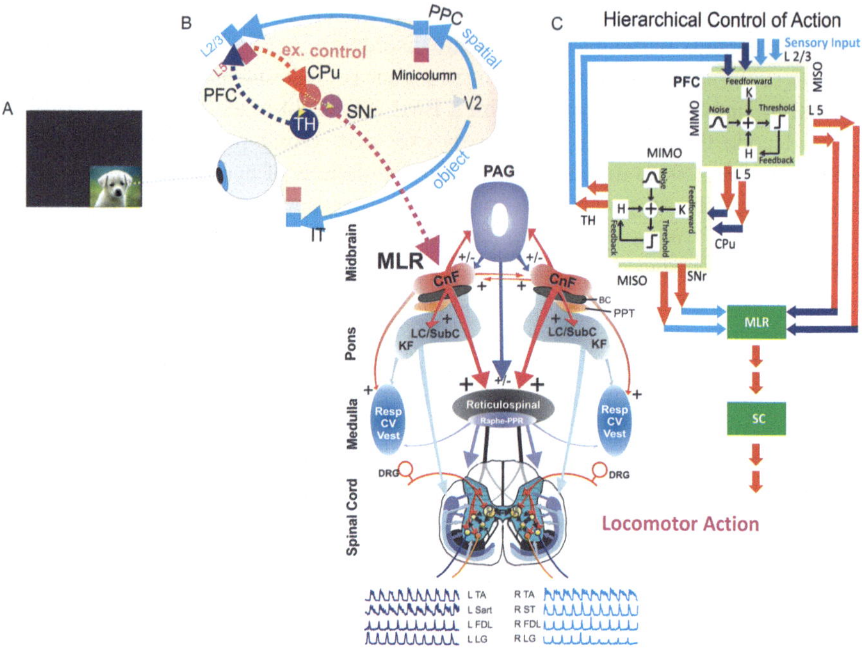

Fig. 2 Diagram for a hierarchical prosthesis for locomotion. (**a**) A pet is seen by the owner. (**b**) The brain with the visual information integrated in the prefrontal cortex and the executive command is sent to the mesencephalic locomotor region (MLR). The information in the brainstem is further sent to the effectors via spinal cord. (**c**) Hierarchical control of locomotion from cortex to spinal cord. (Modified with permission from Noga and Opris (2017), Springer)

3.3 *Augmentation of Cognition*

Cognition may be defined, according to the Oxford Dictionary, as the "mental process of acquiring knowledge and understanding through thought, experience, and the five senses." It involves attention, perception, memory, decision making, reasoning, intuition, and language. The complementary function dealing with mental and feeling reactions is emotion. Let us have a look into how some of these functions may be augmented.

Sensing, assessing, and augmenting threat detection is crucial for survival (Parasuraman and Galster 2013). Recently, a proof of principle for a real-time BCI to decode covert attention signals in higher order visual area was provided by Ekanayake et al. (2018). These researchers implemented a "winner-takes-all approach" to achieve single-block classification of brain activations (with fMRI). By utilizing higher order mental processes, the "cognitive BCIs" are providing a platform for communication in patients who are unable to speak or move due to brain injury. In addition, the cortical cholinergic system is useful in enhancing attention allowing for the potential benefits accrued to improved memory and focus (*vide infra*) (Demeter and Sarter 2013).

Our cerebral cortex integrates the signals and mediates the interactions between the environment and the perceptual–executive systems of the brain (Opris et al. 2013). At the top of the executive hierarchy, prefrontal cortical microcircuits integrate perceptual and executive control information to guide optimal decision making (Opris et al. 2013). For example, spontaneous actions are preceded by brain signals that may sometimes be detected a few hundreds of milliseconds in advance of a subject's intention to act (Rigato et al. 2014). These signals raised doubts about the causal role of the same in conscious free will. However, Rigato et al. (2014) rejected the claim that such "predictive signals" imply unconscious decisions, suggesting instead the relationship to free will. Better decisions are made by groups when undertaking a visual-search task via a hybrid BCI (Valeriani et al. 2017). To demonstrate augmentation, Valeriani et al. (2017) used a hybrid BCI (hBCI) to improve the performance of groups undertaking a realistic visual-search task. This hBCI extracts neural information from EEG signals and combines it with response times to build an estimate of the decision confidence. The hBCI provides significant advantages over non-BCI decision methods in all cases (Valeriani et al. 2017).

Memory prostheses represent a recent "trend" in the stimulation-based BMIs (Berger et al. 2011). This trend covers several of the memory-augmenting approaches. To implement neuroprosthetic memory there are several approaches, including (1) nootropics (Madan 2014); (2) tDCS (Bennabi et al. 2014); (3) memory training games (Deveau et al. 2015); (4) Monte Carlo simulation (Moreau 2014); and (5) memory training that improves emotional states (Takeuchi et al. 2014). A memory prosthesis (for restoring and enhancing memory), where memory content is extracted from hippocampal activity using a MIMO nonlinear dynamical model, was demonstrated in rodents (Song et al. 2014) and nonhuman primates (Deadwyler et al. 2017). In addition to neurostimulation, pharmacological approaches have been developed for augmenting memory and cognition (Lynch et al. 2014). Brain training techniques can also be used to improve memory and cognition (Chapman and Mudar 2014; Haier 2014). In particular, Beatty et al. (2015) investigated several working memory tasks where training could be transferred from one task to another. A multifractal analysis of information processing in hippocampal neural ensembles during working **memory** compared the complexity of firing cells under Δ^9-tetrahydrocannabinol administration vs. control (Fetterhoff et al. 2015).

The neuroprosthetic approaches to language are: (1) BCI with language model–electroencephalography fusion for locked-in syndrome (Oken et al. 2014); (2) language-model assisted and icon-based communication through a BCI with different presentation paradigms (Ahani et al. 2018); (3) language-model assisted BCI for typing (Moghadamfalahi et al. 2015); (4) language model applications to spelling with BCI (Mora-Cortes et al. 2014); (5) integrating language models into classifiers for BCI communication (Speier et al. 2016); (6) high-speed spelling with a noninvasive BCI (Chen et al. 2015). The interface between language and attention recruits a general attention network in spoken language comprehension for prosodic focus marking (Kristensen et al. 2013).

3.4 Augmentation of Emotions

Plutchik defined emotions as "expressive behaviors that had the function of communicating intentions from one animal to another in the presence of conflicts or emergencies" (Plutchik 1997). According to Plutchik's view there are eight basic emotions: joy versus sadness; anger versus fear; trust versus disgust; and surprise versus anticipation. The latter are known as Plutchik' basic emotions or the wheel of emotions. Emotional expressions, from this point of view, regulate interpersonal relations and increase the chances of individual survival (Plutchik 1997).

Emotion dysregulation is an important aspect of many psychiatric disorders (McFarland et al. 2017). BCI technology could be a powerful new approach to facilitating therapeutic self-regulation of emotions (McFarland et al. 2017). One good BCI technique is based on stimulus-specific feedback examining subject-specific responses to emotion-eliciting stimuli by electroencephalography (EEG). To assess the potential of this method, McFarland and colleagues examined the relationships between emotional valence/arousal and the three EEG features: amplitude of alpha activity over frontal cortex; amplitude of theta waves over frontal midline, and also the late positive potential over central and posterior midline areas. For each feature, the ability to predict emotional valence/arousal on both an individual and a group basis was evaluated. EEG data was collected simultaneously and used for developing models to predict subject-based trial ratings. Their results suggest that the frontal midline theta is a better candidate than frontal alpha activity or the late positive potential for use in a BCI-based paradigm designed to modify emotional reactions. Further differentiation of emotional processing and attention has been performed with functional neuroimaging (Kiehl et al. 2017).

Novel paradigms for building BMI involve the use of reward and feedback directly from the brain (Prins et al. 2013). When investigating neuromodulation in the reward center of the brain (i.e., nucleus accumbens), it was shown how to extract a reinforcing signal (in a multi-target reaching task) that could be used to adapt a BMI decoder (Prins et al. 2013). A major challenge was how to "translate" neuromodulator signals into a single binary signal from the distributed representation of neural population, which may encode crucial aspects of reward. To extract these signals, the principal component analysis (PCA) of reward related signals and k-means clustering were used to separate data into two classes (Prins et al. 2013).

4 Clinical Applications

Recent research suggests that executive abilities emerge from cortico-cortical interactions between interlaminar prefrontal cortical microcircuits, while their disruption is involved in a broad spectrum of neurologic and psychiatric disorders such as autism, dementia, depression, anxiety, post-traumatic stress disorder, Alzheimer's disease, and drug addiction. The great promise of neural technology is to provide treatments and possible cures for a broad range of brain disorders.

4.1 Autism

This neurodevelopmental condition is characterized by a bias in the excitatory/inhibitory balance of the cerebral cortex. Neuropathological studies suggest a defect that impairs the coordination of the tangential and radial migration of neuroblasts during the development of the cortical plate (Casanova et al. 2003). Attempts to repair or reestablish the excitatory/inhibitory balance of the cortex have led to trials using repetitive transcranial magnetic stimulation (rTMS). Outcome measures have shown improvements in EEG, ERP, and a reduction of repetitive and maladaptive behaviors (Casanova et al. 2015). Some of these rTMS trials have led to normalization of brainwave oscillations, especially for gamma frequencies, with consequent improvement in executive functions (Sokhadze et al. 2009, 2014). A few studies with a limited number of patients have used tDCS in autism. The physiological effects of this technique have been gauged by changes in the maximum entropy ratio (MER, a reflection of EEG complexity) (Kang et al. 2018) and positive effects have been noted in select behavioral screenings (Amatachaya et al. 2014).

4.2 Depression

As many as two-thirds of patients with depression are not helped by their first course of antidepressant medication. Half of these patients (approximately 1–3% of the US population) may not benefit from pharmacological, psychotherapeutic, or somatic treatment despite adequacy of the trials and multiple therapeutic attempts. In such instances, the feeling of helplessness, sadness, sleep problems, and thoughts of death and/or suicide may persist for years. In these cases, neuromodulatory techniques such as electroconvulsive therapy, rTMS, and VGS may prove of clinical benefit and have gained US Food and Drug Administration approval for treatment. For rTMS, sustained response rates of 50% have been noted after 1 year of treatment (Senova et al. 2019). When rTMS is combined with cognitive behavioral therapy roughly two-thirds of patients with treatment refractory depression experience benefits. Studies with VNS stimulation have reported positive effects but may take 6–12 months for full benefits to occur (Mohr et al. 2011). A small number of open-labeled trials and individual case reports have examined the use of DBS in treatment resistant depression (Mohr et al. 2011). Responsiveness and side effect profiles suggest that neuromodulatory techniques are presently underused and that double-blind and controlled trials are warranted in the case of DBS.

4.3 Alzheimer's Disease

This neurodegenerative condition is marked by neuronal cell loss and the accumulation of beta amyloid and tau proteins with consequent loss of synaptic functions. Depending on implantation site, DBS can potentiate memory circuits involving the hippocampus and mesial temporal lobe circuits. Clinical trials indicate best results for patients in the early stages of AD; however, a recent review concludes that it is premature to suggest using DBS in the treatment of AD (Aldehri et al. 2018). VNS may stabilize or improve cognitive scores in 70% of patients at 6 months but only 41% when longitudinally followed for 1 year (Merrill et al. 2006). The small number of patients in many of these trials and their open-labeled nature warrants further investigations. Few studies have examined the effect of rTMS in AD. Although no cognitive adverse effects have been reported, conclusions are limited given the heterogeneity of cognitive outcome measures applied, stimulation parameters, and the size of the patient population (Limori et al. 2019).

4.4 Post-traumatic Stress Disorder (PTSD)

Approximately 20–30% of patients with PTSD are refractory to conventional treatments. Small case series of trigeminal nerve stimulation in PTSD patients have shown positive changes in response rates, quality of life, and remission after several weeks of treatment (Cook et al. 2016). The trigeminal nerve transfers signal sensations from the face to the brain and has connections to major centers in the mood circuit. More specifically, connections to the amygdala may be involved in fear acquisition and extinction. Defining the salience of a fear stimuli is not only directly related to mood but also alters other cognitive domains like memory, and executive functions. VNS may have a direct effect on the amygdala as it enhances extinction of conditioned fear response by modulating the plasticity of connections linking to the ventromedial prefrontal cortex (Pena et al. 2014). Some of these subcortical centers can be better targeted with DBS or subcranial magnetic stimulation. However, attempts with the latter techniques remain as early phase trials and in need of further investigation.

4.5 Drug Addiction

Following the early success of DBS in movement disorders and its decades-long safety record, the effectiveness of the technique was investigated in other psychiatric disorders including compulsive disorders, eating disorders, and addiction. tDCS over the dorsolateral prefrontal cortex has been used to decrease ambiguous risk-taking behavior and impulsivity related to drug addiction (Bashir and Yoo 2016).

Studies have shown that tDCS modulates intake and craving for addictive substances such as alcohol, nicotine, cocaine, methamphetamine, and cannabis (Lapenta et al. 2018), but placebo response has been high and long-term outcomes understudied (Luigies et al. 2018). These clinical trials have been hampered by small sample sizes, different stimulation protocols, and varying study duration (Lupi et al. 2017). Similar criticisms apply to the use of rTMS and DBS in addiction wherein despite promising case reports, studies tend to be open labeled and lacking controls (Wang et al. 2018).

4.6 *Parkinson's Disease*

A significant portion (28%) of Parkinson's patients suffer from debilitating motor symptoms despite optimal medical therapy. The FDA approved the use of DBS to treat tremor in 1997, and 5 years later approved its use to treat other motor symptoms associated with the condition. DBS reduces the need for levodopa and consequently the side effects of prolonged use of this medication. DBS also helps reduce symptom fluctuation, slow movement, and gait problems. The benefits of DBS have led investigators to study the possible role of neuromodulation in other neurodegenerative conditions like secondary dystonia, Hallervorden–Spatz syndrome, and the chorea of Huntington's disease.

4.7 *Seizures*

Patients with medically refractory seizures who are not candidates for resective surgery may benefit from the implantation of a neuromodulatory device. Vagal nerve stimulation has been approved for use in cases of partial epilepsy regardless of whether a seizure focus has been identified. There is evidence that VNS may be an effective therapy in generalized epilepsy but pivotal trials are missing. The right vagal nerve innervates the sinoatrial node, and for this reason stimulation is performed on the left side. Benefits may depend on device, type of seizure, and other clinical factors like length of treatment. On average patients experience a 25% reduction in seizure frequency after 3 months of therapy and 50– 60% reduction after 2 years (a comprehensive review of the literature is offered by Panebianco et al. 2015). The impact of VNS on mortality and Sudden Unexplained Death in Epilepsy (SUDEP) remains unsettled (Wheless et al. 2018). Preliminary studies using DBS stimulation have shown moderate improvements and the FDA has granted the Medtronics system approval for its use as an add-on treatment for focal epilepsy.

Translating results of brain-augmentation into clinical application has promoted therapeutic solution for conditions such as epilepsy (Höller and Trinka 2014; DeMarse and Carney 2014; Zeitler and Tass 2015), stroke (Grimm et al. 2016),

Parkinson's disease (Lebedev et al. 2014; Lee et al. 2015), Huntington's disease (Nagel et al. 2015), dementia (Garriga et al. 2015; Franco 2014), Alzheimer's disease (Yegla and Parikh 2015), autism spectrum disorders (ASD) (Billeci et al. 2016), traumatic brain injury (Alwis and Rajan 2014; Tajiri et al. 2014), and disorders of consciousness (Bai et al. 2016). Evidence is growing that noninvasive stimulation can be employed to treat a range of neurological conditions (Vicario and Nitsche 2013). Thus, Sokhadze et al. (2009, 2014) report that TMS applied to dorso-lateral prefrontal cortex improves executive functions in ASD, which is evident from the improvements in behavioral reactions and event-related EEG potentials. Additionally, according to the case report by Brem et al. (2014), tDCS can be applied to treat visuospatial neglect. Krawinkel et al. (2015) provide insights on how noninvasive stimulation could treat such conditions as schizophrenia and Parkinson's disease by modulating brain oscillations. Kubera et al. (2015) reviewed the use of low frequency (inhibitory) TMS to the superior temporal gyrus in schizophrenic patients with drug-refractory auditory verbal hallucinations. Moreover, Ayache et al. (2016) report that prefrontal tDCS can decrease pain in patients suffering from multiple sclerosis. Charvet et al. (2015) examined the use of remotely supervised tDCS for clinical trials. The study, which prompted the formulation of guidelines for technology and protocols, is of importance as its implementation removes the need for patients to travel to specialized treatment facilities. Furthermore, Thibeault (2014) argues that efficiency of therapeutic neurostimulation can be improved by neuromorphic components.

5 Technology for Brain Augmentation

Brain augmentation may be achieved using implants for recordings, stimulation, and drug delivery, by employing brain–machine interfaces, and the noninvasive activation of certain brain areas. New technologies provide unprecedented insights into the devastating brain disorders, as well as into the new devices for brain repair. We are interested in applications of neural technology from neuroscience of microcircuits to systems neuroscience of large-scale networks and neural engineering.

5.1 Neural Engineering

Neural engineering is regarded as a discipline within the broad domain of biomedical engineering that uses engineering methods/techniques to investigate, repair, replace, augment/enhance, or otherwise exploit the natural properties of neural systems. Neural engineering integrates elements from robotics, cybernetics, computer engineering, neural tissue engineering, materials science, and nanotechnology with the fields of computational neuroscience, experimental neuroscience, clinical neurology, electrical engineering, and signal processing of living neural tissue. The major goal is restoration and augmentation of human function via direct interactions between the nervous system and artificial devices.

Recent research is focused on the understanding of how coding of stimuli and the processing of information are elaborated in the sensorimotor and cognitive systems, and on quantifying how neural processing is altered in pathological conditions. Research also examines the way neural systems are manipulated by interactions with artificial devices as well as with BCIs and neuroprosthetics. To solve problems at the interface between living neural tissue and nonliving engineering constructs, neural engineers use multi-electrode arrays (MEAs), computer programing for interfaces with the brain with either: machines/actuators (BMI) or computers (BCI).

5.2 *Optogenetic Augmentation of Brain Function*

By combining genetic targeting tool with optical excitation, the recently emerged optogenetics offers the dual possibility to record activity of large ensembles of neurons, and to also manipulate activity of single cells. Recording neuronal activity is obtained by "selective expression of activity sensitive fluorophores" (proteins) into neurons, whose activity can then be "readout by optical imaging" (Knöpfel 2012). Manipulation of neural activity can be achieved via the insertion of photosensitive proteins (opsins) that behave as ion channels or pumps into a neuron's membrane and are preferentially controlled by photons of different wavelengths, allowing temporal control on the milliseconds order. Together, this cell kind specificity and temporal control lead to a tool that may perturb neural circuitry with high precision. Since its first application to neural populations in 2005 (Boyden et al. 2005), it has already had substantial impact as a well-liked technique within the neuroscientific toolkit. In addition to its use as an experimental tool, optogenetic stimulation has been recommended as a novel approach for neuroprosthetics and treatment of brain disorders. While therapies in these domains have traditionally used electrical or pharmacological techniques, optogenetics has one particular advantage over electrical stimulation in being able to target specific cell classes through gene expression. As a consequence, specific neuronal populations may be controlled without potential brain-wide, side effects.

5.3 *Neural Nanotechnology*

Among the promising materials, with applications in biomedical and basic research, we mention colloidal gold nanoparticles (AuNPs), whose wide range of diameters (i.e., 5–400 nm) alters their interaction with visible radiation and yields to a variety of emission spectra. AuNPs can also be coated with molecules and then used as therapeutic-agent delivery systems or as sensors in diagnostic applications.

(a) **Nanoparticles**. Nanoparticle research is currently the most studied branch of science in various fields. The nanoparticles have a wide variety of potential applications in biomedical fields (Vidu et al. 2014; Yue et al. 2012; Pampaloni et al. 2019), including: (1) creation of fluorescent biological labels for important biological markers and molecules in research and diagnosis of diseases; (2) drug delivery systems; (3) gene delivery systems in gene therapy; (4) genetic and tissue engineering.
Nanoparticles are being increasingly used in drug delivery systems. The advantages of using nanoparticles as a drug delivery system include a size and surface characteristic that can be easily manipulated (Vidu et al. 2014). This could be used for each passive and active drug targeting. Nanoparticles can also be made to control and sustain release of the drug during the transportation as well as at the location of the release.
(b) **Nanowire** Nanowire field-effect transistors (NWFETs) were demonstrated that could be used to measure signals from neurons, cardiomyocytes, and heart tissue (Timko et al. 2010). The NWs can be assembled on either planar inorganic or biocompatible flexible plastic surfaces. The NWs-based device built on a flexible surface can measure signals in the millivolts range, which are equal to or substantially greater than those recorded with either planar FETs or MEAs.
(c) **Carbon nanotubes** (CNTs), discovered by Iijima (1991), these cylindrical allotropes of carbon exhibit outstanding mechanical, thermal, and conductive properties. Rolling-up one or additional graphene sheets generates CNTs with good chemical and thermal stability, extreme electric properties, large surface area, and high mechanical strength while carrying ultralight weight (Ajayan 1999). Under well-defined experimental conditions for synthesis, two types of CNTs may be produced: (1) single-wall carbon nanotubes (SWCNTs) and (2) nested multiwall carbon nanotubes (MWCNTs). Being quite close to graphene, CNTs are usually near to atomic-scale perfection making CNTs chemically inert (Enachescu et al. 1999; Vidu et al. 2014). Although the CNTs have 1/6th of the weight of steel, similar to graphene under tension, nanotubes are two orders of magnitude stronger than steel. Computer simulations estimated the melting point of nanotubes around 3700 °C that is higher than that of any metal, but close to that of graphite. SWCNTs scan act as very good conductors of electrons or can show semiconducting behavior, depending on their diameter and the atomic structure of nanotubes. Even the highest thermal conductivity of pure diamond is exceeded by that of CNTs. Additionally, CNTs are biocompatible with many environments (Vidu et al. 2014). Several technological fields are undergoing vigorous developments in CNTs devices due to their unique properties, including conductive nanocomposites, nanometer semiconductor devices, sensors, etc. (Vidu et al. 2014). Functionalization of the CNTs surface has been perfected over the last decade employing various approaches, some being focused on increasing solubility and lowering CNTs' toxic effects in order to fit biomedical conditions. In Fig. 3 a neat illustration of the application of CNTs in neuroscience is shown. As mentioned above, the CNTs are now used (a) in

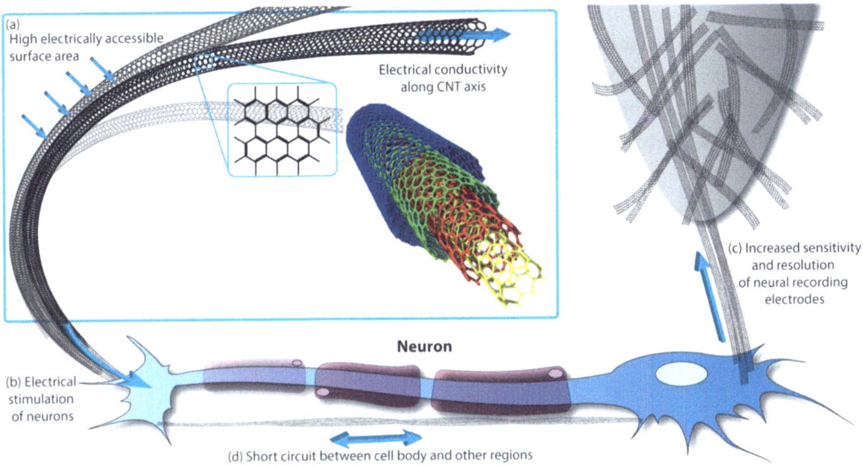

Fig. 3 Application of carbon nanotubes in neuroscience. (**a**) Important aspects of CNTs for use in neural interfaces. Illustration of using CNTs to (**b**) provide electrical input, (**c**) obtain output, and (**d**) alter neuronal behavior. (Modified with permission from Serpell et al. (2016), ACS Cent. Sci)

neural interfaces, (b) to provide electrical input, (c) to obtain output, and (d) to alter neuronal behavior (Serpell et al. 2016).

5.4 *Neural Spintronics*

Spintronics, or "spin electronics," is a branch of nanotechnology dealing with the study of electron (and nuclear) spin and the devices that specifically exploit the properties of the spin (Wolf et al. 2006). For example, "spin relaxation" and "spin transport" in metals and semiconductors are fundamental processes for electronic technology. Spintronic systems are useful in "quantum computing" and "neuromorphic computing" including the data storage and transfer (Bhatti et al. 2017). While the origin of spintronics may be found in the ferromagnet/superconductor tunneling and initial experiments on magnetic tunnel junctions by Julliere within the 1970s (Julliere 1975), the use of semiconductors for spintronics started with the theoretical proposal of a spin field effect transistor by Datta and Das (1990) and of the electric dipole spin resonance by Rashba (1960). Spintronics devices are attracting a lot of attention, due to their noninvasive potential to measure neuronal interactions within the brain, and to modulate neural activity, that enables emergence of brain functions (Torrejon et al. 2017).

1. Magnetic tunneling junctions (MTJ)—an MTJ is a unit consisting of two ferromagnets divided by a thin insulator.

2. Magnetic wall domains and spin current. A magnetic domain may be considered as a region within a magnetic material within which the magnetization is in an exceedingly uniform direction. This means that each individual magnetic moment of the atoms are aligned together and they point within the same direction. A domain wall is an interface separating magnetic domains. The movement of the magnetic wall domain involves a spin current.
3. Racetrack memory uses a spin-coherent current to maneuver magnetic domains along a nanoscopic permalloy wire about 200 nm across and 100 nm thick. As the current is passed through the nanowire, the domains flow by the magnetic read/write heads placed near the wire, which in turn alter the domains to record patterns of bits (Parkin et al. 2008; Parkin and Yang 2015). A racetrack memory device is created by many such wires and read/write components.

5.5 Tools Advances Based on the Interaction with the Brain

1. *Minimally invasive devices*: MEAs are special devices with multiple plates or shanks by which neural signals are obtained and/or delivered to neural tissue. MEAs serve as neural interfaces that connect neurons to electronic circuitry. There are two general types of such arrays: implantable MEAs that are used in vivo and non-implantable MEAs that are used in vitro. MEAs are used for the recording single cell or multi-unit action potentials (or spikes) or local field potentials LFPs/EEGs in which the voltage averages the activity of many cells. MEAs are also used for microstimulation in cortex (ICS), deep (DBS), by MIMO model and by optogenetic stimulation.
2. *Noninvasive devices*: TMS, tDCS, and ultrasound are among the most used non-invasive neurostimulation methods in present use. Noninvasive stimulation methods have gained popularity in recent years as a means to augment brain function. Yet, many unknowns and controversies still remain. Krause and Cohen Kadosh (2014) and Horvath et al. (2014) examine the role of inter-subject differences in responsiveness to tDCS. McKendrick et al. (2015) propose an approach, where wearable devices are used that combine tDCS with a new generation of miniaturized functional near-infrared spectroscopy systems. Blumberg et al. (2015) and Foroughi et al. (2015) report that performance on spatial tasks can be enhanced by tDCS applied to PPC. Younger et al. (2016) demonstrate that tDCS applied to the left inferior parietal lobe can augment reading subskills. Luft et al. (2014) propose that a connectome approach can be combined with brain stimulation. Horschig et al. (2014) discuss neurostimulation methods that could be used to manipulate cortical oscillations. Koganemaru et al. (2015) and Tsagaris et al. (2016) suggest that the efficiency of neurostimulation can be improved if it is combined with the appropriate task patterns.

As mentioned above, Davis and van Koningsbruggen (2013) do not think that the term "noninvasive" is appropriate to describe noninvasive stimulation methods that strongly affect the brain and evoke long-lasting consequences. Therefore, these approaches ought to be used with caution. Among the effects of noninvasive stimulation the researchers name biasing network dynamics (Wokke et al. 2015) and influencing brain hemodynamics (Pulgar 2015; Dutta 2015). Additionally, brain functions can be affected even with transcranial lasers (Gonzalez-Lima and Barrett 2014). Duecker et al. (2014) discuss the potential of noninvasive stimulation as a research tool in the studies of perception, cognition, and behavior. Additionally, Luber (2014) argues that brain augmentation with noninvasive stimulation cannot be explained by a net zero sum proposition, which is the mechanism where brain resources are reallocated: gains in one function are balanced by costs elsewhere.

5.6 Devices for Brain Augmentation

BMI performance critically depends on type of neural signal being recorded and decoded. Invasive and noninvasive BMIs are the two major categories of BMIs (Silva 2018) classified by the recording methodology. Waldert (2009) discusses pros and cons of invasive and noninvasive recordings and the future of these approaches. Among the noninvasive types, EEG is the most popular approach utilized in BMIs. Callan et al. (2015) assess the efficiency of dry EEG recordings and argue that this method is useful for decoding of auditory events from EEG data, even when substantial acoustic noise, and mechanical and physiological artifacts interfere with the recordings from simulated and real flight conditions. Blankerts et al. (2016) review completely different usages of EEG-based BMIs, together with practically oriented applications and using BMIs as experimental tools.

While the BMIs using EEG are easy to implement and safe to use, some limitations arise in their information transfer rate (ITR). ITR can be improved if electrical activity is recorded from the surface of the brain using electrocorticography (ECoG), a minimally invasive method. Kapeller et al. (2014) describe the use of visual evoked potentials in a high-performing ECoG BMI, and Zippo et al. (2015) report a novel epicortical grid tested with wireless recordings, in rhesus monkeys. Functional magnetic resonance imaging is another noninvasive recording technique appropriate for BMI implementations. Caria (2016) discusses the neurophysiological mechanisms involved in self-regulation of blood oxygenation level monitored with fMRI. Liu et al. (2016) found that manipulating pre-rTMS neural activity could augment antidepressant effects to rTMS treatment.

Neural networks with nonmonotonic activation functions can generate greater storage capacity than the ones with Mexican-hat-type activation function (Morita 1996; Nishimori and Opris 1993). Recently, Nie et al. (2016) demonstrated the coexistence of multiple equilibrium points for memristive neural networks with nonmonotonic piecewise linear activation functions and unbounded time-varying delays. Enhanced associative memories should be inevitably linked with the augmentation of memory ability.

6 Ethical Issues

The impact of intelligent neuroprostheses is likely to bear a strong impact in the health sciences, due to their novel therapeutic perspectives for a number of diseases (Nagel 2014). Unavoidably, they will raise basic ethical questions on the "intermingling between man and machine" and, more specifically, on how "deeply" to allow brain processing alterations by implanted "intelligent" artificial systems (Vassanelli and Mahmud 2016). The emerging ethical issues related to brain-augmenting methods (Glannon 2014a, b; Attiah and Farah 2014; Clark 2014; Maslen et al. 2014) deal with the relationship between the diminishment and enhancement following the application of brain-augmenting technologies (Earp et al. 2014), the problem of "mind control" with BMI technologies (Koivuniemi and Otto 2014), free will (Glannon 2014a, b), the duty to use cognitive enhancers in high-responsibility professions (Santoni de Sio et al. 2014), determining the population of people in need of brain enhancement (Schleim 2014), informed public policy (Shook et al. 2014), cognitive biases (Caviola et al. 2014), and the hype caused by the development of brain-augmenting approaches (Rusconi and Mitchener-Nissen 2014). Moreover, Farah (2015) has presented a "toolbox" of concepts to help neurotechnologists analyze these issues across differences of ethical intuition. It is very important that potential military applications be examined from a neuroethical standpoint (Munyon 2018).

7 Conclusion

In this review we discussed recent technological, physiological, and pathological findings involving the augmentation of cortical microcircuit functions and the interactions with subcortical networks that interface cognition and emotion. Knowledge as to the role of these technological advances holds the promise of new treatments for patients with neurological and psychiatric disorders such as neural prosthetics for executive control of behavior and emotions.

Timeliness: In the last few years progress has been made regarding the interface of cognition and neural technology that deserves to be reviewed and synthesized for a broad audience encompassing students, medical researchers, and clinicians.

The use of optogenetics, nanotechnology, nanomagnetism, and the MEAs to record simultaneously from adjacent cortical mini- or microcolumns in supra- and infragranular layers has provided unprecedented insight into the functioning of cortical microcircuits and cortical–subcortical interactions. These findings offer the promise of new treatment avenues for patients with neurological and psychiatric disorders.

The highlights of the review are:

- BCIs
- Noninvasive neural prosthetics

- Nanotechnology and nanochips
- Transfer of memory between two brains
- Microcircuit vs. macrostructure disruption in brain disorders: autism, schizophrenia, dementia, anxiety, depression, post-traumatic stress disorder, and drug addiction
- The use of optogenetics, nanotechnology, nanomagnetism to augment and repair brain function

References

Ahani A, Moghadamfalahi M, Erdogmus D (2018) Language-model assisted and icon-based communication through a brain computer interface with different presentation paradigms. IEEE Trans Neural Syst Rehabil Eng 26:1835–1844. https://doi.org/10.1109/TNSRE.2018.2859432

Ajayan PM (1999) Nanotubes from carbon. Chem Rev 99:1787–1799. https://doi.org/10.1021/cr970102g

Aldehri M, Termel Y, Alnaami I, Jahanshahi A, Hescham S (2018) Deep brain stimulation for Alzheimer's disease: an update. Surg Neurol Int 9:58

Alwis DS, Rajan R (2014) Environmental enrichment and the sensory brain: the role of enrichment in remediating brain injury. Front Syst Neurosci 8:156. https://doi.org/10.3389/fnsys.2014.00156

Amatachaya A, Auvichayapat N, Patjanasoontorn N, Suphakunpinyo C, Ngernyam N, Aree-Uea B, Keeratitanont K, Auvichayapat P (2014) Effect of anodal transcranial direct current stimulation on autism: a randomized double-blind crossover trial. Behav Neurol 2014:173073. https://doi.org/10.1155/2014/173073

Astrand E, Wardak C, Ben Hamed S (2014) Selective visual attention to drive cognitive brain–machine interfaces: from concepts to neurofeedback and rehabilitation applications. Front Syst Neurosci 8:144. https://doi.org/10.3389/fnsys.2014.00144

Atcherson SR, Moreland C, Zazove P, McKee MM (2015) Hearing loss: hearing augmentation. FP Essent 434:18–23

Attiah MA, Farah MJ (2014) Minds, motherboards, and money: futurism and realism in the neuroethics of BCI technologies. Front Syst Neurosci 8:86. https://doi.org/10.3389/fnsys.2014.00086

Ayache SS, Palm U, Chalah MA, Al-Ani T, Brignol A, Abdellaoui M, Dimitri D, Sorel M, Créange A, Lefaucheur J-P (2016) Prefrontal tDCS decreases pain in patients with multiple sclerosis. Front Neurosci 10:147. https://doi.org/10.3389/fnins.2016.00147

Bai Y, Xia X, Kang J, Yin X, Yang Y, He J, Li X (2016) Evaluating the effect of repetitive transcranial magnetic stimulation on disorders of consciousness by using TMS-EEG. Front Neurosci 10:473. https://doi.org/10.3389/fnins.2016.00473

Barolat G, Oakley J, Law JD et al (2001) Epidural spinal cord stimulation for failed back surgery syndrome. Neuromodulation 4:1

Bashir S, Yoo W-K (2016) Neuromodulation for addiction by transcranial direct current stimulation: opportunities and challenges. Ann Neurosci 23(4):241–245

Beatty EL, Jobidon M-E, Bouak F, Nakashima A, Smith I, Lam Q, Blackler K, Cheung B, Vartanian O (2015) Transfer of training from one working memory task to another: behavioural and neural evidence. Front Syst Neurosci 9:86. https://doi.org/10.3389/fnsys.2015.00086

Berger TW, Hampson RE, Song D, Goonawardena A, Marmarelis VZ, Deadwyler SA (2011) A cortical neural prosthesis for restoring and enhancing memory. J Neural Eng 8(4):046017. https://doi.org/10.1088/1741-2560/8/4/046017

Bennabi D, Pedron S, Haffen E, Monnin J, Peterschmitt Y, Van Waes V (2014) Transcranial direct current stimulation for memory enhancement: from clinical research to animal models. Front Syst Neurosci 8:159. https://doi.org/10.3389/fnsys.2014.00159

Bhatti S et al (2017) Spintronics based random access memory: a review. Mater Today 20(9):530–548. https://doi.org/10.1016/j.mattod.2017.07.007

Billeci L, Tonacci A, Tartarisco G, Narzisi A, Di Palma S, Corda D, Baldus G, Cruciani F, Anzalone SM, Calderoni S, Pioggia G, Muratori F, Michelangelo Study Group (2016) An integrated approach for the monitoring of brain and autonomic response of children with autism spectrum disorders during treatment by wearable technologies. Front Neurosci 10:276. https://doi.org/10.3389/fnins.2016.00276

Blankerts B, Acqualagna L, Dähne S, et al (2016) The berlin brain-computer interface: progress beyond communication and control. Front Neurosci 10:530. Published 2016 Nov 21. https://doi.org/10.3389/fnins.2016.00530

Blumberg EJ, Peterson MS, Parasuraman R (2015) Enhancing multiple object tracking performance with noninvasive brain stimulation: a causal role for the anterior intraparietal sulcus. Front Syst Neurosci 9:3. https://doi.org/10.3389/fnsys.2015.00003

Bouyarmane K, Vaillant J, Sugimoto N, Keith F, Furukawa J, Morimoto J (2014) Brain-machine interfacing control of whole-body humanoid motion. Front Syst Neurosci 8:138. https://doi.org/10.3389/fnsys.2014.00138

Boyden ES, Zhang F, Bamberg E, Nagel G, Deisseroth K (2005) Millisecond-timescale, genetically targeted optical control of neural activity. Nat Neurosci 8(9):1263–1268

Brem A-K, Unterburger E, Speight I, Jäncke L (2014) Treatment of visuospatial neglect with biparietal tDCS and cognitive training: a single-case study. Front Syst Neurosci 8:180. https://doi.org/10.3389/fnsys.2014.00180

Burchiel K, Anderson V et al (1996) Prospective, multicenter study of spinal cord stimulation for relief of chronic back and extremity pain. Spine 21:2786–2794

Callan DE, Durantin G, Terzibas C (2015) Classification of single-trial auditory events using dry-wireless EEG during real and motion simulated flight. Front Syst Neurosci 9:11. https://doi.org/10.3389/fnsys.2015.00011

Caria A (2016) Self-regulation of blood oxygenation level dependent response: primary effect or epiphenomenon? Front Neurosci 10:117. https://doi.org/10.3389/fnins.2016.00117

Carr KD (2002) Augmentation of drug reward by chronic food restriction: behavioral evidence and underlying mechanisms. Physiol Behav 76(3):353–364

Casanova MF, Buxhoeveden D, Gomez J (2003) Disruption in the inhibitory architecture of the cell minicolumns: implications for autism. Neuroscientist 9(6):496–507

Casanova MF, Sokhadze E, Opris I, Wang Y, Li X (2015) Autism spectrum disorders: linking neuropathological findings to treatment with transcranial magnetic stimulation. Acta Paediatr 104(4):346–355

Caviola L, Mannino A, Savulescu J, Faulmüller N (2014) Cognitive biases can affect moral intuitions about cognitive enhancement. Front Syst Neurosci 8:195. https://doi.org/10.3389/fnsys.2014.00195

Chapman SB, Mudar RA (2014) Enhancement of cognitive and neural functions through complex reasoning training: evidence from normal and clinical populations. Front Syst Neurosci 8:69. https://doi.org/10.3389/fnsys.2014.00069

Charvet LE, Kasschau M, Datta A, Knotkova H, Stevens MC, Alonzo A, Loo C, Krull KR, Bikson M (2015) Remotely-supervised transcranial direct current stimulation (tDCS) for clinical trials: guidelines for technology and protocols. Front Syst Neurosci 9:26. https://doi.org/10.3389/fnsys.2015.00026

Chen X, Wang Y, Nakanishi M, Gao X, Jung TP, Gao S (2015) High-speed spelling with a noninvasive brain-computer interface. Proc Natl Acad Sci U S A 112(44):E6058–E6067. https://doi.org/10.1073/pnas.1508080112

Clark VP (2014) The ethical, moral, and pragmatic rationale for brain augmentation. Front Syst Neurosci 8:130. https://doi.org/10.3389/fnsys.2014.00130

Cone JJ, Ni AM, Ghose K, Maunsell JHR (2018) Electrical microstimulation of visual cerebral cortex elevates psychophysical detection thresholds. eNeuro 5(5) pii: ENEURO.0311-18.2018. https://doi.org/10.1523/ENEURO.0311-18.2018

Convento S, Bolognini N, Fusaro M, Lollo F, Vallar G (2014) Neuromodulation of parietal and motor activity affects motor planning and execution. Cortex 57:51–59. https://doi.org/10.1016/j.cortex.2014.03.006
Cook IA, Abrams M, Leuchter AF (2016) Trigeminal nerve stimulation for comorbid posttraumatic stress disorder and major depressive disorder. Neuromodulation 19(3):299–305
Datta S, Das B (1990) Electronic analog of the electrooptic modulator. Appl Phys Lett 56(7):665–667. https://doi.org/10.1063/1.102730
Davis NJ, van Koningsbruggen MG (2013) "Non-invasive" brain stimulation is not non-invasive. Front Syst Neurosci 7:76. https://doi.org/10.3389/fnsys.2013.00076
Deadwyler SA, Berger TW, Sweatt AJ, Song D, Chan RHM, Opris I, Gerhardt GA, Marmarelis VZ, Hampson RE (2013) Donor/recipient enhancement of memory in rat hippocampus. Front Syst Neurosci 7:120. https://doi.org/10.3389/fnsys.2013.00120
Deadwyler SA, Hampson RE, Song D, Opris I, Gerhardt GA, Marmarelis VZ, Berger TW (2017) A cognitive prosthesis for memory facilitation by closed-loop functional ensemble stimulation of hippocampal neurons in primate brain. Exp Neurol 287(Pt 4):452–460. https://doi.org/10.1016/j.expneurol.2016.05.031
DeMarse TB, Carney PR (2014) Augmentation of cognitive function in epilepsy. Front Syst Neurosci 8:147. https://doi.org/10.3389/fnsys.2014.00147
Demeter E, Sarter M (2013) Leveraging the cortical cholinergic system to enhance attention. Neuropharmacology 64:294–304. https://doi.org/10.1016/j.neuropharm.2012.06.060
Deng M, Li Z, Kang Y, Chen CLP, Chu X (2018) A learning-based hierarchical control scheme for an exoskeleton robot in human-robot cooperative manipulation. IEEE Trans Cybern 50(1):112–125. https://doi.org/10.1109/TCYB.2018.2864784
Deveau J, Jaeggi SM, Zordan V, Phung C, Seitz AR (2015) How to build better memory training games. Front Syst Neurosci 8:243. https://doi.org/10.3389/fnsys.2014.00243
Di Pino G, Maravita A, Zollo L, Guglielmelli E, Di Lazzaro V (2014) Augmentation-related brain plasticity. Front Syst Neurosci 8:109. https://doi.org/10.3389/fnsys.2014.00109
Diekelmann S (2014) Sleep for cognitive enhancement. Front Syst Neurosci 8:46. https://doi.org/10.3389/fnsys.2014.00046
Duecker F, de Graaf TA, Sack AT (2014) Thinking caps for everyone? The role of neuro-enhancement by non-invasive brain stimulation in neuroscience and beyond. Front Syst Neurosci 8:71. https://doi.org/10.3389/fnsys.2014.00071
Dutta A (2015) Bidirectional interactions between neuronal and hemodynamic responses to transcranial direct current stimulation (tDCS): challenges for brain-state dependent tDCS. Front Syst Neurosci 9:107. https://doi.org/10.3389/fnsys.2015.00107
Earp BD, Sandberg A, Kahane G, Savulescu J (2014) When is diminishment a form of enhancement? Rethinking the enhancement debate in biomedical ethics. Front Syst Neurosci 8:12. https://doi.org/10.3389/fnsys.2014.00012
Ekanayake J, Hutton C, Ridgway G, Scharnowski F, Weiskopf N, Rees G (2018) Real-time decoding of covert attention in higher-order visual areas. Neuroimage 169:462–472. https://doi.org/10.1016/j.neuroimage.2017.12.019
Enachescu M, Schleef D, Ogletree DF (1999) Integration of point-contact microscopy and atomic-force microscopy: application to characterization of graphite/Pt(111). Phys Rev B 60:16913–16919. https://doi.org/10.1103/PhysRevB.60.16913
Farah MJ (2015) An ethics toolbox for neurotechnology. Neuron 86(1):34–37. https://doi.org/10.1016/j.neuron.2015.03.038
Fetterhoff D, Opris I, Simpson SL, Deadwyler SA, Hampson RE, Kraft RA (2015) Multifractal analysis of information processing in hippocampal neural ensembles during working memory under Δ^9-tetrahydrocannabinol administration. J Neurosci Methods 244:136–153. https://doi.org/10.1016/j.jneumeth.2014.07.013
Foroughi CK, Blumberg EJ, Parasuraman R (2015) Activation and inhibition of posterior parietal cortex have bi-directional effects on spatial errors following interruptions. Front Syst Neurosci 8:245. https://doi.org/10.3389/fnsys.2014.00245
Franco R (2014) Enhancing cognition before clinical symptoms of dementia. Front Syst Neurosci 8:240. https://doi.org/10.3389/fnsys.2014.00240

Garriga M, Milà M, Mir M, Al-Baradie R, Huertas S, Castejon C, Casas L, Badenes D, Giménez N, Font MA, Gonzalez JM, Ysamat M, Aguilar M, Slevin M, Krupinski J (2015) 123I-FP-CIT SPECT imaging in early diagnosis of dementia in patients with and without a vascular component. Front Syst Neurosci 9:99. https://doi.org/10.3389/fnsys.2015.00099

Glannon W (2014a) Ethical issues with brain-computer interfaces. Front Syst Neurosci 8:136. https://doi.org/10.3389/fnsys.2014.00136

Glannon W (2014b) Prostheses for the will. Front Syst Neurosci 8:79. https://doi.org/10.3389/fnsys.2014.00079

Gonzalez-Lima F, Barrett DW (2014) Augmentation of cognitive brain functions with transcranial lasers. Front Syst Neurosci 8:36. https://doi.org/10.3389/fnsys.2014.00036

Grau C, Ginhoux R, Riera A, Nguyen TL, Chauvat H, Berg M, Amengual JL, Pascual-Leone A, Ruffini G (2014) Conscious brain-to-brain communication in humans using non-invasive technologies. PLoS One 9(8):e105225. https://doi.org/10.1371/journal.pone.0105225

Graziano MSA (2016) Ethological action maps: a paradigm shift for the motor cortex. Trends Cogn Sci 20(2):121–132. https://doi.org/10.1016/j.tics.2015.10.008

Graziano MS, Aflalo TN, Cooke DF (2005) Arm movements evoked by electrical stimulation in the motor cortex of monkeys. J Neurophysiol 94(6):4209–4223

Grimm F, Naros G, Gharabaghi A (2016) Compensation or restoration: closed-loop feedback of movement quality for assisted reach-to-grasp exercises with a multi-joint arm exoskeleton. Front Neurosci 10:280. https://doi.org/10.3389/fnins.2016.00280

Haier RJ (2014) Increased intelligence is a myth (so far). Front Syst Neurosci 8:34. https://doi.org/10.3389/fnsys.2014.00034

Hampson RE, Gerhardt GA, Marmarelis V, Song D, Opris I, Santos L, Berger TW, Deadwyler SA (2012) Facilitation and restoration of cognitive function in primate prefrontal cortex by a neuroprosthesis that utilizes minicolumn-specific neural firing. J Neural Eng 9(5):056012. This paper provided the first evidence for feasibility of a cognitive neural prosthetics to improve executive function and implementation in a primate brain

Herschkowitz D, Kubias J (2018) Wireless peripheral nerve stimulation for complex regional pain syndrome type I of the upper extremity: a case illustration introducing a novel technology. Scand J Pain 18(3):555–560. https://doi.org/10.1515/sjpain-2018-0014

Höller Y, Trinka E (2014) What do temporal lobe epilepsy and progressive mild cognitive impairment have in common? Front Syst Neurosci 8:58. https://doi.org/10.3389/fnsys.2014.00058

Horschig JM, Zumer JM, Bahramisharif A (2014) Hypothesis-driven methods to augment human cognition by optimizing cortical oscillations. Front Syst Neurosci 8:119. https://doi.org/10.3389/fnsys.2014.00119

Horvath JC, Carter O, Forte JD (2014) Transcranial direct current stimulation: five important issues we aren't discussing (but probably should be). Front Syst Neurosci 8:2. https://doi.org/10.3389/fnsys.2014.00002

Huss DS, Dallapiazza RF, Shah BB, Harrison MB, Diamond J, Elias WJ (2015) Functional assessment and quality of life in essential tremor with bilateral or unilateral DBS and focused ultrasound thalamotomy. Mov Disord 30(14):1937–1943. https://doi.org/10.1002/mds.26455

Iijima S (1991) Helical microtubules of graphitic carbon. Nature 354:56–58. https://doi.org/10.1038/354056a0

Irastorza-Landa N, Sarasola-Sanz A, Lopez-Larraz E, Bibian C, Shiman P, Birbaumer N, Ramos-Murguialday A (2017) Design of continuous EMG classification approaches towards the control of a robotic exoskeleton in reaching movements. IEEE Int Conf Rehabil Robot 2017:128–133. https://doi.org/10.1109/ICORR.2017.8009234

Jahanshahi M, Torkamani M, Beigi M, Wilkinson L, Page D, Madeley L, Bhatia K, Hariz M, Zrinzo L, Limousin P, Ruge D, Tisch S (2014) Pallidal stimulation for primary generalised dystonia: effect on cognition, mood and quality of life. J Neurol 261(1):164–173. https://doi.org/10.1007/s00415-013-7161-2

Julliere M (1975) Tunneling between ferromagnetic films. Phys Lett A 54(3):225–226. https://doi.org/10.1016/0375-9601(75)90174-7

Kang J, Cai E, Han J, Tong Z, Li X, Sokhadze EM, Casanova MF, Li X (2018) Transcranial direct current stimulation (tDCS) can modulate EEG complexity of children with autism spectrum disorder. Front Neurosci 12:201

Kapeller C, Kamada K, Ogawa H, Prueckl R, Scharinger J, Guger C (2014) An electrocorticographic BCI using code-based VEP for control in video applications: a single-subject study. Front Syst Neurosci 8:139. https://doi.org/10.3389/fnsys.2014.00139

Kemler MA, Henrica CW et al (2004) The effect of spinal cord stimulation in patients with chronic reflex sympathetic dystrophy: two years' follow-up of the randomised controlled trial. Ann Neurol 55:13–18

Kern DS, Picillo M, Thompson JA, Sammartino F, di Biase L, Munhoz RP, Fasano A (2019) Interleaving stimulation in Parkinson's disease, tremor, and dystonia. Stereotact Funct Neurosurg 96(6):379–391. https://doi.org/10.1159/000494983

Kiehl EL, Parker AM, Matar RM, Gottbrecht MF, Johansen MC, Adams MP, Griffiths LA, Dunn SP, Bidwell KL, Menon V, Enfield KB, Gimple LW (2017) C-GRApH: a validated scoring system for early stratification of neurologic outcome after out-of-hospital cardiac arrest treated with targeted temperature management. J Am Heart Assoc 6(5):e003821. https://doi.org/10.1161/JAHA.116.003821

Kimura N, Rekimoto J (2018) ExtVision: augmentation of visual experiences with generation of context images for a peripheral vision using deep neural network. 2018 CHI Conference. https://doi.org/10.1145/3173574.3174001

Knöpfel T (2012) Genetically encoded optical indicators for the analysis of neuronal circuits. Nat Rev Neurosci 13(10):687–700. https://doi.org/10.1038/nrn3293

Koganemaru S, Fukuyama H, Mima T (2015) Two is more than one: how to combine brain stimulation and rehabilitative training for functional recovery? Front Syst Neurosci 9:154. https://doi.org/10.3389/fnsys.2015.00154

Koivuniemi A, Otto K (2014) When "altering brain function" becomes "mind control". Front Syst Neurosci 8:202. https://doi.org/10.3389/fnsys.2014.00202

Krack P, Batir A, Blercome N et al (2003) Five-year follow-up of bilateral stimulation of the subthalamic nucleus in advanced Parkinson's disease. N Engl Med J 349(13):1925–1934

Krause B, Cohen Kadosh R (2014) Not all brains are created equal: the relevance of individual differences in responsiveness to transcranial electrical stimulation. Front Syst Neurosci 8:25. https://doi.org/10.3389/fnsys.2014.00025

Krawinkel LA, Engel AK, Hummel FC (2015) Modulating pathological oscillations by rhythmic non-invasive brain stimulation—a therapeutic concept? Front Syst Neurosci 9:33. https://doi.org/10.3389/fnsys.2015.00033

Kristensen LB, Wang L, Petersson KM, Hagoort P (2013) The interface between language and attention: prosodic focus marking recruits a general attention network in spoken language comprehension. Cereb Cortex 23(8):1836–1848. https://doi.org/10.1093/cercor/bhs164

Krucoff MO, Rahimpour S, Slutzky MW, Edgerton VR, Turner DA (2016) Enhancing Nervous System Recovery through Neurobiologics, Neural Interface Training, and Neurorehabilitation. Front Neurosci 10:584. https://doi.org/10.3389/fnins.2016.00584

Kubera KM, Barth A, Hirjak D, Thomann PA, Wolf RC (2015) Noninvasive brain stimulation for the treatment of auditory verbal hallucinations in schizophrenia: methods, effects and challenges. Front Syst Neurosci 9:131. https://doi.org/10.3389/fnsys.2015.00131

Kumar K, Toth C, Nath RK et al (1998) Epidural spinal cord stimulation for treatment of chronic pain—some predictors of success. A 15 year experience. Surg Neurol 50:110–121

Lapenta OM, Marques LM, Rego GG, Confort WE, Boggio PS (2018) tDCS in addiction and impulse control disorders. J ECT 34(3):182–192

Lebedev MA, Tate AJ, Hanson TL, Li Z, O'Doherty JE, Winans JA, Ifft PJ, Zhuang KZ, Fitzsimmons NA, Schwarz DA, Fuller AM, An JH, Nicolelis MA (2011) Future developments in brain-machine interface research. Clinics (Sao Paulo) 66(Suppl 1):25–32

Lebedev AV, Westman E, Simmons A, Lebedeva A, Siepel FJ, Pereira JB, Aarsland D (2014) Large-scale resting state network correlates of cognitive impairment in Parkinson's disease and related dopaminergic deficits. Front Syst Neurosci 8:45. https://doi.org/10.3389/fnsys.2014.00045

Lebedev MA, Opris I, Casanova MF (2018) Editorial: augmentation of brain function; facts, fiction and controversy. Front Syst Neurosci 12:45. https://doi.org/10.3389/fnsys.2018.00045. eCollection.2018

Lee S, Kim DJ, Svenkeson D, Parras G, Oishi MMK, McKeown MJ (2015) Multifaceted effects of noisy galvanic vestibular stimulation on manual tracking behavior in Parkinson's disease. Front Syst Neurosci 9:5. https://doi.org/10.3389/fnsys.2015.00005

Li Z (2014) Decoding methods for neural prostheses: where have we reached? Front Syst Neurosci 8:129. https://doi.org/10.3389/fnsys.2014.00129

Limori T, Nakajima S, Miyazaki T et al (2019) Effectiveness of the prefrontal repetitive transcranial magnetic stimulation on cognitive profiles in depression, schizophrenia, and Alzheimer's disease: a systematic review. Prog Neuropsychopharmacol Biol Psychiatry 88(31-40):2019

Lisi G, Noda T, Morimoto J (2014) Decoding the ERD/ERS: influence of afferent input induced by a leg assistive robot. Front Syst Neurosci 8:85. https://doi.org/10.3389/fnsys.2014.00085

Liu TC, Chuang MC, Chu CY, Huang WC, Lai HY, Wang CT, Chu WL, Chen SY, Chen YY (2016) Implantable graphene-based neural electrode interfaces for electrophysiology and neurochemistry in in vivo hyperacute stroke model. ACS Appl Mater Interfaces 8(1):187–196. https://doi.org/10.1021/acsami.5b08327

Luber B (2014) Neuroenhancement by noninvasive brain stimulation is not a net zero-sum proposition. Front Syst Neurosci 8:127. https://doi.org/10.3389/fnsys.2014.00127

Luft CDB, Pereda E, Banissy MJ, Bhattacharya J (2014) Best of both worlds: promise of combining brain stimulation and brain connectome. Front Syst Neurosci 8:132. https://doi.org/10.3389/fnsys.2014.00132

Luigies J, Segrave R, de Joode N, Figee M, Denys D (2018) Efficacy of invasive and non-invasive brain modulation interventions for addiction. Neuropsychol Rev 29(1):116–138. https://doi.org/10.1007/s11065-018-9393-5

Lupi M, Martinotti G, Santacroce R et al (2017) Transcranial direct current stimulation in substance use disorders: a systematic review of scientific literature. J ECT 33(3):203–209

Lynch G, Cox CD, Gall CM (2014) Pharmacological enhancement of memory or cognition in normal subjects. Front Syst Neurosci 8:90. https://doi.org/10.3389/fnsys.2014.00090

Madan CR (2014) Augmented memory: a survey of the approaches to remembering more. Front Syst Neurosci 8:30. https://doi.org/10.3389/fnsys.2014.00030

Mandonnet E, Duffau H (2014) Understanding entangled cerebral networks: a prerequisite for restoring brain function with brain-computer interfaces. Front Syst Neurosci 8:82. https://doi.org/10.3389/fnsys.2014.00082

Marchesotti S, Martuzzi R, Schurger A, Blefari ML, Del Millán JR, Bleuler H, Blanke O (2017) Cortical and subcortical mechanisms of brain-machine interfaces. Hum Brain Mapp 38(6):2971–2989. https://doi.org/10.1002/hbm.23566

Maslen H, Faulmüller N, Savulescu J (2014) Pharmacological cognitive enhancement—how neuroscientific research could advance ethical debate. Front Syst Neurosci 8:107. https://doi.org/10.3389/fnsys.2014.00107

McFarland DJ, Parvaz MA, Sarnacki WA, Goldstein RZ, Wolpaw JR (2017) Prediction of subjective ratings of emotional pictures by EEG features. J Neural Eng 14(1):016009. https://doi.org/10.1088/1741-2552/14/1/016009

McKendrick R, Parasuraman R, Ayaz H (2015) Wearable functional near infrared spectroscopy (fNIRS) and transcranial direct current stimulation (tDCS): expanding vistas for neurocognitive augmentation. Front Syst Neurosci 9:27. https://doi.org/10.3389/fnsys.2015.00027

Merrill CA, Jonsson MA, Minthon L, Ejnell H, C-son Silander H, Blennow K et al (2006) Vagus nerve stimulation in patients with Alzheimer's disease: additional follow-up results of a pilot study through 1 year. J Clin Psychiatry 67:1171–1178. https://doi.org/10.4088/JCP.v67n0801

Mitchell T (2017) Augmented reality solution for color vision deficiency. http://arinmed.com/augmented-reality-solution-for-color-vision-deficiency/

Mobbs RJ et al (2007) Peripheral nerve stimulation for the treatment of chronic pain. J Clin Neurosci 14(3):216–222

Moghadamfalahi M, Orhan U, Akcakaya M, Nezamfar H, Fried-Oken M, Erdogmus D (2015) Language-model assisted brain computer interface for typing: a comparison of matrix and rapid serial visual presentation. IEEE Trans Neural Syst Rehabil Eng 23(5):910–920. https://doi.org/10.1109/TNSRE.2015.2411574

Mohr P, Rdriguez M, Slavickova A, Hanka J (2011) The application of vagus nerve stimulation and deep brain stimulation in depression. Neuropsychobiology 64(3):170–181

Mora-Cortes A, Manyakov NV, Chumerin N, Van Hulle MM (2014) Model applications to spelling with brain-computer interfaces. Sensors (Basel) 14(4):5967–5993. https://doi.org/10.3390/s140405967

Moreau D (2014) Making sense of discrepancies in working memory training experiments: a Monte Carlo simulation. Front Syst Neurosci 8:161. https://doi.org/10.3389/fnsys.2014.00161

Morita M (1996) Memory and learning of sequential patterns by nonmonotone neural networks. Neural Netw 9(8):1477–1489

Munyon CN (2018) Neuroethics of non-primary brain computer interface: focus on potential military applications. Front Neurosci 12:696. https://doi.org/10.3389/fnins.2018.00696

Nagel SK (2014) Enhancement for well-being is still ethically challenging. Front Syst Neurosci 8:72. https://doi.org/10.3389/fnsys.2014.00072

Nagel SJ, Machado AG, Gale JT, Lobel DA, Pandya M (2015) Preserving cortico-striatal function: deep brain stimulation in Huntington's disease. Front Syst Neurosci 9:32. https://doi.org/10.3389/fnsys.2015.00032

Najarpour Foroushani A, Pack CC, Sawan M (2018) Cortical visual prostheses: from microstimulation to functional percept. J Neural Eng 15(2):021005. https://doi.org/10.1088/1741-2552/aaa904

Nie X, Zheng WX, Cao J (2016) Coexistence and local μ-stability of multiple equilibrium points for memristive neural networks with nonmonotonic piecewise linear activation functions and unbounded time-varying delays. Neural Netw 84:172–180. https://doi.org/10.1016/j.neunet.2016.08.006

Niketeghad S, Pouratian N (2018) Brain machine interfaces for vision restoration: the current state of cortical visual prosthetics. Neurotherapeutics 16:134–143. https://doi.org/10.1007/s13311-018-0660-1

Nimmagadda K, Weiland JD (2018) Retinotopic responses in the visual cortex elicited by epiretinal electrical stimulation in normal and retinal degenerate rats. Transl Vis Sci Technol 7(5):33. https://doi.org/10.1167/tvst.7.5.33

Nishimori H, Opris I (1993) Retrieval process of an associative memory with general input-output function. Neural Netw 6:1061–1067

Noga BR, Opris I (2017) The hierarchical circuit for executive control of movement. Chapter 5. In: Opris I, Casanova MF (eds) Physics of the mind and brain disorders: advances in electrostimulation therapies. Springer series in neural systems. Springer Nature, Berlin, pp 95–128

Noga BR, Sanchez FJ, Villamil LM, O'Toole C, Kasicki S, Olszewski M, Cabaj AM, Majczyński H, Sławińska U, Jordan LM (2017) LFP oscillations in the mesencephalic locomotor region during voluntary locomotion. Front Neural Circuits 11:34. https://doi.org/10.3389/fncir.2017.00034.eCollection.2017

O'Doherty JE, Lebedev MA, Ifft PJ, Zhuang KZ, Shokur S, Bleuler H et al (2011) Active tactile exploration using a brain-machine-brain interface. Nature 479:228–231. https://doi.org/10.1038/nature10489

Oken BS, Orhan U, Roark B, Erdogmus D, Fowler A, Mooney A, Peters B, Miller M, Fried-Oken MB (2014) Brain-computer interface with language model-electroencephalography fusion for locked-in syndrome. Neurorehabil Neural Repair 28(4):387–394. https://doi.org/10.1177/1545968313516867

Onose G, Cârdei V, Crăciunoiu ŞT, Avramescu V, Opriş I, Lebedev MA, Constantinescu MV (2016) Mechatronic wearable exoskeletons for bionic bipedal standing and walking: a new synthetic approach. Front Neurosci 10:343. https://doi.org/10.3389/fnins.2016.00343

Opris I (2013) Inter-laminar microcircuits across neocortex: repair and augmentation. Front Syst Neurosci 7:80. https://doi.org/10.3389/fnsys.2013.00080

Opris I, Casanova MF (2014) Prefrontal cortical minicolumn: from executive control to disrupted cognitive processing. Brain 137(7):1863–1875

Opris I, Barborica A, Ferrera VP (2005a) Microstimulation of dorsolateral prefrontal cortex biases saccade target selection. J Cogn Neurosci 17:893–904

Opris I, Barborica A, Ferrera VP (2005b) Effects of electrical microstimulation in monkey frontal eye field on saccades to remembered targets. Vision Res 45:3414–3429

Opris I, Fuqua JL, Huettl P, Gerhardt GA, Berger TW, Hampson RE, Deadwyler SA (2012a) Closing the loop in primate prefrontal cortex: inter-laminar processing. Front Neural Circuits 6:88

Opris I, Hampson RE, Gerhardt GA, Berger TW, Deadwyler SA (2012b) Columnar processing in primate prefrontal cortex: evidence for executive control microcircuits. J Cogn Neurosci 24(12):2334–2347

Opris I, Santos L, Gerhardt GA, Song D, Berger TW, Hampson RE, Deadwyler SA (2013) Prefrontal cortical microcircuits bind perception to executive control. Sci Rep 3:2285. https://doi.org/10.1038/srep02285

Opris I, Gerhardt GA, Hampson RE, Deadwyler SA (2014) Prefrontal cortical recordings with biomorphic MEAs reveals complex columnar-laminar microcircuits for BMI implementation. J Neurosci Methods 244:104–113

Ordikhani-Seyedlar M, Lebedev MA, Sorensen HBD, Puthusserypady S (2016) Neurofeedback therapy for enhancing visual attention: state-of-the-art and challenges. Front Neurosci 10:352. https://doi.org/10.3389/fnins.2016.00352

Oyama G, Okun MS, Schmidt P, Tröster AI, Nutt J, Go CL, Foote KD, Malaty IA (2014) Deep brain stimulation may improve quality of life in people with Parkinson's disease without affecting caregiver burden. Neuromodulation 17(2):126–132

Pais-Vieira M, Lebedev M, Kunicki C, Wang J, Nicolelis MA (2013) A brain-to-brain interface for real-time sharing of sensorimotor information. Sci Rep 3:1319. https://doi.org/10.1038/srep01319

Pais-Vieira M, Chiuffa G, Lebedev M, Yadav A, Nicolelis MA (2015) Building an organic computing device with multiple interconnected brains. Sci Rep 5:11869. https://doi.org/10.1038/srep11869

Pampaloni NP, Giugliano M, Scaini D, Ballerini L, Rauti R (2019) Advances in nano neuroscience: from nanomaterials to nanotools. Front Neurosci 12:953. https://doi.org/10.3389/fnins.2018.00953.eCollection.2018

Panebianco M, Rigby A, Weston J, Marson AG (2015) Vagus nerve stimulation for partial seizures. Cochrane Database Syst Rev 3(4):CD002896

Parasuraman R, Galster S (2013) Sensing, assessing, and augmenting threat detection: behavioral, neuroimaging, and brain stimulation evidence for the critical role of attention. Front Hum Neurosci 7:273. https://doi.org/10.3389/fnhum.2013.00273.eCollection.2013

Parkin S, Yang SH (2015) Memory on the racetrack. Nat Nanotechnol 10(3):195–198. https://doi.org/10.1038/nnano.2015.41

Parkin SS, Hayashi M, Thomas L (2008) Magnetic domain-wall racetrack memory. Science 320(5873):190–194. https://doi.org/10.1126/science.1145799

Pena DF, Childs JE, Willett S, Vital A, McIntyre CK, Kroener S (2014) Vagus nerve stimulation enhances extinction of conditioned fear and modulates plasticity in the pathway from the ventromedial prefrontal cortex to the amygdala. Front Behav Neurosci 18(8):327

Pezzulo G, D'Ausilio A, Gaggioli A (2016) Predictive technologies: can smart tools augment the brain's predictive abilities? Front Neurosci 10:186. https://doi.org/10.3389/fnins.2016.00186

Pigarev IN, Pigareva ML (2014) Partial sleep in the context of augmentation of brain function. Front Syst Neurosci 8:75. https://doi.org/10.3389/fnsys.2014.00075

Plutchik R (1997) The measurement of emotions. Acta Neuropsychiatr 9(2):58–60. https://doi.org/10.1017/S0924270800036802

Popa I, Donos C, Barborica A, Opris I, Mălîia MD, Ene M, Ciurea J, Mîndruţă I (2016) Intrusive thoughts elicited by direct electrical stimulation during stereo-electroencephalography. Front Neurol 7:114

Prins NW et al (2013) Reward signals for reinforcement based BMI. Otolaryngol Clin North Am 51(4):705–723. https://doi.org/10.1016/j.otc.2018.03.002
Pulgar VM (2015) Direct electric stimulation to increase cerebrovascular function. Front Syst Neurosci 9:54. https://doi.org/10.3389/fnsys.2015.00054
Rashba EI (1960) Cyclotron and combined resonances in a perpendicular field. Sov Phys Solid State 2:1109–1122
Rigato J, Murakami M, Mainen Z (2014) Spontaneous decisions and free will: empirical results and philosophical considerations. Cold Spring Harb Symp Quant Biol 79:177–184. https://doi.org/10.1101/sqb.2014.79.024810
Rusconi E, Mitchener-Nissen T (2014) The role of expectations, hype and ethics in neuroimaging and neuromodulation futures. Front Syst Neurosci 8:214. https://doi.org/10.3389/fnsys.2014.00214
Santoni de Sio F, Faulmüller N, Vincent NA (2014) How cognitive enhancement can change our duties. Front Syst Neurosci 8:131. https://doi.org/10.3389/fnsys.2014.00131
Sarasola-Sanz A, Irastorza-Landa N, Lopez-Larraz E, Bibian C, Helmhold F, Broetz D, Birbaumer N, Ramos-Murguialday A (2017) A hybrid brain-machine interface based on EEG and EMG activity for the motor rehabilitation of stroke patients. IEEE Int Conf Rehabil Robot 2017:895–900. https://doi.org/10.1109/ICORR.2017.8009362
Schicktanz S, Amelung T, Rieger JW (2015) Qualitative assessment of patients' attitudes and expectations toward BCIs and implications for future technology development. Front Syst Neurosci 9:64. https://doi.org/10.3389/fnsys.2015.00064
Schleim S (2014) Whose well-being? Common conceptions and misconceptions in the enhancement debate. Front Syst Neurosci 8:148. https://doi.org/10.3389/fnsys.2014.00148
Seier M, Hiller A, Quinn J, Murchison C, Brodsky M, Anderson S (2018) Alternating thalamic deep brain stimulation for essential tremor: a trial to reduce habituation. Mov Disord Clin Pract 5(6):620–626. https://doi.org/10.1002/mdc3.12685.eCollection
Senova S, Cotovio G, Pascual-Leone A, Oliveira-Maia AJ (2019) Durability of antidepressant response to repetitive transcranial magnetic stimulation: systematic review and meta-analysis. Brain Stimul 12(1):119–128
Serpell CJ, Kostarelos K, Davis BG (2016) Can carbon nanotubes deliver on their promise in biology? Harnessing unique properties for unparalleled applications. ACS Cent Sci 2(4):190–200. https://doi.org/10.1021/acscentsci.6b00005
Shanechi MM, Hu RC, Powers M, Wornell GW, Brown EN, Williams ZM (2012) Neural population partitioning and a concurrent brain-machine interface for sequential motor function. Nat Neurosci 15(12):1715–1722. https://doi.org/10.1038/nn.3250
Shook JR, Galvagni L, Giordano J (2014) Cognitive enhancement kept within contexts: neuroethics and informed public policy. Front Syst Neurosci 8:228. https://doi.org/10.3389/fnsys.2014.00228
Silva GA (2018) A new frontier: the convergence of nanotechnology, brain machine interfaces, and artificial intelligence. Front Neurosci 12:843. https://doi.org/10.3389/fnins.2018.00843b
Sokhadze EM, El-Baz A, Baruth J, Mathai G, Sears L, Casanova MF (2009) Effects of low frequency repetitive transcranial magnetic stimulation (rTMS) on gamma frequency oscillations and event-related potentials during processing of illusory figures in autism. J Autism Dev Disord 39(4):619–634. https://doi.org/10.1007/s10803-008-0662-7
Sokhadze EM, Baruth JM, Sears L, Sokhadze GE, El-Baz AS, Casanova MF (2012) Prefrontal neuromodulation using rTMS improves error monitoring and correction function in autism. Appl Psychophysiol Biofeedback 37(2):91–102. https://doi.org/10.1007/s10484-012-9182-5
Sokhadze EM, El-Baz AS, Sears LL, Opris I, Casanova MF (2014) rTMS neuromodulation improves electrocortical functional measures of information processing and behavioral responses in autism. Front Syst Neurosci 8:134. https://doi.org/10.3389/fnsys.2014.00134
Solopova IA, Selionov VA, Sylos-Labini F, Gurfinkel VS, Lacquaniti F, Ivanenko YP (2015) Tapping into rhythm generation circuitry in humans during simulated weightlessness conditions. Front Syst Neurosci 9:14. https://doi.org/10.3389/fnsys.2015.00014

Song D, Harway M, Marmarelis VZ, Hampson RE, Deadwyler SA, Berger TW (2014) Extraction and restoration of hippocampal spatial memories with non-linear dynamical modeling. Front Syst Neurosci 8:97. https://doi.org/10.3389/fnsys.2014.00097

Song W, Cajigas I, Brown EN, Giszter SF (2015) Adaptation to elastic loads and BMI robot controls during rat locomotion examined with point-process GLMs. Front Syst Neurosci 9:62. https://doi.org/10.3389/fnsys.2015.00062

Speier W, Arnold C, Pouratian N (2016) Integrating language models into classifiers for BCI communication: a review. J Neural Eng 13(3):031002. https://doi.org/10.1088/1741-2560/13/3/031002

Sydow O, Thobois S, Alesch F, Speelman JD (2003) Multicentre European study of thalamic stimulation in essential tremor: a six year follow-up. J Neurol Neurosurg Psychiatry 74:1387–1391

Tajiri N, Duncan K, Antoine A, Pabon M, Acosta SA, de la Pena I, Hernadez-Ontiveros DG, Shinozuka K, Ishikawa H, Kaneko Y, Yankee E, McGrogan M, Case C, Borlongan CV (2014) Stem cell-paved biobridge facilitates neural repair in traumatic brain injury. Front Syst Neurosci 8:116. https://doi.org/10.3389/fnsys.2014.00116

Takakusaki K (2013) Neurophysiology of gait: from the spinal cord to the frontal lobe. Mov Disord 28(11):1483–1491. https://doi.org/10.1002/mds.25669

Takakusaki K (2017) Functional neuroanatomy for posture and gait control. J Mov Disord 10(1):1–17. https://doi.org/10.14802/jmd.16062

Takeuchi H, Taki Y, Nouchi R, Hashizume H, Sekiguchi A, Kotozaki Y, Nakagawa S, Miyauchi CM, Sassa Y, Kawashima R (2014) Working memory training improves emotional states of healthy individuals. Front Syst Neurosci 8:200. https://doi.org/10.3389/fnsys.2014.00200

Tariq M, Trivailo PM, Simic M (2018) EEG-based BCI control schemes for lower-limb assistive-robots. Front Hum Neurosci 12:312. https://doi.org/10.3389/fnhum.2018.00312

Taya F, Sun Y, Babiloni F, Thakor N, Bezerianos A (2015) Brain enhancement through cognitive training: a new insight from brain connectome. Front Syst Neurosci 9:44. https://doi.org/10.3389/fnsys.2015.00044

Thibeault CM (2014) A role for neuromorphic processors in therapeutic nervous system stimulation. Front Syst Neurosci 8:187. https://doi.org/10.3389/fnsys.2014.00187

Timko BP, Cohen-Karni T, Qing Q, Tian BZ, Lieber CM (2010) Design and implementation of functional nanoelectronic interfaces with biomolecules, cells, and tissue using nanowire device arrays. IEEE Trans Nanotechnol 9:269–280

Toleikis JR, Metman LV, Pilitsis JG, Barborica A, Toleikis SC, Bakay RA (2012) Effect of intraoperative subthalamic nucleus DBS on human single-unit activity in the ipsilateral and contralateral subthalamic nucleus. J Neurosurg 116(5):1134–1143. https://doi.org/10.3171/2011.12.JNS102176

Torrejon J, Riou M, Araujo FA, Tsunegi S, Khalsa G, Querlioz D, Bortolotti P, Cros V, Yakushiji K, Fukushima A, Kubota H, Yuasa S, Stiles MD, Grollier J (2017) Neuromorphic computing with nanoscale spintronic oscillators. Nature 547(7664):428–431. https://doi.org/10.1038/nature23011

Tsagaris KZ, Labar DR, Edwards DJ (2016) A framework for combining rTMS with behavioral therapy. Front Syst Neurosci 10:82. https://doi.org/10.3389/fnsys.2016.00082

Urban KR, Gao W-J (2014) Performance enhancement at the cost of potential brain plasticity: neural ramifications of nootropic drugs in the healthy developing brain. Front Syst Neurosci 8:38. https://doi.org/10.3389/fnsys.2014.00038

Usami K, Kano R, Kawai K, Noda T, Shiramatsu TI, Saito N, Takahashi H (2013) Modulation of cortical synchrony by vagus nerve stimulation in adult rats. Conf Proc IEEE Eng Med Biol Soc 2013:5348–5351. https://doi.org/10.1109/EMBC.2013.6610757

Vaisbuch Y, Santa Maria PL (2018) Age-related hearing loss: innovations in hearing augmentation. Otolaryngol Clin North Am 51(4):705–723

Valeriani D, Cinel C, Poli R (2017) Group augmentation in realistic visual-search decisions via a hybrid brain-computer interface. Sci Rep 7(1):7772. https://doi.org/10.1038/s41598-017-08265-7

van den Brand R, Mignardot JB, von Zitzewitz J, Le Goff C, Fumeaux N, Wagner F, Capogrosso M, Martin Moraud E, Micera S, Schurch B, Curt A, Carda S, Bloch J, Courtine G (2015) Neuroprosthetic technologies to augment the impact of neurorehabilitation after spinal cord injury. Ann Phys Rehabil Med 58(4):232–237. https://doi.org/10.1016/j.rehab.2015.04.003

Vassanelli S, Mahmud M (2016) Trends and challenges in neuroengineering: toward "Intelligent" neuroprostheses through brain—"Brain Inspired Systems" communication. Front Neurosci 10:438. https://doi.org/10.3389/fnins.2016.00438

Vicario CM, Nitsche MA (2013) Non-invasive brain stimulation for the treatment of brain diseases in childhood and adolescence: state of the art, current limits and future challenges. Front Syst Neurosci 7:94. https://doi.org/10.3389/fnsys.2013.00094

Vidailhet M, Vercueil L, Houeto JL, Krystkowiak P, Lagrange C, Yelnik J et al (2007) Bilateral, pallidal, deep-brain stimulation in primary generalised dystonia: a prospective 3 year follow-up study. Lancet Neurol 6:223–229

Vidu R, Rahman M, Mahmoudi M, Enachescu M, Poteca TD, Opris I (2014) Nanostructures: a platform for brain repair and augmentation. Front Syst Neurosci 8:91. https://doi.org/10.3389/fnsys.2014.00091

Waldert S, Pistohl T, Braun C, Ball T, Aertsen A, Mehring C (2009) A review on directional information in neural signals for brain-machine interfaces. J Physiol Paris 103(3-5):244–254. https://doi.org/10.1016/j.jphysparis.2009.08.0

Wall A, Borg J, Palmcrantz S (2015) Clinical application of the Hybrid Assistive Limb (HAL) for gait training—a systematic review. Front Syst Neurosci 9:48. https://doi.org/10.3389/fnsys.2015.00048

Wang L, Loh KJ, Chiang WH, Manna K (2018) Micro-patterned graphene-based sensing skins for human physiological monitoring. Nanotechnology 29(10):105503. https://doi.org/10.1088/1361-6528/aaa709

Wheless JW, Gienapp AJ, Ryvlin P (2018) Vagus nerve stimulation (VNS) therapy update. Epilepsy Behav 88(Supplement):2–10

Wokke ME, Talsma LJ, Vissers ME (2015) Biasing neural network dynamics using non-invasive brain stimulation. Front Syst Neurosci 8:246. https://doi.org/10.3389/fnsys.2014.00246

Wolf SA, Chtchelkanova AY, Treger DM (2006) Spintronics—a retrospective and perspective. IBM J Res Dev 50:101. https://doi.org/10.1147/rd.501.0101

Yegla B, Parikh V (2015) Rejuvenating procholinergic treatments for cognition enhancement in AD: current challenges and future prospects. Front Syst Neurosci 8:254. https://doi.org/10.3389/fnsys.2014.00254

Younger JW, Randazzo Wagner M, Booth JR (2016) Weighing the cost and benefit of transcranial direct current stimulation on different reading subskills. Front Neurosci 10:262. https://doi.org/10.3389/fnins.2016.00262

Yue K, Guduru R, Hong J, Liang P, Nair M, Khizroev S (2012) Magneto-electric nano-particles for non-invasive brain stimulation. PLoS One 7(9):e44040. https://doi.org/10.1371/journal.pone.0044040

Zarka D, Cevallos C, Petieau M, Hoellinger T, Dan B, Cheron G (2014) Neural rhythmic symphony of human walking observation: upside-down and Uncoordinated condition on cortical theta, alpha, beta and gamma oscillations. Front Syst Neurosci 8:169. https://doi.org/10.3389/fnsys.2014.00169

Zeitler M, Tass PA (2015) Augmented brain function by coordinated reset stimulation with slowly varying sequences. Front Syst Neurosci 9:49. https://doi.org/10.3389/fnsys.2015.00049

Zippo AG, Romanelli P, Torres Martinez NR, Caramenti GC, Benabid AL, Biella GEM (2015) A novel wireless recording and stimulating multichannel epicortical grid for supplementing or enhancing the sensory-motor functions in monkey (Macaca fascicularis). Front Syst Neurosci 9:73. https://doi.org/10.3389/fnsys.2015.00073

Part II
Brain-Computer Interfaces

Brain Machine Interfaces Within a Critical Perspective

Antonio G. Zippo and Gabriele E. M. Biella

"Que sçay-je?" (ancient French—What do I know?)
Michel de Montaigne

1 Introduction

The concept of prosthesis comes from ancient Greek *prothesis* (pro: before; tizemi (from which thesis): to place) that means to place before. In ancient times, it had a variety of meanings from the grammatical use such as the insertion of a letter or a syllable before a word to the architectural definition of a small room within a temple or a church to collect pilgrims' offers. The current use (the only version remaining in the common language) refers to the medical/surgical substitution of failing parts of the body. The impressive technical and surgical advancements in the fields of surgery techniques, material science, electronics, and electromechanics have deeply changed the context of brute substitution of body pieces with objects (for instance, glass eyes or wooden limbs) with true anatomofunctional supports, integrated at most within the operative ranges of the missing or weakened body regions, have reshaped the whole field of neural prostheses. For example, the recent efficient exploitation of exoskeletal frames allowing for walking in paralyzed patients, responding to brain commands translated from brain-implanted probes, represents an impressive example (Benabid et al. 2019). Just to restrict our discussion to brain prostheses, a wide variety of brain external or internal probes has been recently developed generating a paired overabundance of ethical, moral, and technical questions.

A. G. Zippo
Institute of Neuroscience, National Research Council, Milan, Italy

G. E. M. Biella (✉)
Former Head of Research, Consiglio Nazionale delle Ricerche, Segrate, Milan, Italy

I. Opris et al. (eds.), *Modern Approaches to Augmentation of Brain Function*, Contemporary Clinical Neuroscience,
https://doi.org/10.1007/978-3-030-54564-2_5

Both internal (Laiwalla and Nurmikko 2019; Jin et al. 2019; Li et al. 2019) and external (Jansen et al. 2018; Fisahn et al. 2016; del Ama et al. 2012) brain prostheses must have a direct connection to the brain targets by interfaces that can receive/transmit signals to become either movement- or sensory-associated expressions, or, again, cognitive central activations. All these supportive tasks are meant to ignite coherent responses at the brain level able to generate environment-modulated responses or again sensory inputs able to generate "images" of the external world or to relate with behaviorally organized responses.

2 Roles in Cognition and Behavior

Depending on the fact that the prosthesis may play a passive or an active role (Buch et al. 2018; Sahin and Pikov 2011), the optimization of brain implants often depends from the complex pattern of the involved brain areas composing the neural substrate of brain-derived phenomenologies. For instance, the so-called Cognitive Neural Prostheses (CNP) (Pesaran et al. 2006; Andersen et al. 2010) allow for recordings from microelectrode arrays placed in the motor cortex, then transmitting these signals by opportune wiring or wireless connections to external devices, often electronic miniaturized or electromechanical apparatuses substituting failing functions (e.g., arm movements). This "simplistic" view has to obviously meet with the complexity of the network architecture subserving a reaching or goal-directed movement that requires the articulate cooperation of many brain areas (Irwin et al. 2017; Ramakrishnan et al. 2015). Concurrent serious issues rise with signal encoding at their source in the brain and their subsequent transmission to the external assistive devices. Many experiments on primates have shown the potential efficiency of these techniques, letting the monkeys move a precisely directed external artificial arm to a target and completing the action into correct trajectories (for instance, drinking a juice from a glass). It emerges from this simple example the articulate composition of a directed movement that cannot be merely regulated by the activation of the motor cortex. The cognitive control (Buch et al. 2018; Pesaran et al. 2006; Andersen et al. 2010), in fact, involves the operativity of other areas (cortical areas), in which brain signaling from regions not directly involved with movement is anyway required for complete complex movement schemes. Schemes are often complex trajectories that must be completed along cognitive co-operating functions and resumable in a cohort of coordinated motor planning and imagery activations, their integrated planning into the final motor path, their accompanying decisional expression often under a visual control. It appears as obvious that signals necessary to compose a motor (as well as a sensory or composite cognitive) act is an assortment of very diverse signals, as it appears also in human applications (Bundy et al. 2018; Talakoub et al. 2017; Irwin et al. 2017; van Dijk et al. 2016) and that the final artificial modulation could easily outperform whichever current technological device. If current technology isn't able to complete with natural schemes, as for the current possibilities, all these functions but for a restricted selection of them, it is, however, to be considered that the applied artificial devices process and take advantage of the global still unimpaired dynamics preceding the device operations and its downstream cohort of sensory-motor and "cognitive" designs. In short, the still "healthy" or preserved natural networks may act as kinds of master drivers/effectors delivering or receiving an integrated input/output to the final goal.

3 Sensory-Motor Functions

In cases of motor deficiencies, for instance, anatomofunctional studies may help in indicating the placing of the neuroprosthetic probes toward the best or most efficient result. Going on in considering the sole motor complementation, (implicitly coinvolving equal notes on the other compartments) the motor cortex is a kind of repository of gross movement schemes where their refinements are known to reside in the premotor (PM) and in the posterior parietal cortex (PPC) able to select and refine the opportune scheme of the neuronal motor network scheme.

More than the simple movement, these areas are in fact related to intentional planning of the animal and are strictly goal-related regions generating a cluster of signals able to operate the quoted selection of the necessary scheme on the primary Motor cortex (M1). Due to the many discoveries, in the nineties and the first decade of this century, the group of these motor-related areas became more complex. For instance, the dynamic relations between cortices appeared richer and richer: the reaching activity as observable in the parietal reach region (PRR) is driven mainly by visual coordinates and in much lesser degree by limb movement planning (Mirabella et al. 2008, 2011). The great difference between movement planning and intention versus the true movement command made the situation even more difficult for the machine interface in complex clinical conditions. For instance, the electrical stimulation of the posterior parietal cortex (together with other areas) showed strong relations with the intentional awareness of movement even in its absence. This, as it is shown also in stimulations in humans, declared an intention of movement with no true movement expressed. To complicate this picture, the group of Andersen showed that in experiments on monkeys when the animals instructed to withhold the movement thought planning the movement itself after a short stimulus, they were able to decode by simultaneous recordings a continuous activity in PRR, PMC, and in the dorsal premotor area (PMd) a very fast chain of activations: the decoding could happen in a no more than 100-ms lag time. Other confirmation of these fast successions of movements could be obtained from frontal areas. In PRR two tasks (but not more) have, on the other side, been observed as simultaneously processable.

The motor-stimulating devices face, thus, the problem of the complexity of the cortical motor commands distributed over a large front of frontoparietal cortical regions composing a "cognitive" thread of imagery, planning, attention, and several other features completing the scheme of a movement. All signals must be incorporated within the final motor mainframe that requires a coordinated decoding-encoding for the final output in the shortest time windows. In the strategy of a recording and stimulating device, however, a serious decoding problem is added in, namely, what has to be recorded for a significant reconstruction of the final motor scheme in a common behavior context? A second issue rising from these short notes is related to the mutual timing of these recorded signals.

4 Ontological Criticalities

Due to the most diverse pathological conditions, every specific condition should be evaluated to granting for a significant reconstruction, i.e., for its integrity recovery and potential completeness. In other words, we don't have enough knowledge for satisfying all these requirements. (Just as complementary note, a great number of the experimental data have been obtained from healthy animals with intact sensory-motor networks where the implant represents an apposition interfering with an assumed "copy" of the activated area the implant has been placed in). The necessity of a collective participation to the final output composition has been partially reconstructed and it's still not well understood. A number of models have been exploited in the years leaving intact a very serious issue: by stimulating a brain area, we simply throw a stone in the lake. The activation paths followed by our stimulus reach surely the desired goal of activating that spot, but we will never be sure of the diffusion and distribution on the other paths involved in the intricate networks intermingled with the desired target region and the consequent chain of the neural events. Getting further this critical "exercise," we must also interrogate the meaning of the word "stimulation" that will be discussed at the macroscopic and low-scale levels. As for the gross (or coarse-grain) stimulation, we get stimulations by various means, the most common being the electrical, magnetic, or optical interfaces with fixed or floating electrodes or with either on-board incorporated complex digital and analog circuits hosted on the implanted device. The flagrant strategy is expressed in order to store the stimulus metrics and features and deliver the stimuli in a specific pattern or by parting the stimulus-generative elements from the transmitting probes. This has the precise goal to solving also the problem of energy transfer to achieve a correct output over the transmitting probes. Geometrical and spatial issues represent an additional problem. The last are meant to host the stimulatory device in the narrow intracranial space and its motion adaptation to the intracranial movements of brain and truncal structures (often measurable in terms of many millimeters, displacements conflicting with the need of anatomical stability of the BMI interfaces) (Wagshul et al. 2011). All these factors influence heavily the efficiency of the probes themselves. Letting aside all the technical questions (such as the various internal external stimulating device connecting wires, their damages a.s.o.), this question is also tightly related to the signal producing features of the stimulatory devices and on the instructions the device itself must receive.

The inadequacy of the current models for a practical diffuse clinical use is also due to the implicit need of a weighted efference coupled to coregulating sensory inputs, always nested in every motor constellation.

Namely, due to the intricate and unexplorable fine networks involved in even simple movements their sensory counterpart implicates an integrated and modulating predictor representing the incoming background varieties wherein the movements go and realize.

These predictive possibilities are integrated into the output responses. Many algorithms have been provided to refine the predictive power of diverse movement disorders. For instance, one of the most used filters for efficient sensory-guided prediction has been the Kalman filter that has shown interesting results in motor reaching anomalies when used in forward models of guided movements (Li et al. 2009). Kalman filters have been used also in decoding tasks summating the forward approximations and the motor-instructed signals. As well, other features enriching these predicting forward models are represented by the blocking of planned movements and by an important regulation of the generated movement in smoothing or dissolving the delays inherently associated to the coupling between device-guided motor plans and sensory feedbacks, assuring fluidity to the efferences.

On the other hand, both pro- and antimovements sensorimotor transformations have a distributed functional map for realizing the abstract rules for these transformations that (supposedly) are contained in the prefrontal, premotor, and posterior-parietal cortices. Not surprisingly, also in subcortical regions, such as basal ganglia, sensorimotor transforming rule traces are contained (Boettiger and D'Esposito 2005; White and Wise 1999; Stoet and Snyder 2004; Pasupathy and Miller 2005; Grol et al. 2006; Nixon et al. 2004; Petrides 1982; Toni and Passingham 1999; Wallis et al. 2001).

All said, let's try to rise some conceptual questions that are sometimes quoted and often blinded or hidden in the literature. As already quoted, the stimulus onto neurons may be achieved by different means, i.e., electrical, magnetic, or light (infrared) inputs. Neurons, as widely recognized, are excitable elements providing responses that are far from representing a cause-effect sequence. What do we know about the responses of neurons undergoing the different types of stimulations? Are activations due to electrical or infrared stimuli really reciprocally homogeneous? And, concurrently, are the axonal outputs of the stimulation paths the same if delivered on the dendritic tree (its apices or their scaled compartments) and on other regions of the cell? In common depolarizations due to electrical stimuli, the ignition point at the axon hillock is sometimes (particularly in medium or small sized neurons) so powerful to deliver antidromic depolarizing waves traveling backwards into the big dendrites, especially in the soma proximities. The retrograde currents interfere with the orthodromic incoming inputs and the whole game of ortho- and antidromic waves is also modulated by local dendritic active ionic channels that finally shape the signal wave toward the neurite (or axon). An external stimulation device has its terminals placed randomly (and unpredictably!) onto diverse neuron subsets or again, completely outside the anatomic bounds of neurons, interfacing glial elements (let's remember that the ratio of glia to neuron subsets has been estimated in the range of 1.2 up to 1.5) (Herculano-Houzel 2009).

5 Interfacing With the Neural Code

If these simple observations are reasonable origin of concern, there are other important issues to be raised (see Fig. 1). As it is known from old and extremely suggesting data, the structure timing of spike trains or their distribution in the spike-stream is often a discriminating factor to achieve a synaptic activation or not (Bialek et al. 1999). In addition, single neurons after a constant, recurring, repetitive input respond with diverse spike distribution (and potentially of coding strength). This appears due both to the innate variability of the neuronal behavior under stimulation but in significant amount on the activation milieu or background featuring in those brain activity epochs. The ongoing activity, from a mere cellular reference, heavily interferes with every new incoming input that must cope with the thousands of

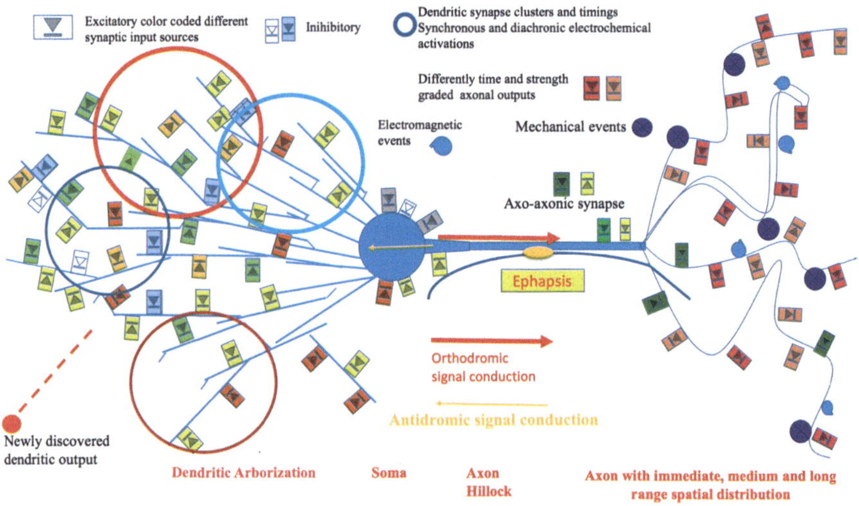

Fig. 1 A «neuron» sketched with major input and output spatiotemporal features. Excitoinhibitory synapses as well as epihaptic connections are indicated. Excitatory inputs come (as a rule) from arrays of different sources with different timings and strength. The dendritic universe is thus transformed in a complex spatiotemporal machine able to convey a hypercomplex input image to the soma, weighting its own inputs with the dendritic input flux. The axon-hillock (or the soma itself) responds with a thresholded activation either characterized by local electrochemical perturbations with no other effect or, reaching a firing threshold, by getting an orthodromic output along the axon and, in many cases, delivering an antidromic backward current invading proximal dendrites matching with the incoming information flow. The extreme complexity of these operations generates electrical, electrochemical, magnetic effects that are magnified in the overcomplex synaptic compartments with a plethora of diverse effects (electrotonic, electrochemical, magnetic, mechanical, and a potential number of still unknown effects). When the whole delicate, synchronic, and diachronic picture roughly imaged here goes under the influence of external stimuli, it is transformed into a stereotypic, «deterministic» selection and repetition of input sequels, warping the extremely delicate thread of the activating events and of the output. A forced selection of stimulus-evoked pathways collapses the implicit variability of the unit outputs and reduces its short-to-medium and long activating/inactivating input ranges to the pertinence networks

spontaneous inputs from the basal activity range that features the overall brain "spontaneous" activity. The differential modality of response of a neuron involved in a spontaneous activity context may present profound differences depending on the immediate activity "history" of the neuron itself (Ruyter Van Steveninck et al. 1994). The artificial stimulation by external devices or probes may result in a complex array of responses. As it is known, neurons are able to respond to simultaneous inputs with finely regulated outputs interspersed on home-keeping activities in the background and with running signaling from the external or internal "worlds" (roughly resumable as esteroceptive and enteroceptive spheres). These inputs may be identified and absorbed at diverse hot-point sources and are mainly generated at the dendritic tree, the soma, axon hillocks, axon en-passant synapses by the synaptic concert of activities and, in the resting state, by spontaneous depolarizing events at the synaptic terminals) with a deep functional semantics. All these source inputs are integrated in the final output and can't be accurately "dissected" to highlight its true origin. This has strict relationships with the coding problem. Is the concept of coding a true sense and complete range of signal interpretation? We don't know currently but some class of spikes (such as sodium/potassium spiking, activities dictated by pure ionotropic channels, metabotropic generated conditions with long-range effects involving the nucleus functions and DNA regulation mechanisms, longer calcium spiking activated in hyperpolarized conditions inhabited by toppings of very fast "combs" of sodium depolarizations, the so-called Low Threshold Calcium Spikes (LTS), through T-type Calcium channels and so on, just to mention a few of spiking modes) expand the coding concept complexity. In addition, the outflow is routed through branching of thousands of outflow ways (not necessarily paritheti-cally involved) distributed over wide and far receiving fields as well as to nearest and narrow spotted regions (just, for instance, a frontoparietal path or a parallel cerebellar fiber onto a single synaptic front on the dendritic arborization of a Purkinjé cell). The heavy changes of the circuitry morphologies induced by the incoming exogenous inputs would be expressed in the profound changes of the graph-like configurations of the circuits under scrutiny.

The synaptic output may assume very different intensities and efficiency configurations. Specifically, in multielectrode recordings, this can be well recognized when in multiple region recordings all those regions known to be connected to a hub region, the diverse local activation patterns become recognizable, both at single neuron and at population level, sharing a common input. These activities have been recognized and identified in high-range-frequency bands neuronal recordings (e.g., within the high-low pass bands of 300 to 6 kHz) as well as in the more recently collective integrations of depolarizations summing in much longer time windows recognized at low-frequency thresholding bands (with high-low pass windows from 0 up to 150 Hz) such as LFP (Local Field Potentials). These are customarily recognized as the true elements summing up in the Electroencephalographic recordings. Surprisingly, high-range EEG oscillatory profiles (reaching even 500–600 Hz profiles, whose physiological significance is still far to be clarified) have recently been evaluated and have not been studied in the BMI contexts. Equally important are the less studied but fundamental electrotonic ephaptic intercellular transmission (unevenly distributed in the brain but particularly relevant in regions needing spiking coordinations in narrowest time windows (such as in the thalamus)). As final remark, the recognition of this wide variety of signals can't be certified as the only set of signal transmission modalities, and, potentially may be thought again as an incomplete set of signaling. From sparse papers in the literature, we are aware that many other "modes" could be envisaged in the future (Faber and Pereda 2018).

Recently (El Hady and Machta 2015), mechanical displacement of the axonal membrane has been shown to be concurrent to the electrical pulse of the action potential (AP). The mechanical membrane displacement could originate from the electrical potentials as arising from wave mode fluxes over the neuronal membranes in which potential energy is cumulated in specific membrane elastic properties, while structures such as the axoplasmic fluid would appear the means to carry and re-express the kinetic energy. In this model, these surface perturbations could emerge from the flowing waves generated by the electrochemical depolarizations of action potentials traveling along the axonal paths. The perturbations could act as a changing contextual "compressive electrostatic force" acting through the neuronal membranes. These electromechanical perturbations have been labeled as Action Waves (AWs), with a still unknown role in the signaling conveyance from neuron to neuron and with an even more enigmatic role in the distribution of signals in local networks.

Another possibility of signal transmission has been very recently identified in slow periodic activities propagation with low-speed profiles around 0.1 m s^{-1} that are prone to weak electric fields. These endogenous fields previously thought to be too weak to start a cell electrical activation appear now to be efficiently modulating the electrical activities of networks as it has been shown in the hippocampus (Chiang et al. 2019). A final note revolves about the propagation of signals in nerve fibers. Notwithstanding the appearing cause-effect direct relationships, the propagation of signals exhibits clear signs of complexity and nonlinearities: the intermingling of electrical, magnetic, mechanical driving forces and their interactions with biomembrane dynamics must match the exceedingly complex balances of ionic currents and their mutual arrangements (Engelbrecht et al. 2018).

6 Cascades of Complex Factors

With these premises, the application of externally applied stimulations by bioinspired devices becomes a multifaceted problem that needs a deep reconversion in the conceptual context that generates both the output of a "stimulus," its configuration, and again the collection of sensory recordings (see Fig. 2). The first question takes in the issue if all the variables involved (some of which have been just discussed together with the many that assumedly elude our current observations and measures) may be causal, necessary, and active or passive effectual events. The brain is the acknowledged Complexity object for excellence and this suggests that whichever scrutiny can't be answered or enlightened in a linear "additive"

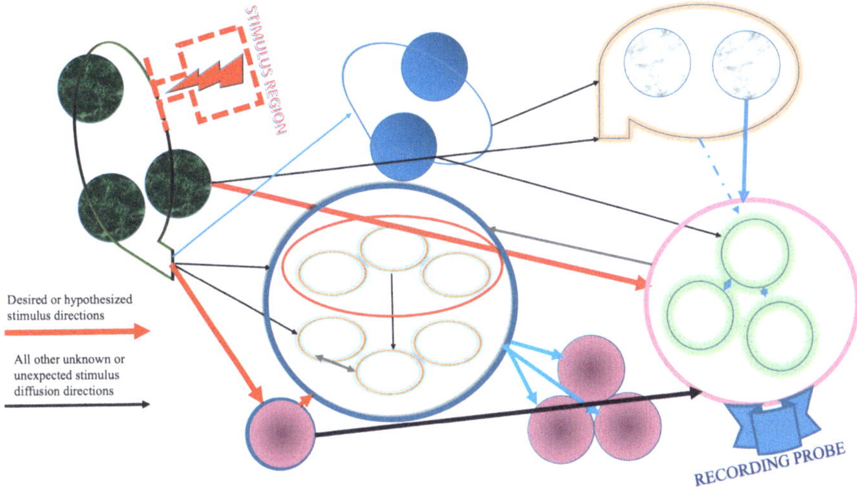

Fig. 2 Each circle (with different color labels) represents a single unit as in Fig. 1 or a strict homogeneous micropopulation. The rich complexity of the networks that each stimulated brain region (stimulating probe in red) belongs to, shows the unpredictable external diffusion waves involving previously unknown or unexpected neural pathways got involved in the stimulus delivery (depending also on many parameters, frequency, and current intensity or patterns included). The recording probes (in blue) may be placed in regions of interest for the stimuli or for the natural activities acting as driving engines to the next regions, or, again, to other potentially undesired regional activations

perspective. A second issue revolves around the consequent complexity of the stimulus structure in the axonal transportation. Recent data have shown a frequency-partitioned differential signal transmission condition, and, namely, that in the visual pathways incoming and outgoing paths, i.e., feedforward and feedback signaling use distinct frequency bands, suggesting that they subserve different communication requirements (Bastos et al. 2015). The deep meaning being that signal transmission in the brain is richly distributed on diverse oscillatory bands selected as preferential for diversely modulated signals (for instance, in feed-forward and feedback signal transfers). The finding highlights how the signal transmission (concurrently to the axonal signal transport complexity quoted earlier) isn't a monotonal process of signal frequencies (Canolty et al. 2012) but requires the intervention of a variety of oscillation bands tuned to the task (and, potentially to the type of task with the associated variety of the cognitive and emotional collections). The third issue regards the intensity, diffusion, and mimesis of the external interfering inputs within the texture of the brain regions involved. Because the three questions (or better the three question sets) can't be afforded straight on the field, i.e., direct application of more and more technically advanced elements on or in a brain that can't encompass the needed matching complexity leaving untouched the root questions that would still lie untouched in the mere try-after-try chain. An effective strategy is, potentially identifiable, in a "come back" or turning approach. In brief, we are obliged to reconsider what is the excitability of a cell from a less ingenuous standpoint. This implies a fractioned answer to questions such as the variety and intensity each electrical, electrochemical, magnetic, mechanical, sonic events a cell is modulated by, in the completion of a response and the evaluation of their necessity and sufficiency for each class of event. The approach is obviously charged by the logic discrepancy inside all the issues and problems of this type and, namely, as just quoted, a complex system can't be reconstructed by an arbitrary partition in an assumed list of identifiable simpler problems. Only a preliminary assessment of the nonlinearities nested in this intricate picture can (or, supposedly may) generate a more pushing and encompassing solution profile. This proposed turning back to the neuron and the search for all the details we are called to take into consideration is a necessary step to understand the very core of a complete or supplementary application of a functional probe. When a neuron loses inputs from afferences, either it goes into an apoptotic path or changes profoundly its input mapping, with newly formed afferences coming from nearby regions and taking the place of the decaying previously established synapses. A typical example is the facial projection on the sensory cortex hand fields in amputated subjects. These newly formed synapses are generated by novel synaptic contacts of sensory facial inputs taking the place of the natural limb innervation. All the borders between two different innervation fields appear to be slightly variable in time, a kind of natural fluctuation if innervation dominance, but sufficiently stable in the final outcome of the synaptic competitions. In case of deciding innervation to a confining field (Sotgiu et al. 1995; Biella and Sotgiu 1995), the still vital and active neighboring regions would take advantage of the progressive regression both of the decaying local neuron apoptotic events and the consequent fiber neurodynamic functions (for instance, due to retrograde wallerian degeneration). This is well illustrated in the quite common finding in amputated patients with phantom limb experience when light stimuli on the face elicit anomalous sensations as originating from the missing hand suggesting that something persists of the original wiring through the missing source (Wittenburg 2010). To underline the extreme exemplary complexity of the disorder, we observe that the "secondary" innervation (in this case, from the face sensory projection cortical fields that lie in the nearbies of the hand fields) is, however, not sufficient to cancel the memorized mappings from the hand itself. The finding suggests that many further elements are involved in the artificial input modulation onto a brain target beyond the target specific responsivity and that of the wider networks it has been nested, an interplay of the ongoing or home-keeping activities with the exogenous stimuli. Local (potential) intracellular memories, with the huge cohort of accompanying biochemical and genic events, are involved too that may induce the surge of late unexpected effects. A stimulus and its input to a cell target must probably match with learnt responses of the target itself, an established condition governed by embryological shaping of the networks' connectivity, and by the shaping from basal or ongoing spontaneous activity (that, as just said, in absence of exogenous stimuli, is most probably related to an overall home-keeping task). All these situations address to a conceivable need of change in the strategy of approaching BMI architecture and dynamics. A turn to the single neuron dynamics appears a necessary step. As it emerges from these considerations, the signal transmission between even a sole couple of neurons is still unclear in its composite complexity. We know only some tile of the complete design and we have only sporadic clues on how electrical, electrochemical, magnetic, mechanical, and metabolic signals may compose a signal. If this is one of the two horn problem, on the other side, neuronal collective behaviors have been read as "electrical" events (see the Local Field Potentials or even the EEG waves). The former EEG frequencies have been dissected in subsections in a neuronal population oscillations whose roles have been progressively enriched, charging delta, theta, alpha, beta, gamma, etc. frequencies of a number of tasks all the oscillations appear involved in (Canolty et al. 2010). The picture has been recently made more and more intriguing. Namely, a most recent paper on sleep-synchronized non-REM cycles has shown that these collective ups and downs of entire neuronal populations interact dramatically with the liquoral currents in the brain and are possibly involved in the cleansing of protein and other "debris" products of daily brain activity such as amyloid or tau proteins (Fultz et al. 2019). Namely, during deep sleep, waves of cerebrospinal fluid, in alternate pulses with the blood flow oscillations concurrent to synchronized ups and downs of the collective neuronal activities, would coincide with the momentary decreases in blood flow. These blood flow decreases would leave more room for the cerebrospinal fluid to push off toxins, including those associated with pathology-linked molecules (e.g., amyloid and Tau proteins seemingly involved in Alzheimer's disease). These intermingled relationships among the neuronal activities, the blood flow, and the cerebrospinal fluid must act as strong alerts of the application of BMIs. None of the current devices is able to keep pace within these intricate dynamics. Single neurons have thus inherent abilities to transmit a signaling manifold and not a mere spike sequence and, at the same time, to be coordinated to thousands of others to generate highly complex ensembles of regional activities. The glial, vascular, and liquoral compartments all are coinvolved in this combined picture, very far from a rigid cause-effectual electrical signal transmission. Thus, a double-track perspective is open and necessary to achieve an enough composite of artificial inputs well organized in the mimicking challenge of the biological background: the single neuron complete transmission grammar and the syntactic approach of local/distant connected networks, let aside the unpredictable input dispersion of anomalous activations along naturally uninvolved regions.

7 Neuroprothesis and Historical Attempts

The already-quoted turn-back to the neuron implies a renovated interest of what happens in the synaptic constellation between two neurons keeping into play the correlated glial intervention with the blood flux, the cerebrospinal flux, the electromagnetic effects, and the fluidic and membrane mechanical noises.

In this environment, old works show a relation of neuron microstimulations with behavioral responses (Salzman et al. 1990), but precise measures and estimates of the microstimuli directed to single neurons need to be made. In the past, some tentative trial has shown with juxtaneuronal stimuli delivered on the somatosensory cortex to be able to modulate in a small fraction of cases the behavioral response where, however, the reaction times for single-cell stimulation were long and variable (Houweling and Brecht 2008), and though the proximity of the stimulating probe to the neurons increased the probability of reduced stimulated volumes, no hint is possible for detection and measurements of signal activation dispersal.

These more recent research using the justappositive stimulations appeared more successful in defining the single neuron responses to micro- or nanostimulations showing new data requiring even stronger need for this return to single-unit studies to better understand the neuronal micro- or larger neuronal populations. Indeed, even these novel approaches, with the meritorious analyses in the turn back to single-cell strategy, miss the many variables of nonelectrochemical nature involved in these experiments. These pitfalls may be obstructive to a wider and deeper comprehension also of BMI construction (Carmena 2013). A multiple view must be coherently active in the balance between single neuron and neuronal population coding. It has been shown that particular stimulus conditions in juxtacellular stimulations may interfere or modulate the population outputs. And this proposal of a novel "backward physiology" is not meant to counteract or invalidate the vast literature on population coding and stimulus-feature completeness but enrich this side. Just this double view should run simultaneously in any approach to both recording and stimulation experiments. Sporadic observations focused on the importance of single-unit activity in the larger context. As it has been shown, single-neuron sensory detection was dramatically poorer in sensory detection as described, for instance, in a series of papers in past years: indeed, responses of individual neurons in sensory cortex in response to repeated presentations of physically identical stimuli display high trial-by-trial variability a factor potentially compromising the detection precision of stimuli. Neurons in the whisker area of primary somatosensory barrel cortex (Nicolelis and Chapin 1994) delivered reduced amounts of information on the identity of a whisker as well as the frequency of stimulation (Reyes-Puerta et al. 2015). On this same line, the group of Nicolelis (Ghazanfar et al. 2000) was able to estimate the difference of stimulus weighting of population vs. single units, where the population coding was sevenfold more accurate in the transmitted information than the single neurons. A review paper on whisker encoding and decoding movements has highlighted the complexity of the regulating networks already in the rat (Mohan et al. 2018). Yet, a role of single units is clearly envisaged in some papers. In some previous studies (Bair and Koch 1996), there were suggestions of this mutual interference of unit vs. ensemble in specific neurophysiological contexts (extrastriate cortex precision on stimulus detection with comparisons of fixed toward moving stimuli) and in comparable experiments a conspicuous number of neurons in the temporal visual region have been shown to perform better in comparison to monkeys' behavioral responses in discrimination visual tasks (Britten et al. 1992). From an information theory perspective, it has been shown in past years that the rich correlation schemes (Zohary et al. 1994) among neurons in small sparse populations act more efficiently than densely packed neurons (Olshausen and Field

2004). The microstimulation of small groups of neurons and the analyses on their effects on behavior while highlighting the effectiveness of the small ensembles on the surrounding network activities and on behavior didn't and doesn't solve, however, the problem of the contingent single-neuron activity roles (Histed et al. 2009). One of the first single unit nanostimulation studies (Houweling et al. 2010) used glass, low impedance pipette with a DC resistance of 4–7 MW brought to contact the neuronal membrane with sudden increases of the impedance value up to 20 MΩ and injecting a current in the nanoampère range. With further technical improvements, Pinault and coworkers planned a novel staining technique allowing for simultaneous recordings of the spiking activity of multiple neurons across the cortical layers. On this line, very recently, Knauer and Stüttgen have been, meanwhile, able, by using a multielectrode array, to record many single neurons in close proximity to the array and to record and stimulate the single cells reliably for hundreds of trials, with stable statistics on the evidences of single-cell neuron, on the spiking background of the network the neuron belonged to, in comparison of the measurable effects on network activities. Again, through the admirable technical progress, the complete fingerprint a single neuron influences aren't again pinned down. As final remark, a very recent research on a worm, often used in neurophysiological research, Caenorhabditis elegans, showed the really revolutionary finding that a single neuron can transmit simultaneously two different signals. Namely, (Tao et al. 2019) the signals are represented by external mechanical touch and proprioceptive body movement. The two signaling pathways involve two metabolic pathways. These two stimuli are detected by distinct mechanosensitive channels and activate distinct cellular outputs linked to mechanonociception and proprioception. Mechanonociception depends on command interneurons through its axon, while proprioception depends on local dendritic Ca^{2+} increases with release of a specific neuropeptide called neuropeptide NLP-12. NLP-12 acts in complex ways on neuromuscular junction on motor axons, regulating muscle tone and movement strength expression. This means that the same neuron simultaneously uses both its axon and dendrites as output pathways to drive distinct responses. While it's given for understanding that in more evoluted species and in Chordates these events may be absent, potential recovery of comparable mechanisms would be a definite blow to too simple doctrines resumable in the conventional directionality of signals in neurons and would make impressively complex the idea of an external artificial stimulation of set of neuronal stimulations.

8 Conclusions

Global recordings taking into account all the variables discussed here and involved in the single and in the network activities are slowly becoming clearer and clearer. What are the relationships between these collective features of a single unit and their influence on the network and in second perspective what are the influences that an active network can project over the single-neuron dynamics? These are two fundamental questions that the BMI device programming would have to include in their functions. However, well aware of the technical primitivity the current devices have in face of the overcomplex tasks, they would be required to grant a true dynamic integrity. We don't aim to diminish or reduce the efforts done until now in the field but to pose the crucial question that is pervading these few lines and that represent a philosophical heritage for a western country. What do we know? That, in the original ancient French, sounds like *"Que sçay-je?"* (what do I know?) by the French renaissance philosopher and essayist Michel de Montaigne. With no particular flight into philosophy and just remaining into the constraints of the theme of BMI, the unpleasant but hard and realistic answer is "too few." The greatly resounding approaches on the sustaining and allowing for purposeful and organized behaviors, the sensation-like producing devices (e.g., for tactile or visual severe deficits) deserve great praise and meanwhile underline the still immature dimension of the field. The seeming mimicking of movement schemes or of the seeing backgrounds and objects in motor or of the visual impairment are produced with intrinsic and hidden cause-effectualistic "linear" logic and not within a nonlinear, highly variable probabilistic environment, too hard to handle biophysically. The dramatic question of the ancient philosopher has an even more dramatic answer: we do know "something" but not enough. This Something is traced in its purpose-inherent design, but the final output (motor, sensorial, or cognitive) appears extremely incomplete in its results. The impressive complexity of the overall world of sensorimotor functions deserves thus a deepest appraisal of what resumes *in se* an active neuron, within the cohort of the neurodynamical events. This flares up a crucial question: is all the concerted integration of these events (being they synchronic or diachronic, being they causalistically related or less) really necessary to the completion of a single neuron expression and, as a consequence, necessary to networks as well? In other words, are they partially influential events (but the spiking activities), i.e., mere epiphenomena or true expression of (as for now) hidden pieces of a coding still far to be understood? The question appears, now, as naturally addressing the requirement of turning back to understand neuronal activity in conjunction to their network dynamics. And for BMI? All the queries still pending on the strict physiological issues inhabiting the brain physiology amplify their weight on the external interventions on brain lesions and the vicarious means drawn up or planned to sustain, substitute, or assist the lacking functions. The technical improvements of the interfaces to brain tissues will need of now far to be reached solutions to become real integrations to brain local dynamics. We must take into account also that the elementary hypothesis of a mere electrical activity substitutive of the brain tissue electrical activity would be characterized by a great variability related to the topographic areas involved: a vicarious BMI for visual functions can't be equivalent to a sensory, motor, or cognitive device in terms of current intensities, resistances, impedances, spatiotemporal unevenness of inputs, and so on; let's keep aside the biological reactions of tissues both in terms of encapsulating immune-like reactions of recording or stimulating probes and for the device function-related consequences such as the thermal effects heating the surrounding tissues and the local signal dispersion as already quoted. The patient, well-weighted, precise measures of the collective properties of the neuron dynamics appear now a sound way out to achieve a correct planning of the future interfaces to the brain taking into account the impressive and cogent obligation to grant for the single units a dynamic syntax joining the single neurons within the concert of regional, local, or larger networks organizing the brain activity.

References

Andersen RA, Hwang EJ, Mulliken GH (2010) Cognitive neural prosthetics. Annu Rev Psychol 61(1):169–190. https://doi.org/10.1146/annurev.psych.093008.100503

Bair W, Koch C (1996) Temporal precision of spike trains in extrastriate cortex of the behaving macaque monkey. Neural Comput 8(6):1185–1202. https://doi.org/10.1162/neco.1996.8.6.1185

Bastos AM, Vezoli J, Bosman CA, Schoffelen J-M, Oostenveld R, Dowdall JR, De Weerd P, Kennedy H, Fries P (2015) Visual areas exert feedforward and feedback influences through distinct frequency channels. Neuron 85(2):390–401. https://doi.org/10.1016/j.neuron.2014.12.018

Benabid AL, Costecalde T, Eliseyev A, Charvet G, Verney A, Karakas S, Foerster M et al (2019) An exoskeleton controlled by an epidural wireless brain-machine Interface in a Tetraplegic patient: a proof-of-concept demonstration. Lancet Neurol 18(12):1112–1122. https://doi.org/10.1016/S1474-4422(19)30321-7

Bialek W, de Ruyter Van Steveninck RR, Rieke F, Warland D (1999) Spikes: exploring the neural code. MIT Press, Cambridge, MA

Biella G, Sotgiu ML (1995) Evidence that inhibitory mechanisms mask inappropriate Somatotopic connections in the spinal cord of Normal rat. J Neurophysiol 74(2):495–505

Boettiger CA, D'Esposito M (2005) Frontal networks for learning and executing arbitrary stimulus-response associations. J Neurosci 25(10):2723–2732. https://doi.org/10.1523/JNEUROSCI.3697-04.2005

Britten KH, Shadlen MN, Newsome WT, Movshon JA (1992) The analysis of visual motion: a comparison of neuronal and psychophysical performance. J Neurosci 12(12):4745–4765. https://doi.org/10.1523/JNEUROSCI.12-12-04745.1992

Buch VP, Richardson AG, Brandon C, Stiso J, Khattak MN, Bassett DS, Lucas TH (2018) Network brain-computer Interface (NBCI): an alternative approach for cognitive prosthetics. Front Neurosci 12(November):790. https://doi.org/10.3389/fnins.2018.00790

Bundy DT, Szrama N, Pahwa M, Leuthardt EC (2018) Unilateral, 3D arm movement kinematics are encoded in Ipsilateral human cortex. J Neurosci 38(47):10042–10056. https://doi.org/10.1523/JNEUROSCI.0015-18.2018

Canolty RT, Ganguly K, Kennerley SW, Cadieu CF, Koepsell K, Wallis JD, Carmena JM (2010) Oscillatory phase coupling coordinates anatomically dispersed functional cell assemblies. Proc Natl Acad Sci U S A 107(40):17356–17361. https://doi.org/10.1073/pnas.1008306107

Canolty RT, Ganguly K, Carmena JM (2012) Task-dependent changes in cross-level coupling between single neurons and oscillatory activity in multiscale networks. PLoS Comput Biol 8(12):e1002809. https://doi.org/10.1371/journal.pcbi.1002809

Carmena JM (2013) Advances in Neuroprosthetic learning and control. PLoS Biol 11(5):1–4. https://doi.org/10.1371/journal.pbio.1001561

Chiang C-C, Shivacharan RS, Wei X, Gonzalez-Reyes LE, Durand DM (2019) Slow periodic activity in the longitudinal hippocampal slice can self-propagate non-synaptically by a mechanism consistent with ephaptic coupling. J Physiol 597(1):249–269. https://doi.org/10.1113/JP276904

del Ama AJ, Koutsou AD, Moreno JC, de-los-Reyes A, Gil-Agudo A, Pons JL (2012) Review of hybrid exoskeletons to restore gait following spinal cord injury. J Rehab Res Dev 49(4):497. https://doi.org/10.1682/JRRD.2011.03.0043

El Hady A, Machta BB (2015) Mechanical surface waves accompany action potential propagation. Nat Commun 6(1):6697. https://doi.org/10.1038/ncomms7697

Engelbrecht J, Peets T, Tamm K, Laasmaa M, Vendelin M (2018) On the complexity of signal propagation in nerve fibres. Proc Estonian Acad Sci 67(1):28. https://doi.org/10.3176/proc.2017.4.28

Faber DS, Pereda AE (2018) Two forms of electrical transmission between neurons. Front Mol Neurosci 11:427. https://doi.org/10.3389/fnmol.2018.00427

Fisahn C, Aach M, Jansen O, Moisi M, Mayadev A, Pagarigan KT, Dettori JR, Schildhauer TA (2016) The effectiveness and safety of exoskeletons as assistive and rehabilitation devices in the treatment of neurologic gait disorders in patients with spinal cord injury: a systematic review. Global Spine J 6(8):822–841. https://doi.org/10.1055/s-0036-1593805

Fultz NE, Bonmassar G, Setsompop K, Stickgold RA, Rosen BR, Polimeni JR, Lewis LD (2019) Coupled electrophysiological, hemodynamic, and cerebrospinal fluid oscillations in human sleep. Science 366(6465):628–631. https://doi.org/10.1126/science.aax5440

Ghazanfar AA, Stambaugh CR, Nicolelis MA (2000) Encoding of tactile stimulus location by somatosensory thalamocortical ensembles. J Neurosci 20(10):3761–3775. https://doi.org/10.1523/JNEUROSCI.20-10-03761.2000

Grol MJ, de Lange FP, Verstraten FAJ, Passingham RE, Toni I (2006) Cerebral changes during performance of overlearned arbitrary visuomotor associations. J Neurosci 26(1):117–125. https://doi.org/10.1523/JNEUROSCI.2786-05.2006

Herculano-Houzel S (2009) The human brain in numbers: a linearly scaled-up primate brain. Front Hum Neurosci 3(November):31. https://doi.org/10.3389/neuro.09.031.2009

Histed MH, Bonin V, Clay Reid R (2009) Direct activation of sparse, distributed populations of cortical neurons by electrical microstimulation. Neuron 63(4):508–522. https://doi.org/10.1016/J.NEURON.2009.07.016

Houweling AR, Brecht M (2008) Behavioural report of single neuron stimulation in somatosensory cortex. Nature 451(7174):65–68. https://doi.org/10.1038/nature06447

Houweling AR, Doron G, Voigt BC, Herfst LJ, Brecht M (2010) Nanostimulation: manipulation of single neuron activity by Juxtacellular current injection. J Neurophysiol 103(3):1696–1704. https://doi.org/10.1152/jn.00421.2009

Irwin ZT, Schroeder KE, Vu PP, Bullard AJ, Tat DM, Nu CS, Vaskov A et al (2017) Neural control of finger movement via Intracortical brain–machine Interface. J Neural Eng 14(6):066004. https://doi.org/10.1088/1741-2552/aa80bd

Jansen O, Grasmuecke D, Meindl RC, Tegenthoff M, Schwenkreis P, Sczesny-Kaiser M, Wessling M, Schildhauer TA, Fisahn C, Aach M (2018) Hybrid assistive limb exoskeleton HAL in the rehabilitation of chronic spinal cord injury: proof of concept; the results in 21 patients. World Neurosurg 110(February):e73–e78. https://doi.org/10.1016/j.wneu.2017.10.080

Jin Y, Chen J, Zhang S, Chen W, Zheng X (2019) Invasive brain machine interface system. In: Neural interface: frontiers and applications. Springer, Singapore, pp 67–89. https://doi.org/10.1007/978-981-13-2050-7_3

Laiwalla F, Nurmikko A (2019) Future of neural interfaces. Adv Exp Med Biol 1101:225–241. https://doi.org/10.1007/978-981-13-2050-7_9

Li Z, O'Doherty JE, Hanson TL, Lebedev MA, Henriquez CS, Nicolelis MAL (2009) Unscented Kalman Filter for brain-machine interfaces. PLoS One 4(7):e6243. https://doi.org/10.1371/journal.pone.0006243

Li C, Jia T, Xu Q, Ji L, Yu P (2019) Brain-computer interface channel-selection strategy based on analysis of event-related Desynchronization topography in stroke patients. J Healthc Eng 2019(August):1–12. https://doi.org/10.1155/2019/3817124

Mirabella G, Pani P, Ferraina S (2008) Context influences on the preparation and execution of reaching movements. Cogn Neuropsychol 25(7–8):996–1010. https://doi.org/10.1080/02643290802003216

Mirabella G, Pani P, Ferraina S (2011) Neural correlates of cognitive control of reaching movements in the dorsal premotor cortex of rhesus monkeys. J Neurophysiol 106(3):1454–1466. https://doi.org/10.1152/jn.00995.2010

Mohan H, de Haan R, Mansvelder HD, de Kock CPJ (2018) The posterior parietal cortex as integrative hub for whisker sensorimotor information. Neuroscience 368(January):240–245. https://doi.org/10.1016/j.neuroscience.2017.06.020

Nicolelis MA, Chapin JK (1994) Spatiotemporal structure of somatosensory responses of many-neuron ensembles in the rat ventral posterior medial nucleus of the thalamus. J Neurosci 14(6):3511–3532. https://doi.org/10.1523/JNEUROSCI.14-06-03511.1994

Nixon P, Lazarova J, Hodinott-Hill I, Gough P, Passingham R (2004) The inferior frontal Gyrus and phonological processing: an investigation using RTMS. J Cogn Neurosci 16(2):289–300. https://doi.org/10.1162/089892904322984571

Olshausen BA, Field DJ (2004) Sparse coding of sensory inputs. Curr Opin Neurobiol 14(4):481–487. https://doi.org/10.1016/J.CONB.2004.07.007

Pasupathy A, Miller EK (2005) Different time courses of learning-related activity in the prefrontal cortex and striatum. Nature 433(7028):873–876. https://doi.org/10.1038/nature03287

Pesaran B, Musallam S, Andersen RA (2006) Cognitive neural prosthetics. Curr Biol 16(3):R77–R80. https://doi.org/10.1016/j.cub.2006.01.043

Petrides M (1982) Motor conditional associative-learning after selective prefrontal lesions in the monkey. Behav Brain Res 5(4):407–413. https://doi.org/10.1016/0166-4328(82)90044-4

Ramakrishnan A, Ifft PJ, Pais-Vieira M, Byun YW, Zhuang KZ, Lebedev MA, Nicolelis MAL (2015) Computing arm movements with a monkey Brainet. Sci Rep 5(1):10767. https://doi.org/10.1038/srep10767

Reyes-Puerta V, Kim S, Sun J-J, Imbrosci B, Kilb W, Luhmann HJ (2015) High stimulus-related information in barrel cortex inhibitory interneurons. PLoS Comput Biol 11(6):e1004121. https://doi.org/10.1371/journal.pcbi.1004121

Ruyter Van Steveninck RR d, Bialek W, Potters M, Carlson RH (1994) Statistical adaptation and optimal estimation in movement computation by the blowfly visual system. In: Proceedings of IEEE International Conference on systems, man and cybernetics, vol 1. IEEE, Piscataway, NJ. Accessed 6 Dec 2019, pp 302–307. https://doi.org/10.1109/ICSMC.1994.399855

Sahin M, Pikov V (2011) Wireless microstimulators for neural prosthetics. Crit Rev Biomed Eng 39(1):63–77. https://doi.org/10.1615/critrevbiomedeng.v39.i1.50

Salzman CD, Britten KH, Newsome WT (1990) Cortical microstimulation influences perceptual judgements of motion direction. Nature 346(6280):174–177. https://doi.org/10.1038/346174a0

Sotgiu ML, Biella G, Riva L (1995) Poststimulus afterdischarges of spinal WDR and NS units in rats with chronic nerve constriction. Neuroreport 9(6):1021–1024

Stoet G, Snyder LH (2004) Single neurons in posterior parietal cortex of monkeys encode cognitive set. Neuron 42(6):1003–1012. https://doi.org/10.1016/J.NEURON.2004.06.003

Talakoub O, Marquez-Chin C, Popovic MR, Navarro J, Fonoff ET, Hamani C, Wong W (2017) Reconstruction of reaching movement trajectories using Electrocorticographic signals in humans. PLoS One 12(9):e0182542. https://doi.org/10.1371/journal.pone.0182542

Tao L, Porto D, Li Z, Fechner S, Lee SA, Goodman MB, Shawn Xu XZ, Lu H, Shen K (2019) Parallel processing of two mechanosensory modalities by a single neuron in *C. elegans*. Dev Cell 51(5):617–631.e3. https://doi.org/10.1016/J.DEVCEL.2019.10.008

Toni I, Passingham RE (1999) Prefrontal-basal ganglia pathways are involved in the learning of arbitrary visuomotor associations: a PET study. Exp Brain Res 127(1):19–32. https://doi.org/10.1007/s002210050770

van Dijk L, van der Sluis CK, van Dijk HW, Bongers RM (2016) Task-oriented gaming for transfer to prosthesis use. IEEE Trans Neural Syst Rehabil Eng 24(12):1384–1394. https://doi.org/10.1109/TNSRE.2015.2502424

Wagshul ME, Eide PK, Madsen JR (2011) The pulsating brain: a review of experimental and clinical studies of intracranial pulsatility. Fluids Barriers CNS 8(1):5. https://doi.org/10.1186/2045-8118-8-5

Wallis JD, Anderson KC, Miller EK (2001) Single neurons in prefrontal cortex encode abstract rules. Nature 411(6840):953–956. https://doi.org/10.1038/35082081

White IM, Wise SP (1999) Rule-dependent neuronal activity in the prefrontal cortex. Exp Brain Res 126(3):315–335. https://doi.org/10.1007/s002210050740

Wittenburg P (2010) Archiving and accessing language resources. Concurr Comput Pract Exp 22(17):2354–2368. https://doi.org/10.1002/cpe.1605

Zohary E, Shadlen MN, Newsome WT (1994) Correlated neuronal discharge rate and its implications for psychophysical performance. Nature 370(6485):140–143. https://doi.org/10.1038/370140a0

An Implantable Wireless Device for ECoG and Cortical Stimulation

Pantaleo Romanelli

1 Introduction

ECoG provides the direct recording of brain electrical activity from the cerebral cortex. ECoG acquisition requires the application of recording electrodes directly on the cortex in order to detect and subsequently resect an epileptic focus. This technique was developed by Penfield and Jasper in the 1940s in order to guide the resection of epileptogenic tissue in patients with drug-refractory epilepsy but proved also useful in brain tumor resection. ECoG acquisition is recently emerging as an essential tool for BCI application aiming to restore neurological functions, for example, by providing guide and feedback to robotic prostheses or exoskeletons. DCS is usually associated to the ECoG to provide direct mapping of eloquent cortex associated with essential neurological functions such as movement, vision, sensation, language comprehension, and production. The resection of an epileptic focus or a brain tumor needs to be tailored to spare eloquent cortex in order to avoid severe neurological damage. DCS is used to guide the resection so that eloquent cortex is spared.

The main limit of current ECoG systems is the need of cables feeding the signal collected by the array of platinum or stainless steel electrodes placed over the cortex to an external amplifier.

Cerebrospinal fluid leaks and meningeal infection are the most common complication following prolonged EcoG recording, restricting the duration of a typical monitoring session to less than a week. Patients with drug-refractory epilepsy undergoing invasive monitoring are confined into a dedicated recording room under careful observation. At the end of the monitoring period, which can be sometimes

P. Romanelli (✉)
Neurosurgery and Brain Radiosurgery, Cyberknife Center, Centro Diagnostico Italiano (CDI), Milan, Italy
e-mail: pantaleo.romanelli@cdi.it

I. Opris et al. (eds.), *Modern Approaches to Augmentation of Brain Function*, Contemporary Clinical Neuroscience,
https://doi.org/10.1007/978-3-030-54564-2_6

too short to identify the seizure focus, the wound is reopened and the grid removed. A short duration of the ECoG acquisition due to leaks or infections can fail to detect the location of the seizure focus with consequent inability to perform the focus resection. Keeping in mind that the most significant shortcoming of current ECoG techniques is the need of connecting cables, a wireless fully implantable ECoG device was developed. This device is made by a silicon grid of submillimetric thickness hosting 128 platinum-iridium contacts, which is connected by intracranial cables to a case made of polyetheretherketone (PEEK) providing wireless transmission. The functions of this implantable neuroprosthesis include wireless data transmission, wireless DCS, and wireless recharge by electromagnetic induction. A set of 64 electrodes is dedicated to wireless DCS, while the other 64 electrodes are used for signal acquisition only. This novel neuroprosthesis was designed to allow prolonged ECoG recording during invasive monitoring. Wireless ECoG offers the possibility to prolong the monitoring for weeks or months, with marked diagnostic gain in selected patients. Also, the monitoring can be performed in an outpatient setting, with substantial reduction of the hospitalization time and of the related costs. ECoG signals can be monitored real time by the involved physician and personnel, providing a novel development of telemedicine. The device can also be used for close-loop responsive stimulation and for a wide range of brain computer interface (BCI) clinical applications. Data transmission exploits the Medical Implant Communication System (MICS), a low-power, short-range (2 m), high-data-rate, 401–406 MHz communication network accepted worldwide for the transmission of biomedical data supporting diagnostic or therapeutic functions associated with implantable devices. The device has been granted the following international patents: EP2699145 B1; US9031657 B2; JP6082942 B2; CA2832520 A; AU2012245942 B2; CN103648367 B. The patents belong to AB Medica (Cerro Maggiore, MI, Italy).

2 Preliminary Experience in Primates

A custom-made dedicated device hosting 16 contacts was developed to perform wireless ECoG and DCS in freely moving primates (Piangerelli et al. 2014; Zippo et al. 2015). The hermetically sealed PEEK enclosure enabled wireless battery recharge through a special cage designed to facilitate the recharge process in monkeys. The cage was developed in accordance with guidelines for accommodation of animals by Council of Europe (ETS123).The inductively recharging cage provides a novel experimental setting where chronic ECoG recording is performed under no restriction of the motor activity of the monkeys. The cage (patent numbers EP2755469 B1, US2015180267, AU2013278927 B2, and CN104427864 A) is a nylon structure containing coils for electromagnetically induced high-frequency recharge in the *x*, *y*, and *z* directions of the space. The coils, generating a constant magnetic field inside the volume they encompass, provide wireless recharging of the device, allowing freedom of movement and avoiding any constraint to the monkey. The innovative combination of wireless ECoG and external seamless battery recharge solves the problems and shortcomings caused by the presence of cables

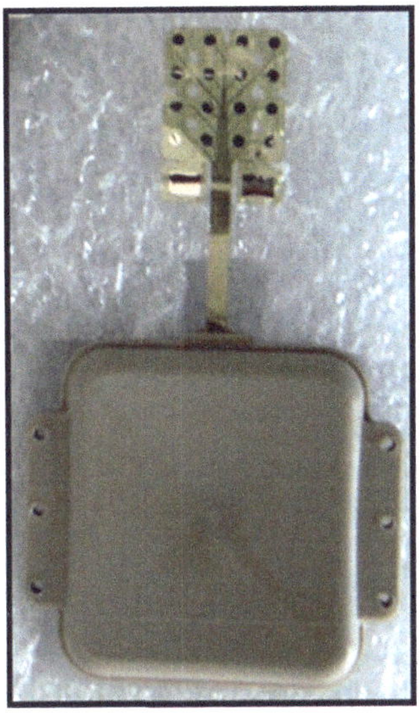

Fig. 1 A wireless implantable device for prolonged ECoG recording and DCS in primates

leaving the skull, providing a safer and easier way to monitor patients and to perform ECoG recording on primates, providing everyday life ECoG signals. In the common experimental settings studying sensorimotor cortex functions or higher executive functions in primates using wired systems, the cables leaving the skull need to be protected from the hands of the animals because they typically try to strip them away. To avoid this problem, behavioral experiments require that the monkey sit in a special chair that does not allow the animal's arms to reach the cables. Our cable-free system therefore opens a new window of opportunities to observe the neural activities in unrestrained, freely moving animals. Using the smart wireless recharge cage, it is possible to recharge the implanted device during the physiological rest period while recording the ECoG signals at the same time.

The wireless ECoG device here described (Fig. 1) was implanted on a male macaque monkey (Macaca fascicularis) weighing 6.95 kg (Romanelli et al. 2018). The experimental protocol was approved by the regional committee (Cometh [Committee on Ethics] Grenoble) and registered to the national committee under the number 12/136 ClinatecNTM-01. The experiment complied with the EU (European Union) directive approved on September 22, 2010 (2010/63/EU), on the care and use of laboratory animals. A 4.7-T MRI study was performed before the craniotomy (BioSpec; Bruker BioSpin) to provide image guidance for the placement of the grid above the sensorimotor cortex. The animal was anesthetized using a loading dose of 5 mg/ kg xylazine and 20 mg/kg ketamine hydrochloride administered intramuscularly, and then a maintenance dose of 1.25 mg/kg and 5 mg/kg xylazine and ketamine, respectively. Physiological parameters were monitored by the veterinary staff during the surgical procedure: heart rate, blood pressure, respiratory depth, and body temperature. Standard aseptic conditions were guaranteed during the surgical procedure. When deep anesthesia was achieved, the animal was

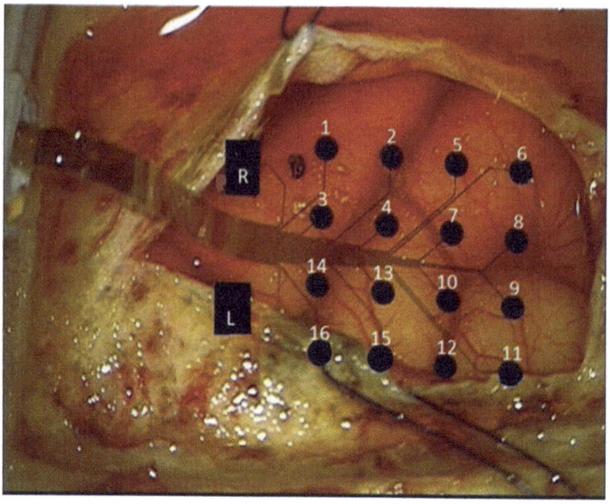

Fig. 2 The cortical grid implanted on primate sensorimotor cortex. The 16 recording contacts are visible. Reference contacts as labeled as R and L. Contacts 1 and 3 are placed on S1 (primary sensory cortex), contacts 2 and 14 are above the sulcus of Rolandus, the other contacts are above M1 (primary motor cortex)

secured to a stereotactic frame and a 3 × 2.5–cm square craniotomy was performed over the left sensorimotor cortex, aiming to center the grid placement over Brodmann area 4. The dura mater was cut in a Y fashion, and the flaps were retracted and sutured on the sides to expose the central sulcus and the surfaces of the primary motor (M1) and sensory (S1) cortex. Radiographic images were acquired during surgery to guide device placement. Electrophysiological confirmation of motor cortex localization was obtained through cortical stimulation with bipolar wand intraoperative neural monitoring (ISIS; INOMED Medizintechnik GmbH). The grid was centered above the hand knob of the left motor cortex (Fig. 2), and then the dura was extended over the grid without suturing it, to avoid excessive pressure over the stem in the subdural penetration point. A silicone adhesive (KWIK-SIL—a translucent silicone elastomer with medium viscosity) was applied over the dura to facilitate the closure and to avoid contact between bone cement and cortex. The bone flap was placed over the dura mater overlying the implanted grid, while the case was placed posteriorly to the craniotomy and fixed to the skull. The wound was then closed. The postoperative course was uneventful. The monkey recovered immediately and was able to resume all normal motor activities (walking and climbing into the cage and feeding unassisted) within a few hours. The ECoG signals of the 16 electrodes were recorded at a 512-Hz sampling rate and a software-imposed band-pass filter from 0.008 to 400 Hz. The ECoG signals were acquired every day for 6 months and remained stable during the study. Electrode impedance values remained at 40 KW, below the acceptable threshold of 100 KW. Spectral analysis of the averaged signal showed the expected power reduction in all frequencies between the first and the last month of observation, due most probably to the thickening of fibrotic tissue surrounding the implant. Overall, the signal quality remained excellent for the entire 6 months of recording, except for a modest degradation during the first week after the procedure (probably due to postoperative debris), as well as during the last

3 weeks, when the implant duration over the cortex drew close to 6 months (Romanelli et al. 2018).

Somatosensory potentials were also recorded and fine finger movements as well as movements of the limbs were elicited by DCS. The cortical responsiveness to light sensory stimuli delivered to somatotopically competent peripheral areas was assessed and the interference of DCS with the peripheral stimuli was assessed. The ECoG recording showed over 6 months the preservation of excellent signal quality while the impedance over the single electrodes remained low. At the end of the experiment, the neurological status of the monkey was excellent, with no sign of motor deficits. Histological examination was performed postmortem 30 weeks after implantation. The cortex was inspected carefully to assess macroscopic signs of damage caused by the prolonged contact of the grid; visual inspection failed to show macroscopic signs of tissue defect. Observation of the implantation site showed a slight brown pigmentation of the brain surface under the grid, probably due to resorbed bleeding. The brain site underlying the electrode array showed its integration into a newly formed dural layer encapsulating the grid. This neoformed tissue was continuous to the constitutive dura mater. We observed no adhesion between the newly formed connective tissue covering the grid and the cortical surface; the grid was easily removed without damaging the cortex. Histological and immunohistochemical evaluation showed no damage done by the prolonged grid placement on the cortex (Romanelli et al. 2018). Observation of the capsule transversal section revealed a connective tissue formation on two sides of the electrode array. The thickness of the reactive tissue ranged between 450 μm (inner layer) and 800 μm (outer layer), while the thickness of the constitutive dura mater was approximately 100 μm. Despite this dural encapsulation of the grid, signal quality remained excellent for several months. However, some loss of signal quality was detected during the last 3 weeks before explantation. Overall, this experiment demonstrated the feasibility and tolerance of chronic ECoG recording and DCS in nonhuman primates, paving the way for a similar experience in humans (Romanelli et al. 2018, 2019).

3 Clinical Applications of Wireless ECoG: Preliminary Experience in Epilepsy Surgery

In patients with refractory epilepsy or brain tumors encroaching eloquent cortex, ECoG and DCS are widely used intraoperative techniques to tailor the resection aiming to spare eloquent cortex while maximizing the removal of epileptic or neoplastic tissue. DCS consists of application of an electrical stimulus directly to the cortex, to assess the contralateral muscle contraction or the related electromyography discharge, in anesthetized patients. Moreover, it can be used to generate transient language and behavioral effects in awake patients while they perform motor or cognitive tasks. DCS allows precise mapping of cortical organization in patients undergoing a resective procedure, providing a valuable instrument to avoid the resection of eloquent cortex and the related neurological sequelae. Intraoperative DCS is especially useful in neurosurgical procedures involving the resection of gliomas involving eloquent cortex. ECoG offers the additional opportunity to record the remote effect of electrocortical stimulation without distortion within a limited distance of a few millimeters, and can provide further details about the functional reorganization caused by the individual's brain pathology. ECoG offers excellent spatial resolution (around 1 mm) with a temporal resolution going down to the timescale of neural activity, thus allowing to identify spatially the cortical region generating the seizures (seizure focus). ECoG signal analysis is also emerging as a valuable for both brain mapping and BCI due to its high signal-to-noise ratio. It allows the examination of high-frequency bands (unavailable for scalp electroencephalography [EEG] recordings) and the use of spectral analysis from sensorimotor cortex. Currently, ECoG provides the most effective tool for brain-computer interface (BCI) applications, i.e., to drive robotic prostheses and to enhance neurorehabilitation. As discussed, the available commercial ECoG systems require cables connecting the subdural grid with an external recording system. The cables connecting the electrodes placed on the cortex with the external apparatus leave the skull through a subcutaneous channel, thus providing a path for CSF leakage and consequent meningeal infection. The use of cables connecting the epicortical grid with an external device is therefore the most significant shortcoming of current ECoG techniques allowing only relatively short recordings.

A proof-of-concept study to show the feasibility of wireless ECoG acquisition in patients undergoing epilepsy surgery procedures was developed in cooperation with the Department of Neurosurgery of the University of Toronto (Romanelli et al. 2019). Short-term wireless ECoG recording was performed in three patients during resective epilepsy surgery procedures. The study was reviewed and approved by the Surgical Innovations Committee of the University Health Network of Toronto University. Informed consent to participate in the study was signed by the patients. The study protocol required the placement of the wireless grid on the surface of the cortex to provide wireless ECoG recording over a 5 min time. After this period, the wireless grid was removed and the operation performed as usual.

The device tested was the model ECoGI-64D developed by AB Medica. This device is an implantable brain diagnostic/stimulating neuroprosthesis designed to localize and monitor the activity of seizure foci in daily life for a prolonged time (up to 30 days, with continuous recording for 24 h a day) and to map the functional activity of the brain by DCS (Figs. 3 and 4). ECoG is recorded through an epicortical grid and transmitted by connecting cables to a radio base station implanted on the skull (Fig. 5).The radio base station originates the wireless transmission of ECoG signals. An external computer with dedicated software control of data acquisition and visualization was used. The external computer has also dedicated audio/video hardware and software providing synchronization with video recordings, thus achieving a video-ECoG. The ECoGI-64D grid is composed of 64 acquisition and 64 stimulation electrodes (plus four reference electrodes), arranged in two separable sections (Fig. 4).The grid can be split into two symmetric parts with 32 acquisition

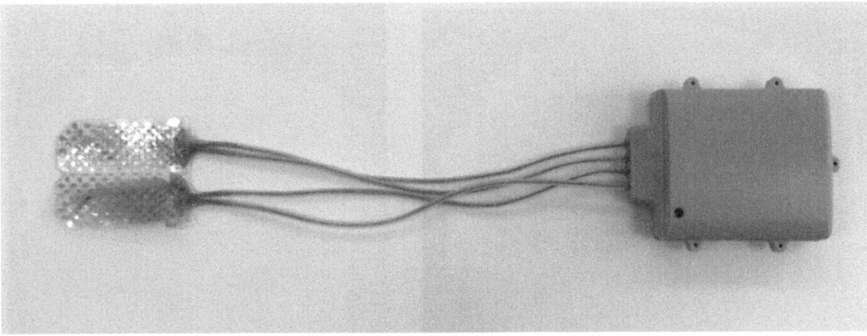

Fig. 3 The clinical-grade wireless neuroprosthesis with a 128-contact grid, 4 connecting cables, and the wireless case providing wireless ECoG transmission and DCS

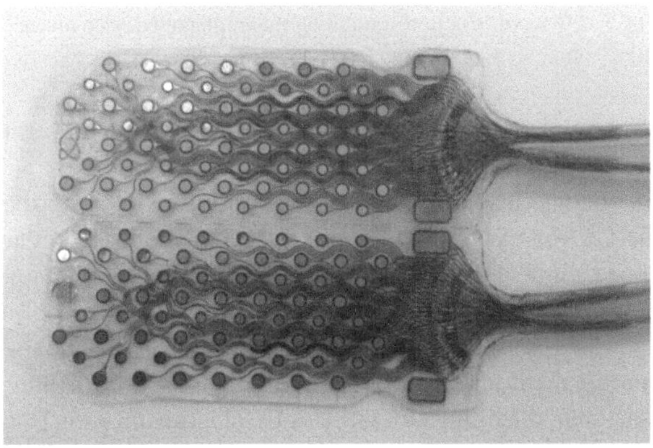

Fig. 4 A detail of the epicortical 128 contacts grid. The grid is made of silicone and the electrodes of platinum-iridium. The grid measures are 43 × 35 mm. The grid thickness is 0.5 mm

and 32 stimulation contacts each plus two reference contacts. Electrodes are connected to an electronic board placed in an airtight case housed in a compact biocompatible sealed PEEK enclosure. The electronic board allows control, acquisition, and data communication. The antenna for communication, the coil for wireless recharge, and the battery for power supply are also placed inside the airtight case. The electronic board contains an Analog-to-Digital front-end and a multiplexing module allowing to choose the stimulation electrodes pair and a microcontroller for device management. A radio link module sends the data to the external unit (Radio Base Station) by MICS radio signals (402 to 405 MHz band). The microcontroller software can be upgraded remotely, providing further flexibility to the system. Wireless recharge is provided by magnetic induction through a dedicated device according to the international WPC (Wireless Power Consortium) standard. Overheating during the recharge procedure is prevented by the ECoG proprietary algorithm, which receives continuous feedback about the skin temperature above the PEEK case. The device is compliant with all the requirements of European Directives for Active Implantable Medical Devices, applicable to the cortical recording for presurgical evaluation of epilepsy in humans, considering all the constraints

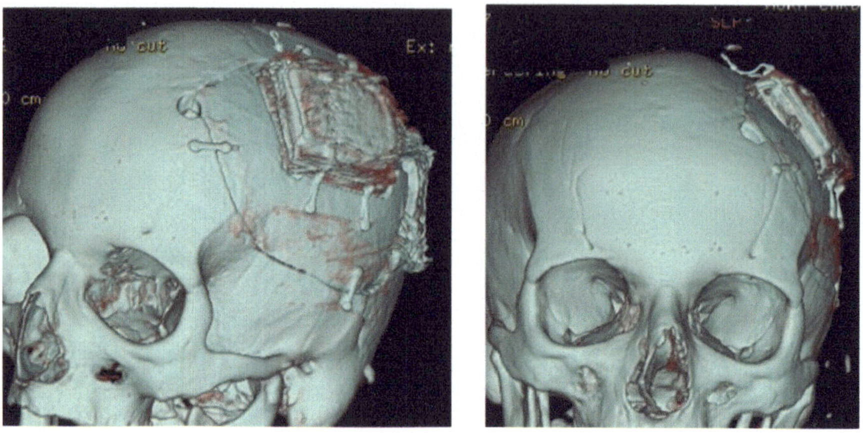

Fig. 5 CT reconstruction simulating the implanted device on cadaver head: lateral view on the left side, frontal view on the right side

of an implantable medical device like ultra-low power, miniaturization, safety, reliability, and regulatory requirements applicable.

The wireless ECoG thus obtained during the described proof-of-concept study at the University of Toronto was submitted to blind review performed by an experienced clinical neurophysiologist. The suitability of the wireless data for clinical interpretation and ability to support the localization of interictal epileptiform activity was assessed. A comparison with the EcoG signals obtained later using a conventional intracranial grid was performed. In all three cases, the grid was easily placed on the brain and wireless ECoG transmission performed. The wireless ECoG data were found to be fully interpretable by the blinded expert. Localization of interictal EEG abnormalities was concordant with that provided by the traditional grid placed later on and could have effectively directed the resection if used alone. The wireless ECoG quality was equivalent to that of traditional wired ECoG. In all three cases, wireless ECoG localized interictal spikes concordant with presurgical data and with subsequent wired recordings. No device-related complication was observed (Romanelli et al. 2019).

4 Conclusions

The development and preliminary clinical experience of a novel neuroprosthesis providing wireless ECoG and DCS have been here presented. Wireless ECoG acquisition offers several innovations and clinical advantages over conventional recording techniques, including: (1) reduced risk of CSF leaks and intracranial infections; (2) prolonged recordings; (3) possibility for remote home monitoring minimizing the hospital stay; (4) responsive closed-loop stimulation; (5) ECoG-driven BCI applications.

References

Piangerelli M et al (2014) A fully integrated wireless system for intracranial direct cortical stimulation, real-time electrocorticography data transmission, and smart cage for wireless battery recharge. Front Neurol 5:156. https://doi.org/10.3389/fneur.2014.00156

Romanelli P et al (2018) A novel neural prosthesis providing long-term electrocorticography recording and cortical stimulation for epilepsy and brain-computer interface. J Neurosurg 130:1–14. https://doi.org/10.3171/2017.10.JNS17400

Romanelli P et al (2019) A wireless Neuroprosthesis for patients with drug-refractory epilepsy: a proof-of-concept study. Cureus 11(10):e5868. https://doi.org/10.7759/cureus.5868

Zippo AG et al (2015) A novel wireless recording and stimulating multichannel epicortical grid for supplementing or enhancing the sensory-motor functions in monkey (Macaca fascicularis). Front Syst Neurosci 9:73. https://doi.org/10.3389/fnsys.2015.00073

BCI Performance Improvement by Special Low Jitter Quasi-Steady-State VEP Paradigm

Ibrahim Kaya, Jorge Bohorquez, and Özcan Özdamar

1 Introduction

Following Hans Berger's introduction of recording brain's electrical activity over the scalp in 1929 to scientific community, electrophysiology has been a critical clinical tool to investigate electrical brain phenomena (Haas 2003). In addition to the clinical use of brain waves as diagnostic indicators for various neurological pathologies, remote communication through brain waves has been at the center of science fiction. Today, recent advancements in the field of neural engineering and electronics technology open new horizons and move us toward reaching such visions once a fantasy. One major product of these developments is the Brain–Computer Interface (BCI) technology.

Brain–computer Interface (BCI) or Brain–Machine Interface (BMI) refers to a technology that converts the recorded brain signals or electrical activity to output actions in order to improve natural brain outputs and the quality of the life of the users (Wolpaw and Wolpaw 2012). BCI technology is a multidisciplinary field and becoming more and more popular each day. The advancements in other fields reflect themselves in BCI systems. The application of BCI to daily life activities are trending with the neural engineering research worldwide. Recent BCI technology involves sophisticated signal processing techniques, advanced sensors, and brain electrophysiology. Since brain research has a vast potential that can transform both technology and medicine, BCIs possess vital importance for those seeking assistance. The population that can benefit from these studies covers a broad range of

I. Kaya (✉)
Department of Biomedical Engineering, İzmir Katip Çelebi University, Çigli, İzmir, Turkey
e-mail: ibrahim.kaya@ikc.edu.tr

J. Bohorquez · Ö. Özdamar
Neurosensory Engineering Laboratory, Department of Biomedical Engineering, University of Miami, Coral Gables, FL, USA

I. Opris et al. (eds.), *Modern Approaches to Augmentation of Brain Function*, Contemporary Clinical Neuroscience,
https://doi.org/10.1007/978-3-030-54564-2_7

neurological, sensory, and muscular disorders. Locked-in syndrome (LIS) and Amyotrophic Lateral Sclerosis (ALS) are two clinical conditions at the focus of BCI research (Hochberg and Anderson 2012). BCI as a very promising tool provides its users with the ability to control and communicate with the environment while they are unable to do so with their own neural pathways due to particular diseases.

BCIs or BMIs can be grouped into three, as motor BMIs for motor control function recovery, sensory BMIs for sensory perception function recovery, and cognitive BMI for improvement of cognitive functions (Lebedev and Opris 2015; Levitskaya and Lebedev 2016). In addition to this classification, BCIs can be divided into two main categories depending on the recording brain levels (invasive and noninvasive) and signal types (Lebedev and Nicolelis 2006). The most well-known noninvasive recording method is Electroencephalogram (EEG). Invasive recording methods are mainly Electrocorticography (ECoG) and Intracortical recordings (LFPs, spikes) (Daly and Wolpaw 2008). Due to minimal risk, EEG is more common and preferred, yet invasive methods have their own advantages. For example, ECoG has higher SNR and better spatial and temporal resolution compared to EEG, in addition, it has less susceptibility for artifacts (Lebedev 2014). ECoG grids covering cortical areas, with its high spatial and frequency resolution, can be used to find out cortical maps for motor function recovery or detection of epileptic focus (Volkova et al. 2019). On the other hand, intracortical electrode arrays record highest SNR signal, however, are risky and more unstable compared to ECoG. BCIs are either dependent or independent (Levitskaya and Lebedev 2016). Independent BCIs do not require the activity of effector organs, peripheral nerves or muscles to generate the brain signal or control. Dependent BCIs rely on the effector activities such as eye muscles and gaze direction to generate the brain signal or control (EEG-VEP) (Wolpaw et al. 2002).

EEG has been widely utilized in various applications such as brain electrical activity recording in clinical studies, lab experiments, diagnosis, health monitoring, and BCI, due to being both noninvasive and advantageous in terms of cost–benefit (Hirsch and Brenner 2010). Most of the EEG-based BCI research focuses on how to improve the Information Transfer Rate (ITR); however, it should be not only ITR but also ease of use, viability for the target population, and the variety of possible output actions. EEG is capable of various signatures for BCI applications. In EEG-based BCIs, Slow Cortical Potentials (SCP), Sensorimotor Rhythms (SMR), P300 Evoked Potentials, and Visual Evoked Potentials (VEP) form the major control signals (Nicolas-Alonso and Gomez-Gil 2012).

VEPs are the most common evoked potentials used in BCI applications. VEPs are recorded with electrodes placed over scalp and reflect the electrical activity in the visual cortex recorded in response to a visual stimulus (Odom et al. 2016). Compared to other EEG signals, VEPs provide relatively high ITR, simple system configuration, and minimal training in BCI systems (Wang et al. 2006). VEP responses are labeled with the stimulation paradigms. There are mainly three VEP types: flash VEP in response to brief flash stimulus, pattern-onset/offset VEP in response to pattern (Odom et al. 2016). Pattern-reversal VEP is more stable in terms of the waveform and the latencies and preferred more in clinical testing (Fahle and Bach 2006). Two common VEP types in literature are transient VEP (TR-VEP) and Steady-State VEP (SSVEP). When the contrast reversal stimulation rate is low

(<3–4 reversals per second (rps)), a transient VEP response decaying before the presentation of the next stimulus can be obtained from the electrical activity on the occipital scalp (Zemon and Gordon 2018). However, if the stimulus is presented uniformly at a higher rate, then a 'Steady-State VEP' response which contains frequency components constant in amplitude and phase is obtained (Özdamar and Bohórquez 2006; Capilla et al. 2011).Transient VEPs are more common in clinical practice to diagnose and monitor neural pathway problems compared to SSVEPs.

VEP-based BCIs can be grouped as time-modulated VEP (t-VEP) BCIs, frequency-modulated VEP (f-VEP) or SSVEP BCIs, and pseudorandom code-modulated VEP (c-VEP) BCIs (Bin et al. 2009). SSVEPs BCIs are widely used and dominate the field. SSVEP BCIs yielded high-speed BCI systems by sophisticated signal processing and classification methods. For instance, Nakanishi et al. (2018) utilized the Task-Related-Component Analysis (TRCA) spatial filtering method in an SSVEP BCI paradigm and achieved higher accuracy and ITR values, such as a mean online ITR of 325.33 ± 38.17 bpm (Nakanishi et al. 2018).

In an SSVEP BCI, targets are encoded with distinct frequencies and the detection is based on the analysis of the frequency content. Three main advantages of SSVEP-based BCIs are easy implementation with common displays, minimal training, and no synchronization. However, the number of targets in SSVEP-based frequency tagging is limited with the display refresh rates. Another disadvantage is that the SSVEP phenomenon cannot be observed at high frequencies, the SNR is very low at some frequency bands due to adaptation.

Early utilization of c-VEP in a BCI paradigm with the ECoG recording method was done by Sutter (Sutter 1992). He applied m-sequences and deconvolution methods to investigate the Electroretinograms as well (Sutter 2001). Later on, Bin et al. (2011) brought the attention to c-VEP BCI paradigm by developing a high-speed EEG-based BCI using m-sequences (Bin et al. 2011). The targets in a c-VEP BCI are encoded with different orthogonal codes and many targets are available with sufficiently long codes. The detection is based on template matching where in a training step templates for each target are obtained by one by one recording, or obtaining a generic template and using it as a basis for other templates.

Another type of c-VEP is called Quasi-Steady-State VEP (QSS-VEPs) which is obtained by low jittered special stimulation sequences. QSS-VEP contains the advantages of both SSVEP and c-VEP paradigms. Moreover, due to the jitter introduced, QSS-VEPs can be observed at higher frequencies as well. Another advantage of QSS-VEPs is that it enables the extraction of clinically important TR-VEPs from QSS-VEPs. These TR-VEPs can be used in monitoring of neural states and progress of BCI users. So far, stimulus specificity of f-VEP BCIs are well studied and it has been known that if one stimulus property such as the size and color of a fixated stimulus, or the number and separation of surrounding stimuli is changed, the neurons and cortex activities may lead to different VEP waveforms (Wei et al. 2016). This affects the classification performance of BCIs. QSS-VEPs with low jitter characteristics can modulate neural activity in an advantageous manner to be used in a BCI system effectively. In this research, we aimed to develop a simple dual target BCI paradigm using QSS-VEPs. The design allows us to observe clinically significant transient VEPs (TR-VEP) as well.

2 Materials and Methods

The main objective is to analyze the feasibility of a BCI switch with four VEP signatures for each stimulation rate (10, 32, 50, and 70 reversal per second), Left QSS-VEP, Left TR-VEP, Right QSS-VEP, Right TR-VEP in a dual target display experiment. Moreover, a comparison of QSS-VEP paradigm and SSVEP paradigm in a template-matching BCI system is provided. Three female and five male healthy subjects, between 18 and 31 years old (mean: 25.75 ± 4.1), participated in the study. All the subjects had normal vision and were right eye dominant. An informed consent form that was approved by the University of Miami Institutional Review Board (IRB) was signed by each subject.

2.1 Continuous Loop Averaging Deconvolution (CLAD)

The CLAD method was developed with the assumption that the low jittered special stimulation sequences can cancel out adverse effects of adaptation and thus better estimate the deconvolved unitary transients generating the steady-state evoked responses compared to broadband spectrum sequences such as m-sequences (Delgado and Özdamar 2004). By the superposition principle, each stimulus leads to the identical per stimulus unitary response which can be provided by the CLAD method. To achieve this, the amount of jitter must be restricted, and the generated responses from these low jitter sequences are called Quasi-Steady-State (QSS) responses due to their resemblance to iso-chronic steady-state responses. In a loop manner, stimulation sequence wraps itself continuously and responses to each stimulus can be represented as an equation. A matrix of equations in time domain can be obtained from the convolved responses. If the stimulation sequence is non-singular, it has an inverse matrix, thus the per stimulus unitary transient response can be obtained. With the help of frequency domain approach, the same procedure can be transformed into a simpler form. Deconvolution is the division of the convolved response with the stimulation sequence in the frequency domain and time domain transient response can be obtained with inverse Fourier Transform (Özdamar and Bohórquez 2006). Thus, the frequency domain approach allows an efficient yet simple way to extract the transient responses.

2.2 Stimulator and Stimuli Design

In this research, a fast switching on-demand LED-based 5 region stimulator display was used to create a black and white horizontal bar pattern stimulus and pattern reversal VEP response. The display was similar to the one developed in (Toft-Nielsen et al. 2011, 2014). The state change or rising and falling times of the LED display were less than 105 μs. Toft-Nielsen et al. (2014) recorded 78rps responses with this display technology (Toft-Nielsen et al. 2014). Odd and even horizontal bars (total 12 bars), that correspond to 1.03 cycles/degree from 1 m viewing distance, were driven simultaneously to make the pattern reversal synchronous. Each of the two target display fields covered a visual field of 5.7° × 5.7° at 1 m. Target displays were placed in the foveal field of view. Separation between targets or fields was 5.7° that was close to the optimum separation of at least 4.8° suggested by (Wei et al. 2016). The luminance of the display and the contrast were set to 580 Cd/m^2 and 95%, respectively.

The two specially designed stimulus sequences of slightly different lengths and jitter were applied to the two targets (right and left fields of display) (Kaya 2019) (Fig. 1). Low jitter and low-noise amplification characteristic sequences that were constructed similarly to (Toft-Nielsen et al. 2014; Bohórquez et al. 2013) triggered reversals of bar displays. The orthogonality of the right and left sequences allowed retrieval of individual unitary transient responses to each display field by deconvolution (See Özdamar and Bohórquez 2006 for deconvolution). Each transient response was possible to be extracted from the same QSS data using the non-singular sequences. Visual cues were provided to the subjects to guide them through the target displays during the experiments. A blue LED for the left gaze, a green

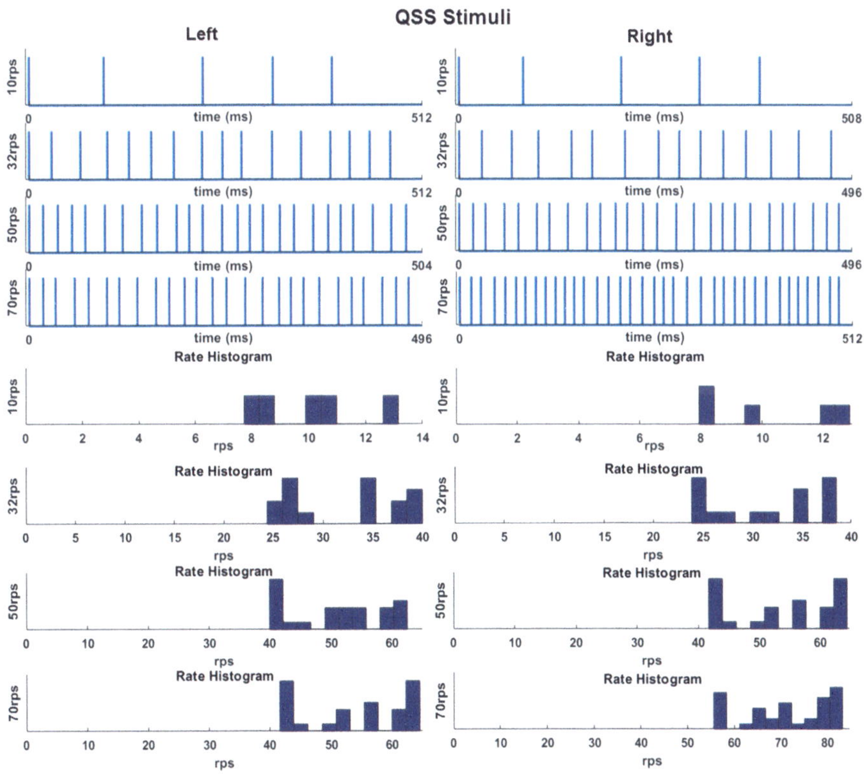

Fig. 1 10rps, 32rps, 50rps, and 70rps QSS-VEP stimuli and rate histograms for both left and right targets are shown. The histograms are composed of multiple rates with a mean rate of nominal 10, 32, 50, and 70rps

LED for center gaze, and a red LED for right gaze were used as indicators. Subjects were directed by these cue lights to attend on the target displays, left or right or none (center).

In the ERP signal acquisition and neurophysiological system identification, pseudo-random binary sequences play a critical role. It has been shown that using some deconvolution approaches, especially CLAD, it is possible to fasten the sensory ERP or evoked potentials (Delgado and Özdamar 2004; Bohórquez and Özdamar 2006). These CLAD sequences can be employed in a BCI target encoding and recognition design. However, they should satisfy unique properties in order to properly stimulate the neural populations and produce characteristics transient-evoked potentials. Although there are such common Pseudo Random Binary Sequence (PRBS) codes for system identification such as m-sequences, these have broadband spectrum and high jitter ratios causing adverse effects due to high adaptation. In this research, we employed CLAD sequences designed with an efficient differential evolution (DE) approach with a cost function optimization for consistent noise attenuation and uniformly distributed low jitter (Kohl et al. 2019). The low jitter enabled us to model the system as in a steady-state mode of operation and it was added to the inter-stimulus-intervals in order to minimize the adaptation related effects. The excellence in the design of stimulus sequence is a key factor in the performance of the sequence and deconvolution.

In this study, there are four stimulation rates, each representing a low (10rps), medium (32rps), high (50rps), and very high (70rps) pattern reversal stimulation rates. For each rate two stimulus sequences were prepared for each target, left and right. The rate histograms for each target sequence are given in Fig. 2

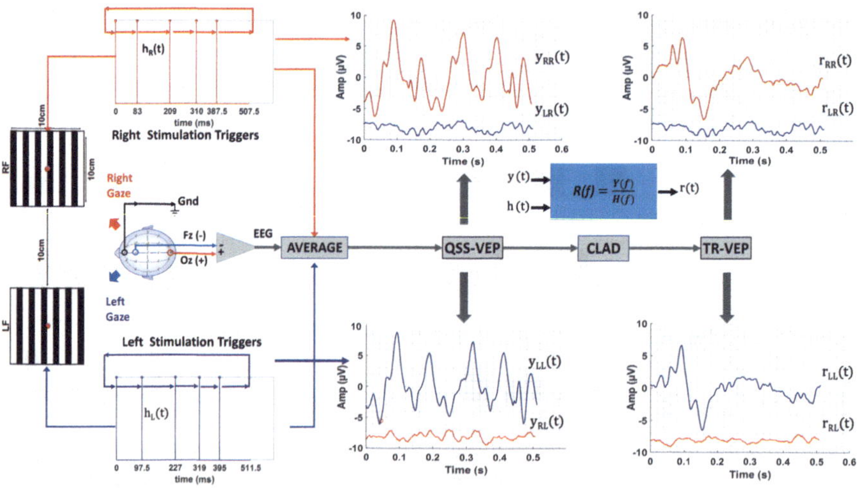

Fig. 2 Experiment design and two LED pattern reversal stimulators at the left. Two stimulation sequences hL(t) (blue) and hR(t) (red) are presented in a circular loop manner. The recorded averaged QSS-VEP signals for each display region and the crosstalk responses are in the middle. The CLAD method in the frequency domain and the resulting deconvolved transient TR-VEPs are shown at the right. (Adapted from Kaya et al. 2019)

(Kaya et al. 2019). As seen from this figure, in contrary to the common m-sequences, the jitter is limited with a threshold and the stimulation rates stay around the nominal stimulation rates. Moreover, the left and right stimulus sequences exhibit similar characteristics, therefore similar neural excitations are expected for both target displays at the same stimulation rate.

2.3 EEG Recording and Signal Processing

For practicality, we utilized from single channel bipolar electrode montage from Oz (+), Fz (−), and forehead (gnd). In the experiments, three Ag/AgCl electrodes were used, and impedances were kept below 7 kΩ. 16 bit medical-grade data acquisition module recorded signals at a sampling rate of 2 kHz. A bandpass filter with cutoff frequencies around 1 to 300 Hz (6 dB/octave) was applied. With a simple yet efficient preprocessing step, the EEG epochs or sweeps with peak to peak amplitudes above 40 μV were labeled as artifacts and discarded in the signal processing stage.

Two experiments were devised in order to find an optimal stimulation rate, and two stimulation paradigms namely QSS and SSVEPs at both low and medium rates were compared. For c-VEP BCIs, template matching is a common method for classification. In this study, three classifiers based on correlation with templates were used.

For the classification purposes, we required two sets of data, one for training and one for testing. Since in this research the classification was based on template matching, QSS-VEP and TR-VEP templates were obtained with the training data. For 10rps example in Fig. 2 for each gaze, training EEG data set was obtained by subject attending to each target for 3 trials while both displays were stimulated simultaneously. For each QSS stimulation rate (10rps, 32rps,50rps, and 70rps) and target or gaze direction (left and right), around 124 sweeps were recorded to generate VEP templates. We aimed to record a similar sweep number for the templates. TR-VEP templates were extracted from QSS-VEPs by the CLAD method. A 2rps stimulation paradigm was carried out and traditional transient VEPs were obtained for each subject for comparison. For the detailed stimulation paradigm and signal processing steps, please refer to (Kaya et al. 2019).

The VEPs in the test data to be classified as left or right target were computed by conventional ensemble averaging. By time-locked (stimulus length) averaging, noise is tapered and the synchronous evoked potential is enhanced. The EEG data were averaged with left stimulus length to get the left gaze QSS-VEP response and with right stimulus length to get the right gaze QSS-VEP response as shown in Fig. 2. SNR was computed as the ratio of the VEP signal power to the noise power in dB. The noise was computed by the +/− averaging method, where the odd sweeps (+) were summed and the even sweeps (−) were subtracted and then averaged. Within the QSS rate comparison, we used 124 to 128 sweeps in template generation. For the 10rps SS/QSS comparison, we selected 117 to 127 sweeps; for 32rps QSS/SS comparison, 93 to 96 sweeps were used.

2.4 Classification

In the target detection stage, a cross-correlation-based template matching method was used. The correlation-coefficient was computed by using the Pearson's Correlation Coefficient (CL or CR) that is given in Eq. (1) (Puth et al. 2014). It provides a measure of the similarity between two signals. N is the data size for the left target class template. T_L is the left reference template which is obtained by averaging the left gazed signal in the training stage. The average EEG data under inspection to be categorized as left or right is S. C_L or C_R is the resultant correlation-coefficient for test left condition and test right condition, respectively.

$$C_L = \frac{\sum_{n=1}^{N} T_L(n) \times S(n)}{\left[\sum_{n=1}^{N} \left(T_L(n)\right)^2 \times \sum_{n=1}^{N} S^2(n)\right]^{1/2}} \tag{1}$$

Three classifiers were utilized; left target template matching, right target template matching, combined detection method, see Fig. 3 (Kaya 2019). First, the signal of interest for each target template matching was computed by an averaging step. For the left target, the average of the EEG data segment (sweeps/epochs) to be classified was found by averaging the total signal of interest with the left template length. Similarly, the signal of interest for right template matching was computed using the right target template length. In the left classifier-based detection processes, the average signal was cross-correlated with the reference template of left. The resulting correlation coefficient was then compared to a predefined left template matching threshold value that was obtained by the training stage. If it was greater than the threshold, the EEG data were classified as left gaze class. Similarly, the right target-based detection could be used with right template threshold. The third detection method is called the 'combined method', which compares both cross correlation coefficients from left and right template matching steps and then assigns the signal of interest class to the highest coefficient's class.

Receiver Operational Characteristic (ROC) curve is a means to analyze the efficiency of a detection method. The error probability or false positive ratio (FPR) and true positive ratio (TPR) for each threshold value was computed. ROC curve plots the FPR (1 − specificity) at the x-axis and TPR (sensitivity) at the y-axis. Pearson's Product-Moment correlation coefficient between the subject's VEP template and the actual data was compared to the threshold and ROC was computed by sweeping threshold from −1 to 1 and finding TPR and FPR for each threshold. The area under the ROC (AUC) is representative of the accuracy of the detection and widely used in BCI research as an indicator of the classifier performance.

Information Transfer Ratio (ITR) is proposed as the optimal metric to evaluate a BCI system performance in terms of throughput, categorical output, unbiased/biased, and practicality (Thompson et al. 2013). ITR can be calculated by Eq. (2),

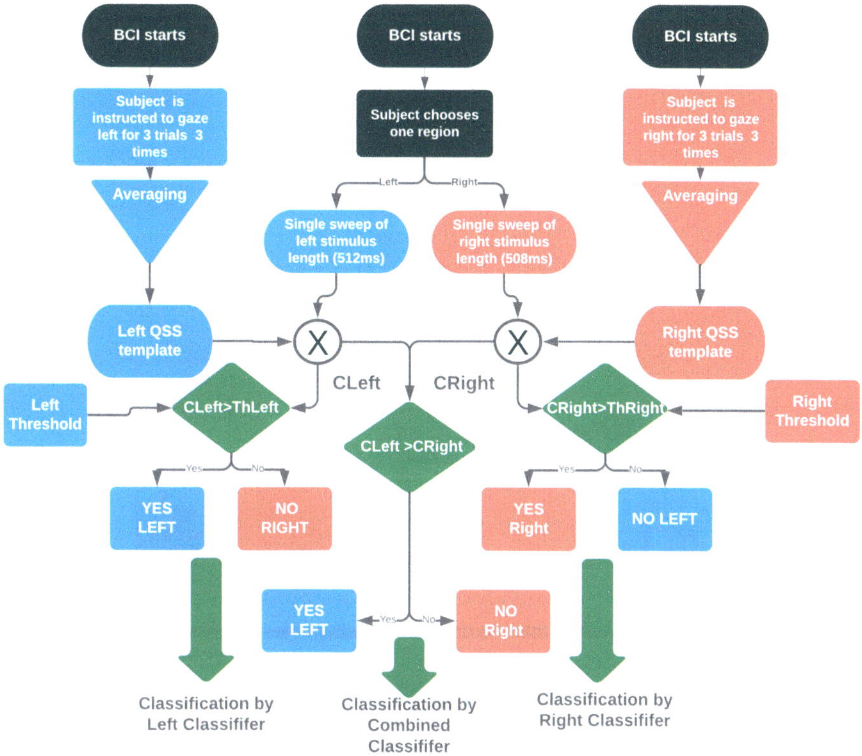

Fig. 3 Three types of classification and command/target detection processes based on 10rps left and right QSS signals type are shown using two template comparison

where the probability of correct selection is P, the number of targets is N, and T (*s*/selection) is the time for selection (Thompson et al. 2013). Its unit is given in bits per minute (bpm). The ITR simply follows the accuracy of the classification, the higher the accuracy, the bigger is the ITR.

$$\mathrm{ITR} = \left[\log_2(N) + P\log_2(P) + (1-P)\log_2\left(\frac{1-P}{N-1}\right)\right]\frac{60}{T} \tag{2}$$

3 Results

For the classification performance of the BCI system operation with single sweep (0.5 s) or three sweep data (1.5 s), EEG data from a gaze switch trial are shown in Fig. 4. Subject-2 started gazing to the left display at 10rps stimulation and then

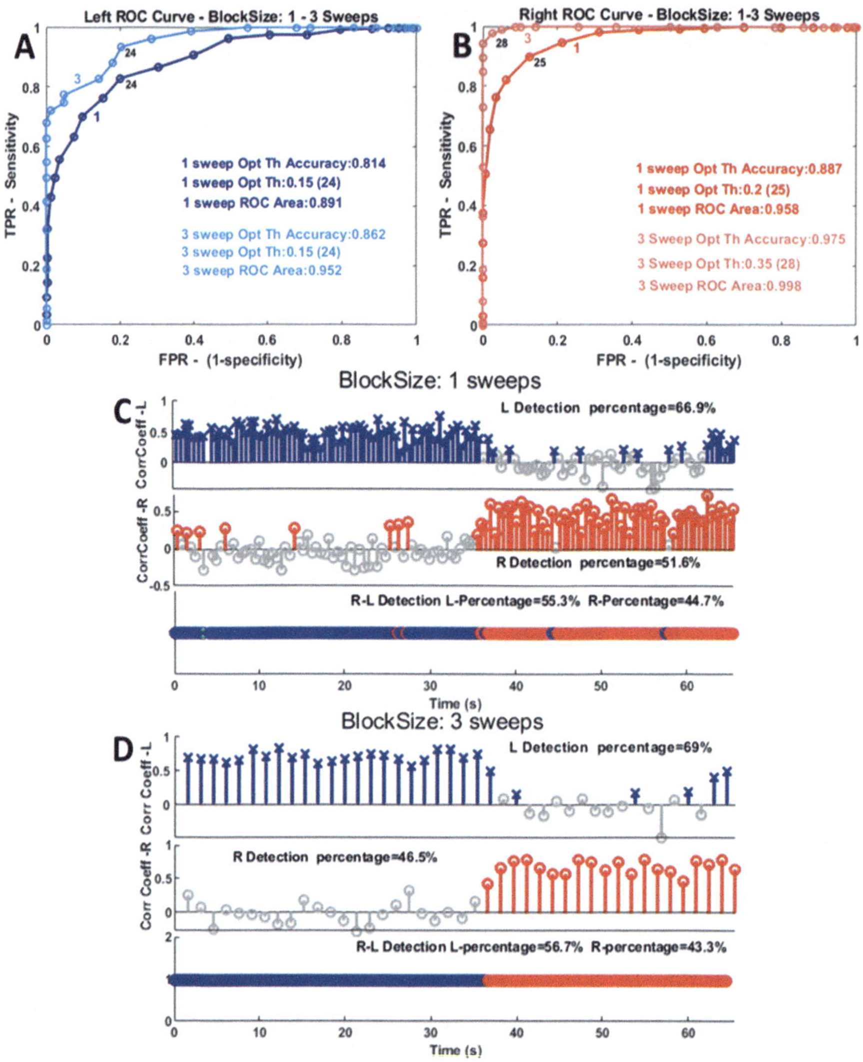

Fig. 4 For S2 the left to right gaze switch was detected by 10rps QSS-VEP template matching and using single sweep or three sweeps data. (**a**) The ROC curves for left gaze detection for single sweep and three sweeps are shown. Selected operational correlation coefficient thresholds for single sweep and three sweeps left gaze detection are 0.15 and 0.15. (**b**) Similar to the **a**, right gaze ROCs, operational thresholds are shown for single and 3 sweeps. (**c**) For single sweep data-based detection, top blue shows the left gaze detected instants (sweep blocks) using only left QSS-VEP threshold parameter for detection, middle (red) shows the right gaze detected instants (sweeps blocks) using only right QSS-VEP ROC threshold parameter for detection, and finally the bottom plot is the detection based on combination of both parameters and comparison of left and right cross correlation coefficients. (**d**) Similar to the **c**, three sweeps data-based detections are shown. It can be seen that three sweeps data lead to better classification, less false positives, and false negatives

switched gaze to the right display by a verbal command in the middle of a trial. This can be seen from Fig. 4c, d panels (Kaya 2019). In the a and b panels, the ROC plots are drawn for each left and right single and three sweep data durations. From these ROC plots, operational threshold values are selected. As sweep number is increased to three, ROC area, thus classification performance or accuracy, increases. The three classifiers, left template matching, right template matching, or combined method, are drawn from top to bottom in panels c and d. The blue color depicts the left gaze-detected EEG sweeps while the red color for the right gaze detection-based sweeps. And in the third plots, the combined classifier assigns the sweep of interest class to the highest correlation yielding side, left or right. While single sweep is used in the gaze detection, there are some false positives and false negatives in the EEG segments. However, when three sweeps' average signal is used, since SNR is increased, the correlation yields better results and classification performance increases significantly.

ROC areas are indicators of the accuracy of the classification. In Fig. 5, ROC curves for each rate (10rps, 32rps, 50rps, and 70rps) and gaze (left, right) for single-sweep QSS-VEP data are shown for subject 6. The 32rps ROC areas are found to be significantly higher than the others, which can be seen from the Subject 6 in Fig. 5 (Kaya 2019).

Table 1 summarizes the accuracy and ITR distribution over all the subjects for single-sweep classification. The 'ave' is the average of the single left, and right classifier performance, whereas 'comb' is the performance of the combined classifier. The combined classifier has significantly higher accuracy than the average of the two classifiers. It can be seen that QSS signal types performed better than TR except for 70rps, when mean accuracy values are compared. The ITR values follow the same relations. There is high variance in between subjects, while Subject 7 has low accuracies and ITR values, S6 particularly has higher accuracy and ITRs. The low performance of Subject 7 can be attributed to the frequently blinking of the subject during the experiment. The mean ITR performances for each rate QSS signal and combined method are 47.82 bpm, 72.26 bpm, 52.40 bpm, and 36.90 bpm for 10rps, 32rps, 50rps, and 70rps, respectively. 32rps QSS has the highest performance. For all four rates and both QSS and TR signal types, the combined classifier performed significantly better ($p < 0.05$) than the average of the single left and right classifiers. 32rps QSS is found to yield the highest accuracy compared to the other rates ($p < 0.05$). Except for the 70rps, QSS-VEP-based template-matching classifiers performed significantly ($p < 0.05$) better than the TR-VEPs.

3.1 ANOVA Results

ANOVA was applied to figure out the sources of variations and the effects of independent variables on dependent accuracy. There were mainly four sources or independent variables, namely gaze (left, right), rate (10rps, 32rps, 50rps, and 70rps), data duration (1 sweep-0.5 s, 2 sweep-1 s, 3 sweep-1.5 s), and signal types (QSS,

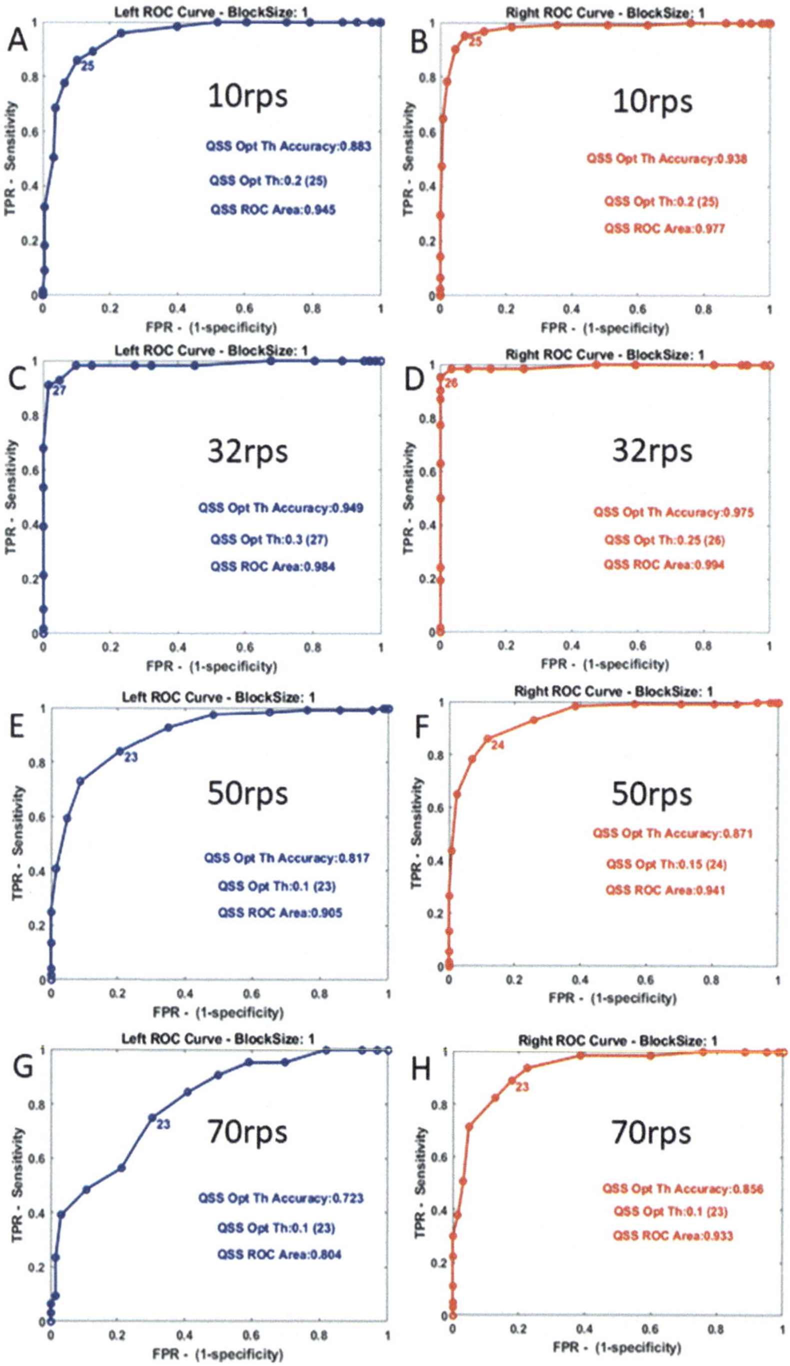

Fig. 5 Subject 6 ROC curve, areas, and operational threshold values are given. (**a**) 10rps QSS ROC curve and operational thresholds to be utilized in 10rps left gaze detection are shown. (**b**) Similar ROC curve and threshold values for 10rps right gaze detection. (**c**, **d**) ROC curves and thresholds for 32rps. (**e**, **f**) ROC curves and thresholds for 50rps. (**g**, **h**) ROC curves and thresholds for 70rps

Table 1 Single sweep (0.5 s) accuracy and ITR table for average classification (Ave) performance and combined classifier performances (Comb)

Subject ID	10rps QSS-Ave	QSS-Comb	TR-Ave	TR-Comb	32rps QSS-Ave	QSS-Comb	TR-Ave	TR-Comb	50rps QSS-Ave	QSS-Comb	TR-Ave	TR-Comb	70rps QSS-Ave	QSS-Comb	TR-Ave	TR-Comb
Accuracy																
S1	0.80	0.82	0.75	0.80	0.81	0.97	0.66	0.79	0.83	0.93	0.81	0.91	0.81	0.90	0.82	0.93
S2	0.91	0.95	0.79	0.85	0.92	0.98	0.88	0.96	0.84	0.93	0.83	0.90	0.89	0.94	0.87	0.89
S3	0.72	0.74	0.65	0.67	0.80	0.86	0.75	0.82	0.70	0.76	0.67	0.69	0.63	0.69	0.61	0.64
S4	0.81	0.90	0.69	0.75	0.82	0.86	0.81	0.84	0.81	0.88	0.73	0.82	0.72	0.81	0.73	0.80
S5	0.73	0.83	0.64	0.70	0.83	0.93	0.72	0.72	0.78	0.81	0.70	0.69	0.69	0.72	0.69	0.71
S6	0.91	0.96	0.76	0.84	0.96	1.00	0.78	0.83	0.84	0.90	0.76	0.86	0.79	0.85	0.78	0.85
S7	0.66	0.74	0.62	0.67	0.72	0.74	0.66	0.69	0.68	0.70	0.62	0.62	0.60	0.62	0.60	0.65
S8	0.66	0.73	0.60	0.66	0.85	0.90	0.77	0.87	0.85	0.93	0.82	0.89	0.81	0.80	0.79	0.84
Mean	0.77	0.83	0.69	0.74	0.84	0.90	0.75	0.82	0.79	0.85	0.74	0.80	0.74	0.79	0.74	0.79
STD	0.09	0.09	0.07	0.07	0.07	0.08	0.07	0.08	0.06	0.08	0.07	0.11	0.09	0.10	0.09	0.10
ITR																
S1	32.54	38.79	23.13	32.65	36.71	97.28	9.08	30.23	41.90	74.32	34.83	66.83	35.70	63.72	39.05	76.99
S2	68.23	87.18	31.95	47.42	73.02	105.09	57.34	89.56	43.74	73.88	41.21	63.72	58.22	80.23	54.10	60.74
S3	17.35	21.34	8.02	9.67	33.73	48.49	23.22	37.74	13.66	25.20	10.46	12.34	5.56	13.45	4.54	7.33
S4	35.20	62.59	13.31	22.74	38.00	48.34	36.96	43.45	35.95	56.48	18.51	37.74	17.51	35.82	18.77	33.97
S5	19.11	39.72	7.18	14.54	41.63	73.88	16.78	17.67	27.80	35.70	14.54	13.45	12.61	16.62	13.03	15.14
S6	67.82	92.60	24.00	43.03	92.03	120.00	28.34	42.18	45.04	64.68	24.50	49.42	30.91	47.88	29.00	46.07
S7	8.46	21.34	4.74	10.71	17.51	21.06	8.46	12.89	11.41	14.84	5.29	5.38	3.63	4.78	3.35	7.59
S8	9.31	19.02	3.81	9.08	45.48	63.91	26.74	54.27	45.78	74.10	37.22	59.65	36.71	32.65	31.02	44.60
Mean	32.25	47.82	14.52	23.73	47.26	72.26	25.87	41.00	33.16	52.40	23.32	38.57	25.11	36.90	24.11	36.56
STD	22.52	27.73	9.85	14.50	22.27	31.25	14.98	22.37	13.13	22.39	12.44	23.49	17.44	24.31	16.43	23.80

Combined method checking the cross-correlation coefficients for left gaze and right gaze and classifying based on the higher coefficient significantly improves the accuracy and ITRs

Table 2 ANOVA results for tests of Between-Subject Effects

Source	Type III sum of squares	df	Mean square	F	Sig.
Gaze	4.167E-6	1	4.167E-6	0.000	0.983
Data duration	0.700	2	0.350	39.834	0.000
Rate	0.266	3	0.089	10.094	0.000
Signal	0.287	1	0.287	32.678	0.000
Gaze * Data duration	0.002	2	0.001	0.116	0.891
Gaze * Rate	0.008	3	0.003	0.297	0.827
Gaze * Signal	0.001	1	0.001	0.080	0.777
Data Duration * Rate	0.001	6	0.000	0.026	1.000
Data Duration * Signal	0.001	2	0.001	0.064	0.938
Rate * Signal	0.108	3	0.036	4.081	0.007

Data duration, Rate, and Signal are three main independent factors significantly affecting the dependent variable accuracy. There is a significant interaction between rate (10rps, 32rps, 50rps, and 70rps) and signal type (QSS or TR)

TR). Before the application of ANOVA, a homogeneity test, the Levene's Test of Equality of Error Variances, was used. Since no significance was found, ANOVA could be applied .The ANOVA table is given in Table 2.

The findings of the ANOVA can be summarized as follows. The gaze (left or right target) has no significant effect on the accuracy. Increasing data duration significantly increases the accuracy ($p < 0.01$). Another factor is the rate; it also significantly affects the accuracy. Among these, 32rps has the highest accuracy followed by the 50rps. The signal type whether QSS or TR has significant effect ($p < 0.01$) on the accuracy so that QSS yields higher accuracies compared to TR-VEP signal types in the classification.

3.2 SSVEP Vs. QSS-VEP Comparison

It is known that SSVEPs are not prominent at higher stimulation rates (Diez et al. 2013). This is mainly due to the adaptation of the visual cortex at high rates. High stimulation rates can be better in VEP-based BCIs due to low background EEG noise at those frequency bands, and less proneness to seizures. In this experiment step, we investigated if there are any characteristic differences between SSVEP and QSS-VEP stimulation paradigms in terms of template-matching BCI performance. Two rates were selected as 10rps and 32rps for stimulation. For 10rps, the same number of sweeps was used in training and test data sets of QSS- and SS-based template matching classification. As seen from Fig. 6 (Kaya 2019), for 10rps, similar accuracies and ITRs are observed for both paradigms. On the other hand, for 32rps, since QSS-VEPs have broader spectrum and stimulate a wider frequency bandwidth, they produced high SNR-enhanced responses at these rates.

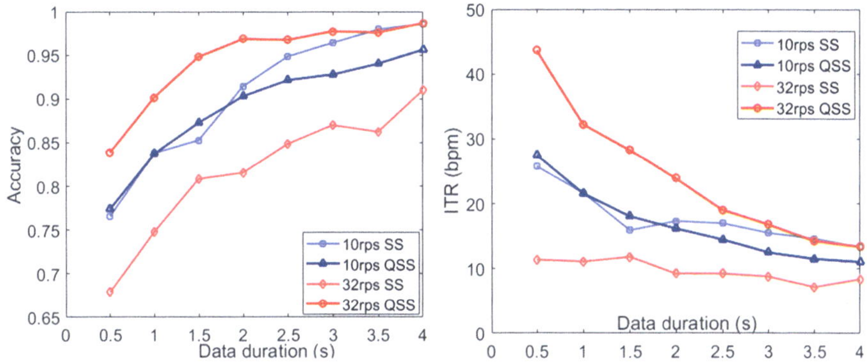

Fig. 6 QSS-VEP and steady-state visual evoked potentials (SSVEP) performances, (accuracies (left) and ITRs (right), are compared. 32rps QSS-VEP performed better than SSVEP for short data durations whereas 10rps performed similar

This resulted in remarkable higher accuracies and ITRs compared to SSVEPs. QSS paradigm clearly improves the template-matching classification in a BCI design at medium-high rates.

4 Discussions

The c-VEP paradigms utilizing unitary generator transient VEPs to predict random codes have been efficient in various BCI designs in (Thielen et al. 2015; Nagel and Spüler 2018). By using broad band VEPs (BBVEPs) generated by Gold codes and a 6 × 6 target matrix, Thielen et al. (2015) developed a novel BCI speller (Thielen et al. 2015). They used linear regression methods to obtain the transient VEPs, then for the test data the transients were re-convolved to generate BBVEP templates for detection. With this method based on regression of VEPs and re-convolution, BBVEPs in response to random codes composed of long and short pulses of visual stimuli were predicted accurately. Accuracy of 0.86 and ITR of 48 bpm (9 symbols per min) were achieved using 3.21 s data. On the other hand, in another regression-based study, Nagel and Spüler (2018) modeled the VEPs in response to random code modulations by using windows of 250 ms and regression method (pattern prediction by EEG2Code and VEP-EEG prediction by the Code2EEG models) (Nagel and Spüler 2018). The system reached a mean ITR of 108 bpm and theoretical limit of 470 bpm. In 2019, by combining these models with deep learning methods, Nagel and Spüler (2019) reached an average ITR of 701 bpm (Nagel and Spüler 2019). However, a deconvolution-based method could also be used to extract transient VEPs more efficiently. With QSS-VEP paradigm and CLAD method, it would be possible to reach high-speed BCIs as demonstrated by the predictive models developed by (Thielen et al. 2015; Nagel and Spüler 2018, 2019).

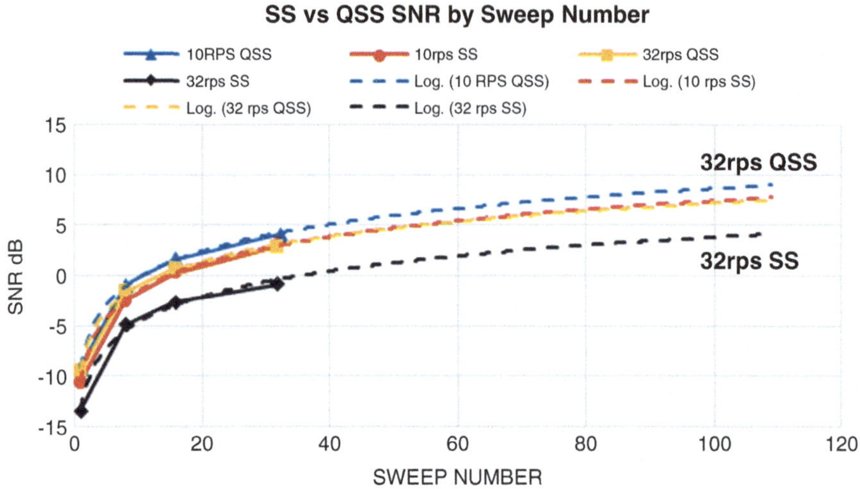

Fig. 7 QSS-VEP and steady-state visual evoked potential (SSVEP) performances, (accuracies (left) and ITRs (right), are compared. 32rps QSS-VEP performed better than SSVEP for short data durations whereas 10rps performed similar

QSS-VEP paradigm achieved mean ITR of 72 bpm with 32rps rate and single sweep (0.5 s); this is remarkable for a two-target system. The low jittered QSS paradigm significantly improves the VEP SNR. This can be seen from Fig. 7 (Kaya 2019), where QSS and SS paradigms are compared in terms of SNR by sweep number. It can be clearly seen that QSS reaches the same SNR by less sweeps than the SS paradigm. For BCI applications the time required to detect a control signal is critical in determination of the system performance. Since QSS has high SNR with less sweep, the possible application of this method in BCI design will promote the speed of operation. Another advantage of QSS paradigm is the extraction of TR-VEPs from the convolved responses. With the help of the CLAD method, fast acquisition of TR-VEPs in a BCI could help monitoring of diseases such as glaucoma or other neural origin pathologies. Moreover, by using CLAD, transient VEPs can be obtained by smaller window than 250 ms which is used in (Nagel and Spüler 2018, 2019). The ceiling limit mentioned in their study can be leveled up.

5 Conclusion

VEP has been used extensively in BCI research due to the fact that VEP-based BCIs provided significantly higher accuracy and ITR values compared to other EEG signal modalities. SSVEP-based BCIs have been leading this competition; however, recent advancements in the c-VEP-based BCIs enabled very high ITRs. These are mainly due to the generative frameworks developed to model the single unitary transient VEPs to single pulse stimulus and predictive models for VEPs in response to random stimuli.

In our study, we utilized the QSS-VEP stimulation paradigm in a dual target BCI system. Four different stimulation rates were investigated, namely 10rps, 32rps, 50rps, and 70rps. Different sizes of EEG data, single sweep (0.5 s)—3 sweeps (1.5 s), were used in the classification. For the target identification, right and left QSS-VEP template-matching-based detection was applied for classification for each rate. In these two methods, cross-correlation coefficients were compared to the threshold values and classification was performed. And another method combining these two classification correlation coefficients is developed in order to improve accuracy and ITR. This method significantly improved the mean accuracies for all the rates; from 0.77 to 0.83 for 10rps QSS, from 0.84 to 0.90 for 32rps QSS, 0.79 to 0.85 for 50rps, and 0.74 to 0.79 for 70rps QSS. One subject achieved a maximum of 100% accuracy by combined classification and 120 bpm in offline ITR performance at 32rps QSS and 0.5 s (single sweep) data duration. The developed system provided remarkable performance.

In the last experiment, Quasi-Steady-State (QSS) and Steady-State (SS) paradigms were compared in a template-matching BCI scenario. QSS paradigm significantly improved the SNR values for the same number of sweeps. The low jitter introduced by the QSS method prevents the adverse adaptation effects on the template matching performance. This method can be integrated with other signal processing and machine learning methods to develop better VEP-based BCI systems.

References

Bin G, Gao X, Wang Y et al (2009) VEP-based brain-computer interfaces: time, frequency, and code modulations [research frontier]. IEEE Comput Intell Mag 4(4):22–26

Bin G, Gao X, Wang Y (2011) A high-speed BCI based on code modulation VEP. J Neural Eng 8(2):025015

Bohórquez J, Özdamar Ö (2006) Signal to noise ratio analysis of maximum length sequence deconvolution of overlapping evoked potentials. JASA 119(5):2881–2888

Bohórquez J, Lozano S, Kao A et al (2013) Deconvolution and modeling of overlapping visual evoked potentials. In: 29th Southern Biomedical Engineering Conference (SBEC). IEEE, Piscataway, NJ, pp 31–32

Capilla A, Pazo-Alvarez P, Darriba A et al (2011) Steady-state visual evoked potentials can be explained by temporal superposition of transient event-related responses. PLoS One 6(1):e14543

Daly JJ, Wolpaw JR (2008) Brain–computer interfaces in neurological rehabilitation. Lancet Neurol 7(11):1032–1043

Delgado RE, Özdamar Ö (2004) Deconvolution of evoked responses obtained at high stimulus rates. JASA 115(3):1242–1251

Diez PF, Müller SMT, Mut VA et al (2013) Commanding a robotic wheelchair with a high-frequency steady-state visual evoked potential based brain–computer interface. Med Eng Phys 35(8):1155–1164

Fahle M, Bach M (2006) Origin of the visual evoked potentials. In: Heckenlively JR, Arden GB (eds) Principles and practice of clinical electrophysiology of vision. MIT Press, Cambridge, MA, pp 207–234

Haas LF (2003) Hans Berger (1873–1941), Richard Caton (1842–1926), and electroencephalography. J Neurol Neurosurg Psychiatry 74(1):9–9

Hirsch L, Brenner R (2010) Atlas of EEG in critical care. Wiley, New York, pp 187–216

Hochberg LR, Anderson K (2012) BCI users and their needs. In: Brain–computer interfaces: principles and practice. Oxford University Press, New York, p 317. ch. 19

Kaya I (2019) High rate quasi-steady-state pattern visual evoked potentials for brain-computer Interface. Dissertation. University of Miami

Kaya I, Bohórquez J, Özdamar Ö (2019) A BCI gaze sensing method using low jitter code modulated VEP. Sensors 19(17):3797

Kohl MC, Schebsdat E, Schneider EN et al (2019) Fast acquisition of full-range auditory event-related potentials using an interleaved deconvolution approach. JASA 145(1):540–550

Lebedev M (2014) Brain-machine interfaces: an overview. Transl Neurosci 5(1):99–110

Lebedev MA, Nicolelis MA (2006) Brain–machine interfaces: past, present and future. Trends Nerosci 29(9):536–546

Lebedev M, Opris I (2015) Brain-machine interfaces: from macro-to microcircuits. In: Recent advances on the modular organization of the cortex. Springer, Dordrecht, pp 407–428

Levitskaya O, Lebedev MA (2016) Brain-computer interface: the future in the present. Bull RSMU 2

Nagel S, Spüler M (2018) Modelling the brain response to arbitrary visual stimulation patterns for a flexible high-speed brain-computer Interface. PLoS One 13(10):e0206107

Nagel S, Spüler M (2019) World's fastest brain-computer interface: combining EEG2Code with deep learning. PLoS One 14(9):e0221909

Nakanishi M, Wang Y, Chen X (2018) Enhancing detection of SSVEPs for a high-speed brain speller using task-related component analysis. IEEE Trans Biomed Eng 65(1):104–112

Nicolas-Alonso LF, Gomez-Gil J (2012) Brain computer interfaces, a review. Sensors 12(2):1211–1279

Odom JV, Bach M, Brigell M et al (2016) ISCEV standard for clinical visual evoked potentials:(2016 update). Doc Ophthalmol 133(1):1–9

Özdamar Ö, Bohórquez J (2006) Signal-to-noise ratio and frequency analysis of continuous loop averaging deconvolution (CLAD) of overlapping evoked potentials. JASA 2006(119):429–438. https://doi.org/10.1121/1.2133682

Puth MT, Neuhäuser M, Ruxton GD (2014) Effective use of Pearson's product–moment correlation coefficient. Anim Behav 93:183–189

Sutter EE (1992) The brain response interface: communication through visually-induced electrical brain responses. J Microcomput Appl 15(1):31–45

Sutter EE (2001) Imaging visual function with the multifocal m-sequence technique. Vision Res 41(10–11):1241–1255

Thielen J, van den Broek P, Farquhar J et al (2015) Broad-band visually evoked potentials: re (con) volution in brain-computer interfacing. PLoS One 10(7):e0133797

Thompson DE, Blain-Moraes S, Huggins JE (2013) Performance assessment in brain-computer interface-based augmentative and alternative communication. Biomed Eng 12(1):43

Toft-Nielsen J, Bohórquez J, Özdamar Ö (2011) Innovative pattern reversal displays for visual electrophysiological studies. In: 2011 Annual International Conference of the IEEE Engineering in Medicine and Biology Society. IEEE, Piscataway, NJ, pp 2009–2012

Toft-Nielsen J, Bohórquez J, Özdamar Ö (2014) Unwrapping of transient responses from high rate overlapping pattern electroretinograms by deconvolution. Clin Neurophysiol 125(10):2079–2089

Volkova K, Lebedev MA, Kaplan A et al (2019) Decoding movement from electrocorticographic activity: a review. Front Neuroinform 13:74

Wang Y, Wang R, Gao X et al (2006) A practical VEP-based brain-computer interface. IEEE Trans Neural Syst Rehabil Eng 14(2):234–240
Wei Q, Feng S, Lu Z (2016) Stimulus specificity of brain-computer interfaces based on code modulation visual evoked potentials. PLoS One 11(5):e0156416
Wolpaw JR, Wolpaw EW (2012) Brain-computer interfaces: something new under the sun. In: Brain–computer interfaces: principles and practice. Oxford University Press, New York
Wolpaw JR, Birbaumer N, McFarland DJ et al (2002) Brain–computer interfaces for communication and control Clin. Neurophysiology 113:767–791
Zemon VM, Gordon J (2018) Quantification and statistical analysis of the transient visual evoked potential to a contrast-reversing pattern: a frequency-domain approach. Eur J Neurosci 48:1765–1788

Communication with Brain–Computer Interfaces in Medical Decision-Making

Walter Glannon

1 Introduction

Brain–computer interfaces (BCIs) can restore or augment motor functions that have been impaired or lost from traumatic brain injury, amputation or neurodegenerative disease. BCIs produce these effects in bypassing affected brain regions through real-time direct connections between functionally intact regions and a computer (Wolpaw and Wolpaw 2012; Lebedev 2014). Electrodes placed on the scalp or implanted in the brain can decode electrical signals in the motor cortex that provide informational input to the computer. Informational output from the computer completes a sensory feedback loop in which a person can modulate brain activity through brain-mediated mental acts. BCIs have enabled people with tetraplegia to move a computer cursor, robotic arm and other objects. People who cannot communicate verbally or gesturally can use BCIs to communicate in other ways to confirm a diagnosis or express attitudes about their condition (Birbaumer et al. 2008; Naci et al. 2012; Linden 2014: 19–26). In these and other respects, BCIs can allow neurologically compromised patients to form and execute intentions in actions. They can enable them to regain some degree of voluntary motor control functional independence and improve their quality of life.

In this chapter, I discuss the potential of BCIs to help cognitively capable patients communicate decisions about medical care when their motor functions are impaired. While communication requires both cognitive and motor capacities, it is motor impairment that BCIs are designed to overcome. BCIs can serve as assisted communication systems. I focus mainly on patients with locked-in syndrome (LIS) and those with advanced amyotrophic lateral sclerosis (ALS). The first condition typically results from a primary brainstem stroke. The second condition results from

W. Glannon (✉)
Department of Philosophy, University of Calgary, Calgary, AB, Canada
e-mail: wglannon@ucalgary.ca

I. Opris et al. (eds.), *Modern Approaches to Augmentation of Brain Function*, Contemporary Clinical Neuroscience,
https://doi.org/10.1007/978-3-030-54564-2_8

degeneration of motor neurons. Although they have a distinct pathophysiology, both conditions cause profound paralysis, including the inability to communicate through normal means. I examine how some of these patients could use BCIs to communicate in expressing consequential decisions about therapy that could extend or end their lives. These decisions would include whether to continue or discontinue artificial nutrition and hydration (ANH) in the LIS and ALS, whether to initiate or end mechanical ventilation in ALS, and whether to request organ donation or assisted dying in both conditions. I also explore the possibility that some patients in the minimally conscious state (MCS) may have enough cognitive capacity to use a BCI to express similar decisions about life-sustaining therapy.

I address specific empirical and ethical questions. How does a BCI decode electrical signals in the motor cortex and adjacent brain regions? How does this decoding enable the motor functions necessary for behaviorally nonresponsive patients to communicate? What constitutes clear, consistent, and reliable communication? What can a patient communicate by using a BCI? Even if a BCI enabled patients to give more than 'yes' or 'no' binary responses to questions from investigators, would it "allow them to fully exercise their right to actively participate in consequential medical decisions about their own medical care" (Bernat 2020: 1)? How can a patient benefit from or be harmed by using a BCI in wanting and trying to express these decisions?

I explain differences between decoding neural states and mental states and argue that the latter can at best be decoded only indirectly. These differences explain why communication involves more than neural decoding. After describing different types of communication, I point out that they all require cognitive and motor functions and mental and neural causation. Physical events in the brain generate and sustain mental acts like forming and executing an intention to communicate a thought. The subject's mental states of wanting, intending, and trying to use a BCI to communicate activate the neural signals that are transmitted to the computer for language output. BCI-communication thus involves interacting neural and mental events in an integrated brain-mind-machine network. This includes both extracortical and intracortical systems.

I then discuss potential benefit and harm from using a BCI to express decisions about life-sustaining medical care. The most likely patient to use this technology for this purpose would be one with LIS or ALS whose cognitive functions were intact. Patients with complete locked-in syndrome (CLIS) present a greater challenge because they are unable to make voluntary eyelid movements and cannot use eye-tracking or BCI systems involving visual feedback. More controversial is whether patients in the MCS have enough cognitive capacity to use a BCI in executing the motor functions necessary to communicate decisions about continuing or discontinuing ANH. I describe a case of an MCS patient where her use of a BCI might have resolved a legal conflict between a court and the patient's family. The ethical stakes in BCI-mediated communication about life-sustaining therapy are high. Failure in trying to express a decision by patients, or misinterpretation by caregivers of what is expressed, could defeat patients' intentions. They could be harmed by continuing or discontinuing life-sustaining therapy when either of these actions was contrary to their best interests. Yet BCIs could enable some patients to clearly and reliably express these decisions and thereby benefit them. I conclude the chapter by summarizing the main points and with some general thoughts on the potential of current and more advanced BCIs to enable neurologically compromised patients to communicate.

2 Decoding and Communicating

BCIs enable communication by decoding neural signals and converting them into letters and words as synthesized speech. Electrodes placed on the scalp, epidural subdural implants, or intracortical microelectrode arrays record electrical signals in motor and language-processing areas, which are then transmitted to a receiver and computer for language output. At a brain-systems level, this type of decoding refers to the recording of specific areas of neural signal activation in these areas (Saur et al. 2008). The neural signatures can be detected by EEG or functional neuroimaging such as positron emission tomography (PET) and functional magnetic resonance imaging (fMRI). At a mental-systems level, the decoding refers to the identification of intentions and other mental states associated with wanting and trying to communicate.

Underscoring the connection between decoding and communication, Mariska Vansteensel and coinvestigators describe a 58-year-old ventilator-dependent patient with ALS who used a fully implanted BCI system consisting of a computer typing program to communicate by typing two letters per minute (Vansteensel et al. 2016). Electrodes were implanted over the patient's sensorimotor and dorsolateral prefrontal cortex. Electrical cortical signals passed through a transmitter implanted under her collarbone. An antenna sent the signals to a receiver and then to a wheelchair-mounted computer tablet from which she selected the letters. The authors write that, "mental acts, such as an attempt to move a limb, lead to reproducible signals in corresponding cortical regions. The detection of signals from the cortex requires computational processing to separate ("decode") them from background noise. Decoding of sufficient quality provides the input for a computer system that directs typing software, thereby enabling communication" (p. 2060). Adrian Owen and Lorina Naci note that fMRI and other imaging modalities have been used in research on consciousness to "'decode' mental decisions or thoughts in healthy participants" (Owen and Naci 2016: 101). They explain the process further: "The principle often employed capitalizes on the fact that certain types of thought are associated with a unique brain activation pattern that can be used as a signature for that specific thought. If a statistical classifier is trained to recognize these characteristic signatures, a volunteer's thoughts can be decoded (within the constraints of the experimental design) using their brain activity alone" (Owen and Naci 2016: 101; Naci).

Communication roughly consists in expressing one's thoughts to others. These include wishes, intentions, decisions, and other conscious mental states and acts. Communication is reliable when the letters, words, or other signs one produces are coherent and can be understood and interpreted by others as an accurate reflection of one's thoughts. Communication is consistent when one clearly expresses the same mental states and acts at different times. In normal circumstance, this occurs verbally or gesturally. BCIs can enable behaviorally nonresponsive patients to communicate by activating motor areas and producing letters and words through a computer. The patient uses the BCI to communicate through a combination of mental acts and neural events.

Neural decoding does not provide a complete explanation of communication. Recording electrical signals in cortical regions constitutes only one component of the sequence of events that extends from a person's intention and decision to use a BCI to the outcome of successfully expressing them to others. As Vansteensel and coauthors point out, the patient's mental acts generate the signals in the sensorimotor and prefrontal cortex that are transmitted to the word-producing software. Contrary to what Owen and Naci claim at the end of the passage cited above, the use of "mental acts" and "mental decisions" in these passages suggests that communicating through an fMRI or BCI involves more than brain activity alone. Although intending or trying to perform certain speech or bodily acts depends on normal cortical function, these mental acts are not identical or reducible to neural events. The phenomenology and content of mental states can only be indirectly inferred from recording electrical signals in the brain. Decoding the neural correlates of thoughts is not equivalent to decoding the content of the thoughts themselves. This content includes cognitive, affective, and volitional properties shaped by factors inside and outside the brain. Specifically, neural decoding cannot capture a patient's experience of being locked-in (Vidal, 2020).

Producing letters and words through a BCI is an artificial speech act (Searle 1969). The ability of a behaviorally nonresponsive patient to use the interface to perform this act cannot be explained in terms of dualist model in which mental events are independent of neural events (Lee 2016). Nor can it be explained as a completely physical sequence of events extending from the brain to the computer and back again. A satisfactory explanation of how one uses the interface to produce words in expressing decisions requires both mental and physical events as interacting components of an integrated brain-mind-machine system. While the informational input to and informational output from the computer is a physical process, the patient activates these signals through the mental acts of wanting and trying to activate them. The patient also activates the computer typing program by trying to move her hand as if to type. Electrical cortical signals and the computer software allow the necessary motor functions for successful communication. But it is the patient, not signals or software, who communicates through the combined effects of her mental acts and the interface. By modulating activity in language-processing regions of the brain, the patient can make and express decisions about medical care through the BCI. Her mental acts of initiating and sustaining this process indicate that she has a causal role in expressing these decisions. There is shared control between the patient and the BCI in producing the desired result.

EEG-based electrodes on the scalp, epidural subdural electrocorticography, or intracortical microelectrode arrays do not activate and transmit these signals entirely on their own. The patient's mental acts of intending to use the BCI and responding to computer feedback in trying to communicate are not epiphenomenal but have causal efficacy in initiating and sustaining the neural events that provide input to the computer typing program (cf. Pernu 2019). There is both mental and physical causation in this process (Davidson 1992, 2001; Mele 2009). In the case of the ALS patient described earlier, her thinking about moving her hand as if to type initiated the sequence of transmitting cortical signals to the typing software. By trying to move her hand in this way, she caused the computer tablet to display a row of letters. When the letter she wanted was highlighted, she mentally selected it, and the computer typing program produced the letters. The patient's mental acts of thinking about and trying to move her hand as if to type letters and the neural events that provide the input to the program are both necessary to explain her ability, or inability, to communicate wishes and decisions in using, or failing to use, the BCI.

In normal brain function, activation of electrical signals in the motor cortex for limb movement and in the superior temporal lobe and premotor cortex of the frontal lobe for speech can occur without having to think about performing these actions. We can execute these motor and language skills without conscious effort. To communicate with a BCI, a patient who cannot speak must put considerable conscious effort into learning how to activate electrical signals in the brain and transmit them to the computer to produce selected letters and words. In addition to an initial desire and intention to use the BCI to communicate, the patient must also have the capacity for sustained motivation, attention, concentration, mental effort, and patience in using it. These cognitive, emotional, and volitional capacities in turn are influenced by interaction between the patient and the practitioner in the process of learning how to use the BCI. They are also influenced by the patient's values and how they shape his interest in continuing or discontinuing life-sustaining therapy. While these mental states emerge from and are sustained by normal cortical and subcortical brain function, their content and subjective quality is more than a function of physical events involving electrical cortical signals, a brain implant, or input to and output from the computer.

EEG, fMRI, and implanted microelectrode arrays can decode brain signals associated with wanting, intending, and trying to use a BCI to communicate. But these mental acts cannot be decoded in the same way as neural signals because they are not located in the brain. The presence or absence of mental content is typically inferred and confirmed by observing a patient's behavior rather than by recording brain activity. This supports the idea that an integrated brain-mind-machine network is necessary for behaviorally nonresponsive patients to make and communicate decisions about their medical care. It is also necessary for health-care providers to confirm and act on these decisions.

In a study led by Owen, a patient initially diagnosed as being in a vegetative state was able to activate her supplementary motor cortex in response to commands to imagine moving around her house and playing tennis. She thereby communicated to the investigators that she was in fact aware and at least minimally conscious (Owen et al. 2006; Owen and Coleman 2008). In a different study led by Martin Monti, a minimally conscious patient was able to give 'yes' and 'no' responses to questions by activating certain brain regions and confirming the ability to understand spoken language and sustain focus on a given task (Monti et al. 2010; Naci et al. 2012). Each of these studies used fMRI to record brain activation in response to motor commands. Although it was unclear to what extent the patients understood what the investigators were saying, the brain's response confirmed a basic form of communication. A more recent EEG study of ICU patients with acute brain injury also suggests the potential for communication despite cognitive-motor dissociation (Claassen et al. 2019).

The ability to activate brain regions by consciously responding to instructions or answer affirmatively or negatively to questions is not adequate for more complex and robust communication. Specifically, these responses do not provide evidence of the neural and mental capacities necessary to make and express value-laden decisions about continuing or discontinuing life-sustaining therapy. These capacities are necessary for a patient's decision to be autonomous and meet standard criteria of decisional capacity and informed consent. A BCI with a word-processing program, like the one used by the ALS patient described earlier, may allow this type of communication. Yet some patients wanting and attempting to use a BCI may not be able to learn how to use it. Or they may not be able to use it to produce the letters and words necessary to express these momentous decisions clearly and reliably (Naci and Owen 2013). This may allow misinterpretation of these decisions by caregivers and attending physicians. It may lead caregivers to conclude that the patient lacks decisional capacity. The magnitude of harm to conscious patients from inability or failure to communicate these decisions underscores the significance of this problem. It could result in discontinuing therapy and ending a patient's life against his interest in staying alive. On the other hand, it could result in continuing therapy and prolonging a different patient's life against her interest in dying. If a patient experienced pain and suffering from her condition, then this second action could cause her to experience more pain and suffering and experience greater harm.

Ned Block distinguishes "phenomenal consciousness" from "access consciousness" (Block 1995, 2007). He describes the first as "experience" and claims that the "phenomenally" conscious aspect of a state is what it is like to be in that state" (Block 1995: 227). Access consciousness consists in information processing and its "availability for use in reasoning and rationally guiding speech and action" (Block 1995: 227). This second type of consciousness includes the ability to report our experiences to others and make decisions based on the information we process. Access consciousness requires cognitive and motor functions to communicate verbally or gesturally. These functions and the ability to communicate are impaired in LIS, ALS, and the MCS. Most patients in the first two groups have cognitive control but impaired motor control. Patients in the third group are impaired in both cognitive and motor control. They are impaired in both phenomenal and access consciousness. These deficits come in degrees, however. Some patients at the higher end of the MCS spectrum may have a level of cognitive function allowing for different forms of agency. BCIs might augment their access consciousness, partly restore motor functions, and enable them to communicate their thoughts about care to family and caregivers. As assisted communication systems, they can help some patients translate intentions into actions. Whether or to what extent patients can use these systems for this purpose depends on the level of their cognitive and other mental capacities.

3 Extracortical and Intracortical BCIs for Communication

Extracortical EEG-based BCIs utilizing electrodes on the scalp have enabled some locked-in patients to activate slow cortical potentials, sensorimotor rhythm, and the P300 event-related potential and respond affirmatively or negatively to questions about their quality of life (Birbaumer et al. 2008, 2014) These responses have had a high level of accuracy (around 70%) in assessing patients' attitudes about living with their condition. There is a statistically significant correlation between support from family and caregivers and patients' affirmative responses. This shows that the type and content of patients' responses reflect the influence of contextual factors in addition to their neurological condition. This type of support, as well as support from investigators and practitioners teaching patients how to operate a BCI, can sustain patients' conscious effort and increase the probability of expression and accurate interpretation of their decisions.

EEG-based BCIs are limited in their ability to allow subjects to translate an intention to speak into a communicative act. This is mainly because the cranium can deflect and thus interfere with the transmission of neural signals from the motor and premotor cortex to the computer. Subdural electrocorticography (ECoG) avoids this problem because the electrodes are placed on the surface of the brain (Leuthardt et al. 2004). In a recent study, Gopala Anumanchipalli, Josh Chartier, and Edward Chang used an ECoG BCI in a dual-stage decoding process to reconstruct speech (Anumanchipalli et al. 2019). They "first transformed neural signals into representations of movements of the vocal-tract articulators, and then transformed the decoded movements into spoken sentences" (Pandarinath and Ali 2019: 466). In contrast to the typing-based BCI used by the patient described by Vansteensel et al., this system may enable the production of more words per minute. It may enable more fluid communication of natural speech, which corresponds to 150 words per minute (Pandarinath and Ali 2019: 466). Even if the technique could not reach this level, approximating it would be a major advance in BCI-based communication.

Significantly, the five volunteers in this study were all capable of speaking. LIS, advanced ALS, and most MCS patients are unable to speak. These noncommunicative patients are the group that would benefit from an ECoG BCI. They are the subjects on whom it must be tested to demonstrate its communicative capacity. Fluid communication of speech involves more than translating an intention to speak directly into a speech act. The subject must learn how to activate language-processing areas in the frontal and temporal cortex. It is not obvious that this system would enable speech if patients and research subjects had limited activity in these areas. One cannot infer that the effects of reconstructing speech in subjects who are capable of speaking would be replicated in those who are incapable of speaking.

Moreover, the fact that ECoG does not directly record localized brain activity in cortical regions associated with language may limit its ability to produce words with the clarity and reliability necessary for robust communication.

Neural interfaces consisting of implantable intracortical microelectrode arrays can overcome this limitation. Because the implant has direct contact with motor areas, it can record neural signals more directly for transmission to the software. The array can also record signals in a wider area of cortical networks. Noninvasive scalp-based and less-invasive subdural systems involve cumbersome and time-consuming electrode setups. The wires connecting the electrodes to the EEG can limit a patient's mobility as well. A wireless implantable system allows greater patient mobility and can operate continuously. Still, the ability of an implant to record and transmit neural signals to a computer typing program depends on the specific features of the neurological deficit. The nature and extent of the brain injury or neurodegeneration could limit a patient's ability to operate the BCI. In addition, invasive systems entail more risks associated with intracranial surgery. These include bleeding, infection, and other possible deleterious effects on neural tissue surrounding the implant. There may not be perfect biocompatibility of the microelectrode array with tissue in the brain (Kennedy et al. 2011). Also, the implant could move over time and require adjustment. Nor is it clear how long the implant would function. It might have to be replaced by another implant after a certain period.

In 2014, Leigh Hochberg and Merit Cudkowicz stated that, "though both scalp-based EEG and corticography signals have been recorded in people with ALS and total LIS, we are aware of no reports of restoring communication using a neural signal-based BCI in this most severely affected population" (Hochberg and Cudkowicz 2014: 1852). Presumably, the authors are referring to a more complex process than binary responses to questions. An implantable BCI could allow patients with a high level of cognitive function to make and express autonomous and verifiable decisions about medical care. The implanted microelectrode array that decoded the electrical signals in the ALS patient's sensorimotor and dorsolateral prefrontal cortex when she tried to move her hand and finger as if to type allowed input to the computer typing program through which she spelled out letters. Because the patient was able to use an eye-tracker, arguably the implanted system was not necessary. Vansteensel et al. point out that the implanted BCI offered "autonomous communication that supplemented and at times supplanted the patient's eye-tracking device" (p. 2060). If this patient had lost voluntary eye movement, then she would not have been able to use an eye-tracking system. In that case, only the implantable BCI would have been effective. Also, the implant probably facilitated her ability to select letters by recording motor cortical signals and transmitting them more directly to the computer. These connections might not have been possible with an eye-tracking device even if the patient had preserved voluntary eye movement.

BCIs typically utilize visual feedback. Patients rely on this type of sensory feedback from computer output to modulate brain activity. CLIS patients are unable to respond to visual stimuli in learning how to operate these systems. Alternatively, auditory or tactile feedback could combine with sensorimotor processing to enable some form of letter and word production for communication (Kubler 2009;

Hochberg and Cudkowicz 2014). Still, it is questionable whether anything other than visual feedback would be able to activate a computer typing program. One can only speculate on whether advanced BCIs not relying on vision would allow expression of the cognitive functions associated with decision-making. Systems relying on auditory or tactile feedback are more cognitively and emotionally demanding for patients to use than those relying on visual feedback, which can be demanding enough. This could strain a patient's ability to execute a word-processing function to communicate. These challenges could preclude many CLIS patients from expressing wishes and decisions about medical care and leave them to suffer in silence.

In a study of BCI using functional near-infrared spectroscopy (fNIRS), Ujwal Chaudhary and coinvestigators reported that four completely locked-in patients with advanced ALS were able to respond to questions requiring a 'yes' or no' response to questions with an above-chance level accuracy rate of 70% (Chaudhary et al. 2017). Yet the authors report that "electroencephalogram (EEG) changes in the theta-frequency band correlated with inferior communication performance, probably because of decreased vigilance and attention" (2017: 1). Inferior communication could also be attributed to deflection or smearing of electrical signals from the scalp-based electrodes by the cranium. Decreased vigilance and attention would preclude a level of communication necessary for decisions about and consent to maintaining or altering their medical care. A system like the one used by the ALS patient would be more likely to enable this level of communication than an EEG-based system. Still, the fact that completely locked-in patients would lack the visual feedback necessary for them to operate a computer typing program may limit what they could communicate with an implantable system. Chaudhary et al. conclude from their study that "these positive results could indicate the first step towards abolition of complete locked-in states, at least for ALS" (p. 1). But it is unclear how this result could be achieved if patients were capable of only simple affirmative or negative responses to questions. This would fall short of evidence for the cognitive and motor capacity to make and express decisions with BCIs about interventions that would abolish these states. Like the fMRI studies mentioned earlier, functional neuroimaging displaying brain activity in response to commands can only determine a minimal sense of communication. It cannot provide the level of communication necessary for consequential decision-making.

Not all ALS or LIS patients with preserved cognitive and visual functions could learn to use a BCI to communicate wishes and decisions to caregivers. A BCI with a computer writing tablet might overcome the dysphagia, anarthria, and aphonia from pseudobulbar palsy if the patient had intact consciousness and cognition (Bernat, 2020, 231). There would, however, be variation among patients in the cognitive capacity necessary to use the system. Advanced ALS, for example, may involve cognitive impairment and loss of rational and decisional capacity (Phukan et al. 2007; Schnakers et al. 2008). This could limit the ability to follow instructions in executing the motor functions necessary to successfully operate the interface.

Patients in the MCS face greater challenges in communicating with BCIs because they have both cognitive and motor impairment. The MCS is defined as the presence of inconsistent but clearly discernable behavioral signs of consciousness (Giacino et al. 2002). Some patients with a higher level of awareness and some preserved cognitive functions may be able to use a BCI to make and express decisions about the care they receive. Communication in these cases would likely require an implantable BCI. The more direct connections between the implant and motor cortical electrical signals may compensate for motor and cognitive deficits associated with brain injury. This would depend on the extent of axonal injury in cortical regions. MCS patients may be aphasic and unable to understand or use language with the aid of a computer (Massimini and Tononi 2018: 39). They may not even have the capacity for minimal communication with 'yes' and 'no' responses to questions. Others may have apraxia and be impaired in motor planning and unable to perform motor tasks when asked. They may be willing to perform a task but unable to execute it. Although implantable systems may have more communicative potential because of stronger connections between motor neural signals and the computer, the extent of brain injury may limit the patient's ability to interact with the practitioner and learn how to use the system.

Nevertheless, because awareness and cognition come in degrees, a patient at the higher end of the minimally conscious spectrum may have enough cognitive capacity for some degree of decision-making about therapy. She may have enough preserved neural networks supporting cognitive functions for communicating decisions despite profound behavioral impairment (Chennu et al. 2014). Recovery from the MCS is defined by the reemergence of functional communication and/or functional use of objects (Giacino et al. 2002). Activating cortical regions in response to commands and engaging in this behavior could serve a diagnostic purpose in confirming that a person had recovered from the MCS. But this primitive form of communication would not imply that a patient had a level of cognitive function high enough to use a BCI to make and express decisions about life-sustaining medical care. Anything less than accurate communication of these decisions could lead to misinterpretation of the patient's wishes. It could result in an outcome that could be just as harmful to the patient as a situation in which the patient was unable to learn how to use a BCI.

There may be cases where there is enough clarity and reliability in language production to meet criteria of decisional capacity and informed consent to continue or discontinue therapy. Despite traumatic brain injury resulting in the MCS, there may be areas of preserved axonal connectivity in thalamo-cortical and cortico-cortical networks to support some degree of reasoning and decision-making. This may also support using a BCI to express decisions about medical care. The fact that ALS and LIS patients generally have more preserved cognitive capacities than MCS patients suggests that the probability of successful use of a BCI to communicate would be higher in the first group than in the second. If a patient had the cognitive capacity to make and express decisions, then the considerable time and effort required to produce enough words through the interface would suggest that they were deliberative rather than spontaneous. This could contribute to the reliability of these decisions and confirm that they were sufficiently informed and autonomous.

4 Ethical Issues in BCI-Based Communication

Studies indicate that the main source of frustration and suffering among locked-in patients and those at the higher end of the MCS spectrum is their inability to communicate with family and caregivers (Birbaumer et al. 2014; Fins 2015: 179–195). They are frustrated by the disconnection between their phenomenal consciousness and access consciousness and the disconnection between their cognitive capacity and motor incapacity. This negative emotional experience may be greater in those with CLIS than those with LIS and ALS because they cannot use voluntary eye-blinks or eye-tracking systems to respond to questions. The frustration and suffering of patients with these three disorders presumably would be greater than in MCS patients. Unlike the latter, CLIS, LIS, and ALS patients are fully aware of the gap between their cognitive function and motor dysfunction and their inability to close this gap. Yet if an MCS patient is aware of her cognitive impairment in addition to her motor impairment and its role in limiting communication, then she may also experience frustration and suffer because of it. Among patients deemed cognitively capable of operating a BCI, mental fatigue from constantly intending and trying to produce letters and words through a computer typing program or speech synthesizer could cause not only those with MCS but also those with ALS and LIS/CLIS to fail to communicate. In that case, patients within all these groups could continue to suffer in silence. Still, some may have the necessary capacities to effectively express decisions through a BCI.

Samuel Maiser and coauthors spell our four abilities that LIS patients must demonstrate for decision-making capacity: (1) the ability to receive information; (2) to understand the relevant information presented; (3) to reason with the information to logically weigh options, risks, and benefits of alternative treatments; and (4) express a choice that is generally consistent with their past expressed values and that had been reached free of coercion (Maiser et al. 2016). These four abilities are consistent with Paul Appelbaum and Thomas Grisso's four criteria of competence in medical decision-making. The "legal standards for competence include the four related skills of (1) communicating a choice, (2) understanding relevant information, (3) appreciating the current situation and its consequences, and (4) manipulating information rationally" (Appelbaum and Grisso 1998: 1635; Beauchamp and Childress 2019, Ch. 4). Most of the conditions in these two models of competence or capacity fall within Block's broader concept of access consciousness. Some behaviorally nonresponsive patients have enough cognitive function to meet these conditions. They can appreciate their current situation and the consequences of their decisions, even if they are limited in weighing options beyond continuing or discontinuing therapy. Critically, though, their ability to successfully use a BCI would be necessary to express these decisions and for caregivers to act accordingly.

James Bernat states: 'In discussions about continuing or discontinuing life-sustaining therapy, it is imperative to maintain a system of effective communication between the physician and patient so that the patient's decisions are fully informed about rehabilitation options, palliative care options, treatment of depression, and other available treatments (Bernat 2020: 234). He states further that "communication systems allowing only 'yes' or 'no' binary responses by a LIS patient simply are inadequate to permit consequential decision-making" (Bernat 2020: 234). These responses would be "crude" and fall short of the robust communication necessary to confirm these decisions as clear and reliable (234). In specifying their conditions for medical decision-making capacity among LIS patients, Maiser and coauthors argue that there should be a higher standard of this capacity when there is greater gravity to the decision. Supporting their position, Bernat writes, "To this reasonable point, I would add that a higher standard also should be set for establishing effective and nuanced communication to allow the patient's decision to be actuated" (Bernat 2020: 234).

Would it be reasonable and fair to hold patients using a BCI as an assisted communication system to such a standard? If a behaviorally nonresponsive patient can use the interface to express a decision about medical care clearly, reliably, and consistently to caregivers, then this could be enough to meet a reasonable standard of decisional capacity. This could include a decision about continuing or discontinuing life-sustaining therapy. Presumably, what Bernat calls "nuanced communication" would also be necessary to confirm that the patient had this capacity. This would involve discussion between the patient and caregivers (Jox 2013). Patients who are locked-in or minimally conscious are unable to do this. But some of these patients may have the capacity to process information, make decisions, and use a BCI to express them. It would be unfair to hold these patients to a higher standard that included interlocution and discussion with caregivers if the words they produced through the BCI constituted an informed decision that accurately reflected their mental states. If a patient could produce two or more letters per minute with the computer tying program and could sustain this output long enough to produce a coherent set of words, then this may be enough to confirm the ability to process information and express a decision to continue or discontinue care. What she expressed through a BCI could reflect an understanding of the gravity of her decision and make it informed, rational, and autonomous. A less-stringent standard of decision-making capacity than the one proposed by Maiser et al. and Bernat could be enough to make and express consequential medical decisions (Glannon 2016).

Not allowing BCI-mediated word production as evidence of this capacity and these decisions because patients could not discuss them with caregivers could unfairly discriminate against them. It would impose an unnecessary and unreasonable burden on them if they had the same mental capacity to reason and make decisions as other patients but, because of their motor impairment, could not demonstrate that they had this capacity verbally or gesturally. A patient's ability to bypass nonfunctional motor regions through the interface shows that motor impairment does not necessarily preclude effective communication. If a patient intended to discontinue care and end his life, then not acknowledging BCI-based communication as an expression of an informed decision would harm him by preventing the realization of his intention.

There is an ethically significant asymmetry between behaviorally nonresponsive patients who wanted and those who did not want to continue life-sustaining therapy.

Continuing therapy would be the default position for a patient who was unable to communicate. If he wanted to initiate or maintain mechanical ventilation, ANH, or other interventions to stay alive, then the ability to communicate his wish would benefit him by realizing it. But he could benefit even if he could not communicate if his wish was consistent with the default position. In contrast, a patient who wanted to forego or discontinue care could be harmed by the defeat of his wish if he could not communicate it to caregivers. This underscores the need for BCIs that would enable patients to express decisions about medical care that would be in their best interests.

Maiser et al.'s consistency condition underscores the importance of knowledge of the patient's values and previously expressed wishes about the types of medical care she would accept or refuse. Consistency in having these wishes over time would be necessary to confirm that the patient's decision was informed, rational, and autonomous. Together with the patient's behavior and any evidence of pain and suffering, this knowledge would minimize the potential for misinterpreting the patient's intention and decision about the type of medical care she would or would not want. This information would have to supplement the output from the computer typing or synthesized speech program to clarify any ambiguity in what the patient expressed. The combination of these factors could objectively confirm her state of mind. What the patient expressed through the BCI, and knowledge of her standing values and wishes, could ensure that any actions by caregivers in response to what she expressed would benefit rather than harm her.

The stakes in continuing therapy against a behaviorally nonresponsive patient's wish to die are just as high as they are in discontinuing therapy against a different patient's wish to live. For competent patients who wanted to discontinue therapy, caregiver refusal to accept computer-generated words as evidence of informed decisions could harm patients further by preventing them from communicating and exacerbating their loss of motor control. Given the default position, continuing therapy for patients would not be problematic for those who wanted it. But it would be problematic for patients who did not want it.

When MCS or advanced ALS patients are too cognitively impaired to make decisions, families may be allowed to give proxy consent for continuing or discontinuing medical interventions such as ANH (Buchanan and Brock 1990: 122–148). They may also give proxy consent for the patient to participate in research on BCIs. But if a patient's wishes are unclear, or if the burden in caring for such a patient unduly influences their decisions about her medical care, then families may not always act in the patient's best interests. Patients with the cognitive capacity to process information and make decisions should be the sole decision-makers about their care. This would exclude decisions by proxy and uphold patient autonomy. However, a patient may have decisional capacity but not be able to report it to others if she could not successfully use a BCI as an assisted communication system. There would still be a gap between the patient's cognitive capacity and her motor incapacity. The patient could have one component of access consciousness in processing information but lack the other component necessary to report her mental states to others. If she could not use the BCI to communicate, and could not communicate in any other way, then there would be no objectively verifiable way of knowing whether she had or lacked decisional capacity. In that case, the only ethically justifiable action would be to continue therapy despite the possibility of harming the patient by doing something she did not want. Court decisions typically support families making proxy decisions to continue care for noncompetent patients. But the fact that not all patients want to continue care, or continue the care they are receiving, underscores the critical need for BCIs that could facilitate the translation of thoughts into actions for behaviorally nonresponsive patients.

To minimize the probability of harm to patients and research subjects who may use BCIs, practitioners and investigators should adopt strict selection criteria and include only those patients with preserved neural and cognitive capacities necessary to operate the interface. They must explain the aim, potential outcomes, and limitations of these systems. This would include disabusing patients and research subjects of unreasonable expectations about the potential of these systems to restore motor functions necessary for communication. The selection criteria may seem unfair to those with impaired cognition who would not be candidates for BCI training and application. Yet they could be harmed if they wanted to use a BCI but could not meet the cognitive demands in learning how to use it. Based on the patient's neurological and psychological status, if an investigator judged that the patient could not meet these demands, then the investigator would not be obligated to select the patient simply because he wanted to use it to communicate. Excluding him because he lacked the necessary level of cognitive function would be a form of fair discrimination and thus medically and ethically justified (Glannon 2014). One could support this claim even if the interface offered the only possible means of expressing his thoughts. Indeed, the investigator or clinician would have an ethical obligation of nonmaleficence to exclude patients from training who lacked the capacity to successfully operate a BCI (Beauchamp and Childress 2019, Ch. 5). Rather than causing harm, selecting patients on these grounds would prevent it. It would protect them from harm by preventing frustration and despair in failed attempts at trying to use the system, which would defeat their hope or expectation to communicate.

Having a high level of cognitive function alone would not entail success in operating the interface. It may not protect patients from negative psychological effects from failing to operate it to achieve their communicative goal. A patient may have the cognitive capacity to respond to sensory feedback from the computer and activate brain signals in motor and language areas. But she may lack the cognitive, emotional, and volitional capacity for the sustained motivation, attention, and concentration to produce the letters and words necessary to communicate. The mental workload in using the interface to produce these items at a very slow rate would require considerable time and mental effort. It could test her patience and perseverance. She may experience mental fatigue from repeated failed attempts to express her wishes and intentions. Operating a BCI can be challenging even for people with intact cognitive and motor functions. The failure rate of healthy controls in operating BCIs in research may be as high as 30% (Birbaumer et al. 2008, 2014).

In some cases, harm to patients from failing to produce words through a computer typing or similar program may not be preventable. Investigators may not be able to assess all relevant mental capacities when they consider a potential research subject as a suitable candidate for training to use a BCI. Critical mental incapacities may not manifest themselves until after the training had begun. It would not be unfair but tragic that they would suffer from failed attempts due to motor impairment caused by their neurological condition. What *would* be unfair would be to hold them to a higher standard of competence and consent requiring interlocution with investigators and caregivers if they could demonstrate decisional capacity through a BCI. This would unjustifiably prevent release from their noncommunicative state and realization of their intentions in appropriate actions.

One group of investigators has explained how patients at the higher end of the minimally conscious spectrum could be trained to use fMRI-guided BCIs to communicate decisions about life-sustaining therapy involving more than affirmative or negative responses to questions or commands (Peterson et al. 2013). These could include not only expressions about one's quality of life but also decisions about how caregivers should act in response to them. Alternatively, speech-synthesis programs involving EEG-based, ECoG or an intracortical microelectrode array activating regions mediating reasoning and language processing could allow some of these patients to express decisions. Such a system could enable a patient to produce the words necessary for caregivers to ascertain these mental acts. This could resolve uncertainty regarding crude responses correlating with cortical brain activity displayed by neuroimaging. A BCI could enable a deliberative decision about medical care the patient wanted or did not want. Still, it is important to emphasize that this would require a high level of cognitive capacity. It is likely that only a small number of patients with this disorder of consciousness could do this.

Kathrine Bendtsen points out that "in order for the MCS patient to understand and provide consent, he or she must be able to retain the information in short-term memory long enough to evaluate and assign value to the information presented and the possible outcomes" (Bendtsen 2013: 49). "Assigning value" suggests a fairly high level of cognitive and emotional processing, a level that many MCS patients probably lack. Allowing a less stringent criterion of consent in which values are implicitly assumed to be a property of patient decisions may be justifiable. If the patient knows that continuing ANH or other life-sustaining therapy will extend her life, that discontinuing therapy will end it, and she has an interest in realizing or avoiding one of these outcomes, then this would be enough for her to make an informed and voluntary decision and consent to continue or discontinue it. In cases where a competent patient requests discontinuation of chemotherapy for cancer or mechanical ventilation, there is often sustained discussion between the patient and caregivers. This may not be possible for an ALS, LIS, or MCS patient who could only communicate by producing a few letters per minute with a BCI. But if a patient can use the computer to produce enough words to clearly express a decision to stop therapy, if the decision is consistent with her previously expressed wishes, and if caregivers observe that the patient is in pain or suffering, then a single instance of communicating the decision with the BCI may be enough to meet a reasonable standard of consent (Glannon 2014, 2016).

Crucially, though, the patient must communicate her decision for caregivers to translate it into the appropriate action. Not all neurologically compromised patients can do this. Yet some of them retain enough cognitive, emotional, and volitional capacity to make and express consequential decisions about their medical care with a BCI. The gravity of these decisions lies not just in the possibility of discontinuing therapy for those who wanted it, but also the possibility of continuing therapy for those who did not want it. Although it has its limitations, BCI-mediated communication can promote patient autonomy and avoid these harmful outcomes.

5 A Case Study

A 2011 legal case in the United Kingdom highlights medical, ethical, and legal issues surrounding life-sustaining therapy for a behaviorally nonresponsive MCS patient with some preserved cognition. It raises the question of whether a fully implanted BCI could have enabled the patient to express a decision about continuing or withdrawing ANH. If she had the neurological and mental capacity to process information about her condition, and if she could successfully use a BCI, then she might have produced enough words to express her decision to her caregivers, her family, and the Court. Still, it is unclear whether even a patient with many preserved cognitive functions would have the necessary neural and mental capacities to activate electrical signals in her motor cortex and use a computer typing program to clearly express a decision to continue or discontinue ANH. Given accounts of the patient's behavior, I will assume that, in principle, she could have used a BCI to do this.

In *Re M*, the English Court of Protection ruled that it would be unlawful to withdraw ANH from a woman who had been in an MCS for 8 years (*Wv M* 2011). Her family's request that life-support be withdrawn was based on a statement by the patient before her injury that she would not want to remain alive in a completely dependent condition. This was rejected by the Court, which ruled that the patient's welfare justified continuation of ANH. Expert evidence indicated that the patient was regularly in pain. This was consistent with studies indicating that neural networks of pain processing are active in MCS patients (Boly et al. 2008; Demertzi and Laureys 2012). Depending on her level of awareness and the affective processing in these networks, she may have been suffering both from her experience of pain and feeling of dependence on others. Analgesia may have relieved her pain. But this would not have resolved her suffering from the feeling of dependence. If she had the cognitive and emotional capacity for an interest in how her life should go, and she wanted to end her life, then continued ANH would have harmed her by defeating this interest. The problem is that she was unable to communicate this wish and a decision to end life-sustaining therapy. This resulted in conflicting interpretations of the patient's neurological and mental status by the Court and the family and the Court's ruling to continue ANH on the presumption that it benefited the patient.

There are lingering questions about whether the patient in this case had decision-making capacity and could have used a BCI to communicate her decision about life-sustaining therapy to others. This is hypothetical because this technology was not considered in this case. If she had these capacities, then a BCI that allowed her to clearly and reliably express a decision about continuing or discontinuing ANH and her life could have resolved the dispute between the Court and her family about which course of action she wanted. This would have established her as the sole decision-maker about her care. By compensating for motor impairment enabling this and other patients to communicate when they cannot do this on their own, BCI-mediated communication can help patients to be autonomous decision-makers and ensure that other parties act to promote rather than interfere with their interests (Glannon 2013).

As Hochberg and Cudkowicz point out, BCIs or other technologies that allow patients to communicate decisions about life-sustaining treatment "by proclamation rather than proxy will be revolutionary" (2014: 1853). While they make this point regarding patients with LIS, it could also apply to ALS patients and MCS patients with higher levels of awareness and cognitive function. Some of these patients could use a BCI in meeting a less stringent but adequate standard of decision-making and consent. By allowing them to express decisions about medical care, the BCI could restore some degree of control over their lives. This would apply both to patients who wanted to continue life-sustaining therapy, and those who wanted to discontinue it. Nevertheless, researchers training patients to use these systems are obligated to protect them from the harm resulting from failed attempts and unmet expectations. This would involve careful selection of patients, informing them of the potential and limitations of the technique, and monitoring their behavior during the training and use of the interface.

6 Conclusion

Extracortical and intracortical BCIs may enable paralyzed and behaviorally nonresponsive patients to turn thoughts into actions. These actions include moving prosthetic limbs and other objects. They also include communicating affirmative and negative responses to questions and the more complex process of expressing decisions about medical care. For patients with the cognitive and other mental functions necessary to perform these mental acts, BCIs can restore or augment motor functions that have been impaired or lost from brain injury amputation or neurodegenerative disease. The feedback between the brain and the computer may allow patients to produce words in computer-generated speech. Expressing decisions in this way would involve shared control between the patient and the interface. These systems shed light on the brain-mind relation. They show that the ability to use BCIs to communicate depends on brain-mind-machine interaction in the activation and transmission of brain signals as computer input, and in the production of letters, words, and sentences as computer output. They also raise ethical questions about the potential benefit or harm to patients who successfully or unsuccessfully try to use them. I have focused on the potential of implantable BCIs to allow ALS, LIS/CLIS, and some MCS patients to communicate consequential medical decisions about continuing or discontinuing life-sustaining therapy.

New techniques may increase the likelihood of successful communication in behaviorally nonresponsive patients. One such technique is the subdural ECoG system discussed earlier, where decoding neural signals of movement of vocal-tract articulation can result in spoken sentences. This system would have to be tested in neurologically compromised subjects who were unable to communicate by normal means. To benefit them, a BCI would have to enable motor functions necessary to produce coherent sentences that accurately reflected their intentions and could be interpreted and assessed correctly by others. Patients would need neurological, cognitive, emotional, and volitional capacities to be trained to use a BCI to communicate in a clear, consistent, and sustained way. Another technique is fNIRS, which was also discussed earlier. The head-mounted infrared laser in this technique may decode and activate electrical signals in the brain and transmit them to the computer more effectively than scalp-based EEG or ECoG models. Still, the effects of head-mounted systems may be limited because of the possibility of cranial deflection.

Even if implantable systems are more effective in facilitating transmission of neural signals to the computer, there would still be the possibility of failed attempts by the subject in trying to execute motor tasks through the machine. This shows that the mental capacities of patients wanting and trying to operate BCIs to communicate are as important as neural events and technical features of the software in producing the desired result. Given the risk of harm from failed attempts at using BCIs, investigators training patients and research subjects to use them have a medical and ethical obligation to select only those deemed capable of successfully using them. They also have an obligation to explain the limitations of interface technology so that patients have reasonable expectations about its potential to restore their ability to communicate.

In 2013, Davina Fernandez-Espejo and Adrian Owen acknowledged that simple "Yes" or "No" binary responses to questions with existing interface technology could not establish that a minimally conscious patient could decide whether she wanted to continue living. They pointed out that fMRI-based BCIs could not confirm whether such a patient had the "cognitive and emotional capacity to make such a complex decision" (Fernandez-Espejo and Owen 2013: 808). But they claimed that "it is only a matter of time before all of these obstacles are overcome" (2013: 808; Peterson et al. 2013: 10). At the time, this may have seemed an exaggerated claim. But current and future BCI technology may enable patients who cannot speak to clearly express their thoughts. It may enable them to meet reasonable criteria of decisional capacity and informed consent to continue or discontinue life-sustaining therapy. This would validate Fernandez-Espejo and Owen's claim. It would also validate Hochberg and Cudkowicz's point about the potential of artificial communication systems to enable medical decision-making by proclamation from patients rather than by proxy from surrogates. Developing this technology further is critical for maximizing benefit and minimizing harm among the large population of behaviorally nonresponsive conscious patients who suffer in silence because they cannot communicate.

Questions remain as to whether BCI-based typing or synthesized speech programs could enable neurologically compromised patients to express decisions about life-sustaining therapy as clearly and reliably as patients who can do this verbally or gesturally. The limitations of BCIs, combined with motor and language-processing impairment, may preclude effective communication in many cases. These factors could result in health-care providers failing to discern or misinterpreting patients' thoughts. Nevertheless, for patients with preserved cognitive, emotional, and volitional capacities, BCIs can bypass damaged motor and other cortical brain regions and enable them to confirm decision-making capacity. BCI-mediated communication may allow them to express informed consequential decisions about medical care that is in their best interests. It can restore some degree of functional independence and allow them to regain some control of their lives.

References

Anumanchipalli G, Chartier J, Chang E (2019) Speech synthesis from neural decoding of spoken sentences. Nature 568:493–498

Appelbaum P, Grisso T (1998) Assessing patients' capacity to consent to treatment. N Eng; J Med 319:1635–1638

Beauchamp T, Childress J (2019) Principles of biomedical ethics, 8th edn. Oxford University Press, New York

Bendtsen K (2013) Communicating with the minimally conscious: ethical implications in end-of-life care. Am J Bioethics Neurosci 4:46–51

Bernat J (2020) Medical decision making by patients in the locked-in syndrome. Neuroethics 13:229–238. https://doi.org/10.1007/s12152-018-9358-7

Birbaumer N, Murguialday A, Cohen L (2008) Brain-computer interface in paralysis. Curr Opin Neurol 21:634–638

Birbaumer N, Gallegos-Ayala G, Wildgruber M, Silvoni S, Soekadar S (2014) Direct brain control and communication in paralysis. Brain Topogr 27:4–11

Block N (1995) On a confusion about a function of consciousness. Behav Brain Sci 18:227–287

Block N (2007) Consciousness, accessibility and the mesh between psychology and neuroscience. Behav Brain Sci 30:481–499

Boly M, Faymonville M-E, Schnakers C, Peigneux P, Lambermont B, Phillips C et al (2008) Perception of pain in the minimally conscious state with PET activation: an observational study. Lancet Neurol 7:1013–1020

Buchanan A, Brock D (1990) Deciding for others: the ethics of surrogate decision making. Cambridge University Press, New York

Chaudhary U, Xia B, Silvoni S, Cohen L, Birbaumer N (2017) Brain-computer interface-based communication in the completely locked-in state. PLoS Biol 15:e1002593. https://doi.org/10.1371/journal.pbio.1002593

Chennu S, Finoia P, Kamau E, Allanson J, Williams G, Monti M et al (2014) Spectral signatures of reorganized brain networks in disorders of consciousness. PLoS Comput Biol 10:e1003887. https://doi.org/10.1371/journal/pcbi.1003887

Claassen J, Doyle K, Matory A, Couch C, Burger K, Velasquez A et al (2019) Detection of brain activation in unresponsive patients with acute brain injury. N Engl J Med 380:2497–2505

Davidson D (1992) Thinking causes. In: Mele A, Heil J (eds) Mental causation. Oxford University Press, New York, pp 3–18

Davidson D (2001) Essays on actions and events, 2nd edn. Oxford University Press, New York

Demertzi A, Laureys S (2012) In: Richmond S, Rees G, Edwards SJ (eds) Where in the brain is pain? Evaluating painful experiences in non-communicative patients. Oxford University Press, Oxford, pp 89–98

Fernandez-Espejo D, Owen A (2013) Detecting awareness after severe brain injury. Nat Rev Neurosci 14:801–809

Fins J (2015) Rights come to mind: brain injury, ethics and the struggle for consciousness. Cambridge University Press, New York

Giacino J, Ashwal S, Childs N, Cranford R, Jennett B, Kelly J et al (2002) The minimally conscious state: definition and diagnostic criteria. Neurology 58:349–353

Glannon W (2013) Burdens of ANH outweigh benefits in the minimally conscious state. J Med Ethics 39:551–552. http://www.jme.bmj.com. https://doi.org/10.1136/medethics-2012-100882

Glannon W (2014) Ethical issues with brain-computer interfaces. Front Syst Neurosci 8:136. https://doi.org/10.3389/fnsys.2014.00136

Glannon W (2016) Brain-computer interfaces in end-of-life decision-making. Brain Comput Interfaces 3:133–139. http://www.tandfonline.com. https://doi.org/10.1080/2326263X.2016.1207496

Heil, J, Mele, A (eds.) (1992), Mental Causation. Clarendon Press, Oxford

Hochberg L, Cudkowicz M (2014) Locked in, but not out? Neurology 82:1852–1853

Jox R (2013) Interface cannot replace interlocution. Why the reductionist concept of neuroimaging-based capacity determination fails. Am J Bioethics Neurosci 4:15–17

Kennedy P, Andreasen D, Bartels J, Ehiring P, Mao H, Velliste M et al (2011) Making the lifetime connection between brain and machine for restoring and enhancing function. Prog Brain Res 194:1–25

Kubler A (2009) Brain-computer interfaces for communication in paralyzed patients and implications for disorders of consciousness. In: Laureys S, Tononi G (eds) The neurology of conciousness, pp 217–233

Laureys S, Tononi G (eds) (2009) The neurology of consciousness: cognitive neuroscience and neuropathology. Elsevier, Amsterdam

Lebedev M (2014) Brain-machine interfaces: an overview. Transl Neurosci 5:99–110

Lee J (2016) Brain-computer interfaces and dualism: a problem of brain, mind and body. AI Soc 31:29–40

Leuthardt E, Schalk G, Wolpaw J, Ojemann J, Moran D (2004) A brain computer interface using electrocortigraphic signals in humans. J Neural Eng 1:63–71

Linden D (2014) Brain control: developments in therapy and implications for society. Palgrave Macmillan, Basingstoke

Maiser S, Kabir A, Sabsevitz D, Peltier W (2016) Locked-in syndrome. Case report and discussion of decisional capacity. J Pain Symptom Manage 51:789–793

Massimini M, Tononi G (2018) Sizing up consciousness: towards an objective measure of the capacity for experience, trans. F. Anderson. Oxford University Press, Oxford

Mele A (2009) Effective intentions: the power of conscious will. Oxford University Press, New York

Monti M, Vanhaudenhuyse A, Coleman M, Boly M, Pickhard J, Tshibanda L et al (2010) Willful modulation of brain activity in disorders of consciousness. N Engl J Med 362:579–598

Naci L, Owen A (2013) Making every word count for nonresponsive patients. JAMA Neurol 70:1235–1241

Naci L, Cusack R, Jia V, Owen A (2012) The brain's silent messenger: using selective attention to decode human thought for brain-based communication. J Neurosci 33:9385–9393

Owen A, Coleman M (2008) Functional imaging in the vegetative state. Nat Rev Neurosci 9:235–243

Owen A, Naci L (2016) Decoding thoughts in behaviorally nonresponsive patients. In: Sinnott-Armstrong W (ed) Finding consciousness. Oxford University Press, New York, pp 100–121

Owen A, Coleman M, Boly M, Davis M, Laureys S, Pickard J (2006) Detecting awareness in the vegetative state. Science 313:1402

Pandarinath C, Ali Y (2019) Brain implants that let you speak your mind. Nature 568:466–467

Pernu T (2019) Mental causation via neuroprosthetics? A critical analysis. Synthese 195:5159–5174

Peterson A, Naci L, Weijer C, Cruse D, Fernandez-Espejo D, Graham M et al (2013) Assessing decision-making capacity in the behaviorally nonresponsive patient with residual covert awareness. AJOB Neurosci 4:3–14

Phukan J, Pender N, Hardiman O (2007) Cognitive impairment in amyotrophic lateral sclerosis. Lancet Neurol 6:994–1003

Richmond S, Rees G, Edwards S (eds) (2012) I know what you're thinking: brain imaging and mental privacy. Oxford University Press, Oxford

Saur D, Kreher B, Schnell S, Kummerer D, Kellmeyer P, Vry M-S et al (2008) Ventral and dorsal pathways for language. Proc Natl Acad Sci 105:18035–18040

Schnakers C, Majerus S, Goldman S, Boly M, Van Eeckhout P, Gay S et al (2008) Cognitive function in the locked-in syndrome. J Neurol 255:323–330

Searle J (1969) Speech acts: an essay in the philosophy of language. Cambridge University Press, Cambridge

Sinnott-Armstrong W (ed) (2016) Finding consciousness: the neuroscience, ethics and law of severe brain damage. Oxford University Press, New York

Vansteensel M, Pels E, Bleichner M, Branco M, Denison T, Freudenburg Z et al (2016) Fully implanted brain-computer interface in a locked-in patient with ALS. N Engl J Med 375:2060–2066

Vidal, F. (2020). Phenomenology of the locked-in syndrome: an overview and some suggestions.
Neuroethics 13: 119–143. https://doi.org/10.1007/s12152-018-9388-1

Wolpaw J, Wolpaw E (eds) (2012) Brain-computer interfaces: principles and practice. Oxford University Press, New York

Part III
Augmenting Cognitive Function

Neuroprotection and Neurocognitive Augmentation by Photobiomodulation

Francisco Gonzalez-Lima

1 Introduction

This chapter discusses studies with laboratory animals and human volunteers, showing that specific interventions using transcranial light stimulation can cause neuroprotection and neurocognitive augmentation. These neural effects are part of a more general phenomenon known as photobiomodulation, which primarily involves stimulation with red to near-infrared wavelengths of light. Photobiomodulation or low-level light therapy refers to the application of such light to biological systems with the purpose of providing a beneficial effect. The absorption of incident photon energy by tissue chromophores induces electronically excited states that modify their energy state and can promote oxygen consumption and the generation of reactive oxygen species. These two biochemical events have been proposed to activate a variety of intracellular signaling and metabolic pathways that modify cellular functions (Karu 1989, 2000), which in turn modify brain function and behavior because brain cells are critically dependent on oxygen consumption for energy production by mitochondria (Gonzalez-Lima et al. 2014). Photobiomodulation has been found to influence various biological processes in cell cultures, animal models, and clinical conditions (Yu et al. 1994; Eells et al. 2003; Wong-Riley et al. 2005). For example, photobiomodulation is being used clinically in humans to accelerate wound healing (Conlan et al. 1996), to relieve inflammation (Whelan et al. 2002), and to relieve neurogenic pain (Iijima et al. 1991).

Wong-Riley et al. (2001) reported that photobiomodulation using light-emitting diodes (LEDs) was able to reverse the neurotoxic effects of tetrodotoxin on neuronal cultures by a process involving the mitochondrial respiratory enzyme cytochrome

F. Gonzalez-Lima (✉)
Department of Psychology and Institute for Neuroscience, The University of Texas at Austin, Austin, TX, USA
e-mail: gonzalezlima@utexas.edu

I. Opris et al. (eds.), *Modern Approaches to Augmentation of Brain Function*, Contemporary Clinical Neuroscience,
https://doi.org/10.1007/978-3-030-54564-2_9

oxidase (also called cytochrome c oxidase or Complex IV of the Electron Transport Chain). A number of additional in vitro effects of photobiomodulation on nervous tissue have been described. These include, among others, the following: (1) increased expression of the antiapoptotic protein Bcl-2 and reduced expression of the proapoptotic protein Bax (Shefer et al. 2002; Liang et al. 2006); (2) decreased numbers of apoptotic cells after exposure to the amyloid beta protein (Duan et al. 2003); (3) improved function of cortical neurons inactivated by toxins (Wong-Riley et al. 2005; Liang et al. 2006); (4) increased survival and ATP content of striatal neurons after rotenone- and MPP^+-induced toxicity and decreased oxidative stress and nitric oxide production (Liang et al. 2008); (5) increased neurite outgrowth (Wollman and Rochkind 1998); (6) regulation of gene encoding for DNA repair proteins, antioxidant enzymes, and molecular chaperones (Eells et al. 2004); (7) increased proliferation of olfactory ensheathing stem cells (Byrnes et al. 2005a), Schwann cells (Van Breugel and Bar 1993), astrocytes, and oligodendrocytes (Rochkind et al. 1990). In vivo, photobiomodulation induces peripheral and central nerve regeneration after trauma (Byrnes et al. 2005b), reduces neuroinflammation (Byrnes et al. 2005b), and prevents methanol-induced photoreceptor degeneration (Eells et al. 2003).

In photobiomodulation, a target tissue or organism is exposed to a low power, high fluency source of directional and monochromatic light, which delivers energy doses that are too low to cause heating, yet high enough to modulate cell functions (Sommer et al. 2001). Photobiomodulation is best done using wavelengths at the end of the visible electromagnetic spectrum (λ = 610–1100 nm). This wavelength range is used because light energy at wavelengths shorter than 600 nm is generally absorbed by the skin pigment melanin, whereas water significantly absorbs energy at wavelengths higher than 1150 nm. This implies the existence of an "optical window" in tissue that covers the red and near-infrared wavelengths, where there is room for a biological effect due to excitation of intracellular photoacceptors (Hamblin and Demidova 2006).

This chapter describes our animal and human research using red to near-infrared lasers and LEDs, a noninvasive and relatively inexpensive intervention, which we are investigating for the augmentation of cognitive brain functions and the treatment of neurodegenerative disorders. Part 1 deals with animal studies on the prevention of neurodegeneration and behavioral deficits by photobiomodulation. Part 1 shows light parameters that prevent optic neuropathy in vivo. These parameters include fractionated exposure to far-red to near-infrared light wavelengths, primarily corresponding to the absorption spectrum of the mitochondrial respiratory enzyme cytochrome oxidase. We also show that such light stimulates cell respiration and upregulates cytochrome oxidase activity and the antioxidant enzyme superoxide dismutase. The translation of this knowledge for human studies in Part 2 shows that this research may have a broad impact, not only relevant for cognitive enhancement in healthy people but also potential for the noninvasive treatment of neurodegenerative and neurocognitive disorders.

2 Part 1: Animal Studies—Prevention of Neurodegeneration and Behavioral Deficits by Photobiomodulation

2.1 Objectives of Animal Studies of Neuroprotection by Photobiomodulation

Targeted stimulation of mitochondrial function with photobiomodulation or low-level light therapy constitutes a sound, yet largely unexplored strategy with great potential for neurocognitive enhancement and neuroprotection from neurodegeneration. Mitochondrial dysfunction, concomitant free radical damage, and decrease in energy production are believed to play a central role in neurodegeneration. Despite the acknowledged central role of mitochondrial dysfunction in the pathophysiology of neurodegeneration, stimulation of mitochondrial function has not been thoroughly considered in the design of neuroprotective interventions (Gonzalez-Lima et al. 2014). Photobiomodulation is a potential noninvasive and safe method to effectively counteract these early events in neurodegeneration induced by mitochondrial dysfunction. Cytochrome oxidase is a potential target for mitochondrial neuroprotective interventions since it is an inducible respiratory enzyme that plays a limiting step role in energy production. Cytochrome oxidase expression is tightly linked to neuronal energetic demands and is mediated through a genome-wide activation of transcription factors with the potential to impact neuronal homeostasis (Gonzalez-Lima and Cada 1998). Being the major photoacceptor in neurons, cytochrome oxidase is an ideal target to promote neuroprotection with photobiomodulation. Although photobiomodulation has been shown to be highly effective at preventing neuronal death in vitro, its in vivo effects and mechanism of action are still poorly characterized (Rojas and Gonzalez-Lima 2011).

Our *long-term research goal* is to develop metabolic interventions that effectively prevent or treat neurodegenerative disorders in humans. The *objective* of Part 1 is to describe beneficial effects of photobiomodulation, which prevent neurodegeneration and behavioral deficits in an in vivo animal model of optic neuropathy induced by mitochondrial dysfunction, and to identify the mechanism of photobiomodulation action. Part 1 is primarily based on our data showing that photobiomodulation reduces retinal damage and associated functional visual deficits induced by mitochondrial dysfunction in vivo (Rojas et al. 2008b). Our *overarching hypothesis* is that photobiomodulation delivered *via* light-emitting diode (LED) arrays will upregulate the mitochondrial respiratory enzyme cytochrome oxidase and the antioxidant enzyme superoxide dismutase, which will be highly effective against mitochondrial optic neuropathy in vivo. Our *rationale* is that determining the parameters to prevent mitochondrial optic neuropathy with photobiomodulation in vivo would establish a strong basis for the design of clinical trials in humans, which we discuss in Part 2 of this chapter. We are especially well prepared to undertake this demonstration because, in addition to our strong data on photobiomodulation, we have developed an efficient in vivo animal model of optic neuropathy induced by mitochondrial dysfunction (Zhang et al. 2002, 2006), and have used it to characterize neuroprotective effects with the antiexcitotoxic agent memantine (Rojas et al. 2008a) and the metabolic enhancer methylene blue (Rojas et al. 2009). We will explain how using our animal model of optic neuropathy can address the following behavioral, anatomical, and mechanistic specific aims:

Aim 1: Demonstrate an effective photobiomodulation protocol to prevent behavioral deficits induced by mitochondrial dysfunction in vivo. Our hypothesis, based on our in vitro and in vivo data, is that a fractionated photobiomodulation protocol over 6 days with wavelengths aimed at the copper absorption peaks of cytochrome oxidase, will be the most effective in preventing functional visual deficits induced by mitochondrial failure in vivo.

Aim 2: Demonstrate the efficacy of photobiomodulation to prevent neurodegeneration induced by mitochondrial dysfunction in vivo. Our hypothesis, based on our in vitro and in vivo data, is that the most effective photobiomodulation protocol will be able to prevent structural degenerative damage in optic neuropathy induced by mitochondrial failure in vivo.

Aim 3: Demonstrate the mechanism of action of photobiomodulation that prevents neurodegeneration induced by mitochondrial dysfunction. Our hypothesis is that photobiomodulation will exert its neuroprotective effects via upregulation of the mitochondrial respiratory enzyme cytochrome oxidase and the antioxidant enzyme superoxide dismutase.

2.2 Significance of Animal Studies of Neuroprotection by Photobiomodulation

Exposure of neuronal cultures to red or near-infrared light (photobiomodulation) is known to increase survival after neurotoxin exposure in vitro (Wong-Riley et al. 2001). Such intervention might represent a potential preventive strategy to avoid neurodegeneration, an otherwise irreversible process. Despite positive evidence, skepticism about the potential clinical use of photobiomodulation still persists because the scientific rationale for in vivo treatment protocols for particular responses is lacking, and current photobiomodulation use is based on empirical choice of wavelength, total dose, dose fractionation, and other parameters. Nevertheless, the fact remains that photobiomodulation exerts a wide variety of neuronal effects. For example, prophylactic photobiomodulation in vitro has proved to be very effective at protecting neurons from neurodegeneration induced by mitochondrial toxins (Wong-Riley et al. 2001, 2005; Liang et al. 2008). Photobiomodulation has also been successfully employed for nerve repair and reduction of neural injury after stroke in animal models (Lapchak et al. 2004; Byrnes et al. 2005b), and it is clinically used to relieve pain in humans (Iijima et al. 1991). Rojas et al. (2008b) was the first in vivo animal study to use photobiomodulation in the treatment of optic neuropathy. Optic neuropathy is found in conditions with high morbidity for which no effective treatments are available (Abu-Amero et al. 2006; Iseri et al. 2006; He et al. 2008; Carelli et al. 2009). Furthermore, no previous studies have addressed the effectiveness of photobiomodulation in any model of optic neuropathy and it is also uncertain how energy from photobiomodulation works and what are the optimal light parameters for specific applications in vivo. Our research validated the concept that the mechanism of action of photobiomodulation in vivo involves direct photonic stimulation of the respiratory enzyme cytochrome oxidase. With this animal research, we expect to set the preclinical basis for the design of an effective, noninvasive, inexpensive, and safe intervention against optic neuropathy, which could potentially have a large clinical impact worldwide. We also expect to stimulate research on the applicability of photobiomodulation in other neurodegenerative disorders linked to mitochondrial dysfunction such as Alzheimer's and Parkinson's diseases (Rojas and Gonzalez-Lima 2017).

2.3 *Cytochrome Oxidase as a Molecular Target of Photobiomodulation*

Virtually, all biological tissues feature photoacceptor molecules adapted to maximize the assimilation of noncoherent electromagnetic radiation from the environment (Alberts 2002). This adaptation reaches its highest degree of efficiency in the thylakoid membrane of chloroplasts in plants, but it is conserved in animal tissue as well. In mammalian tissue, the known photoacceptors are mainly heme-containing metalloproteins. The three most important metalloproteins are hemoglobin, myoglobin, and cytochrome oxidase, but others such as superoxide dismutase, cytochrome *c*, cytochrome *b*, nitric oxide synthase, catalase, guanylate cyclase, flavoproteins, and cryptochromes may also play a role as photoacceptors. Although all these molecules may be involved in the photobiomodulation effects (Wong-Riley et al. 2005), cytochrome oxidase is by far the most abundant photoacceptor in neurons and its absorption peaks have been shown to coincide with the action spectra of photobiomodulation in vitro. For these reasons, *cytochrome oxidase is regarded as the primary photoacceptor of light in the red to near-infrared region of the light spectrum in neural tissue* (Yamanaka et al. 1988; Pastore et al. 2000; Fan et al. 2006).

Cytochrome oxidase is the terminal complex of the mitochondrial electron transport chain and catalyzes the reduction of more than 95% of the oxygen taken up by aerobic organisms. Thus, cytochrome oxidase constitutes an efficient energy-transducing device, acting as a redox-linked proton pump that creates a transmembrane electrochemical gradient and as a rate-limiting step for the synthesis of the energy-storing molecule adenosine triphosphate (ATP) (Hatefi 1985). Cytochrome oxidase is a bigenomically regulated enzyme, whose expression is tightly coupled to neuronal energy demands and glutamatergic activation (Liang et al. 2006; Dhar et al. 2009; Dhar and Wong-Riley 2009). Its activity has been extensively used as a marker of neuronal metabolic activity (Wong-Riley 1989; Sakata et al. 2005).

Cytochrome oxidase is a heme-containing enzyme and the most abundant metalloprotein in neurons. Cytochrome oxidase contains four redox metal centers: Cu_A, Cu_B, Heme *a*, and Heme a_3. The redox state of the enzyme can vary from fully reduced to fully oxidized, with intermediate states that include oxidation of one, two, or three metal centers. These metal centers determine different light absorption peaks for the enzyme: 613.5–623.5 nm, 667.5–683.7 nm, 750.7–772.3 nm, and 812.5–846 nm, which correspond to Cu_A reduced, Cu_A oxidized, Cu_B reduced, and Cu_B oxidized, respectively (Hamblin and Demidova 2006). These absorption peaks have been shown to coincide with its peaks of catalytic activity and with ATP content (Eells et al. 2004) and, as mentioned above, with the action spectra of photobiomodulation in vitro (Wong-Riley et al. 2005).

2.4 In Vitro Neuroprotective Effects of Cytochrome Oxidase Stimulation by Photobiomodulation

In vitro evidence supports that photobiomodulation directly stimulates the catalytic activity of cytochrome oxidase and further regulates mitochondrial function. For example, photobiomodulation has been shown to increase cytochrome *c* oxidation in the presence of cytochrome oxidase (Pastore et al. 2000). It also modulates the production of nitric oxide (Liang et al. 2008) and modifies the interactions between cytochrome oxidase and nitric oxide, promoting the generation of free radicals (Karu 1999). Photobiomodulation also increases the rate of oxygen consumption in hepatic mitochondria as well as the mitochondrial phosphate potential, energy charge, catalytic activity of mitochondrial complexes I, II, III, and IV (Yu et al. 1997), and ATP cellular content (Liang et al. 2008; Ying et al. 2008). Further in vitro evidence also supports that besides its immediate effects on neuronal cytochrome oxidase catalytic activity, photobiomodulation induces upregulation of the neuronal cytochrome oxidase pool, implicating an indirect effect of photobiomodulation on gene expression (Wong-Riley and Liang 2017). Rat neuronal cultures exposed to photobiomodulation showed a 12% increase in cytochrome oxidase activity. A similar enhancing effect was also observed when delivered after exposure to tetrodotoxin, which blocks electrical neural activity and indirectly inhibits cytochrome oxidase activity (Wong-Riley et al. 2001). In addition, photobiomodulation partially restored enzyme activity blocked by potassium cyanide, a cytochrome oxidase inhibitor, and significantly reduced neuronal cell death (Liang et al. 2006). Similar protective effects were observed when striatal and cortical rat neuronal cultures were exposed to rotenone and MPP^+, toxins that inhibit mitochondrial complex I. Photobiomodulation significantly increased cellular ATP content, decreased the number of neurons undergoing cell death, and reduced the expressions of reactive oxygen species and reactive nitrogen species in rotenone- or MPP^+-exposed neurons as compared to untreated ones (Liang et al. 2008). Furthermore, prophylactic use of photobiomodulation also suppressed rotenone- or MPP^+-induced apoptosis in both striatal and cortical neurons, and pretreatment plus photobiomodulation treatment during neurotoxin exposure was significantly better than photobiomodulation treatment alone during exposure to neurotoxins (Ying et al. 2008). Despite this large body of in vitro compelling evidence supporting a protective role of photobiomodulation against mitochondrial toxins via induction of cytochrome oxidase, there is little evidence regarding its effects against mitochondrial inhibitors in in vivo systems, which prompted the design of our animal studies presented in this chapter.

2.5 *In Vivo Light Delivery and Dosing Considerations*

A number of parameters are known to influence the efficacy, feasibility, and safety of photobiomodulation in vivo (Rojas and Gonzalez-Lima 2011). These should be taken into account since they are expected to vary depending on the specific therapeutic purpose. The main parameters include (1) dosage, (2) wavelength, (3) dosing schedules or fractionation, and (4) source of radiation.

1. *Dose*. There is essentially a very poor characterization of photobiomodulation dosing responses for neuroprotective applications in vivo. Photobiomodulation doses are expressed as radiant energy in Joules per surface area (energy density). Incident energy on a given surface is the product of power density (or light intensity, in watts) and time of exposure (in seconds). Thus, for achieving a desired dose, either power or time of exposure can be varied. Photobiomodulation dosimetry should consider the dose-response phenomenon of *hormesis*. A hormetic dose-response (also known as U-shaped, biphasic, or bell-shaped dose response) involves stimulation of a biological process at a low dose and inhibition of that process at a high dose. Hormetic models are superior to linear threshold models in their capacity to accurately predict responses below a pharmacological threshold (Calabrese and Baldwin 2003; Kaiser 2003). This is important because the stimulatory responses to photobiomodulation are usually modest, being only about 30–60% greater than control values. This contrasts with the several-fold increase in a specific variable expected according to traditional dose-response models (Calabrese 2008). Dose-response hormesis is well documented in photobiomodulation applications since photostimulatory or photoinhibitory effects are obtained with low (0.001–10 J/cm^2) and high (>10 J/cm^2) energy densities, respectively (Brondon et al. 2005). Positive response is expected to vary within this dose range for a particular desired outcome (Hamblin and Demidova 2006). For example, Liang et al. (2008) demonstrated that photobiomodulation daily doses within the stimulatory range (4 and 8 J/cm^2) are more effective than higher doses (12 and 16 J/cm^2) at preventing neurotoxicity induced by mitochondrial complex I inhibitors rotenone and 1-methyl-4-phenylpyridinium in vitro. Therefore, there exist optimal doses of photobiomodulation for neuroprotection in vivo. Doses lower than this optimum value, and more importantly,

doses larger than the optimum value will a have a diminished treatment response. High doses might even produce a negative effect.

2. *Wavelength.* Most photobiomodulation protocols use wavelengths in the range of *620–870 nm* to coincide with maximal absorbance profiles of endogenous chromophores, mainly cytochrome oxidase (Karu 1999). Near-infrared wavelengths offer acceptable penetration into tissues, as they overlap the "optical window". Wavelengths inferior to 600 nm are usually absorbed by abundant chromophores such as melanin and hemoglobin, whereas wavelengths above 1150 nm are usually absorbed by water. This means that the major penetration occurs in a window that includes the red and near-infrared spectrum. Such infrared wavelengths include the absorption peaks of the most important photoacceptor in neurons: the respiratory enzyme cytochrome oxidase. As mentioned before, cytochrome oxidase contains four redox metal centers: Cu_A, Cu_B, Heme *a*, and Heme a_3. The redox state of the enzyme can vary from fully reduced to fully oxidized, with intermediate states that include oxidation of one, two, or three metal centers that have different absorption spectra. The absorption peaks of the two copper centers in cytochrome oxidase (Cu_A and Cu_B) have been shown to coincide with the action spectra of photobiomodulation. In addition, the endogenous antioxidant enzyme superoxide dismutase (SOD) shows absorption peaks at 670–680 nm (Kubota and Yang 1984). SOD scavenges superoxide anions generated during successive reduction of molecular oxygen by single electrons, in electron transport redox reactions. The absorption peaks of SOD fall in the near-infrared window, which makes it an alternate photoacceptor relevant for neuroprotective actions. Hence, near-infrared wavelengths that stimulate cytochrome oxidase might also boost the antioxidant activity of SOD.
3. *Fractionation protocol.* Delivering large total doses of photobiomodulation in a single session produces less favorable outcomes than giving the same doses over several sessions. This dose fractionation has been tested in vitro and in vivo, and it has been shown to be highly effective at preventing neurodegeneration (Liang et al. 2008; Rojas et al. 2008b). Fractionation protocols that deliver the total dose over 6 days or more have demonstrated favorable results, including those that split the daily dose into two sessions (Liang et al. 2006, 2008; Ying et al. 2008). In addition, photobiomodulation fractionation protocols including prophylactic doses given before the neurotoxic lesion are also effective at preventing neurodegeneration (Rojas et al. 2008b; Ying et al. 2008).
4. *Photobiomodulation source.* Light-emitting diode (LED) arrays as well as lasers can produce photobiomodulation. Laser sources allow better penetration than LED sources, and so we prefer lasers for human brain photobiomodulation, as shown in Part 2. However, some lasers have limitations in beam width, wavelength capabilities, and areas of tissues that can be treated. In addition, some lasers are associated with heat production that can induce tissue damage (Eells et al. 2004). In contrast, LED arrays generate negligible amounts of heat reducing the risk of thermal injury. For example, a photobiomodulation source using LEDs with a wavelength of 633 nm, an intensity of less than 10 mW, and divergent beams is considered a safe source because, owing to its divergence, it cannot

damage the eye (Rojas et al. 2008b). In addition, LED arrays are compact and portable and have achieved nonsignificant risk status for human trials by the FDA (Wong-Riley et al. 2005).

In addition to these four variables, the response to photobiomodulation also depends on a number of intrinsic biological properties of the target tissue, including photoacceptor content, cell viability, redox status, susceptibility to induction of genetic expression, metabolic capacity, location in the body with reference to tissues with high absorbances, and conditions of the interstitial milieu. Hence, an optimal neuroprotective response depends on photobiomodulation delivery variables that should be tailored to the properties of the target tissue, and finding an effective protocol requires a careful modification of the relevant parameters.

2.6 *Mitochondrial Dysfunction in Neurodegenerative Disorders and Therapeutic Role of Photobiomodulation*

Mitochondria play a central role in neuronal physiology. These organelles integrate cell respiration, energy metabolism, and ionic balance into a homeostatic coherent adaptation for energy maintenance and cell survival (Kann and Kovacs 2007). The retina contains neurons with extremely high energy demands that rely mostly on mitochondrial-derived ATP to meet these requirements (Astrup et al. 1981; Ames III et al. 1992; Kann and Kovacs 2007). Thus, similar to other neuronal populations, retinal neurons are very vulnerable to events that lead to oxidative stress and energy depletion, including oxygen or glucose deprivation and dysfunction of the mitochondrial machinery that uses both of them to generate ATP (Beretta et al. 2006; Levin 2007). Evidence accumulated in the last 25 years suggests that mitochondrial dysfunction, induced by both genetic and environmental factors, plays a key role in the pathogenesis of retinal neurodegeneration, the main morphological feature of optic neuropathy.

Common eye disorders, such as glaucoma and age-related macular degeneration, and even neurodegenerative disorders, such as Alzheimer's disease, have been increasingly recognized to feature optic neuropathy induced by mitochondrial dysfunction. Patients with Alzheimer's disease show a reduction in the number of retinal ganglion cells and axons, compared to healthy individuals (Hinton et al. 1986; Valla et al. 2001; Danesh-Meyer et al. 2006; Iseri et al. 2006). In addition, the most common primary mitochondrial disorder, Leber's optic neuropathy, is responsible for approximately 2% of all cases of blindness (Chalmers and Schapira 1999). All these conditions feature retinal ganglion cell degeneration, optic nerve atrophy, and blindness that severely decrease the quality of life of affected individuals and represent a major public health problem. A number of genetic and acquired conditions featuring blindness secondary to degeneration of the retina and optic nerve have been linked to mitochondrial dysfunction. Besides Leber's hereditary optic neuropathy, genetic conditions featuring optic neuropathy and mitochondrial failure include Leigh syndrome (Borit 1971; DiMauro 1999), Friedreich's ataxia (Carelli et al. 2002), myoclonic epilepsy ragged-red-fibers (MERRF) (Chinnery et al. 1997), mitochondrial encephalomyopathy-lactic acidosis and stroke-like syndrome (MELAS) (Hwang et al. 1997), hereditary spastic paraplegia (Casari et al. 1998), and the deafness-dystonia-optic atrophy syndrome (Tranebjaerg et al. 2000). Similarly, acquired diseases featuring optic neuropathy with an association with mitochondrial dysfunction include the tobacco-alcohol amblyopia and intoxication with chloramphenicol, ethambutol, carbon monoxide, clioquinol, cyanide, hexachlorophene, isoniazid, lead, methanol, plasmocid, or triethyltin (Carelli et al. 2002).

We developed and tested a toxicological model of mitochondrial optic neuropathy with a favorable response to methylene blue and photobiomodulation (Rojas and Gonzalez-Lima 2010). Light tissue penetration depends on both the types of target tissues, wavelength and the source of photobiomodulation. Besides being safe, at red to near-infrared wavelengths, light penetration to the eye is maximal, where absorbance by the cornea and lens is negligible (<10%) and high refractive indices favor low light scattering and a high degree of focusing on the retina (Jester et al. 1999). In addition, spectroscopic measures have shown that photons at wavelengths between 630 and 800 nm are able to travel approximately 23 cm even in layers of tissues such as skin, dense connective tissue, muscle, bone, and spinal cord, with about 6% of the total energy density being detectable at the ventral surface of a living rat, when the photobiomodulation source is located on the dorsal surface (Byrnes et al. 2005b; Wong-Riley et al. 2005). Thus, the studies in Part 1 are not only relevant for the treatment of a number of neuro-ophthalmological conditions but also pursue a more general goal of developing therapeutic approaches against neurodegeneration, which focus on counteracting the immediate consequences of mitochondrial failure.

Animal studies relevant for our aims 1, 2, and 3 are briefly presented below. These studies used fixed daily photobiomodulation doses, wavelengths, and power density (3.6 J/cm^2, 633 nm, and 2 mW, respectively) and tested three different photobiomodulation fractionation protocols. The results showed that these variables allow effective neuroprotection in the model of optic neuropathy (Rojas et al. 2008b).

2.7 *Photobiomodulation Prevents Impairment of Visual Function in a Rat Model of Optic Neuropathy Induced by Mitochondrial Dysfunction*

Rationale. Animals were male Long-Evans rats (40 days-old). Visual function was assessed behaviorally before and after treatment, using a descending method of limits in a 2-choice visual task apparatus designed to determine minute changes in the dark-adapted illuminance sensitivity threshold. This was accomplished to test the in vivo protective effects of photobiomodulation at the functional level.

Model of optic neuropathy. Animals were anesthetized with 1.5% isoflurane and received bilateral intravitreal injections (total volume 3 μL). Subjects were divided into control and experimental groups. In the control group ($n = 7$), the vitreous body in both eyes of each subject was injected with the vehicle dimethyl sulfoxide (DMSO). Subjects in the experimental group ($n = 23$) received bilateral intravitreal injections of 200 μg/kg rotenone in DMSO. Rotenone is a mitochondrial complex I inhibitor shown to induce oxidative stress and energy depletion-mediated neurotoxicity (Sherer et al. 2003; Beretta et al. 2006) and retinal degeneration after a single intravitreal injection (Zhang et al. 2002). Sixteen days after the rotenone injection, the subjects were decapitated, and the eyeballs and brains were rapidly removed and frozen in isopentane (−40 °C).

Photobiomodulation delivery. Photobiomodulation was delivered via two R30-123 narrow-angle light-emitting diode (LED) arrays (radius = 4.4 cm) (LEDtronics, Inc., Torrance, CA) located 3.8 cm above the subjects' head. All photobiomodulation sessions were given at peak $\lambda = 633$ nm, with a power density of 2 mW/cm^2, during 30 min for an energy density dose of 3.6 J/cm^2. This single-session energy density was based on previous studies showing beneficial photobiomodulation effects in in vitro and in vivo models of neurodegeneration induced by mitochondrial dysfunction (Liang et al. 2006). Three different photobiomodulation protocols were assessed in subjects receiving bilateral rotenone intravitreal injections (i.e., the experimental group) for their effectiveness to prevent rotenone-induced retinotoxicity. Protocol photobiomodulation 1 (NILT 1, $n = 7$) consisted of a total dose of 10.8 J/cm^2 fractionated in three sessions of photobiomodulation treatment, each one occurring at 10 min, 24 h, and 48 h after rotenone injections. Protocol photobiomodulation 2 (NILT 2, $n = 5$) consisted of six sessions of photobiomodulation treatment occurring at 10 min, 24 h, 48 h, 72 h, 96 h, and 120 h after rotenone injections, for a total dose of 21.6 J/cm^2. Protocol photobiomodulation 3 (NILT 3, $n = 5$) also consisted of a total dose of 21.6 J/cm^2, similar to protocol photobiomodulation 2, but in this case, treatment sessions occurred at 48 and 24 h before the rotenone injections and continued at 10 min, 24 h, 48 h, and 72 h, after the injections. These dose fractionation schedules were implemented based on studies reporting that multiple photobiomodulation treatment sessions over days are more beneficial than the administration of a single treatment (Brondon et al. 2005; Liang et al. 2008; Ying et al. 2008). Six subjects in the experimental group received bilateral rotenone injections but no photobiomodulation treatment (i.e., rotenone group). Regardless of whether they received rotenone or DMSO injections or whether they were treated with photobiomodulation or not, all subjects were anesthetized with 1.5% isoflurane during similar periods of time at the same intervals.

Results. Our data support that photobiomodulation is able to prevent the visual function deficits induced by mitochondrial dysfunction in vivo. Effective parameters were a daily dose of 3.6 J/cm^2, a wavelength of 633 nm, and a protocol including six sessions given postinjection (protocol NILT 2 in Fig. 1). This protocol prevented the rotenone-induced increase in the illuminance threshold (Fig. 1a), the increase in latency (Fig. 1b), and the decrease in correct choices (Fig. 1c) in the

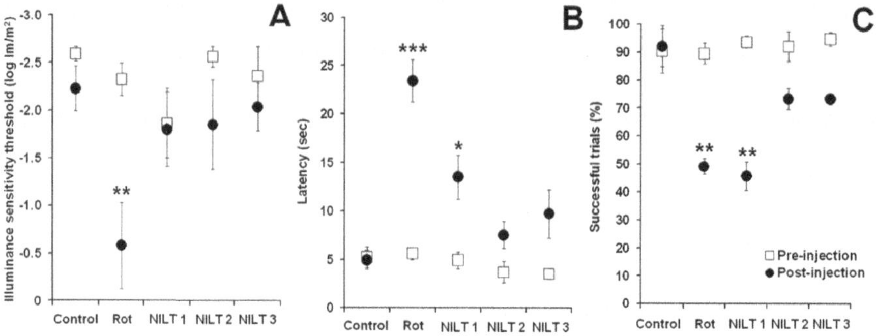

Fig. 1 Effects of photobiomodulation on visually guided behavior. (**a**) Rotenone (Rot) induced a significant decrease in the illuminance sensitivity threshold, but no significant changes postinjection were observed in subjects receiving rotenone plus photobiomodulation. (**b**) Subjects in the Rot group displayed mean postinjection escape latencies across illuminance levels that were significantly higher than control and baseline. A total dose of 21.6 J/cm² photobiomodulation at 630 nm given in six daily session postrotenone of 3.6 J/cm² (NILT 2) prevented this change, whereas a total dose of 10.8 J/cm² photobiomodulation (NILT 1) over three daily sessions of 3.6 J/cm² using the same wavelength was not as effective. A third protocol consisted of a total dose of 21.6 J/cm² given in six daily doses of 3.6 J/cm² at 630 nm, including prophylactic sessions before rotenone injection and four sessions after rotenone injection (NILT 3). This third protocol was also effective at preventing the increase in escape latency. (**c**) Rotenone-treated subjects displayed mean postinjection escape performances across illuminance levels that were not different from chance. This performance was significantly worse than control and baseline. The dose-response effects on performance corresponded to those on escape latency. Asterisks indicate a significant difference compared to control. * = $p < 0.05$, ** = $p < 0.01$, and *** = $p < 0.001$

visually guided behavioral task. In addition, two prophylactic sessions of 3.6 J/cm² photobiomodulation at 633 nm followed by four sessions of postlesion photobiomodulation were also effective at preventing the visual deficits induced by rotenone (NILT 3 in Fig. 1).

2.8 Photobiomodulation Prevents Structural Retinal Damage in the Model of Optic Neuropathy

Rationale. This study tested whether functional effects observed in the behavioral experiment were associated with actual preservation of retinal structure. Measures focus on the retinal nerve fiber layer + ganglion cell layer (RNFL + GCL) because these retinal layers are known to be affected by the intravitreal administration of rotenone and because they constitute the structural basis for the major output pathway from the retina to the brain.

Quantification of retinal structure. Estimates of RNFL + GCL and inner plexiform layer (IPL) thickness were obtained by systematic analyses of NADH dehydrogenase activity-stained retinal sections from animals used in behavioral studies.

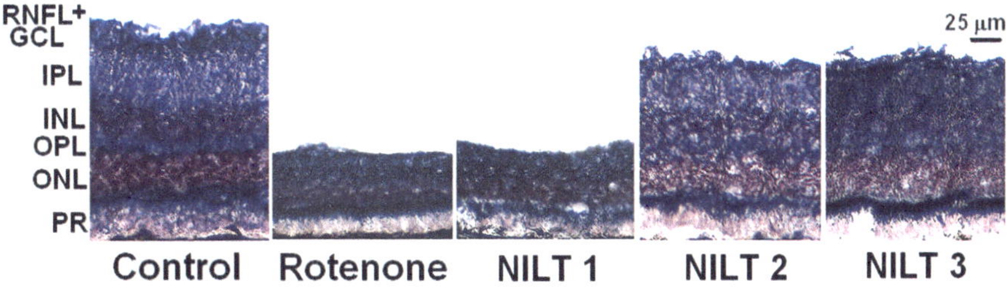

Fig. 2 Neuroprotective effects of photobiomodulation on retinal structure. Layer thinning is a histopathological feature of rotenone-induced neurotoxicity. This is more drastic in the innermost layers including the retinal nerve fiber layer (RNFL), ganglion cell layer (GCL), inner plexiform layer (IPL), and inner nuclear (INL) layer. Whereas structural defects of similar severity were observed in the group treated with the photobiomodulation NILT 1 protocol, photobiomodulation protocols NILT 2 and NILT 3 were effective at preventing the retinotoxic effects of rotenone. NILT 1 = rotenone + 10.8 J/cm^2 photobiomodulation postinjection, NILT 2 = rotenone + 21.6 J/cm^2 postinjection, and NILT 3 = rotenone + 10.8 J/cm^2 preinjection + 10.8 J/cm^2 postinjection. *OPL* outer plexiform layer, *ONL* outer nuclear layer, *PR* photoreceptor layer. Light microscopy, 10×

Results. Photobiomodulation displayed a dose-response effect on the integrity of the retinal layers after rotenone injections (Fig. 2). This is a clear indication that variations in the dose protocol have a large influence on the neuroprotective effects of photobiomodulation, independent of the daily dose and the used wavelength. This supports the need to investigate the protocols that will be most effective at preventing neurodegeneration in different models and disorders.

Mechanistic data relevant for aim 3 are presented below. These studies were aimed at determining the effects of photobiomodulation at effective parameters on cytochrome oxidase activity and oxidative stress. The results support the hypothesis that the mechanism of action of photobiomodulation is mediated through enhancement of the activity of the respiratory enzyme cytochrome oxidase and of the antioxidant enzyme superoxide dismutase.

2.9 *Protective Effects of Photobiomodulation Are Not Related to Photodegradation of Rotenone*

Rationale. To rule out the possibility that the observed effects of photobiomodulation on visual behavior and retinal structure are due to photodegradation of rotenone by light, we exposed rotenone solutions to 3.6 J/cm^2 633 nm photobiomodulation and characterized the chemical properties of the solutions after the light treatment.

Results. The results showed that rotenone solutions exposed to photobiomodulation do not lose their prooxidant potential and are thus still able to potently induce oxidative stress in brain homogenates, as quantified by generation of lipid peroxides in vitro (Fig. 3).

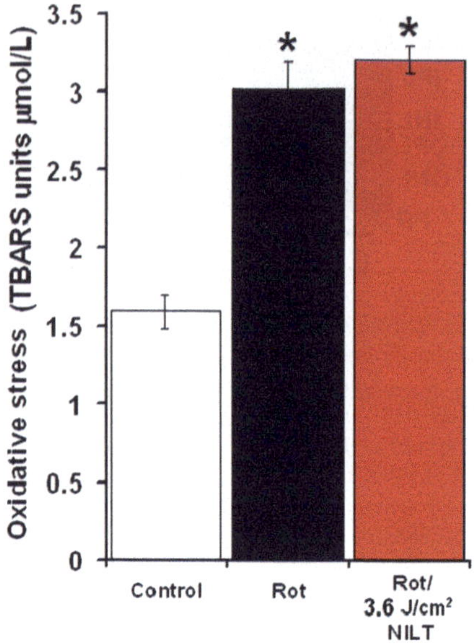

Fig. 3 Photobiomodulation does not decrease the pro-oxidant potential of rotenone. Rotenone solutions exposed to 3.6 J/cm^2 induce similar levels of lipid peroxides in brain homogenates. This shows that light effects are not mediated through photodegradation of rotenone. * = $p < 0.05$

2.10 Preservation of Visual Function and Retinal Structure Are Not Mediated by Isoflurane Exposure

Rationale. We ruled out the possibility that the behavioral and structural effects of photobiomodulation are unrelated to a photobiomodulatory effect and are secondary to a neuroprotective effect of isoflurane instead. Subjects received inhaled anesthesia concomitant to photobiomodulation treatments. This concern was based on suggestion that isoflurane and related anesthetics can prevent neuronal damage secondary to a series of insults. We compared the visually guided behavior (escape latency) of animals intravitreally infused with bilateral rotenone with that of animals given intravitreal rotenone plus 1.5% isoflurane in three 30 min sessions (one session per day). We expected that if isoflurane exposure was responsible for the preservation of visual function after rotenone infusion, isoflurane alone (no photobiomodulation) would prevent the abnormal visually guided behavior induced by rotenone.

Results. Compared to control ($n = 5$) (no rotenone, no isoflurane), rotenone-infused subjects ($n = 5$) showed an overall increase in mean escape latencies across all illuminance levels postlesion (3.2 ± 0.3 s, control vs. 15.6 ± 2 s, rotenone, $p < 0.05$). This effect resulted in a right displacement of the escape latency curve compared to control, which maintained low escape latencies even at low illuminance levels. Rotenone-treated subjects that received isoflurane ($n = 6$) also showed increased escape latencies compared to control across all illuminance levels. In fact, mean escape latencies in the isoflurane group were higher than those of the rotenone-only group (23.4 ± 3, $p < 0.05$) (Fig. 4). These results demonstrate that isoflurane does not prevent the impairments in visually guided

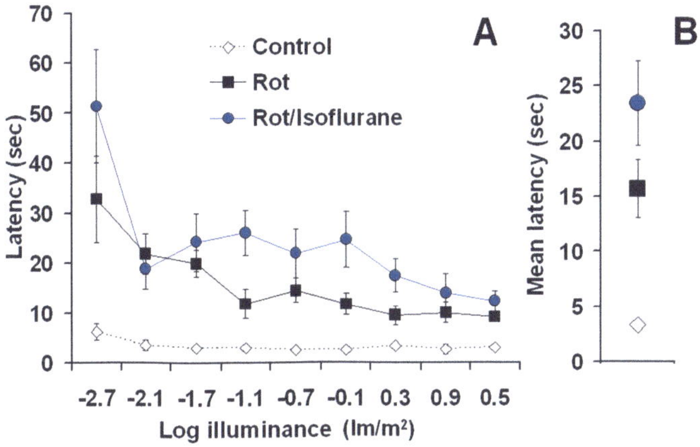

Fig. 4 Isoflurane does not prevent rotenone-induced visual deficits. (**a**) Psychometric curve of escape latencies as a function of platform illuminance. Rotenone increased escape latency compared to control and isoflurane did not prevent this increase. (**b**) Post-lesion mean group latencies pooled across illuminance levels

behavior induced by rotenone and support the fact that the effects of photobiomodulation are due to a photobiomodulatory interaction of light with neural tissue.

2.11 Photobiomodulation Prevents Decreases in Cell Respiration in Brain Homogenates In Vitro

Rationale. We tested the hypothesis that photobiomodulation directly stimulates cell respiration in vitro. Cell respiration is expected to increase if cytochrome oxidase activity is enhanced, since this enzyme is not only the rate limiting step for ATP synthesis but also catalyzes the synthesis of water from molecular oxygen. However, an in vitro system consisting of membrane isolates is uncoupled and oxygen consumption occurs at the maximal possible rate. Inhibition of this system through blockade of the respiratory chain (e.g., addition of the complex I inhibitor rotenone) can provide the conditions to avoid a ceiling effect for oxygen consumption and yet reveal the respiration-enhancing effects of photobiomodulation.

Results. Cell respiration measures in vitro showed that a single session of 0.1 J/cm^2 or 1 J/cm^2 photobiomodulation at 633 nm was able to increase the rate of oxygen consumption in the presence of rotenone (Fig. 5).

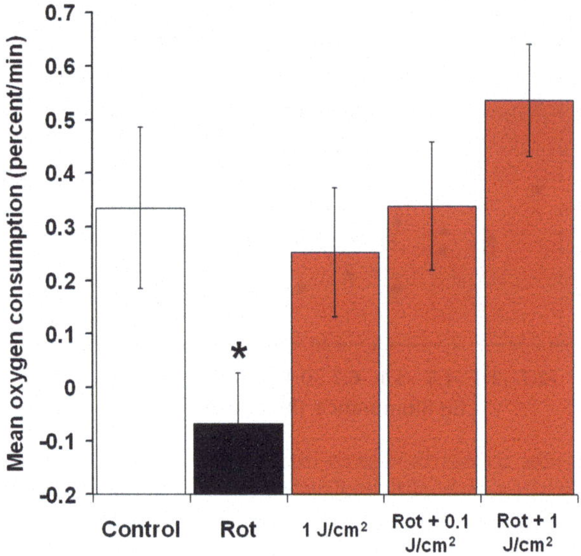

Fig. 5 Photobiomodulation effect on cell respiration in vitro. Rotenone decreased the rate of oxygen consumption. Small (0.1 J/cm^2) and large (1 J/cm^2) light doses reversed the inhibitory effect of rotenone. Rot = 10 μM rotenone, * = significant difference vs. control at $p < 0.05$

2.12 Photobiomodulation Increases Brain Antioxidant Capacity in a Dose-Response Manner In Vivo

Rationale. The effect of photobiomodulation on the antioxidant capacity of neuronal tissue was first shown by Rojas et al. (2008b). Although photobiomodulation is expected to directly stimulate the respiratory enzyme cytochrome oxidase, activation of the latter is expected to allow the expression of enzymatic systems that will coordinate a response for cell survival. Enzymatic systems anticipated to respond to photobiomodulation include inducible endogenous antioxidants such as superoxide dismutase (SOD). In addition, SOD shows absorption peaks at wavelengths similar to cytochrome oxidase. For this reason, SOD expression and activity are anticipated to be affected by photobiomodulation and mediate its neuroprotective effects. In our study, we tested the dose-response effects of 10.8 and 21.6 J/cm^2, 633 nm photobiomodulation on brain SOD activity in vivo. Total doses were delivered in daily sessions of 3.6 J/cm^2 photobiomodulation.

Results. High dose photobiomodulation induced a 50% increase in the whole-brain SOD activity in vivo in a dose-response manner (Fig. 6). The effect reached significance on the cytosolic fraction of the enzyme, but the mitochondrial fraction showed a similar trend. The results support that photobiomodulation increases SOD activity in the brain in vivo and indicate further testing of photobiomodulation to prevent oxidative stress in vivo.

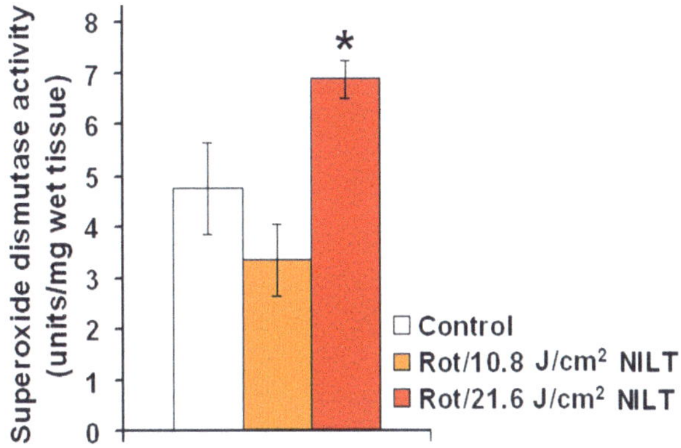

Fig. 6 Photobiomodulation increases brain superoxide dismutase (SOD) activity in vivo. A high dose of 633 nm light given in daily doses of 3.6 J/cm^2 induced a 50% increase in SOD capacity, implicating a transcranial effect. * = $p < 0.05$

2.13 A Single Dose of Photobiomodulation Enhances Brain Cytochrome Oxidase Activity in a Hormetic Fashion In Vivo

Rationale. Previous in vivo experiments tested the effects of a fixed daily dose of photobiomodulation and different fractionation protocols on behavioral and structural variables. This experiment was conducted to test the effects of different daily doses of photobiomodulation delivered in a single session on levels of whole-brain cytochrome oxidase activity in vivo. Unanesthetized animals (male, Sprague-Dawley rats) were exposed to 660 nm light at either 10.9 J/cm^2 ($n = 5$), 21.6 J/cm^2 ($n = 4$), 32.9 J/cm^2 ($n = 4$) or no photobiomodulation (control, $n = 5$) in polycarbonate home cages. Treatments were delivered via four LED arrays with a power density of 9 mW/cm^2 for total treatment times of 20 min, 40 min, and 60 min for each dose, respectively. Twenty-four hours after the single treatment session, animals were decapitated and their brains were extracted, frozen, sectioned, and histochemically stained for quantitative cytochrome oxidase activity (Gonzalez-Lima et al. 1997).

Results. A single session of photobiomodulation at different doses showed enhancement of whole-brain cytochrome oxidase capacity in vivo following a hormetic dose-response pattern. A single dose of 10.9 J/cm^2 photobiomodulation resulted in a 13.6% increase in cytochrome oxidase activity ($p < 0.05$). In turn, a single dose of 21.6 J/cm^2 resulted in an increase of only 10.3%, whereas the highest dose induced no significant change in cytochrome oxidase activity (3%) (Fig. 7). These results suggest that responses of whole-brain cytochrome oxidase levels to photobiomodulation in vivo are characterized by a hormetic pattern, with a low dose given in a single day having a stimulatory effect, while higher doses are less

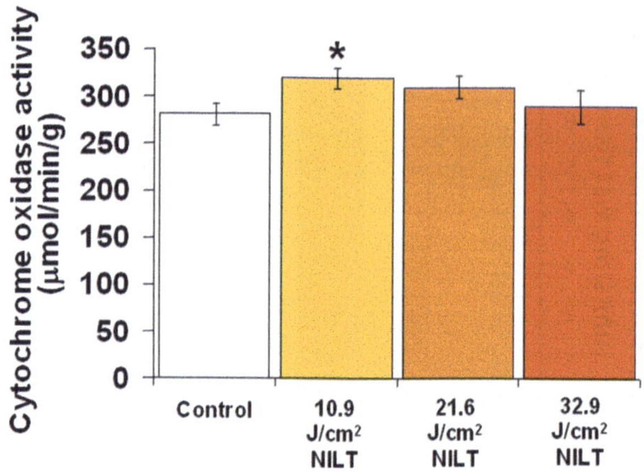

Fig. 7 A single light dose increases whole-brain cytochrome oxidase capacity in unanesthetized rats. Low-dose 660 nm photobiomodulation boosted cytochrome oxidase activity by 13.6%, whereas higher doses were not as effective. This response to a low dose with an absent response to a high dose is typical of hormesis. These data also support a transcranial effect of photobiomodulation. * = $p < 0.05$

effective. Thus, we anticipate that effective photobiomodulation protocols will include daily doses lower than 10.9 J/cm^2.

2.14 Fractionated Photobiomodulation Increases Brain Cytochrome Oxidase Activity in a Dose-Response Manner In Vivo

Rationale. This experiment tested the effects of a fractionated protocol of daily photobiomodulation delivered in vivo to rats on whole-brain cytochrome oxidase activity. Daily doses of 3.6 J/cm^2 633 nm photobiomodulation were given at a power density of 2 mW/cm^2.

Results. With a fractionated protocol, we observed dose-response effects of photobiomodulation on the activity of cytochrome oxidase. While a single-session dose of 21.6 J/cm^2 was ineffective in the previous study, the same cumulative dose of 21.6 J/cm^2 at 633 nm fractionated into six daily sessions of 3.6 J/cm^2 each is sufficient to induce an approximately 20% increase in cytochrome oxidase activity, which is a reflection of increased global energy metabolism (Fig. 8). The results suggest that low daily doses of photobiomodulation provide long-term enhancing effects on whole-brain metabolic capacity. These results further support a transcranial effect of photobiomodulation in living rats. Taken together with the results using 660 and 633 nm photobiomodulation, these data suggest that it is possible that noneffective daily doses trigger a favorable cumulative response in vivo, if wavelengths and power densities are modified.

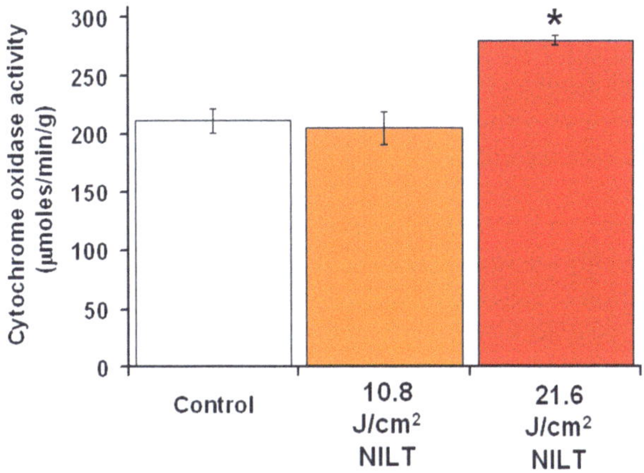

Fig. 8 Fractionated photo biomodulation increased brain cytochrome oxidase in a dose-response fashion in vivo. Rats treated with the low-dose (10.8 J/cm^2) protocol showed no difference to control. Rats treated with the higher-dose protocol (21.6 J/cm^2) showed a 20% increase in cytochrome oxidase activity compared to control. * = $p < 0.01$

In conclusion, the animal experiments reviewed characterized neuroprotective effects of photobiomodulation at three different levels of analysis: *functional, structural, and neurochemical.* The experiments demonstrated optimal photobiomodulation parameters for the treatment of optic neuropathy and its mechanism of action. They served to improve the understanding of the mechanisms mediating the effects of photobiomodulation in neural tissue in vivo. Using in vivo animal models to characterize the effects of photobiomodulation helped us address important questions on how light interacts with neural tissue in a more translational-relevant fashion in the studies in Part 2.

3 Part 2: Human Studies—Augmentation of Neurocognitive Functions by Photobiomodulation

3.1 Introduction and Objectives of the Human Studies

We have conducted randomized controlled trials (RCTs) to test the safety and efficacy of a new noninvasive photobiomodulation intervention, *transcranial infrared laser stimulation (TILS)*, for neurocognitive enhancement in younger and middle-aged adults. Considering that cognition is one of the most important determinants of quality of life and functional ability in middle and older age (Gaugler et al. 2009), it is critical to seek new treatments to prevent or delay cognitive impairment in vulnerable populations. Adults showing early preclinical signs of cognitive decline are prime candidates for interventions intended to enhance cognitive function.

TILS penetrates approximately 40 mm to the brain and improves cognitive functions (Rojas and Gonzalez-Lima 2013; Gonzalez-Lima and Barrett 2014; Tedford et al. 2015). Specifically, TILS of the human prefrontal cortex with a wavelength of 1064 nm and a power density of 0.25 W/cm^2 upregulates the levels of oxidized *cytochrome oxidase*, the conformation of the enzyme that has the highest oxygen consumption activity, which leads to improved cerebral oxygenation in a nonthermal manner (Tian et al. 2016; Wang et al. 2017). This unique neurophotonics in vivo mechanism is highly relevant for cognitive enhancement because neurons are critically dependent on cytochrome oxidase-mediated oxygen consumption to sustain electrophysiological activity (Ojaimi et al. 1999; Wong-Riley et al. 2005). Since cerebral physiology is critically dependent on oxygen metabolism, the mechanistic action of TILS on cytochrome oxidase has strong potential for cognitive enhancement.

We have shown that TILS is safe and effective for increasing cognitive functions in young adults in five controlled studies using photobiomodulation of the right prefrontal cortex (Barrett and Gonzalez-Lima 2013; Blanco et al. 2015, 2016; Hwang et al. 2016; Disner et al. 2016). Based on our successful TILS studies with young and middle-aged adults, our ongoing project is to determine if TILS may also enhance cognitive performance by improving prefrontal oxygen metabolism in older adults. For TILS to become a mainstream intervention for older adults, it is essential to evaluate quantitatively its cognitive and neurophysiological effects in an older population. However, there are no completed RCTs of TILS in older adults.

The *long-term research goal* of our human studies is to quantify the cognitive and physiological effects of TILS on the human prefrontal cortex using prefrontal-based cognitive tasks and multimodal in vivo evaluation of neurophysiological mechanisms mediating cognitive enhancement, using near infrared spectroscopy (NIRS), electroencephalography (EEG), and functional magnetic resonance imaging (fMRI) techniques.

Aim 1: Demonstrate the neurophysiological effects of transcranial infrared laser stimulation in humans. Our hypothesis, based on our animal studies, is that TILS of the prefrontal cortex will enhance cerebral oxygen metabolism and neurocognitive network function in humans. *Methods*: Using a placebo-controlled design, we were the first to determine the cerebral effects of TILS using noninvasive techniques (NIRS, EEG, and fMRI), including in vivo measures of oxidized cytochrome oxidase, oxygenated hemoglobin (HbO), electrophysiological power spectral density (PSD), cerebral blood flow (CBF), and blood oxygen level-dependent (BOLD) response. We compared cerebral responses with attention/memory tasks, before and after treatment. *Results*: We found that TILS-mediated cytochrome oxidase upregulation promoted better cerebral oxygen metabolism and neurocognitive network function and more efficient prefrontal response posttreatment, relative to participants in the placebo condition.

Aim 2: Demonstrate the cognitive-enhancing effects of transcranial infrared laser therapy in humans. Our hypothesis, based on the animal studies, is that TILS of the prefrontal cortex will enhance cognitive performance in humans. *Methods*:

Using a randomized placebo-controlled design, we are comparing participants' cognitive performance before and after 4–5 week active or placebo treatments and at 1-week and 8-weeks posttreatment. A psychomotor vigilance task for sustained attention and a delayed-match-to-sample working memory task were chosen for weekly evaluation because they engage prefrontal-based attention-memory-executive domains particularly vulnerable to aging, and our data demonstrate that performance in those domains is improved by TILS in younger adults (Barrett and Gonzalez-Lima 2013; Blanco et al. 2015, 2016; Hwang et al. 2016; Disner et al. 2016; Vargas et al. 2017). Weekly simultaneous bbNIRS/EEG and cognitive measures will give us cytochrome oxidase, electrophysiological and behavioral changes to allow us to directly compare how these neural metrics are related to each other and to the cognitive performance. Standard neuropsychological tests will also be administered at baseline and after the treatment. *Results*: We expect that the active treatment group will exhibit better cognitive performance posttreatment, relative to participants in the placebo condition.

This research provided an important translational step between human studies and a large literature utilizing animal models establishing that photobiomodulation improves neuronal oxygenation and upregulates mitochondrial respiration in a way that is safe, noninvasive, and therapeutically beneficial (Anders et al. 2014; Wong-Riley et al. 2005; Rojas and Gonzalez-Lima 2011; de la Torre 2017; Rojas et al. 2008a, b, 2012; Barrett and Gonzalez-Lima 2013; Blanco et al. 2015, 2016; Hwang et al. 2016; Disner et al. 2016; Vargas et al. 2017; Fulop et al. 2010; Lampl et al. 2007; Zivin et al. 2009; Hacke et al. 2014; Wang et al. 2016). In particular, middle-aged and older adults as the study population will provide relevant translational data for early intervention before onset of clinical impairment. The continuous success of this project will also help develop a new way to alleviate or stabilize the cognitive deficits common to aging, Alzheimer's disease, and related dementias.

3.2 Significance of Human Cognitive Enhancement by Photobiomodulation

The potential impact of this research increases as the overall population ages. It is estimated that by 2050, there will be nearly two billion people aged over 60 (He et al. 2016). Along with a rapidly aging population comes the health problems associated with it, including cognitive decline and dementia. While aerobic exercise is effective to support oxygen metabolism, associated with a range of health benefits including cognitive enhancement (Etnier et al. 1997), compliance with exercise is abysmal and only 5% of US adults report the recommended daily physical activity of 30 min (Buford et al. 2013). Aerobic exercise is not a viable option for older patients with severe cardiovascular disease or orthopedic injuries. There is a pressing need for alternatives to aerobic exercise to preserve and enhance cognitive function for all. Although significance is high for older adults, impairments in attention-memory-executive function are not unique to this population. The success of this project will have far-reaching significance because it will provide relevant data for early intervention before onset of clinical impairment. This may help alleviate or stabilize the cognitive deficits common in aging, Alzheimer's disease, and related dementias. This is also significant to other patients who would benefit from cognitive enhancement, e.g., in traumatic brain injury, metabolic, cardiovascular, and many other neurological and mental disorders.

3.3 Cytochrome Oxidase as Molecular Target for Human Cognitive Enhancement

Aging-dependent and neurodegeneration-dependent decline in cytochrome oxidase activity has been well documented, ranging from studies in drosophila (Ren et al. 2010) to the human brain (Ojaimi et al. 1999), and cytochrome oxidase decline is more pronounced in mild cognitive impairment and Alzheimer's disease (Valla et al. 2006) (PubMed lists 1071 reports on 'cytochrome oxidase and aging'). Cytochrome oxidase is the enzyme responsible for utilizing over 95% of the oxygen we breathe for cellular respiration. Nerve cell function is critically dependent on oxygen utilization (i.e., aerobic metabolism). Animal studies from our lab have shown that cytochrome oxidase activity is critical for learning and memory performance (Gonzalez-Lima et al. 2014). For example, partial inhibition of cytochrome oxidase activity is sufficient to cause cognitive impairment, whereas improvement of cytochrome oxidase activity by metabolic interventions causes cognitive enhancement (Callaway et al. 2002, 2004). For the past 20 years, our lab has been searching for a noninvasive, safe, and effective method to upregulate cytochrome oxidase that could be translated to humans for the purpose of cognitive enhancement. We think that photobiomodulation is the answer because it increases cellular oxygen metabolism by delivering photons that oxidize cytochrome oxidase, the main intracellular photon acceptor at red-to-near-infrared wavelengths (Anders et al. 2014; Wong-Riley et al. 2005; Rojas and Gonzalez-Lima 2011; de la Torre 2017). High bioavailability to brain tissue in vivo is supported by preclinical evidence of increases in brain cytochrome oxidase activity and oxygen consumption that improved behavioral outcome in animal models (Rojas et al. 2008a, b, 2012) and by near-infrared light penetration of approximately 40 mm through the human head (Tedford et al. 2015). Recent findings from our team demonstrated that TILS at 1064 nm upregulates cytochrome oxidase and produces beneficial effects on human prefrontal cortex oxygenation in young adults (Tian et al. 2016; Wang et al. 2017). We found that TILS of the right prefrontal cortex at 1064 nm, 0.25 W/cm^2 is safe and effective for increasing cognitive functions (sustained attention, working memory, executive skills, and category learning) in young adults in controlled studies (Barrett and Gonzalez-Lima 2013; Blanco et al. 2015, 2016; Hwang et al. 2016; Disner et al. 2016), and we are extending this approach to older adults.

3.4 Need to Investigate How TILS Affects Human Neurocognitive Function

In 2002, the FDA approved low-level light/laser therapy with red-to-near-infrared wavelengths for pain relief in head and neck pain, arthritis, and carpal tunnel syndrome (Fulop et al. 2010). FDA-cleared laser diodes and light-emitting diodes (LEDs) delivering low-power density (low irradiance) but high-energy density (high fluence) light are a highly promising, affordable, and safe alternative for improving human cognitive function (Gonzalez-Lima and Barrett 2014). However, this intervention has not been adopted in spite of safe, portable, effective, and promising outcomes from animal and human studies. One major reason is that the action of TILS in the human brain in vivo has not been quantitatively studied and understood using optical methods to reveal light-tissue interactions. One important example was a large clinical study (*"Neurothera Effectiveness and Safety Trials"; NEST*) sponsored by PhotoThera (Lampl et al. 2007; Zivin et al. 2009; Hacke et al. 2014). NEST had three clinical trials, lasted a total of 7 years, and recruited a total of more than 1000 acute stroke patients. During the studies, an 810-nm laser was applied over the entire surface of the head (20 locations in the 10/20 EEG system) without targeting any particular cortical region regardless of stroke and at a very low laser energy density (1 J/cm^2 over the entire cortical surface) for only 2 min without repeated treatment. While this intervention was found to be safe, without any side effects, and worked well in the first two trials for moderate and moderate-severe acute stroke patients, it did not achieve the expected statistical significance for severe stroke patients. Given our recent results from 1064-nm laser stimulation of the human forearm (Wang et al. 2016) and forehead (Wang et al. 2017) as well as human cognition measures (Barrett and Gonzalez-Lima 2013; Blanco et al. 2015, 2016; Hwang et al. 2016; Disner et al. 2016), it is clear to us that the laser parameters and stimulation setup used in the NEST studies were not chosen effectively, while many other factors could have contributed to the unsuccessful clinical trial. This negative example strongly demonstrates the necessity to investigate how TILS affects neurophysiology both in time and space in our target population, so it can become an effective, noninvasive tool for treating prefrontal-based cognitive decline. We envision a near future with scientifically validated, safe portable devices to conveniently administer TILS in homes, nursing homes, and healthcare facilities.

We propose TILS as an exciting new intervention to enhance neurocognitive function in adults at risk for developing cognitive impairment. Building upon a body of evidence demonstrating increases in brain cytochrome oxidase activity and brain oxygenation, which improve behavioral outcomes and memory in animal models (Rojas et al. 2008a, b; Rojas and Gonzalez-Lima 2017) and healthy humans (Barrett and Gonzalez-Lima 2013; Blanco et al. 2015, 2016; Hwang et al. 2016; Disner et al. 2016; Tian et al. 2016; Wang et al. 2017), we conducted the first randomized clinical trials (RCTs) to test if TILS improves neurocognitive function in humans. We were the first to quantify TILS effects on the human brain in vivo using state-of-the-art neurophysiological techniques (fNIRS, qEEG, and fMRI) and a new broadband NIRS (bbNIRS) apparatus to quantify cytochrome oxidase in the human brain in vivo. We were enabled in this innovation in the new field of laser-mediated cognitive enhancement by strong interdisciplinary collaborations using all these techniques, including established experts in cerebral metabolism, cognition, and biomedical optics who are coauthors of the studies cited below.

3.5 Overview of Our Approach for Human Cognitive Enhancement by Photobiomodulation

Neurophysiological effects were investigated using a combination of noninvasive NIRS, EEG, and MRI techniques. Cognitive-enhancing effects were investigated in tests of global and prefrontal cortex (PFC)-based cognitive function, with a special emphasis on sustained attention-working memory-executive functions, which show a decline in older adults and patients with chronic neurodegenerative and mental disorders.

Figure 9a shows that the left part of the laser unit has on/off controls and multiple safety interlocks, including key and emergency stop. The center has a screen display and keypad to program output power, number of treatment counts, and exposure time. Output is programmable between 0.1 and 20 W, and optimal parameters for TILS are 3.4–3.9 W and 8 counts of 60-s each (8 min). On screen messages confirm correct handling, calibration, and use of laser. The right side of the laser unit has a calibration port that securely locks the handpiece in place, while the laser is being calibrated before each use. Beam output characteristics are continuously monitored, while laser is active. Since the 1064-nm laser is invisible, a red beam area provides visual confirmation for tissue targeting (Fig. 9b). During operation, the laser is locked into position near the skin and participants are instructed to sit still and keep their eyes closed. Experimenters and participants wear dark safety glasses that block the infrared light from reaching the eyes, as required by the laser manufacturer and the University of Texas Laser Safety Program. Dr. Fenghua Tian and Dr. Hanli Liu are pioneers in the field of optical brain imaging who introduced a realistic brain model and numerical simulation of light propagation through the tissues (Tian and Liu 2014). Based on this model and diffusion theory, Dr. Tian estimated the cortical region of the TILS as shown in Fig. 9c. The model supports that TILS penetrates through the cortical gray matter (Tedford et al. 2015) and effectively targets the dorsolateral PFC, not the ventrolateral or medial PFC regions. We have not had any report of discomfort or adverse events caused by TILS, as low-intensity laser (0.25 W/cm^2) produces negligible heat that most people cannot detect. As shown in Fig. 9d, we have demonstrated that thermal stimulation that matches precisely the heat produced by our laser treatment has no significant effect on cytochrome oxidase in the PFC (Wang et al. 2018). Therefore, TILS-specific molecular action on cytochrome oxidase cannot be explained by thermal effects and cannot be due to a more trivial effect of water heating and increased circulation. In summary, TILS causes 1064-nm photons to oxidize cytochrome oxidase in the dorsolateral PFC and heating would not be a possible explanation for TILS effects (Wang et al. 2018).

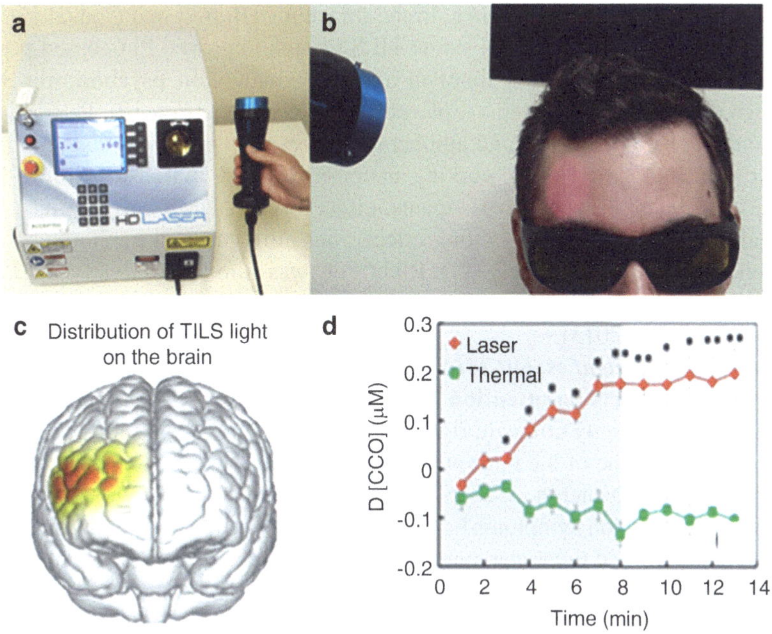

Fig. 9 Transcranial infrared laser stimulation (TILS). (**a**) Laser device for TILS. The FDA-cleared Class IV laser device (HD Laser, Cell Gen Therapeutics, Dallas, Texas) consists of a control unit (16″ × 14″ × 13″) with a fiber optic cable coupled to a handpiece. (**b**). Laser delivery. For illustration purposes, the largest laser aperture aims at the forehead using an internal red diode aiming light. The laser aperture can be adjusted to a desired spot size from 1 to 45 mm diameter. We used a 4-cm diameter laser beam size to match the size of the prefrontal cortex (PFC) area we aim to stimulate. (**c**) Cortical target. Model of intensity distribution of treatment light on the right PFC (orange indicates effective light intensity) made by Dr. Fenghua Tian. (**d**) Molecular target. Oxidized [cytochrome oxidase] increases during laser (red; $n = 11$) but not thermal (green; $n = 11$) stimulation measured in vivo by bbNIRS. "*" $p < 0.05$ and "**" $p < 0.01$ laser vs. thermal stimulation (mean ± SE) (Wang et al. 2018)

Beginning in 2013, we have published twelve studies of TILS neurocognitive effects involving 432 participants, which demonstrated TILS feasibility in humans. First, we briefly review the six studies demonstrating cognitive enhancement and then the six studies demonstrating the neurophysiological effects of TILS.

3.6 Cognitive-Enhancing Effects of TILS of the Human Prefrontal Cortex

We published the first six placebo-controlled studies (333 participants, 177 females, ages 17–40) demonstrating that TILS of the right PFC produces beneficial effects on PFC-modulated attention/memory/executive functions such as sustained attention and working memory, executive skills, attention bias modification, and rule-based category learning (Barrett and Gonzalez-Lima 2013; Blanco et al. 2015, 2016; Hwang et al. 2016; Disner et al. 2016).

1. ***Attention and working memory***. In the first study (Barrett and Gonzalez-Lima 2013) ($n = 40$, ages 18–35), a single TILS session improved PFC-based performance (decreased reaction time) in a sustained-attention **psychomotor vigilance task** (PVT) in treated vs. placebo control groups (Fig. 10). Performance (memory retrieval latency and number of correct responses) was also improved in the treated group on a working memory **delayed match-to-sample task** (DMS). In healthy young participants, effect sizes of DMS correct trials were small, about 2–6% due to ceiling effects, but differences in memory retrieval latency were about 20% (Fig. 11). Right PFC was stimulated because PVT and DMS cognitive tasks are predominantly mediated by the right PFC (Barrett and Gonzalez-Lima 2013).
2. ***Comparison of right vs. left TILS***. The second study ($n = 51$, ages 18–40) compared TILS effects on **attention bias modification** (ABM) learning. We used this study to directly compare right vs. left PFC stimulation. Participants were randomized to one of the three stimulation conditions: right forehead, left forehead, or sham (Disner et al. 2016). ABM is a cognitive intervention designed to improve depression symptoms by learning to decrease negative attentional bias. The right TILS led to greater symptom improvement among participants whose attention was directed away from negative stimuli. Minimal change was observed in the left and sham TILS, suggesting that the beneficial effects of ABM learning on depression symptoms may be enhanced when paired with right but not left PFC stimulation. This is consistent with the *right PFC being more dominant* in the human brain's attention-memory-executive network, which supports our rationale to continue with stimulation of the right PFC.

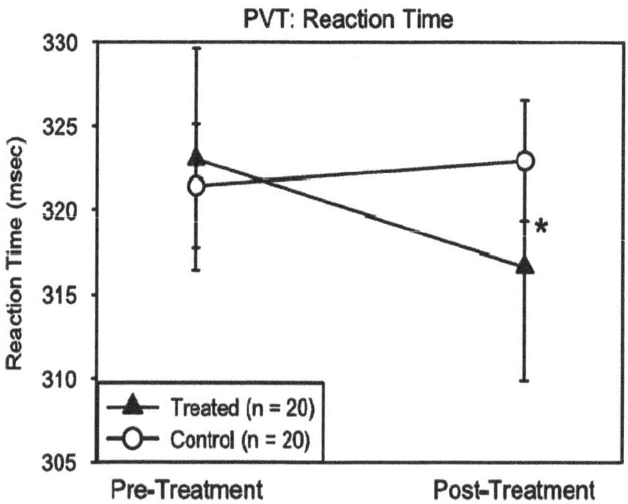

Fig. 10 Improved cognitive performance in the psychomotor vigilance test (PVT) (shorter reaction times) post TILS vs. placebo control group in healthy young adults. ($n = 40$, mean ± SE), $^*p < 0.05$ (Barrett and Gonzalez-Lima 2013)

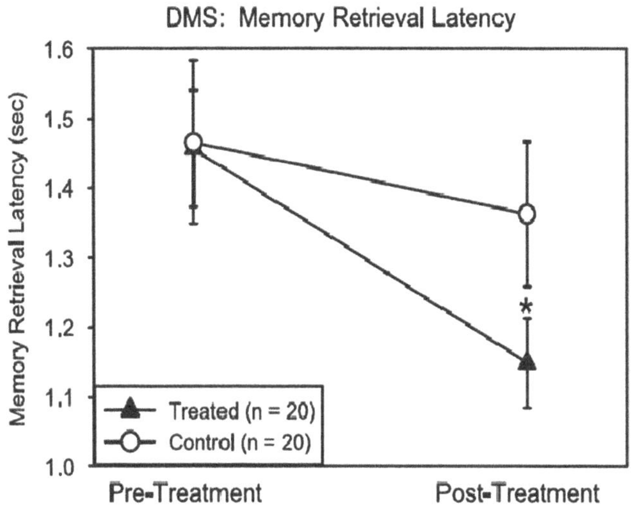

Fig. 11 Improved cognitive performance in the delayed match to sample test (DMS) of working memory (shorter memory retrieval latency) post-TILS vs. placebo control group in the same subjects used for Fig. 10

3. ***Executive skills***. In the third study (Blanco et al. 2015), we built on this research by studying the effects of TILS on executive function (n = 30, ages 18–40). Executive function is modulated by the PFC and plays a role in several cognitive skills including selective and divided attention, manipulation, task switching, and inhibition of interfering stimuli. Executive dysfunction is characteristic of cognitive aging and mild cognitive impairment (MCI) with deficits worsening as they develop into Alzheimer's disease and related dementias (Kirova et al. 2015). Executive function deficits can be detected by the **Wisconsin Card Sorting Task** (WCST), a gold-standard neuropsychological measure of executive function, and previous studies have shown poor performance in this task by individuals with healthy aging (Salthouse et al. 2003) as well as aging-related diseases such as Parkinson's disease (Monchi et al. 2004) and Alzheimer's disease (Binetti et al. 1996). Participants who received laser treatment aimed at the right PFC made significantly fewer errors and showed improved set-shifting ability relative to placebo controls (Blanco et al. 2015) (Fig. 12). These findings demonstrated the ability of TILS to enhance executive function in healthy people and its potential to improve executive function deficits in elderly and clinical populations. Thus, we will continue to use WCST to evaluate pre- vs. posttreatment effects of TILS on executive skills.
4. ***Comparison of TILS with exercise***. The fourth study (n = 60, ages 18–30) revealed that compared to placebo, **vigorous aerobic exercise and TILS resulted in the same improvement in cognitive performance in PVT and DMS tasks**. However, the effective vigorous aerobic exercise protocol involved 20 min running in a treadmill at 85–90% maximal oxygen consumption

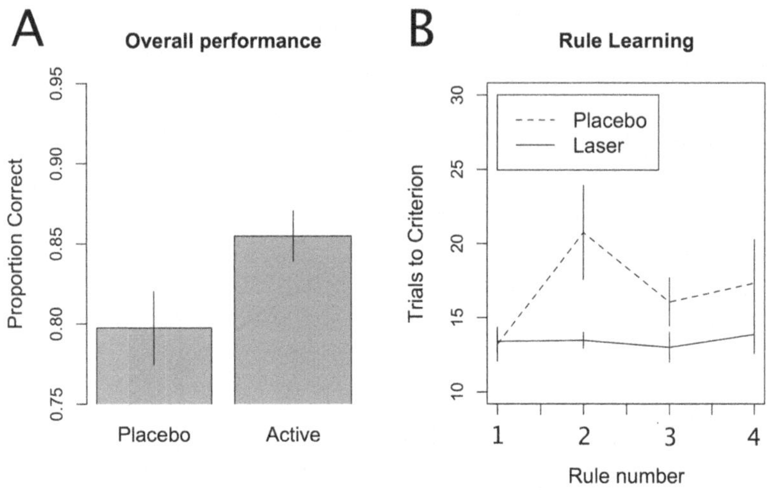

Fig. 12 (**a**) Overall WCST accuracy. The laser group correctly sorted the cards more often than the placebo group. (**b**) Trials to criterion for each of the first four rules learned. The placebo group took significantly longer to reach criterion on the second rule than the laser group, suggesting a benefit in set shifting ability in the laser group. Error bars represent standard errors (Blanco et al. 2015)

(VO_2max), while the laser treatment involved 8-min TILS to the right PFC while seating quietly (Hwang et al. 2016). Thus, it is expected that older adults and patients who may be unable to engage in vigorous aerobic exercise to improve cognition could benefit from TILS as an alternative intervention. We intend to answer this question in future studies.

5. ***Category learning***. In the fifth study (n = 118, ages 17–35) (Blanco et al. 2016), we tested the effects of TILS on category learning, an essential function of everyday life. The study tested two different types of category structures: rule-based and information-integration. **Rule-based category learning** is optimized by a reflective system of learning associated with PFC processing. Information-integration category learning is optimized by a reflexive system of learning associated with processing in the striatum. As hypothesized, participants in the PFC-TILS group had higher learning rates and improved performance in rule-based category learning tasks, but there was no significant effect in information-integration category learning tasks (Fig. 13). These results had two major implications: (1) additional evidence that application of TILS to the right PFC can improve PFC-mediated cognitive functions and (2) different forms of learning can be modulated by TILS without affecting others.
6. ***Overall cognitive rate correct score***. In the sixth study, 34 healthy adults (16 males, 18 females; average age: 31, and standard error: 2.5) were recruited (Holmes et al. 2019). The 18 experimental group participants (9 males, 9 females) received full laser stimulation and completed all tasks using a within-subject control design. They performed the PVT and DMS cognitive tasks before and after TILS, with concomitant fNIRS recordings, to reflect the hemodynamic

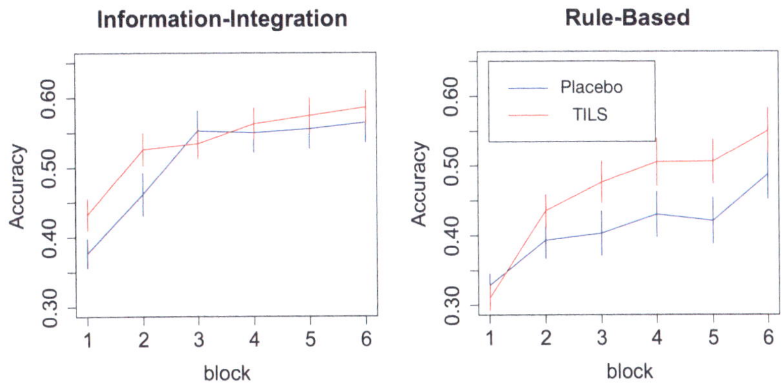

Fig. 13 Accuracy (proportion correct classification). In rule-based learning, the TILS group performed significantly ($p < 0.05$) better than the placebo group in each block after the first block, while information-integration did not show significant group differences ($n = 118$, mean ± SE) (Blanco et al. 2016)

effects of TILS on cognitive performance. Another 16 participants (7 males, 9 females) were matched blind to treatment as sham controls without photobiomodulation (TILS procedure used with light off). Cognitive data were measured in seven of the sham participants (3 males, 4 females) as previously described (Barrett and Gonzalez-Lima 2013). Hemodynamic data were measured in nine of the sham participants (four males, five females) as previously described (Wang et al. 2017). The hemodynamic results are shown below in the discussion of neurophysiological effects. For each subject's cognitive performance in both sessions of the PVT and DMS, an overall cognitive score was calculated, incorporating both speed (reaction time) and accuracy (number of correct responses), using the rate correct score or RCS (Woltz and Was 2006). This overall cognitive score was equal to the number of correct responses divided by the sum of all reaction times. Cognitive results from the PVT and DMS were obtained before and after TILS and sham. As expected, overall cognitive processing improved after TILS, as indicated by the significantly higher rate correct score (Fig. 14a), whereas there were no significant differences after sham (Fig. 14b). This score reflects the speed and accuracy of cognitive processing. The overall rate correct score effect size (Cohen's d) after TILS was $|d| = 0.62$, indicating a medium effect size, given that $|d| < 0.2$ = small effect; $0.2 < |d| < 0.8$ = medium effect; and $|d| > 0.8$ = large effect (Cohen 1988).

The behavioral sham data comparing pre-post tasks without TILS served to confirm the lack of learning/placebo effects for repeated testing of our specific DMS and PVT tasks and replicated the same sham findings from our published studies. The cognitive effects are consistent with our previous two studies in which we compared TILS with sham/placebo groups in the same PVT and DMS tasks (Barrett and Gonzalez-Lima 2013; Hwang et al. 2016). Indeed, pre-post testing of these PVT and DMS tasks in a single session without laser stimulation does not lead to

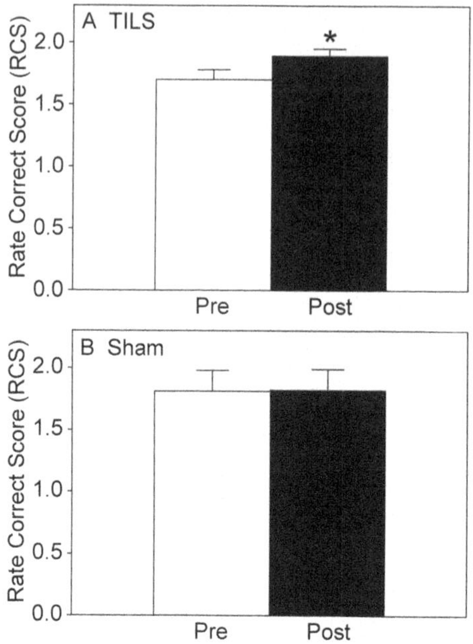

Fig. 14 Rate correct score (RCS) for cognitive performance before and after TILS (**a**) and Sham (**b**). Mean ± S.E., * = Significant mean difference between Pre- vs. Post- TILS scores, $p \leq 0.01$. * No randomization occurred for sham, since we used a within-subject design in which the same subjects are their own control by directly comparing pre- vs post- measures statistically. There were no significant pre-post differences in the sham subjects (Holmes et al. 2019)

improved cognitive scores (Barrett and Gonzalez-Lima 2013). Therefore, placebo or practice effects cannot explain the cognitive improvement produced by TILS. Overall, the results from these six placebo-controlled trials demonstrate the potential of TILS as a cognitive-enhancing approach.

3.6.1 Cognitive Benefits of TILS in Older Participants

Further, preliminary data with older participants also suggested beneficial neurocognitive effects of repeated TILS aimed at the right PFC. In a proof-of-concept open-label pilot study ($n = 21$, 11 females, ages 49–90), we investigated the feasibility of studying older participants using the proposed cognitive protocols ($n = 12$), with qEEG ($n = 6$) and fMRI ($n = 6$) (Vargas et al. 2017). PVT (sustained attention) and DMS (working memory) were conducted for 5 weeks. The first session measured baseline scores, and the second session investigated acute effects of TILS. Chronic effects were investigated after four additional weekly treatments with the proposed TILS. Repeated measure ANOVA using percent change from baseline showed significant ($p < 0.05$) effects: reduced PVT reaction time (Fig. 15a) and failures to respond (Fig. 15b), while increased correct memory retrieval responses in DMS (Fig. 15c). Detailed inspection of the data verified that all older participants improved cognitive performance after single and repeated treatments.

Younger participants had reaction times and correct responses that were consistently better than those of the older cohort, and placebo-treated subjects showed no improvements. However, the relative effect size of TILS in this older group was larger than that in our previous controlled studies with younger participants. Effect sizes (Cohen's D) for younger (18–35) vs. older (49–90) were 0.21 vs. 0.65 for PVT

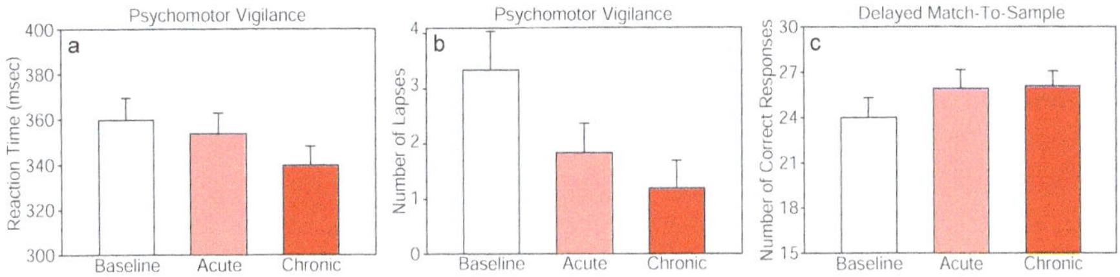

Fig. 15 Cognitive effects of TILS on attention and memory in older adults. (**a**) Reaction time; (**b**) number of lapses (trials in which the subject failed to respond within 500 ms of the stimulus) for the PVT; and (**c**) number of correct responses for the DMS. Bars show group means plus standard errors, $n = 12$ (Vargas et al. 2017)

reaction time, 0.67 vs. 1.05 for PVT lapses, and 0.26 vs. 0.53 for DMS correct responses. For example, the number of trials in which older subjects failed to respond within 500 ms of the stimulus reduced by 55% after a single laser treatment and by 74% after four additional laser treatments. Laser-induced memory improvement was also found in the DMS test. But several participants below 56 (median age of cohort) reached ceiling effects in the correct number of responses after three weeks of treatment. In contrast, the oldest participants (one man and one woman, each 90 years old) continued to improve over weeks, with a mean 15% increase in their number of correct memory retrieval responses as compared to 11% improvement for participants in their 50s. Based on these results, we plan to avoid ceiling effects in future studies by administering two DMS tests, one easier (4-s delay) and one more difficult (8-s delay). The longer the delay before memory retrieval, the more the difficulty involved in this working memory task. In this way, the more difficult task will allow us to evaluate if there are further memory improvements in participants showing ceiling effects in the easier task after repeated laser treatments.

3.7 Neurophysiological Effects of TILS of the Human Prefrontal Cortex

We published the first six placebo-controlled neurophysiological studies (99 participants, 44 females, ages 18–45) demonstrating the effects of TILS on metabolic, hemodynamic, and electrophysiological indices using fNIRS (Tian et al. 2016; Holmes et al. 2019), bbNIRS (Wang et al. 2017), qEEG (Wang et al. 2017a, 2019), and fMRI (Vargas et al. 2017).

1. ***Near-infrared spectroscopy (NIRS) studies***. Transcranial infrared laser stimulation of the human prefrontal cortex causes cognitive enhancement. To investigate the hemodynamic effects in prefrontal cortex by which this cognitive enhancement occurs, we used near-infrared spectroscopy (NIRS), which is a safe, noninvasive method of monitoring hemodynamics. We measured concentration changes in oxygenated and deoxygenated hemoglobin, total hemoglobin, and differential effects in healthy adults during resting conditions and during sustained attention and working memory performance, before and after laser of the

right prefrontal cortex. We also measured sham controls without photobiomodulation. NIRS involves optical imaging methods using two-wavelength light (**fNIRS**) or broadband light (**bbNIRS**), which have been developed for mapping human brain functions and functional networks (Hoshi 2007; Ferrari and Quaresima 2012; Boas et al. 2014; Zeff et al. 2007; Eggebrecht et al. 2014; Niu et al. 2012; Niu and He 2014). We conducted two studies using fNIRS and two studies using bbNIRS. The principle of fNIRS is that light absorption of oxy- and deoxy-hemoglobin (HbO and HHb) will vary with cerebral functions, similar to the principle for blood-oxygen-level-dependent (BOLD) signals used in fMRI, but BOLD-fMRI cannot separate HbO and HHb. Utilization of two wavelengths in fNIRS is adequate for quantification of changes (Δ) in concentrations of HbO and HHb (and then $\Delta HbT = \Delta HbO + \Delta HHb$ and $\Delta HbD = \Delta HbO - \Delta HHb$) (Hoshi 2007; Ferrari and Quaresima 2012; Boas et al. 2014; Zeff et al. 2007; Eggebrecht et al. 2014; Niu et al. 2012; Niu and He 2014). TILS is not due to water heating (Wang et al. 2018) or increased circulation (i.e., both HbO and HHb increase). Tian et al. (2016) employed fNIRS and for the first time reported that **TILS increased human cerebral oxygenation (i.e., more HbO but less HHb)** in healthy adults under resting conditions, $n = 18$ (Tian et al. 2016).

In a second fNIRS study, we investigated another 18 participants under cognitive activation. fNIRS revealed large effects on prefrontal oxygenation during cognitive enhancement postlaser and provided the first demonstration that cognitive enhancement by transcranial photobiomodulation is associated with cerebrovascular oxygenation of the prefrontal cortex. Sham control data served to rule out that the laser effects were due to pre-post task repetition or other nonspecific effects. The effects of TILS on the fNIRS signal are illustrated in Figs. 16 and 17, in terms of the temporal and spatial effects of TILS, respectively. The level of oxygenated hemoglobin rose significantly after TILS and maintained a higher

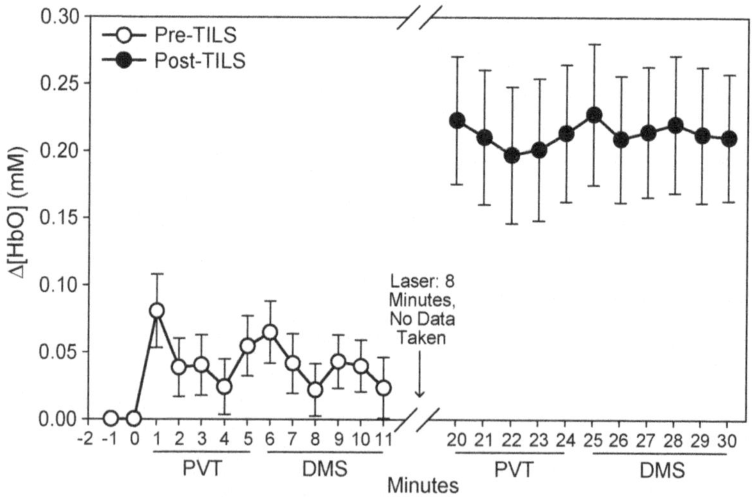

Fig. 16 Temporal effects of TILS on human cerebral oxygenation. Temporal sequence of changes in oxygenated hemoglobin (μM), pre- and postTILS during cognitive processing in the PVT and DMS tasks. Mean ± S.E (Holmes et al. 2019)

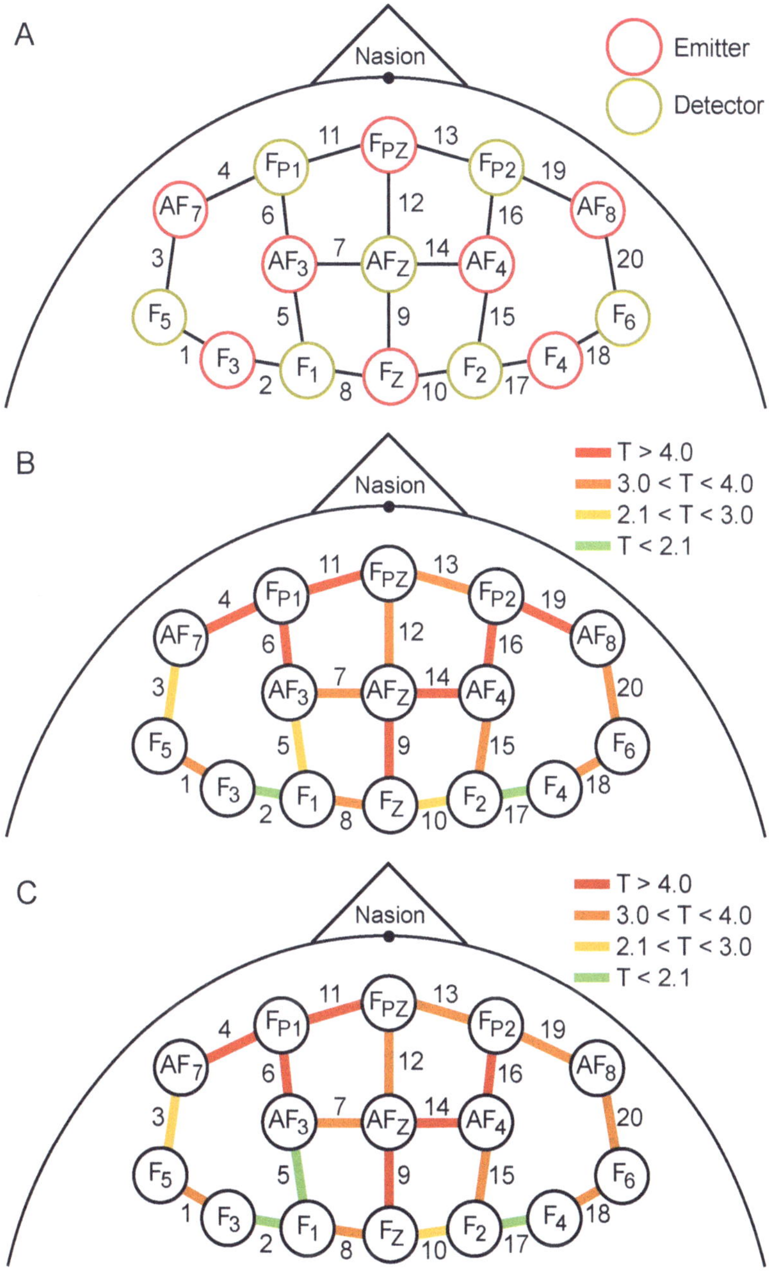

Fig. 17 Spatial effects of TILS on human prefrontal cortex. (**a**) Schematic spatial layout of the frontal montage used for fNIRS recording. Red and yellow circles correspond to emitters and detectors, respectively. The labels inside the circles correspond to the 10–20 EEG system locations used for each emitter and detector. Each adjacent emitter/detector pair formed a channel that gathered fNIRS data (20 labeled lines). (**b**) Colored lines represent the magnitude of *T*-scores from the analysis of mean differences between pre- and postTILS PVT epochs. (**c**) The magnitude of *T*-scores comparing the pre- and postTILS DMS epochs. The affected channels are almost identical, with two exceptions (channels 5 and 19) (Holmes et al. 2019)

level throughout the postTILS period (Fig. 16). The laser treatment resulted in a large increase in oxygenated hemoglobin in the anterior frontal region measured by fNIRS, corresponding to the prefrontal cortex region engaged during the PVT and DMS tasks (Nieder and Miller 2004; Drummond et al. 2005). This large hemodynamic response was about five times greater than that in the preTILS condition, and it was sustained for more than 10 min of cognitive processing after TILS.

Comparing differences between pre- and postTILS in terms of spatial effects, the anterior channels in the fNIRS frontal montage (Fig. 17a) registered higher levels of oxygenated hemoglobin than the posterior channels during cognitive processing of PVT (Fig. 17b) and DMS (Fig. 17c).

It has also become evident that bbNIRS is able to accurately quantify changes in cerebral oxidized cytochrome oxidase (Tachtsidis et al. 2010; Papademetriou et al. 2012; Kolyva et al. 2012, 2014). In a third study, Wang et al. (2017) implemented a bbNIRS system (Fig. 18) and showed in vivo that **cytochrome oxidase increase by 1064-nm laser precedes by minutes the increase in HbO and the decrease in HHb** in forearms (n = 11) (Wang et al. 2016) and PFC (n = 11) (Wang et al. 2017). In addition, oxidized cytochrome oxidase in PFC is greater than placebo by 2 min after onset of TILS and continues to increase until 8 min of stimulation (Fig. 19) (Wang et al. 2017).

In conclusion, TILS at 1064 nm wavelength enhanced cognitive performance and both fNIRS and bbNIRS responses in human prefrontal cortex. Altogether, the results of the NIRS studies and our previous cognitive studies suggest that TILS is a form of photobiomodulation that can successfully augment cerebrovascular oxygenation and thereby improve human cognitive brain functions.

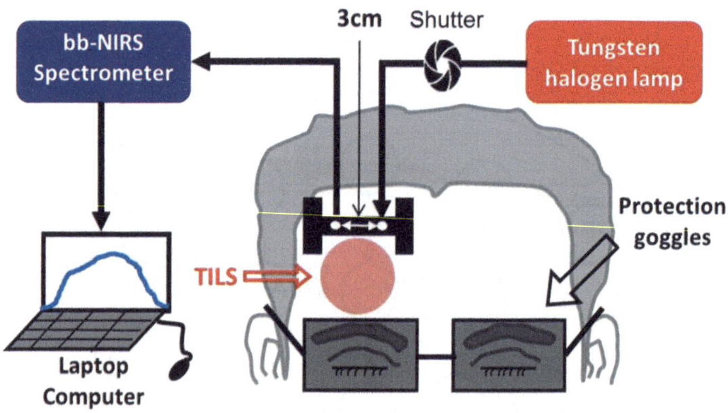

Fig. 18 Schematic of bbNIRS setup. This consisted of a tungsten halogen lamp as the light source and a miniature back-thinned CCD spectrometer as the detector. TILS (red circle) was administered underneath the black probe holder. A laptop computer was used to acquire, display, and save the data from the spectrometer. The shutter controlled the on and off function for the white light from the tungsten-halogen lamp to the subject's forehead. A pair of protection goggles was worn during the whole procedure (Wang et al. 2017)

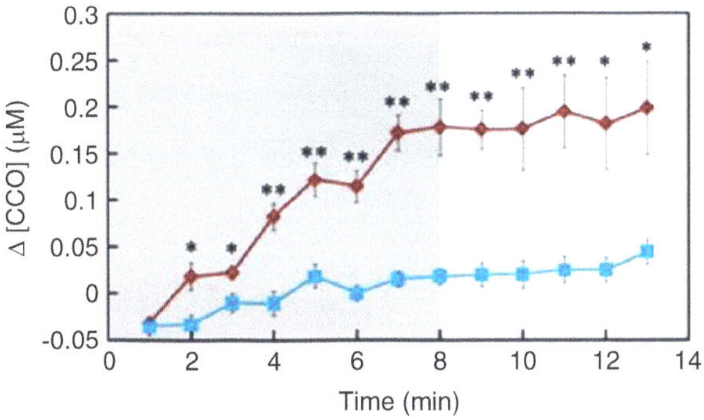

Fig. 19 TILS increases cytochrome oxidase in the human prefrontal cortex. Subject-averaged time courses of TILS (red) vs. placebo (blue) changes (Δ) of oxidized [cytochrome oxidase = CCO] recorded from human foreheads in vivo (mean ± SE, $n = 11$). Time zero ($t = 0$) is the onset of TILS. "*" $0.01 < p < 0.05$ and "**" $p < 0.01$, two sample t-test between TILS and placebo (Wang et al. 2017)

2. ***Electroencephalographic (EEG) studies***. We conducted EEG studies using a 64-channel qEEG analysis before, during, and after TILS or placebo, $n = 20$ (Wang et al. 2017a, 2019). A gradual and strong increase of power density in the Alpha band (8–13 Hz) and smaller effects in Beta (13–30 Hz) and Gamma (32+ Hz) bands were observed across time during TILS to right PFC (Fig. 20). Using a novel analysis method and open-source software, eLORETA (Aoki et al. 2015), Dr. Xinlong Wang and Dr. Hanli Liu developed a new spatiotemporal image processing algorithm to estimate TILS-induced EEG responses in time and spectral frequency and focused on specific frequency bands to estimate and reconstruct source localizations induced by TILS (Fig. 21). The results are very promising because they revealed that (a) it takes minutes for the electrophysiological signals to respond (much different from transcranial direct current or magnetic stimulation, tDCS or TMS) and this change manifested after approximately 2 min and peaked by 8 (matching cytochrome oxidase upregulation), (b) the neuromodulation propagated from frontal to posterior cortical regions, (c) TILS modulated ipsilateral activity at the right frontal cortex, while also modulating bilateral activity at the posterior cortex, and (d) TILS modulated an ipsilateral, fronto-parieto-occipital network and a contralateral, parieto-occipital network at the Alpha frequency.
3. ***fMRI studies***. BOLD fMRI signal is a combined output of *decreased* oxygen consumption and increased cerebral blood flow (CBF) and blood volume. Since TILS *increases* oxygen consumption in the PFC, as shown in our previous studies (Rojas et al. 2008a, b, 2012; Wang et al. 2017), a reduced BOLD-fMRI signal was expected. BOLD-fMRI responses to a working memory 2-back task were measured in six participants before and after TILS (Vargas et al. 2017). As predicted, TILS-induced cognitive improvement was accompanied by reduced task-evoked BOLD-fMRI signal in right PFC relative to its own baseline, suggesting that TILS-facilitated oxygen consumption promotes more *efficient* PFC function

Fig. 20 TILS increases EEG power density. Time-frequency heat plot of TILS-induced power density difference with respect to placebo. Data were taken from an electrode near the TILS site and averaged over 20 subjects. TILS started at time = 0 and ended at 11 (red line on top) (Wang et al. 2017a)

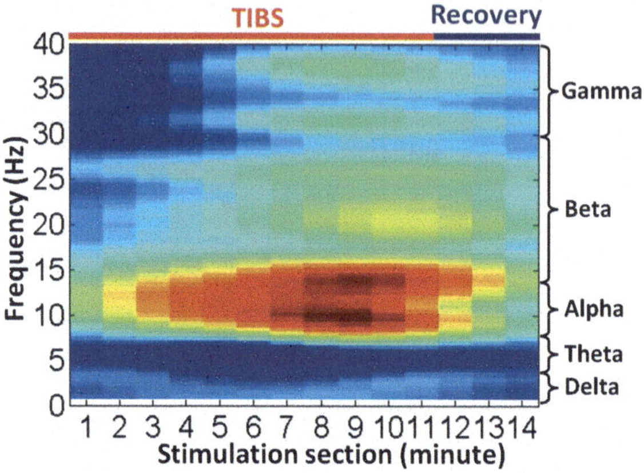

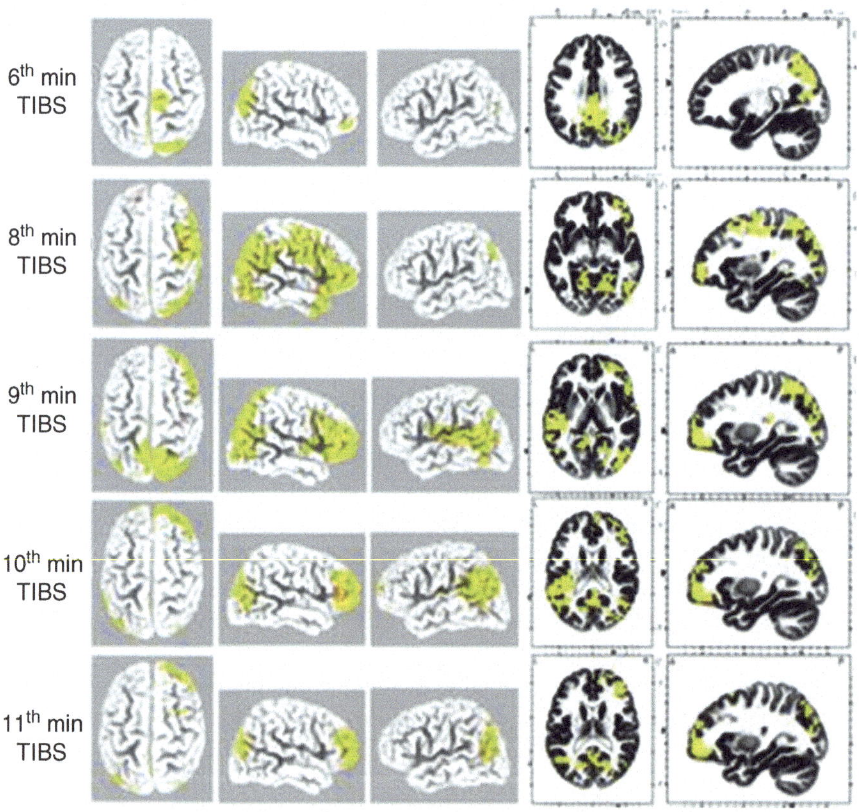

Fig. 21 TILS modulates a large-scale fronto-parieto-occipital network. 3D views of t-maps (yellow = $p < 0.01$) between TILS and placebo conditions based on Alpha-band (8–13 Hz) power density at 6, 8, 9, 10, and 11 min after right PFC TILS started ($n = 20$) (Wang et al. 2017a)

that enhances cognitive processing. This is consistent with higher cognitive performing older adults showing reduced CBF in right PFC and bilateral neural plasticity (Cabeza et al. 2002). We will conduct more fMRI studies to develop a high-resolution spatial map of the network effects of TILS in the brain. Future fMRI studies of TILS will follow the same PVT and DMS protocols of two recent fMRI studies of methylene blue that we coauthored using BOLD, CBF, and functional connectivity (Rodriguez et al. 2016a, b).

In conclusion, the neurophysiological studies identified a large-scale attention-memory-executive network that corresponds to the network modulated by TILS, including associative areas of frontal, posterior parietal, and occipital cortices, reliably associated with tasks that require high-sustained attention-working memory-executive demands among younger and older adults (Haley et al. 2007a, b, 2008, 2010; Sweet et al. 2008). Our combined cognitive and neurophysiological studies support the hypothesis that TILS of the right PFC modulates a large-scale fronto-parieto-occipital network that promotes cognitive enhancement of attention-memory-executive functions.

4 Conclusions

The studies presented demonstrate that photobiomodulation may be translated from animals to humans to become a noninvasive, safe, nonpharmacological, and cost-effective method for augmentation of brain functions. Noninvasive methods such as fNIRS, bbNIRS, and EEG may be able to monitor hemodynamic and electrophysiological effects on human brains in real time. Combining photobiomodulation with such neurophysiological monitoring allows for calibration of the laser dose for individuals based on their own particular cerebrovascular response, as a new form of "precision medicine". This approach might also permit identifying potential candidates for photobiomodulation before clinical presentation of neurodegenerative effects. This approach is particularly promising for populations with prefrontal hypometabolism, such as in cognitive aging, mild cognitive impairment, Alzheimer's dementia, and many other neurological and psychiatric conditions that lead to neurocognitive decline (Rojas and Gonzalez-Lima 2013).

Acknowledgements I thank all my students and collaborators who contributed to the studies presented in this chapter. They are listed as my coauthors in the cited references to our original papers. In particular, I thank Dr. Douglas W. Barrett, my lab manager, who worked with me in the human studies. I thank my former trainee Dr. Julio C. Rojas for his leading role in the animal studies. I thank Dr. Nathaniel J. Blanco and Dr. Jungyun Hwang for their contributions to the cognitive studies. I thank my engineering collaborators for their critical contributions, especially Dr. Hanli Liu, Dr. Fenghua Tian, Dr. Xinlong Wang, and Dr. Lida Huang. I also thank my current trainees who are advancing this exciting field in ongoing studies. This work was supported by grants from the National Institutes of Health and from the Oskar Fischer Project Fund.

References

Abu-Amero KK, Morales J, Bosley TM (2006) Mitochondrial abnormalities in patients with primary open-angle glaucoma. Invest Ophthalmol Vis Sci 47:2533–2541

Alberts B (2002) Molecular biology of the cell, 4th edn. Garland Science, New York

Ames A III, Li YY, Heher EC, Kimble CR (1992) Energy metabolism of rabbit retina as related to function: high cost of Na^+ transport. J Neurosci 12:840–853

Anders JJ, Moges H, Wu X, Erbele ID, Alberico SL, Saidu EK et al (2014) In vitro and in vivo optimization of infrared laser treatment for injured peripheral nerves. Lasers Surg Med 46(1):34–45

Aoki Y, Ishii R, Pascual-Marqui RD, Canuet L, Ikeda S, Hata M et al (2015) Detection of EEG-resting state independent networks by eLORETA-ICA method. Front Hum Neurosci 9:31

Astrup J, Sorensen PM, Sorensen HR (1981) Oxygen and glucose consumption related to Na^+-K^+ transport in canine brain. Stroke 12:726–730

Barrett DW, Gonzalez-Lima F (2013) Transcranial infrared laser stimulation produces beneficial cognitive and emotional effects in humans. Neuroscience 230:13–23

Beretta S, Wood JP, Derham B, Sala G, Tremolizzo L, Ferrarese C, Osborne NN (2006) Partial mitochondrial complex I inhibition induces oxidative damage and perturbs glutamate transport in primary retinal cultures. Relevance to Leber Hereditary Optic Neuropathy (LHON). Neurobiol Dis 24:308–317

Binetti G, Magni E, Padovani A, Cappa SF, Bianchetti A, Trabucchi M (1996) Executive dysfunction in early Alzheimer's disease. J Neurol Neurosurg Psychiatry 60(1):91–93

Blanco NJ, Maddox WT, Gonzalez-Lima F (2015) Improving executive function using transcranial infrared laser stimulation. J Neuropsychol 11(1):14–25

Blanco NJ, Saucedo CL, Gonzalez-Lima F (2016) Transcranial infrared laser stimulation improves rule-based, but not information-integration, category learning in humans. Neurobiol Learn Mem 139:69–75

Boas DA, Elwell CE, Ferrari M, Taga G (2014) Twenty years of functional near-infrared spectroscopy: introduction for the special issue. NeuroImage 85:1–5

Borit A (1971) Leigh's necrotizing encephalomyelopathy. neuro-ophthalmological abnormalities. Arch Ophthalmol 85(4):438–442

Brondon P, Stadler I, Lanzafame RJ (2005) A study of the effects of phototherapy dose interval on photobiomodulation of cell cultures. Lasers Surg Med 36:409–413

Buford TW, Roberts MD, Church TS (2013) Toward exercise as personalized medicine. Sports Med 43(3):157–165

Byrnes KR, Wu X, Waynant RW, Ilev IK, Anders JJ (2005a) Low power laser irradiation alters gene expression of olfactory ensheathing cells in vitro. Lasers Surg Med 37:161–171

Byrnes KR, Waynant RW, Ilev IK, Wu X, Barna L, Smith K, Heckert R, Gerst H, Anders JJ (2005b) Light promotes regeneration and functional recovery and alters the immune response after spinal cord injury. Lasers Surg Med 36:171–185

Cabeza R, Anderson ND, Locantore JK, McIntosh AR (2002) Aging gracefully: compensatory brain activity in high-performing older adults. NeuroImage 17(3):1394–1402

Calabrese EJ (2008) Hormesis: principles and applications for pharmacology and toxicology. Am J Pharmacol Toxicol 3:59–71

Calabrese EJ, Baldwin LA (2003) Toxicology rethinks its central belief. Nature 421:691–692

Callaway NL, Riha PD, Wrubel KM, McCollum D, Gonzalez-Lima F (2002) Methylene blue restores spatial memory retention impaired by an inhibitor of cytochrome oxidase in rats. Neurosci Lett 332(2):83–86

Callaway NL, Riha PD, Bruchey AK, Munshi Z, Gonzalez-Lima F (2004) Methylene blue improves brain oxidative metabolism and memory retention in rats. Pharmacol Biochem Behav 77(1):175–181

Carelli V, Ross-Cisneros FN, Sadun AA (2002) Optic nerve degeneration and mitochondrial dysfunction: genetic and acquired optic neuropathies. Neurochem Int 40:573–584

Carelli V, La Morgia C, Valentino ML, Barboni P, Ross-Cisneros FN, Sadun AA (2009) Retinal ganglion cell neurodegeneration in mitochondrial inherited disorders. Biochim Biophys Acta 1787(5):518–528
Casari G, De Fusco M, Ciarmatori S, Zeviani M, Mora M, Fernandez P, De Michele G, Filla A, Cocozza S, Marconi R, Dürr A, Fontaine B, Ballabio A (1998) Spastic paraplegia and OXPHOS impairment caused by mutations in paraplegin, a nuclear-encoded mitochondrial metalloprotease. Cell 93(6):973–983
Chalmers RM, Schapira AH (1999) Clinical, biochemical and molecular genetic features of Leber's hereditary optic neuropathy. Biochim Biophys Acta 1410:147–158
Chinnery PF, Howell N, Lightowlers RN, Turnbull DM (1997) Molecular pathology of MELAS and MERRF. The relationship between mutation load and clinical phenotypes. Brain 120(Pt 10):1713–1721
Cohen J (1988) Statistical power analysis for the behavioural sciences, 2nd edn. Lawrence Erlbaum Associates, New York
Conlan MJ, Rapley JW, Cobb CM (1996) Biostimulation of wound healing by low-energy laser irradiation. A review. J Clin Periodontol 23:492–496
Danesh-Meyer HV, Birch H, Ku JY, Carroll S, Gamble G (2006) Reduction of optic nerve fibers in patients with Alzheimer disease identified by laser imaging. Neurology 67:1852–1854
de la Torre JC (2017) Treating cognitive impairment with transcranial low level laser therapy. J Photochem Photobiol B Biol 168:149–155
Dhar SS, Wong-Riley MT (2009) Coupling of energy metabolism and synaptic transmission at the transcriptional level: role of nuclear respiratory factor 1 in regulating both cytochrome c oxidase and NMDA glutamate receptor subunit genes. J Neurosci 29(2):483–492
Dhar SS, Liang HL, Wong-Riley MT (2009) Nuclear respiratory factor 1 co-regulates AMPA glutamate receptor subunit 2 and cytochrome c oxidase: tight coupling of glutamatergic transmission and energy metabolism in neurons. J Neurochem 108(6):1595–1606
DiMauro S (1999) Mitochondrial encephalomyopathies: back to Mendelian genetics. Ann Neurol 45(6):693–694
Disner SG, Beevers CG, Gonzalez-Lima F (2016) Transcranial laser stimulation as neuroenhancement for attention bias modification in adults with elevated depression symptoms. Brain Stimul 9(5):780–787
Drummond S, Bischoff-Grethe A, Dinges DF, Ayalon L, Mednick SC, Meloy MJ (2005) The neural basis of the psychomotor vigilance task. Sleep 28:1059–1068
Duan R, Zhu L, Liu TC, Li Y, Liu J, Jiao J, Xu X, Yao L, Liu S (2003) Light emitting diode irradiation protect against the amyloid beta 25–35 induced apoptosis of PC12 cell in vitro. Lasers Surg Med 33:199–203
Eells JT, Henry MM, Summerfelt P, Wong-Riley MT, Buchmann EV, Kane M, Whelan NT, Whelan HT (2003) Therapeutic photobiomodulation for methanol-induced retinal toxicity. Proc Natl Acad Sci U S A 100:3439–3444
Eells JT, Wong-Riley MT, VerHoeve J, Henry M, Buchman EV, Kane MP, Gould LJ, Das R, Jett M, Hodgson BD, Margolis D, Whelan HT (2004) Mitochondrial signal transduction in accelerated wound and retinal healing by near-infrared light therapy. Mitochondrion 4:559–567
Eggebrecht AT, Ferradal SL, Robichaux-Viehoever A, Hassanpour MS, Dehghani H, Snyder AZ et al (2014) Mapping distributed brain function and networks with diffuse optical tomography. Nat Photonics 8(6):448–454
Etnier JL, Salazar W, Landers DM, Petruzzello SJ, Han M, Nowell P (1997) The influence of physical fitness and exercise upon cognitive functioning: a meta-analysis. J Sport Exerc Psychol 19(3):249–277
Fan W, Agarwal N, Cooper NG (2006) The role of CaMKII in BDNF-mediated neuroprotection of retinal ganglion cells (RGC-5). Brain Res 1067:48–57
Ferrari M, Quaresima V (2012) A brief review on the history of human functional near-infrared spectroscopy (fNIRS) development and fields of application. NeuroImage 63(2):921–935
Fulop AM, Dhimmer S, Deluca JR, Johanson DD, Lenz RV, Patel KB et al (2010) A meta-analysis of the efficacy of laser phototherapy on pain relief. Clin J Pain 26(8):729–736

Gaugler JE, Yu F, Krichbaum K, Wyman JF (2009) Predictors of nursing home admission for persons with dementia. Med Care 47(2):191–198
Gonzalez-Lima F, Barrett DW (2014) Augmentation of cognitive brain functions with transcranial lasers. Front Syst Neurosci 8:36
Gonzalez-Lima F, Cada A (1998) Quantitative histochemistry of cytochrome oxidase activity: theory, methods, and regional brain vulnerability. In: Gonzalez-Lima F (ed) Cytochrome oxidase in neuronal metabolism and Alzheimer's disease. Plenum Press, New York, pp 55–90
Gonzalez-Lima F, Valla J, Matos-Collazo S (1997) Quantitative cytochemistry of cytochrome oxidase and cellular morphometry of the human inferior colliculus in control and Alzheimer's patients. Brain Res 752:117–126
Gonzalez-Lima F, Barksdale BR, Rojas JC (2014) Mitochondrial respiration as a target for neuroprotection and cognitive enhancement. Biochem Pharmacol 88:584–593
Hacke W, Schellinger PD, Albers GW, Bornstein NM, Dahlof BL, Fulton R et al (2014) Transcranial laser therapy in acute stroke treatment: results of neurothera effectiveness and safety trial 3, a phase III clinical end point device trial. Stroke 45(11):3187–3193
Haley AP, Sweet LH, Gunstad J, Forman DE, Poppas A, Paul RH et al (2007a) Verbal working memory and atherosclerosis in patients with cardiovascular disease: an fMRI study. J Neuroimaging 17(3):227–233
Haley AP, Forman DE, Poppas A, Hoth KF, Gunstad J, Jefferson AL et al (2007b) Carotid artery intima-media thickness and cognition in cardiovascular disease. Int J Cardiol 121(2):148–154
Haley AP, Gunstad J, Cohen RA, Jerskey BA, Mulligan RC, Sweet LH (2008) Neural correlates of visuospatial working memory in healthy young adults at risk for hypertension. Brain Imaging Behav 2(3):192–199
Haley AP, Tarumi T, Gonzales MM, Sugawara J, Tanaka H (2010) Subclinical atherosclerosis is related to lower neuronal viability in middle-aged adults: a 1 H MRS study. Brain Res 1344:54–61
Hamblin MR, Demidova TN (2006) Mechanisms of low level light therapy. Proc SPIE 6140:1–12
Hatefi Y (1985) The mitochondrial electron transport and oxidative phosphorylation system. Annu Rev Biochem 54:1015–1069
He Y, Leung KW, Zhang YH, Duan S, Zhong XF, Jiang RZ, Peng Z, Tombran-Tink J, Ge J (2008) Mitochondrial complex I defect induces ROS release and degeneration in trabecular meshwork cells of POAG patients: protection by antioxidants. Invest Ophthalmol Vis Sci 49:1447–1458
He W, Goodkind D, Kowal P (2016) An aging world: 2015 US Census Bureau, international population reports. US Government Publishing Office, Washington, DC
Hinton DR, Sadun AA, Blanks JC, Miller CA (1986) Optic-nerve degeneration in Alzheimer's disease. N Engl J Med 315:485–487
Holmes E, Barrett DW, Saucedo CL, O'Connor P, Liu H, Gonzalez-Lima F (2019) Cognitive enhancement by transcranial photobiomodulation is associated with cerebrovascular oxygenation of the prefrontal cortex. Front Neurosci 13:1129
Hoshi Y (2007) Functional near-infrared spectroscopy: current status and future prospects. J Biomed Opt 12(6):062106
Hwang JM, Park HW, Kim SJ (1997) Optic neuropathy associated with mitochondrial tRNA[Leu(UUR)] A3243G mutation. Ophthalmic Genet 18:101–105
Hwang J, Castelli DM, Gonzalez-Lima F (2016) Cognitive enhancement by transcranial laser stimulation and acute aerobic exercise. Lasers Med Sci 31(6):1151–1160
Iijima K, Shimoyama N, Shimoyama M, Mizuguchi T (1991) Evaluation of analgesic effect of low-power He:Ne laser on postherpetic neuralgia using VAS and modified McGill pain questionnaire. J Clin Laser Med Surg 9:121–126
Iseri PK, Altinas O, Tokay T, Yuksel N (2006) Relationship between cognitive impairment and retinal morphological and visual functional abnormalities in Alzheimer disease. J Neuroophthalmol 26:18–24
Jester JV, Moller-Pedersen T, Huang J, Sax CM, Kays WT, Cavangh HD, Petroll WM, Piatigorsky J (1999) The cellular basis of corneal transparency: evidence for 'corneal crystallins'. J Cell Sci 112(Pt 5):613–622
Kaiser J (2003) Hormesis. Sipping from a poisoned chalice. Science 302:376–379

Kann O, Kovacs R (2007) Mitochondria and neuronal activity. Am J Physiol Cell Physiol 292:C641–C657
Karu T (1989) Laser biostimulation: a photobiological phenomenon. J Photochem Photobiol B 3:638–640
Karu T (1999) Primary and secondary mechanisms of action of visible to near-IR radiation on cells. J Photochem Photobiol B 49:1–17
Karu T (2000) Mechanisms of low-power laser light action on cellular level. Proc SPIE 4159:1–19
Kirova AM, Bays RB, Lagalwar S (2015) Working memory and executive function decline across normal aging, mild cognitive impairment, and Alzheimer's disease. Biomed Res Int 2015:1–9
Kolyva C, Tachtsidis I, Ghosh A, Moroz T, Cooper CE, Smith M, Elwell CE (2012) Systematic investigation of changes in oxidized cerebral cytochrome c oxidase concentration during frontal lobe activation in healthy adults. Biomed Opt Express 3(10):2550–2566
Kolyva C, Ghosh A, Tachtsidis I, Highton D, Cooper CE, Smith M, Elwell CE (2014) Cytochrome c oxidase response to changes in cerebral oxygen delivery in the adult brain shows higher brain-specificity than haemoglobin. NeuroImage 85:234–244
Kubota S, Yang JT (1984) Bis[cyclo(histidylhistidine)]copper(II) complex that mimicks the active center of superoxide dismutase has its catalytic activity. Proc Natl Acad Sci U S A 81:3283–3286
Lampl Y, Zivin JA, Fisher M, Lew R, Welin L, Dahlof B et al (2007) Infrared laser therapy for ischemic stroke: a new treatment strategy: results of the NeuroThera Effectiveness and Safety Trial-1 (NEST-1). Stroke 38(6):1843–1849
Lapchak PA, Wei J, Zivin JA (2004) Transcranial infrared laser therapy improves clinical rating scores after embolic strokes in rabbits. Stroke 35:1985–1988
Levin LA (2007) Mechanisms of retinal ganglion specific-cell death in Leber hereditary optic neuropathy. Trans Am Ophthalmol Soc 105:379–391
Liang HL, Whelan HT, Eells JT, Meng H, Buchmann E, Lerch-Gaggl A, Wong-Riley M (2006) Photobiomodulation partially rescues visual cortical neurons from cyanide-induced apoptosis. Neuroscience 139:639–649
Liang HL, Whelan HT, Eells JT, Wong-Riley MT (2008) Near-infrared light via light-emitting diode treatment is therapeutic against rotenone- and 1-methyl-4-phenylpyridinium ion-induced neurotoxicity. Neuroscience 153:963–974
Monchi O, Petrides M, Doyon J, Postuma RB, Worsley K, Dagher A (2004) Neural bases of set-shifting deficits in Parkinson's disease. J Neurosci 24(3):702–710
Nieder A, Miller E (2004) A parieto-frontal network for visual numerical information in the monkey. Proc Natl Acad Sci U S A 101:7457–7462
Niu H, He Y (2014) Resting-state functional brain connectivity: lessons from functional near-infrared spectroscopy. Neuroscientist 20(2):173–188
Niu H, Wang J, Zhao T, Shu N, He Y (2012) Revealing topological organization of human brain functional networks with resting-state functional near infrared spectroscopy. PLoS One 7(9):e45771
Ojaimi J, Masters CL, Opeskin K, McKelvie P, Byrne E (1999) Mitochondrial respiratory chain activity in the human brain as a function of age. Mech Ageing Dev 111(1):39–47
Papademetriou MD, Tachtsidis I, Elliot MJ, Hoskote A, Elwell CE (2012) Multichannel near infrared spectroscopy indicates regional variations in cerebral autoregulation in infants supported on extracorporeal membrane oxygenation. J Biomed Opt 17(6):067008
Pastore D, Greco M, Passarella S (2000) Specific helium-neon laser sensitivity of the purified cytochrome c oxidase. Int J Radiat Biol 76:863–870
Ren JC, Rebrin I, Klichko V, Orr WC, Sohal RS (2010) Cytochrome c oxidase loses catalytic activity and structural integrity during the aging process in Drosophila melanogaster. Biochem Biophys Res Commun 401(1):64–68
Rochkind S, Vogler I, Barr-Nea L (1990) Spinal cord response to laser treatment of injured peripheral nerve. Spine 15:6–10
Rodriguez P, Zhou W, Barrett DW, Altmeyer W, Gutierrez JE, Li J et al (2016a) Multimodal randomized functional MR imaging of the effects of methylene blue in the human brain. Radiology 281(2):516–526

Rodriguez P, Singh AP, Malloy KE, Zhou W, Barrett DW, Franklin CG et al (2016b) Methylene blue modulates functional connectivity in the human brain. Brain Imaging Behav 11(3):640–648
Rojas JC, Gonzalez-Lima F (2010) Mitochondrial optic neuropathy: in vivo model of neurodegeneration and neuroprotective strategies. Eye Brain 2:21–37
Rojas JC, Gonzalez-Lima F (2011) Low-level light therapy of the eye and brain. Eye Brain 3:49–67
Rojas JC, Gonzalez-Lima F (2013) Neurological and psychological applications of transcranial lasers and LEDs. Biochem Pharmacol 86(4):447–457
Rojas JC, Gonzalez-Lima F (2017) Transcranial low-level laser light therapy for neurocognitive enhancement. In: Hamblin MR, Sousa MV, Agrawal T (eds) Handbook of low-level laser therapy. Pan Stanford Publishing, Singapore, pp 1057–1076
Rojas JC, Saavedra JA, Gonzalez-Lima F (2008a) Neuroprotective effects of memantine in a mouse model of retinal degeneration induced by rotenone. Brain Res 1215:208–217
Rojas JC, Lee J, John JM, Gonzalez-Lima F (2008b) Neuroprotective effects of near-infrared light in an in vivo model of mitochondrial optic neuropathy. J Neurosci 28:13511–13521
Rojas JC, John JM, Lee J, Gonzalez-Lima F (2009) Methylene blue provides behavioral and metabolic neuroprotection against optic neuropathy. Neurotox Res 15:186–200
Rojas JC, Bruchey AK, Gonzalez-Lima F (2012) Low-level light therapy improves cortical metabolic capacity and memory retention. J Alzheimers Dis 32(3):741–752
Sakata JT, Crews D, Gonzalez-Lima F (2005) Behavioral correlates of differences in neural metabolic capacity. Brain Res Brain Res Rev 48:1–15
Salthouse TA, Atkinson TM, Berish DE (2003) Executive functioning as a potential mediator of age-related cognitive decline in normal adults. J Exp Psychol Gen 132(4):566
Shefer G, Partridge TA, Heslop L, Gross JG, Oron U, Halevy O (2002) Low-energy laser irradiation promotes the survival and cell cycle entry of skeletal muscle satellite cells. J Cell Sci 115:1461–1469
Sherer TB, Betarbet R, Testa CM, Seo BB, Richardson JR, Kim JH, Miller GW, Yagi T, Matsuno-Yagi A, Greenamyre JT (2003) Mechanism of toxicity in rotenone models of Parkinson's disease. J Neurosci 23:10756–10764
Sommer AP, Pinheiro AL, Mester AR, Franke RP, Whelan HT (2001) Biostimulatory windows in low-intensity laser activation: lasers, scanners, and NASA's light-emitting diode array system. J Clin Laser Med Surg 19:29–33
Sweet LH, Paskavitz JF, Haley AP, Gunstad JJ, Mulligan RC, Nyalakanti PK, Cohen RA (2008) Imaging phonological similarity effects on verbal working memory. Neuropsychologia 46(4):1114–1123
Tachtsidis I, Gao L, Leung TS, Kohl-Bareis M, Cooper CE, Elwell CE (2010) A hybrid multi-distance phase and broadband spatially resolved spectrometer and algorithm for resolving absolute concentrations of chromophores in the near-infrared light spectrum. In: Takahashi E, Bruley D (eds) Oxygen transport to tissue XXXI, Advances in experimental medicine and biology, vol 662. Springer, Boston, MA
Tedford CE, DeLapp S, Jacques S, Anders J (2015) Quantitative analysis of transcranial and intraparenchymal light penetration in human cadaver brain tissue. Lasers Surg Med 47(4):312–322
Tian F, Liu H (2014) Depth-compensated diffuse optical tomography enhanced by general linear model analysis and an anatomical atlas of human head. NeuroImage 85:166–180
Tian F, Hase SN, Gonzalez-Lima F, Liu H (2016) Transcranial laser stimulation improves human cerebral oxygenation. Lasers Surg Med 48(4):343–349
Tranebjaerg L, Hamel BC, Gabreels FJ, Renier WO, Van Ghelue M (2000) A de novo missense mutation in a critical domain of the X-linked DDP gene causes the typical deafness-dystonia-optic atrophy syndrome. Eur J Hum Genet 8:464–467
Valla J, Berndt JD, Gonzalez-Lima F (2001) Energy hypometabolism in posterior cingulate cortex of Alzheimer's patients: superficial laminar cytochrome oxidase associated with disease duration. J Neurosci 21:4923–4930
Valla J, Schneider L, Niedzielko T, Coon KD, Caselli R, Sabbagh MN et al (2006) Impaired platelet mitochondrial activity in Alzheimer's disease and mild cognitive impairment. Mitochondrion 6(6):323–330

Van Breugel HH, Bar PR (1993) He-Ne laser irradiation affects proliferation of cultured rat Schwann cells in a dose-dependent manner. J Neurocytol 22:185–190
Vargas E, Barrett DW, Saucedo CL, Huang LD, Abraham JA, Tanaka H, Haley AP, Gonzalez-Lima F (2017) Beneficial neurocognitive effects of transcranial laser in older adults. Lasers Med Sci 32(5):1153–1162
Wang X, Tian F, Soni SS, Gonzalez-Lima F, Liu H (2016) Interplay between up-regulation of cytochrome-c-oxidase and hemoglobin oxygenation induced by near-infrared laser. Sci Rep 6(1):30540
Wang X, Tian F, Reddy DD, Nalawade SS, Barrett DW, Gonzalez-Lima F, Liu H (2017) Up-regulation of cerebral cytochrome-c-oxidase and hemodynamics by transcranial infrared laser stimulation: a broadband near-infrared spectroscopy study. J Cereb Blood Flow Metab 37(12):3789–3802
Wang X, Dmochowski J, Husain M, Gonzalez-Lima F, Liu H (2017a) Transcranial infrared brain stimulation modulates EEG alpha power. Brain Stimul 10:e46–e83
Wang X, Reddy DD, Nalawade SS, Pal S, Gonzalez-Lima F, Liu H (2018) Impact of heat on metabolic and hemodynamic changes in transcranial infrared laser stimulation measured by broadband near-infrared spectroscopy. Neurophotonics 5(1):011004
Wang X, Dmochowski J, Zeng L, Kallioniemi E, Husain M, Gonzalez-Lima F, Liu H (2019) Transcranial photomodulation with 1064-nm laser modulates brain electroencephalogram rhythms. Neurophotonics 6(2):025013
Whelan HT, Connelly JF, Hodgson BD, Barbeau L, Post AC, Bullard G, Buchmann EV, Kane M, Whelan NT, Warwick A, Margolis D (2002) NASA light-emitting diodes for the prevention of oral mucositis in pediatric bone marrow transplant patients. J Clin Laser Med Surg 20:319–324
Wollman Y, Rochkind S (1998) In vitro cellular processes sprouting in cortex microexplants of adult rat brains induced by low power laser irradiation. Neurol Res 20:470–472
Woltz DJ, Was CA (2006) Availability of related long-term memory during and after attention focus in working memory. Mem Cogn 34:668–684
Wong-Riley MT (1989) Cytochrome oxidase: an endogenous metabolic marker for neuronal activity. Trends Neurosci 12:94–101
Wong-Riley M, Liang HL (2017) Cytoprotective effect of low-level light therapy using LEDs on neurons. In: Hamblin MR, Sousa MV, Agrawal T (eds) Handbook of low-level laser therapy. Pan Stanford Publishing, Singapore, pp 185–206
Wong-Riley MT, Bai X, Buchmann E, Whelan HT (2001) Light-emitting diode treatment reverses the effect of TTX on cytochrome oxidase in neurons. Neuroreport 12:3033–3037
Wong-Riley MT, Liang HL, Eells JT, Chance B, Henry MM, Buchmann E, Kane M, Whelan HT (2005) Photobiomodulation directly benefits primary neurons functionally inactivated by toxins: role of cytochrome c oxidase. J Biol Chem 280:4761–4771
Yamanaka T, Fukumori Y, Numata M, Yamazaki T (1988) The variety of molecular properties of bacterial cytochromes containing heme a. Ann N Y Acad Sci 550:39–46
Ying R, Liang HL, Whelan HT, Eells JT, Wong-Riley MT (2008) Pretreatment with near-infrared light via light-emitting diode provides added benefit against rotenone- and MPP+-induced neurotoxicity. Brain Res 1243:167–173
Yu W, Naim JO, Lanzafame RJ (1994) The effect of laser irradiation on the release of bFGF from 3T3 fibroblasts. Photochem Photobiol 59:167–170
Yu W, Naim JO, McGowan M, Ippolito K, Lanzafame RJ (1997) Photomodulation of oxidative metabolism and electron chain enzymes in rat liver mitochondria. Photochem Photobiol 66:866–871
Zeff BW, White BR, Dehghani H, Schlaggar BL, Culver JP (2007) Retinotopic mapping of adult human visual cortex with high-density diffuse optical tomography. Proc Natl Acad Sci U S A 104(29):12169–12174
Zhang X, Jones D, Gonzalez-Lima F (2002) Mouse model of optic neuropathy caused by mitochondrial complex I dysfunction. Neurosci Lett 326:97–100
Zhang X, Rojas JC, Gonzalez-Lima F (2006) Methylene blue prevents neurodegeneration caused by rotenone in the retina. Neurotox Res 9:47–57
Zivin JA, Albers GW, Bornstein N, Chippendale T, Dahlof B, Devlin T et al (2009) Effectiveness and safety of transcranial laser therapy for acute ischemic stroke. Stroke 40(4):1359–1364

Avoiding Partial Sleep: The Way for Augmentation of Brain Function

Ivan N. Pigarev and Marina L. Pigareva

1 Sleep and "Human Factor"

The modern industrial style of life and necessity to live in non-natural environments often leads to dramatic errors in human behavior. Usually, these errors are included in the general term "human factor" and are attributed to reduced attention, inability to predict consequences of an action in complex conditions, and, finally, to reduced brain ability. Thus, the intention to augment the brain function can be considered a challenging goal for Neuroscience.

It was noticed that industrial anthropogenic disasters, traffic accidents, and medical errors often occur during the night, suggesting a probable link between sleepiness and reduced quality of brain functioning (Mitler et al. 1988; Dinges 1995; Dinges et al. 1997; Cajochen et al. 1999; Barger et al. 2006; Akerstedt et al. 2011). Given the importance of this issue, many special studies devoted to investigation of the possible link between sleep disturbances and mental abilities were undertaken.

It was demonstrated that short-term total sleep deprivation results in cognitive impairments, especially in learning, and memory tasks (Manganotti et al. 2001; Stickgold 2005; Born et al. 2006; Walker 2008, 2009; Diekelmann and Born 2010; McCoyand and Strecker 2011; Diekelmann et al. 2012, 2013).

In some studies, authors concluded that periods of slow-wave sleep (SWS) played a particular role in these processes (Fowler et al. 1973; Plihal and Born 1997; Diekelmann et al. 2011). Others maintained a rapid eye movement (REM) sleep

I. N. Pigarev (✉)
Institute for Information Transmission Problems (Kharkevich Institute), Russian Academy of Sciences, Moscow, Russia
e-mail: pigarev@iitp.ru

M. L. Pigareva
Institute of Higher Nervous Activity and Neurophysiology, Russian Academy of Sciences, Moscow, Russia

I. Opris et al. (eds.), *Modern Approaches to Augmentation of Brain Function*, Contemporary Clinical Neuroscience,
https://doi.org/10.1007/978-3-030-54564-2_10

dependency (Empson and Clarke 1970), although in some studies, REM sleep deprivation had no effect on different aspects of memory function (Hornung et al. 2007; Saxvig et al. 2008).

In addition to the function of memory consolidation, sleep has been proposed to benefit the encoding of new information during succeeding periods of wakefulness (McDermott et al. 2003; Yoo et al. 2007; Mander et al. 2011).

2 Peculiarities of Sleep Deprivation Effects

In their review of the sleep deprivation literature, Harrison and Horne (2000) concluded that sleep deprivation has little effect on simple rule-following tasks, but it obstructs decision-making in complex integration tasks requiring flexibility, innovation, or plan revision. However, other researchers observed performance decrements in relatively simple tasks such as identification and vigilance tasks (Chee et al. 2008; Ratcliff and Van Dongen 2009). Sleep-deprived subjects exhibited decreased performance in the taking advantage task (Glass et al. 2011).

The impact of sleep deprivation on memory formation did not appear to be universal, but instead differed on the basis of many factors such as the types of tasks used for learning, and personal characteristics of subjects, including their emotionality. It has been shown that sleep-dependent motor skill memory improvement was dependent on the nature of the skill to be learned (Cohen et al. 2005; Cohen and Robertson 2007; Siengsukon and Boyd 2008). In another study, perception, attention, and memory were impaired by sleep deprivation, but visual search and logical reasoning tasks were not (Williamson et al. 2000).

Gensel et al. (2009) used different deprivation conditions throughout the experiment and did not find that an intense decrease in the total amount of REM sleep or slow-wave sleep of their volunteers led to the inhibition of learning. They even proposed that sleep-dependent memory consolidation did not rely only on intact amounts of SWS or REM sleep across a night, but required different EEG microstructures, e.g., sleep spindles, delta waves, and PGO waves.

It was proposed that reduced visual short-term memory after sleep deprivation may be connected, not with impairment of memory consolidation mechanisms but, rather, with a decline in visual attention and/or visual processing (Chuah and Chee 2008).

An important factor was the length of time between training and test or amount of skill practice, independent of whether there was sleep or not (Shadmehr and Brashers-Krug 1997; Robertson et al. 2004; Keisler et al. 2007; Song et al. 2007; Criscimagna-Hemminger and Shadmehr 2008; Doyon et al. 2009; Debas et al. 2010; Borich et al. 2011; Borich and Kimberley 2011; Voderholzer et al. 2011; Reis et al. 2015). In a nap study (Mander et al. 2011), no differences were observed between the sleep and no sleep groups in a specific alertness control task.

Results of sleep deprivation experiments were often rather contradictory. According to Lo et al. (2012), influence of sleep deprivation depends on the task domain, prior sleep debt, circadian phase at which performance is assessed, and genetically determined subject characteristics. Degree of task difficulty and the subject's emotionality influence the outcome of experiments concerning the connection between sleep and memory (Smith 2001; Walker and Stickgold 2006). Small differences in test design can cause large discrepancies in the studies of sleep dependency of memory processes. This may explain why some results are regularly found only by the same groups of scientists (Genzel et al. 2009).

Even in those experiments where positive effects of sleep on memory consolidation were demonstrated, these effects were minimal. It seemed unlikely to us that the only function of sleep was simply to provide such a modest improvement in memory. On the other hand, the probable connection of sleepiness with the rare but dramatic consequences of anthropogenic disasters, as well as the fantastic pictures of dreams, supports the general belief that the first function of sleep is for efficient functioning of the brain.

3 Modern Theories of Sleep

Several modern theories concerning the function of sleep offer hypothetical mechanisms, which could be used by the brain for this purpose, e.g., the theory of neuronal groups (Krueger and Obál 1993; Krueger et al. 2008) or the theory of synaptic homeostasis (Tononi and Cirelli 2003, 2006). All theories, which were based on the assumption that sleep first of all is important for efficient brain function, referred to numerous studies which demonstrated, both in humans and animals, "local use-dependent processes." It was shown that delta power during the first hours of sleep is higher in those cortical areas, which were more active immediately before sleep (Kattler et al. 1994; Rector et al. 2005; Huber et al. 2006). Delta power was considered therefore as an indicator and measure of recuperative processes in the brain.

However, if to accept an idea that function of sleep was to keep the efficient brain function, it would be logical to expect that the brain would be the organ most vulnerable to sleep deprivation. However, the results of fundamental studies of A. Rechtshaffen and his colleagues (Everson et al. 1989; Cirelli et al. 1999; Rechtshaffen and Bergmann 2002) do not confirm this suggestion. Their experiments demonstrated that total sleep deprivation led first of all to multiple visceral disorders (hair loss, skin, and gastrointestinal ulcerations and so on) and, finally, to unavoidable death of animals. A striking finding was that in rats that died after several days of total sleep deprivation, the organ, which did not have obvious degenerative changes, was the brain (Cirelli et al. 1999). This observation was surprising, since the negative effects of sleep deprivation on mental ability were well known. However, investigation of neuronal activity in the cerebral cortex in the sleep-wake cycle offers insights into the other sleep-dependent mechanisms which can explain the reduction in mental abilities for complex problem-solving, even in the normally working brain.

4 Phenomenon of Partial (Local) Sleep

It was generally assumed for a long time that sleep develops synchronously in all areas of the mammalian cortex. This conclusion was based on the observations obtained in the first studies of sleep in humans and animals. In these studies, experimenters tried to ensure optimal conditions for sleep, and sleep could develop in all cortical areas nearly simultaneously. The only exception has been reported for dolphins, whose EEG shows periods of deep slow-wave sleep in either the right or left hemisphere alone (Mukhametov et al. 1977; Mukhametov 1984, 1987; Oleksenko et al. 1992). After this discovery, features of the unilateral sleep were noticed in other sea mammals and some birds (Rattenborg et al. 2012; Mascetti 2016). However, even in dolphins, EEG activity in different areas of one hemisphere was found to be synchronized or desynchronized always simultaneously.

We came to investigation of sleep from physiology of vision. In our experiments, the goal was opposite—to have awake cats as long as possible. Nevertheless, animals became tired and drowsy after a couple of hours of these experiments. We first used various tricks to keep animals active but soon noticed that even on the background of still clear visually driven behavior neuronal responses to visual stimuli in extrastriate areas strongly diminished and finally disappeared. Neurons began to fire in a typical sleep mode—with bursts and pauses. It was obvious that in still behaving animals extrastriate areas were sleeping. This situation led us to the idea to start combining visual and sleep studies on the same animal in the same experiment. However, phenomenon of local sleep in extrastriate visual area was described later, when one of us (IP) studying with German colleagues, neuronal properties in the visual area V4 in monkeys observed and documented partial sleep in this area (Pigarev 1997; Pigarev et al. 1997). In this work, it was demonstrated that when a monkey had to perform long and monotonic visual discrimination task, neurons became less responsive to the same visual stimuli and finally stopped responding at all, while the monkey continued to work in the task. If the task was interrupted, the monkey fell asleep for 10–20 min. After the nap, neuronal responses to visual stimuli often recovered. Neuronal background firing during such periods of temporal inactivity resembled that which these neurons demonstrated during periods of natural sleep of the animal. The monkey's performance in the visual task during periods of local sleep in the area V4 was rather high, although it was slightly reduced in comparison with that at the beginning of the experiment. Thus, one could conclude that at least visual area V1 still was working. In the same study, it was noticed that, even within the area V4, local sleep developed not simultaneously but started from the periphery of the visual field. The last neurons, which were recruited in sleep, were neurons in the region of the fovea representation. These experiments demonstrated that at least in visual system development of sleep started from the "higher order" sensory areas most likely crucial for the most complicated behavioral situations. Taking all those considerations into account, one could expect to observe local sleep more often in organisms with better expressed multiple sensory representations. Indeed, the fronto-occipital trend in delta power, most pronounced during beginning of sleep, was discovered in human subjects by the group of

A. Borbely (Werth et al. 1996, 1997). Extensive studies of sleep-spread dynamics are presented and discussed in the review of M. Ferrara and L. De Gennaro (2011). It seemed less likely that local sleep would be found in animals with more simple cortical organization and a limited number of sensory areas. However, recently, local sleep was described in behaviorally active rats, and again in the frontal cortical area (Vyazovskiy et al. 2011).

Recently, collecting material for our essay devoted to the history of first sleep studies in Russia (Pigarev and Pigareva 2019), we found that actually effect of partial sleep was discovered much earlier, in the laboratory of Ivan Pavlov. In physiological investigations of conditioned reflexes, they faced the problems very similar to ours. In the long behavioral experiments dogs, after sufficient food reward, became drowsy and had tendency to fall asleep while experimenters wanted to receive more data and forced animals to work longer. In such cases, they noticed that conditioned reflexes to somatosensory stimuli diminished and finally disappeared while reflexes to visual or auditory stimuli could be still preserved (see in (Rozjanskiy 1954)). They interpreted this behavior as development of local sleep in somatosensory cortex. Pavlov on his clinical seminars also used the possibility of partial sleep of only limited areas of the cerebral cortex explaining some observations in psychiatry (see in (Pavlov 1954)). Local sleep at that time looked so natural that it never was considered, as the physiological phenomena deserved publication.

At present, local sleep attracts more and more attention of scientists, and observations, demonstrating the local features of sleep become more diverse (e.g., (Vyazovskiy et al. 2011; Borbely 1982; Siclari and Tononi 2017; Fernandez et al. 2018)).

We would like to remind here that since 1993 (Tononi and Cirelli 2003; Huber et al. 2006; Krueger et al. 2019), the mentioned above idea concerning local use-dependent sleep was widely discussed in sleep literature. Although experiments demonstrated local use-dependent sleep, and observations of local sleep in behaviorally awake animals were mutually supportive, they were not absolutely identical. In the first case, it was shown that in sleeping brain, the amplitudes of slow waves could locally vary from region to region dependent on previous history of activation. In the second case, sleep in some cortical areas appeared during behavioral wakefulness. Namely, this second case we will have in mind using the term "partial sleep" in future.

5 Partial Sleep and Cognitive Impairments After Sleep Deprivation

Taking into account the phenomenon of partial sleep, the cognitive impairments after sleep deprivation can be explained not by the general deterioration of brain efficiency, but by switching of several cortical areas from those functions, which they have to perform in wakefulness.

Such partial sleep might be especially dangerous, because neither the person himself nor other individuals could notice its appearance and development. At the same time, dangerous consequences of temporal disengagement of some cortical areas from the control of behavior potentially can be rather dramatic (Maric et al. 2017; Andrillon et al. 2019). That is why understanding the physiological mechanisms involved in initiation of sleep, and particularly of partial sleep, can be considered an important element in attempts to augment brain functionality.

At first sight, the hypothesis that partial sleep is responsible for a reduction in brain functionality looks similar to the proposal that after a long period of wakefulness, the brain needs recuperation during sleep. The difference is only in the temporal sequence of events. After prolonged wakefulness, all regions of the brain may fall asleep for recovery simultaneously, or certain areas may do so at first, while others remain awake for some time.

However, it is possible to consider another, fundamentally different, scenario. What if the brain, as all other visceral organs of an organism, does not need any special recuperative rest connected with total interruption of functionality? What if the brain, like a computer, can work efficiently for long periods of time? What if observed "sensory isolation" of the brain during sleep just reflects switching over of the normally working brain for processing of another flow of incoming information?

We should not forget that sleep-deprived animals die not because they become blind, deaf, have forgotten the ways to a food tray or because of serious problems with decision-making. They die mainly because of multiple visceral disorders in virtually all life supporting systems, including the immune system (Rechtshaffen and Bergmann 2002). At the same time, the brain appears to be the most resistant organ.

Our observations of the neuronal activity in the sleep-wake cycle also did not convince us that during long periods of wakefulness, there were crucial pathological changes of the neuronal state, which forced a brain to switch into a sleeping "restorative" mode. Nevertheless, one can argue that it is generally recognized that the pattern of slow-wave sleep EEG is very specific for this state, and differs from the pattern of EEG in wakefulness. However, interpretation of this observation also is equivocal. The EEG pattern of natural deep slow-wave sleep was usually compared with the EEG pattern in a rather artificial state of very passive wakefulness, when human subjects or animals were immobile and without intensive sensory stimulation.

6 Whether Cortical EEG Reflects Peculiarity of Brain Activity in Wakefulness and Sleep, or Just Pattern of the Cortical Afferent Flow?

We proposed that difference of cortical afferentation in wakefulness and sleep might define the observed difference of neuronal activity in slow-wave sleep and wakefulness. To check this proposal, general EEG and eye movements were recorded in cats

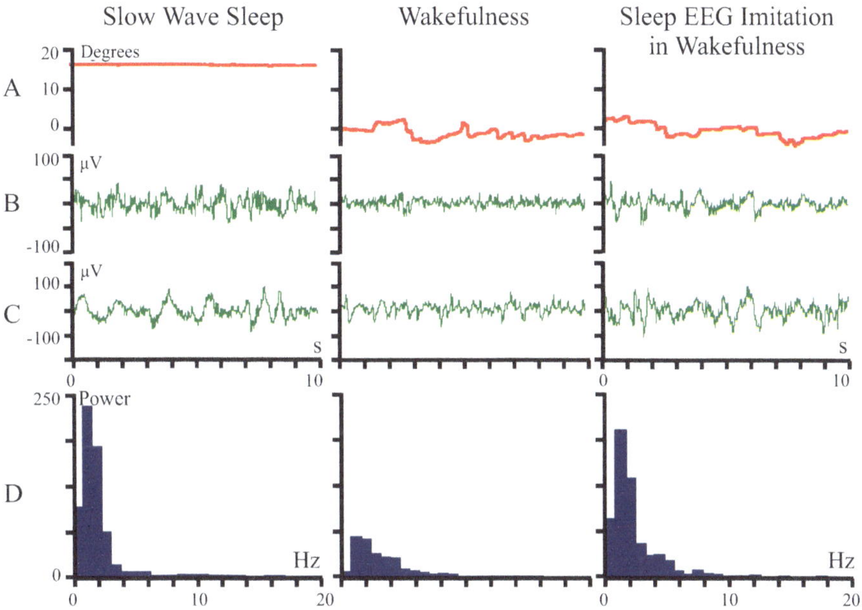

Fig. 1 Rhythmic stimulation by the optimal visual stimulus during wakefulness evokes sleep-like slow-wave activity in the cat visual cortex. The rows: A—vertical component of eye movements which helps to distinguish sleep (upward deviation) from wakefulness (downward deviation); B—general EEG recorded between electrodes over temporal and frontal cortical areas of the cat; C—local field potentials (local EEG) recorded between two tungsten microelectrodes located 300 μm one from another within the cortical visual area V1; D—power spectrum of the local EEG presented above in C. All parameters were collected during slow-wave sleep (left column), during passive wakefulness (middle column), and during the procedure of "sleep EEG imitation" by visual stimulation in wakefulness, which produced strong excitation of the cortical neurons recorded by the microelectrodes (right column). (Technical details of the study in Pigarev et al. 2013)

during slow-wave sleep and active wakefulness between electrodes located over parietal and frontal cortical areas. In addition, we recorded neuronal activity and local field potentials (local EEG) from visual (Fig. 1) and somatosensory (Fig. 2) cortical areas using bipolar tungsten microelectrodes with distances of about 300 μm between the tips of the electrodes.

During slow-wave sleep, the animal eyes were deviated upward, and this allowed us to distinguish periods of sleep from active wakefulness in the obtained recordings. For every group of recorded neurons, we applied the optimal parameters of stimulation (either visual or somatosensory) and delivered these stimuli in a rhythmic manner. We called this procedure "sleep EEG imitation in wakefulness." Using this procedure (right column), in actively awake cats, we got burst neuronal firing (not shown) and EEG slow waves, which were indistinguishable from, or even higher than, those which we had observed during the periods of natural slow-wave sleep (left column). These sleep-like waves were especially well visible in the chan-

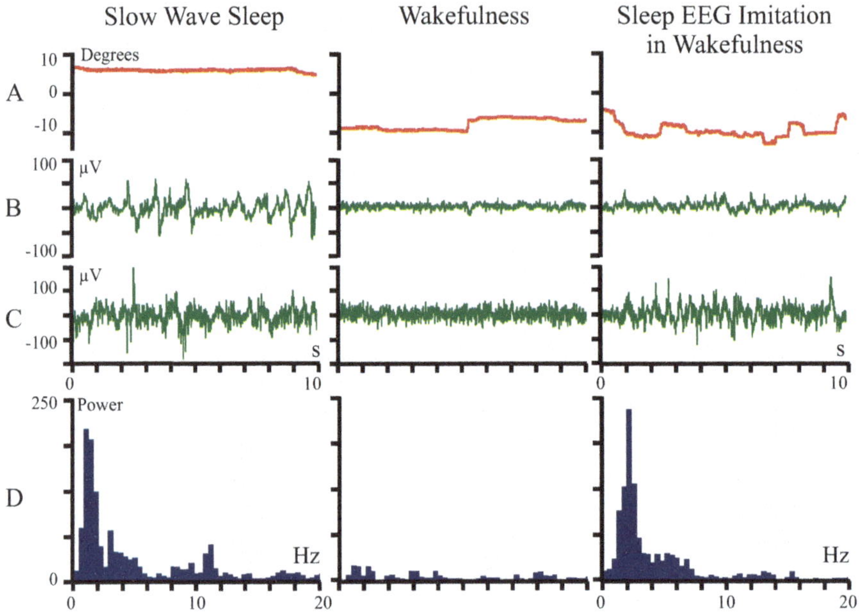

Fig. 2 Periodic optimal somatic stimulation during wakefulness evokes sleep-like slow-wave activity in the cat somatosensory cortex. Designation the same as in Fig. 1. All parameters were collected during slow-wave sleep (left column), during passive wakefulness (middle column) and during the procedure of "sleep EEG imitation" in wakefulness by rhythmic stimulation of the contralateral forepaw, which produced strong excitation of the cortical neurons recorded by the microelectrodes. (Technical details of the study in Pigarev et al. 2013)

nel of the local field potentials (C) because the local EEG reflected activity of the neurons for which we used the optimal stimulation. The general EEG reflected averaged activity collected from the large cortical territory, including those neurons for which applied stimuli were not optimal. Nevertheless, some sleep-like waves were seen even in the general EEG (B). In row D in Figs. 1 and 2, we present power spectrums calculated for the 10 s fragments of the local EEG shown in row C of the corresponding column. It is seen that spectral compositions of the local EEG for slow-wave sleep (left column) and imitation of sleep EEG in wakefulness (right column) were rather similar, and both differed from the usual spectrum of quiet wakefulness (central column).

The presented observations supported an idea that patterns of the cortical afferentation, rather than the state of vigilance, determine the pattern of cortical activity. All abovementioned considerations inclined us to conclude that switching to analysis of another flow of incoming information would be able to better explain the phenomenology of transition from wakefulness to sleep. The temporal organization of these incoming signals, specific for the state of sleep, will define the pattern of cortical activity during sleep.

7 Which Signals Could Provide Periodic and Synchronous Afferentation During Sleep?

Animal physiology offered the answer on this question, that it could be periodic activity of various visceral systems, e.g., gastrointestinal peristalsis, heart, and respiratory activities. We proposed that during sleep, the same brain neurons that in wakefulness process exteroceptive information of various modalities switch over to the analysis of interoceptive information coming from visceral systems. Rhythmic activities of different visceral systems define this periodic afferent flow toward the cortical areas, which is reflected in cortical slow-wave sleep activity. Thus, the central nervous system during sleep might be involved in the process of visceral regulation (Pigarev 2014).

According to this proposal, periods of partial sleep are not the periods when "tired" brain areas stop processing of exteroceptive information in favor of self-recuperation. During periods of partial sleep, normally working brain areas respond to messages from the internal organs and switch to the processing of the visceral afferentation. Within the frame of this hypothesis, we should "think differently" about the nature of sleep and local sleep.

This suggestion may be too fantastic for the brain paradigm generally accepted at present. This paradigm was established mainly on the basis of data collected for the state of wakefulness. On the other hand, our "fantastic" proposal opened the way for its experimental validation in simple experiments, which could not be conducted without this theoretical background. Below, we offer a short review of the experiments performed to investigate such nontrivial predictions of the visceral hypothesis of sleep.

8 Experimental Validation of the Visceral Hypothesis of Sleep

First of all, to check this hypothesis, responses of different cortical regions to extero- and interoceptive stimulation during sleep and wakefulness were compared. These experiments were started from the visual cortical areas. Visual areas were selected for these experiments because they were well studied, and it was generally recognized that in behaviorally awake animals, neurons of these areas were responding exclusively to visual stimulation. In addition, one of us (I.P.) had considerable experience in investigation of various visual areas in behaviorally awake animals. Later, similar experiments were conducted with neurons not only in occipital, but also in frontal and parietal cortical areas.

In Fig. 3 (adopted from (Pigarev 1994; Pigarev and Pigareva 2012)), we show responses of complex neurons in visual cortical area V1 (panel a) and somatosensory area 5 (panel b) of cats to electrical intraperitoneal stimulation delivered in slow-wave sleep and in wakefulness. These neurons in the state of wakefulness

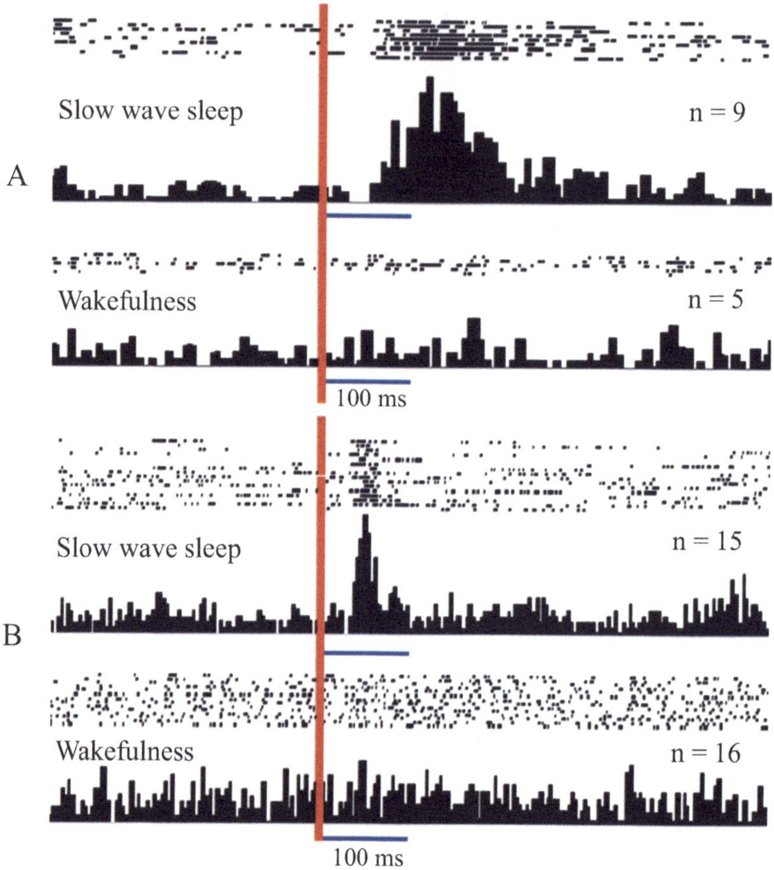

Fig. 3 Responses of neurons in the cat cortical visual area V1 (**a**) and somatosensory area 5 (**b**) to intraperitoneal electrical stimulation delivered in sleep and in wakefulness. Responses presented as rasters where every line corresponds to single stimulation trial. Dots represent single spikes. Below are averaged histograms. Vertical red line—moment of intraperitoneal stimulation, *N*—number of averaged trials. (Technical details of the study in Pigarev 1994)

responded to visual and somatosensory stimuli, respectively. During slow-wave sleep, both neurons responded to electrical stimulation of the area of the small intestine, and these apparent responses immediately disappeared in REM sleep and after awakening. About 40% of neurons out of 49 recorded in these experiments demonstrated such properties.

Similar experiments were conducted with monkeys, where evoked responses to intraperitoneal electrical stimulation were recorded above the cortical visual area V1 (Pigarev et al. 2006). Evoked responses were again obtained only during slow-wave sleep and disappeared during REM sleep and in the state of wakefulness.

In experiments with one monkey, we used magnetic stimulation, with the coil located close to the surface of the monkey's stomach. In response to magnetic

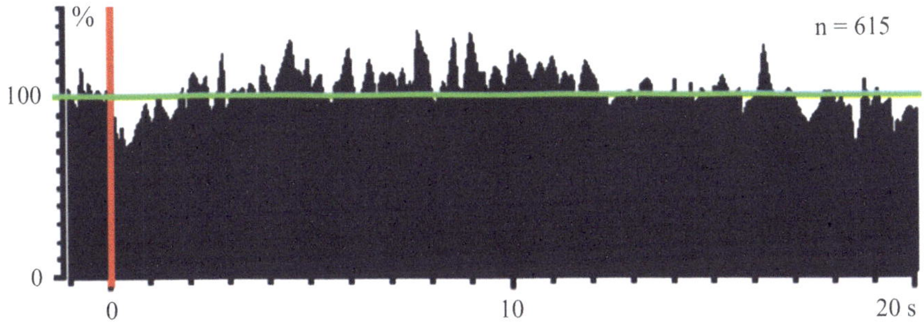

Fig. 4 Population response of 61 neurons (615 trials) in monkey's cortical visual area V4 to magnetic stimulation of stomach during slow-wave sleep. Vertical red line—moment of stimulation. Horizontal green line indicates mean background firing rate before stimulation taken as 100%. (Technical details of the study in Pigarev et al. 2008)

pulses, which did not wake the animal, we obtained cortical evoked responses, recorded by the electrodes above the occipital pole of the skull. These responses were observed again only during slow-wave sleep (Pigarev et al. 2008). Simultaneous recording of the neuronal activity in the visual area V4 revealed a strong short latency inhibition in response to these magnetic pulses, which was obviously visible even without averaging, in the population response of 61 neurons (Fig. 4). After this short latency inhibition, the delayed (5–15 s) activation of the background firing took place. This result deserves attention because receptive fields in visual area V4 had small central excitatory areas and huge inhibitory surrounds. The applied magnetic pulses could activate those parts of visceral organs, which projected to this huge inhibitory periphery of the studied receptive fields. On the other hand, after some delay, peristaltic waves provoked by the stimulation could reach regions, which projected to the central excitatory part of the receptive fields, causing the observed delayed activation. All these responses to magnetic stimulation again disappeared in wakefulness.

In experiments with rabbits (Pigarev et al. 2004), we also recorded evoked responses to electrical intraperitoneal stimulation in visual and somatosensory cortical areas, which appeared again exclusively during slow-wave sleep.

It has been argued that electrical and magnetic stimuli are not natural, and that observed effects could have a nonspecific origin. Although the main information concerning organization of the nervous system was obtained using method of electrical stimulation, we agreed that it would be much more important to demonstrate a functional link between visceral organs and cortical areas during sleep in natural conditions, without any artificial stimulation.

Such experiments were conducted with the help of our colleagues from the Pavlov Institute of Physiology (St. Petersburg), prof. V. A. Bagaev and I. I. Busigina. Recording electrodes were implanted in the walls of the small intestine and stomach of cats, together with stomach fistula. With this approach, in addition to cortical neuronal activity, EEG, and ocular movements, we could record myoelectrical activity of small intestine and stomach, and to change intragastric contents.

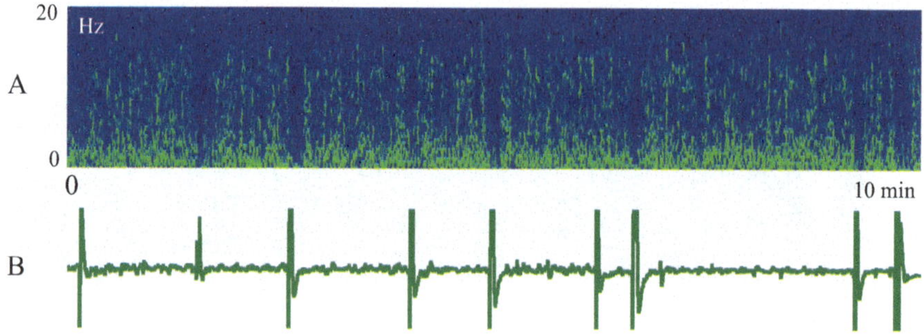

Fig. 5 Episode of slow-wave sleep. (**a**) Spectrogram of the cat cortical EEG. Yellow color indicates higher power. (**b**) Myoelectrical activity of stomach. (Technical details of the study in Pigarev et al. 2013)

In Fig. 5, we present a spectrogram of cat cortical EEG (a) recorded over occipital cortex simultaneously with myoelectrical activity of the stomach (b) during an episode of slow-wave sleep. In the spectrogram, yellow colors indicate higher power, and periodic vertical blue fragments indicate moments of short desynchronizations connected with lack of low frequency components. These desynchronized intervals are well known to anybody who has recorded EEG during slow-wave sleep. It was previously demonstrated (Oniani et al. 1974) that, behaviorally, sleep was not interrupted during these periods, and thresholds for awakening during such short desynchronizations still were very high. What was new in the presented figure was a surprising coincidence of these EEG desynchronizations with the appearance of periodic migrating myoelectrical complexes in the stomach (short vertical inclinations in Fig. 5b). There was no need for any special analysis in order to notice such coincidence. Simultaneous appearance of the migrating myoelectrical complexes in stomach activity and short desynchronizations in the cortical EEG usually happened during intervals of 10–20 min of slow-wave sleep. The observed coincidence of these effects can disappear for a while and appear again later. This was a very robust effect, observed in most of our sleep recordings, which included the periods of corresponding stomach activity.

More impressive were results of those experiments where we have studied interaction of the neuronal activity in various cortical visual areas and myoelectrical activity from the wall of the duodenum (Pigarev et al. 2013). It was demonstrated that about one-third of more than 200 of the studied cortical neurons during slow-wave sleep established a causal relationship with the activity of the duodenum during slow-wave sleep. Even more, these neurons demonstrated selectivity to particular types of duodenal rhythmicity. Some neurons preferred simple duodenal waves, and others responded only to waves with spike potentials. Such a relationship was never observed in wakefulness.

Finally, it was found that changes of the intragastric medium (water infusion via fistula into the stomach) performed in the period of slow-wave sleep lead to changes

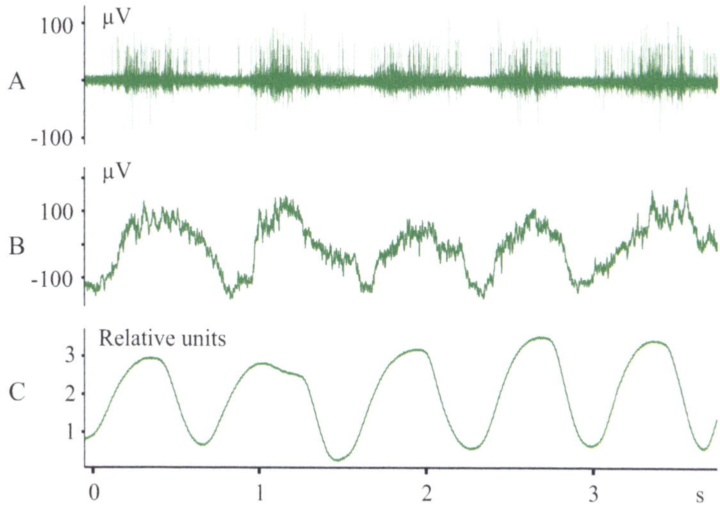

Fig. 6 Neuronal activity in the cat primary visual cortex synchronized with respiration during slow-wave sleep. (**a**) Multiunit activity; (**b**) Local field potentials (local EEG); (**c**) Nasal air flow in relative units. (Technical details of the study in Pigarev et al. 2013)

in the EEG pattern and temporal reorganization of the background neuronal spiking, revealed by Fano factor analysis (Pigarev et al. 2016a).

We do not imply that only the structures of the digestive system are represented in the cerebral cortex during sleep. In other experiments, evoked responses to heartbeats during sleep were recorded (Lavrova et al. 2019). An example of neuronal firing and local field potentials in the visual area V1, which synchronized with respiration during slow-wave sleep, is shown in Fig. 6.

According to the visceral sleep theory, patterns of periodic activation coming from the visceral organs determine the oscillating picture of cortical activity during slow-wave sleep. The desynchronized pattern of cortical EEG during REM sleep can be connected with afferentation coming to the cerebral cortex from visceral systems lacking obvious rhythmic activity, e.g., liver, kidneys, reproductive organs, and, finally, the brain itself. The brain's status within this theory is obviously dual. On the one hand, the brain is the central processor, which controls behavior in the environment during wakefulness and defines recovery of all visceral organs during sleep. However, on the other hand, the brain itself is an enormously complicated visceral organ, which certainly should be in need of service. How and when such brain self-service is realized is a challenging question. It may happen, for example, during particular phases of REM or slow-wave sleep, or it may be organized as a permanent service, e.g., by glial cells. The recently discovered "glymphatic" mechanism may reflect elements of such brain self-servicing (Nedergaard 2013; Xie et al. 2013). Various other options can be offered, but that is a topic for future studies.

9 The Visceral Sleep Theory and Observations of "Slow-Wave" Activity in the Cortical Slabs and Slices

Our approach to sleep function supposes that cortical activity during sleep is defined by the afferent flow coming to the cerebral cortex from various visceral organs. On the other hand, there is substantial evidence that sleep-like activity can be generated in cortical slabs (Timofeev et al. 2000) and isolated cortical slices (Sanchez-Vives and McCormick 2000) without any interoceptive inputs. However, we do not think that these observations are inconsistent with our theoretical proposal. Of relevance is the important discovery of Steriade and his colleagues (2001), who performed the first intracellular recordings of neuronal cortical activity in naturally sleeping cats. They found that waves of hyperpolarization reflected as periodic silent pauses in neuronal firing during slow-wave sleep were connected not with active inhibition, but with disfacilitation caused by the temporal periodic lack of excitatory inputs to these cortical neurons. At the time of their study, it was generally recognized that, during sleep, the cerebral cortex was disconnected from any afferent inputs, and they had to conclude that, "during slow-wave sleep, neocortical neurons may be engaged in information processing of internally generated signals…," which provided such excitatory inputs. As a source of such "internally generated signals," they considered intracortical excitation (Steriade et al. 2001).

The studies performed on the isolated cortical slices demonstrated that, in certain conditions, it was possible to evoke periodic neuronal discharges, which had some features of similarity with real slow-wave sleep oscillations. Later, it was shown in experiments on thalamo-cortical slices that activation of the thalamo-cortical neurons dominates in triggering such cortical oscillations (Contreras and Steriade 1995; Rigas and Castro-Alamancos 2007). In a review (Crunelli and Hughes 2010), it was recognized that most likely several mechanisms might elicit the cortical slow waves. We propose that "internally generated signals," which define cortical waves during slow-wave sleep, are actually coming from various visceral systems using the same thalamo-cortical pathways, which activate cortical neurons during wakefulness.

10 The Pathways for the Visceral Afferentation to the Cerebral Cortex During Sleep

One may inquire about the ways by which the information from various visceral systems may reach "the same thalamo-cortical pathways." For the somatosensory system, it is well investigated. It is known since early anatomical studies in the nineteenth century (e.g., Head 1896), that visceral and somatosensory afferents terminate at the same neurons in the spinal cord, and thus, visceral information may travel to the cerebral cortex through the fibers of the somatosensory columns. The fact of such combined projections was confirmed in many studies (Kuo et al. 1981; Cervero 1983; Cervero et al. 1984; Akeyson and Schramm 1994; Perry and Lawson

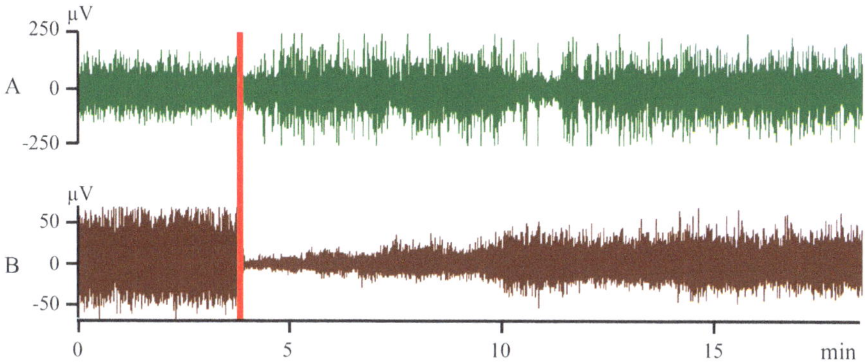

Fig. 7 Rabbit spinal cord fibers activity in wakefulness and slow-wave sleep. (**a**) Cortical EEG. (**b**) Activity of the fibers in the dorsal column of spinal cord at the thoracic level in wakefulness and developing slow-wave sleep. Vertical red line indicates the border between these two states. (Technical details of the study in the text and Pigarev et al. 2011)

1998), and this overlap is regarded as the most probable mechanism of the referred pains (Head 1896; Arendt-Nielsen and Svensson 2001; Peles et al. 2004; Hobson et al. 2010).

However, this overlap created a yet unresolved problem—how the central nervous system manages to distinguish spikes coming by a single fiber from so different sources. Our hypothesis offers a solution of this problem. The transmission of the somatosensory information happens during wakefulness, while visceral information is transmitted to the central nervous system by the same fibers but during sleep, on the background of muscles atonia when their movements are excluded.

Figure 7 illustrates results of a pilot study, undertaken to check this proposal (Pigarev et al. 2016b). Experiments on behaving New Zealand rabbit were performed with Dr. Helena Almirall in her laboratory in Barcelona. In Fig. 7, top record (channel a) presents cortical EEG, and spike activity of many spinal fibers recorded by the electrode located above the dura mater on the dorsal surface of thoracic vertebra is shown below (channel b). Vertical red line indicates moment of transition from active wakefulness to sleep. During wakefulness, rabbit moved and strong firing in spinal fibers reflected these movements. At the moment indicated by the red line, rabbit stopped moving and spike activity in spinal cord also disappeared. After that slow-wave sleep began to develop, and rabbit stayed immobile. However, in parallel with sleep development, in immobile rabbit, spike activity in spinal cord fibers began to grow again. During certain periods of slow-wave sleep, intensity of the recorded spike activity was as high as during active motions (not shown). One could often detect rhythmicity of respiration or heart beats in the activity of spinal fibers during sleep, and this rhythmicity disappeared in wakefulness.

To our knowledge, pathways of the visceral information to the visual cortical areas have never been directly described yet. Here, we offer only some considerations. It is well known that, in the main thalamic visual relay (lateral geniculate nucleus), retinal synaptic terminals form only one-third of all synaptic terminals. Another third of terminals belong to backward cortico-thalamic projections. The remaining one-third of terminals is of nonvisual origin and comes from the pontine and the brain stem regions (Hughes and Mullikin 1984). Activation from pontine nuclei reaches the lateral geniculate nucleus during sleep, especially during REM sleep, and reflects in the visual cortical areas as well-known ponto-geniculo-occipital waves (Brooks and Bizzi 1963). The origin of this pontine activity has not yet been investigated. On the other hand, pontine and brain stem projections to the lateral geniculate nucleus come from the regions of entering and passage of various visceral nerves. Therefore, their link with visceral information is very probable. The widespread connections of the enteric nervous system of duodenum with various brain structures from brain stem to the cerebral cortex were demonstrated in the study (Parker et al. 2020).

11 K-Complexes and Visceral Afferentation, Use Dependency, and Sleep Homeostasis

In discussions of cortical visceral activation during sleep, one often argues that these visceral responses can be of nonspecific origin, resembling sensory-triggered K-complexes. This topic was investigated in detail in our special study (Pigarev et al. 2011). It was shown that visually induced K-complexes had absolutely specific origin. Even more, K-complexes could be induced by sensory stimuli only during short intervals of developing sleep. In contrast, cortical visceral responses could be better recorded during periods of deep slow-wave sleep, when K-complexes could not be induced by any sensory stimulation.

The attempts to link various visceral events with the elements of EEG recorded during sleep, e.g., K-complexes have been undertaken in many studies (Pampiglione and Ackner 1958; Johnson and Karpan 1968; Halász et al. 1985; Heald et al. 1989; Okada et al. 1991; Hornyak et al. 1991; Niiyama et al. 1996; Monstad and Guilleminault 1999; Tank et al. 2003). The same idea was expressed in a study of Cash et al. (2009), where it was suggested that spontaneous K-complexes appearing in the EEG during transition from wakefulness to sleep could be induced by a "sensory stimulus occult to the investigator (e.g., gastric)."

In the middle of the previous century, cortical responses to the stimulation of various visceral nerves were described and intensively investigated in several laboratories (e.g., (Bailey and Bremer 1938; Amassian 1951; Patton and Amassian 1952; Gardner et al. 1955; Chernigovskii 1960)). These studies were performed in acute experiments under anesthesia. However, in later experiments without anesthesia, these results could not be reproduced. In wakefulness, neurons in these areas responded only to visual or somatosensory stimulation. Thus, cortical responses to visceral stimulation were regarded as probable artifacts of anesthesia.

In our studies, we have demonstrated that without any anesthesia, in natural conditions, cortical areas do establish connections with visceral organs, but this link is functionally active only during sleep. Involvement of the high levels of the central nervous system, including the cerebral cortex in mammals, in the processing of visceral information during periods of sleep may be the main, if not the exclusive, function of sleep.

Here, we should come back to the abovementioned effect of "use dependency," which is widely explored now as an experimental argument in favor of the concepts that sleep is necessary for brain recovery (Kattler et al. 1994; Rector et al. 2005; Huber et al. 2006; Krueger et al. 2019). The alternative mechanism of this effect can also be proposed by the visceral sleep theory. It is known that activation of cortical areas by intensive exteroceptive stimulation during wakefulness will lower the thresholds of the neurons responding to this stimulation (e.g., due to the LTP mechanism). As a result, during sleep, these neurons will respond more strong to visceral stimulation, coming to the same neurons through the same synaptic connections. Consequently, the power of the EEG in these areas will grow.

Discussing the visceral sleep theory, we would like to draw attention to the distinguished theory of A. Borbély—the two-process model of sleep regulation (Borbely 1982; Borbely and Achermann 1999). This model proposes that sleep is regulated by the interaction of two processes—homeostatic and circadian. The homeostatic process allows for a constant amount of sleep during 24 h. Visceral sleep theory offers the physiological framework of this homeostatic mechanism. Homeostasis of all visceral systems is supported during sleep due to the involvement of the cerebral cortex in the processing of information from internal organs, evaluation of their states, and recovery of their functionality. These processes define the total length of sleep.

12 Mechanism of Sleep Initiation and Features of Local Sleep

The visceral sleep theory offers the logic behind the initiation of sleep. Indeed, a mismatch between the current parameters of any visceral system and the genetically determined range for these parameters would provide the feeling of tiredness, or sleep pressure. If an environmental situation allows sleep, an organism would transit to normal total sleep in all cortical areas. In the cases when, because of visceral problems, the need for sleep is dramatically increased, but environmental conditions do not allow sleep to occur, sleep may progress only in some cortical areas in still behaviorally active organisms. According to the observations cited above (Pigarev 1997), the development of sleep starts from the most recent "high-order" cortical areas. The proportion of such areas is highest in the frontal pole of the brain. This might underlie the reported fronto-occipital trend in the development of sleep.

It is logical to propose that those behavioral tasks, which do not need engagement of the highest cortical resources, would be normally realized even in conditions when part of the brain is sleeping. However, in the situation when all cortical computational ability is required for decision-making in a complicated problem, partial sleep may lead to severe and often dramatic behavioral errors.

Conditions of partial sleep development indicate that this situation is very probable when it is necessary to remain awake during periods of high sleep pressure. For humans, this might happen during work at the time of the maximal sleepiness. For rats, partial sleep can accompany experiments during the light phase of the day, if the rats are caged in animal houses with non-inverted conditions of illumination.

People with habitual or forced short length of night sleep are at permanent risk of partial sleep development. For some professions, partial sleep provoked by such chronic sleep deprivation may not cause any troubles. However, for professions connected with responsible and complex decision-making, especially during night shifts, the dangerous consequences of local sleep are very high. For these professions, any visceral disorders, especially in the gastrointestinal system, may dramatically increase the risk of wrong decisions.

Besides the negative effects of partial sleep for mental ability due to the disengagement of some cortical areas from the intellectual work, one should also not neglect the possible negative consequences of this phenomenon for the visceral health of an organism. As participation of all cortical areas in information processing is essential for the efficient solution of complex problems in wakefulness, for efficient management of visceral systems, all cortical areas should be involved into the processing of visceral information. Appearance of partial sleep during wakefulness means that the length of total sleep during the nighttime was not sufficient, or may indicate hidden problems in some visceral systems.

13 Conclusion

The detailed investigation of brain involvement in the regulation of the various visceral systems during sleep is a goal for further studies. At present, following the visceral theory of sleep, we can state that efficient and sufficient sleep, together with visceral health, might be the cheapest, safest and most pleasant way to augment brain function.

Acknowledgments We are very thankful to Dr. Denys Garden for critical reading of the manuscript and many helpful comments. We also thank our collaborators listed as our coauthors in the cited references. Preparation of this article was partly supported by Russian Foundation for Basic Researches grants 19-04-00215 and 17-04-00594-A.

Conflict of Interest Statement Authors declare no conflicts of interest.

Author Contributions Both coauthors have equal contribution to all steps of preparation of this article, and both approved the version to be published.

References

Akerstedt T, Philip P, Capelli A et al (2011) Sleep loss and accidents–work hours, life style, and sleep pathology. Prog Brain Res 190:169–188
Akeyson EW, Schramm LP (1994) Splanchnic and somatic afferent convergence on cervical spinal neurons of the rat. Am J Phys 266(Suppl 1):R268–R276
Amassian VE (1951) Cortical representation of visceral afferents. J Neurophysiol 14:435–446
Andrillon T, Windt J, Silk T (2019) Does the mind wander when the brain takes a break? Local sleep in wakefulness, attentional lapses and mind-wandering. Front Neurosci 13:949
Arendt-Nielsen L, Svensson P (2001) Referred muscle pain: basic and clinical findings. Clin J Pain 17:11–19
Bailey P, Bremer F (1938) A sensory cortical representation of the vagus nerve. With a note on the effects of low blood pressure on the cortical electrogramm. J Neurophysiol 1:405–414
Barger LK, Ayas NT, Cade BE et al (2006) Impact of extended-duration shifts on medical errors, adverse events, and attentional failures. PLoS Med 3:e487
Borbely AA (1982) A two process model of sleep regulation. Human Neurobiol 1:195–204
Borbely AA, Achermann P (1999) Sleep homeostasis and models of sleep regulation. J Biol Rhythm 14:557–568
Borich MR, Kimberley TJ (2011) Both sleep and wakefulness support consolidation of continuous, goal-directed, visuomotor skill. Exp Brain Res 214:619–630
Borich M, Furlong M, Holsman D et al (2011) Goal-directed visuomotor skill learning: off-line enhancement and the importance of the primary motor cortex. Restor Neur Neurosci 29:105–113
Born J, Rasch B, Gais S (2006) Sleep to remember. Neuroscientist 12:410–424
Brooks DC, Bizzi E (1963) Brain stem electrical activity during deep sleep. Arch Ital Biol 101:648–665
Cajochen C, Foy R, Dijk DJ (1999) Frontal predominance of a relative increase in sleep delta and theta EEG activity after sleep loss in humans. Sleep Res Online 2:65–69
Cash SS, Halgren E, Dehghani N et al (2009) The human K-complex represents an isolated cortical down-state. Science 324:1084–1087
Cervero F (1983) Somatic and visceral inputs to the thoracic spinal cord of the cat: effects of noxious stimulation of the biliary system. J Physiol 337:51–67
Cervero F, Connell LA, Lawson SN (1984) Somatic and visceral primary afferents in the lower thoracic dorsal root ganglia of the cat. J Comp Neurol 228:422–431
Chee MW, Tan JC, Zheng H et al (2008) Lapsing during sleep deprivation is associated with distributed changes in brain activation. J Neurosci 28:5519–5528
Chernigovskii VN (1960) Interoreceptori (Interoceptors). Medgiz, Moscow
Chuah LY, Chee MW (2008) Cholinergic augmentation modulates visual task performance in sleep-deprived young adults. J Neurosci 28:11369–11377
Cirelli C, Shaw PJ, Rechtschaffen A et al (1999) No evidence of brain cell degeneration after long-term sleep deprivation in rats. Brain Res 4:184–193
Cohen DA, Robertson EM (2007) Motor sequence consolidation: constrained by critical time windows or competing components. Exp Brain Res 177:440–446
Cohen DA, Pascual-Leone A, Press DZ et al (2005) Off-line learning of motor skill memory: a double dissociation of goal and movement. Proc Natl Acad Sci U S A 102:18237–18241
Contreras D, Steriade M (1995) Cellular basis of EEG slow rhythms: a study of dynamic corticothalamic relationships. J Neurosci 15:604–622
Criscimagna-Hemminger SE, Shadmehr R (2008) Consolidation patterns of human motor memory. J Neurosci 28:9610–9618
Crunelli V, Hughes SW (2010) The slow (<1 Hz) rhythm of non-REM sleep: a dialogue between three cardinal oscillators. Nat Neurosci 13:9–17
Debas K, Carrier J, Orban P et al (2010) Brain plasticity related to the consolidation of motor sequence learning and motor adaptation. Proc Natl Acad Sci U S A 107:17839–17844

Diekelmann S, Born J (2010) The memory function of sleep. Nat Rev Neurosci 11:114–126
Diekelmann S, Buchel C, Born J et al (2011) Labile or stable: opposing consequences for memory when reactivated during waking and sleep. Nat Neurosci 14:381–386
Diekelmann S, Biggel S, Rasch B et al (2012) Offline consolidation of memory varies with time in slow wave sleep and can be accelerated by cuing memory reactivations. Neurobiol Learn Mem 98:103–111
Diekelmann S, Wilhelm I, Wagner U et al (2013) Sleep to implement an intention. Sleep 36:149–153
Dinges DF (1995) An overview of sleepiness and accidents. J Sleep Res 4:4–14
Dinges DF, Pack F, Williams K et al (1997) Cumulative sleepiness, mood disturbance, and psychomotor vigilance performance decrements during a week of sleep restricted to 4–5 hours per night. Sleep 20:267–277
Doyon J, Korman M, Morin A et al (2009) Contribution of night and day sleep vs. simple passage of time to the consolidation of motor sequence and visuomotor adaptation learning. Exp Brain Res 195:15–26
Empson JAC, Clarke P (1970) Rapid eye movements and remembering. Nature 227:28–288
Everson CA, Bergmann BM, Rechtschaffen A (1989) Sleep deprivation in the rat. III. Total sleep deprivation. Sleep 12:13–21
Fernandez LMG, Vantomme G, Osorio-Forero A, Cardis R, Béard E, Lüthi A (2018) Thalamic reticular control of local sleep in mouse sensory cortex. eLife 7:e39111
Ferrara M, De Gennaro L (2011) Going local: insights from EEG and stereo-EEG studies of the human sleep-wake cycle. Curr Top Med Chem 11:2423–2437
Fowler M, Sullivan M, Ekstrand B (1973) Sleep and memory. Science 179:302–304
Gardner ED, Thomas LM, Morin F (1955) Cortical projections of fast visceral afferents in the cat and monkey. Am J Physiol 183:438–445
Genzel L, Dresler M, Wehrle R et al (2009) Slow wave sleep and REM sleep awakenings do not affect sleep dependent memory consolidation. Sleep 32:302–310
Glass BD, Maddox WT, Bowen C et al (2011) The effects of 24-hour sleep deprivation on the exploration-exploitation trade-off. Biol Rhythm Res 42:99–110
Halász P, Pál I, Rajna P (1985) K-complex formation of the EEG in sleep: a survey and new examinations. Acta Physiol Hung 65:3–35
Harrison Y, Horne JA (2000) The impact of sleep deprivation on decision making: a review. J Exp Psychol Appl 6:236–249
Head H (1896) On disturbances of sensation with especial reference to the pain of visceral disease. Brain 19:211–276
Heald S, Siebers RW, Maling TJ (1989) K-complex vasoconstrictor response: evidence for central vasomotor downregulation in borderline hypertension. J Hypertens Suppl 7:S28–S29
Hobson AR, Chizh B, Hicks K et al (2010) Neurophysiological evaluation of convergent afferents innervating the human esophagus and area of referred pain on the anterior chest wall. Am J Physiol Gastrointest Liver Physiol 298:G31–G36
Hornung OP, Regen F, Danker-Hopfe H et al (2007) The relationship between REM sleep and memory consolidation in old age and effects of cholinergic medication. Biol Psychiatry 61:750–757
Hornyak M, Cejnar M, Elam M et al (1991) Sympathetic muscle nerve activity during sleep in man. Brain 114:1281–1295
Huber R, Ghilardi MF, Massimini M, Ferrarelli F, Riedner BA, Peterson MJ, Tononi G (2006) Arm immobilization causes cortical plastic changes and locally decreases sleep slow wave activity. Nat Neurosci 9:1169–1176
Hughes HC, Mullikin WH (1984) Brainstem afferents to the lateral geniculate nucleus of the cat. Exp Brain Res 54:253–258
Johnson LC, Karpan WE (1968) Autonomic correlates of the spontaneous K-complex. Psychophysiology 4:444–452
Kattler H, Dijk DJ, Borbély AA (1994) Effect of unilateral somatosensory stimulation prior to sleep on the sleep EEG in humans. J Sleep Res 3:159–164

Keisler A, Ashe J, Willingham DT (2007) Time of day accounts for overnight improvement in sequence learning. Learn Mem 14:669–672
Krueger JM, Obál F (1993) A neuronal group theory of sleep function. J Sleep Res 2:63–69
Krueger JM, Rector DM, Roy S et al (2008) Sleep as a fundamental property of neuronal assemblies. Nat Rev Neurosci 9:910–919
Krueger JM, Nguyen JT, Dykstra-Aiello CJ, Ping T (2019) Local sleep. Sleep Med Rev 43:14e21
Kuo DC, Krauthamer GM, Yamasaki DS (1981) The organization of visceral sensory neurons in thoracic dorsal root ganglia (DRG) of the cat studied by horseradish peroxidase (HRP) reaction using the cryostat. Brain Res 208:187–191
Lavrova VD, Busygina II, Pigarev IN (2019) Otrajenie aktivnosti serdca v electroencefalogramme koshek v periodi medlennogo sna (Heartbeat-evoked responses on EEG in slow wave sleep in cats). Sensornie Systemi 33:70–76
Lo JC, Groeger JA, Santhi N et al (2012) Effects of partial and acute total sleep deprivation on performance across cognitive domains, individuals and circadian phase. PLoS One 7:e45987
Mander BA, Santhanam S, Saletin JM et al (2011) Wake deterioration and sleep restoration of human learning. Curr Biol 21:R183–R184
Manganotti A, Palermo S, Patuzzo G et al (2001) The role of sleep in learning and memory. Science 294:1048–1052
Maric A, Montvai E, Werth E et al (2017) Insufficient sleep: enhanced risk-seeking relates to low local sleep intensity. Ann Neurol 82:409–418
Mascetti GG (2016) Unihemispheric sleep and asymmetrical sleep: behavioral, neurophysiological, and functional perspectives. Nat Sci Sleep 8:221–238
McCoyand G, Strecker RE (2011) The cognitive cost of sleep lost. Neurobiol Learn Mem 96:564–582
McDermott CM, LaHoste GJ, Chen C et al (2003) Sleep deprivation causes behavioral, synaptic, and membrane excitability alterations in hippocampal neurons. J Neurosci 23:9687–9695
Mitler MM, Carskadon MA, Czeisler CA et al (1988) Catastrophes, sleep, and public policy: consensus report. Sleep 11:100–109
Monstad P, Guilleminault C (1999) Cardiovascular changes associated with spontaneous and evoked K-complexes. Neurosci Lett 263:211–213
Mukhametov LM (1984) Sleep in marine mammals. Exp Brain Res 8:227–238
Mukhametov LM (1987) Unihemispheric slow-wave sleep in the Amazonian dolphin, Inia geoffrensis. Neurosci Lett 79(1–2):128–132
Mukhametov LM, Supin AY, Polyakova IG (1977) Interhemispheric asymmetry of the electroencephalographic sleep patterns in dolphins. Brain Res 134:581–584
Nedergaard M (2013) Garbage truck of the brain. Science 340:1529–1530
Niiyama Y, Sato N, Katsuzava O et al (1996) Electrophysiological evidence suggesting that sensory stimuli of unknown origin induced spontaneous K-complexes. Electroencephalogr Clin Neurophysiol 98:394–400
Okada H, Iwase S, Mano T, Sugiyama Y, Watanabe T (1991) Changes in muscle sympathetic nerve activity during sleep in humans. Neurology 411:1961–1966
Oleksenko AI, Mukhametov LM, Polyakova IG et al (1992) Unihemispheric sleep deprivation in bottlenose dolphins. J Sleep Res 1(1):40–44
Oniani TN, Koridze MG, Kavkasidze MG et al (1974) Dinamika vosbudimosti rasnich struktur mosga v rasnii fasyi cikla bodrstvovanie-son (The dynamics in excitability of various brain structures during different phases of wakefulness-sleep cycle). In: Oniani TN (ed) Neirofiziologia emotsii i tsikla bodrstvovanie-son. Metsniereba, Tbilisi, pp 120–159
Pampiglione G, Ackner B (1958) The effects of repeated stimuli upon EEG and vasomotor activity during sleep in man. Brain 81:64–74
Parker CG, Dailey MJ, Phillips H et al (2020) Central sensory-motor crosstalk in the neural gut-brain axis auton, vol 225. Basic and Clin, Neurosci, p 102656. https://doi.org/10.1016/j.autneu.2020.102656

Patton HD, Amassian VE (1952) Cortical projection zone of chorda tympani nerve in cat. J Neurophysiol 15:245–254

Pavlov IP (1954) Klinicheskie sredi Pavlova (Pavlov's clinical Wednesdays) 1931–1933. In: M.-L: Isd-vo AN SSSR, vol 1. Publishing house of Soviet Academy of Sciences, Moscow, pp 104–105, 164–165, 2

Peles SH, Miranda A, Shaker R et al (2004) Acute nociceptive somatic stimulus sensitizes neurons in the spinal cord to colonic distension in the rat. J Physiol 560:291–302

Perry MJ, Lawson SN (1998) Differences in expression of oligosaccharides, neuropeptides, carbonic anhydrase and neurofilament in rat primary afferent neurons retrogradely labelled via skin, muscle or visceral nerves. Neuroscience 85:293–310

Pigarev IN (1994) Neurons of visual cortex respond to visceral stimulation during slow wave sleep. Neuroscience 62:1237–1243

Pigarev IN (1997) Partial sleep in cortical areas. WFSRS Newsl 5:7–8

Pigarev IN (2014) The visceral theory of sleep. Neurosci Behav Physiol 44:421–434

Pigarev IN, Pigareva ML (2012) Sleep and control of visceral functions. Neurosci Behav Physiol 42:948–956

Pigarev IN, Pigareva ML (2019) Historical view on the attempts to understand the function of sleep in the school of Ivan Pavlov and his Russian forerunners and followers. Clin Transl Neurosci 3(1):2514183X1983476

Pigarev IN, Nothdurft H-C, Kastner S (1997) Evidence for asynchronous development of sleep in cortical areas. Neuroreport 8:2557–2560

Pigarev IN, Almirall H, Marimon J et al (2004) Dynamic pattern of the viscero-cortical projections during sleep. Study in New Zealand rabbits. J Sleep Res 13(Suppl.1):574–575

Pigarev IN, Almirall H, Pigareva ML et al (2006) Visceral signals reach visual cortex during slow wave sleep: study in monkeys. Acta Neurobiol Exp 66:69–73

Pigarev IN, Almirall H, Pigareva ML (2008) Cortical evoked responses to magnetic stimulation of macaque's abdominal wall in sleep-wake cycle. Acta Neurobiol Exp 68:91–96

Pigarev IN, Fedorov GO, Levichkina EV et al (2011) Visually triggered K-complexes: a study in New Zealand rabbits. Exp Brain Res 210:131–142

Pigarev IN, Bagaev VA, Levichkina EV et al (2013) Cortical visual areas process intestinal information during the periods of slow-wave sleep. Neurogastroenterol Motil 25:268–e169

Pigarev IN, Bibikov NG, Busygina II (2016a) Changes in the intragastric environment during sleep affect the statistical characteristics of neuron activity in the cerebral cortex. Neurosci Behav Physiol 46:64–72

Pigarev I, Pigareva ML, Lavrova VD et al (2016b) Spinal cord fibers, transmitting somatic information in wakefulness, are engaged in transmission of the visceral information during sleep. 23rd Congr the Eur Sleep Res Soc 140. http://www.esrs-congress.eu/2016.html

Plihal W, Born J (1997) Effects of early and late nocturnal sleep on declarative and procedural memory. J Cogn Neurosci 9:534–547

Ratcliff R, Van Dongen HPA (2009) Sleep deprivation affects multiple distinct cognitive processes. Psychon Bull Rev 16:742–751

Rattenborg NC, Lima SL, Lesku JA (2012) Sleep locally, act globally. Neuroscientist 18(5):533–546

Rechtshaffen A, Bergmann BM (2002) Sleep deprivation in the rat: an update of the 1989 paper. Sleep 25:18–24

Rector DM, Topchiy IA, Carter KM, Rojas MJ (2005) Local functional state differences between rat cortical columns. Brain Res 1047:45–55

Reis J, Fischer JT, Prichard G et al (2015) Time–but not sleep-dependent consolidation of tDCS-enhanced visuomotor skills. Cereb Cortex 25(1):109–117

Rigas P, Castro-Alamancos MA (2007) Thalamo-cortical up states: different effects of intrinsic and extrinsic cortical inputs on persistent activity. J Neurosci 27:4261–4272

Robertson EM, Pascual-Leone A, Press DZ (2004) Awareness modifies the skill-learning benefits of sleep. Curr Biol 14:208–212

Rozjanskiy NA (1954) Materialii fisiologii sna (Materials to sleep physiology). In: M.: Gos. Isd-vo med literaturi. Publishing House of Medical Literature, Moscow
Sanchez-Vives MV, McCormick DA (2000) Cellular and network mechanisms of rhythmic recurrent activity in neocortex. Nat Neurosci 3:1027–1034
Saxvig IW, Lundervold AJ, Gronli J (2008) The effect of a REM sleep deprivation procedure on different aspects of memory function in humans. Psychophysiology 45:309–317
Shadmehr R, Brashers-Krug T (1997) Functional stages in the formation of human long-term motor memory. J Neurosci 17:409–419
Siclari F, Tononi G (2017) Local aspects of sleep and wakefulness. Curr Opin Neurobiol 44:222–227
Siengsukon CF, Boyd LA (2008) Sleep enhances implicit motor skill learning in individuals post-stroke. Top Stroke Rehabil 15:1–12
Smith C (2001) Sleep states and memory processes in humans: procedural versus declarative memory systems. Sleep Med Rev 5:491–506
Song S, Howard JH, Howard DV (2007) Sleep does not benefit probabilistic motor sequence learning. J Neurosci 27:12475–12483
Steriade M, Timofeev I, Grenier F (2001) Natural waking and sleep states: a view from inside neocortical neurons. J Neurophysiol 85:1969–1985
Stickgold R (2005) Sleep-dependent memory consolidation. Nature 437:1272–1278
Tank J, Diedrich A, Hale N et al (2003) Relationship between blood pressure, sleep K-complexes, and muscle sympathetic nerve activity in humans. Am J Physiol Regul Integr Comp Physiol 285:R208–R214
Timofeev I, Grenier F, Bazhenov M (2000) Origin of slow cortical oscillations in deafferented cortical slabs. Cereb Cortex 10:1185–1199
Tononi G, Cirelli C (2003) Sleep and synaptic homeostasis: a hypothesis. Brain Res Bull 62:143–150
Tononi G, Cirelli C (2006) Sleep function and synaptic homeostasis. Sleep Med Rev 10:49–62
Voderholzer U, Piosczyk H, Holz J (2011) Sleep restriction over several days does not affect long-term recall of declarative and procedural memories in adolescents. Sleep Med 12:170–178
Vyazovskiy VV, Olcese U, Hanlon EC, Nir Y, Cirelli C, Tononi G (2011) Local sleep in awake rats. Nature 472:443–447
Walker MP (2008) Cognitive consequences of sleep and sleep loss. Sleep Med 9:S29–S34
Walker MP (2009) The role of sleep in cognition and emotion. Ann N Y Acad Sci 1156:168–197
Walker MP, Stickgold R (2006) Sleep, memory, and plasticity. Annu Rev Psychol 57:139–166
Werth E, Achermann P, Borbely AA (1996) Brain topography of the human sleep EEG: antero-posterior shifts of spectral power. Neuroreport 8:123–127
Werth E, Achermann P, Borbely AA (1997) Fronto-occipital EEG power gradients on human sleep. J Sleep Res 6:102–112
Williamson AM, Feyer A, Mattick RP et al (2000) Developing measures of fatigue using an alcohol comparison to validate the effects of fatigue on performance. Accid Anal Prev 33:313–326
Xie L, Kang H, Xu Q et al (2013) Sleep drives metabolite clearance from the adult brain. Science 342:373–377
Yoo SS, Hu PT, Gujar N et al (2007) A deficit in the ability to form new human memories without sleep. Nat Neurosci 10:385–392

Augmentation of Brain Functions by Nanotechnology

Ioan Opris, Nicholas Preza, Mikhail A. Lebedev, Brian R. Noga, Stephano J. Chang, Manuel F. Casanova, Mircea Lupusoru, Victor M. Pulgar, Sakhrat Khizroev, Jorge Bohorquez, and Aurel I. Popescu

1 Nanotechnologies in Neuroscience

Among the nanotechnologies that emerged recently in neuroscience, we mention the nanoparticles (including magnetic nanoparticles) and their involvement in the blood-brain barrier (BBB) therapy, in nanoelectrical and chemical stimulation, as

I. Opris (✉)
Department of Neurosurgery, University of Miami, Miami, FL, USA

N. Preza · S. Khizroev · J. Bohorquez
Department of Biomedical Engineering, University of Miami, Coral Gables, FL, USA

M. A. Lebedev
Department of Neurobiology, Duke University, Durham, NC, USA

Center for Bioelectric Interfaces of the Institute for Cognitive Neuroscience, National Research University Higher School of Economics, Moscow, Russia

Department of Information and Internet Technologies of Digital Health Institute, I.M. Sechenov First Moscow State Medical University, Moscow, Russia

B. R. Noga
Department of Neurological Surgery, University of Miami, Miami, FL, USA

S. J. Chang
Miami Project to Cure Paralysis, Miller School of Medicine, Miami, FL, USA

Neuroscience Graduate Program, University of Miami Miller School of Medicine, Miami, FL, USA

Division of Neurosurgery, Department of Surgery, University of British Columbia, Vancouver, BC, Canada

M. F. Casanova (✉)
University of South Carolina, Greenfield, SC, USA
e-mail: manuel.casanova@louisville.edu

I. Opris et al. (eds.), *Modern Approaches to Augmentation of Brain Function*, Contemporary Clinical Neuroscience,
https://doi.org/10.1007/978-3-030-54564-2_11

well as recent insights into neuroengineering, involving the characterization of biophysical features of neural cells and the function of neural microcircuits (Vidu et al. 2014). Neural interfacing has been already confirmed feasible for sensors and brain-machine interfaces.

2 Nanoparticles

Nanoparticles (NPs) are ultrafine units in the microscopic domain of few to hundreds of nanometers, but less than a micron in size. NPs appear to be classified with respect to their roles: those with alumina and silver on the toxic range and gold nanoparticles more on the therapeutic spectrum. In tight connection to NPs are the BBB and the exosomes that are relevant for the vital functions of the brain. Other significant nanotechnological advances include a variety of artificial nanomaterials, such as fullerenes (Dunk et al. 2012), carbon nanotubes, CNTs (Iijima 1991), and graphene (Novoselov et al. 2004). Among the promising nanotools with applications in biomedical research, and basic research, are colloidal gold nanoparticles (AuNPs) with their variety of emission spectra encouraged also their adoption in bioimaging. AuNPs can also be coated with molecules and then used as therapeutic-agent delivery or as sensors in diagnostic applications.

Alumina nanoparticles (Al_2O_3-NPs) have been used for almost a century as common immunologic adjuvants in human and veterinary vaccines injected into the muscle (Gherardi et al. 2016). Aluminum oxyhydroxide (Alhydrogel(®)) has been well tolerated in the short term, but the persistence of aluminic granuloma is associated with neurologic problems, including cognitive dysfunction (Gherardi et al. 2016). These recent insights into the aluminic granuloma strongly demand a serious reevaluation of long-term alumina NPs adjuvant safety. Moreover, nano-alumina neurotoxicity induces oxidative stress that accelerates amyloid beta (Aβ) production in humans and Institute of Cancer Research (ICR) female mice (Shah et al. 2015). These NPs induced Alzheimer's disease (AD) neuropathology by enhancing the amyloidogenic pathway of Aβ production and aggregation, and implied the progression of neurodegeneration in the cortex and hippocampus of mice. Overall, prolonged exposure to such NPs may cause developing the AD neurodegenerative disease.

Silver nanoparticles (AgNPs) are thought to create substantial risks to both the environment and the human health. Since the main mechanism of AgNPs toxicity relates to oxidative stress, Dąbrowska-Bouta et al. (2019) showed that oxidative

M. Lupusoru
Carol Davila University of Medicine and Pharmacy, Bucharest, Romania

V. M. Pulgar
Department of Pharmaceutical Sciences, Campbell University, Buies Creek, NC, USA

A. I. Popescu
Department of Electricity, Physics of Solid and Biophysics, Faculty of Physics, University of Bucharest, Bucharest, Romania
e-mail: prof.aurel.popescu@gmail.com

stress-induced alterations, in myelin membranes, may be the cause of ultrastructural disturbances in myelin sheaths. Previous results suggested that the central nervous system (CNS) may be a target of low-level toxicity of AgNPs (Dąbrowska-Bouta et al. 2016). Silver nanoparticle exposure induces neurotoxicity in the rat hippocampus without increasing the BBB permeability (Dan et al. 2018). Ferdous et al. (2019) showed that pulmonary exposure to AgNPs induced lung inflammation that caused cardiac oxidative stress, DNA damage, and apoptosis in the heart, alteration of coagulation markers and thrombosis. AgNPs can produce oxygen free radicals that lead to NO formation. Ag-NPs may protect the morphine-conditioned rats against the naloxone (NLX)-induced withdrawal symptoms due to high-level NO in the carcinoembryonic antigen (Rahimpour et al. 2020).

Gold nanoparticles (AuNPs) with small dimensions and low toxicity possess the desired properties for diagnosis, imaging, control of drug delivery, and the possibility to interact with a variety of substances, largely due to their excellent biocompatibility with the organism (Cabuzu et al. 2015). AuNPs have (1) optical properties, being able to absorb infrared light; (2) large surface and the ability of being coated with a variety of therapeutic agents; (3) a great potential to be used as drug delivery systems; (4) ability to cross the BBB; and (5) capacity to interact with the DNA to produce genotoxic effects. Because of their ability, the principal applications of AuNPs are in amyloid-like fibrillogenesis inhibitors and drug delivery systems. The herpes simplex virus-1 (HSV-1) infection may deregulate the balance between the amyloidogenic and non-amyloidogenic pathways, raising the accumulation of Aβ peptides, one of the hallmarks in the neurodegenerative diseases. Rodriguez-Izquierdo et al. (2020) demonstrated that NPAuG3-S8 crosses the BBB and does not generate cerebral damage to in vivo CD1 mice model. The NPAuG3-S8 could be a promising treatment against neuronal HSV-1 infections and neuronal disorders related to the Aβ peptides.

3 Nanoparticle Formulations in the Diagnosis and Therapy of Alzheimer's Disease

Alzheimer's disease (AD) is one of the most common age-related diseases that occur due to the deposition of amyloid fibrils in the form of extracellular plaques containing Aβ peptide and intracellular tangles of aggregated microtubule-binding proteins (Gupta et al. 2019). Scores of small molecule inhibitors have been designed for the treatment of AD. However, some of these drugs cannot pass through the BBB. To overcome this problem, various NPs or nanomedicines (NMs) have been synthesized. These nanoparticles exploit the existing physiological mechanisms of passing through the BBB, including receptor- and adsorptive-mediated transcytosis that facilitates the transcellular transport of nanoparticles from the blood to the brain. During the last decades, varieties of nanoparticles that differ in their composition have been developed, having potential applications in the diagnostics and therapy of AD. The most common NP formulations that have major impact include in the diagnosis and therapy of AD include polymeric NPs (PPs), gold NPs, gadolinium NPs, selenium NPs, protein-based NPs, and polysaccharide-based NPs.

Manipulating Aβ aggregation provides critical means for the diagnosis and cure of AD and the applications of Aβ-based aggregation systems. Lee et al. (2014) showed that the brain lipid bilayer provides the binding sites for Aβ and drives the fast and efficient formation of Aβ aggregate structures. It is important to emphasize that large Aβ plaque structures (>15 μm in diameter), hallmark for AD, were formed without going through fibril structures when Aβ peptides were co-incubated with AuNPs on the brain supported lipid bilayer (SLB). By simultaneously inducing Aβ aggregation with these two substrates, large Aβ plaque structures (>15 μm in diameter) were formed within a short incubation time, without going through fibril structures that are typically found in a majority of other Aβ aggregation processes. AuNPs were heavily involved with Aβ aggregation, especially in the core part, and the structural features were induced due to the presence of both AuNPs on the brain SLB. Further, this approach could be readily applied for the treatment of Parkinson's disease.

4 Multimodal Nanoparticles Labeling of Neurons

The development of imaging tools for single cells, over long timescales, in intact animals, may depend on the strength of the signal, stability, and specificity of cell labeling (Amirav et al. 2019). Noninvasive imaging techniques, like magnetic resonance imaging (MRI), do not provide cellular resolution or specificity, suffering from low signal-to-noise ratio and, sometimes from low temporal resolution.

To enhance MRI signals, cell-selective bifunctional magneto-fluorescent contrast agents may provide a means for increasing resolution and cellular specificity. Fluorescence provides means for identification of labeled cells and particle location after MRI acquisition, and it can be used to facilitate the design of cell-selective labeling of defined targets. Amirav et al. (2019) reviewed some recent magneto-fluorescent markers and highlighted key differences between them, in terms of durability and relevant approaches.

4.1 Quantum Dots

Quantum dots (QDs), inorganic fluorescent semiconductor nanoparticles, are an attractive alternative with unique optical and electronic properties (i.e., narrow emission spectra, resistance to photobleaching, high quantum yield) and ease of synthesis, and are used both for high-resolution imaging and as probes to label specific molecules or biological tissues (Brus 1983; Jaiswal et al. 2004; Medintz et al. 2005). QDs have high molar extinction coefficient and fluorescence quantum yield, broad absorption, and narrow tunable emission spectra, in addition to excellent tem-

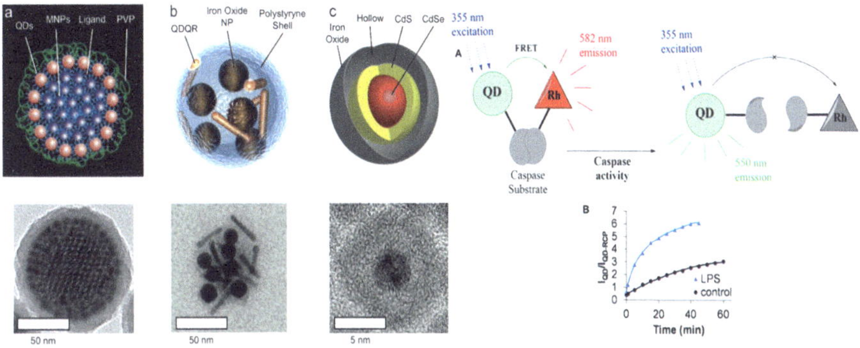

Fig. 1 Left panel. Magneto-Fluorescent Nanoparticles. (**a**) Schematic illustration (top) and TEM micrograph (bottom) of co-assembled super-nanoparticles consisting of closed-packed Fe_3O_4 magnetic nanoparticles "core" surrounded by a "shell" of fluorescent CdSe@CdS quantum dots (QDs). Following coating with silica, the superstructure obtains an average diameter of 100 nm. (Adapted with permission from Chen et al. 2014). (**b**) Schematic illustration and TEM micrograph of magneto-fluorescent nanohybrid produced by two-step co-encapsulation of iron oxide nanoparticles and fluorescent CdSe@CdS quantum dots/quantum rods (QDQRs) into polystyrene shell, with a 50 nm diameter structures. (Adapted with permission from Feld et al. 2015). (**c**) Schematic illustration and TEM micrograph of magneto-fluorescent yolk-shell hybrid structures composed of CdSe@CdS core that is encapsulated within a hollow-Fe_2O_3 shell, with a 10 nm diameter. (Adapted with permission from Pahari et al. 2018). Right panel. QD-based nanosensor for the detection of caspase-1 activity. (**a**) In the absence of caspase activity, FRET between the quantum dot (QD) and rhodamine (Rh) results in the latter's fluorescence emission at 582 nm (shown in red). Active caspase cleaves the substrate peptide linking the QD and rhodamine, and FRET is abolished. This results in the loss of Rh fluorescence and the gain of QD emission (shown in green). (**b**) Representative measures of caspase-1 sensor activity using QD-based sensor. Lipopolysaccharide (LPS) induces caspase activation which leads to rapid cleavage of the peptide link between QD and Rh an increase in the fluorescence ratio of IQD (550 nm)/IQD-RCP (582 nm). *RCP* rhodamine-conjugated peptide. (With permission from Amirav et al. 2019 and Maysinger et al. 2015)

poral stability and resistance to photobleaching. Two prominent examples of such designs are presented in Fig. 1a, b on Left panel. Bawendi and coworkers (Chen et al. 2014) developed colloidal superstructures comprised of close-packed magnetic nanoparticle cores that are fully surrounded by a shell of fluorescent QDs, the core-shell superparticle being coated with a protective silica shell. Furthermore, QDs spectral properties make them ideal candidate to use as donors in Fluorescence Resonance Energy Transfer, FRET (Cooper and Nadeau 2009), which is another important nanotechnique based on the energy transfer from a donor chromophore to an acceptor chromophore, and mostly used to investigate molecular dynamics such as protein-protein interactions. These make QDs particularly advantageous over other fluorescent agents. Hence, alternative synthetic strategies for the fabrication of magneto-fluorescent materials include conjugation of separate nanoconstructs, or co-encapsulation into organic structures or inorganic materials (Cho et al. 2014; Amirav et al. 2019). These conceptual designs are expected to prevent undesirable interactions within the hybrid that could abrogate the respective properties (e.g., see Fig. 1 on Right panel).

5 Nanoparticle-Based Therapeutics for Brain Injury

Brain injuries affect a large patient population with major physical/emotional suffering, at a significant cost to the society (Bharadwaj et al. 2018). Effective diagnostic and therapeutic options available for brain injuries are limited by the complex brain injury pathology involving BBB.

Brain injuries, including ischemic stroke and brain trauma, initiate BBB opening for a short period of time, which is followed by a second reopening for an extended time. The leaky BBB and/or the alterations in the receptor expression on BBB may provide opportunities for therapeutic delivery via nanoparticles (NPs). The approaches for therapeutic interventions via NP delivery are aimed at salvaging the pericontusional/penumbra area for possible neuroprotection and neurovascular unit preservation. Neural injury, particularly brain injury assaults such as stroke, and traumatic brain injury (TBI), provide a unique window of BBB disruption that NP delivery may be able to exploit to enhance passive delivery to injury penumbra. In this section, we highlight studies that exploit passive or active delivery mechanisms following brain injury to achieve drug delivery of various therapeutic molecules, summarized in Fig. 2. The focus of this is to provide a survey of NP strategies

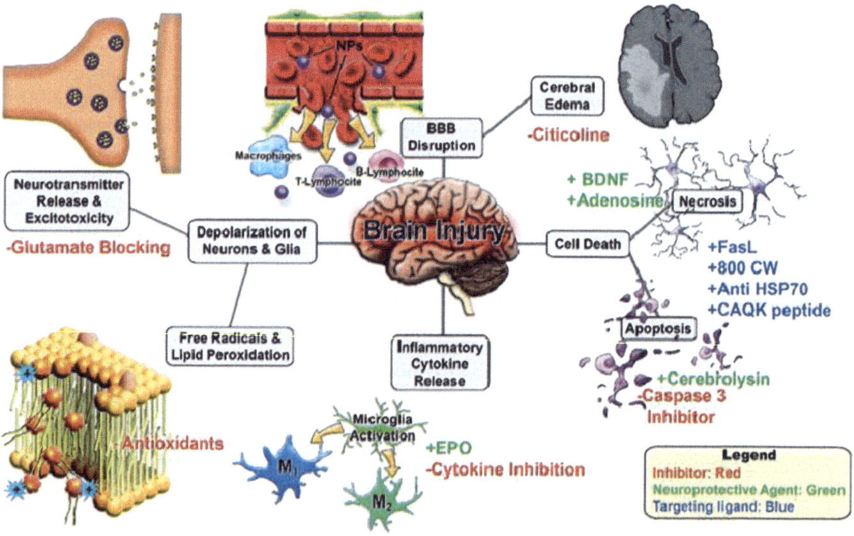

Fig. 2 Therapeutic targets for brain injury for nanoparticle-mediated drug delivery. Brain injury leads to a cascade of secondary damage events (denoted in gray boxes). These secondary injuries are potential targets for therapeutics where NP delivery may be useful. Potential intervention strategies are highlighted with red/green/blue font where inhibitors/blocking agents are in red, neuroprotective agents via pathway modulation in green, and targeting ligands in blue. *EPO* erythropoietin, *BDNF* brain-derived neurotrophic factor, *FasL* Fas ligand, *800CW* nanoparticles with/without coatings, *HSP* heat shock protein, *CAQK* conjugated with targeting peptide. (With permission from Bharadwaj et al. 2018)

employed in cerebral ischemia and brain trauma and, finally, provide insights for improved NP-based diagnostic/treatment approaches.

6 Nanotherapeutic Approaches

In the last few years, a series of therapeutically novel approaches with potential applications emerged. We acknowledge here the ones involving: (1) nanoparticles, (2) carbon nanomaterials, such as graphene, nanotubes, nano-onions, or multilayered fullerenes for therapy, (3) biosensing and imaging approaches that have antioxidant action, (4) intrinsic photoluminescence, (5) the ability to cross the BBB, carrying oligonucleotides and cells, and (6) the ability to induce cell differentiation. Fernandes et al. (2018) used liposomes and carbon nanomaterials, in recent diagnosis and therapies, in acute ischemic stroke. Liposomes represent a biomimetic system with composition, structural organization, and properties very similar to biological membranes. Carbon nanomaterials, not being naturally parts of the human body, reveal new modes of interaction and integration with biological molecules and systems, resulting in completely unique pharmacological properties.

Novel genetic neuroprotective cell therapeutics are bringing promising approaches for the regenerative functions in the eye. Nafissi and Foldvari (2015) discussed these genetic nanotechnology neuroprotective therapies in glaucoma. The development of highly specific gene delivery methods, with safe noninvasive in the eye, is crucial. Nadeau (2015) reported initial photophysical characterization of a new genetically encoded voltage sensor (based upon the fluorescence of rhodopsins), namely the "proteorhodopsin optical proton sensor" (PROPS). This is the first sensor capable of indicating the changes in membrane voltage by means of changes in fluorescence. Nadeau reported in two strains of *Escherichia coli*, a nanosecond time-resolved emission of this protein, before and after membrane depolarization. Infante (2018), proposed to examine the "affinity and nanoparticle-based strategies" for the delivery of neurotrophic factors to the spinal cord in an adequate, tunable, and safe therapeutic manner.

d'Amora and Giordani (2018) showed that the zebrafish is an adequate animal model for "high-throughput" screening of chemicals, due to their small size, low-price, and transparency. The zebrafish has emerged as a powerful tool for screening developmental neurotoxicity. Convertino et al. (2018) elaborated on the graphene's potential for nerve tissue regeneration hinting to novel approaches of active nerve conduits for peripheral neuron survival and outgrowth. Moldovan et al. (2018) carried out experiments in adult transgenic mice with fluorescent-tagged liposomes that provided insight into the local anesthetic effect of nanomedicines in postoperative pain. The effect of local anesthetic nanomedicines has important implications for humans.

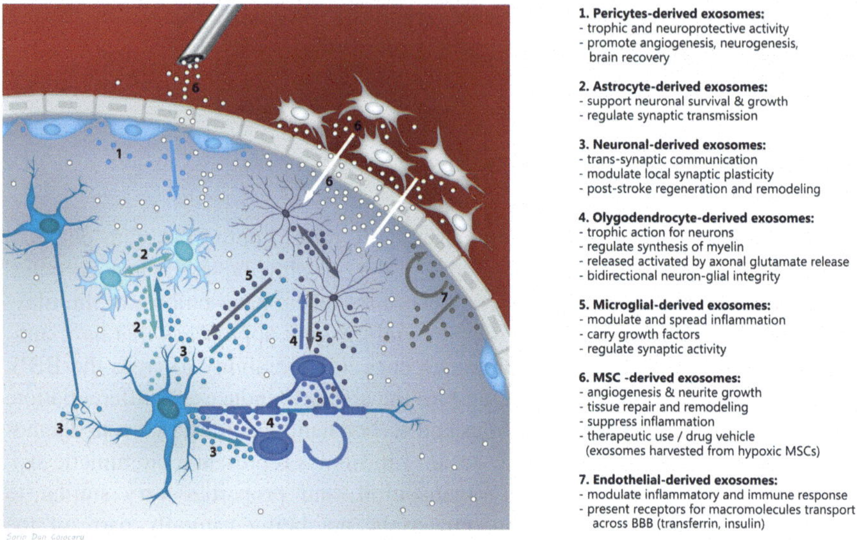

Fig. 3 Exosomes (represented by differently colored dots) as fundamental participants to the mechanism of communication between components of the neurovascular unit. *MSC* mesenchymal stem cell. (With permission from Zagrean et al. 2018)

6.1 Exosomes as a Communication Tool

Exosomes are defined as 30- to 100-nm-sized membrane vesicle derivatives of the endosomal compartment and correspond to the intraluminal vesicles of multivesicular bodies (MVBs) that, upon fusion of the MVBs with the plasma membrane, are released as exosomes into the extracellular environment, where they act as signaling organelles for intercellular communication (Fig. 3).

From the extracellular milieu, exosomes may contact target cells by (1) receptor-mediated adhesion to the cellular plasma membrane, followed by endocytic uptake and internalization, and (2) direct fusion of the exosome membrane with the target cell membrane and subsequent exosomal content release into the recipient cell (Bang and Thum 2012). Emerging research has revealed that the nano-sized exosomes could be used for neurovascular unit (NVU) remodeling after stroke, due to their ability to mediate cell-to-cell communication. Exosomes are becoming effective therapeutic tools for many brain disorders. These vesicles can bypass biological barriers, including BBB, and can play the role of drug or gene therapy transporters. Herein, Betzer et al. (2017) provided a noninvasive method based on glucose-coated gold nanoparticle (GAuNP) labeling and computed tomography imaging, for in vivo neuroimaging and tracking of exosomes. Betzer et al. (2017) demonstrated that 5 nm AuNPs enabled improved exosome labeling. Also, when comparing the intranasal to intravenous administration, it led to higher brain accumulation and thus "enhanced" in vivo neuroimaging. Next, Perets et al. (2019) perfected the neuroimaging tool to track the migration and homing patterns of intranasally administrated exosomes derived from bone marrow mesenchymal stem cells (MSC-exo) in several brain pathologies, including AD, PD, stroke, and autism. The neuroinflammatory signal in pathological states appears highly correlated with MSC-exo accumulation, suggesting that the "homing" mechanism is inflammatory-driven. Overall, these findings show that golden exosomes "selectively target brain pathologies" in neurodegenerative and neurodevelopmental disorders.

7 Overcoming BBB to Treat Neurodegenerative Diseases

The BBB is the unique and protective feature of the CNS microvasculature controlling brain homeostasis as well as ion and molecule movement (Saraiva et al. 2016). This highly specialized endothelial structure lines the lumen of brain microvessels and modulates the transport of blood proteins and cells into the CNS, also providing an efficient removal of potentially neurotoxic substances from brain to the blood. Thus, brain micro-vessel endothelial cells (MVECs) control the functions of the BBB; their intercellular connections formed by tight and adherent junctions control paracellular diffusion whereas active and passive receptors and channels control the flux of macromolecules in and out of the neural milieu. MVECs are highly polarized cells with distinct apical and basolateral components where efflux transporters and nutrient's transporters are integral components of the BBB. Failure in maintaining any of these components results in the breakdown of this specialized multicellular structure and consequently promotes neuro-inflammation and neurodegeneration. In several pathologies such as MS, epilepsy, traumatic brain injury, stroke, AD, and PD, the BBB is impaired. For example, the structure of brain endothelium as well as BBB transport is altered in AD patients, and several reports showed that the expression of the GLUT1 glucose transporter is diminished in AD patients and AD animal models, suggesting an energy impairment in the AD brain.

Restorative strategies of brain function after stroke are centered on the repairing of cerebral endothelial and parenchymal cells. Communication between the cells and signaling within the neurovascular unit, including the multicellular brain-vessel-blood interface, with its highly selective BBB, are crucial to the homeostasis and function of the central nervous system. The work of Zagrean et al. (2018) highlights the important role of exosomes in mediating the crosstalk of the cells within the neurovascular unit. It further reveals the restorative therapeutic potential of exosomes in ischemic stroke, a frequent neurologic condition, in need of an efficient therapy.

MVECs display transcytosis at a lower rate compared to other non-neural ECs; however, BBB transcytosis seems to be upregulated during injury. Pulgar draws our attention to the physiological operation of "receptor-mediated transcytosis" (RMT) to carry load across the brain endothelial cells toward brain parenchyma, exemplifying critical advances in RMT-mediated brain drug delivery (Pulgar 2019). However, even a damaged, and a more permeable BBB, can pose serious challenges to drug delivery into the brain. The use of NP formulations able to encapsulate molecules with therapeutic value, while targeting specific transport processes in the brain vasculature, may enhance drug transport through the BBB in neurodegenerative/ischemic disorders and target relevant regions in the brain for regenerative processes.

The more common types of NPs formulations used to deliver drugs into the brain involve polymers, biomimetics, and inorganic NPs. Several of these groups of NPs have been formulated into therapeutic strategies for brain diseases. For example, PEG-PLA NPs containing paclitaxel, liposomes containing doxorubicin, and lactoferrin NPs containing temozolomide represent an example of each group and have been tested in the treatment of glioma.

Oxytocin is a large-molecule neuropeptide with therapeutic promise for social-deficit disorders. In order to improve oxytocin's brain penetrance and duration of action, even with intranasal administration, Oppong-Damoah et al. (2019) designed a NP formulation to increase its brain bioavailability through active transport and increase its duration of action through encapsulation and sustained release. The transport of oxytocin-like large molecules was evaluated in a cell culture BBB model, and it was found that NP formulation increased BBB transport both in vitro and in vivo. Moreover, nanoparticle-encapsulated oxytocin, intranasally administered, exhibited greater pro-social effects both acutely and 3 days after administration, in comparison with oxytocin alone, in mouse social-interaction experiments.

8 Neuromodulation of the Brain

Nanomagnetic and nanoelectric particles are endowed with valuable interactive abilities for neuronal cells. Pinkernelle et al. (2015) assessed the "bio-functionality" of "functionalized" growth factors using appropriate biological models. Thus, successful "functionalization" of magnetic nanoparticles with growth factors seems dependent on the "binding" chemistry. These magnetic nanoparticles support the regeneration idea in the nervous system (Fig. 4).

9 Noninvasive Neuromodulation by Magneto-Electric Nanoparticles

Khizroev (2019) and his team are using magneto-electric nanoparticles (MENPs) to artificially stimulate the neural activity deep in the brain. The new technology provides a unique way to couple electric signals in the neural network to the magnetic dipoles in the nanoparticles with the purpose to enable a noninvasive approach. Simulations of the effect of MENPs for noninvasively stimulating the brain of a patient with PD to bring the pulsed sequences of the electric field to the levels

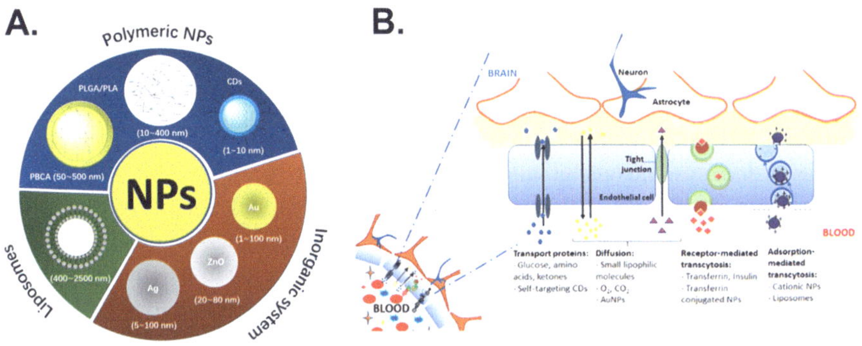

Fig. 4 (**a**) Types of NPs and the BBB. Different types of NPs used to cross the BBB are shown with sizes indicated. (**b**) Main components of the BBB showing the MVECs (light blue), astrocytes (white), and neurons (dark blue). Mechanisms of BBB crossing indicated include transporter- and diffusion-mediated, as well as receptor- and adsorption-mediated transcytosis (RMT). (With permission from Zhou et al. 2018)

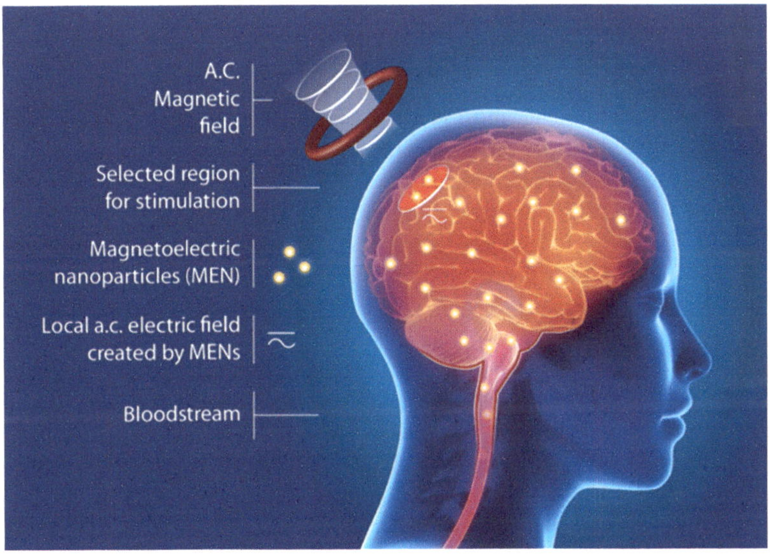

Fig. 5 Illustration of a wireless electric stimulation with magneto-electric nanoparticles (MENPs) via application of A.C. magnetic fields. (With permission from Khizroev 2019)

comparable to those of healthy people show that the optimized values for the concentration of the 20 nm nanoparticles (with the magneto-electric coefficient of 10^6 V cm/T in the aqueous solution) are 3×10^6 particles/cm^3, and the frequency of the externally applied 3×10^{-2} T magnetic field is 80 Hz (Fig. 5).

To enable patient- and disease-specific diagnosis and treatment, at the intracellular level, there needs to be a way to locally stimulate specific neurons to navigate and dispense cargo to definitive locations. Significant progress has been made in biotechnology in discovering techniques to improve diagnosis of many diseases, most recently, the use of magneto-electric nanoparticles. MENPs are a class of multiferroic nanostructure that has unique properties that permit them to couple the wireless control efficiency of magnetic fields and the complex access to intrinsic molecular level information by local electric fields. MENPs have the potential to open new doors for nanotechnology capabilities due to the magneto-electric effect. Regarding diagnosis, MENPs can enhance image sites with relatively high efficiency and adequate resolution. MENPs can be used for image-guided navigation through organs to intensify resolution through the application of external magnetic field gradients. Imaging modalities like magnetic resonance imagining (MRI) and the relatively new magnetic particle imaging (MPI) can greatly enhance diagnosis when combined with MENPs. Recent studies have indicated that MENPs will heighten the image contrast in MRI and MPI while providing an image not only containing structural information but also revealing local intrinsic electric fields.

MENPs have significant capabilities to provide wireless stimulation to specific regions deep in the brain. Improved treatment of disabilities related to motor and sensory impairment or damage and curing patients that suffer from neurodegenerative diseases is feasible with the introduction of MENP therapies. Most neurodegenerative diseases like PD and epilepsy result from one or several electrical subcircuits being damaged or completely broken. Electric field stimulation approaches are current methods used today, but most proved to be inefficient. However, MENPs, on the other hand, have recently proven to enhance repair of damaged neurons affected by disease to regain their communication skills.

Another way, MENPs have proven to improve the field of nanomedicine through its effective capability to enhance controlled drug delivery and release. MENPs make it possible to deliver therapeutic loads across the BBB with a high efficiency and allow adequate release to specific target sites at any time. The novel concept to use MENPs to access selective regions deep in the brain for mapping starts with the NPs being guided into the brain across the BBB through the application of a d.c. magnetic field gradient. When the particles are in the brain, they are dispersed over the entire brain or to a specific target region via the application of varying d.c. magnetic field gradients. The particles create a new microenvironment in the brain by strongly coupling the intrinsic signals generated by neural activity at the subneuronal level with the external magnetic field. The coupling allows for improved neural stimulation compared to current therapies. The use of MENPs for ultimate patient treatment and diagnosis is a milestone in the development of personalized nanomedicine.

10 Nanoelectrical and Chemical Stimulation

A developing generation of minimally invasive or noninvasive neural stimulation techniques is supported by nanotechnology to reach at a high spatial resolution. In these approaches of neural stimulation, as pointed out by Wang and Guo (2016), a nanomaterial transforms a faraway transmitted primary stimulus (like a magnetic or ultrasonic signal), into a localized secondary stimulus, such as an electric field, to stimulate neurons. Stimulating neural systems with "applied" electric field (EF) is a common tool for testing neural network responses. For example, Tang-Schomer et al. (2018) examined the effects of "applied" EFs on neuronal networks of in vitro cultures. The interface cultures were "set up" with extensions of the silk protein film-embedded gold wires, connected to an electrical stimulator. Cortical cultures displayed large-scale oscillations, synchronized by EF at specified frequencies. These effects of EF on random neuronal networks have significant implications for studies of brain function and neuromodulation. Furthermore, Goldental et al. (2016) mimicking the collective firing patterns of connected neurons proved the emergence of cooperative phenomena like synchronous oscillations, the coexistence of fast γ and slow δ oscillations, and other dynamical phenomena within large-scale neuronal networks.

Novel nanofluidic mechanisms like hydrophobic gating, suggested by Jones and Stelzle (2016), may support the control of chemical release appropriate for mimicking neurotransmission. Nanofluidic chemical release facilitates fast high-resolution neurotransmitter-based neurostimulation that could bring improvements over electrical neurostimulation.

11 Neuroengineering

Nanotechnology is a fast-developing field that provides simple and efficient tools to study the brain in health and disease. Of particular importance are the biosensors, multielectrode arrays, memory resistive devices, and the brain-machine interfaces.

12 Sensors

Moretti et al. (2018) demonstrated the biocompatibility of a magnetic sensor array for the detection of neuronal signals in the in vitro culture.

12.1 Magnetic Tunneling Junctions Sensor

First chip based on magnetic tunnel junction (MTJ) technology for cell culture studies, that shows the biocompatibility of these sensors, was reported by Moretti et al. (2018) (Fig. 6).

Moretti's team obtained a full biocompatibility of the system through the planarization of the sensors and the use of a three-layer capping of $SiO_2/Si_3N_4/SiO_2$. They grew primary neurons up to 20 days on the top of the devices and obtained

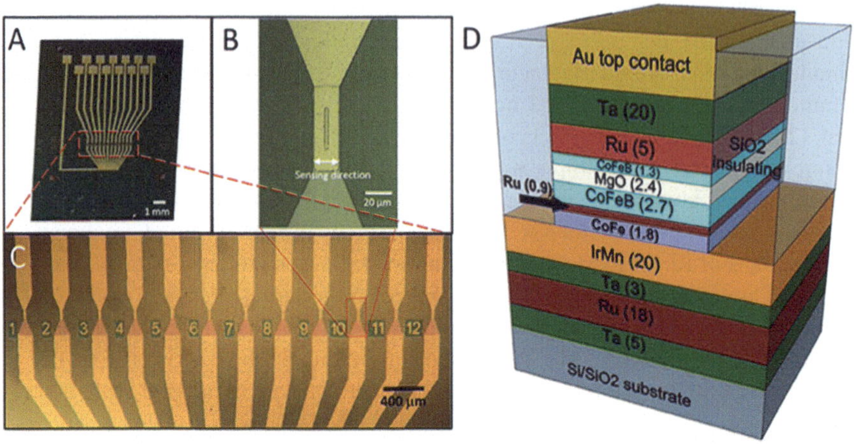

Fig. 6 (**a**) Optical image of the chip featuring the 12 sensors, each provided with an independent contact and a common ground contact; (**b**) Zoom image of a single sensor area (4 × 30) μm^2; (**c**) Zoom image of the sensor active region, composed of 12 magnetic tunneling junctions; (**d**) Structure of a magnetic tunneling junction (thickness of the layers in nm). (With permission by Moretti et al. 2018)

proper functionality and viability of the overlying neuronal networks. At the same time, MTJ sensors kept their performances unchanged for several weeks in contact with neurons and neuronal medium. These results pave the way to the development of high performing biomagnetic sensing technology for the electrophysiology of in vitro systems, in analogy with Multielectrode Arrays.

Cellular processes such as the deformation of membrane, the transport of organelles, or the migration of cells are sensitive to mechanical forces, operating through the "chaperoning" force-inducing nanoparticles in electrical/magnetic field gradients, with spatial precision in the range of sub-micrometers. Gahl and Kunze (2018) used force-mediating magnetic nanoparticles to generate neuronal cell function. Moretti et al. (2018) produced the first bio-magnetic chip using a novel technology based on MTJ for cell culture, demonstrating that these sensors are biocompatible. Such advancements of nanomagnetic field in cellular organization/communication/signaling and intracellular trafficking can be used in the next neurotherapeutic devices.

12.2 Optical Probes for Neurobiological Sensing and Imaging

The biochemical signaling, both inside and among cells, can be imaged with the advent of new generation tools: fluorescent nanosensors and molecular probes. Until now, the single-cell electrophysiological techniques were considered as "the

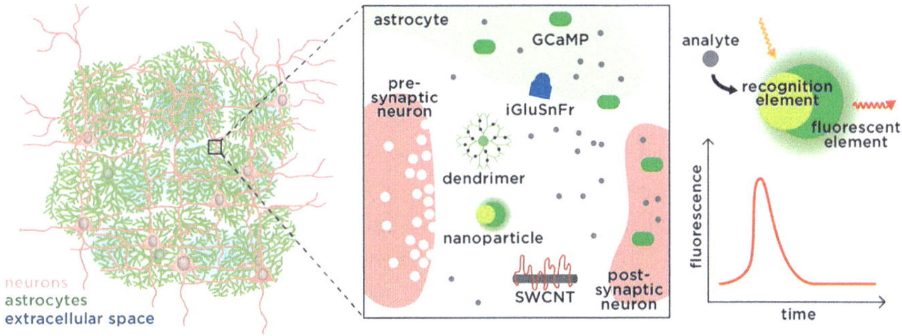

Fig. 7 Fluorescent nanosensors. Optical probes for neurobiological sensing and imaging. *GCaMP* genetically encoded Calmodulin protein, *iGluSnFr* glutamate-sensitive fluorescent reporters, *SWCNT* single-wall carbon nanotube. (With permission of Kim et al. 2018a, 2018b; Acc. Chem. Res.)

gold standard" able to elucidate the neural network dynamics with a high spatiotemporal resolution (Fig. 7).

However, these electrophysiological techniques are confronted with serious challenges in clarifying, with an adequate space and temporal precision, the molecules involved in neuronal network activation. But the imaging of molecule dynamics is extremely important for the understanding the synergic interactions between neurons themselves, as well as between neurons and non-neural cells. Fortunately, in the last years, important technological progress, both in optical probes and imaging techniques, have allowed to approach, at molecular level, the single neurons, as well as their circuitry promising the acquisition of biochemical, temporal, and spatial information. These new techniques, complementary to the classical physiological ones, will certainly allow to unravel the intricate neuronal interactions, at different levels of the brain morphology, in order to obtain the spatiotemporal image of neuronal dynamics. Until now, just a few of probe candidates have provided detailed brain biochemical information. For instance, the intracellular Ca^{2+} dynamics technique was the main method to follow the biochemical mapping of neuron activities, by the aid of fluorescent molecular reporters.

Advances for monitoring arrays of chemicals in the brain will make possible the mapping of neuronal circuitry/units with a high spatiotemporal accuracy. In this respect, the optical probes can sense, with minimal injury to brain tissue, a large variety of molecule arrays, providing very accurate spatiotemporal information on neuronal dynamics (Kim et al. 2018a, b). Optical customized nanosensors, dynamically monitoring the neuron microenvironment with a high spatiotemporal resolution, will allow not only new insights into brain cell function, but also the understanding how various signaling systems act in conjunction with the neighbor neural and non-neural cells.

13 Biosensors

Many challenges of sensor development, including the bioengineered probes and sensors, arise when the physiological and pathological biomarkers are tested in neural cells (Maysinger et al. 2015). The nanoparticle-based sensors have the ability to detect properties (biochemical and physiological) of neurons and glia and to generate signals proportional to the changes (physical, chemical or electrical) in these cells (Maysinger et al. 2015). Among the most used nanostructures are the carbon-based structures (such as C-dots, graphene, and nanodiamonds), the QDs, and AuNPs. They are capable to detect and measure the activity of proteases (e.g., metalloproteinases, caspases), ions, and other biomolecules under physiological or pathological conditions, in neuronal cells. Such genetically manipulated probes and sensors are useful to reveal the changes in protease activities or calcium ion concentrations.

14 Neuronal Recording

Conventional neural recording electrodes, such as the Michigan-type silicon probe, are orders of magnitude stiffer than the brain tissue. By taking advantage of the dependence of bending stiffness on probe thickness, mesh electronics is the only neural probe to date with effective bending stiffness comparable to that of neural tissue (Fig. 8).

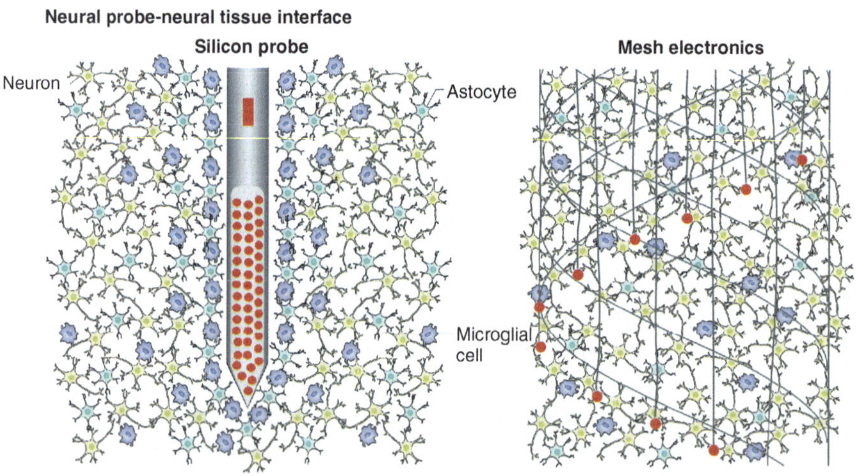

Fig. 8 Left panel: Neural (silicon) probe-neural tissue interface. Right panel: Mesh electronic neural probes. The probes are designed to have the largest feature sizes similar to individual neuron soma, bending stiffness values on par with the brain tissue. (With permission of Dai et al. 2018)

A stiff implant causes repeated chronic damage to the much softer surrounding brain tissue, generating various chronic immune responses, including, most prominently, death and damage of neurons near the probe, and the proliferation of astrocytes and microglia to form a ~100 μm thick glial sheath that insulates the recording electrodes from neurons (Dai et al. 2018). As a result of the chronic immune response, the natural distribution of neurons and glial cells is considerably perturbed at the electrode-brain interface. Therefore, a chronically stable electrode-brain interface should minimize disturbing the natural distribution of neurons by making tissue-like electronics, using materials with similar mechanical properties to brain tissue.

Neural recording electrode technologies have contributed considerably to neuroscience by enabling the extracellular detection of low-frequency local field potential oscillations and high-frequency action potentials of single units. Nevertheless, several longstanding limitations exist, including low multiplexity, deleterious chronic immune responses, and long-term recording instability. Driven by initiatives encouraging the generation of novel neurotechnologies and the maturation of technologies to fabricate high-density electronics, novel electrode technologies are emerging. Here, we provide an overview of recently developed neural recording electrode technologies with high spatial integration, long-term stability, and multiple functionalities. We describe how these emergent neurotechnologies can approach the ultimate goal of illuminating chronic brain activity with minimal disruption of the neural environment, thereby providing unprecedented opportunities for neuroscience research in the future.

15 Multisite Attenuated Intracellular Recordings by Extracellular Multielectrode Arrays

Multielectrode arrays (MEA) have been developed and extensively used in basic and applied research in neuronal and cardiomyocyte networks, both in vivo and in vitro (Spira et al. 2018).

The two panels of Fig. 9 depict the basic principles and physical constraints of electrode technologies. Schematic depicting the basic physical principles of measuring bioelectrical signals by recording electrodes in neural tissue (left), with representative raw, low-pass-filtered and high-pass-filtered neural recording traces (middle) and sorted spikes (right) measured by an electrode. (With permission from Spira et al. 2018).

Right panel: The MEA platforms consisting of thousands of sensors (with high-density, small diameter, and low impedance) use vertical nanowires that pass through the cultured cell's membrane and record the action potentials in a similar manner to that of a sharp intracellular microelectrode. Spira's team developed a "bioinspired approach" in which cell's energetic resources are utilized with extracellular gold microelectrodes to record attenuated synaptic and action potentials

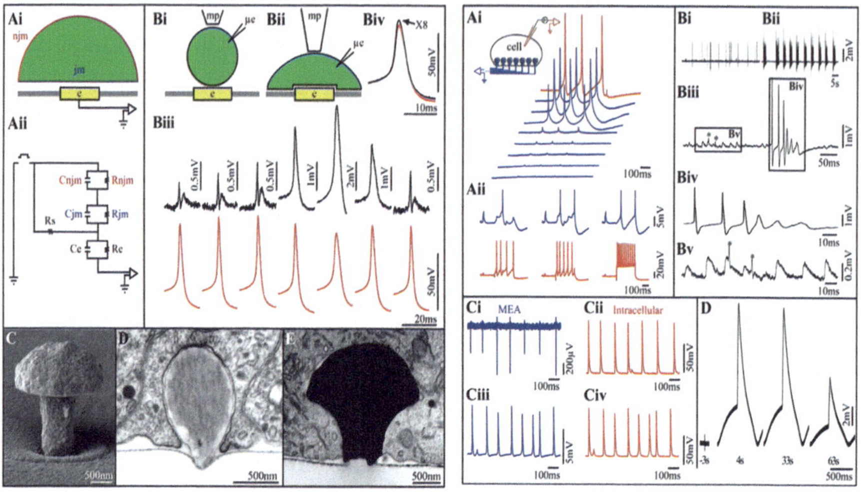

Fig. 9 Left panel: (Ai) Schematic drawing of a cell (green) residing on a planar sensing electrode (yellow) and the space between them (white). (Aii) An analog electrical circuit of the cell-electrode interface. In the model, the cell's surface area is subdivided into a non-junctional membrane (njm, red) that faces the grounded culture medium, and a junctional membrane (jm, blue) that faces the electrode. Each of these membrane compartments is represented by a resistor and capacitor in parallel R_{njm}, C_{njm}, R_{jm}, and C_{jm}, respectively. The cleft between the cell and the sensor is represented by a resistor (the seal resistance—R_s). The electrode is represented by a resistor and capacitor (R_e, C_e, respectively). The electrical coupling coefficient (CC) between a cell and a recording device is defined as the ratio between the voltage recorded by the device (electrode-amplifier) and the voltage generated across the plasma membrane of an excitable cell (V_{elect}/V_{cell}). The square pulse in between the ground jm and njm is a voltage calibration pulse. (Bi, Bii) Schematic illustrations of downward displacement of a cell to increase the seal resistance (mp—a fire-polished pipette to exert the pressure, μe—intracellular recording electrode). (Biii) Concomitant alterations in the extracellularly recorded FPs (upper traces—black) and intracellular APs (lower traces—red) during the displacement of the neuron's cell body toward the planar electrode. From left to right, initially the increased FPs amplitude is not associated with changes in the intracellular AP amplitude. Thereafter (fourth trace), the extracellular FP recorded by the planar electrode transformed to intracellular recordings of an AP. This is associated by reduction in the AP amplitude recorded by the μE. Release of the mechanical pressure led to the reversal of the process (traces 6 and 7). (Biv) Super positioning of the intracellular recorded APs with the sharp electrode and the extracellular planar electrode (multiplied by 8). Note that although the amplitudes of the two APs are different, the shapes of the APs are similar. (Biii, Biv modified with permission from Cohen et al. 2008). (C) A scanning electron microscope image of a gMμE. (D) A latex bead engulfed by a cultured Aplysia neuron. (E) Electron microscope image of a thin section showing a gMμE (black) tightly engulfed by a cultured PC12 cell. Right panel: Differences in the levels of the seal resistance formed between a single Aplysia neuron residing on eight gMμEs (insert) lead to differences in the IN-CELL recorded APs amplitudes. A cultured Aplysia neuron was intracellularly stimulated to fire three consecutive APs (red trace). Simultaneous recordings of these APs by eight gMμE (blue traces) revealed differences in the IN-CELL recorded amplitudes. (Aii) Synaptic and action potentials recorded by an extracellular gMμE. Stimulation of a presynaptic neuron by an intracellular sharp electrode (red) leads to the generation of excitatory postsynaptic potentials (blue) recorded by a gMμE. Note the summation of the EPSPs to reach firing threshold. (Ai, Aii modified with permission from Hai et al. 2010). (B) Spontaneous activity recorded by a gMμEs from a cultured hippocampal neuron. (Bi) Control spontaneous APs firing recorded by a gMμE. (Bii) Ten minutes

with characteristic features resembling those of intracellular recordings. Moreover, the approach allowed to record intracellular potentials by an array of extracellular electrodes. Intracortical microelectrodes (IME) are tools that allow neuroscience researchers to examine the nervous system. Recent strategies to enhance interfacing with the brain's systems have been suggested by methods that mimic the biological tissue. Kim et al. (2018a, b) review focused on nano-architecture, a concept that considers the surface of the implant. Different nano-architectural approaches have been discussed to enhance the "biocompatibility" of IMEs, increase the recording quality, and augment the longevity of the implant.

Microelectrode material together with cell culture medium plays important roles in cell's health during in vitro electrophysiological studies. Ryynänen et al. (2018) reported an "ion beam assisted e-beam deposition" (IBAD)-based process as being an alternative to titanium nitride (TiN) method of deposition by "sputtering" in the fabrication of TiN microelectrode arrays. The developed IBAD TiN process enables the MEA manufacturers with more choices as to which method to use in order to deposit TiN electrodes. In addition to electrode material, also the insulator layer and cell culturing medium keep a crucial role in successful long-term MEA measurements.

16 Interface Microelectrodes for Ultrasensitive Monitoring of Alzheimer's Disease

Ultrasensitive detection of monomeric Aβ peptides is of fundamental significance when studying the pathological progression of AD. Ding et al. (2020) developed a novel electrochemical biosensor for sensitive and selective monitoring Aβ peptides (monomers in cerebrospinal fluid, CSF).

Fig. 9 (continued) after the application of 10 μm GABAzin the firing pattern was changed. (Biii) Enlargements of the records in (Bi). (Biv) Enlargement of the right box in (Biii). (Bv) Enlargement of the left box in (Biii). The low amplitude long duration potentials in (Biii) (left box) have the features of excitatory postsynaptic potentials. The fast spikelets indicated by asterisk could be either dendritic spikes or the firing of electrically coupled neurons. As the dynamics and amplitudes of the potentials shown in (Bv) are not altered by GABAzin, it is unlikely that they reflect barrage of FPs generated by remote neurons. (C) Comparison of intracellular recorded potentials to IN-CELL recordings from cultured myotubes obtained by gMμE electroporation. (Ci, Cii) Simultaneous extracellular FPs recordings by a gMμE (Ci) and intracellular recordings by a sharp electrode (Cii). The recordings (Ci, Cii) revealed identical firing patterns and similar qualitative alterations in the amplitudes of the recorded action potentials. (Ciii) Electroporation of the myotube changed the recording mode by the gMμE from extracellular to the IN-CELL. Note that although the IN-CELL recorded amplitude of the APs is about an order of magnitude lower than that of the intracellular electrode APs, the shape of the recorded potentials is identical. Also, note that both the gMμE (Ciii) and the intracellular sharp electrode (Civ) recorded subthreshold potentials in between the APs. (C modified with permission from Rabieh et al. 2016). (D) IN-CELL recordings from cultured human cardiomyocytes: 3 s before electroporation and 4, 33, and 63 s after electroporation. (With permission from Spira et al. 2018)

Through specific Cu^{2+}-Aβ-hemin coordination, Aβ directs the assembly of Cu^{2+}-PEI/AuNPs-hemin nanoprobes into network aggregates on a microelectrode interface, thus promoting the enrichment of Aβ monomers on the microelectrode. The AuNP aggregate promotes the deposition of AgNPs, which are then utilized for the electrochemical stripping analysis of the Aβ monomers. The proposed method displayed ultrasensitivity for Aβ monomers with the detection limit down to 0.2 pM. The exposed method also displayed a high selectivity toward Aβ monomers. This remarkable analytical performance renders the electrochemical biosensor useful for evaluating the dynamic change of Aβ monomer levels in the CSF of live mice with AD, prompting the investigation of the role that Aβ monomers play in brain events. In Fig. 10, we present a schematic illustration of the electrochemical biosensor for the measurement of Aβ monomers. The process of Aβ-directed aggregation of Cu^{2+}-PEI/AuNPs-hemin nanoprobes was transferred to the gold electrode surface and converted colorimetric assay into enhanced electrochemical analysis.

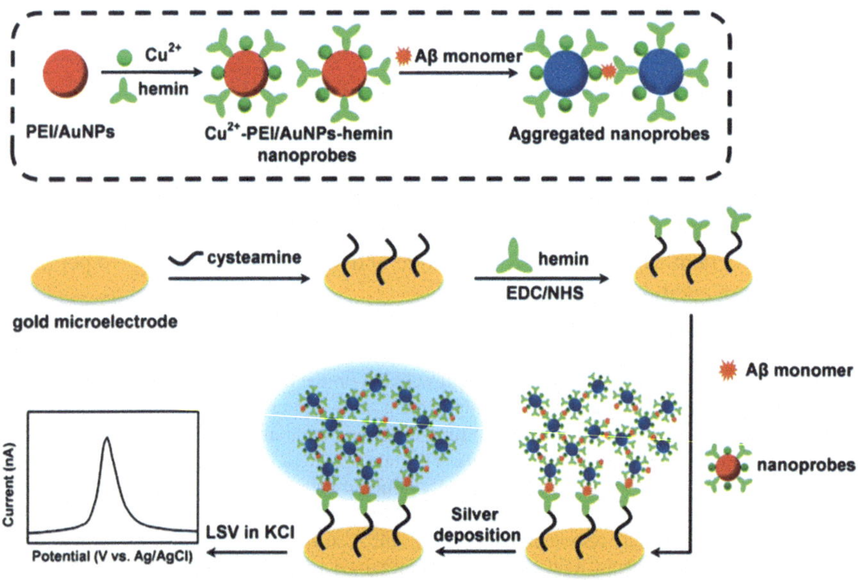

Fig. 10 Ultrasensitive detection of monomeric Aβ peptides has a fundamental significance for studying the pathological progression of AD. *PEI* polyethyleneimine, *EDC/NHS* (1-ethyl-3-[3-dimethylami-nopropyl]-carbodiimide/N-hydroxysuccinimide, *LSV* linear sweep voltammetry. (With permission from Ding et al. 2020)

17 Resistive Memory Devices

A pioneering computational method, derived from studies in brain computing, is gaining momentum in recent years. Resistive random-access memory (RRAM) arrays use little energy and hold a potential for enormous densities. An interesting type of RRAM was recently demonstrated to have alternating (dynamic switching) current rectification properties, like those of Complementary Metal Oxide Semiconductors (CMOS) transistors (Berco 2018). Such artificial synaptic devices can be switched between two modes (excitatory and inhibitory) to double the array density and to significantly reduce the peripheral circuit complexity.

Gavrilov et al. (2018) discussed the "associative spatial-temporal memories" based on neuromorphic networks with restricted connectivity, termed "CrossNets." Such networks have the capability to be implemented naturally in nanoelectronic hardware with hybrid memristive circuits (a memristor is a nanoelectric element of circuitry used in parallel computing memory technology). This may allow extremely high energy efficiency, comparable to that of the biological cortical circuits, functioning at a much higher operation speed. Numerical simulations performed by Gavrilov et al. (2018), and confirmed by analytical calculations, show that the characteristics significantly depend on the method of information recording into the memory. Most importantly, CrossNet memories provide a capacity higher than that of "Ternary Content-Addressable Memories" with the same number of nonvolatile memory cells (e.g., memristors, while the input noise immunity of the CrossNet memories is lower).

18 Brain-Machine Interfaces

A brain-machine interface (BMI) may be defined as a direct communication line between an augmented brain and an external device (Lebedev et al. 2018; Lebedev and Nicolelis 2006). Silva (2018) reviewed the recent technological capabilities for machine learning and artificial intelligence (AI) to implement "smart" nanobrain machine interfaces (nBMI). Silva's view consists of novel technologies that will "communicate" with the brain using approaches that allow contextual learning and adaptation to dynamic functional demands. It applies to both technologies: (1) invasive (e.g., neural prosthesis) and (2) noninvasive (e.g., electroencephalography, EEG). Advances in computation, hardware, and software (such as algorithms that learn and adapt in a contextually dependent way) will have the ability to enhance the capabilities that nanotechnology provides to the design and functionality of nBMI.

The opportunity to optically connect/interface with the mammalian/human brain in vivo has favored an unparalleled investigation of functional connectivity in the brain's neuronal circuitry. Pisanello et al. (2016a, b) reviewed the role of nanotechnology for optical-neuronal interfaces, focusing on the new devices and methods for optogenetic control of neuronal firing, and on the "detection" and "triggering of action potentials" using "optically active colloidal nanoparticles."

Future nanotechnology will allow us to interface the cloud with a human brain (Saniotis et al. 2018). Martins et al. (2019) labeled this as a "human brain/cloud interface" ("B/CI"), based on the nanotechnologies referred to here as "neural nanorobotics." The latter may endow a "B/CI" with "controlled" connectivity to coordinate neuronal firing with external storage and processing of data, via the direct "monitoring" of the brain's ~86 billion neurons and ~200 trillion synapses. A neural nanorobotically that allows human "B/CI" might serve as a "personalized conduit," enabling subjects to get a direct, instantaneous access to each aspect of human knowledge.

19 Conclusion

This chapter had briefly discussed neural interfacing that has been already confirmed feasible for sensors and brain-machine interfaces. Advances in nanotechnology give hopes of the augmentation of brain functions in the near future and, at the same time, for precocious diagnostic, monitoring, and recovery of neurodegenerative brain diseases (Alzheimer, Parkinson, etc.).

The nanotechnological methods, described in this chapter, are able to offer efficient ways to overcome the "impenetrable" brain-blood barrier to the benefit of the affected subjects. However, the current nanotechnological techniques are still limited as they:

1. Either depend on permanent implants
2. Require invasive procedures
3. Are not cell-type specific
4. Involve slow pharmacokinetics, or
5. Have a restricted penetration depth, thus making it difficult to stimulate regions deep within the brain
6. Their side effects (e.g., of metallic nanoparticles and quantum dots) are not enough known and thus require careful long-term researches

The present and future nanotechnologies will certainly contribute to a great momentum in medicine and will allow the expected access to the customized medicine of the brain.

20 Future Directions

As the current trend of research points toward the bioinspired neuron-like electronics (Yang et al. 2019), the current technological capabilities create optimal conditions for artificial intelligence and machine learning to devise and realize the nanobrain-machine interfaces (nBMI), which will enable us to communicate, in real time, with the brain and to change some of its functional requirements. This communication is possible in the case of neural prosthesis implant (an invasive procedure) as well as in the case of electroencephalography (EEG) recording, a noninvasive technique. The latest progress is in computation (both in hardware and software) and in versatile algorithms that will endow the nanotechnology with capacity to assure the BMI design and functionality. In order to achieve these objectives, a great number of challenging problems need to be solved. We indicate several promising directions:

1. Brain devices for future diagnosis and treatment in psychiatry (Costa et al. 2017).
2. (a) Thought-controlled nanoscale robots in a living host (Arnon et al. 2016).
   (b) Remote control of cells (Knöpfel and Akemann 2010).
3. (a) Transcriptome profiles of neurons interfacing adjacent cells (Baranes et al. 2019).
   (b) Intracortical extracellular matrix-microelectrodes for improving neural interfaces (Shen et al. 2018).
   (c) Vertical nanowire electrode arrays as a scalable platform for intracellular interfacing to neuronal circuits (Robinson et al. 2012).
   (d) Bio-hybrid interfaces to study neuromorphic functionalities with memristive properties (Juarez-Hernandez et al. 2016).
4. Patterned neuronal networks using nanodiamonds and the effect of varying nanodiamond properties on neuronal adhesion and outgrowth (Edgington et al. 2013).
5. Long-term, multisite, parallel, intracell recording/stimulation by an array of extracellular microelectrodes (Hai et al. 2010).
6. (a) Nanoscale transistors for interfacing with brain (El-Atab et al. 2019).
   (b) Low-voltage synaptic transistor array (Molina-Lopez et al. 2019).
7. Chitosan as a suitable nanocarrier material for anti-Alzheimer drug delivery (Sarvaiya and Agrawal 2015).
8. (a) Neuromodulation by the aid of magnetic nanoparticles (Roet et al. 2019; Khizroev 2019).
   (b) Polymer-based nanomedicine using polymeric nanoparticles, micelles, dendrimers, polymersomes, polyplexes, polymer-lipid hybrids, etc. for improving cancer therapy (Prabhu et al. 2015).

References

Amirav L, Berlin S, Olszakier S, Pahari SK, Kahn I (2019) Multi-modal nano particle labeling of neurons. Front Neurosci 13:12. https://doi.org/10.3389/fnins.2019.00012

Arnon S, Dahan N, Koren A, Radiano O, Ronen M, Yannay T, Giron J, Ben-Ami L, Amir Y, Hel-Or Y, Friedman D, Bachelet I (2016) Thought-controlled nanoscale robots in a living host. PLoS One 11(8):e0161227. https://doi.org/10.1371/journal.pone.0161227. eCollection 2016

Bang C, Thum T (2012) Exosomes: new players in cell-cell communication. Int J Biochem Cell Biol 44(11):2060–2064. https://doi.org/10.1016/j.biocel.2012.08.007

Baranes K, Hibsh D, Cohen S, Yamin T, Efroni S, Sharoni A, Shefi O (2019) Comparing transcriptome profiles of neurons interfacing adjacent cells and nanopatterned substrates reveals fundamental neuronal interactions. Nano Lett 19(3):1451–1459. https://doi.org/10.1021/acs.nanolett.8b03879

Berco D (2018) Rectifying resistive memory devices as dynamic complementary artificial synapses. Front Neurosci 12:755. https://doi.org/10.3389/fnins.2018.00755

Betzer O, Perets N, Angel A, Motiei M, Sadan T, Yadid G, Offen D, Popovtzer R (2017) In vivo neuroimaging of exosomes using gold nanoparticles. ACS Nano 11(11):10883–10893. https://doi.org/10.1021/acsnano.7b04495

Bharadwaj VN, Nguyen DT, Kodibagkar VD, Stabenfeldt SE (2018) Nanoparticle-based therapeutics for brain injury. Adv Healthc Mater 7(1). https://doi.org/10.1002/adhm.201700668

Brus R (1983) Drugs and factors antagonizing the renin-angiotensin system. Postepy Hig Med Dosw 37(3):271–285

Cabuzu D, Cirja A, Puiu R, Grumezescu AM (2015) Biomedical applications of gold nanoparticles. Curr Top Med Chem 15:16. https://doi.org/10.2174/1568026615666150414144750

Chen O, Riedemann L, Etoc F, Herrmann H, Coppey M, Barch M et al (2014) Magneto-fluorescent core-shell supernanoparticles. Nat Commun 5:5093. https://doi.org/10.1038/ncomms6093

Cho M, Contreras EQ, Lee SS, Jones CJ, Jang W, Colvin VL (2014) Characterization and optimization of the fluorescence of nanoscale iron oxide/quantum dot complexes. J Phys Chem C 118:14606–14616. https://doi.org/10.1021/jp502194z

Cohen A, Shappir J, Yitzchaik S, Spira ME (2008) Reversible transition of extracellular field potential recordings to intracellular recordings of action potentials generated by neurons grown on transistors. Biosens Bioelectron 23(6):811–819

Convertino D, Luin S, Marchetti L, Coletti C (2018) Peripheral neuron survival and outgrowth on graphene. Front Neurosci 12:1. https://doi.org/10.3389/fnins.2018.00001

Cooper DR, Nadeau JL (2009) Nanotechnology for in vitro neuroscience. Nanoscale 1(2):183–200. https://doi.org/10.1039/b9nr00132h

Costa E, Silva JA, Steffen RE (2017) The future of psychiatry: brain devices. Metabolism 69S:S8–S12. https://doi.org/10.1016/j.metabol.2017.01.010

d'Amora M, Giordani S (2018) The utility of Zebrafish as a model for screening developmental neurotoxicity. Front Neurosci 12:976. https://doi.org/10.3389/fnins.2018.00976

Dąbrowska-Bouta B, Zięba M, Orzelska-Górka J, Skalska J, Sulkowski G, Frontczak-Baniewicz M, Talarek S, Listos J, Strużyńska L (2016) Influence of a low dose of silver nanoparticles on cerebral myelin and behavior of adult rats. Toxicology 363–364:29–36. https://doi.org/10.1016/j.tox.2016.07.007

Dąbrowska-Bouta B, Sulkowski G, Strużyński W, Strużyńska L (2019) Prolonged exposure to silver nanoparticles results in oxidative stress in cerebral myelin. Neurotox Res 35(3):495–504. https://doi.org/10.1007/s12640-018-9977-0

Dai X, Hong G, Gao T, Lieber CM (2018) Mesh nanoelectronics: seamless integration of electronics with tissues. Acc Chem Res 51(2):309–318. https://doi.org/10.1021/acs.accounts.7b00547

Dan M, Wen H, Shao A, Xu L (2018) Silver nanoparticle exposure induces neurotoxicity in the rat Hippocampus without increasing the blood-brain barrier permeability. J Biomed Nanotechnol 14(7):1330–1338. https://doi.org/10.1166/jbn.2018.2563

Ding S, Xu Y, Liu Q, Gu H, Zhu A, Shi G (2020) Interface engineering of microelectrodes toward ultrasensitive monitoring of β-amyloid peptides in cerebrospinal fluid in Alzheimer's disease. Analyst 145:2331. https://doi.org/10.1039/c9an02285f

Dunk PW, Kaiser NK, Hendrickson CL, Quinn JP, Ewels CP, Nakanishi Y, Sasaki Y, Shinohara H, Marshall AG, Kroto HW (2012) Closed network growth of fullerenes. Nat Commun 3:855. https://doi.org/10.1038/ncomms1853

Edgington RJ, Thalhammer A, Welch JO, Bongrain A, Bergonzo P, Scorsone E, Jackman RB, Schoepfer R (2013) Patterned neuronal networks using nanodiamonds and the effect of varying nanodiamond properties on neuronal adhesion and outgrowth. J Neural Eng 10(5):056022. https://doi.org/10.1088/1741-2560/10/5/056022

El-Atab N, Shaikh SF, Hussain MM (2019) Nano-scale transistors for interfacing with brain: design criteria, progress and prospect. Nanotechnology 30(44):442001. https://doi.org/10.1088/1361-6528/ab3534
Feld A, Merkl JP, Kloust H, Flessau S, Schmidtke C, Wolter C et al (2015) A universal approach to ultrasmall magneto-fluorescent nanohybrids. Angew Chem Int Ed Engl 54:12468–12471. https://doi.org/10.1002/anie.201503017
Ferdous Z, Al-Salam S, Greish YE, Ali BH, Nemmar A (2019) Pulmonary exposure to silver nanoparticles impairs cardiovascular homeostasis: effects of coating, dose and time. Toxicol Appl Pharmacol 367:36–50. https://doi.org/10.1016/j.taap.2019.01.006
Fernandes LF, Bruch GE, Massensini AR, Frézard F (2018) Recent advances in the therapeutic and diagnostic use of liposomes and carbon nanomaterials in ischemic stroke. Front Neurosci 12:453. https://doi.org/10.3389/fnins.2018.00453
Gahl TJ, Kunze A (2018) Force-mediating magnetic nanoparticles to engineer neuronal cell function. Front Neurosci 12:299. https://doi.org/10.3389/fnins.2018.00299
Gavrilov D, Strukov D, Likharev KK (2018) Capacity, fidelity, and noise tolerance of associative spatial-temporal memories based on memristive neuromorphic networks. Front Neurosci 12:195. https://doi.org/10.3389/fnins.2018.00195
Gherardi RK, Aouizerate J, Cadusseau J, Yara S, Authier FJ (2016) Aluminum adjuvants of vaccines injected into the muscle: normal fate, pathology and associated disease. Morphologie 100(329):85–94. https://doi.org/10.1016/j.morpho.2016.01.002
Goldental A, Sabo P, Sardi S, Vardi R, Kanter I (2016) Mimicking collective firing patterns of hundreds of connected neurons using a single-neuron experiment. Front Neurosci 9:508. https://doi.org/10.3389/fnins.2015.00508
Gupta J, Fatima MT, Islam Z, Khan RH, Uversky VN, Salahuddin P (2019) Nanoparticle formulations in the diagnosis and therapy of Alzheimer's disease. Int J Biol Macromol 130:515–526. https://doi.org/10.1016/j.ijbiomac.2019.02.156
Hai A, Shappir J, Spira ME (2010) In-cell recordings by extracellular microelectrodes. Nat Methods 7(3):200–202. https://doi.org/10.1038/nmeth.1420
Iijima S (1991) Helical microtubules of graphitic carbon. Nature 354:56–58. https://doi.org/10.1038/354056a0
Infante JC (2018) Nanoparticle-based systems for delivery of protein therapeutics to the spinal cord. Front Neurosci 12:484. https://doi.org/10.3389/fnins.2018.00484
Jaiswal JK, Goldman ER, Mattoussi H, Simon SM (2004) Use of quantum dots for live cell imaging. Nat Methods 1(1):73–78
Jones PD, Stelzle M (2016) Can nanofluidic chemical release enable fast, high resolution neurotransmitter-based neurostimulation? Front Neurosci 10:138. https://doi.org/10.3389/fnins.2016.00138
Juarez-Hernandez LJ, Cornella N, Pasquardini L, Battistoni S, Vidalino L, Vanzetti L, Caponi S, Dalla Serra M, Iannotta S, Pederzolli C, Macchi P, Musio C (2016) Bio-hybrid interfaces to study neuromorphic functionalities: new multidisciplinary evidences of cell viability on poly(anyline) (PANI), a semiconductor polymer with memristive properties. Biophys Chem 208:40–47. https://doi.org/10.1016/j.bpc.2015.07.008
Khizroev S (2019) Technobiology's enabler: the magnetoelectric nanoparticle. Cold Spring Harb Perspect Med 9(8) pii: a034207. https://doi.org/10.1101/cshperspect.a034207
Kim EH, Chin G, Rong G, Poskanzer KE, Clark HA (2018a) Optical probes for neurobiological sensing and imaging. Acc Chem Res 51(5):1023–1032. https://doi.org/10.1021/acs.accounts.7b00564
Kim Y, Meade SM, Chen K, Feng H, Rayyan J, Hess-Dunning A, Ereifej ES (2018b) Nanoarchitectural approaches for improved intracortical interface technologies. Front Neurosci 12:456. https://doi.org/10.3389/fnins.2018.00456
Knöpfel T, Akemann W (2010) Remote control of cells. Nat Nanotechnol 5:560–561
Lebedev MA, Nicolelis MA (2006) Brain-machine interfaces: past, present and future. Trends Neurosci 29(9):536–546

Lebedev MA, Opris I, Casanova MF (2018) Editorial: augmentation of brain function: facts, fiction and controversy. Front Syst Neurosci 12:45. https://doi.org/10.3389/fnsys.2018.00045

Lee H, Kim Y, Park A, Nam J-M (2014) Amyloid-β aggregation with gold nanoparticles on brain lipid bilayer. Small 10(9):1779–1789

Martins NRB, Angelica A, Chakravarthy K, Svidinenko Y, Boehm FJ, Opris I, Lebedev MA, Swan M, Garan SA, Rosenfeld JV, Hogg T, Freitas RA Jr (2019) Human brain/cloud interface. Front Neurosci 13:112. https://doi.org/10.3389/fnins.2019.00112

Maysinger D, Ji J, Hutter E, Cooper E (2015) Nanoparticle-based and bioengineered probes and sensors to detect physiological and pathological biomarkers in neural cells. Front Neurosci 9:480. https://doi.org/10.3389/fnins.2015.00480

Medintz IL, Uyeda HT, Goldman ER, Mattoussi H (2005) Quantum dot bioconjugates for imaging, labelling and sensing. Nat Mater 4(6):435–446

Moldovan M, Alvarez S, Rothe C, Andresen TL, Urquhart A, Lange KHW, Krarup C (2018) An in vivo mouse model to investigate the effect of local anesthetic nanomedicines on axonal conduction and excitability. Front Neurosci 12:494. https://doi.org/10.3389/fnins.2018.00494

Molina-Lopez F, Gao TZ, Kraft U, Zhu C, Öhlund T, Pfattner R, Feig VR, Kim Y, Wang S, Yun Y, Bao Z (2019) Inkjet-printed stretchable and low voltage synaptic transistor array. Nat Commun 10(1):2676. https://doi.org/10.1038/s41467-019-10569-3

Moretti D, DiFrancesco ML, Sharma PP, Dante S, Albisetti E, Monticelli M, Bertacco R, Petti D, Baldelli P, Benfenati F (2018) Biocompatibility of a magnetic tunnel junction sensor array for the detection of neuronal signals in culture. Front Neurosci 12:909. https://doi.org/10.3389/fnins.2018.00909

Nadeau JL (2015) Initial photophysical characterization of the proteorhodopsin optical proton sensor (PROPS). Front Neurosci 9:315. https://doi.org/10.3389/fnins.2015.00315

Nafissi N, Foldvari M (2015) Neuroprotective therapies in glaucoma: II. Genetic nanotechnology tools. Front Neurosci 9:355. https://doi.org/10.3389/fnins.2015.00355

Novoselov KS, Geim AK, Morozov SV, Jiang D, Zhang Y, Dubonos SV, Grigorieva IV, Firsov AA (2004) Electric field effect in atomically thin carbon films. Science 306(5696):666–669

Oppong-Damoah A, Zaman RU, D'Souza MJ, Murnane KS (2019) Nanoparticle encapsulation increases the brain penetrance and duration of action of intranasal oxytocin. Horm Behav 108:20–29. https://doi.org/10.1016/j.yhbeh.2018.12.011

Pahari SK, Olszakier S, Kahn I, Amirav L (2018) Magneto-fluorescent yolk–shell nanoparticles. Chem Mater 30:775–780. https://doi.org/10.1021/acs.chemmater.7b04253

Perets N, Betzer O, Shapira R, Brenstein S, Angel A, Sadan T, Ashery U, Popovtzer R, Offen D (2019) Golden exosomes selectively target brain pathologies in neurodegenerative and neurodevelopmental disorders. Nano Lett 19(6):3422–3431. https://doi.org/10.1021/acs.nanolett.8b04148

Pinkernelle J, Raffa V, Calatayud MP, Goya GF, Riggio C, Keilhoff G (2015) Growth factor choice is critical for successful functionalization of nanoparticles. Front Neurosci 9:305. https://doi.org/10.3389/fnins.2015.00305

Pisanello F, Sileo L, De Vittorio M (2016a) Corrigendum: micro- and nanotechnologies for optical neural interfaces. Front Neurosci 10:468. https://doi.org/10.3389/fnins.2016.00468

Pisanello F, Sileo L, De Vittorio M (2016b) Micro- and nanotechnologies for optical neural interfaces. Front Neurosci 10:70. https://doi.org/10.3389/fnins.2016.00070

Prabhu RH, Patravale VB, Joshi MD (2015) Polymeric nanoparticles for targeted treatment in oncology: current insights. Int J Nanomedicine 10:1001–1018

Pulgar VM (2019) Transcytosis to cross the blood brain barrier, new advancements and challenges. Front Neurosci 12:1019. https://doi.org/10.3389/fnins.2018.01019

Rabieh N, Ojovan SM, Shmoel N, Erez H, Maydan E, Spira ME (2016) On-chip, multisite extracellular and intracellular recordings from primary cultured skeletal myotubes. Sci Rep 6:36498. https://doi.org/10.1038/srep36498

Rahimpour M, Karami M, Haeri Rohani A (2020) Silver nanoparticles (Ag-NPs) in the central amygdala protect the rat conditioned by morphine from withdrawal attack due to naloxone via high-level nitric oxide. Arch Pharmacol 393:857. https://doi.org/10.1007/s00210-019-01784-2

Robinson JT, Jorgolli M, Shalek AK, Yoon M-H, Gertner RS, Park H (2012) Vertical nanowire electrode arrays as a scalable platform for intracellular interfacing to neuronal circuits. Nat Nanotechnol 7:180–184

Rodriguez-Izquierdo I, Serramia MJ, Gomez R, De La Mata FJ, Bullido MJ, Muñoz-Fernández MA (2020) Gold nanoparticles crossing blood-brain barrier prevent HSV-1 infection and reduce herpes associated amyloid-β secretion. J Clin Med 9(1). pii: E155. https://doi.org/10.3390/jcm9010155

Roet M, Heschama S-A, Jahanshahia A, Rutten BPF, Anikeeva PO, Temel Y (2019) Neuromodulation of the brain: a role for magnetic nanoparticles? Prog Neurobiol 177:1–14. https://doi.org/10.1016/j.pneurobio.2019.03.002

Ryynänen T, Toivanen M, Salminen T, Ylä-Outinen L, Narkilahti S, Lekkala J (2018) Ion beam assisted E-beam deposited TiN microelectrodes—applied to neuronal cell culture medium evaluation. Front Neurosci 12:882. https://doi.org/10.3389/fnins.2018.00882

Saniotis A, Henneberg M, Sawalma AR (2018) Integration of nanobots into neural circuits as a future therapy for treating neurodegenerative disorders. Front Neurosci 12:153. https://doi.org/10.3389/fnins.2018.00153

Saraiva C, Praça C, Ferreira R, Santos T, Ferreira L, Bernardino L (2016) Nanoparticle-mediated brain drug delivery: overcoming blood-brain barrier to treat neurodegenerative diseases. J Control Release 235:34–47. https://doi.org/10.1016/j.jconrel.2016.05.044

Sarvaiya J, Agrawal YK (2015) Chitosan as a suitable nanocarrier material for anti-Alzheimer drug delivery. Int J Biol Macromol 72:454–465. https://doi.org/10.1016/j.ijbiomac.2014.08.052

Shah SA, Yoon GH, Ahmad A, Ullah F, Ul Amin F, Kim MO (2015) Nanoscale-alumina induces oxidative stress and accelerates amyloid beta (Aβ) production in ICR female mice. Nanoscale 7(37):15225–15237. https://doi.org/10.1039/c5nr03598h

Shen W, Das S, Vitale F, Richardson A, Ananthakrishnan A, Struzyna LA, Brown DP, Song N, Ramkumar M, Lucas T, Cullen DK, Litt B, Allen MG (2018) Microfabricated intracortical extracellular matrix-microelectrodes for improving neural interfaces. Microsyst Nanoeng 4:30. https://doi.org/10.1038/s41378-018-0030-5. eCollection 2018

Silva GA (2018) A new frontier: the convergence of nanotechnology, brain machine interfaces, and artificial intelligence. Front Neurosci 12:843. https://doi.org/10.3389/fnins.2018.00843

Spira ME, Shmoel N, Huang S-HM, Erez H (2018) Multisite attenuated intracellular recordings by extracellular multielectrode arrays, a perspective. Front Neurosci 12:212. https://doi.org/10.3389/fnins.2018.00212

Tang-Schomer MD, Jackvony T, Santaniello S (2018) Cortical network synchrony under applied electrical field in vitro. Front Neurosci 12:630. https://doi.org/10.3389/fnins.2018.00630

Vidu R, Rahman M, Mahmoudi M, Enachescu M, Poteca TD, Opris I (2014) Nanostructures: a platform for brain repair and augmentation. Front Syst Neurosci 8:91. https://doi.org/10.3389/fnsys.2014.00091

Wang Y, Guo L (2016) Nanomaterial-enabled neural stimulation. Front Neurosci 10:69. https://doi.org/10.3389/fnins.2016.00069

Yang X, Zhou T, Zwang TJ, Hong G, Zhao Y, Viveros RD, Fu TM, Gao T, Lieber CM (2019) Bioinspired neuron-like electronics. Nat Mater 18(5):510–517. https://doi.org/10.1038/s41563-019-0292-9

Zagrean A-M, Hermann DM, Opris I, Zagrean L, Popa-Wagner A (2018) Multicellular crosstalk between exosomes and the neurovascular unit after cerebral ischemia. Therapeutic implications. Front Neurosci 12:811. https://doi.org/10.3389/fnins.2018.00811

Zhou Y, Peng Z, Sevena ES, Leblanc RM (2018) Crossing the blood-brain barrier with nanoparticles. J Control Release 270:290–303

The Impact of Aging and Age-Related Comorbidities on Stroke Outcome in Animal Models and Humans

Aurel Popa-Wagner, Mircea Popescu-Driga, and Daniela Glavan

1 Introduction

Globally, cerebrovascular diseases are the second cause of death being surpassed by ischemic heart disease by just 1.5%. The consequences of stroke are often devastating and extend well beyond the patients to impact not only on families but also employers, caregivers, and social networks. Around 60–83% survivors will fully recover and achieve independence in self-care by 1 year after a stroke. However, a significant proportion may require a temporary or lifelong assistance. Thus, between 10% and 15% of survivors require assistance and primary care in a specialized institution at 1 year following stroke (Appelros et al. 2002; Bonita and Beaglehole 1988).

In western aging societies, the incidence of stroke events increases abruptly at the age of about 73 years. Advances in stroke medicine and the adaptation to stroke risk factors were successful in decreasing stroke incidence and increasing the number of stroke survivors in western societies. However, due to the increased in the number of elderly, the incidence of stroke has increased again paralleled by an

A. Popa-Wagner (✉)
Griffith University Menzies Health Institute of Queensland, Gold Coast, QLD, Australia

Center of Experimental and Clinical Medicine, University of Medicine and Pharmacy, Craiova, Romania

Department of Neurology Chair of Vascular Neurology and Dementia, University of Medicine, Essen, Germany
e-mail: aurel.popa-wagner@geriatrics-healthyageing.com

M. Popescu-Driga
Center of Experimental and Clinical Medicine, University of Medicine and Pharmacy, Craiova, Romania

D. Glavan
Department of Psychiatry, University of Medicine and Pharmacy, Craiova, Romania

I. Opris et al. (eds.), *Modern Approaches to Augmentation of Brain Function*, Contemporary Clinical Neuroscience,
https://doi.org/10.1007/978-3-030-54564-2_12

increase in the number of stroke survivors, many with severe disabilities, that has led to an increased economic and social burden in society.

2 The Risk of Cerebral Ischemia Increases with Age

Age is the only one non-modifiable risk of cerebral ischemia. Advances in emergency medicine and primary, secondary, and tertiary medical care have led to a 12% decrease in first stroke incidence in western societies while in lo- and middle-income countries has increased by a similar number (Bonita and Beaglehole 1988).

Stroke patients are at the highest risk of death in the first week after the event, mostly due to inflammation and cerebral edema. Thereafter, between 20% and 50% die within the first month depending on age, brain localization, severity, comorbidities, and general response to therapeutic approaches. Patients who survive may be left with no disability or with mild, moderate, or severe disability. In younger patients spontaneous recovery is not uncommon and occurs mostly within 6 months (Lloyd-Jones 2010). However, advances in medicine led to an increased number of survivors with disabilities so that stroke burden as measured by the disability-adjusted life-years lost has risen by 24% from 1990 to 2013 (Burn et al. 1994). We can postpone the first stroke event by appropriate measures such as changes in life-style and primary prevention (Tacutu et al. 2010; Appelros et al. 2002). However, patients with a first stroke are at risk (10%) of being hit by a second in the first year and 5% per year thereafter (Burn et al. 1994; Roger et al. 2012).

3 Cerebral Ischemia and Comorbidities

Elderly with a low-risk lifestyle behavior (physical exercise, nonsmoking, healthy nutritious diet, healthy weight during midlife, moderate alcohol consumption) have a significantly lower risk of stroke among women and men than individuals without a low-risk lifestyle. However, at least four criteria must be met for a good outcome after stroke (Lim and Kang 2015). Conversely, elderly with comorbidities including stress, obesity, metabolic syndrome, high cholesterol, diabetes, hypertension, alcoholism, smoking, and a sedentary lifestyle will have a higher incidence of stroke (Donnan and Davis 2008; Goldstein et al. 2011; Rewell et al. 2010).

Numerous studies conducted in the economically advanced countries have documented an "obesity epidemic." The American Heart Association (AHA) guidelines indicate that obesity characterized by a body mass index (BMI) of at least 30 kg/m^2 as well as physical inactivity are risk factors for stroke (Goldstein et al. 2011). In numbers, for overweight patients, every 5 kg/m^2 is associated with 40% increased mortality in stroke patients (Goldstein et al. 2011). In fact, obesity is never described as a single abnormality as it by itself leads to other pathological complications such as diabetes, hypercholesterolemia, and as a consequence, hypertension (Towfighi and Ovbiagele 2009; Strazzullo et al. 2010; Miname and Santos 2019). For example, simultaneous presence of two comorbidities like chronic diabetes and hypertension has an additive, negative effect by increasing the level of ischemic damage in animals, probably due to vascular diabetic complications (Rewell et al. 2010). Indeed, cell treatment in comorbid animals was associated with a reduced efficacy in terms of infarct size and behavioral outcome when compared with those obtained in healthy animals (Cui et al. 2019).

Still we are left with the unexplained "obesity paradox": after the acute event, the highest mortality risk was described in young patients with increased BMI, while in obese elderly patients the risk decreased in a linear fashion. The authors claim that elderly obesity may induce a protective effect (Towfighi and Ovbiagele 2009). However, this finding is poorly understood and is yet to be verified in more experimental and clinical studies. Further since BMI reflects also total lean mass, it is possible that a certain magnitude of body mass is necessary to decrease mortality and disability in stroke survivors. For example, recently it has been reported that low uric acid is associated with functional outcome after stroke in low BMI, suggesting one of the mechanisms underlying the obesity-stroke paradox of the outcome in ischemic stroke patients (Tang et al. 2019). Indeed, several studies suggest that thrombolytic therapy is less successful in obese patients and in patients with metabolic syndrome (Arenillas et al. 2008; Deguchi et al. 2012).

In animal models, poststroke, aged calorie-restricted Sprague–Dawley rats showed improved behavioral recovery that requires complex sensorimotor skills, such as the rotating rod and inclined plane tasks, or cutaneous sensitivity and sensorimotor integration or spatial memory compared with ad libitum-fed, overweighed rats. Likewise, CR aged rats showed significant poststroke increases in serum glucose, insulin, and IGF1 levels as well as CR-specific changes in gene expression including downregulation of genes involved in the ubiquitin proteasome degradation system (Ciobanu et al. 2017).

Cellular stress and hyperglycemia are known to accelerate the aging process. In stroke patients, hyperglycemia was also shown to accelerate infarct progression in cortical areas, while glycemic variability and hyperglycemia during acute ischemic stroke were identified as novel predictors for poststroke cognitive decline (Martin et al. 2006; Lim and Kang 2015). However, the mechanisms of hyperglycemia-associated infarct progression remain unclear. It could be that hyperglycemia-associated neuroinflammation and oxidative stress promote hemorrhagic transformation leading to blood–brain barrier (BBB) disruption and neuronal cell death, and aggravating brain dysfunction (Kumari et al. 2012; Soejima et al. 2013a, b).

Our knowledge about the molecular and cellular mechanisms underlying accelerated infarct progression in subjects with metabolic syndrome is still poor. One attractive hypothesis is that obesity, hyperinsulinemia, and hypercholesterolemia activate several proinflammatory signaling pathways, leading to a condition of chronic low-grade inflammation, especially in the elderly (Catrysse and van Loo 2017). Zhang and colleagues suggested that a crucial pathway in the context of obesity and the metabolic syndrome is the IKKβ/NF-κB signaling pathway (Zhang et al. 2008).

Some studies report a strong connection unhealthy diet and increased oxidative stress (NADPH oxidase dependent), mitochondrial dysfunction, or proinflammatory changes in the brain, both promoting neuronal network imbalance and elevated glucose level (Cai and Liu 2012). In light of this, a better understanding of the contribution of comorbidities to stroke-induced pathological sequelae as well as identifying molecular factors and signaling pathways underlying the metabolic syndrome may be exploited to develop prevention strategies as well as more successful therapies against age-associated diseases which, in turn, would improve quality of life and extend lifespan.

4 Stroke Models Using Aged Animals Are Clinically More Relevant

Animal models of stroke often ignore age and comorbidities frequently associated with senescence. Since stroke afflicts mostly the elderly comorbid patients, it is highly desirable to test the efficacy of stroke therapies in an appropriate animal stroke model. Ignoring the strong relationship between increased age and the incidence of stroke could be one of the explanations for unsuccessful bench-to-bedside translation of successful therapies in young animal models and humans. Just to mention, of some 1000 drugs tested to cure or diminish the devastating effects of stroke, none has worked in the clinic. The only therapeutic option is to recanalize the blocked artery using tissue plasminogen activator or removing the thrombus surgically (thrombectomy). Studies of stroke have demonstrated an age and gender effects on incidence, functional recovery and mortality, not only in humans but also in animal models (Bergerat et al. 2011; Gokcay et al. 2011). Therefore, studies on physiologically complex organisms like rats, mice, or nonhuman primates are required to investigate the molecular mechanisms of aging in humans or to predict human responses to age-related diseases or the response of aged organisms to drugs. Further, since epidemiological studies have documented that the incidence of stroke in humans is higher in late middle age (50–70 years) than in the more aged (more than 70 years) (Feigin et al. 2003a, b), we are using middle aged instead of geriatrics animals as stroke model (Popa-Wagner et al. 2007a, b). In this context, the aged rodent model is highly recommended to investigate the molecular pathways underlying the effects of therapeutic drugs and functional outcome in preclinical studies. Further recommendations for a translational roadmap have been made recently by Boltze and coworkers (Modo et al. 2018).

All stroke models in aged rats are based on the middle cerebral artery occlusion (MCAO). The occlusion can be permanent (Futrell 1991; Wang et al. 1995; Brown et al. 2003; Zhang et al. 2005) or for a limited period of time of 30–120 min (transient) using a thrombus (Wang et al. 1995), intraluminal filament occlusion (Lindner et al. 2003; Sutherland et al. 1996; Rosen et al. 2005; Trueman et al. 2011), or a hook attached to a micromanipulator (Popa-Wagner et al. 1998). Still other models have used the occlusion of distal branches of the MCA. Likewise, long-term hypoxia–ischemia could also be induced by unilateral common carotid artery occlusion (Macri et al. 2006).

The thrombus model, which mimics more accurate the clinical stroke, has been developed by Charles Rosen and his coworkers (2005), achieves MCA occlusion by placing the blood clot with the aid of a micro-catheter, and uses laser Doppler flowmetry to monitor the drop in the blood flow during the procedure. This method is more accurate because it avoids the nonselective thrombus injection into the internal carotid artery. Using this method, smaller sized clots are needed to achieve successful, reliable MCA occlusion with a higher efficiency. Moreover, the infarct size is more uniform and the location is more predictable resulting in substantially larger infarctions than those seen in aged animals following nonselective embolization, or young animals subjected to filamentous occlusion.

Comorbidities aggravate the outcome after cerebral ischemia. Currently, there are several different rodent models with comorbidities available including the streptozocin rat model for diabetes, high-sugar diet Sprague–Dawley rats, spontaneously high-fat diet, hypertensive rat (SHR) model, or spontaneously hypertensive rat stroke-prone (SHRSP). For example, magnetic resonance imaging of the ischemic penumbra in SHRSP showed increased infarct size after 60 min MCAO was twice as larger than in SHR or WKY normotensive rats (McCabe et al. 2009). However, obesity fed a controversial paradox. Thus, a large study conducted on 12,617 stroke Japanese patients registered in the Kyoto Stroke Registry found that hyperlipidemic had a favorable early outcome in terms of remaining sequelae and hazard ratio for death. A similarly large Asian study conducted in China on 10,905 eligible acute ischemic stroke patients reported favorable outcome (Rankin scale) in overweight patients (Zhao et al. 2014). Another large study done on 4782 acute ischemic stroke patients concluded that the risks of death at 12 months and death or high dependency at 3 and 12 months were higher in underweight patients and not significantly different in overweight vs. normal-weight patients. A study done on Korean patients reported that advanced age (over 65) and extreme obesity are predictors of a favorable outcome at 6 months poststroke. Similar favorable outcome for obese American patients was reported by Burke et al. (2014). However, no obesity paradox was observed in patients after intravenous thrombolysis, suggesting that the favorable odds of recovery are related to better spontaneous recanalization in obese patients (Oesch et al. 2017). Finally, a recent systematic review of data for behavior- and medication-based weight loss and weight loss maintenance interventions on the incidence or prevalence of obesity-related conditions among adults who are overweight or have obesity noted that weight loss interventions resulted in a decreased risk of developing diabetes, particularly among those with prediabetes. This study emphasizes that prevention remains the major target for social behavior (LeBlanc et al. 2018). Indeed, recently we reported that recovery from stroke is enhanced in aged rats by calorie restriction that reduces body weight prior to infarct.

5 Age-Dependent Recovery from Cerebral Ischemia

Aging is accompanied by both functional and cognitive decline which are closely related to morphological changes in the brain during aging. In addition, the increased susceptibility of the aged brain to cerebral ischemia is associated with cognitive decline and incomplete behavioral recovery (Manwani et al. 2011). Indeed, post-stroke studies conducted on aged (18- to 20-month-old) rats have shown that middle aged rats start recovery with a delay of 4 days and there was poorer behavioral outcomes and diminished functional recovery after ischemia in aged rats as compared with young animals (Andersen et al. 1999; Sutherland et al. 1996; Brown et al. 2003; Wang et al. 2003; Badan et al. 2003; Zhang et al. 2005; Won et al. 2006; Chen and Sun 2007; Buchhold et al. 2007; DiNapoli et al. 2008). Similar findings have been reported for poststroke recovery of senescence-accelerated prone mice (Lee et al. 2006). It should be noted that the extent of recovery was also dependent on the complexity and difficulty of the test. For example, aged rats had difficulties in mastering complex tasks such as rotating pole that requires a complexity of motor, sensory, reflex, and balance outcome, or somatosensory function tested by the adhesive removal test, or spatial memory as measured in the Morris water maze (Zhang et al. 2005; Badan et al. 2003; Zhao et al. 2005; Buchhold et al. 2007; Popa-Wagner et al. 2010). However, more simple tasks such as foot-fault test score and corner test score were equally mastered by the aged rats. Of note, functional impairments were largely dependent on the infarct size, i.e., rats with the larger infarcts (20% tissue loss) were more severe impaired than animals with 4% tissue loss (Lindner et al. 2003).

A recent study has pointed out differences in study design between preclinical experimental studies and clinical trials, and recommends the use of aged female and male comorbid (hypertension, diabetes) animal models. Further recommendations include more sensitive and reliable behavioral tests, including modified neurological severity score (mNSS), adhesive removal and foot-fault test obtained from the same subjects. Finally, a wider therapeutic time window for clinical trials is highly recommended (Cui et al. 2019). Likewise, the use of imaging shall be widely employed to assess tissue and behavioral outcome (Seehafer et al. 2010; Jiang et al. 2012).

Similarly, recent advances in stroke connectome can be used to uncover the interaction between changes in functional connectivity and behavioral outcome. In particular, for therapeutic regeneration studies it is highly advisable to monitor functional improvements for an extended period of time. For example, brain imaging revealed that restoration of interhemispheric functional connectivity in rat model correlated with enhanced sensorimotor function (van Meer et al. 2012; Lim and Kang 2015). Recuperation and regeneration is further complicated due to the inherent, age-related reduction in the functional brain networks strength in the rodent brain (Egimendia et al. 2019).

Recommendations for a translational roadmap have been made recently by Boltze and coworkers (Modo et al. 2018). They note that rodent stroke models currently in use are homogenous in behavioral outcome, allowing the use of relatively low sample sizes to achieve therapeutic significance. However, the animals are young, healthy subjects that do not reflect comorbidities doctors have to deal with in daily clinical practice. Indeed, we found large differences in genomic, cellular, and behavioral responses to cerebral ischemia in aged rats as compared with young animals (Badan et al. 2003; Buga et al. 2008, 2012; Schmidt et al. 2011). To summarize, translating therapeutic strategies from young animals to aged humans could lead to false conclusions.

6 Spontaneous Stroke Recovery in Aged Patients and Animals

In young mice or rats, functional recovery starts by day 2 poststroke and complete spontaneous recovery may occur, depending on age, location, and size of the ischemic lesion. However, in aged rats stroke neurological recovery starts with a delay of 1 week to reach maximum 75% of the functional improvement observed in young rats (Buchhold et al. 2007). Indeed, the age-dependent increase in the evolution of ischemic tissue into infarction suggests that age is a biological marker for the variability in tissue outcome in acute human stroke (Ay et al. 2005). Nevertheless, stroke patients do regain some of their lost neurological functions during the first weeks or months after the stroke and physical therapy is effective in stimulating poststroke recovery (Honmou et al. 2012; Liepert et al. 2004).

Mechanistically, recovery is thought to occur via recruitment of neighboring neuronal circuitries (Hallett 2001) or exploiting synaptic plasticity in the contralesional hemisphere (Qin et al. 2014). For example, recovery occurs more easily if the infarct is located in striatum, a subcortical structure that is important for motor learning by activity-dependent plasticity. This is why patients with subcortical stroke are more likely to regain control of movement early after stroke (Bejot et al. 2008; Rothrock et al. 1995).

7 Anti-neuroinflammatory Therapies to Enhance Poststroke Recovery

The immune status of the CNS is strictly regulated. In the young brain, microglia are kept in a quiescent state by interactions between the microglia receptor CD200 and its ligand, which is expressed on neurons (Matsumoto et al. 2007). During normal brain aging, that neuronal inhibition of microglial activation may become compromised due to the age-related upregulation of microglial cell surface markers MHCII which is paralleled by a gradual decrease in the expression of CD200 (Wang et al. 2011; Cox et al. 2012; Shrivastava et al. 2012). As a result, with increasing age microglia in the brain becomes increasingly activated in an attempt to remove damaged structures generated by oxidative, proteotoxic, and metabolic stresses (Salminen et al. 2011).

Recent experimental evidence suggests that the microenvironment of the normal aged brain is characterized by chronic low-level inflammation and increased microglia reactivity. Activated or reactive microglia or primed microglia is characterized by morphological changes and upregulation of MHCII, but minimal basal production of proinflammatory phenotype of heightened reactivity is often referred to as "sensitized," due to a more rapid induction and massive cytokine release upon activation when compared to normal microglia (Perry 2007; Salminen et al. 2011). Systemically, with increasing age the acquired immune system fails to mount appropriate responses to new pathogens. At the same time, however, plasma levels of inflammatory cytokines increase with age due the conversion to a proinflammatory, "senescence-associated secretory phenotype," (Fyhrquist et al. 2013; Tchkonia et al. 2013).

Following stroke, the aged brain mounts a strong inflammatory response which could interfere with regenerative events like axonal growth (Popa-Wagner et al. 2007a, b). Therefore, many studies have focused on those pathways that are associated with inflammation after cerebral ischemia (Rawlinson et al. 2020). The effects of cerebral ischemia on the neuroimmune functions are complex. There is evidence of both beneficial (neuroprotection and negative feedback on microglial activation) and detrimental to the neurovascular unit (Iadecola and Anrather 2011; Jayaraj et al. 2019). The lesion rapidly activates resident cells and opening of the BBB allows infiltration of large numbers of inflammatory cells into the ischemic region causing tissue edema and exacerbating brain damage (Popa-Wagner et al. 2007a, b; Jayaraj et al. 2019). The aging brain reacts strongly to ischemia-reperfusion injury with an early inflammatory response (Badan et al. 2003; Popa-Wagner et al. 2007a, b). This inflammatory response is characterized by increased chemokine expression and increased cell death. At transcriptional level, five inflammation-related genes in the peri-infarcted region (Buga et al. 2008).

Clinical data strongly implicate chronic systemic inflammatory conditions, which are associated with metabolic inflammation caused by obesity, atherosclerosis, and hyperlipidemia, in creating a "primed" inflammatory environment in the brain and increasing the vulnerability of the aged brain to stroke and exacerbating poststroke inflammation (Amor et al. 2010; Drake et al. 2011; Eldahshan et al. 2019; Dhungana et al. 2013).

Higher body temperatures before or after endovascular thrombectomy were associated with poorer clinical outcomes (Diprose et al. 2020). Therefore, to minimize the incapacitating sequelae of stroke during the acute and post-acute phase of stroke, a promising focus of research is physical cooling, or hypothermia, either confined to the head or including the entire body (Esposito et al. 2014; Hennerici et al. 2013; Kollmar et al. 2007; Wu and Grotta 2013; Kurisu and Yenari 2017). The feasibility of hypothermia to reduce sequelae after stroke and improve functional recovery has been addressed by several studies in stroke patients. Stroke patients were exposed for 6–24 h to mild hypothermia (in the range 33–35.5 °C). Hypothermia was well tolerated but its clinical benefits are limited, especially in the long term as measured by the NIHSS (National Institutes of Health Stroke Scale) scores (Hemmen et al. 2010; Kammersgaard et al. 2000; Wan et al. 2014). However, one study reported that the infarct volume was lower in hypothermia patients than in non-hypothermia patients (De Georgia et al. 2004). More recently, it was shown that patients with ischemic stroke, who underwent mild hypothermia after recanalization, showed improved clinical outcome, possibly due to a reduction in cerebral edema (Hong et al. 2014; Piironen et al. 2014).

In clinical trials, hypothermia has been applied for 2–48 h with various beneficial effects (Clark et al. 2009; Wei et al. 2008). Therefore, the optimal time range of exposure to hypothermia yielding the most efficacious neuroprotection is not clear. While most authors agree that mild hypothermia (i.e., hypothermia range 32–34 °C) provides optimal neuroprotection (Johansen et al. 2014; Goossens and Hachimi-Idrissi 2014), the ideal hypothermic therapy in terms of depth and duration of mild hypothermia has to be established yet.

The feasibility and clinical outcome after longer exposure to systemic hypothermia (34–35 °C and maintained for 12 or 24 h), achieved by intravenous infusion of 20 mL/kg refrigerated normal saline (4 °C), were assessed in a large multicenter phase 3 randomized trial. Systemic cooling started within 6 h of symptom onset improves functional outcome at 3 months in awake patients with acute ischemic stroke. The results of this trial provided no evidence of mild hypothermia applied for a longer period of time has an impact on functional outcome at 3 months (van der Worp et al. 2019).

In an aged rat model of stroke, exposure of animals to an atmosphere containing a low dose of H_2S after stroke led to a gradual decrease in whole body temperature, which stabilized at 32 ± 0.5 °C after 12 h. After 24 h, animals were returned to normal atmospheric conditions. The animals recovered within minutes and did not show any signs of neurological or physiological deficits related to H_2S-induced hypothermia (Vintilescu et al. 2016; Sandu et al. 2015; Joseph et al. 2012).

The beneficial effects of hypothermia have been attributed to diminished neuroinflammation, apoptosis excitotoxicity, or free radical production (Chopp et al. 1989). Other neuroprotective mechanisms could include increased neurogenesis and vascular density after stroke. Recently, we reported that the microenvironment of the injured, aged brain is fully capable of neurorestorative processes (Balseanu et al. 2014; Tatarishvili et al. 2014). However, whether stroke stimulates endogenous neurogenesis is still debated, especially in aged subjects in whom neurogenesis is normally decreased (Hermann and Chopp 2012; Jinno 2011).

Recent research has revealed that young-adult animals exposed to mild hypothermia show an increased endogenous repair capacity in the brain. In these subjects, the number of newly formed neurons in the dentate gyrus is higher compared to normothermic animals (Bregy et al. 2012; Silasi and Colbourne 2011, Silasi et al. 2012). Likewise, the number of newly born striatal neurons in aged rodents after stroke was similar to that in young-adult rodents (Ahlenius et al. 2009; Darsalia et al. 2005) despite a 50% decline in neurogenesis in the SVZ of older animals compared with young-adult animals (Enwere et al. 2004).

Adipose-derived mesenchymal stem cells markedly attenuated brain infarct size and improved neurological function in rats. Quite recently, it was shown that in vivo administration of human adipose derived mesenchymal stem cells (hADMSCs) loaded with p5, a 24-residue peptide derived from p35 increased the number of surviving transplanted cells in the peri-infarcted area of animals treated with hADMSCs+P5 than that in hADMSCs-treated or control animals, concomitant with reduced number of phagocytic, annexin3-positive cells in the peri-infarcted region.

8 Cell Therapies

Although rehabilitation is important for improving functional recovery in the early stages after stroke, it does not provide a replacement of lost tissue. One strategy to replenish the lost cells is the transplantation of multipotent fetal brain tissue. However, if and how the aged brain responds to grafted neural tissue is largely unknown because the survival of grafted cells in an inflammatory environment causing cell necrosis and glial scar formation is questionable (Liu et al. 2014). The neurogenic potential of the host tissue is probably the most important factor for a successful integration of transplanted cells. Radial glia-like neural stem cells (RGLs) in the dentate gyrus sub-region of the hippocampus give rise to dentate granule cells (DGCs) and astrocytes throughout life, a process referred to as adult hippocampal neurogenesis (Vicidomini et al. 2020). Understanding how the niche performs its functions may guide strategies to maintain its potency throughout the lifespan and provide a permissive milieu for cell replacement after injuries in the adult hippocampus. For example, mouse fetal hippocampal NSCs expanded in vitro as neurospheres and implanted into the injured Ca3 region of the hippocampus of 24-month-old rats, survived but exhibited limited directed migration and neuronal plasticity (Shetty et al. 2008). Yet another study using NSCs transplanted into young-adult (3-month-old) and aged (24-month-old) rat brains at 1 day after stroke reportedly reduced ischemic brain injury in aged rats (Liu et al. 2014).

Pluripotent stem cells can be derived from blastocyst (embryonic stem cells, ESCs) or through reprogramming of postmitotic somatic cells, most commonly fibroblasts, generating induced pluripotent stem cells (iPSCs) (Kokaia et al. 2018). The induced pluripotent stem cells (iPSCs) offer new prospects for stroke treatment. iPSCs can be generated from a patient, avoiding both ethical problems and immune rejection and a limited differentiation potential of adult stem cells (Yuan et al. 2013). In stroke models, hiPSC-lt-NES cells derived from a young-adult male have the potential to survive, differentiate into immature and mature neurons, and migrate to the peri-infarct area of aged rats. The treated aged rats showed improved behavioral recovery after implantation into the stroke-injured striatum and cortex of adult rats (Oki et al. 2012; Tornero et al. 2013). In a recent study, it could be shown that human iPSC survived and differentiated into neurons after intracortical transplantation in aged rats with cortical stroke and also improved functional recovery in cylinder tests at 4 and 7 weeks (Tatarishvili et al. 2014). Recent studies indicate that iPSCs can also be generated from aged humans and differentiate into specific cell types (Tatarishvili et al. 2014; Mohamad et al. 2013; Phanthong et al. 2013). Moreover, it seems that the redifferentiation efficiency of human fibroblasts via iPSCs into functional motor neurons is the same as in 29–82-year-old individuals (Boulting et al. 2011). However, comorbidities such as diabetes type II (T2DM) impair stroke-induced neurogenesis and parvalbumin (PV) + interneurons-mediated neuroplasticity in a mouse model fed for 12 months with high-fat diet (Pintana et al. 2019). Another study showed that T2DM impedes poststroke WM recovery by suppressing both oligodendrogenesis and beneficial microglia/macrophage responses in wild-type, homozygous diabetic db/db, and heterozygous db/+ mice subjected to distal middle cerebral artery occlusion. Behaviorally, T2DM was associated with poor long-term neurobehavioral outcomes for spatial working memory (Morris water maze), cutaneous sensitivity (adhesive tape removal), and exploratory behavior (open field) (Ma et al. 2018).

Advanced age is associated with a strong poststroke inflammatory response, and is associated with increased risk of stroke and poor poststroke outcomes (Popa-Wagner et al. 2020). In preclinical models of systemic inflammatory disease, the proinflammatory cytokine interleukin-1 (IL-1) is a major player of inflammation, being associated with deleterious effects on neurobehavioral recovery in cerebral ischemia (McColl et al. 2007). Early subcutaneous administration of the IL-1 antagonist, IL-1Ra, to poststroke young and aged animals has been shown to be neuroprotective and increases poststroke neurogenesis for 4 weeks in young/healthy (200–250 g) animals and for at least 7 days in aged/comorbid (corpulent, 800–1000 g animals) (Pradillo et al. 2017).

Most clinical studies conducted so far used neural cells derived from human fetal donors. The techniques to achieve effective survival and growth of neuronal tissues transplanted into the CNS are meanwhile well established (Dunnett 2013). Even though effective, neural grafting has, however, not become a standard treatment for several reasons, including the limited supply of fetal tissue of human origin, and the beneficial effects have been controversial (Morizane et al. 2008). Of the various options, stem cell therapy presents us with a viable alternative (Stoll 2014).

In order to enable the replacement of lost tissue, cell replacement strategies were used in human stroke patients (Stoll 2014; Strazzullo et al. 2010). However, these early clinical studies lacked appropriate control groups. The RECOVER-Stroke trial conducted by Savitz et al. (2019) may well serve as a role model for future early-stage cell therapy clinical trials in stroke. The study featured an impressive array of safety endpoints and stratified patients according to NIHSS scores ($\leq$15 versus $\geq$16) and whether the patients suffered from a lacunar versus a cortical stroke. Importantly, this study also represents the first serious attempt to assess the safety of cell delivery by the intra-arterial route in stroke patients. In this study, a special subpopulation of commercially available bone marrow cells that express CD34+ and CD133+ stem and progenitor cell surface markers and high levels of aldehyde dehydrogenase (ALDH) was administered.

Stroke itself stimulates neurogenesis in the hippocampus and subventricular zone both in young and aged rodent models of stroke. However, the proportion of surviving neurons is discouragingly low (Arvidsson et al. 2002; Parent et al. 2002;

Lindvall and Kokaia 2015). In animal models, the number of new striatal neurons in aged rodents after stroke was similar to that in young animals (Ahlenius et al. 2009; Darsalia et al. 2005), despite a 50% decline in neurogenesis in the SVZ of aged rodents compared to young-adult animals (Tropepe et al. 1997; Enwere et al. 2004).

Cell-based therapy augments this endogenous response. Thus, human iPSCs implanted into the striatum of young-adult animals at 1 week after MCAO protected substantia nigra from atrophy, probably through a trophic effect via the release of survival-promoting growth factors (Polentes et al. 2012). However, how cells are transplanted and where they are placed after stroke are important issues in graft survival and efficacy in promoting behavioral recovery. Data from many groups have shown that stroke increases proliferation of neuronal progenitors in the ipsilateral subventricular region of young-adult rodents with a maximum at 1–2 weeks, and the newly generated neuroblasts migrate to the damaged area in the peri-infarcted striatum over a period of several months. Eventually, the neuroblasts differentiate into medium-sized spiny neurons and may become part of the neuronal network (Arvidsson et al. 2002; Jin et al. 2006; Thored et al. 2006; Hou et al. 2008; Mine et al. 2013). It seems that the injected cells themselves can also stimulate neurogenesis in the SVZ (Zhang et al. 2011; Mine et al. 2013).

Earlier studies on postmortem human brains provided evidence that there might be SVZ cell proliferation and neuroblast formation after stroke even in aged patients (Jin et al. 2006; Macas et al. 2006; Minger et al. 2007). The finding that new neurons are continuously added in the adult human striatum (Ernst et al. 2014), along with the presence of an increased number of putative neuroblasts in the human striatum after stroke, lends support to this hypothesis (Macas et al. 2006). However, whether endogenous neurogenesis contributes to spontaneous recovery after stroke has not yet been established. In addition, age, comorbidities, physical condition of the patient, and severity of disease could substantially influence these steps and, therefore, the outcome of the healing process.

9 Genetic Conversion Therapy

After cerebral ischemia, the ratio between astroglial cells and neurons in the neurovascular unit is disrupted in the perilesional area of aged rodents. It has been hypothesized that restoring the balance within the neurovascular unit may lead to an improved neurorestoration after focal ischemia (Gresita et al. 2019).

Genetic methods using used transcription factor reprogramming have been developed to convert nonneuronal cells into neurons. For example, three different combinations of transcription factors, NEUROGENIN 2 (NGN2) only, NGN2 plus Forebrain Embryonic Zinc Finger-Like Protein 2 (FEZF2), and NGN2 plus Special AT-Rich Sequence-Binding Protein 2 (SATB2), were delivered to human ES cells by lentiviral vectors. All combinations gave rise to pyramidal-shaped hES-iNs, morphologically resembling adult human cortical neurons expressing cortical projection neuron (PN) markers and with mature electrophysiological properties. Using ex vivo transplantation to human organotypic cultures, it was found that the hES-iNs could integrate into adult human cortical networks. However, there was no evidence that the hES-iNs had acquired a distinct cortical layer phenotype. Instead, the hES-iNs, similar to fetal human cortical neurons, expressed both upper and deep layer cortical neuronal markers (Miskinyte et al. 2018).

Human fibroblasts can be directly converted to several subtypes of neurons by using three transcription factors, BRN2, MYT1L, and FEZF2. The conversion efficiency was increased to about 16% by treatment with small molecules and microRNAs. When transplanted ex vivo to organotypic cultures of adult human cerebral cortex, the iCtx cells exhibited morphological and electrophysiological properties of mature neurons, integrated structurally into the cortical tissue, and received synaptic inputs from adult human neurons (Miskinyte et al. 2017).

More recently, a retroviral delivery system encoding the transcription factor Ngn2 alone or in combination with the antiapoptotic factor Bcl-2 was used to target proliferating astrocytes in the neocortex of young and aged mice after cerebral ischemia. Successful direct in vivo reprogramming of reactive glia into neuroblasts and mature neurons was assessed by cellular phenotyping and used to assess the conversion efficacy of proliferating astrocytes into neurons after cerebral ischemia in young and aged mice. However, the conversion efficacy was disappointingly low, most likely because the therapeutic vectors carrying the conversion gene are engulfed by phagocytes shortly after intracortical administration (Gresita et al. 2019).

10 Conclusion

We conclude that data translatability from animal models to humans is poor for cerebral ischemia, most likely because ignoring the strong relationship between increased age and the incidence of stroke could be one of the explanations for unsuccessful bench-to-bedside translation of successful therapies in young animal models and humans.

Acknowledgments This work was supported by UEFISCDI, Ministry of Education and Scientific Research, Romania [research grant code PN-II-ID-PCE-2012-4-0133 to MS; project numbers PN-III-P4-ID-PCE-2016-0340 to DH; PN-III-P2-2.1-PED-2016-1013 and PN-III-P4-ID-PCE-2016-0215 to APW].

Disclosure/Conflict of Interest: We have no conflict of interest to declare.

References

Ahlenius H, Visan V, Kokaia M, Lindvall O, Kokaia Z (2009) Neural stem and progenitor cells retain their potential for proliferation and differentiation into functional neurons despite lower number in aged brain. J Neurosci 29:4408. https://doi.org/10.1523/JNEUROSCI.6003-08.2009

Amor F, Puentes P, Baker D, van der Valk P (2010) Inflammation in neurodegenerative diseases. Immunology 129:154. https://doi.org/10.1111/j.1365-2567.2009.03225.x

Andersen MB, Zimmer J, Sams-Dodd F (1999) Specific behavioral effects related to age and cerebral ischemia in rats. Pharmacol Biochem Behav 62:673. https://doi.org/10.1016/S0091-3057(98)00204-4

Appelros P, Nydevik I, Viitanen M (2002) Poor outcome after first-ever stroke: predictors for death, dependency, and recurrent stroke within the first year. Stroke 34:122. https://doi.org/10.1161/01.str.0000047852.05842.3c

Arenillas J, Ispierto L, Millán M, Escudero D, Herrero N, Dorado L, Guerrero C, Serena J, Castillo J, Dávalos A (2008) Metabolic syndrome and resistance to IV thrombolysis in middle cerebral artery ischemic stroke. Neurology 71:190. https://doi.org/10.1212/01.wnl.0000317092.21210.e6

Arvidsson A, Collin T, Kirik D, Kokaia Z, Lindvall O (2002) Neuronal replacement from endogenous precursors in the adult brain after stroke. Nat Med 8(9):963–970

Ay H, Koroshetz WJ, Vangel M, Benner T, Melinosky C, Zhu M et al (2005) Conversion of ischemic brain tissue into infarction increases with age. Stroke 36:2632. https://doi.org/10.1161/01.STR.0000189991.23918.01

Badan I, Buchhold B, Hamm A, Gratz M, Walker LC, Platt D, Kessler C, Popa-Wagner A (2003) Accelerated glial reactivity to stroke in aged rats correlates with reduced functional recovery. J Cereb Blood Flow Metab 23:845. https://doi.org/10.1097/01.WCB.0000071883.63724.A7

Balseanu AT, Buga AM, Catalin B, Wagner DC, Boltze J, Zagrean AM et al (2014) Multimodal approaches for regenerative stroke therapies: combination of granulocyte colony-stimulating factor with bone marrow mesenchymal stem cells is not superior to G-CSF alone. Front Aging Neurosci 6. https://doi.org/10.3389/fnagi.2014.00130

Bejot Y, Catteau A, Caillier M, Rouaud O, Durier J, Marie C et al (2008) Trends in incidence, risk factors, and survival in symptomatic lacunar stroke in Dijon, France, from 1989 to 2006: a population-based study. Stroke 39:1945. https://doi.org/10.1161/strokeaha.107.510933

Bergerat A, Decano J, Wu CJ, Choi H, Nesvizhskii AI, Moran AM, Ruiz-Opazo N, Steffen M, Herrera VL (2011) Prestroke proteomic changes in cerebral microvessels in stroke-prone, transgenic[hCETP]-hyperlipidemic, Dahl salt-sensitive hypertensive rats. Mol Med 17:588. https://doi.org/10.2119/molmed.2010.00228

Bonita R, Beaglehole R (1988) Recovery of motor function after stroke. Stroke 19:1497. https://doi.org/10.1161/01.str.19.12.1497

Boulting GL, Kiskinis E, Croft GF, Amoroso MW, Oakley DH, Wainger BJ et al (2011) A functionally characterized test set of human induced pluripotent stem cells. Nat Biotechnol 29(3):279–286

Bregy A, Nixon R, Lotocki G, Alonso OF, Atkins CM, Tsoulfas P et al (2012) Posttraumatic hypothermia increases doublecortin expressing neurons in the dentate gyrus after traumatic brain injury in the rat. Exp Neurol 233:821. https://doi.org/10.1016/j.expneurol.2011.12.008

Brown AW, Marlowe KJ, Bjelke B (2003) Age effect on motor recovery in a post-acute animal stroke model. Neurobiol Aging 24:607. https://doi.org/10.1016/S0197-4580(02)00129-X

Buchhold B, Mogoanta L, Suofu Y, Hamm A, Walker L, Kessler C, Popa-Wagner A (2007) Environmental enrichment improves functional and neuropathological indices following stroke in young and aged rats. Restor Neurol Neurosci 25(5–6):467–484

Buga AM, Sascau M, Pisoschi C, Herndon JG, Kessler C, Popa-Wagner A (2008) The genomic response of the ipsilateral and contralateral cortex to stroke in aged rats. J Cell Mol Med 12:2731. https://doi.org/10.1111/j.1582-4934.2008.00252.x

Buga AM, Scholz CJ, Kumar S, Herndon JG, Alexandru D, Cojocaru GR, Dandekar T, Popa-Wagner A (2012) Identification of new therapeutic targets by genome-wide analysis of gene expression in the ipsilateral cortex of aged rats after stroke. PLoS One 7:e50985. https://doi.org/10.1371/journal.pone.0050985

Burke DT, Al-Adawi S, Bell RB, Easley K, Chen S, Burke DP (2014) Effect of body mass index on stroke rehabilitation. Arch Phys Med Rehabil 95(6):1055–1059. https://doi.org/10.1016/j.apmr.2014.01.019

Burn J, Dennis M, Bamford J, Sandercock P, Wade D, Warlow C (1994) Long-term risk of recurrent stroke after a first-ever stroke. The Oxfordshire Community Stroke Project. Stroke 25:333. https://doi.org/10.1161/01.STR.25.2.333

Cai D, Liu T (2012) Inflammatory cause of metabolic syndrome via brain stress and NF-kappaB. Aging (Albany NY). https://doi.org/10.18632/aging.100431

Catrysse L, van Loo G (2017) Inflammation and the metabolic syndrome: the tissue-specific functions of NF-κB. Trends Cell Biol 27:417. https://doi.org/10.1016/j.tcb.2017.01.006

Chen Y, Sun F-Y (2007) Age-related decrease of striatal neurogenesis is associated with apoptosis of neural precursors and newborn neurons in rat brain after ischemia. Brain Res 1166:9. https://doi.org/10.1016/j.brainres.2007.06.043

Chopp M, Knight R, Tidwell CD, Helpern JA, Brown E, Welch KM (1989) The metabolic effects of mild hypothermia on global cerebral ischemia and recirculation in the cat: comparison to normothermia and hyperthermia. J Cereb Blood Flow Metab 9:141. https://doi.org/10.1038/jcbfm.1989.21

Ciobanu O, Sandu R, Balseanu A, Zavaleanu A, Andrei G, Petcu E, Uzoni A, Popa-Wagner A (2017) Caloric restriction stabilizes body weight and accelerates behavioral recovery in aged rats after focal ischemia. Aging Cell 16:1394. https://doi.org/10.1111/acel.12678

Clark DL, Penner M, Wowk S, Orellana-Jordan I, Colbourne F (2009) Treatments (12 and 48 h) with systemic and brain-selective hypothermia techniques after permanent focal cerebral ischemia in rat. Exp Neurol 220:391. https://doi.org/10.1016/j.expneurol.2009.10.002

Cox FF, Carney D et al (2012) CD200 fusion protein decreases microglial activation in the hippocampus of aged rats. Brain Behav Immun 26:789. https://doi.org/10.1016/j.bbi.2011.10.004

Cui L, Golubczyk DT, Boltze AM, Jukka JJ (2019) Cell therapy for ischemic stroke: are differences in preclinical and clinical study design responsible for the translational loss of efficacy? Ann Neurol. https://doi.org/10.1002/ana.25493

Darsalia V, Heldmann U, Lindvall O, Kokaia Z (2005) Stroke-induced neurogenesis in aged brain. Stroke 36:1790. https://doi.org/10.1161/01.STR.0000173151.36031.be

De Georgia MA, Krieger DW, Abou-Chebl A, Devlin TG, Jauss M, Davis SM et al (2004) Cooling for acute ischemic brain damage (COOL AID): a feasibility trial of endovascular cooling. Neurology 63:312. https://doi.org/10.1212/01.WNL.0000129840.66938.75

Deguchi I, Ohe Y, Fukuoka T, Dembo T, Nagoya H, Kato Y, Maruyama H, Horiuchi Y, Tanahashi N (2012) Relationship of obesity to recanalization after hyperacute recombinant tissue-plasminogen activator infusion therapy in patients with middle cerebral artery occlusion. J Stroke Cerebrovasc Dis 21:161. https://doi.org/10.1016/j.jstrokecerebrovasdis.2011.11.003

Dhungana H, Malm T et al (2013) Aging aggravates ischemic stroke-induced brain damage in mice with chronic peripheral infection. Aging Cell. https://doi.org/10.1111/acel.12106

DiNapoli VA, Huber JD, Houser K, Li X, Rosen CL (2008) Early disruptions of the blood–brain barrier may contribute to exacerbated neuronal damage and prolonged functional recovery following stroke in aged rats. Neurobiol Aging 29:753. https://doi.org/10.1016/j.neurobiolaging.2006.12.007

Diprose W, Liem B, Wang M, Sutcliffe J, Brew S, Caldwell J, Mcguinness B, Campbell D, Barber P (2020) Impact of body temperature before and after endovascular thrombectomy for large vessel occlusion stroke. Stroke. https://doi.org/10.1177/10.1161/STROKEAHA.119.028160

Donnan GA, Davis SM (2008) Breaking the 3 h barrier for treatment of acute ischaemic stroke. Lancet Neurol 7:981. https://doi.org/10.1016/S1474-4422(08)70230-8

Drake C, Boutin H, Jones M, Dene A, McColl B, Selvarajah J, Hulme S, Georgiou R, Hinz R, Gerhard A, Vail A, Prenant C, Julyan P, Maroy R, Brown G, Smigova A, Herholz K, Kassiou M, Crossman D, Allan S (2011) Brain inflammation is induced by co-morbidities and risk factors for stroke. Brain Behav Immun 25:1113. https://doi.org/10.1016/j.bbi.2011.02.008

Dunnett SB (2013) Neural tissue transplantation, repair, and rehabilitation. Handb Clin Neurol. https://doi.org/10.1016/B978-0-444-52901-5.00004-6

Egimendia A, Minassian A, Diedenhofen M, Wiedermann D, Ramos-Cabrer P, Hoehn M (2019) Aging reduces the functional brain networks strength—a resting state fMRI study of healthy mouse brain. Front Aging Neurosci 11. https://doi.org/10.3389/fnagi.2019.00277

Eldahshan W, Fagan S, Ergul A (2019) Inflammation within the neurovascular unit: focus on microglia for stroke injury and recovery. Pharmacol Res 147:104349. https://doi.org/10.1016/j.phrs.2019.104349

Enwere E, Shingo T, Gregg C, Fujikawa H, Ohta S, Weiss S (2004) Aging results in reduced epidermal growth factor receptor signaling, diminished olfactory neurogenesis, and deficits in fine olfactory discrimination. J Neurosci 24:8354. https://doi.org/10.1523/JNEUROSCI.2751-04.2004

Ernst A, Alkass K, Bernard S, Salehpour M, Perl S, Tisdale J et al (2014) Neurogenesis in the striatum of the adult human brain. Cell 156(5):1072–1083

Esposito E, Ebner M, Ziemann U, Poli S (2014) In cold blood: intraarteral cold infusions for selective brain cooling in stroke. J Cereb Blood Flow Metab 34:743. https://doi.org/10.1038/jcbfm.2014.29

Feigin V, Anderson N, Gunn A, Rodgers A, Anderson C (2003a) The emerging role of therapeutic hypothermia in acute stroke. Lancet Neurol 2:529. https://doi.org/10.1016/S1474-4422(03)00500-3

Feigin VL, Lawes CMM, Bennett DA, Anderson CS (2003b) Stroke epidemiology: a review of population based studies of incidence, prevalence, and case-fatality in the late 20th century. Lancet Neurol 2:43. https://doi.org/10.1016/S1474-4422(03)00266-7

Futrell N (1991) An improved photochemical model of embolic cerebral infarction in rats. Stroke 22:225

Fyhrquist F, Saijonmaa O et al (2013) The roles of senescence and telomere shortening in cardiovascular disease. Nat Rev Cardiol 10:274. https://doi.org/10.1038/nrcardio.2013.30

Gokcay F, Arsava EM, Baykaner T, Vangel M, Garg P, Wu O, Singhal AB, Furie KL, Sorensen AG, Ay H (2011) Age-dependent susceptibility to infarct growth in women. Stroke 42:947. https://doi.org/10.1161/STROKEAHA.110.603902

Goldstein LB, Bushnell CD, Adams RJ, Appel LJ, Braun LT, Chaturvedi S et al (2011) Guidelines for the primary prevention of stroke: a guideline for healthcare professionals from the American Heart Association/American Stroke Association. Stroke 42:517. https://doi.org/10.1161/str.0b013e3181fcb238

Goossens J, Hachimi-Idrissi S (2014) Combination of therapeutic hypothermia and other neuroprotective strategies after an ischemic cerebral insult. Curr Neuropharmacol 12:399. https://doi.org/10.2174/1570159X12666140424233036

Gresita A, Glavan D, Udristoiu I, Catalin B, Hermann DM, Popa-Wagner A (2019) Very low efficiency of direct reprogramming of astrocytes into neurons in the brains of young and aged mice after cerebral ischemia. Front Aging Neurosci 11:334. https://doi.org/10.3389/fnagi.2019.00334. eCollection 2019

Hallett M (2001) Plasticity of the human motor cortex and recovery from stroke. Brain Res Brain Res Rev 36:169. https://doi.org/10.1016/S0165-0173(01)00092-3

Hemmen TM, Raman R, Guluma KZ, Meyer BC, Gomes JA, Cruz-Flores S et al (2010) Intravenous thrombolysis plus hypothermia for acute treatment of ischemic stroke (ICTuS-L): final results. Stroke 41:2265. https://doi.org/10.1161/strokeaha.110.592295

Hennerici MG, Kern R, Szabo K (2013) Non-pharmacological strategies for the treatment of acute ischaemic stroke. Lancet Neurol 12:572. https://doi.org/10.1016/S1474-4422(13)70091-7

Hermann DM, Chopp M (2012) Promoting brain remodeling and plasticity for stroke recovery: therapeutic promise and potential pitfalls of clinical translation. Lancet Neurol 11:369. https://doi.org/10.1016/S1474-4422(12)70039-X

Hong JM, Lee JS, Song HJ, Jeong HS, Choi HA, Lee K (2014) Therapeutic hypothermia after recanalization in patients with acute ischemic stroke. Stroke 45:134. https://doi.org/10.1161/STROKEAHA.113.003143

Honmou O, Onodera R, Sasaki M, Waxman SG, Kocsis JD (2012) Mesenchymal stem cells: therapeutic outlook for stroke. Trends Mol Med 18:292. https://doi.org/10.1016/j.molmed.2012.02.003

Hou SW, Wang YQ, Xu M, Shen DH, Wang JJ, Huang F et al (2008) Functional integration of newly generated neurons into striatum after cerebral ischemia in the adult rat brain. Stroke 39(10):2837–2844

Iadecola C, Anrather J (2011) The immunology of stroke: from mechanisms to translation. Nat Med 17:796. https://doi.org/10.1038/nm.2399

Jayaraj RL, Azimullah S, Beiram R et al (2019) Neuroinflammation: friend and foe for ischemic stroke. J Neuroinflammation 16:142. https://doi.org/10.1186/s12974-019-1516-2

Jiang Q, Thiffault C, Kramer C, Ding L, Zhang L, Nejad-Davarani S, Li L, Arbab A, Lu M, Navia B, Victor S, Hong K, Li JQ, Wang SY, Li Y, Chopp M (2012) MRI detects brain reorganization after human umbilical tissue-derived cells (hUTC) treatment of stroke in rat. PLoS One 7:e42845. https://doi.org/10.1371/journal.pone.0042845

Jin K, Wang X, Xie L, Mao XO, Zhu W, Wang Y et al (2006) Evidence for stroke-induced neurogenesis in the human brain. Proc Natl Acad Sci U S A 103(35):13198–13202

Jinno S (2011) Decline in adult neurogenesis during aging follows a topographic pattern in the mouse hippocampus. J Comp Neurol 519:451. https://doi.org/10.1002/cne.22527

Johansen FF, Hasseldam H, Rasmussen RS, Bisgaard AS, Bonfils PK, Poulsen SS et al (2014) Drug-induced hypothermia as beneficial treatment before and after cerebral ischemia. Pathobiology 81:42. https://doi.org/10.1159/000352026

Joseph C, Buga AM, Vintilescu R, Balseanu A, Moldovan M, Junker H, Lary W, Lotze M, Popa-Wagner A (2012) Prolonged gaseous hypothermia prevents the upregulation of phagocytosis-specific protein Annexin 1 and causes low-amplitude EEG activity in the aged rat brain after cerebral ischemia. J Cereb Blood Flow Metab 32:1632. https://doi.org/10.1038/jcbfm.2012.65

Kammersgaard LP, Rasmussen BH, Jorgensen HS, Reith J, Weber U, Olsen TS (2000) Feasibility and safety of inducing modest hypothermia in awake patients with acute stroke through surface cooling: a case-control study: the Copenhagen stroke study. Stroke 31:2251. https://doi.org/10.1161/01.str.31.9.2251

Kokaia Z, Llorente IL, Carmichael ST (2018) Customized brain cells for stroke patients using pluripotent stem cells. Stroke 49(5):1091–1098. https://doi.org/10.1161/STROKEAHA.117.018291

Kollmar R, Blank T, Han JL, Georgiadis D, Schwab S (2007) Different degrees of hypothermia after experimental stroke: short- and long-term outcome. Stroke 38:1585. https://doi.org/10.1161/STROKEAHA.106.475897

Kumari S, Anderson L, Farmer S, Mehta SL, Li PA (2012) Hyperglycemia alters mitochondrial fission and fusion proteins in mice subjected to cerebral ischemia and reperfusion. Transl Stroke Res 3:296. https://doi.org/10.1007/s12975-012-0158-9

Kurisu K, Yenari M (2017) Therapeutic hypothermia for ischemic stroke; pathophysiology and future promise. Neuropharmacology 134:302. https://doi.org/10.1016/j.neuropharm.2017.08.025

LeBlanc EL, Patnode CD, Webber EM, Redmond N, Rushkin M, O'Connor EA (2018) Behavioral and pharmacotherapy weight loss interventions to prevent obesity-related morbidity and mortality in adults: an updated systematic review for the U.S. Preventive Services Task Force [Internet]. Agency for Healthcare Research and Quality (US), Rockville, MD. Available from http://www.ncbi.nlm.nih.gov/books/NBK532379/

Lee JC, Cho GS, Choi BO, Kim HC, Kim YS, Kim WK (2006) Intracerebral hemorrhage-induced brain injury is aggravated in senescence-accelerated prone mice. Stroke 37:216. https://doi.org/10.1161/01.STR.0000195151.46926.7b

Liepert J, Hamzei F, Weiller C (2004) Lesion-induced and training-induced brain reorganization. Restor Neurol Neurosci 22:269–277

Lim JS, Kang DW (2015) Stroke connectome and its implications for cognitive and behavioral sequela of stroke. J Stroke 17:256. https://doi.org/10.5853/jos.2015.17.3.256

Lindner MD, Gribkoff VK, Donlan NA, Jones TA (2003) Long-lasting functional disabilities in middle-aged rats with mall cerebral infarcts. J Neurosci 23:10913. https://doi.org/10.1523/JNEUROSCI.23-34-10913.2003

Lindvall O, Kokaia Z (2015) Neurogenesis following stroke affecting the adult brain. Cold Spring Harb Perspect Biol 7(11):a019034

Liu X, Ye R, Yan T, Yu SP, Wei L, Xu G, Fan X, Jiang Y, Stetler RA, Liu G, Chen J (2014) Cell based therapies for ischemic stroke: from basic science to bedside. Prog Neurobiol 115:92–115. https://doi.org/10.1016/j.pneurobio.2013.11.007

Lloyd-Jones DM (2010) Cardiovascular risk prediction: basic concepts, current status, and future directions. Circulation 121:1768. https://doi.org/10.1161/circulationaha.109.849166

Ma S, Wang J, Wang Y, Dai X, Xu F, Gao X, Johnson J, Xu N, Leak RK, Hu X, Luo Y, Chen J (2018) Diabetes mellitus impairs white matter repair and long-term functional deficits after cerebral ischemia. Stroke 49(10):2453–2463. https://doi.org/10.1161/STROKEAHA.118.021452

Macas J, Nern C, Plate KH, Momma S (2006) Increased generation of neuronal progenitors after ischemic injury in the aged adult human forebrain. J Neurosci 26(50):13114–13119

Macri MA, D'Alessandro N, Di GC, Di IP, Di LS, Giuliani P, Bianchi G, Esposito E (2006) Regional changes in the metabolite profile after long-term hypoxia-ischemia in brains of young and aged rats: a quantitative proton MRS study. Neurobiol Aging 27:98. https://doi.org/10.1016/j.neurobiolaging.2005.01.007

Manwani B, Liu F, Xu Y, Persky R, Li J, McCullough L (2011) Functional recovery in aging mice after experimental stroke. Brain Behav Immun 25:1689. https://doi.org/10.1016/j.bbi.2011.06.015

Martin A, Rojas S, Chamorro A, Falcon C, Bargallo N, Planas AM (2006) Why does acute hyperglycemia worsen the outcome of transient focal cerebral ischemia? Role of corticosteroids, inflammation, and protein O-glycosylation. Stroke 37:1288. https://doi.org/10.1161/01.str.0000217389.55009.f8

Matsumoto H, Kumon Y et al (2007) Expression of CD200 by macrophage-like cells in ischemic core of rat brain after transient middle cerebral artery occlusion. Neurosci Lett 418:44. https://doi.org/10.1016/j.neulet.2007.03.027

McCabe C, Gallagher L, Gsell W, Graham D, Dominiczak AF, Macrae IM (2009) Differences in the evolution of the ischemic penumbra in stroke-prone spontaneously hypertensive and Wistar-Kyoto rats. Stroke 40:3864. https://doi.org/10.1161/strokeaha.109.559021

McColl BW, Allan SM, Rothwell NJ (2007) Systemic inflammation and stroke: aetiology, pathology and targets for therapy. Biochem Soc Trans 35:1163–1165

Miname M, Santos R (2019) Reducing cardiovascular risk in patients with familial hypercholesterolemia: risk prediction and lipid management. Prog Cardiovasc Dis 62:414. https://doi.org/10.1016/j.pcad.2019.10.003

Mine Y, Tatarishvili J, Oki K, Monni E, Kokaia Z, Lindvall O (2013) Grafted human neural stem cells enhance several steps of endogenous neurogenesis and improve behavioral recovery after middle cerebral artery occlusion in rats. Neurobiol Dis 52:191–203

Minger SL, Ekonomou A, Carta EM, Chinoy A, Perry RH, Ballard CG (2007) Endogenous neurogenesis in the human brain following cerebral infarction. Regen Med 2(1):69–74

Miskinyte G, Devaraju K, Grønning Hansen M, Monni E, Tornero D, Woods NB, Bengzon J, Ahlenius H, Lindvall O, Kokaia Z (2017) Direct conversion of human fibroblasts to functional excitatory cortical neurons integrating into human neural networks. Stem Cell Res Ther 8(1):207. https://doi.org/10.1186/s13287-017-0658-3

Miskinyte G, Grønning Hansen M, Monni E, Lam M, Bengzon J, Lindvall O, Ahlenius H, Kokaia Z (2018) Transcription factor programming of human ES cells generates functional neurons expressing both upper and deep layer cortical markers. PLoS One 13(10):e0204688. https://doi.org/10.1371/journal.pone.0204688. eCollection 2018

Modo M, Jolkkonen J, Zille M, Boltze J (2018) Future of animal modeling for poststroke tissue repair. Stroke 49:1099. https://doi.org/10.1161/STROKEAHA.117.018293

Mohamad O, Drury-Stewart D, Song M, Faulkner B, Chen D, Yu SP et al (2013) Vector-free and transgene-free human iPS cells differentiate into functional neurons and enhance functional recovery after ischemic stroke in mice. PLoS One 8(5):e64160

Morizane A, Li JY, Brundin P (2008) From bench to bed: the potential of stem cells for the treatment of Parkinson's disease. Cell Tissue Res 331(1):323–336

Oesch L, Tatlisumak T, Arnold M, Sarikaya H (2017) Obesity paradox in stroke - myth or reality? A systematic review. PLoS One 12(3):e0171334. https://doi.org/10.1371/journal.pone.0171334. eCollection 2017. Review

Oki K, Tatarishvili J, Wood J, Koch P, Wattananit S, Mine Y et al (2012) Human-induced pluripotent stem cells form functional neurons and improve recovery after grafting in stroke-damaged brain. Stem Cells 30(6):1120–1133

Parent JM, Vexler ZS, Gong C, Derugin N, Ferriero DM (2002) Rat forebrain neurogenesis and striatal neuron replacement after focal stroke. Ann Neurol 52(6):802–813

Perry VH (2007) Stress primes microglia to the presence of systemic inflammation: implications for environmental influences on the brain. Brain Behav Immun 21:45. https://doi.org/10.1016/j.bbi.2006.08.004

Phanthong P, Raveh-Amit H, Li T, Kitiyanant Y, Dinnyes A (2013) Is aging a barrier to reprogramming? Lessons from induced pluripotent stem cells. Biogerontology 14(6):591–602

Piironen K, Tiainen M, Mustanoja S, Kaukonen KM, Meretoja A, Tatlisumak T et al (2014) Mild hypothermia after intravenous thrombolysis in patients with acute stroke: a randomized controlled trial. Stroke 45:486. https://doi.org/10.1161/STROKEAHA.113.003180

Pintana H, Lietzau G, Augestad IL, Chiazza F, Nyström T, Patrone C, Darsalia V (2019) Obesity-induced type 2 diabetes impairs neurological recovery after stroke in correlation with decreased neurogenesis and persistent atrophy of parvalbumin-positive interneurons. Clin Sci (Lond) 133(13):1367–1386. https://doi.org/10.1042/CS20190180

Polentes J, Jendelova P, Cailleret M, Braun H, Romanyuk N, Tropel P et al (2012) Human induced pluripotent stem cells improve stroke outcome and reduce secondary degeneration in the recipient brain. Cell Transplant 21(12):2587–2602

Popa-Wagner A, Schroder E, Walker LC, Kessler C (1998) Beta-amyloid precursor protein and ss-amyloid peptide immunoreactivity in the rat brain after middle cerebral artery occlusion: effect of age. Stroke 29:2196. https://doi.org/10.1161/01.str.29.10.2196

Popa-Wagner A, Badan I, Walker L, Groppa S, Patrana N, Kessler C (2007a) Accelerated infarct development, cytogenesis and apoptosis following transient cerebral ischemia in aged rats. Acta Neuropathol 113:277. https://doi.org/10.1007/s00401-006-0164-7

Popa-Wagner A, Carmichael ST, Kokaia Z, Kessler C, Walker LC (2007b) The response of the aged brain to stroke: too much, too soon? Curr Neurovasc Res 4:216. https://doi.org/10.2174/156720207781387213

Popa-Wagner A, Dumitrascu DI, Capitanescu B, Petcu EB, Surugiu R, Fang WH, Dumbrava DA (2020) Dietary habits, lifestyle factors and neurodegenerative diseases. Neural Regen Res 15(3):394–400. https://doi.org/10.4103/1673-5374.266045

Popa-Wagner A, Stocker K, Balseanu AT, Rogalewski A, Diederich K, Minnerup J et al (2010) Effects of granulocyte-colony stimulating factor after stroke in aged rats. Stroke 41:1027. https://doi.org/10.1161/STROKEAHA.109.575621

Pradillo JM, Murray KN, Coutts GA, Moraga A, Oroz-Gonjar F, Boutin H, Moro MA, Lizasoain I, Rothwell NJ, Allan SM (2017) Reparative effects of interleukin-1 receptor antagonist in young and aged/co-morbid rodents after cerebral ischemia. Brain Behav Immun 61:117–126. https://doi.org/10.1016/j.bbi.2016.11.013

Qin L, Jing D, Parauda S, Carmel J, Ratan RR, Lee FS et al (2014) An adaptive role for BDNF Val66Met polymorphism in motor recovery in chronic stroke. J Neurosci 34:2493. https://doi.org/10.1523/jneurosci.4140-13.2014

Rawlinson C, Jenkins S, Thei L, Dallas ML, Chen R (2020) Post-ischaemic immunological response in the brain: targeting microglia in ischaemic stroke therapy. Brain Sci 10. https://doi.org/10.3390/brainsci10030159

Rewell SS, Fernandez JA, Cox SF, Spratt NJ, Hogan L, Aleksoska E et al (2010) Inducing stroke in aged, hypertensive, diabetic rats. J Cereb Blood Flow Metab 30:729. https://doi.org/10.1038/jcbfm.2009.273

Roger VL, Go AS, Lloyd-Jones DM, Benjamin EJ, Berry JD, Borden WB et al (2012) Heart disease and stroke statistics—2012 update: a report from the American Heart Association. Circulation 125:e2. https://doi.org/10.1161/cir.0b013e31823ac046

Rosen CL, Dinapoli VA, Nagamine T, Crocco T (2005) Influence of age on stroke outcome following transient focal ischemia. J Neurosurg 103:687. https://doi.org/10.3171/jns.2005.103.4.0687

Rothrock JF, Clark WM, Lyden PD (1995) Spontaneous early improvement following ischemic stroke. Stroke 26:1358. https://doi.org/10.1161/01.str.26.8.1358

Salminen A, Ojala J et al (2011) Astrocytes in the aging brain express characteristics of senescence-associated secretory phenotype. Eur J Neurosci 34:3. https://doi.org/10.1111/j.1460-9568.2011.07738.x

Sandu R, Buga AM, Balseanu A, Moldovan M, Popa-Wagner A (2015) Twenty four hours hypothermia has temporary efficacy in reducing brain infarction and inflammation in aged rats. Neurobiol Aging 38:127. https://doi.org/10.1016/j.neurobiolaging.2015.11.006

Savitz SI, Yavagal D, Rappard G, Likosky W, Rutledge N, Graffagnino C, Alderazi Y, Elder JA, Chen PR, Budzik RF Jr, Tarrel R, Huang DY, Hinson JM Jr (2019) A phase 2 randomized, sham-controlled trial of internal carotid artery infusion of autologous bone marrow-derived ALD-401 cells in patients with recent stable ischemic stroke (RECOVER-Stroke). Circulation 139(2):192–205

Schmidt H, Zeginigg M, Wiltgen M, Freudenberger P, Petrovic K, Cavalieri M, Gider P, Enzinger C, Fornage M, Debette S, Rotter J, Ikram M, Launer L, Schmidt R (2011) Genetic variants of the NOTCH3 gene in the elderly and magnetic resonance imaging correlates of age-related cerebral small vessel disease. Brain 134:3384. https://doi.org/10.1093/brain/awr252

Seehafer JU, Kalthoff D, Farr TD, Wiedermann D, Hoehn M (2010) No increase of the blood oxygenation level-dependent functional magnetic resonance imaging signal with higher field strength: implications for brain activation studies. J Neurosci 30:5234. https://doi.org/10.1523/JNEUROSCI.0844-10.2010

Shetty AK, Rao MS, Hattiangady B (2008) Behavior of hippocampal stem/progenitor cells following grafting into the injured aged hippocampus. J Neurosci Res 86(14):3062–3074. https://doi.org/10.1002/jnr.21764

Shrivastava K, Gonzalez P et al (2012) The immune inhibitory complex CD200/CD200R is developmentally regulated in the mouse brain. J Comp Neurol 520:2657. https://doi.org/10.1002/cne.23062

Silasi G, Colbourne F (2011) Therapeutic hypothermia influences cell genesis and survival in the rat hippocampus following global ischemia. J Cereb Blood Flow Metab 31:1725. https://doi.org/10.1038/jcbfm.2011.25

Silasi G, Klahr AC, Hackett MJ, Auriat AM, Nichol H, Colbourne F (2012) Prolonged therapeutic hypothermia does not adversely impact neuroplasticity after global ischemia in rats. J Cereb Blood Flow Metab 32:1525. https://doi.org/10.1038/jcbfm.2012.38

Soejima H, Ogawa H, Morimoto T, Nakayama M, Okada S, Sakuma M et al (2013a) Aspirin possibly reduces cerebrovascular events in type 2 diabetic patients with higher C-reactive protein level: subanalysis from the JPAD trial. J Cardiol 62:165. https://doi.org/10.1016/j.jjcc.2013.03.015

Soejima Y, Hu Q, Krafft PR, Fujii M, Tang J, Zhang JH (2013b) Hyperbaric oxygen preconditioning attenuates hyperglycemia-enhanced hemorrhagic transformation by inhibiting matrix metalloproteinases in focal cerebral ischemia in rats. Exp Neurol 247:737. https://doi.org/10.1016/j.expneurol.2013.03.019

Stoll EA (2014) Advances toward regenerative medicine in the central nervous system: challenges in making stem cell therapy a viable clinical strategy. Mol Cell Ther 2:12

Strazzullo P, D'Elia L, Cairella G, Garbagnati F, Cappuccio F, Scalfi L (2010) Excess body weight and incidence of stroke meta-analysis of prospective studies with 2 million participants. Stroke 41:e418. https://doi.org/10.1161/STROKEAHA.109.576967

Sutherland GR, Dix GA, Auer RN (1996) Effect of age in rodent models of focal and forebrain ischemia. Stroke 27:1663. https://doi.org/10.1161/01.STR.27.9.1663

Tacutu R, Budovsky A, Fraifeld VE (2010) The NetAge database: a compendium of networks for longevity, age-related diseases and associated processes. Biogerontology 11:513. https://doi.org/10.1007/s10522-010-9265-8

Tang SC, Lee LJ-H, Jeng J-S, Hsieh S-T, Chiang M-C, Yeh S-J, Hsueh HW, Chao C-C (2019) Pathophysiology of central poststroke pain: motor cortex disinhibition and its clinical and sensory correlates. Stroke 50:2851. https://doi.org/10.1161/STROKEAHA.119.025692

Tatarishvili J, Oki K, Monni E, Koch P, Memanishvili T, Buga AM et al (2014) Human induced pluripotent stem cells improve recovery in stroke-injured aged rats. Restor Neurol Neurosci 32:547. https://doi.org/10.3233/RNN-140404

Tchkonia T, Zhu Y et al (2013) Cellular senescence and the senescent secretory phenotype: therapeutic opportunities. J Clin Invest 123:966. https://doi.org/10.1172/JCI64098

Thored P, Arvidsson A, Cacci E, Ahlenius H, Kallur T, Darsalia V et al (2006) Persistent production of neurons from adult brain stem cells during recovery after stroke. Stem Cells 24(3):739–747

Tornero D, Wattananit S, Gronning Madsen M, Koch P, Wood J, Tatarishvili J et al (2013) Human induced pluripotent stem cell-derived cortical neurons integrate in stroke-injured cortex and improve functional recovery. Brain 136(Pt 12):3561–3577

Towfighi A, Ovbiagele B (2009) The impact of body mass index on mortality after stroke. Stroke. https://doi.org/10.1161/STROKEAHA.109.550228

Tropepe V, Craig CG, Morshead CM, van der Kooy D (1997) Transforming growth factor-alpha null and senescent mice show decreased neural progenitor cell proliferation in the forebrain subependyma. J Neurosci 17(20):7850–7859

Trueman RC, Harrison DJ, Dwyer DM, Dunnett SB, Hoehn M, Farr TD (2011) A critical re-examination of the intraluminal filament MCAO model: impact of external carotid artery transection. Transl Stroke Res 2:651. https://doi.org/10.1007/s12975-011-0102-4

van der Worp HB, Macleod MR, Bath PM, Bathula R, Christensen H, Colam B, Schwab S (2019) Therapeutic hypothermia for acute ischaemic stroke. Results of a European multicentre, randomised, phase III clinical trial. Eur Stroke J 4:254. https://doi.org/10.1177/2396987319844690

van Meer MPA, Otte WM, van der Marel K, Nijboer CH, Kavelaars A, van der Sprenkel JWB, Viergever MA, Dijkhuizen RM (2012) Extent of bilateral neuronal network reorganization and functional recovery in relation to stroke severity. J Neurosci 32:4495. https://doi.org/10.1523/JNEUROSCI.3662-11.2012

Vicidomini C, Guo N, Sahay A (2020) Communication, cross talk, and signal integration in the adult hippocampal neurogenic niche. Neuron 105(2):220–235. https://doi.org/10.1016/j.neuron.2019.11.029

Vintilescu R, Uzoni A, Ciobanu O, Moldovan M, Anghel A, Radu E, Coogan A, Popa-Wagner A (2016) Post-stroke gaseous hypothermia increases vascular density but not neurogenesis in the ischemic penumbra of aged rats. Restor Neurol Neurosci 34:401. https://doi.org/10.3233/RNN-150600

Wan YH, Nie C, Wang HL, Huang CY (2014) Therapeutic hypothermia (different depths, durations, and rewarming speeds) for acute ischemic stroke: a meta-analysis. J Stroke Cerebrovasc Dis 23:2736. https://doi.org/10.1016/j.jstrokecerebrovasdis.2014.06.017

Wang LC, Futrell N, Wang DZ, Chen FJ, Zhai QH, Schultz LR (1995) A reproducible model of middle cerebral infarcts, compatible with long-term survival, in aged rats. Stroke 26:2087. https://doi.org/10.1161/01.STR.26.11.2087

Wang R-Y, Wang S-G, Yang Y-R (2003) Effect of age in rats following middle cerebral artery occlusion. Gerontology 49:27. https://doi.org/10.1159/000066505

Wang XJ, Zhang S et al (2011) Impaired CD200-CD200R-mediated microglia silencing enhances midbrain dopaminergic neurodegeneration: roles of aging, superoxide, NADPH oxidase, and p38 MAPK. Free Radic Biol Med 50:1094. https://doi.org/10.1016/j.freeradbiomed.2011.01.032
Wei G, Hartings JA, Yang X, Tortella FC, Lu XC (2008) Extraluminal cooling of bilateral common carotid arteries as a method to achieve selective brain cooling for neuroprotection. J Neurotrauma 25:549. https://doi.org/10.1089/neu.2007.0498
Won SJ, Xie L, Kim SH, Tang H, Wang Y, Mao X (2006) Influence of age on the response to fibroblast growth factor-2 treatment in a rat model of stroke. Brain Res 1123:237. https://doi.org/10.1016/j.brainres.2006.09.055
Wu TC, Grotta JC (2013) Hypothermia for acute ischaemic stroke. Lancet Neurol 12:275. https://doi.org/10.1016/s1474-4422(13)70013-9
Yuan T, Liao W, Feng NH, Lou YL, Niu X, Zhang AJ et al (2013) Human induced pluripotent stem cell-derived neural stem cells survive, migrate, differentiate, and improve neurologic function in a rat model of middle cerebral artery occlusion. Stem Cell Res Ther 4(3):73
Zhang L, Zhang RL, Wang Y, Zhang C, Zhang ZG, Meng H, Chopp M (2005) Functional recovery in aged and young rats after embolic stroke: treatment with a phosphodiesterase type 5 inhibitor. Stroke 36:847. https://doi.org/10.1161/01.STR.0000158923.19956.73
Zhang X, Zhang G, Zhang H, Karin M, Bai H, Cai D (2008) Hypothalamic IKKbeta/NF-kappaB and ER stress link overnutrition to energy imbalance and obesity. Cell 135:61. https://doi.org/10.1016/j.cell.2008.07.043
Zhang RL, Chopp M, Roberts C, Jia L, Wei M, Lu M et al (2011) Ascl1 lineage cells contribute to ischemia-induced neurogenesis and oligodendrogenesis. J Cereb Blood Flow Metab 31(2):614–625
Zhao CS, Puurunen K, Schallert T, Sivenius J, Jolkkonen J (2005) Effect of cholinergic medication, before and after focal photothrombotic ischemic cortical injury, on histological and functional outcome in aged and young adult rats. Behav Brain Res 156:85. https://doi.org/10.1016/j.bbr.2004.05.011
Zhao L, Du W, Zhao X, Liu L, Wang C, Wang Y, Wang A, Liu G, Wang Y, Xu Y (2014) Favorable functional recovery in overweight ischemic stroke survivors: findings from the China National Stroke Registry. J Stroke Cerebrovasc Dis 23(3):e201–e206. https://doi.org/10.1016/j.jstrokecerebrovasdis.2013.10.002

Diagnostic Markers of Subclinical Depression Based on Functional Connectivity

Yunkai Zhu, Jorge Bohorquez, and Ioan Opris

1 Introduction

Major depression disorder is a chronic, relapsing illness affecting 3–6% of the entire world population. Depression is defined as a mood disorder that induces a persistent feeling of sadness and loss of interest. Also known, as major depressive disorder or clinical depression, it affects how one feels, thinks, and behaves and can lead to a variety of emotional and physical problems (Johnson et al. 2013). In the next, we show some anatomical structures that play a key role in depression.

The lateral habenula (LHb) is a small epithelial structure located between the medial thalamus and the third ventricle, near its posterior union (Fig. 1) (Lawson et al. 2013; Boulos et al. 2016; Pauli et al. 2018). It receives input from the basal ganglia and limbic system and sends projections to the inhibitory interneurons in the midbrain (Wang et al. 1977; hou et al. 2009; Jennifer et al. 2009; Stamatakis et al. 2012; Hikosaka 2010). There is increasing clinical evidence that the LHb is one of the key regions regulating the reward circuit of the midbrain (Matsumoto and Hikosaka 2007; Hikosaka 2010; Stopper and Floresco 2014; Lawson et al. 2014; Ely et al 2016; Benarroch 2015; Proulx et al. 2014). Several studies in recent years have shown that LHb plays a significant role in depression: in experiments on depression in rats, the depressed rats have been observed to discharge their lateral habenular nucleus much more frequently than normal rats (Fig. 2) (Li et al. 2011, 2013; Yang et al. 2010, 2018). Changes in neuronal activity in the brain can lead to changes in local brain blood flow. Therefore, fMRI could be expected to provide, to

Y. Zhu (✉) · J. Bohorquez
Department of Biomedical Engineering, University of Miami, Coral Gables, FL, USA
e-mail: yxz1585@miami.edu

I. Opris
Department of Neurosurgery, University of Miami, Miami, FL, USA

I. Opris et al. (eds.), *Modern Approaches to Augmentation of Brain Function*, Contemporary Clinical Neuroscience,
https://doi.org/10.1007/978-3-030-54564-2_13

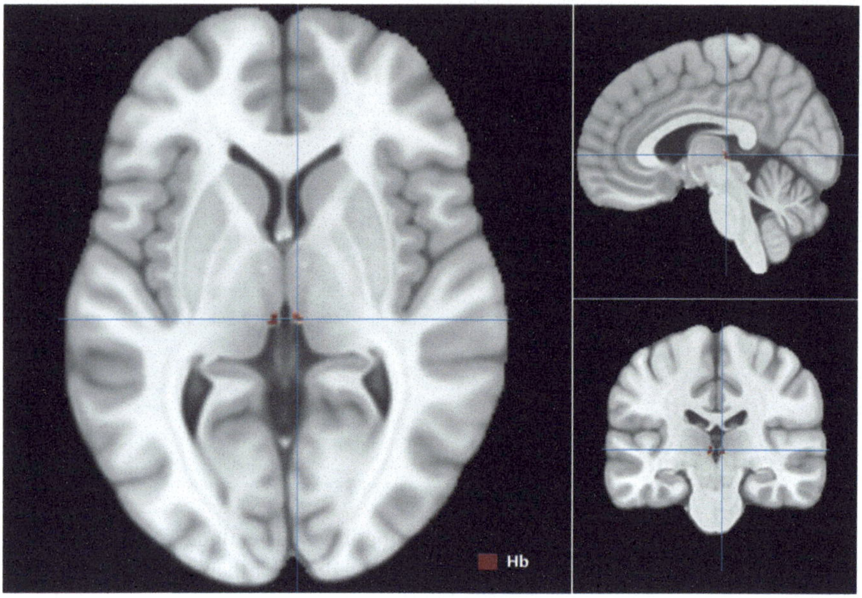

Fig. 1 The red marker on the MRI image shows the anatomical location of Hb. It locates in the iliac triangle at the head of the pineal gland and inside the dorsal thalamus. Hb can be divided into medial habenula and lateral habenula. The content of myelin in Hb is high, so it can show a high signal intensity in T1-weighted imaging. Hb shows an obvious boundary with the surrounding tissues and cerebrospinal fluid space

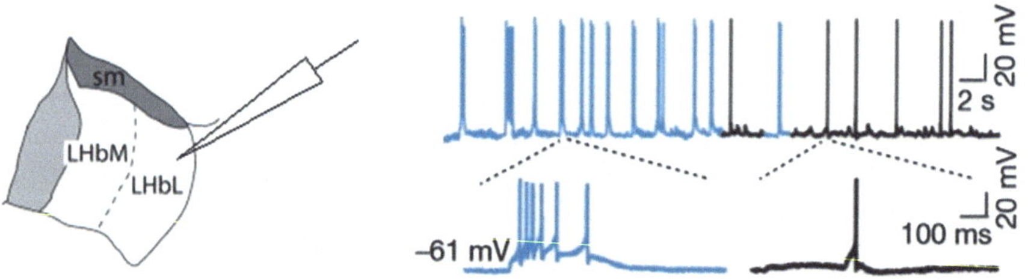

Fig. 2 Blue represents the habenula action potential in the depressed rats. Black represents a change after dosing of the drug to eliminate depressive symptoms. The discharge frequency of the habenula nucleus before dosing is higher than the frequency in rats treated with drugs to eliminate depressive symptoms

a certain extent, a biomarker that is capable of distinguishing depressed patients from normal people.

Subclinical depression refers to the presence of clinical depressive symptoms, which are not enough to meet the diagnostic criteria for depression (Fergusson et al. 2005; Judd 1994). Numerous studies have shown both that subclinical depression may eventually lead to major depression disorder and that subclinical depression is potentially associated with brain activity (Enns et al. 2001; Lewinsohn et al. 2003; Shankman et al. 2009).

Many studies have confirmed that by extracting characteristics of neuroimaging, such as the degree of activation in specific areas detected from functional magnetic resonance imaging (fMRI), the results can be used as a reliable indicator to evaluate the health of the subject (Woo et al. 2017). These characteristics can be used as an aid in diagnosing a disease or assessing the severity of symptoms (Mayberg et al. 1999; Keedwell et al. 2005). The connection between the brain activity provided by functional imaging and symptoms can provide a basis for redefining the diagnosis category, determining neuropathological characteristics, and evaluating healthy brain function beyond the current clinical diagnosis category (Bassett and Sporns 2017). In addition, neuroimaging analysis can provide a basis for the intervention of clinical or subclinical patients, which may help to prevent or even reverse the course of the disease. The goal of most neuroimaging research is to understand what functions and processes are encoded in the target brain area. For example, through the standard parameter mapping scale, many individual tests can be performed on local areas or voxels to create a whole-brain map, which is currently the most popular neuroimaging analysis method. Characteristic studies based on the whole brain also found isolated brain regions to be of importance for the expression of clinical disease and symptoms. In this regard, previous studies have reported on the relationship between the anterior cingulate cortex (ACC) and depression (Mayberg et al. 1999), the gray matter around the aqueduct of both the thalamus and midbrain and chronic pain (Hamani et al. 2006), basal ganglia and obsessive-compulsive disorder (Welter et al. 2011), subthalamic nucleus and Parkinson's disease (Singh et al. 2015; Krack et al. 2003).

Functional connectivity refers to the functional integration between spatially separated areas of the brain (Biswal et al. 1995). Unlike structural connections that look for physical connections in the brain, functional connections are similar to the patterns of activation in different areas of the brain, regardless of their physical connections (Fornito et al. 2015). Functional connections are usually measured using functional magnetic resonance imaging (fMRI) in the resting state and analyzed in terms of correlation, consistency, and spatial grouping based on temporal similarity (Cohen et al. 2017). These methods have been used to demonstrate that functional connectivity is related to behavior in a variety of different tasks. According to the correlation between the changes of the cerebral blood flow in different brain regions at the same time, functional connection strength was calculated to model the functional brain network of functional connections in different brain regions (Fig. 3). A great number of brain functional network analyses, based on resting-state functional magnetic resonance imaging, have been applied to the study of mental illness (Lynal et al. 2008; Guo et al. 2014; Woo et al. 2017). Network-based connection strength analysis and network parameter analysis can be used to search for functional links in the brain regions and functional links between regions that are significantly associated with the disease (Horstmann et al. 2010).

The value of machine learning-assisted predictive modeling has grown rapidly in the field of neuroimaging, and there are currently more than 500 papers using this multivariate predictive model (Koutsouleris et al. 2009; Orrù et al. 2012; Woo et al. 2017). At the same time, the sample size also increased rapidly. Relevant research organizations are working on gathering and sharing data in small groups to make large-scale research possible. These include the Alzheimer's disease neuroimaging initiative (ADNI), autism brain imaging data exchange (ABIDE), Parkinson's series

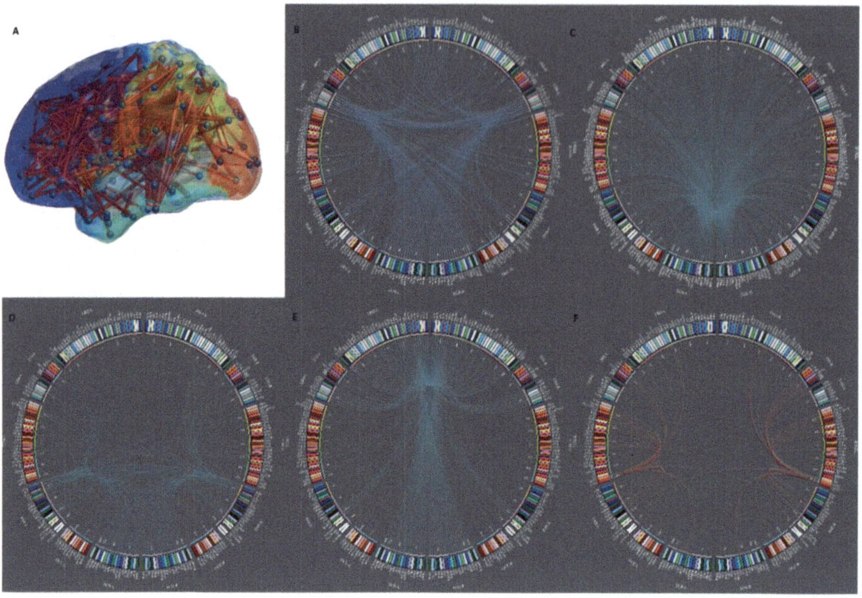

Fig. 3 (**a**) Connection model based on BrainNetome Atlas. (**b–f**) Examples that show the connection patterns of different brain regions, which include (**b**) Anterior cingulate cortex, (**c**) Thalamus, (**d**) Superior parietal lobule, (**e**) Superior frontal gyrus, (**f**) Posterior superior temporal sulcus

marking initiative (PPMI), and human connectome project (HCP). This is a significant aim for this field, which promotes the development of large sample models and greatly improves statistical capabilities; while at the same time, it promotes the development of multi-sample pattern recognition.

In the research reviewed in this chapter, the fMRI of patients with subclinical depression and normal subjects were used to extract the parameters of cerebral blood flow in the corresponding brain regions and model the brain network. Through network connection analysis, network node analysis, and network parameter analysis, combined with machine learning, the brain regions with significant influence on subclinical depression were extracted. The aim is to provide obvious biomarkers for the diagnosis of subclinical depression and effective therapeutic targets for treatment via electrical stimulation, e.g., transcranial magnetic stimulation.

2 Experimental Paradigm

All participants were recruited from volunteers who underwent health screening at Guangzhou Medical University from 2012 to 2014. The Baker depression scale (BDI-II) was used to assess the severity of depressive symptoms (Beck et al. 1996). Thirty-four subjects with a BDI-II score > 13 (11 men and 23 women) were included in the SD group (average BDI score SD: 22.58 6.92), and 40 healthy controls (21 men and 19 women) were selected to match the age, sex, and education level of the

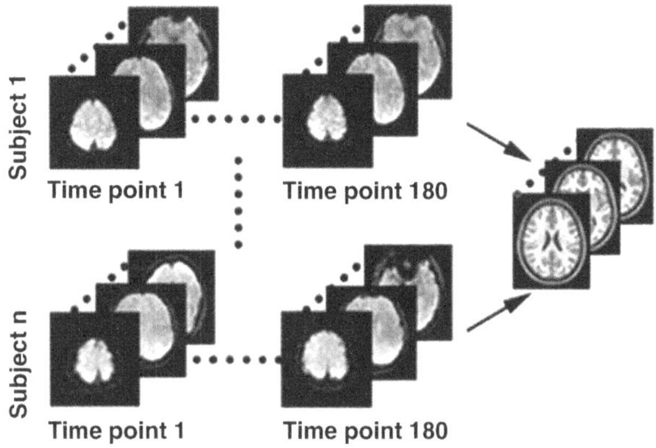

Fig. 4 Functional MRI (fMRI) image preprocessing. The steps include eliminating the first 20-time points to ensure the subject is in a resting state; Head movement correction to erase movement interference; Slice time correction to erase the interference of blood flow changes between different layers; Co-registration to normalize images into a standard structure space; Filtering to denoise

SD group. According to the two-sample, *t*-test, there was no significant difference in age (years) between the SD group and the HC group (mean SD: 19.91 1.64 vs. 19.70 0.85, $p = 0.50$), and there was no significant difference in education (years) (mean SD: 13.18 0.58 vs. 13.08 0.62, $p = 0.47$). Chi-square test showed no significant difference in gender ($p = 0.07$). None of the participants met the MDD criteria based on the diagnostic and statistical manual of mental Illness (DSM-IV). Other inclusion criteria for all participants included: age 19–25 years, right-handedness, no MRI scans showing lesions, no neurological disease, and no alcohol or drug dependence. This study was approved by the Medical Ethics Committee of the First People's Hospital of Guangzhou Medical University and meets the 1964 Helsinki Declaration and its subsequent amendments or similar ethical standards (Zhu et al. 2019).

After the fMRI data were obtained, using DPABI toolbox, the first, 10-time points were removed to ensure that the subjects entered a resting state. After that, Slice Timing was performed to eliminate the influence of blood flow on different layers of fMRI. To eliminate the disturbance of the subject's head movement during the test, head movement correction was applied. In the next step, we performed filtering with a bandwidth of (0.01–0.1) Hz. Finally, spatial domain correction was carried out to register the different shapes of the brain to the standard MNI coordinate system (Fig. 4) (Uddin et al. 2009; Yang et al. 2010).

3 Modeling Functional Brain Network

Blood-oxygen level-dependent (BOLD) response parameters of different brain regions were extracted using BrainNetome Atlas as a standard brain segmentation template (Fan et al. 2016). BOLD of time series was used to calculate the Pearson correlation coefficient of the brain region to obtain the functional connection strength of different brain regions to model the functional brain network of the whole brain (Fig. 5).

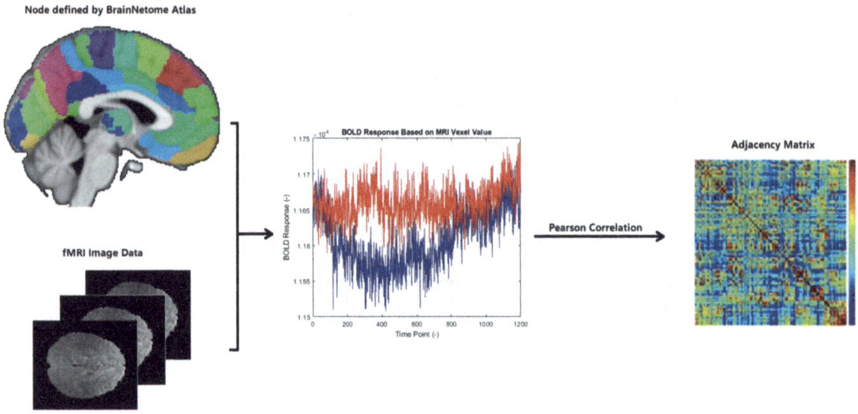

Fig. 5 Construction of functional brain networks. BOLD response of different brain regions was extracted with BrainNetome Atlas. Pearson correlation coefficients based on BOLD response of the brain regions were calculated. After fisher Z-transformation, the adjacency matrix was inserted by the correlation value to construct the functional brain network

4 Identification of Dysfunctional Connections

The experimental results showed that the discharge frequency of LHb in the depressed rats was much higher than in the control group (Yang et al. 2018). From this conclusion, we can speculate that the functional connection strength of the nodes of the habenula nucleus in the subclinical depression group should be stronger than that in the control group. Furthermore, the nucleus accumbens is associated with the brain's reward system, and if subclinical depression is associated with a lack of reward in the brain, then the level of activity in the nucleus accumbens should differ between the depressed and normal groups. Previous studies have demonstrated that the activity of the anterior cingulate cortex in the depressed patients is different from that in the normal controls (Gao et al. 2016; Drevets et al. 2008). Therefore, we should also be able to detect significant differences in the anterior cingulate cortex in subclinical patients. In this study, we use leave-one-out two-sample t-test to extract the functional connections with a significant level of $p < 0.05$ to generate multiple functional connection masks (Fig. 6).

5 Validating the Model with Machine Learning

In order to verify whether the functional connection mask is effective, we input the extracted functional connection into the support vector machine (SVM) model. SVM uses the RBF kernel function. The accuracy rate, ROC curve, and AUC were obtained through cross-validation to evaluate the classification efficiency of the above model (Adankon et al. 2009; Chang et al. 2011). By analyzing the model with

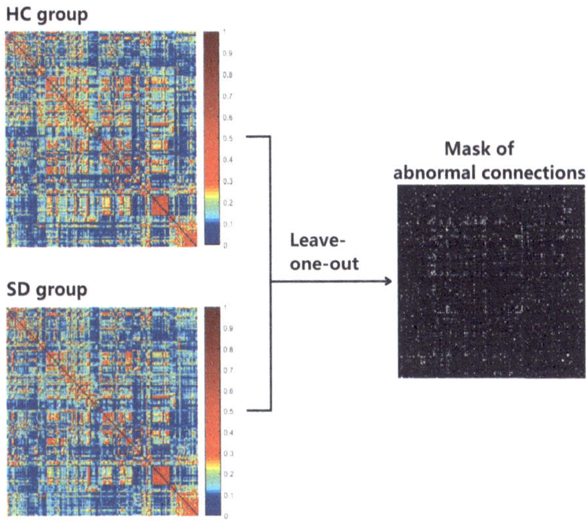

Fig. 6 Identification of dysfunctional connections. Double *t*-test was used to screen the functional connections that might have great differences. We generated multiple masks to extract the functional connection then input SVM. Functional connections with the significant difference were detected by evaluating SVM results

the highest classification efficiency, the functional connection mask with the greatest difference between subclinical depression patients and normal people can be found (Zhu et al. 2019).

After finding the most effective mask, we can extract the subnetwork functional connections located in different brain regions and input the SVM model. Accuracy, ROC curve, and AUC will be calculated to evaluate the classification efficiency of the functional connections of the subnetwork. In this way, we will be able to identify the brain regions with the most significant differences between the subclinical depression patients and the control group (Zhu et al. 2019).

6 Reliable Biomarkers for Subclinical Depression Prediction

Functional connections were plotted based on BrainNetome Atlas (Fig. 6) and extracted functional connections were inputted to a machine learning classifier to determine whether it can effectively work for HC and SD classification (Zhu et al. 2019). It can be observed that under the SVM-based classification, anterior cingulate cortex, thalamus, superior parietal lobule, superior frontal gyrus, posterior superior temporal sulcus achieved the highest classification effect (Fig. 7) (Xia et al. 2013). This confirms the feasibility of using functional connection strength as a biomarker for the diagnosis of subclinical depression.

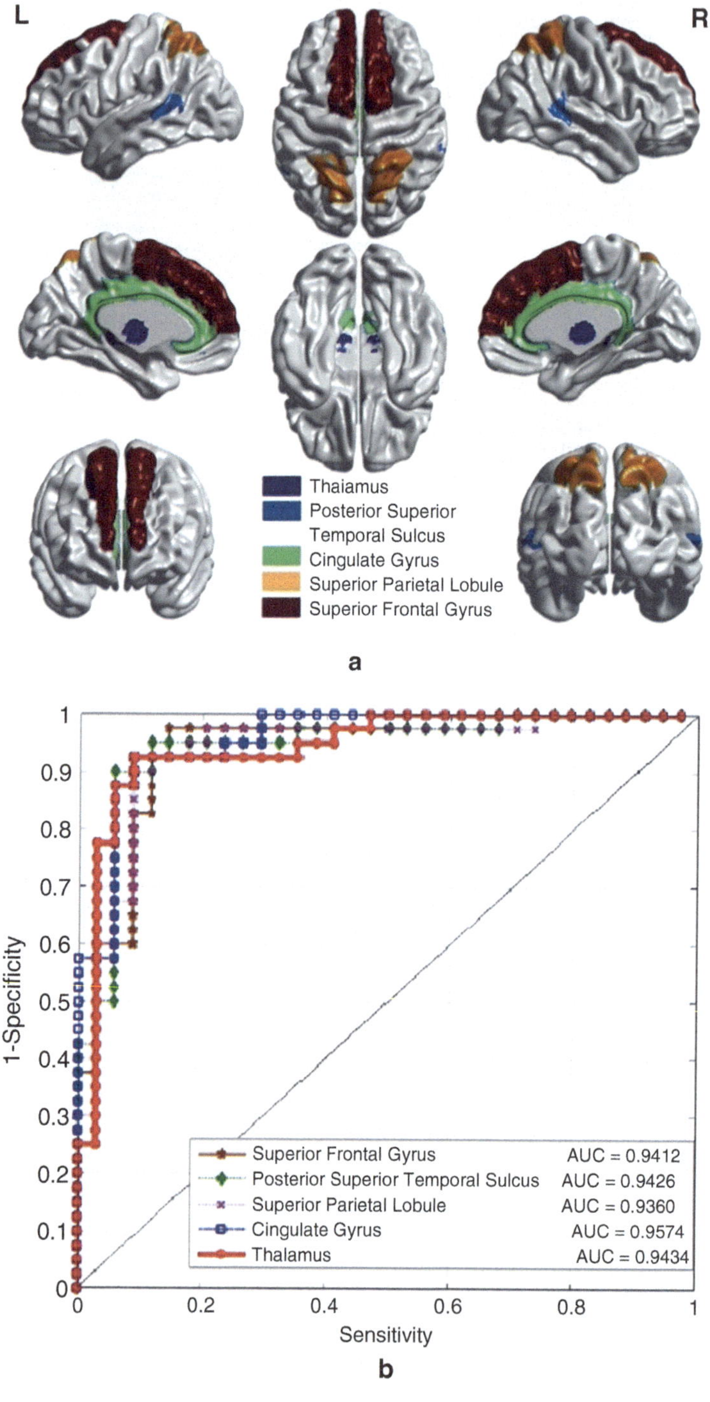

Fig. 7 Five brain regions (**a**) with significant functional connectivity differences were identified. Based on SVM, the functional connection of each brain region was tested. Classification results were evaluated by ROC curves and AUC (**b**)

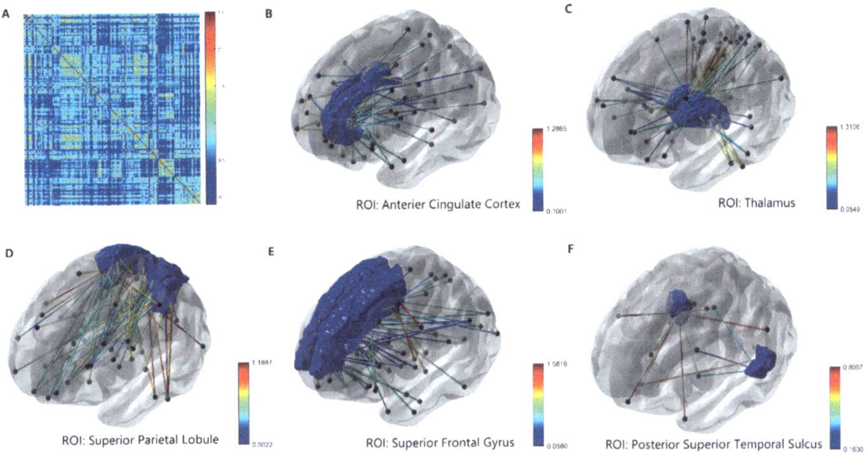

Fig. 8 (**a**) Functional connection matrix based on BrainNetome Atlas. The color shows the strength of the functional connections used to predict preclinical depression. (**b–f**) Significantly different functional connections in the five brain regions

Activity in the habenula inhibits activity in its downstream brain regions. We can observe functional connections with significant differences between LHb and NA, LHb and VTA, LHb and SPL in the SD group. This may indicate that the reward circuitry in the brain is affected in subclinical depression (Fig. 8).

In the node degree analysis of the network, there were 18 abnormal functional connections with the right Hb, which is the highest in the thalamus subregions. Meanwhile, the node degree of ten thalamus subregions in the SD group showed significant differences from the HC group (Fig. 9). Using these models, we were able to identify biomarkers that effectively diagnosed sub-clinical depression.

7 Potential for Recovery of Brain Function in Patients with Depression by Transcranial Magnetic Stimulation (TMS)

Due to the widespread nature of depression, noninvasive interventions are of great significance in the treatment of depression. Transcranial magnetic stimulation is used for emergency intervention and treatment of depression due to its non-invasive characteristics (George et al. 2013). Studies have shown that the activation of the prefrontal cortex (PFC) is low in some patients with depression (George et al. 1994), and that PFC-based stimulation can activate regulatory circuits connected to it such as movement, perception, and emotional management circuits (George 1998, Alexander et al. 1986). Nowadays, the "5 cm rule" is often used in clinical practice to determine the stimulation site of TMS, which is the area 5 cm in the front of the

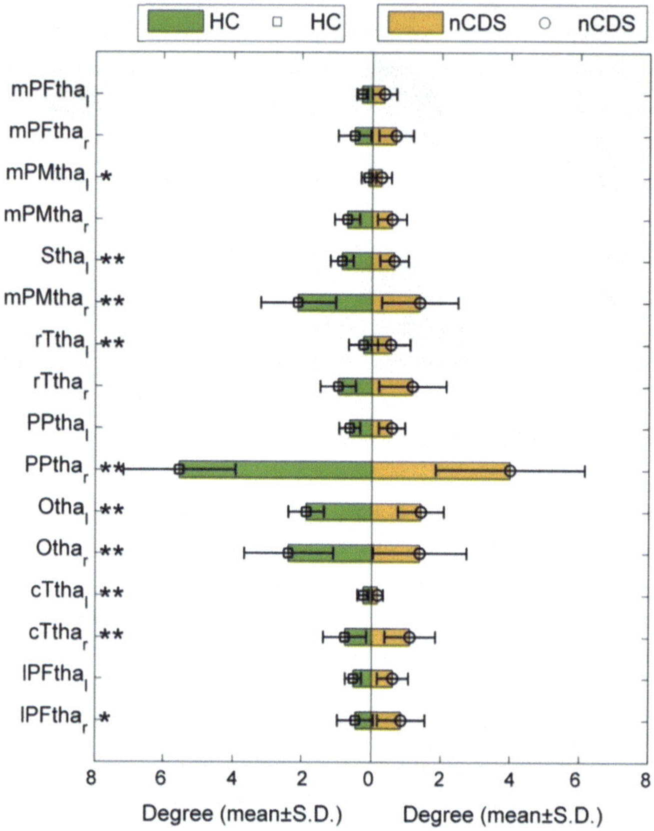

Fig. 9 Degree analysis of the thalamus subregions. The right posterior parietal thalamus, where LHb is located, has the most abnormal functional connections, the number of which is 18, and the left LHb has only 3 (an unusual functional connection). At a significant level $p < 0.01$, the network node degree of the ten subregions of the thalamus in the SD group showed a significant difference from the corresponding node degree in the HC group

motor cortex. This area was identified as a stimulus site for transcranial magnetic stimulation (Johnson et al. 2013). Clinically-based double-blind experiments show that subjects who received low-field synchronized transcranial magnetic stimulation (sTMS) for a period of time showed improvements in depressive symptoms, especially in patients with adverse reactions to drugs or a history of drug intolerance (Andrew F. Leuchter A et al. 2015). There is also evidence showing that TMS has the same therapeutic effect in the treatment of major depression as standard drug therapy (George 2010; Schutter 2008; Lam et al. 2008).

In the functional connectivity analysis of subclinical depression, it can be seen that a large number of abnormal functional connectivity exists in PFC of the SD patients (Fig. 8) (Zhu et al. 2019), but there is no evidence that TMS can affect the functional connectivity between LHb and PFC and ACC. We can look for the areas of the brain that are more closely related to the lateral habenula (Lecourtier et al. 2008). If the parts of the brain are greatly inhibited from the lateral habenula in subclinical depression or MDD, then electrical stimulation based on these areas, such as transcranial magnetic stimulation, may serve as a therapeutic modality by reshaping the feedback loop in the brain of the depressed patients. In this regard, we should stress that the force of a magnetic field falls off as the cube of the distance from the magnetic field. Deep TMS often relies on the use of special coils (H), which allow the magnetic field to penetrate a couple of inches away from the skull. Given the size and location of the habenula, deep within the brain, it is evident that any functional changes to the same will be mediated through connectivity from the main site of stimulation. The '5 cm rule' (*vide supra*) helps to identify the dorsolateral prefrontal cortex (DLPC). The tremendous output territory of DLPC makes it a connection hub within the small-world network of corticocortical connectivity. It is thought that by modulating the output of DLPC, a beneficial cascading effect is procreated with many areas associated with the same.

8 Comparison to Other Studies

In recent years, HN and its surrounding areas have become the focus of depression research. For example, in a controlled study of rats, ketamine was used to inhibit the activity of LHb, confirming that the abnormal cluster discharge in LHb causes depression (Yang et al. 2018). In an analysis based on fMRI, the cerebral blood flow difference of HN was confirmed between the SD group and HC group (Ely et al. 2016). In the study that is reviewed in this chapter, based on the analysis of functional connections around the whole brain, a model for predicting subclinical depression was constructed by looking for abnormal functional connections (Zhu et al. 2019). In the treatment of major depression, applying direct current stimulation to HN through deep brain stimulation can quickly relieve or even eliminate depressive symptoms (Sartorius et al. 2010). However, this approach is not noninvasive. TMS for PFC requires a long time to achieve the same therapeutic effect (George et al. 2013). In the study that is reviewed in this chapter, through the analysis of functional connections, the brain regions with strong functional connections to HN can be found as the targets of TMS, and it is expected that the stimulation of these targets can play a role in the rapid relief of depressive symptoms in future studies.

9 Conclusion

This section reviews a method that uses machine learning models to extract functional connections as features to predict subclinical depression. It could provide neuroimaging evidence for finding the brain regions that cause subclinical depression. Meanwhile, it is expected to provide effective targets for bioelectromagnetic treatment of depression disorder.

Acknowledgments The authors thank Profs. Aurel Popescu and Manuel F Casanova for their valuable help in editing and reading of the manuscript. Also, the authors are grateful to the former mentors of student Yunkai Zhu for his support.

References

Adankon MM, Cheriet M, Biem A (2009) Semisupervised least squares support vector machine. IEEE Trans Neural Netw 20(12):1858–1870

Alexander GE, DeLong MR, Strick PL (1986) Parallel organization of functionally segregated circuits linking basal ganglia and cortex. Annu Rev Neurosci 9:357–381

Andrew F. Leuchter A, Ian A, Cook A et al (2015) Efficacy and safety of low-field synchronized transcranial magnetic stimulation (sTMS) for treatment of major depression. Brain Stimul 8(4):787–794

Bassett DS, Sporns O (2017) Network neuroscience. Nat Neurosci 20:353–364. https://doi.org/10.1038/nn.4502

Beck AT, Steer RA, Ball R, Ranieri WF (1996) Comparison of Beck depression inventories-IA and-II in psychiatric outpatients. J Pers Assess 67(3):588–597

Benarroch EE (2015) Habenula: recently recognized functions and potential clinical relevance. Neurology 85(11):992–1000

Biswal B, Yetkin FZ, Haughton VM, Hyde JS (1995) Functional connectivity in the motor cortex of resting human brain using echo-planar MRI. Magn Reson Med 34(4):537–541

Boulos LJ, Darcq E, Kieffer BL (2016) Translating the habenula – from rodents to humans. Biol Psychiatry 81(4):296–305

Chang CC, Lin CJ (2011) LIBSVM: a library for support vector machines. 2(3):1–27

Cohen JD, Daw N, Engelhardt B, Hasson U, Li K, Niv Y et al (2017) Computational approaches to fMRI analysis. Nat Neurosci 20:304–313

Drevets WC et al (2008) The subgenual anterior cingulate cortex in mood disorders. CNS Spectr 13(8):663–681

Ely BA et al (2016) Resting-state functional connectivity of the human Habenula in healthy individuals: associations with sub-clinical depression. Human Brain Mapp 37:2369–2384

Enns MW, Cox BJ, Borger SC (2001) Correlates of analogue and clinical depression: a further test of the phenomenological continuity hypothesis. J Affect Disord 66(2):175–183

Fan L, Li H, Zhuo J, Zhang Y, Wang J, Chen L et al (2016) The human brainnetome atlas: a new brain atlas based on connectional architecture. Cereb Cortex 26(8):3508–3526

Fergusson DM, Horwood LJ, Ridder EM, Beautrais AL (2005) Subthreshold depression in adolescence and mental health outcomes in adulthood. Arch Gen Psychiatry 62(1):66–72

Fornito A, Zalesky A, Breakspear M (2015) The connectomics of brain disorders. Nat Rev Neurosci 16:159–172. https://doi.org/10.1038/nrn3901

Gao C, Wenhua L, Liu Y, Ruan X, Chen X, Liu L, Yu S, Chan RCK, Wei X, Jiang X (2016) Decreased subcortical and increased cortical degree centrality in a nonclinical college student sample with sub-clinical depressive symptoms: a resting-state fMRI study. Front Human Neurosci 10:1–9

George MS (1994) Introduction: the emerging neuroanatomy of depression. Psychiatr Ann 24(12):635–636

George MS (2010) Transcranial magnetic stimulation for the treatment of depression. Expert Rev Neurotherap 10(11):1761–1772

George MS (2013) The expanding evidence base for rTMS treatment of depression. Curr Opin Psychiatry 26(1):13–18
George MS, Huggins T, Mcdermut W, Parekh PI, Rubinow D, Post RM (1998) Abnormal Facial Emotion Recognition in Depression: Serial Testing in an Ultra-Rapid-Cycling Patient. Behavior Modification 22(2):192–204
George MS, Ketter TA, Robert M (1994) Prefrontal cortex dysfunction in clinical depression. Depression 2(2):59–72
George MS, Taylor JJ, Short B (2013) Treating the depressions with superficial brain stimulation methods. In: Handbook of clinical neurology, Brain stimulation, 3rd series, vol 116. Elsevier, Amsterdam
Guo H, Cheng C, Xiaohua C et al (2014) Resting-state functional connectivity abnormalities in first-onset unmediated depression. Neural Regen Res 9(2):153–163
Hamani C et al (2006) Deep brain stimulation for chronic neuropathic pain: long-term outcome and the incidence of insertional effect. Pain 125:188–196
Hikosaka O (2010) The habenula: from stress evasion to value-based decision making. Nat Rev Neurosci 11(7):503–513
Horstmann MT, Bialonski S, Noenning N, Lehnertz K (2010) State dependent properties of epileptic brain networks. Clin Neurophysiol 121(2):172–185
Jennifer K, Veinante P, Pawlowski SA, Freund-Mercier M-J, Barrot M (2009) Afferents to the GABAergic tail of the ventral tegmental area in the rat. J Comp Neurol 513:597–621
Johnson KA, Baig M et al (2013) Prefrontal rTMS for treating depression: location and intensity results from the OPT-TMS multi-site clinical trial. Brain Stimul 6(2):108–117
Judd LL (1994) Subsyndromal symptomatic depression. CNS Drugs 1(6):399–404
Keedwell PA, Andrew C, Williams SC, Brammer MJ, Phillips ML (2005) The neural correlates of anhedonia in major depressive disorder. Biol Psychiatry 58:843–853
Koutsouleris N et al (2009) Use of neuroanatomical pattern classification to identify subjects in at-risk mental states of psychosis and predict disease transition. Arch Gen Psychiatry 66:700–712
Krack P et al (2003) Five-year follow-up of bilateral stimulation of the subthalamic nucleus in advanced Parkinson's disease. N Engl J Med 349:1925–1934
Lam RW, Chan P, Wilkins-Ho M, Yatham LN (2008) Repetitive transcranial magnetic stimulation for treatment-resistant depression: a systematic review and metaanalysis. Can J Psychiatr 53:621–631
Lawson RP, Drevets WC, Roisera JP (2013) Defining the habenula in human neuroimaging studies. Neuroimage 64(100):722–727
Lawson RP, Seymour B, Loh E et al (2014) The habenula encodes negative motivational value associated with primary punishment in humans. Proc Natl Acad Sci U S A 111(32):11858–11863
Lecourtier L, Defrancesco A, Moghaddam B (2008) Differential tonic influence of lateral habenula on prefrontal cortex and nucleus accumbens dopamine release. Eur J Neurosci 27(7):1755–1762
Lewinsohn PM, Klein DN, Durbin EC, Seeley JR, Rohde P (2003) Family study of subthreshold depressive symptoms: risk factor for MDD? J Affect Disord 77(2):149–157
Li B et al (2011) Synaptic potentiation onto habenula neurons in the learned helplessness model of depression. Nature 470:535–539
Li K, Zhou T, Liao L, Yang Z, Wong C, Henn F, Malinow R, Yates JR III, Hu H (2013) βCaMKII in lateral habenula mediates core symptoms of depression. Science 341(6149):1016–1020
Lynal M-E, Bassett DS, Kerwin R, McKenna PJ, Kitzbichler M, Muller U (2008) Functional connectivity and brain networks in schizophrenia. J Neurosci 30(28):9477–9487
Matsumoto M, Hikosaka O (2007) Lateral habenula as a source of negative reward signals in dopamine neurons. Nature 447(7148):1111–1115
Mayberg HS et al (1999) Reciprocal limbic-cortical function and negative mood: converging PET findings in depression and normal sadness. Am J Psychiatry 156:675–682
Orrù G, Petterssonyeo W, Marquand AF, Sartori G, Mechelli A (2012) Using support vector machine to identify imaging biomarkers of neurological and psychiatric disease: a critical review. Neurosci Biobehav Rev 36:1140–1152

Pauli W, Nili A, Tyszka J (2018) A high-resolution probabilistic in vivo atlas of human subcortical brain nuclei. Sci Data 5:180063. https://doi.org/10.1038/sdata.2018.63

Proulx CD, Hikosaka O, Malinow R (2014) Reward processing by the lateral habenula in normal and depressive behaviors. Nat Neurosci 17:1146–1152

Sartorius A, Kiening KL, Kirsch P, von Gall CC, Haberkorn U, Unterberg AW, Henn FA, Meyer-Lindenberg A (2010) Remission of major depression under deep brain stimulation of the lateral habenula in a therapy-refractory patient. Biol Psychiatry 67:e9–e11

Schutter DJLG (2008) Antidepressant efficacy of high-frequency transcranial magnetic stimulation over the left dorsolateral prefrontal cortex in double-blind sham-controlled designs: a meta-analysis. Psychol Med 39:65–75

Shankman SA, Lewinsohn PM, Klein DN, Small JW, Seeley JR, Altman SE (2009) Subthreshold conditions as precursors for full syndrome disorders: a 15-year longitudinal study of multiple diagnostic classes. J Child Psychol Psychiatry 50(12):1485–1494

Singh G, Samavedham L (2015) Unsupervised learning based feature extraction for differential diagnosis of neurodegenerative diseases: a case study on early-stage diagnosis of Parkinson disease. J Neurosci 256:30–40

Stamatakis AM, Stuber GD (2012) Activation of lateral habenula inputs to the ventral midbrain promotes behavioral avoidance. Nat Neurosci 15:1105–1107

Stopper CM, Floresco SB (2014) What's better for me? Fundamental role for lateral habenula in promoting subjective decision biases. Nat Neurosci 17(1):33–35

Uddin L, Clare-Kelly AB, Xavier-Castellanos F, Milham M (2009) Functional connectivity of default mode network components: correlation, anticorrelation, and causality. Hum Brain Mapp 30(2):625–637

Wang RY, Aghajanian GK (1977) Physiological evidence for habenula as major link between forebrain and midbrain raphe. Science 197(4298):89–91

Welter ML et al (2011) Basal ganglia dysfunction in OCD: subthalamic neuronal activity correlates with symptoms severity and predicts high-frequency stimulation efficacy. Transl Psychiatry 1:e5

Woo C-W, Chang LJ, Linquistand MA, Wager TD (2017) Building better biomarkers: brain models in translational neuroimaging. Nat Neurosci 20(3):365–377

Xia M, Wang J, Yong H (2013) BrainNet viewer: a network visualization tool for human brain connectomics. PLoS One 8(7):e68910

Yan C, Zang Y (2010) DPARSF: a MATLAB toolbox for "pipeline" data analysis of resting-state fMRI. Front Syst Neurosci 4(13):13

Yang LM, Hu B, Xia YH, Zhang BL, Zhao H (2008) Lateral habenula lesions improve the behavioral response in depressed rats via increasing the serotonin level in dorsal raphe nucleus. Behav Brain Res 188(1):84–90

Yang Y, Cui Y, Sang K, Dong Y, Ni Z, Ma S, Hu H (2018) Ketamine blocks bursting in the lateral habenula to rapidly relieve depression. Nature 554(7692):317–322

Zhu Y, Qi S et al (2019) Connectome-based biomarkers predict sub-clinical depression and identify abnormal brain connections with the lateral habenula and thalamus. Front Psychiatry 10:371

Transcranial Magnetic Stimulation in Autism Spectrum Disorders: Modulating Brainwave Abnormalities and Behaviors

Manuel F. Casanova, Ioan Opris, Estate M. Sokhadze, Emily L. Casanova, and Xiaoli Li

1 Introduction

The medical literature characterizes autism as a neurodevelopmental disorder, that is, a condition where the growth and maturation of the brain is impaired and attendant difficulties are manifested from a young age. In autism spectrum disorder (ASD), problems in communication and socialization occur early and tend to last for the lifetime of the individual. Other common manifestations include restricted interest, repetitive behaviors, and sensory abnormalities. These manufestation vary from patient to patient in regards to their presence, age of onset, severity, developmental trajectories, and attendant comorbidities. This variability has led both researchers and clinicians to consider autism as a spectrum of conditions with multiple possible subtypes and even *forme frustes*. Given this heterogeneity, one treatment may not prove to be effective for all patients. This is even more so as

M. F. Casanova (✉)
Division of Developmental Behavioral Pediatrics, Department of Pediatrics, Greenville Health System, Greenville, SC, USA

University of South Carolina School of Medicine Greenville, Greenville, SC, USA
e-mail: manuel.casanova@louisville.edu

I. Opris
Department of Neurosurgery, University of Miami, Miami, FL, USA

E. M. Sokhadze · E. L. Casanova
University of South Carolina School of Medicine Greenville, Greenville, SC, USA

X. Li
State Key Laboratory of Cognitive Neuroscience and Learning & IDG/McGovern Institute for Brain Research, Beijing Normal University, Beijing, China

Center for Collaboration and Innovation in Brain and Learning Sciences, Beijing Normal University, Beijing, China

I. Opris et al. (eds.), *Modern Approaches to Augmentation of Brain Function*, Contemporary Clinical Neuroscience,
https://doi.org/10.1007/978-3-030-54564-2_14

treatment attempts remain symptomatic rather than curative. Indeed, symptomatic or supportive care leaves the underlying proclivity for symptoms unabated and the patient vulnerable to the side effects of the interventions. In the case of neuroleptics, these side effects include sleepiness, weight gain, diabetes, constipation, sexual problems, and stiff jerky movements of the face and body (i.e., tardive dyskinesia). These side effects offer serious handicaps, especially to young children as they attend school. It is, therefore, of importance that future treatment interventions target commonalities in pathological mechanisms as a way of addressing the basic cause(s) of the disorder.

Transcranial magnetic stimulation (TMS) is one of the first treatment interventions for ASD, which is aimed at a core pathological feature of the condition. TMS is meant to stabilize gamma oscillation abnormalities (vide infra), which have been related to executive dysfunction and faulty binding/coherence of information processing. The gamma oscillations are modulated by fast-spiking basket interneurons (inhibitory neurons). Not coincidentally, neuropathological studies in autism claim a significant reduction in basket cells in total numbers across laminae and in all cortical areas examined (Hashemi et al. 2017). It is, therefore, of interest that animal models of autism, whose neuropathological examination indicate a diminished number of basket cells, also provide an autism-like phenotype (Saunders et al. 2013; Wöhr et al. 2015). Pharmacological interventions that rescue parvalbumin (i.e., calcium-binding albumin protein present in both chandelier and basket cells) neuronal deficits relieve the reduction in phase-locked gamma oscillations and ameliorate social behavioral deficits (Nakamura et al. 2015). According to some researchers, the downregulation of parvalbumin represents one point of convergence that provides a "common link between apparently unrelated ASD-associated synapse structure/function phenotypes" (Wöhr et al. 2015, p. 1).

Neuropathological studies in idiopathic and secondary autism have provided findings suggestive of a neuronal migration disorder. The presence of heterotopias, neuronal accumulation in the gray/white matter junction, and a dysplastic cerebral cortex all suggest that periventricular and rhombic lip germinal cell divisions occur in a heterochronous fashion (Bailey et al. 1998; Casanova 2007; Wegiel et al. 2010; Casanova et al. 2013a; Hutsler and Casanova 2016). Daughter cells that radially migrate to the cortex are thus impeded from reaching their final destination and those that arrive cannot make appropriate connections. The end result being that the tangentially migrating cells (traveling perpendicularly to the radial glial cells) stemming from the ganglionic eminences (lateral, medial, and caudal eminences) cannot synchronize their activities with the radially migrating cells originating from the periventricular germinal matrix. Ultimately, the radially migrating neuroblasts will mature into pyramidal (excitatory) cells, while the tangentially migrating cohort will provide interneurons or inhibitory cells. These otherwise disparate cell populations get together in the cerebral cortex and coordinate their function by forming a synergetic functional alliance of cells (pyramidal cell-interneuron dyads, Marin-Padilla 2011). In autism, the lack of homeostasis between these cell populations (e.g., an imbalance in their cell numbers) provides an excitatory/inhibitory cortical bias, which helps to explain the sensory deficits and seizures that are often observed in ASD (Casanova et al. 2013a).

The neuropathological changes that were previously described occur during brain development. Findings like heterotopias and cortical dysplasias vary greatly in topography and severity between patients (Casanova et al. 2013a). Researchers have ascribed this variability to the putative effects of environmental exigencies and a triple hit model of autism causation (Casanova 2007). In autism, a genetic susceptibility interacts with an environmental exigency at certain times during brain development to provide a risk factor for the condition. Variability in any of these factors procreates in turn heterogeneity in the clinical phenotype. In this regard, autism is a multifactorial or complex disorder similar to obesity, diabetes, or heart disease. There are multiple contributing factors and although some familial clusters clearly exist, there does not appear to be a clear-cut pattern of inheritance.

The neuronal migration disorder affects the modular arrangement of the cortex, especially its minicolumns. Lorente de Nó discussed the functional role of these "vertical cylinders" when he said, "All the elements of the cortex are represented in it, and therefore, it may be called an elementary unit, in which theoretically, the whole process of transmission of impulses from the afferent fiber to the efferent axon may be accomplished" (Lorente de Nó 1938). However, minicolumns were first characterized in the pioneering electrophysiological experiments of Vernon Mountcastle. After impaling over 10,000 cortical cells Mountcastle noted that neurons vertically disposed across the width of the cortex would respond in unison after stimulation (Mountcastle 1978, 1997, 1998). The results led Mountcastle to claim that mini- or microcolumns, a translaminar ecosystem of neurons and their connections, were the smallest unit of function within the cortex that reproduced the holistic properties of the brain. More recently, work by Ioan Opris using conformal multielectrode arrays to define the multilayer coding of minicolumns have attributed the genesis of some cognitive functions to this modular structure (Opris et al. 2012; for review see Opris and Casanova 2014).

In autism, both the internal arrangement and connectivity of minicolumns appear to be abnormal. Minicolumns are smaller in size and exhibit a reduction in their peripheral neuropil space while their central compartment (containing pyramidal cells) remains unchanged (Casanova et al. 2010). The peripheral neuropil space is the anatomical compartment that contains the cell somas of interneurons as well as many of their projections. Mountcastle used to describe the function of this compartment as a strong vertical flow of inhibition while other researchers have given it the descriptive appellation of a shower curtain of inhibition (Szentàgothai 1975; Mountcastle 1998). The idea for the name came from the suggestion that the shower curtain of inhibition served to keep processing of stimuli within the confines of the central compartment. The inhibitory surround in this regard was a mechanism that helped to enhance contrast when processing information. In autism, a faulty shower of inhibition allows the information to suffuse among adjacent minicolumns, diminishing contrast, and procreating a cascade of excitation.

Smaller minicolumns provide a bias in connectivity favoring short connections over longer ones. Neuroimaging studies have shown that in autism, the outer radiate matter of the brain (compartment for the short-arcuate corticocortical fibers) is increased in size, while the relative size of the corpus callous (compartment for long-projection fibers linking homologous areas of both hemispheres) is reduced in size. In this regard, the changes in neuronal migration offer a core pathological feature of ASD, while changes in connectivity (including plasticity) are secondary/ compensatory or downstream effects.

Studies in schizophrenia have shown that transcranial magnetic stimulation over the dorsolateral prefrontal cortex (DLPC) normalizes abnormalities in gamma oscillations and provides improvements in some symptoms. Outcome measures show the beneficial effects of this treatment modality in regards to both coherence and binding. This line of thinking suggested using TMS in autism and observing gamma oscillations as well as changes in executive functions as outcome measures.

2 Transcranial Magnetic Stimulation

Transcranial magnetic stimulation works on the principle of magnetic induction of electricity (Faraday's Law). According to this principle, a change in the strength of a magnetic field acting on a conductor induces a voltage across its ends. The induced electromotive force (EMF) or voltage is dependent on the number of turns and cross-section of the coil. The reverse effect is also true. A current through a conductor causes a magnetic field around the same. The strength of the engendered magnetic field depends in turn on the strength of the electric current and the type of core used (e.g. air vs. ferrite). A well-known high school experiment meant to demonstrate Faraday's law by passing current through a coil and showing how the induced magnetic field is able to deflect the needle of a nearby compass. Since the needle of the compass aligns itself along the magnetic flux lines, the experiment serves to detect the presence of a magnetic field near the conductor.

The turns of the insulated wire around a core provide a passive electronic element called an inductor (also called a coil or choke). The turns of the coil serve to concentrate the magnetic flux lines (making them closer together), thus increasing its strength, while simultaneously adding polarity to the magnetic field. The shape of the coil focuses the flux lines in a particular area or target. The strength of the generated magnetic field varies with distance (an inverse square law). This fact is of importance in TMS in two different ways. First, the wand with coil can be held at a glancing angle (or even at a right angle) as sham treatment. Masking the trial in this way allows for a control population that isolates many of the specific effects of the treatment and thus, neutralizes some biases of the placebo effect. Second, for physiological response, the depth of penetration with current TMS equipment is restricted to some 30 mm. The skull is largely invisible to electromagnetic radiation. It provides protection against trauma but lacks the properties of a Faraday's cage or shield. Since the brain lies about 20 mm from the surface of the scalp, TMS targets only the crest of the gyri. Some commercial companies claim that deeper penetration (4 cm) is possible with different coil configurations (e.g., H coil). However, the magnetic field lines splay open with distance making the exclusive targeting of small locations difficult. Overall, the deeper the stimulation, the larger and more nonspecific, the area of stimulation.

In TMS, a kiloampere (Kamp) current is discharged from a bank of capacitors through a coil, thus generating a magnetic field of 1–3 T strength (note: for comparison the strength of the Earth's magnetic field is 25–65 μT). The resistance of the conducting wire is kept to a minimum, a fact that translates into a small-time constant or transient response by the capacitor. The fast response time is, in part, because the potential energy of a capacitor is stored in an electric field. A similar response would have been impossible with a battery given that the rate of discharge is dependent on the chemical reaction that creates the voltage difference between its electrodes. A specialized electronic switch, capable of handling high currents, is used to regulate the firing of the capacitors, thus allowing its users to select different frequencies for their trials. Despite the low resistance of the circuit path linking the capacitors to the coil, the high current creates a voltage drop along resistive elements, which is converted into heat. The power dissipation increases with higher frequencies. It is for this reason that specialized cooling equipment have been employed when using coils at higher frequencies.

The strength of magnetic induction upon a conductor depends on the number of flux lines passing through the same. For most objects, a 90° angle affords the highest number of intersects. In TMS, current through a coil provides a rapidly expanding magnetic field while closely adjacent membrane bags of electrolytes (cells and their projections) provide conductors. We believe that the disposition of interneurons and their projections within the peripheral neuropil space of minicolumns affords an ideal geometrical orientation (90° to the magnetic field) for stimulation (vide supra, the strong vertical flow of inhibition designated by Mountcastle). At low frequencies (<1 Hz), TMS acts preferentially on inhibitory elements with the more favorable geometrical orientation by inducing currents along the axons rather than across the same (Fox et al. 2004). When higher frequencies are used (>5 Hz), all cells are stimulated regardless of their geometrical orientation. Since most cortical cells are pyramidal, the end result is that TMS becomes excitatory (see Wassermann et al. 1998).

Hoffman and Cavus, in their review of slow (low frequency) TMS studies, proposed long-term depression and long-term depotentiation as models for understanding the mechanism of slow TMS (Hoffman and Cavus 2002). Neocortical long-term depression and changes in the cortical excitability induced by slow TMS appear to accumulate in an additive fashion as the number of stimulations is increased over many days. Studies of both slow TMS and long-term depression suggest additive efficacy when higher numbers of stimulations are administered. The reversal, or depotentiation, of previously enhanced synaptic transmission due to long-term potentiation may be the most relevant model for slow TMS when used as a therapeutic tool.

In TMS, the influence of the magnetic field is restricted to an area approximately 3 cm^2. Targeting a particular area of the brain is, therefore, of importance if we are to maximize the effects of treatment. The dorsolateral prefrontal cortex (DLPC) has been implicated in the genesis of executive functions, such as cross-modal and set-shifting attention, that are impaired in autism. DLPFC is also involved in the inhibition of inappropriate responses. It is for this reason that dysfunction of DLPC is claimed to be responsible for the inability to inhibit context-inappropriate/inflexible behaviors that impair adaptive responses. Given the fact that DLPC provides global multisystem connectivity within the brain, correcting the function of this region can be postulated to help normalize the function of its multiple interconnected sites.

TMS is generally considered safe, that is, as having no lasting side effects and is well accepted by patients (Fig. 1). Reported side effects include a mild, transient tension-type headache on the day of stimulation and mild discomfort due to the noise made by the machine as the magnetic pulses are generated. There is a certain risk of inducing a seizure (Lerner et al. 2019) and subjects with epilepsy, a family history of epilepsy, or brain trauma are generally excluded from TMS studies. As a safety precaution, some TMS studies adjust the stimulation intensity below the participant's motor threshold. TMS is generally considered safe for use in pediatric populations, as no significant adverse effects have been reported (Garvey and Gilbert 2004; Quintana 2005).

In general, parameters for our studies have used low-frequency (inhibitory) TMS over DLPC. As previously explained, we have excluded patients with seizures or brain trauma from our subject population. Higher functioning individuals were necessary in order to pay attention and comply with the testing paradigms. For the same reasons, age range was usually 8–18 years. Outcome measures included gamma oscillations, event-related potentials (often using an oddball paradigm), behavioral screening, and autonomic measures. We thought that the best control would be a series of autistic individuals not subjected to the active treatment. In this regard,

Fig. 1 Patient with ASD during the rTMS procedure

most of our studies have used a waitlist control of autistic subjects. After finishing the study and breaking the blind for them, the waitlist controls were offered the active treatment.

3 Outcomes of TMS Studies in Autism

Modern communication systems (e.g., radio and TV) transmit information through voltage frequencies. The brain is similar to these modern communication systems as voltage fluctuations or oscillations recorded from the scalp surface convey intelligence. Brainwaves can be divided into bandwidths each one correlating with different behavioral states. Slower ones are associated to sleep while higher frequencies have been involved in higher cognitive functions. In the brain, the highest frequency and largest bandwidth are devoted to the so-called gamma oscillations. These frequencies provide the phenomenon of "binding", that is, the way we gather different aspects of perception into a whole or a single experience. In autism, binding is impaired, thus making it difficult to learn and generalize experiences.

Gamma oscillations can be analyzed as event-related responses that happen during specific time windows after stimulus onset. An early, phase-locked component usually defined as "evoked" is thought to represent the binding of information within a confined cortical field (Casanova et al. 2013b, 2015; Rippon 2017). It has been proposed that some of the symptoms of autism could result from a reduction in the integration of specialized local networks in the brain caused by a deficit in temporal binding. For gamma oscillations, a later component, labeled as "induced" is believed to represent the binding of feedforward and feedback processing across networks of different cortical regions. Simply stated, induced gamma activity correlates to the binding of widely distributed cell assemblies, which is thought to underlie cohesive stimulus representation in the human brain. According to this assumption changes in gamma, EEG have been considered as indicators for the processing of Gestalt-like patterns. The induced component is not phase-locked and the attendant variability in latency contributes to averaging out potentially significant findings.

Bursts of gamma oscillations can be seen over the occipital lobes during visual processing. When involved in more complex tasks, gamma busts may be seen in other cortical locations that are involved in that undertaking (Brown et al. 2005). Figures of illusory contours (Kanizsa figures) whose edges form or evoke the precept of a shape, produce gamma oscillations during visual cognitive tasks. In autism, EEG recordings during a Kanizsa figure have shown an overall increase in the gamma activity as compared to controls (Brown et al. 2005). The findings have been interpreted as being the result of reduced inhibitory processing.

Ogawa et al. examined the changes in high frequency oscillations (HFOs) of somato-sensory evoked potentials (SEPs) before and after slow TMS over the right primary somatosensory cortex (0.5 Hz, 50 pulses, 80% motor threshold intensity) (Ogawa et al. 2004). HFOs, which represent a localized activity of intracortical inhibitory interneurons, were significantly increased, without notable alterations of SEPs, after slow TMS. The results suggest that slow TMS affects cortical excitability by modulating the activity of the intracortical inhibitory interneurons beyond the time of the stimulation and that TMS may have therapeutic effects on such disorders. This is in agreement with our hypothesis that slow TMS increases the activity of inhibitory cells within minicolumns, which then enhances the spatial contrast that is needed for functional discrimination.

Our laboratory performed the first TMS study in autism ($N = 8$ ASD children, $n = 13$ age-matched controls) using the power of the EEG gamma band during a Kanizsa task as an outcome measure (Sokhadze et al. 2009a, b; for reviews of TMS studies in autism see Casanova et al. 2015; Oberman et al. 2016; Barahona-Corrêa et al. 2018; Casanova et al. 2019; Cole et al. 2019; Finisguerra et al. 2019; Masuda et al. 2019; Ni and Huang 2019; Gómez et al. 2019). Low-frequency stimulation (0.5 Hz) was delivered twice per week for 3 weeks. At the baseline, the power of gamma activity was higher and had a shorter latency in our ASD group. After treatment, the patient group showed a wider difference in power between gamma responses to target as compared to nontarget stimuli. The findings were reproduced in later studies using different populations and number of TMS sessions (Baruth et al. 2010; Casanova et al. 2012; Sokhadze et al. 2014a).

Studies by our group have examined the effects of bilateral DLPC TMS on both gamma phase coherence and event-related potentials (ERP). One study consisted of 18 sessions with 54 participants using two groups of children with ASD (TMS and wait list as controls, 27 individuals in each group). The results indicated a significant post-treatment increase in latency and reduction in the amplitude of frontal and fronto-central N100, N200, and P3a ERP components to nontargets in the treatment group as compared to the control group (Sokhadze et al. 2014a). In another study, 18 sessions of bilateral DLPC TMS was used to examine EEG gamma phase coherence between the frontal and parietal sites (Hensley et al. 2014) in 32 participants (TMS and wait list controls, 16 subjects each). TMS had its most significant effect on induced gamma in the frontal region of our active treatment group as indicated by increased gamma phase coherence in response to target stimuli.

Similar to our previously reported gamma findings (vide supra), ERP studies during a visual novelty processing tasks have indicated that ASD individuals lack stimulus discrimination between target and nontarget stimuli as compared to controls. This is manifested as significantly prolonged and augmented ERP components to irrelevant distracters over frontal and parietal recording sites (Sokhadze et al. 2009a, b). The reported changes are especially salient for early ERP components peaking within the first 100 ms (e.g., P100 and N100). These early components are labeled as "sensory" or "exogenous" as they depend on the physical parameters of the stimulus. These findings are similar to those reported by Grice et al. (2001) where autistic individuals did not show significant differences in the frontal gamma activity when comparing the processing of upright and inverted faces (the latter acting as "physical parameters" of the stimuli) as opposed to clear increase in the control subjects. Our findings are also in agreement with studies showing how the brains of autistic individuals are inappropriately activated. This may be due to a disruption in the ratio between cortical excitation and inhibition. In autism, increased cortical activity is made evident by gamma and ERP responses, which indicate that activity induced by perceptual processes starts earlier and continues longer because the neural networks subserving cognitive processes involved in combining information processing are not functioning normally.

In several of our studies, we have compared clinical, behavioral, and electrocortical measures of outcomes in groups of children with autism and controls (active TMS and wait-list group $N = 27$ per group) using 18 weekly sessions of TMS applied bilaterally over DLPC. Post-TMS evaluation showed decreased irritability and hyperactivity on the aberrant behavior checklist (ABC) and decreased stereotypic behaviors on the repetitive behavior scale (RBS-R). Following the TMS course, we also found a decreased amplitude and prolonged latency in the frontal and fronto-central N100, N200, and P3A ERP components to non-targets in the TMS group as compared to the wait-list group (Fig. 2). TMS also resulted in an increase of the frontal P2a difference wave. These ERP changes, along with increased centro-parietal and parieto-occipital P100 and P300 (P3b) to targets, are indicative of increased efficiency of information processing post-TMS treatment.

We have also investigated executive functions in 14 individuals with ASD and 14 age- and IQ-matched controls by evaluating error monitoring and correction (Fig. 3). The ASD group showed significant evidence of compromised error detection, evaluation, and correction, which may underlie a general impairment in self-monitoring related to behavioral and/or social disturbances in ASD. We then evaluated the effects of 12 sessions of bilateral slow TMS on error monitoring and correction. The active TMS group showed significant improvement in error detection and correction

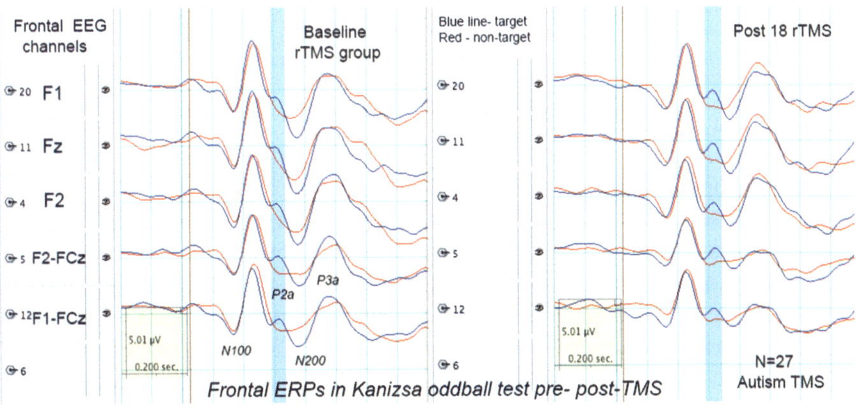

Fig. 2 Frontal event-related potentials (ERP) in the visual oddball test at the baseline and after 18 sessions of rTMS in children with autism ($N = 27$). Changes of the frontal N100 ERP after the rTMS course were significant for nontarget Kanizsa stimuli ($F = 5.47$, $p = 0.023$) as N100 became less negative post-TMS. In a similar manner, the amplitude of the orienting P3a component decreased to nontarget Kanizsa stimuli ($F = 5.40$, $p = 0.024$). On the other hand, the amplitude of the P2a component to target stimuli increased post-TMS reflecting more efficient target identification

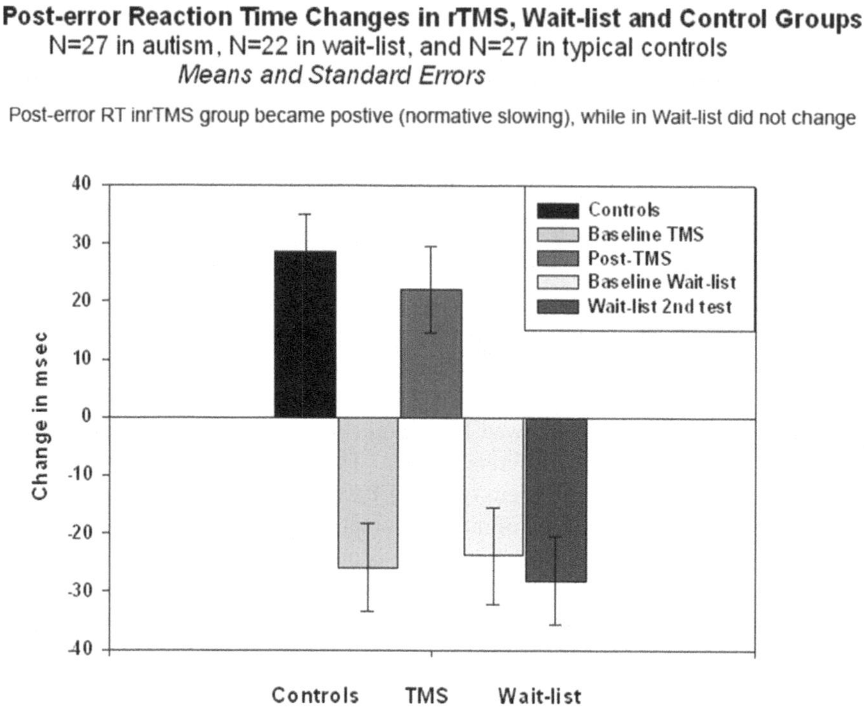

Fig. 3 Post-error reaction time (RT) changes in children in TMS and wait-list ASD groups and typical children in visual oddball test with illusory Kanizsa figures. Both autism groups did not show normative post-error RT slowing. The TMS group post-treatment (18 sessions of rTMS) showed post-error slowing, post-error RT change (post TMS post-error RT change minus baseline post-error RT change) was positive (from −25.8 ± 7.4 ms to 22.1 ± 7.5 ms, $t = 2.24$, df = 26, $p = 0.033$). Wait-list did not show any significant changes in post-error RTs or error rate

compared to a randomized, non-active TMS group; the results point to improved executive functioning and behavioral performance in ASD as a result of TMS.

Another important finding from our TMS studies was a decrease of the latency and an increase of the negative amplitude for the error-related negativity (ERN) component during commission errors that further confirmed the improvement in error monitoring and corrective functions in ASD individuals. Facilitated information processing was also manifested in such behavioral response measures as motor response accuracy. Results from our study indicate that TMS helps in the normalization of ERP responses, reaction time and accuracy during tests of executive function, as well as in improvements in behavioral evaluations.

The combined use of rTMS and EEG neurofeedback (i.e. the use of brain activity parameters as feedback to regulate a brainwave frequency) has been used to operantly condition post-TMS EEG changes (Sokhadze et al. 2014b). The underlying hypothesis being that combining TMS and neurofeedback therapy would prove synergistic when measuring outcomes of executive functions and behaviors in ASD as compared to the wait-list control group (Sokhadze et al. 2014b). The results of the combined treatment supported the initial hypothesis by demonstrating significant improvements in behavioral and ERP measures of executive functions, as well as significant changes in the EEG outcomes of neurofeedback training, such as frontal theta-to-beta ratio and an increase in the relative power of gamma activity (Sokhadze et al. 2014b).

Heart rate variability and electrodermal activity have been used as noninvasive measures of the autonomic nervous system activity during TMS therapy in autism (Casanova et al. 2014; Wang et al. 2016; Sokhadze et al. 2017). The autonomic nervous system (ANS) is directly involved in the manifestations of affect, emotional expression, facial gestures, vocal communication, and social engagement behaviors. The dysfunction of this system is often hypothesized as contributing to abnormalities of cognitive, emotional, and behavioral responses in children with autism. Cardiac under-reactivity during socially engaging situations results in lower behavioral flexibility and reduced attentional capacity to relevant stimuli; a skill critical for social communication development. Our studies have found that ASD individuals have an accelerated heart rate in association with lower heart rate variability (HRV) indexed by low frequency (LF) to high frequency (HF) ratio (LF/HF of HRV, so-called cardiac autonomic balance index) and low standard deviation of HR (SDHR) along with high electrodermal activity (skin conductance level or SCL). These are indicators of excessive sympathetic and reduced parasympathetic activation at the pre-treatment stage, which are normalized by TMS treatment (Sokhadze et al. 2014c).

In summary, TMS, especially at low frequencies, has proven to be a safe intervention for children with ASD. This is the first treatment targeting a core pathological abnormality of autism. Outcome measures reveal improvements in executive functions as shown by the normalization of error monitoring (i.e., detection, evaluation, and correction of errors) and attendant ERP components. Patient assessments reveal improvements in maladaptive behaviors and correlated changes in the autonomic nervous system. Increased heart rate variability after TMS may prove to be preventive in reducing sudden or unexpected death in ASD. Still, the total number of patients examined so far are small and there is a need for large population studies having adequate (sham) controls. In addition, the longevity of benefits needs to be studied along with the possibility of booster TMS sessions in order to maintain therapeutic benefits.

Acknowledgements This article is based on several studies partially supported by a grant from the National Institutes of Health (MH86784) awarded to Manuel F. Casanova.

References

Bailey A, Luthert P, Dean A et al (1998) A clinicopathological study of autism. Brain 121:889–905

Barahona-Corrêa JB, Velosa A, Chainho A, Lopes R, Oliveira-Maia AJ (2018) Repetitive transcranial magnetic stimulation for treatment of autism spectrum disorder: a systematic review and meta-analaysis. Front Integr Neurosci 12:27

Baruth J, Casanova MF, El-Baz A (2010) Low-frequency repetitive transcranialmagnetic stimulation modulates evoked-gamma frequency oscillations in autismspectrum disorders. J Neurother 14:179–9

Brown CC, Gruber T, Boucher J, Rippon G, Brock J (2005) Gamma abnormalities during perception of illusory figures in autism. Cortex 41:364–376

Casanova MF (2007) The neuropathology of autism. Brain Pathol 17(4):422–433

Casanova MF, El-Baz A, Vanbogaert E, Narahari P, Switala A (2010) A topographical study of minicolumnar core width by lamina comparison between autistic subjects and controls: possible minicolumnar disruption due to an anatomical element in-common to multiple laminae. Brain Pathol 20(2):451–458

Casanova MF, Baruth JM, El-Baz A, Tasman A, Sears L, Sokhadze E (2012) Repetitive transcranial magnetic stimulation (rTMS) modulates event-related potential (ERP) indices of attention in autism. Transl Neurosci 3:170–180

Casanova MF, El-Baz A, Kamat SS et al (2013a) Focal cortical dysplasias in autism spectrum disorders. Acta Neuropathol Commun 1:67

Casanova MF, Baruth J, El-Baz AS, Sokhadze GE, Hensley M, Sokhadze ES (2013b) Evoked and induced gamma frequency oscillations in autism. In: Casanova MF, El-Baz AS, Suri JS (eds) Imaging the brain in autism. Springer, New York, pp 87–106

Casanova MF, Hensley MK, Sokhadze EM, El-Baz AS, Wang Y, Li X, Sears L (2014) Effects of weekly low-frequency rTMS on autonomic measures in children with autism spectrum disorder. Front Hum Neurosci 8:851

Casanova MF, Sokhadze E, Opris I, Wang Y, Li X (2015) Autism spectrum disorders: linking neuropathological findings to treatment with transcranial magnetic stimulation. Acta Paediatr 104(4):346–355

Casanova MF, Sokhadze EM, Opris I, Li X (2019) Autism, transcranial magnetic stimulation and gamma frequencies. In: Sokhadze EM, Casanova MF (eds) Autism spectrum disorder: neuromodulation, neurofeedback, and sensory integration approaches to research and treatment. FNNR & BMED Press, Murfreesboro, TN, pp 49–65

Cole EJ, Enticott PG, Oberman LM (2019) rTMS in ASD Consensus Group. Thepotential of repetitive transcranial magnetic stimulation for autism spectrum disorder: a consensus statement. Biol Psychiatry 85(4):e21–2

Finisguerra A, Borgatti R, Urgesi C (2019) Non-invasive brain stimulation for the rehabilitation of children and adolescents with neurodevelopmental disorders: a systematic review. Front Psychol 10:135

Fox PT, Narayana S, Tandon N, Sandoval H, Fox SP, Kochunov P, Lancaster JL (2004) Column-based model of electric field excitation of cerebral cortex. Hum Brain Mapp 22:1–16

Garvey MA, Gilbert DL (2004) Transcranial magnetic stimulation in children. Eur J Paediatr Neurol 8:7–19

Gómez L, Vidal B, Morales L, Berrillo S, Baez M, Maragoto C, Vera H (2019) Non-invasive brain stimulation in children with autism spectrum disorder. In: Sokhadze EM, Casanova MF (eds) Autism spectrum disorder: neuromodulation, neurofeedback, and sensory integration approaches to research and treatment. FNNR & BMED Press, Murfreesboro, TN, pp 89–114

Grice SJ, Spratling MW, Karmiloff-Smith A, Halit H, Csibra G, de Haan M, Johnson MH (2001) Disordered visual processing and oscillatory brain activity in autism and Williams syndrome. Neuroreport 12:2697–2700

Hashemi E, Ariza J, Rogers H et al (2017) The number of parvalbumin-expressing interneurons is decreased in the medial prefrontal cortex in autism. Cereb Cortex 27(3):1931–1943

Hensley MK, El-Baz AS, Sokhadze E, Sears L, Casanova MF (2014) Effects of 18 session TMS therapy on gamma coherence in autism. Psychophysiology 51:S16. (Abstract)

Hoffman RE, Cavus I (2002) Slow transcranial magnetic stimulation, long-term potentiation, and brain hyperexcitability disorders. Am J Psychiatry 159:1093–1102

Hutsler JJ, Casanova MF (2016) Cortical construction in autism spectrum disorder: columns, connectivity and the subplate. Neuropathol Appl Neurobiol 42(2):115–134

Lerner AJ, Wassermann EM, Tamir DI (2019) Seizures from transcranial magnetic stimulation 2012–2016: Results of a survey of active laboratories and clinics. Clin Neurophysiol 130(8):1409–1416

Lorente de Nó R (1938) Architectonics and structure of the cerebral cortex. In: Fulton JF (ed) Physiology of the nervous system. Oxford University Press, New York, pp 291–330

Marin-Padilla M (2011) The human brain: prenatal development and structure. Springer, Berlin

Masuda F, Nakajima S, Miyazaki T et al (2019) Clinical effectiveness of repetitive transcranial magnetic stimulation treatment in children and adolescents with neurodevelopmental disorders: a systematic review. Autism 20:1362361318822502. https://doi.org/10.1177/1362361318822502

Mountcastle VB (1978) An organizing principle for cerebral function: the unit module and the distributed system. In: Edelman GM, Mountcastle VB (eds) The mindful brain: cortical organization and the group-selective theory of higher brain function. MIT Press, Cambridge, MA, pp 7–51

Mountcastle VB (1997) The columnar organization of the neocortex [review]. Brain 120:701–722

Mountcastle VB (1998) Perceptual neuroscience: the cerebral cortex. Harvard University Press, Cambridge, MA

Nakamura T, Matsumoto J, Takamura Y et al (2015) Relationships among parvalbumin-immunoreactive neuron density, phase-locked gamma oscillations, and autistic/schizophrenic symptoms in PDGFR-B knock-out and control mice. PLoS One 10(3). https://doi.org/10.1371/journal.pone.0119258

Ni H-C, Huang Y-Z (2019) Theta burst stimulation in autism. In: Sokhadze EM, Casanova MF (eds) Autism spectrum disorder: neuromodulation, neurofeedback, and sensory integration approaches to research and treatment. FNNR & BMED Press, Murfreesboro, TN, pp 67–87

Oberman LM, Enticott PG, Casanova MF, Rotenberg A, Pascual-Leone A, McCracken JT, TMS in ASD Consensus Group (2016) Transcranial magnetic stimulation in autism spectrum disorder: challenges, promise, and roadmap for future research. Autism Res 9(2):184–203

Ogawa A, Ukai S, Shinosaki K, Yamamoto M, Kawaguchi S, Ishii R, Takeda M (2004) Slow repetitive transcranial magnetic stimulation increases somatosensory high-frequency oscillations in humans. Neurosci Lett 358:193–196

Opris I, Casanova MF (2014) Prefrontal cortical minicolumns: from executive control to disrupted cognitive processing. Brain 137(Pt 7):1863–1875

Opris I, Hampson RE, Gerhardt GA, Berger TW, Deadwyler SA (2012) Columnar processing in primate PFC: evidence for executive control microcircuits. J Cogn Neurosci 24:2334–2337

Quintana H (2005) Transcranial magnetic stimulation in persons younger than the age of 18. J ECT 21:88–95

Rippon G (2017) Gamma abnormalities in autism spectrum disorders, ch. 22. In: Casanova MF, El-Baz A, Suri JS (eds) Autism imaging and devices. CRC Press, Taylor and Francis Group, Boca Raton, FL, pp 457–496

Saunders JA, Tatard-Leitman VM, Suh J et al (2013) Knockout of NMDA receptors in parvalbumin interneurons recreates autism-like phenotypes. Autism Res 6(2):69–77

Sokhadze E, El-Baz A, Baruth J, Mathai G, Sears L, Casanova MF (2009a) Effects of low frequency repetitive transcranial magnetic stimulation (rTMS) on gamma frequency oscillations and event-related potentials during processing of illusory figures in autism. J Autism Dev Disord 39:619–634

Sokhadze EM, Baruth JM, Tasman A, Sears L et al (2009b) Event-related potential study of novelty processing abnormalities in autism. Appl Psychophysiol Biofeedback 34:37–51

Sokhadze EM, El-Baz AS, Sears LL, Opris I, Casanova MF (2014a) rTMS neuromodulation improves electrocortical functional measures of information processing and behavioral responses in autism. Front Syst Neurosci 8:134

Sokhadze EM, El-Baz AS, Tasman A, Sears LL et al (2014b) Neuromodulation integrating rTMS and neurofeedback for the treatment of autism spectrum disorder: an exploratory study. Appl Psychophysiol Biofeedback 39(304):237–257

Sokhadze EM, El-Baz AS, Tasman A, Sears LL, Wang Y, Lamina EV (2014c) Neuromodulation integrating rTMS and neurofeedback for the treatment of autism spectrum disorder: an exploratory study. Appl Psychophysiol Biofeedback 39:237–257

Sokhadze G, Casanova MF, Kelly D, Casanova E, Russell B, Sokhadze EM (2017) Neuromodulation based on rTMS affects behavioral measures and autonomic nervous system activity in children with autism. NeuroRegulation 4(2):65

Szentàgothai J (1975) The "module concept" in cerebral cortex architecture. Brain Res 95(2–3):475–496

Wang Y, Hensley MK, Tasman A, Sears L, Casanova MF, Sokhadze EM (2016) Heart rate variability and skin conductance during repetitive TMS course in children with autism. Appl Psychophysiol Biofeedback 41(1):47–60

Wassermann EM, Wedegaertner FR, Ziemann U et al (1998) Crossed reduction of motor cortex excitability by 1 Hz transcranial magnetic stimulation. Neurosci Lett 250:141–144

Wegiel J, Kuchna I, Nowicki K et al (2010) The neuropathology of autism: defects of neurogenesis and neuronal migration, and dysplastic changes. Acta Neuropathol 11:755–770

Wöhr M, Orduz D, Gregory P et al (2015) Lack of parvalbumin in mice leads to behavioral deficits relevant to all human autism core symptoms and related neural morphofunctional abnormalities. Transl Psychiatry 5(3):e525

Neurofeedback Training with Concurrent Psychophysiological Monitoring in Children with Autism Spectrum Disorder with Comorbid Attention Deficit/Hyperactivity Disorder

Estate M. Sokhadze, Desmond P. Kelly, Eva Lamina, and Manuel F. Casanova

1 Introduction

Autism spectrum disorder (ASD) is characterized by severe disturbances in reciprocal social relations, varying degrees of language and communication difficulty, and restricted, repetitive, and stereotyped behavioral patterns (APA 2013). Standard treatment of ASD relies mostly on behavioral interventions (e.g., Applied Behavioral Analysis [ABA]) and pharmacotherapy for comorbid conditions such as attention deficit and hyperactivity, anxiety, and depression. There is a need to develop and adopt neurotherapeutic approaches that will use applied psychophysiological training methods based on neuromodulation. One such technique is neurofeedback (NFB). This technique, a version of electroencephalogram (EEG) biofeedback, aims at acquiring self-regulation skills over certain brain activity patterns in an operant conditioning paradigm (Sherlin et al. 2011). By operant conditioning of EEG, NFB provides an effective way to train electrophysiological activity of the targeted cortical topography. Neurofeedback training is considered as one of the most effective and salient treatments for children with attention deficit/hyperactivity disorder (ADHD).

Clinical neurophysiological and psychophysiological studies showed that some EEG characteristics of ASD are similar to the reported brain EEG abnormalities in ADHD. It should be noted that NFB is considered in several reviews as an efficacious and even specific treatment for some ADHD subtypes (e.g., inattentive type) (Arns et al. 2009, 2014; Hurt et al. 2014; Sherlin et al. 2011). Neurofeedback training outcomes in ADHD symptom treatment were reported to be very positive and are well-publicized (Arns et al. 2014; Sherlin et al. 2010). Currently, there are controlled randomized clinical trials underway (e.g., design reviewed in Kerson and Collaborative Neurofeedback Group 2013) to further examine the utility of the

E. M. Sokhadze (✉) · D. P. Kelly · E. Lamina · M. F. Casanova
Prisma Health, University of South Carolina School of Medicine, Greenville, SC, USA

I. Opris et al. (eds.), *Modern Approaches to Augmentation of Brain Function*, Contemporary Clinical Neuroscience,
https://doi.org/10.1007/978-3-030-54564-2_15

method and clinical efficacy of neurofeedback training for ADHD treatment. This fact encouraged some research groups to apply protocols typically used in ADHD population in children with ASD, e.g., downregulation of theta-to-low beta ratio at Cz or FCz, and upregulation of sensorimotor rhythm (SMR) in C3 or C4 topography with simultaneous downregulation of frontocentral theta rhythm.

The rationale for adopting the typical ADHD NFB protocols as an intervention of choice for ASD neurotherapy is based on the assumption that neurofeedback protocols successfully applied for treatment of ADHD may also be efficacious to the treatment of children with autism. The evidence that some of the symptoms of ASD can be improved using this approach has been reported in the literature (Jarusiewicz 2002; Coben and Padolsky 2007; Coben 2008, 2013; Kouijzer et al. 2009a, b, 2010). A study conducted by Jarusiewicz (2002) investigating the utility of neurofeedback in autistic children supports the proposition that the theta-to-beta neurofeedback training protocol, which is generally applied to ADHD, can also be of use in autism (Kouijzer et al. 2009b). However, according to Coben, protocols for ASD need to be selected and developed individually since autism has a wide range of symptoms and variable EEG manifestations (Coben and Padolsky 2007; Coben and Myers 2010; Coben 2013). Studies by Coben and his associates reported advantages of using qEEG-guided individualized protocols without limitation of treatment for enhancement/suppression of specific rhythms or using the interventions that target only a preselected topography or are restricted to a specific EEG band (Coben and Padolsky 2007; Coben et al. 2010; Coben and Myers 2010; Coben 2013).

Whether the current protocols that have proven efficacious in ADHD are also effective for ASD, or how to choose the appropriate protocols for core symptoms of ASD, are still open to questions. Such studies must eventually include analyses of the moderators and mediators of neurofeedback-based treatment process and biomarkers predicting intervention response and should consider the mechanisms underlying learning of EEG self-regulation skills in children with ASD and ADHD diagnosis. The placebo-controlled NFB studies should have well-defined training targets, demonstrate ability to engage these targets, and show relevance of functional and behavioral outcomes to improvements in behavioral symptoms of autism. At the same time, there are multiple factors affecting neurofeedback treatment outcomes that require investigation before initiating a randomized clinical trial (RCT). Among such factors are the specific characteristics of the neurofeedback protocol (topography, selected EEG target, length of treatment, etc.) to be used as active treatment in the RCT. According to Arns et al. (2014), evaluation of neurofeedback for ADHD has gone through a long and winding road but still has to travel further in order to cover all grounds related to clinical effectivity and specificity. More details about specific protocols used in ASD can be found in several published studies on neurofeedback training in autism (Coben 2008, 2013; Coben et al. 2010, 2014; Datko et al. 2017; Friedrich et al. 2014, 2015; Kouijzer et al. 2009a, b, 2010; Linden and Gunkelman 2013; Pineda et al. 2012, 2014a, b; Sokhadze et al. 2014; Thompson and Thompson 2013; Wang et al. 2016; Zivoder et al. 2015).

DSM-5 allows clinicians to diagnose ADHD and ASD simultaneously. It is estimated that 30–75% of children diagnosed with ASD have symptoms of ADHD

(Frazier et al. 2011). The prevalence of cases of comorbid ADHD and ASD has been reported to be increasing (Leitner 2014). Reports from both parents and teachers have shown that children with comorbid ADHD and ASD experience greater difficulty in daily life relative to those with only ADHD or only ASD and they have poorer quality of life relative to those with only ASD (Gadow et al. 2009; Leitner 2014). Therefore, it is necessary to develop treatment methods that address ADHD symptoms in children with comorbid ASD and ADHD. In light of findings supporting brain neuroplasticity, the future studies should address both theoretical and practical implications of empowering individuals to self-regulate their brain states (comparing different approaches, topographies, and targets of neurofeedback). There is a need to aim at investigation of feasibility, palatability, acceptance, adherence, preferred frequency and dosage (e.g., number of sessions administered), and other important methodological variables and factors that need to be clarified before proceeding to a larger randomized clinical trial. Refining protocols of the most advanced NFB approaches in ASD using EEG biomarkers will open the road for novel treatments for children with ASD and comorbid ADHD.

2 Rationale for Prefrontal Neurofeedback for Gamma Self-Regulation Training

Our previous neurofeedback and qEEG studies on ASD and ADHD were guided by experimental data and theoretical considerations related to specifications of high frequency EEG activity atypicality in children with autism spectrum disorder and ADHD (Sokhadze 2012; Sokhadze et al. 2014; Wang et al. 2016). Abnormalities of high-frequency EEG oscillations have been associated with binding problems (the coactivation of neural assemblies) present in autism and other psychiatric conditions (Brock et al. 2002; Grice et al. 2001; Loring et al. 1985; Sheer 1989). Oscillatory activity in the gamma band of the EEG (i.e., 30–80 Hz) has been related to cognitive functions such as attention, learning, and memory (Kaiser 2003). Binding of widely distributed cell assemblies by synchronization of their gamma frequency activity is thought to underlie cohesive stimulus representation in the human brain (Kahana 2006). According to this assumption, changes in gamma EEG activity have been considered indicators of processing of Gestalt-like patterns (Herrmann and Mecklinger 2000, 2001; Von Stein et al. 1999). High-frequency EEG oscillations in the gamma range, especially those centered around 40 Hz, are intimately related to mental processes such as consciousness (Llinas and Ribary 1993), binding of sensory features into coherent percepts (Engel and Singer 2001; Tallon-Baudry et al. 1996), object representation (Bertrand and Tallon-Baudry 2000), attention (Fell et al. 2001), perception (Sedley and Cunningham 2013), and memory (Herrmann et al. 2004).

There are only a few EEG studies employing resting-state examinations in individuals with ASD, and practically, all of them report oscillatory anomalies.

Specifically, eyes open resting-state exams have shown greater relative delta and less relative alpha power in 4- to 12-year-old low functioning children with ASD (Cantor et al. 1986) and greater high beta and gamma power in 3- to 8-year-old boys with ASD (Orekhova et al. 2007). Eyes-closed exams have shown greater relative 3–6 Hz and 13–17 Hz power and less 9–10 Hz power in adults with ASD (Murias et al. 2007) and decreased delta and beta power, as well as increased theta power, in children with ASD (Coben 2013). Although the aforementioned results implicate an atypical oscillatory activity in ASD, findings are discrepant and probably due to between-study differences in age, level of functioning, and medication status of the ASD participants.

Cornew et al. (2012) showed that children with ASD exhibited regionally specific elevations in delta, theta, alpha, and high-frequency beta and gamma power, supporting an imbalance of neural excitation/inhibition as a neurobiological feature of ASD (Uzunova et al. 2016). In the auditory domain, reduced entrainment to auditory stimulation at 40 Hz in participants with ASD (Wilson et al. 2007) has been demonstrated. In contrast, during visual perception, there is evidence for both hyperactivity and hypoactivity of gamma-band oscillations (Grice et al. 2001; Brown 2005; Brown et al. 2005; Milne et al. 2009; Stroganova et al. 2012, 2015), raising the question of the link between high-frequency oscillations and perceptual dysfunctions in this disorder. Indeed, gamma-band abnormalities have been reported in many studies of autism spectrum disorders. Gamma-band activity is associated with perceptual and cognitive functions that are compromised in autism. Despite all the evidence, the utility of gamma-band related variables as functional diagnostic biomarkers is currently unexplored, suggesting an urgent need for using gamma oscillation measures as targets of self-regulation training using neurofeedback as well as valuable and informative functional markers of response to interventions such as neurofeedback.

Neurofeedback of gamma frequency, specifically 40 Hz gamma, was first explored in healthy subjects in a series of studies performed in the late 1970s and the early 1980s by a group of researchers headed by Sheer (Bird et al. 1978a, b; Ford et al. 1980). Interest in using gamma activity as a target of neurofeedback-based self-regulation has been renewed during the last decade (Keizer et al. 2010a, b; Kober et al. 2013, 2017; Sedley and Cunningham 2013; Staufenbiel et al. 2014). Though some of these studies showed only moderate changes in gamma and were less associated with improvements in cognitive functions as compared to other protocols, such as sensorimotor rhythm (SMR, 12–15 Hz at motor strip) training (Kober et al. 2013, 2017), other studies did show improvements in cognitive functions and performance, as well as in improved memory retrieval (Keizer et al. 2010a, b; Salari et al. 2014; Staufenbiel et al. 2014).

The series of pilot studies of our group addressed some technical, feasibility, acceptability, and conceptual issues that need to be clarified to create prerequisites for a double-blind randomized clinical trial of neurofeedback training for children with autism spectrum disorder. More specifically, in the series of studies, we planned to develop methodology to monitor EEG and autonomic activity and analyze changes during neurofeedback sessions in high-functioning children with ASD. The pilot studies represented one of the approaches aimed at the understanding of EEG correlates of neurofeedback training in high-functioning ASD population, rather than an attempt at claiming clinical improvements resulting from the prefrontal brainwave training. Before moving to sham neurofeedback-controlled trials, it is proposed that more research studies should be done to understand: (1) whether children with high-functioning autism can control EEG in NFB mode, (2) how EEG characteristics are changing during the training course in an ASD population, (3) what additional efforts are needed to correctly identify specific changes in EEG rhythms known to be abnormal in ASD, specifically gamma activity at the frontal sites, (4) how many neurofeedback sessions are required to observe reliable changes in targeted neurofeedback indexes and monitored EEG bands of interest, (5) how ADHD comorbidity in ASD is affecting training outcomes, and (6) how psychophysiological measures collected during individual training session can be used to determine participants' attention to training task and engagement in self-regulation process. Another important factor is the specifications of neurofeedback protocol.

Our approach includes neurofeedback training in the prefrontal topography, specifically at the midline prefrontal site. Considering the role of the prefrontal cortex in executive functions, including attention and cognitive processes, it was feasible to investigate effects of neurofeedback using training at the anterior, frontal location rather than at the central or posterior (e.g., parietal) sites. This selection of cortical topography was also determined by our previous studies on gamma oscillations in children with autism (Baruth et al. 2010; Sokhadze et al. 2009, 2018), which showed alterations of evoked and induced gamma oscillations during attention tests especially well present in the frontal topographies.

The goal of this study was to conduct neurofeedback in children with ASD using the Peak Brain Happiness Trainer (PBHT, Neurotek, Goshen, USA) neurofeedback device with the "Focus/Neureka!" ("Focused Attention" index and "40 Hz-centered gamma" index) training protocol to investigate relative changes in EEG bands (e.g., theta [4–8 Hz] and beta [13–30 Hz]) and their ratios (e.g., theta-to-beta) throughout the 24 sessions long course of neurofeedback training in ASD with comorbid ADHD and during each individual training session. The aim was to investigate how 40 Hz centered gamma power and EEG theta and beta band ratios change within the course of neurofeedback training in high functioning individuals with ASD with co-occurring ADHD, and whether there are any correlations between EEG measures of interest (i.e., relative gamma power and theta-to-beta ratio) and neurofeedback training indexes such as "Focused Attention" (i.e., "Focus") index and "40 Hz-centered gamma" (i.e., "40 Hz gamma") index.

It was expected that all participants would complete weekly sessions of ~25–30 min long training and learn to increase the "Focused Attention" index and control level of "40 Hz centered gamma" parameter within predefined range of targets in neurofeedback mode. It was similarly expected that an increase in so-called "Focused Attention" measure of the PBHT device protocol would be manifested in a gradual decrease of theta-to-beta EEG ratios, while an increase in the "40 Hz centered gamma" measure would be accompanied by the gradual increase of the relative power of gamma (35–45 Hz) band. We were also interested in exploring whether power of the selected EEG bands and their ratios correspond to and correlate with neurofeedback measures calculated in real-time online mode in BioExplorer-based commercial software application of the PBHT device. Another point of interest was related to investigation of dynamics of psychophysiological measures (heart rate, heart rate variability [HRV] indexes, electrodermal activity, respiration rate, etc.) concurrently recorded during neurofeedback sessions to determine their utility as moderators and mediators of neurofeedback treatment outcomes.

The study explored our hypothesis that the relative power of gamma oscillations and the individual EEG band ratios will be changed and modulated as the predicted tendencies, in particular, (1) higher power of gamma during individual session and across the course of neurofeedback; (2) lower theta-to-beta ratio during individual sessions and across the courses; (3) positive correlation of the relative power of gamma band with the measure of "40 Hz centered gamma" index used as a neurofeedback training target, along with negative correlation of theta-to-beta ratio with a "Focused Attention" index measure, which was also used as a target during the neurofeedback training in this study, (4) the longer course of NFB training (24 instead of previously used 12 and 18-session long training) would have more profound effects showing acquisition of index control, expressed as a gradual increase of target measures during and across the session, and (5) autonomic measures such heart rate, HRV, respiration rate, and skin conductance level measures will show positive dynamics across the course of NFB training.

3 Methods

3.1 *Patient Demographics and Recruitment*

In this pilot study, we screened 18 children, then recruited 14 children and adolescents with ASD and ADHD diagnosis (mean age 10.28 years, SD = 1.93, 11 males, and 3 females), and enrolled them in the 24 session-long course of neurofeedback training. In this pilot study, it was not required for participants to be off medication during the whole course of the neurofeedback trainings. We monitored medication status, dosage, and other variables of pharmacotherapy and kept record, but this information was not used as a part of the patients' demographic descriptive characteristics in this study. Participants with ASD were recruited initially through the Prisma Health-Upstate (formerly Greenville Health System—GHS) Pediatrics Department.

All participants with ASD were diagnosed by experienced pediatricians according to the Diagnostic and Statistical Manual of Mental Disorders (DSM-IV-TR) (APA 2000) or DSM-5 (APA 2013) and further ascertained with the Autism Diagnostic Interview—Revised (ADI-R, Le Couteur et al. 2003) and/or ADOS/ADOS-2 (Lord and Rutter 2005). ADHD diagnosis was confirmed according to standard procedures that included clinical evaluations and parental and teachers' reports. Further medical estimations were made to exclude the participants with a history of seizure, significant hearing or visual impairment, a brain abnormality, or an identified genetic disorder. Participants with severe psychiatric comorbidities were not included in the study. All patients except one participant were naive to neurofeedback training procedures and never participated in any previous neurofeedback study. One of the participants was treated in the past for neurological condition using neurofeedback.

Using previous record data of the Wechsler Intelligence Scale for Children (WISC-IV, Wechsler 2004) or (for adolescents) the Wechsler Abbreviated Scale of Intelligence (WASI) (Wechsler 1999), all participants were determined to have full-scale IQ > 70. Child and adolescent psychiatrists and clinical psychologists at the Prisma Health/GHS performed pre- and postneurofeedback clinical evaluations. Participants enrolled by referral from third party clinical providers presented records confirming ASD and ADHD diagnosis. Neurofeedback sessions with concurrent autonomic activity monitoring were conducted by an experienced applied psychophysiologist certified in neurofeedback (BCN) by the Biofeedback Certification International Alliance (BCIA). All required IRB-approved consent/assent forms were signed by the participants and their parents/guardians.

3.2 Behavioral Measures and Evaluations

In conjunction with the EEG data, we collected the behavioral rating results with the pre- and postneurofeedback data using the *Aberrant Behavior Checklist (ABC)* (Aman and Singh 1994) from the parents of the ASD participants. *Irritability*, *Lethargy/Social Withdrawal*, *Stereotypy*, *Hyperactivity*, and *Inappropriate Speech* were the five problem aspects that were contained in and assessed by the ABC rating scale. In our studies, we focused primarily on the *Hyperactivity*, *Lethargy/Social Withdrawal*, and *Irritability* ratings before and after a course of NFB treatments. *Social Responsiveness Scale (SRS)* (Constantino and Gruber 2005) is a caregiver completed rating scale assessing social interest and interaction in autism. *Repetitive Behavior Scale—R (RBS)* (Bodfish et al. 1999) is a caregiver completed rating scale assessing repetitive and restricted behavior patterns in ASD. *The Achenbach System of Empirically Based Assessment* (ASEBA) checklist will be used for assessing adaptive and maladaptive functioning. The school-age assessment forms are the CBCL/6–18, completed by parents, and the TRF/6–18, completed by teachers. Empirically based syndrome scales scored from the CBCL/6–18, TRF, and YSR are based on factor analyses coordinated across the forms for Anxious/Depressed, Withdrawn/Depressed, Somatic Complaints, Social Problems, Thought Problems, Attention Problems, Rule-breaking, and Aggressive Behavior. The DSM-oriented scales are also scored from all three forms.

Consumer Satisfaction Questionnaire (CSQ, Parent version): Brief consumer satisfaction questionnaire for the parents developed by the OSU (Ohio State University, Columbus) was used to examine participant's views about the helpfulness and practicality of the intervention. The 11-item questionnaires were given in the 18th and 24th sessions. The questionnaire had a different form for Treatment 18 from that for the other administration. The first five questions vary between the forms for treatment 18 and treatment 24. For example, the version for treatment 18th lacks the question of the session 24th questionnaire asking for comparison of the visit frequencies from those who switched (one vs. two vs. three visits per week).

Compliance was checked by attendance and by the computer-collected record of the neurofeedback session, which shows the number of minutes of training conducted, the average amplitudes of frequencies trained, variance within the frequencies trained, and percent rewards (so-called "run-time") provided to the subject during training.

3.3 Neurofeedback Protocol and Data Collection

In our pilot study, ASD participants completed a course of NFB trainings using a "Focus/Neureka!" protocol designed to modulate the "Focused attention" index (FI) using "InhibitAll" type of protocol and "40 Hz-centered gamma" index (GI) using "Neureka!" protocol. The prefrontal neurofeedback training application used in this study was based on the BioExplorer software (CyberEvolution, Seattle, WA, USA) platform. The protocol provided the exercises for each subject to enhance FI and GI throughout the session while maintaining an adequate level of GI measure ("40 Hz gamma" index) within a certain range. During all the treatment sessions, different scenes from the BBC "Planet Earth" and "Life" series, as well as similar nature documentaries from National Geographic DVDs, were shown to maintain the participants' interest. The protocol provided feedback to the subjects in both visual and auditory modalities. Based on the thresholds set, parameters related to visual feedback, such as the brightness, size, and continuation of the video have been modulated, and the sound volume of the video was adjusted simultaneously according to the targeted "FI" and "GI" indexes during the treatment. All EEG signals and training parameters were measured using three electrodes, one active electrode at the prefrontal EEG (AFz) site, the second being a reference on the left ear, and a third sensor serving as ground and located between the above two electrodes. All of the subjects in the study were requested to complete a 25–30 min recording per session and a total of 24 (once or twice per week) neurofeedback sessions, in order to increase the "FI" and "GI" using the "Focus/Neureka!" Peak BrainHappiness Trainer (PBHT) protocol. More than 90% of the sessions met the requirement of a 20-min minimum usable EEG data recording. Eye blink and EMG artifact removal was implemented using the specific BioExplorer application that can be found in the operation manual of the NFB device (Fig. 1).

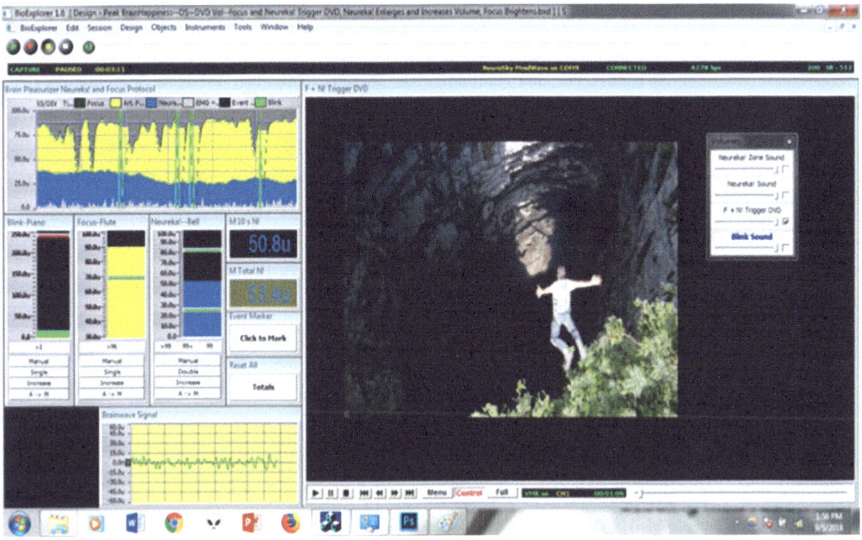

Fig. 1 Example of screenshots of neurofeedback screens presented to participants

3.4 The EEG Signal Processing

The EEG signal collected and recorded by BioExplorer applications during NFB treatments were exported using BioReviewer and further analyzed by a series of customized codes using Matlab software (MathWorks, Inc., Massachusetts). As an extension application of BioExplorer software, BioReview was used to export the raw EEG and to calculate the desired frequency bands of data for each session. By configurations in the BioReview report, along with the raw EEG, the separated delta (2–4 Hz), theta (4–8 Hz), alpha (8–13 Hz), beta (13–30 Hz), and gamma (30–50 Hz) were also acquired using seventh order elliptical bandpass filters. The exported data were arranged in a text file in which the different items were organized into columns and each subsequent row represented the data point in time series between samples.

For the relative power calculation, it was necessary to gain the total power of the band from 2 to 45 Hz (the whole bands from delta to gamma frequencies). A custom band-pass filter application integrated of wavelet transformation and a Harris window configuration were created to filter and separate the 2–45 Hz frequency band from the raw signal that was exported from the BioReview reports. The wavelet analysis was used to provide enhanced temporal resolution of frequency responses of a given signal, and it allowed us to apply a band pass filter to the individual waveform and avoid the distortion when applying the filter to the entire signal. Besides the relative power calculation for each band, the ratio of certain bands was also calculated. The ratios of interest for this study were theta (4–8 Hz) to beta (13–30 Hz).

The primary statistical analyses in the study mainly included linear regression estimation and paired sample Student *t*-test and one-way ANOVA methods. Each EEG-dependent variable over 24 sessions of neurofeedback course was analyzed using linear regression analysis. The mean values of pre- and postNFB behavioral measures using ABC, SRS-2, and ASEBA questionnaire were compared with the paired sample *t*-test and one-way ANOVA method using test time point (pre- vs. postNFB) as a factor. EEG variables and FI and GI NFB training indexes were calculated as well on per minute basis during each training session. Each dependent EEG variable went through the normality distribution analysis using *t*-test to ensure appropriateness for the test.

3.4.1 Baseline EEG Power Spectrum Analysis

A baseline measure of EEG arousal (theta to low beta ratio at the central or fronto-central site) was calculated using the ratio of (4–8 Hz)/(13–21 Hz), so-called Theta-to-beta ratio (TBR). This is Monastra et al. (2001) attention index, and it was derived from the baseline data for exploratory purposes. Before the start of the neurofeedback course, the participants were assessed using SmartMind3 and SmartMind4 (later modification) instruments of BrainTrain (Sheffield, VA) company. This EEG system uses advanced method of eye-movement and facial EMG artifact removal for reliable estimation of EEG recorded from the prefrontal site. This is a professional level miniEEG device that provides accurate artifact-free data, which allows calculation of correct theta-to-low beta ratio. The EEG sensors are attached to AFz site (active) and are referenced to linked earlobes. The electrodes use TEN20 conductive crème, and attachment sites are treated first with NuPrep or OmniPrep gel. The obtained results are useful for estimation of more traditional ranges of theta-to-beta ratio compatible with databases used for complementary functional measures for ADHD diagnostics. Single channel EEG data are collected during a sequence of tasks including baseline inattention and focused processing tasks. The baseline tasks of inattention include 3 min eyes closed and 3 min of eyes open. The attention tasks include 5-min silent reading and the 5-min timed arithmetic test used for an outcome measure. This assessment procedure is adapted from Monastra et al. (1999), where it demonstrated a sensitivity of 86% as an attention index in the diagnosis of ADHD. The power spectrum for purposes of determining arousal level will average the different conditions. Other EEG mapping tests and evaluations were conducted using Discovery 24 system and Neuroguide software.

3.4.2 Measurement of the ANS-Dependent Variables

Electrocardiogram (ECG), pneumogram (PNG), and electrodermal activity (EDA) were acquired (2048 Hz sampling rate for ECG, 128 Hz for PNG and EDA) using a NeXus-10 wireless system (Mind Media, B.V., Netherlands) with BioTrace+ software, Procomp2 Thought Technology device with Infiniti software, and C2 system of J&J Engineering with USE-3 Physiodata software. Three Ag/AgCl electrodes were attached for measurement of Lead II ECG, and PNG was recorded using a strain gauge transducer. EDA was recorded by Ag/AgCl electrodes (with Unibase isotonic gel) attached to the distal phalanx of index and middle fingers to measure skin conductance level (SCL), skin conductance response (SCR), and number of nonspecific SCRs per minute (NS.SCR) when appropriate. *Cardiovascular activity:* Average heart rate (HR), the standard deviation of the HR (SDHR), power of high frequency (HF), low frequency (LF), very low frequency (VLF) components, % power of LF (% LF [of VLF + LF + HF]) and of HF (%), and the ratio of the LF over the HF (LF/HF ratio is used as an indirect autonomic balance index) of HRV are calculated as cardiac activity measures. Artifact-corrected 3-min long recording epochs are analyzed with FFT to assess HRV. Integrals of the spectrum in 0.003–0.04 (VLF of HRV), 0.04–0.15 Hz (LF of HRV), and 0.15–0.40 Hz (HF of HRV) bands are measured (in ms^2) (Berntson et al. 1993, 1997). Furthermore, all HR data were analyzed off-line using Kubios HRV software (Finland). *Respiratory activity*: Respiration rate (RESP) on per minute basis, inspiration wave amplitude (IA), and Peak respiration frequency (PFRQ) are calculated. These measures are used to control HF peak in HRV related to respiratory frequencies in HRV. *Electrodermal activity:* SCL (in μS) and amplitude of the SCR, defined as fluctuation with more than 0.02 μS increment and NS.SCR—number of nonspecific SCR (per minute). The magnitude of EDA and NS.SCR corresponds to the strength of emotion and can be used as an indicator of quantitative aspects of emotional arousal and reactivity (Boucsein 2012). Considering the length of each neurofeedback session, only SCL changes were used as outcome measures. Other measures of interest were analyzed on session-by-session basis, i.e., 20–25 min *N* session and over 24 sessions of neurofeedback course. In addition, selected measures were compared for the first 5 min vs. last 5 min window of the training session.

3.4.3 Statistical Analysis

For each outcome variable, the described statistics, such as mean and standard deviation (SD), for each of the three active treatment groups (i.e., TBR, SMR, and Gamma) at each time point (pre- and post24 NFB) were collected, and a graphical presentation such as histogram and box plot will be presented to examine normality of the data distribution. To examine the treatment effect, following completion of treatment, the changes in the pre- and posttreatment measures of behavior ratings were compared using one-way analyses of variances (ANOVA) and post hoc t-test analysis. One-way ANOVA and post hoc *t*-test analysis were applied to examine the treatment effect on each behavioral outcome variable. Again, normality of the data was checked, and, if necessary, appropriate transformation was applied. In addition, multiple linear regression and graphic presentation of the data were applied to examine the association between behavioral outcome (e.g., ABC, SRS-2, RBS-R, CSQ, etc.), and/or electrophysiological activity outcome variable (such as Theta-to-beta ratio, relative gamma power, HRV measures, SCL, etc.). In addition, during the

NFB training course period, data of each monitored EEG and autonomic variable over 24 sessions of NFB were summarized by the slope of the linear regression model over the 24 time points. As it was mentioned earlier, the emphasis of this study was not on statistical significance, but on collection of sufficient data to make common sense and informed decisions about how to proceed with a larger trial.

4 Results

4.1 EEG Activity Measures Across Sessions of Neurofeedback

Outcomes of 24 NFB sessions: Regression analysis was completed with the whole 24 session-long course for ten participants that completed the study (Fig. 2).

4.2 Neurofeedback Training Indexes

The "Focused Attention" index (FI, i.e., measure of "Inhibit All" protocol in neurofeedback) showed a statistically significant linear increase over 24 sessions of training ($r = 0.52$, $F = 8.97$, and $p = 0.007$, Fig. 3). However, the neurofeedback measure reflecting relative power of "40-Hz centered Gamma" index did not show a linear increase trend over 24 sessions of training.

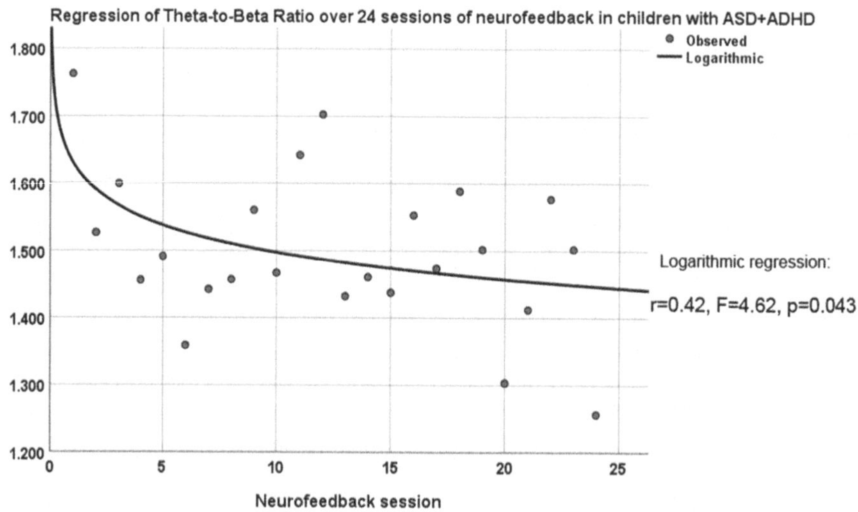

Fig. 2 Regression of Theta-to-beta ratio across 24 sessions of neurofeedback in children with ASD and ADHD ($N = 10$) fits logarithmic regression curve and reached significance level ($r = 0.42$, $F = 4.62$, and $p = 0.043$)

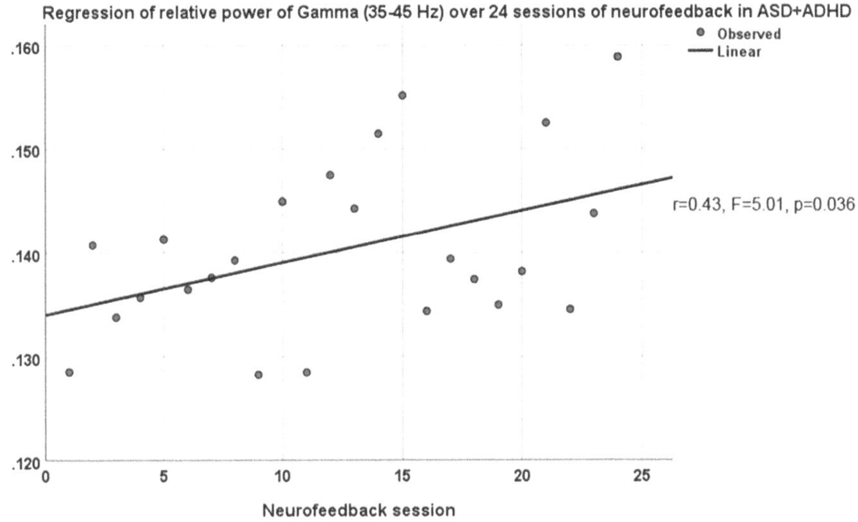

Fig. 3 Relative power of gamma (35–45 Hz) showed statistically significant linear regression over 24 sessions of neurofeedback training in ten children with ASD with comorbid ADHD. Linear regression $r = 0.43$, $F = 5.01$, and $p = 0.036$

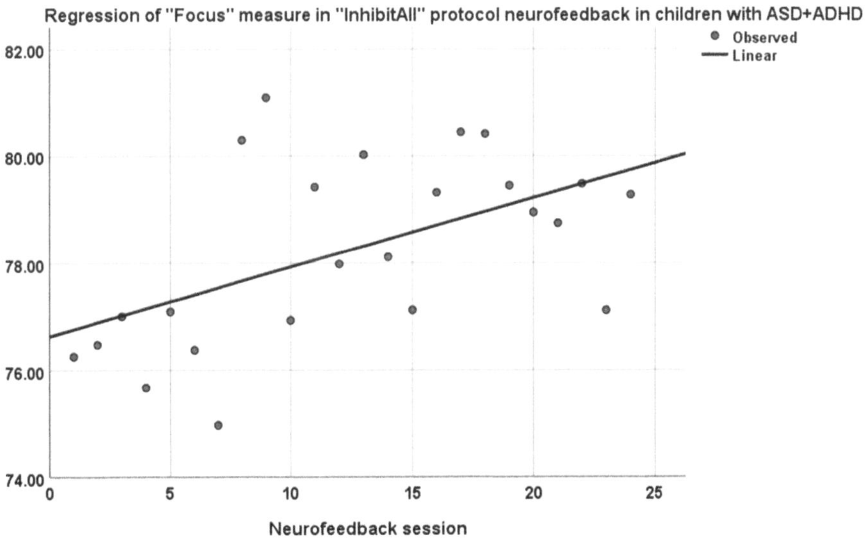

Fig. 4 Linear regression of "Focus" index in a patient J001 enrolled in 24 session-long course of neurofeedback training

Even though the group data did not yield significant changes, several participants did demonstrate linear increase of both "Focus" and "40 Hz-centered gamma" indexes over 24 sessions. Their results are presented in Fig. 4.

4.3 *Autonomic Activity Indexes During Individual Neurofeedback Sessions and Across the Course*

See Figs. 5, 6, 7, 8, 9, 10, 11 and 12.

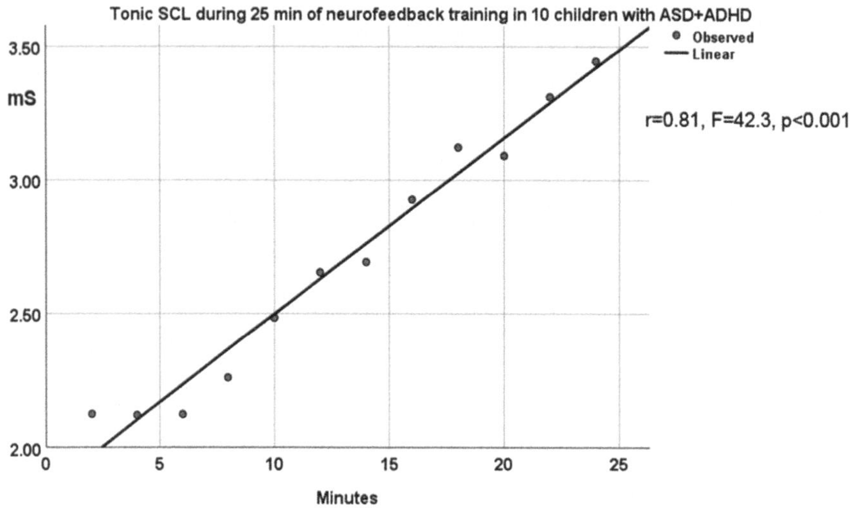

Fig. 5 Skin conductance level during 25 min of neurofeedback training in ten children with ASD with co-occurring ADHD. SCL shows linear increase toward the end of training session ($r = 0.81$, $F = 42.3$, and $p < 0.001$)

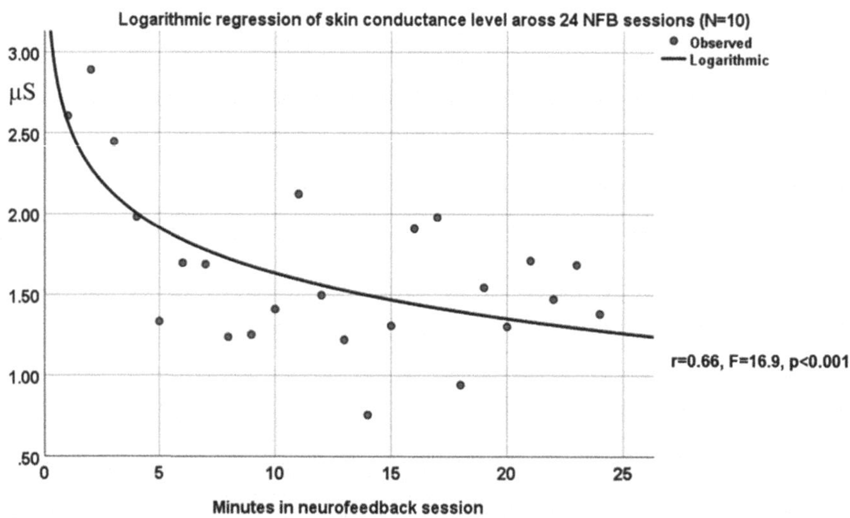

Fig. 6 Logarithmic regression of SCL over 24 session of neurofeedback in children with ASD and ADHD ($r = 0.66$, $F = 16.9$, and $p < 0.001$)

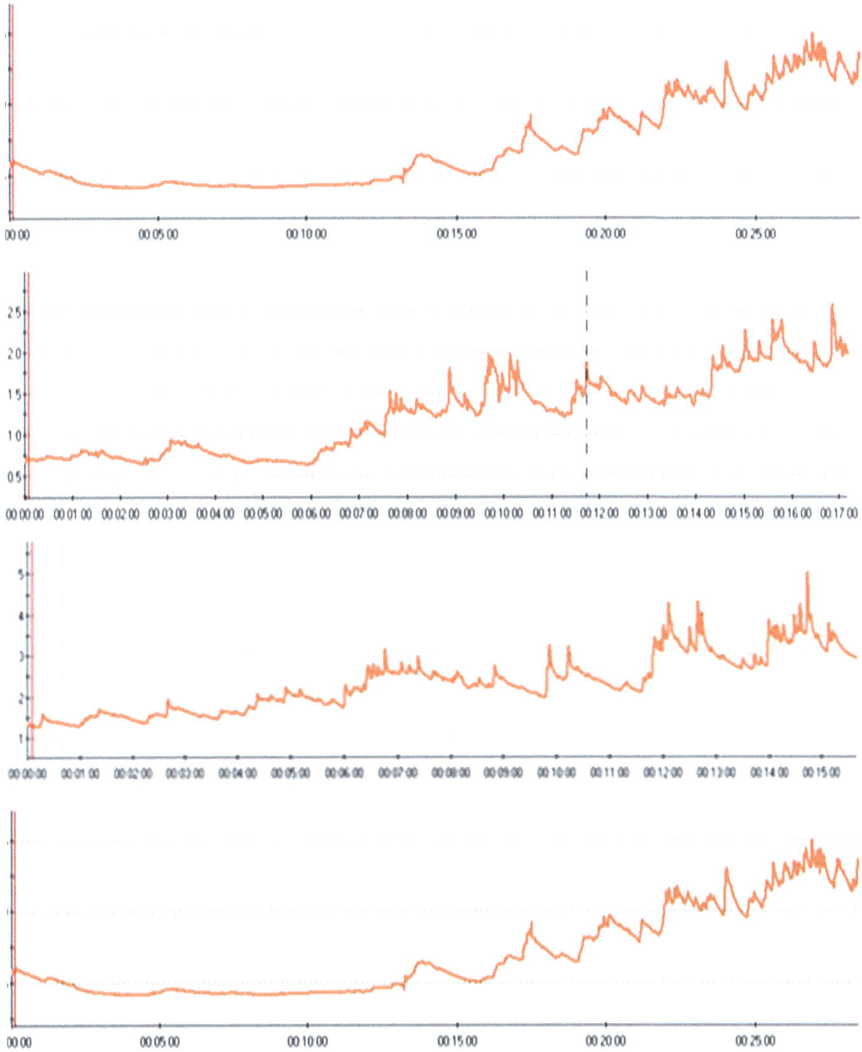

Fig. 7 Examples of SCL changes toward the end of successful neurofeedback training sessions in four participants with ASD with ADHD

4.4 Behavioral Evaluations

ABC scores showed Irritability decrease from 13.20 ± 6.83 down to 7.30 ± 3.80, df = 1.18, $F = 5.61$, and $p = 0.029$ and Hyperactivity decrease from 21.40 ± 8.30 down to 14.10 ± 6.65, df = 1.18, $F = 4.71$, and $p = 0.044$. ASEBA questionnaire

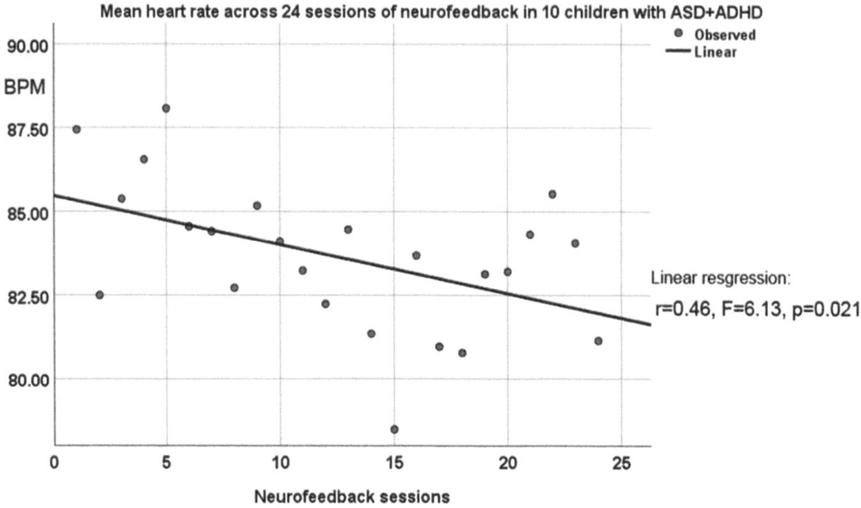

Fig. 8 Liner regression of mean Heart Rate across 24 sessions of neurofeedback training in ten children with ASD and ADHD diagnosis ($r = 0.46$, $F = 6.13$, and $p = 0.021$)

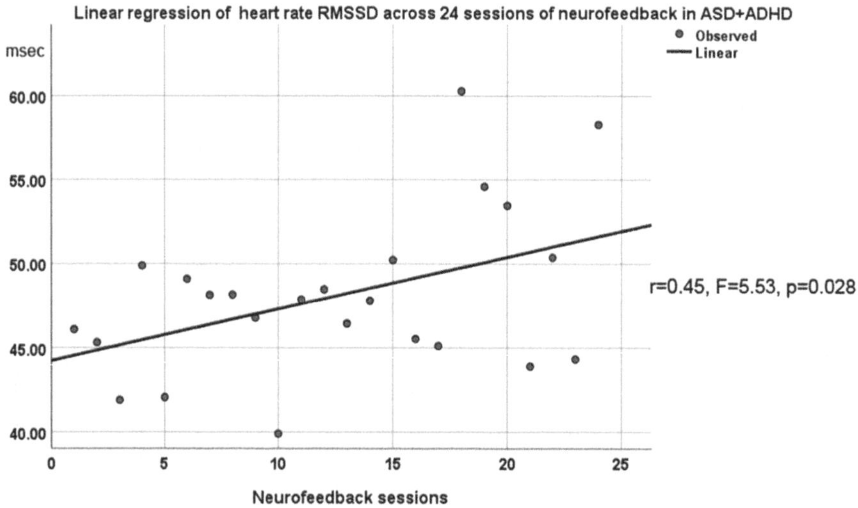

Fig. 9 Linear regression of time domain HRV measure RMSSD across 24 sessions of neurofeedback in children with ASD and ADHD ($r = 0.45$, $F = 5.53$, and $p = 0.028$)

showed that Attention deficit scores decrease from 70.91 ± 5.17 down to 65.36 ± 5.84, df = 1.20, $F = 5.57$, and $p = 0.029$, and Conduct behavior scores decrease from 61.64 ± 6.39 down to 56.18 ± 4.68, df = 1.20, $F = 5.21$, and $p = 0.034$.

SRS-2 questionnaire posttraining showed T-score decrease from 84.50 ± 8.66 down to 4.10 ± 11.13, df = 1.20, $F = 5.46$, and $p = 0.031$.

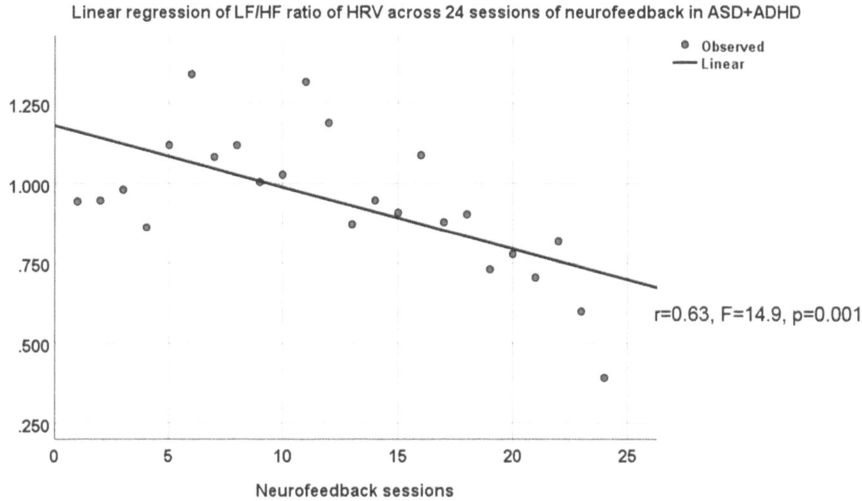

Fig. 10 Linear regression of frequency domain HRV index (LF/HF ratio) across 24 sessions of neurofeedback in children with ASD with ADHD ($r = 0.63$, $F = 14.9$, and $p = 0.001$)

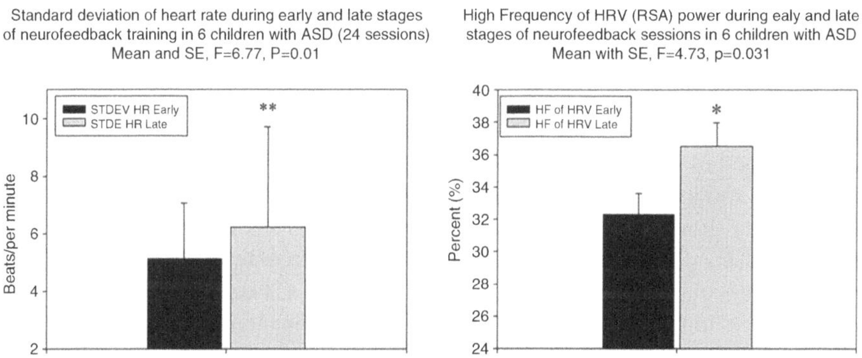

Fig. 11 Left: Standard Deviation of heart Rate during early and late stages of neurofeedback session. HR standard deviation increased at the late stages of neurofeedback ($F = 6.77$ and $p = 0.01$). Right: Changes of High Frequency of HRV during late stages of neurofeedback ($F = 4.73$ and $p = 0.031$)

4.4.1 Consumer Satisfaction Questionnaire (CSQ, Parent Version)

Brief consumer satisfaction questionnaire for the parents of the OSU (Ohio State University, Columbus) was used to examine participant's views about the helpfulness, and practicality of the intervention showed preference of once per week schedule of visits for neurofeedback sessions. This 11-item questionnaire was being given in the 18th and 24th sessions. A majority of parents (85%) selected once per week visit schedule due to several other procedures and extra-curriculum activities of their child.

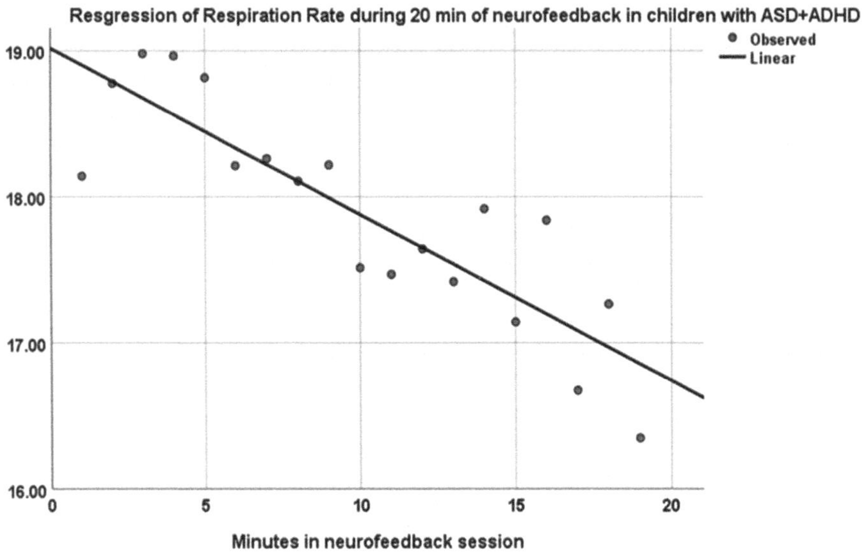

Fig. 12 Linear regression of Respiration Rate during 20 min of neurofeedback in children with ASD with ADHD ($r = 0.59$, $F = 12.1$, and $p = 0.01$)

5 Discussion

The results indicate that the pilot study outcomes closely aligned to those predicted by our hypotheses, especially in regard to regression of the dependent-EEG and autonomic variables across the neurofeedback training sessions. For instance, the theta-to-beta showed a decrement across NFB sessions, while the relative power of the gamma band showed a linear increase over the course of the training in 24 session-long courses. Neurofeedback training indexes ("Focused Attention" index and "40 Hz gamma" index) showed a linear increase over training sessions only for "Focused Attention" measure in whole group of trainees, but only in several subjects, they showed gradual increase of the "40 Hz-centered gamma" index. We found, however, significant trends of the EEG variable changes within 20–25 min of individual neurofeedback sessions. Only one training index (i.e., "Focused Attention") in our previous studies in ADHD patients showed a linear increase over the minutes within individual sessions in the 18 session study, yielding a high negative correlation with both theta-to-low beta and theta-to-beta ratios in the EEG (Hillard et al. 2013). We found a notable decrease in the theta-to-beta proportions from session to session in the 24 session-long neurofeedback courses. These results are in accordance with the goals of NFB treatment described earlier for children with ADHD (Arns et al. 2009; Lofthouse et al. 2010). Even though the theta-to-low beta ratios used in previous ADHD NFB studies were mostly collected from the central or frontocentral cortical sites (e.g., Cz or FCz), our prefrontal and theta-to-beta ratio showed similar trends in an ADHD population (Hillard et al. 2013; Sokhadze et al. 2013).

Due to the improvements in behavioral outcomes indicated by ABC, SRS, and ASEBA questionnaire, it is also possible to discuss whether training of the "Focus/40 Hz gamma" measures of the PBHT protocol accompanied by the decrease of theta-to-beta ratio and increase of the relative power of 40 Hz-centered gamma activity are related to functional behavioral improvements reported by the patients. Determining which of these two measures is more fundamental to the effects of neurofeedback in ASD would increase the efficiency and aid in the delivery of more effective neurofeedback treatment methods.

As mentioned earlier, autism is characterized by an imbalanced inhibitory/excitatory ratio in local cortical network (reviewed in Uzunova et al. 2016), which may cause the disordered gamma oscillations in ASD reflected at the electroencephalographic level. The gamma abnormalities and excessive cortical excitation (E/I ratio) in autism have been considered as important EEG biomarkers for ASD based on recent theoretical reviews and experimental studies (Uzunova et al. 2016; Casanova et al. 2015). Brown et al. (2005) interpreted the abnormal gamma responses in their study on individuals with autism as reflecting decreased "signal to noise" ratio due to decreased inhibitory processing (Grice et al. 2001; Lansbergen et al. 2011). Brock et al. (2002) described the parallels between the psychological model of 'central coherence' (Frith and Happé 1994) in information processing and their neuroscience model of neural integration or "temporal binding". This concept was further elaborated in an "impaired connectivity" hypothesis of autism, which summarized theoretical and empirical advances in research implicating disordered connectivity in autism (Brown 2005). The authors highlighted recent developments in the analysis of the temporal binding of information and the relevance of gamma activity to current models of structural and effective connectivity based on the balance between excitatory and inhibitory cortical activity (Belmonte and Yurgelun-Todd 2003; Casanova et al. 2003, 2013; Rippon et al. 2007; Rojas and Wilson 2014). Based on the minicolumn hypothesis of autism, disrupted patterns of coordinated high-frequency oscillatory output in distributed networks might be associated with cortical "disconnection" in autism according to Casanova et al. (2006).

The current pilot study indicated the potential of prefrontal neurofeedback aimed at modulating the disordered EEG activities associated with ASD and ADHD. Also, from the results of the correlation between "40 Hz Gamma" index and the relative power of gamma calculated in our custom made program, it is found that the "Focus/40 Hz Gamma" protocol provided by the neurofeedback equipment used in our study can effectively help to improve gamma activity along with the decrement of theta-to-beta ratio in prefrontal EEG in children ASD with comorbid ADHD.

Theta-to-beta ratio showed the significant linear decrease over neurofeedback course. Theta-to-beta ratio is one of the classical indexes for characterizing the ability to focus attention and to concentrate. The current study showed that both prefrontal theta-to-beta ratio and power of gamma activity could be modulated positively by operant conditioning during the NFB training in high-functioning children with ASD with comorbid ADHD. It is well known that most ASD and ADHD subjects have difficulties with switching focused attention. The "Focus/40 Hz Gamma" protocol used in the study provided a successful way for positively modulating both gamma activity and focused concentration abilities in ASD. The positive effects of the neurofeedback training further can be manifested by the improvement in the behavioral scores measured by the ABC and ASEBA parental questionnaire. Our results show a significant reduction in the *Irritability and Hyperactivity* subscales of the ABC and *Attention Deficit* problems according to the ASEBA.

Our study showed that compared to previous protocols that required more sessions per subject (>30) and a more frequent training rate (e.g., twice per week), the statistically significant improvement either in EEG or in behavioral measures (Sokhadze et al. 2009) can also be achieved within a shorter number of sessions (i.e., 24 and 18 NFB sessions in ASD and in ADHD) and weekly visits. Probably, 24 sessions rather than 18 sessions might contribute to better consolidation of results of operant conditioning using neurofeedback, and currently, we have studies in progress that will compare outcomes of 12 vs. 18 vs. 24 sessions of neurofeedback using the same protocol in children with autism.

It should be noted that the study has several limitations. The enrollment to the neurofeedback training was open to only high-functioning children with autism with confirmed comorbid ADHD, and thus, results cannot be directly interpolated for low-functioning children with ASD or ASD children without co-occurring ADHD. The study was not designed as a clinical research as it had no control group of participants, and the number of clinical behavioral evaluations was minimal. The focus of current study was directed toward more accurate analysis of the dynamic of target indexes (FI and GI) and theta-to-beta ratio and relative power of 40 Hz gamma changes within and across neurofeedback sessions in children with ASD and primarily toward feasibility and collection of preliminary results for early phase clinical trial. Records of patients' demographic specifications (e.g., social status of families, ASD onset, duration data, etc.) and detail of their medication status were not analyzed. Analysis of some datasets for 24 session-long course was not complete and was pending recruitment of a larger sample. In order to foster the neurofeedback treatment applications for children with ASD with comorbid ADHD and its scientific rationale, further methodological advances are necessary: controlled and randomized study designs, larger sample sizes of patients, a more accurate selection of subjects with ASD and ADHD, and more intensive and rigorous baseline, and posttreatment and follow-up evaluations.

5.1 Consideration of Important Factors During Designing Future Neurofeedback Studies

During last decade, numerous publications have addressed important issues related to the use of neurofeedback, operant conditioning, and the role of learning (Coben and Ricca 2015; Enriquez-Geppert et al. 2017; Gaume et al. 2016; Kerson and Collaborative Neurofeedback Group 2013; Niv 2013; Pigott and Cannon 2014; Pigott et al. 2018; Pineda et al. 2012; Ros et al. 2014; Sherlin et al. 2011; Strehl 2014; Vollebregt et al. 2014; Zuberer et al. 2015). It is very important to have these papers reviewed and considered while developing neurotherapy protocols. Gaume et al. (2016) proposed that designing effective and efficient neurofeedback protocols would benefit from a comprehensive model of the mechanisms of learning during neurofeedback training procedures. Among the key elements relevant to such model, the authors listed perceptibility, autonomy mastery, and learnability. Within the framework of a proposed model, the number of neurofeedback sessions, duration of each session, time interval between sessions, and several other related protocol details are crucial parameters of neurofeedback protocols. Ros et al. (2014) admitted that neurofeedback emerged as a promising technique that enables self-regulation of ongoing brain activity, but pointed out that despite empirical evidence of clinical benefits, a solid theoretical basis is still lacking on the mechanisms of NFB effects on these outcomes. The authors attempted to combine together several concepts from neurobiology, bioengineering, and theory of dynamic systems to put together a theoretical model aimed at describing mechanistic effects of neurofeedback training.

Strehl (2014) noted in her theoretical conceptual paper that even though the majority of definitions of NFB considers it as an operant conditioning method that results in acquisition of brain activity self-regulation skills, the role of classical Pavlov conditioning should also be considered beyond operant conditioning, two-process-theory, as well as role of motivation in neurofeedback skill acquisition process. The model supports the hypothesis that learning aimed at self-regulation of EEG has to be considered within psychotherapeutic, or in other words, within behavioral therapy framework. In behavioral therapy, the therapist assists patient in learning a new behavior focusing on overt behavior, cognition, and emotions, while in neurofeedback, therapist tries to help change the activity of the brain that becomes visible thanks to the biofeedback equipment. However, an important point is that therapist should know the rules of learning and be proficient in designing and applying neurofeedback training to be able to effectively guide patients in this version of behavioral therapy intervention (Strehl 2014).

Zuberer et al. (2015) discussed the role of the analysis of learning and adaptation processes during the course of training. The paper outlines the need to relate these processes to improvements in self-regulated EEG activity across training sessions to behavioral, neuropsychological, and electrophysiological outcomes. At the same time, it proposed that much attention is devoted to the analysis of EEG changes and dynamics in the course of the neurofeedback training (both during the individual session and over the whole course of training) and how these measures impact behavioral and clinical outcomes. These suggestions call for the necessity of improving target analysis and monitoring EEG measures across sessions in the course of training and within individual session, assessment of learning trajectories in population under study, and to provide the best conditions for learning.

Pigott et al. (2018) critically reviewed the methodology from six sham-controlled trials using neurofeedback to treat ADHD and noted that some of the methodologies may have violated established rules of operant conditioning by improperly using either automated or manually adjusted EEG reward thresholds. Some studies had other methodology flaws as well, which may impede neurofeedback subjects from learning EEG self-regulation skills (Pigott et al. 2018). Some of such methodologically flawed studies led Thibault et al. (2018) to argue that these beneficial effects are due to placebo phenomena rather than specific clinical and behavioral effects of neurofeedback. Discussions about NFB specificity should be encouraged to include analyses of the changes of targeted EEG parameters to be able to report learning curve, as well as changes of EEG measures in individual training session and across course (Zuberer et al. 2015). These changes should be correlated with the gains of behavioral and clinical outcomes of the treatment course. It is necessary that for the evaluation of efficacy and specificity of NFB in autism, strict methodological standards should be adhered in the study design along with scientific rationale for the selection of the targets of EEG self-regulation. Several reviews and metastudies (Coben 2013; Hurt et al. 2014; Linden and Gunkelman 2013; Pineda et al. 2014a, b) have demonstrated the potential efficacy of NFB training with regard to the improvement of ASD symptoms. Whether NFB is efficacious and, at the same time, specific in neurotherapy of autism with comorbid ADHD still needs further investigation and rigorous research, which should go beyond analyzing pre-post changes and must include analyses of the dynamic of targeted EEG indexes and monitor EEG and autonomic parameters during the process of EEG self-regulation skill acquisition.

5.2 *Autonomic Nervous System Activity in Autism*

A series of current studies have shown exponentially increasing interest in the investigation of the autonomic nervous system (ANS) activity abnormalities in children with ASD (Cohen et al. 2015; Klusek et al. 2015; Kushki et al. 2013, 2014; Ming et al. 2011, 2016; Sokhadze et al. 2019). Considering that sympathetic activation is often associated with autonomic arousal and anxiety, investigation of impairments of arousal regulation definitely deserves to become one of the main goals of autism research. Stereotyped and repetitive motor behaviors, one of the core features of autism, have been suggested to serve as a self-medicating attempt at reducing hyper-responsive sympathetic activity (Hirstein et al. 2001). Symptoms suggestive of a deficiency in autonomic control in ASD indicate the feasibility for developing a set of biomarkers defining autonomic phenotypes. It is possible to propose that clearly defined autonomic arousal phenotypes may open a new perspective for behavioral, pharmacological, and neuromodulation (e.g., heart rate variability [HRV] biofeedback) interventions in autism. Eventually, as noted by Rees (2014), there is an urgent need to accelerate recognition of the importance of ANS activity investigation in pediatrics, not limiting it only to neurodevelopmental disorders.

5.3 *Connection Between Central and Autonomic Nervous Systems*

Thayer and Lane (2009) reviewed neuroanatomical and neuroimaging studies that implicate inhibitory GABAergic pathways from the prefrontal lobe to the limbic system and inhibitory pathways between the amygdala and the sympathetic and parasympathetic neurons in the medulla known to be involved in the modulation of HRV. This group of authors (Thayer and Lane 2000) described a neurovisceral integration model that is directly involved in the regulation of emotion and proposed a role for dysregulation that may result in various psychopathologies, including anxiety disorder (Friedman 2007) and potentially anxiety symptoms that are typical of children with ASD. Benarroch (1997) used the term "the central autonomic network (CAN)" and proposed connections of the CAN with the sinoatrial node of the heart via the stellate ganglia and the vagus nerve. As it was outlined by Thayer (2015), decreased tonic inhibitory output of the prefrontal cortex (PFC) can lead to disinhibition of the tonically inhibited structures under central autonomic control (i.e., CAN) and result in a simultaneous disinhibition of sympatho-excitatory neurons and an inhibition of parasympatho-excitatory neurons accompanied by an increase in HR and a concomitant decrease of vagally mediated HRV. Thayer et al. (2012) outlined that connections between the amygdala and medial PFC, which show that threat and safety regulate HRV through their connections with the NST. Furthermore, the CAN model proposes that vagally mediated HRV is linked to prefrontal executive functions and that HRV reflects the functional capacity of the PFC support emotional and physiological self-regulation. They hypothesize that parasympathetically mediated HRV is positively correlated with prefrontal cortical performance. In their model, when the prefrontal cortical functioning is decreased, HR increases and HRV decreases. Prolonged prefrontal cortical inactivity can lead to hypervigilance, defensiveness, and social isolation (Thayer and Lane 2005; Thayer et al. 2009). The CAN model predicts reduced HRV and hypofunctional vagal activity in anxiety, as it might be associated with abnormal ANS cardiac control (Friedman 2007). This challenges the sympathetic over-activation model of anxiety that underestimates the role of a hypofunctional parasympathetic system. From his perspective, disorders such as autism presenting with anxiety and dysregulated autonomic control can involve varying degrees of sympathetic over-activation and parasympathetic under-activation. Bal et al. (2010) suggested that children with ASD rely mostly on downregulation of sympathetic arousal as they fail to engage the parasympathetic system for self-regulation. Analysis of both central (e.g., frontal asymmetry indexes) and autonomic (HRV, SCR, respiration, etc.) biomarkers during exposure to effective stimuli and during neurofeedback training allows for a better understanding of their contribution to attentional engagement, emotional reactivity, and anxiety in autism.

6 Conclusions

In a clinical research study with neurofeedback training sessions in children with ASD and comorbid ADHD, we attempted to change the power of the gamma band and theta-to-beta ratio at midline prefrontal site to investigated whether these changes would influence behavioral symptoms. The results of our studies show that children with autism and attention deficit problems are indeed able to alter the power in the gamma band and theta-to-beta ratio if provided with neurofeedback. Our results also provide support for better expression of positive changes in behavioral rating scores following more extended length of neurofeedback course as compared to our previous neurofeedback trials in ASD and ADHD. At the same time, we showed decrease of theta-to-beta ratio during session and across the course of neurofeedback, and the "Focused Attention" index targeted by "InhibitAll" arm of the protocol did show statistically significant increases. Neurofeedback training was accompanied by detectable changes in monitored autonomic activity markers (SCL. HRV, respiration rate, etc.). Some of these psychophysiological indexes can be potentially considered as predictors of successful outcome of neurofeedback training. Neurofeedback had positive effects on ratings of aberrant behavior, attention deficit, and social responsiveness scores following 24 sessions of self-regulation training. Results of these pilot studies will be used to inform design of the controlled trials aimed to evaluate clinical efficacy of the neurofeedback method in ASD and ADHD.

Acknowledgements This study was supported by the GHS HSC 2018 Transformative Pilot Study Grant to Desmond Kelly and Estate Sokhadze.

References

Aman MG, Singh NN (1994) Aberrant behavior checklist - community. Supplementary manual. Slosson Educational Publications, East Aurora, NY

American Psychiatric Association (APA) (2000) Diagnostic and statistical manual of mental disorders (DSM-IV), 4th edn. American Psychiatric Association, Washington, DC

American Psychiatric Association (APA) (2013) Diagnostic and statistical manual of mental disorders (DSM-5), 5th edn. American Psychiatric Association, Washington, DC

Arns M, de Ridder S, Strehl U, Breteler M, Coenen A (2009) Efficacy of neurofeedback treatment in ADHD: the effects on inattention, impulsivity and hyperactivity: a meta-analysis. Clin EEG Neurosci 40(3):180–189

Arns M, Heinrich H, Strehl U (2014) Evaluation of neurofeedback in ADHD: the long and winding road. Biol Psychol 95:108–115

Baruth J, Casanova M, El-Baz A, Horrell T, Mathai G, Sears L, Sokhadze E (2010) Low-frequency repetitive transcranial magnetic stimulation modulates evokedgamma frequency oscillations in autism spectrum disorders. J Neurother 14(3):179–194

Bal E, Harden E, Lamb D, Van Hecke AV, Denver JW, Porges SW (2010) Emotion recognition in children with autism spectrum disorders: relations to eye gaze and autonomic state. J Autism Dev Disord 40(3):358–370

Belmonte MK, Yurgelun-Todd DA (2003) Functional anatomy of impaired selective attention and compensatory processing in autism. Cogn Brain Res 17:651–664

Benarroch EE (1997) The central autonomic network. In: Low PA (ed) Clinical autonomic disorders, 2nd edn. Lippincott-Raven, Philadelphia, pp 17–23

Bernston G, Bigger JT, Eckberg D, Grossman P, Kaufmann PG, Malik M et al (1997) Heart rate variability: origins, methods and interpretive caveates. Psychophysiology 34:623–648

Berntson G, Cacippo JY, Quigley K (1993) Cardiac psychophysiology and autonomic space in humans: empirical perspectives and implications. Psychol Bull 114:296–322

Bertrand O, Tallon-Baudry C (2000) Oscillatory gamma activity in humans: a possible role for object representation. Int J Psychophysiol 38(3):211–223

Bird BL, Newton FA, Sheer DE, Ford M (1978a) Biofeedback training of 40-Hz EEG in humans. Biofeedback Self Regul 3(1):1–11
Bird BL, Newton AD, Sheer E, Ford M (1978b) Behavioral and electroencephalographic correlates of 40 Hz EEG biofeedback training in humans. Biofeedback Self Regul 3(1):13–28
Bodfish JW, Symons FJ, Lewis J (1999) Repetitive behavior scale. Western Carolina Center Research Reports
Boucsein W (2012) Electrodermal activity, 2nd edn. Springer, New York
Brock J, Brown CC, Boucher J, Rippon G (2002) The temporal binding deficit hypothesis of autism. Dev Psychopathol 14(2):209–224
Brown C (2005) EEG in autism: is there just too much going on in there? In: Casanova MF (ed) Recent developments in autism research. Nova Science Publishers, New York, pp 109–126
Brown C, Gruber T, Boucher J, Rippon G, Brock J (2005) Gamma abnormalities during perception of illusory figures in autism. Cortex 41(3):364–376
Cantor DS, Thatcher RW, Hrybyk M, Kaye H (1986) Computerized EEG analysis of autistic children. J Autism Dev Disord 16(2):169–187
Casanova MF, Buxhoeveden D, Gomez J (2003) Disruption in the inhibitory architecture of cell minicolumn: implications for autism. Neuroscientist 9:496–507
Casanova MF, Baruth J, El-Baz AS, Sokhadze GE, Hensley M, Sokhadze EM (2013) Evoked and induced gamma-frequency oscillation in autism. In: Casanova MF, El-Baz AS, Suri JS (eds) Imaging the brain in autism. Springer, New York, pp 87–106
Casanova MF, Sokhadze E, Opris I, Wang Y, Li X (2015) Autism spectrum disorders: linking neuropathological findings to treatment with transcranial magnetic stimulation. Acta Pediatr 104(4):346–355
Casanova MF, van Kooten IA, van Engeland H, Heinsen H, Steinbursch WM, Hof PR et al. (2006) Minicolumnar abnormalities in autism. Acta Neuropathol 112:287–303
Coben R (2008) Autistic spectrum disorder: a controlled study of EEG coherence training focused on social skills deficits. J Neurother 12:57–75
Coben R (2013) Neurofeedback for autistic disorders: emerging empirical evidence. In: Casanova MF, El-Baz AS, Suri J (eds) Imaging the brain in autism. Springer, New York, NY, pp 107–134
Coben R, Myers TE (2010) The relative efficacy of connectivity guided and symptom based EEG biofeedback for autistic disorders. Appl Psychophysiol Biofeedback 35(1):13–23
Coben R, Padolsky I (2007) Assessment-guided neurofeedback for autistic spectrum disorder. J Neurother 11(1):5–23
Coben R, Ricca R (2015) EEG biofeedback for autism spectrum disorder: a commentary on Kouijzer et al. (2013). Appl Psychophysiol Biofeedback 40(1):53–56
Coben R, Linden M, Myers TE (2010) Neurofeedback for autistic spectrum disorder: a review of the literature. Appl Psychophysiol Biofeedback 35(1):83–105
Coben R, Sherlin L, Hudspeth WJ, McKeon K, Ricca R (2014) Connectivity-guided EEG biofeedback for autism spectrum disorder: evidence of neurophysiological changes. NeuroRegulation 1(2):109–130
Cohen S, Masyn K, Mastergeorge A, Hessl D (2015) Psychophysiological responses to emotional stimuli in children and adolescents with autism and fragile X syndrome. J Clin Child Adolesc Psychol 44(2):250–263
Constantino JN, Gruber CP (2005) The Social Responsiveness Scale (SRS) Manual. Los Angeles, CA: Western Psychological Services
Cornew L, Roberts TP, Blaskey L, Edgar JC (2012) Resting-state oscillatory activity in autism spectrum disorders. J Autism Dev Disord 42(9):1884–1894
Datko M, Pineda JA, Müller RA (2017) Positive effects of neurofeedback on autism symptoms correlate with brain activation during imitation and observation. Eur J Neurosci 47(6):579–591
Engel AK, Singer W (2001) Temporal binding and the neural correlates of sensory awareness. Trends Cogn Sci 5(1):16–25
Enriquez-Geppert S, Huster RJ, Herrmann CS (2017) EEG-neurofeedback as a tool to modulate cognition and behavior: a review tutorial. Front Hum Neurosci 11:51. https://doi.org/10.3389/fnhum.2017.00051

Fell J, Klaver P, Lehnertx K, Grunwald T, Schaller C, Elger CE, Fernandez G (2001) Human memory formation is accompanied by rhinal-hippocampal coupling and decoupling. Nat Neurosci 4(12):1259–1264
Ford M, Bird BL, Newton FA, Sheer DE (1980) Maintenance and generalization of 40-Hz EEG biofeedback effects. Biofeedback Self Regul 5(2):193–205
Frazier TW, Shattuck PT, Narendorf SC, Cooper BP, Wagner M, Spitznagel EL (2011) Prevalence and correlates of psychotropic medication use in adolescents with an autism spectrum disorder with and without caregiver-reported attention-deficit/hyperactivity disorder. J Child Adolesc Psychopharmacol 21(6):571–579
Friedman BH (2007) An autonomic flexibility--neurovisceral integration model of anxiety and cardiac vagal tone. Biol Psychol 74(2):185–199
Friedrich EV, Suttie N, Sivanathan A, Lim T, Louchart S, Pineda JA (2014) Brain-computer interface game applications for combined neurofeedback and biofeedback treatment for children on the autism spectrum. Front Neuroeng 7:21
Friedrich EV, Sivanathan A, Lim T, Suttie N, Louchart S, Pillen S, Pineda JA (2015) An effective neurofeedback intervention to improve social interactions in children with autism spectrum disorder. J Autism Dev Disord 45(12):4084–4100
Frith U, Happé F (1994) Autism: beyond theory of mind. Cognition 50:115–132
Gadow KD, DeVincent CJ, Schneider J (2009) Comparative study of children with ADHD only, autism spectrum disorder + ADHD, and chronic multiple tic disorder + ADHD. J Atten Disord 12(5):474–485
Gaume A, Vialatte A, Mora-Sánchez A, Ramdani C, Vialatte FB (2016) A psychoengineering paradigm for the neurocognitive mechanisms of biofeedback and neurofeedback. Neurosci Biobehav Rev 68:891–910
Grice SJ, Spratling MW, Karmiloff-Smith A, Halit H, Csibra G, de Haan M, Johnson MH (2001) Disordered visual processing and oscillatory brain activity in autism and Williams syndrome. Neuroreport 12(12):2697–2700
Herrmann CS, Mecklinger A (2000) Magnetoencephalographic responses to illusory figures: early evoked gamma is affected by processing of stimulus features. Int J Psychophysiol 38(3):265–281
Herrmann CS, Mecklinger A (2001) Gamma activity in human EEG is related to high-speed memory comparisons during object selective attention. Vis Cogn 8(3–5):593–608
Herrmann CS, Munk MHJ, Engel AK (2004) Cognitive functions of gamma-band activity: memory match and utilization. Trends Cogn Sci 8(8):347–355
Hillard B, El-Baz AS, Sears L, Tasman A, Sokhadze EM (2013) Neurofeedback training aimed to improve focused attention and alertness in children with ADHD: a study of relative power of EEG rhythms using custom-made software application. Clin EEG Neurosci 44(3):193–202
Hirstein W, Iversen P, Ramachandran VS (2001) Autonomic responses of autistic children to people and objects. Proc Biol Sci 268(1479):1883–1888
Hurt E, Arnold LE, Lofthouse N (2014) Quantitative EEG neurofeedback for the treatment of pediatric attention-deficit/hyperactivity disorder, autism spectrum disorders, learning disorders, and epilepsy. Child Adolesc Psychiatr Clin N Am 23(3):465–486
Jarusiewicz B (2002) Efficacy of neurofeedback for children in the autistic spectrum. A pilot study. J Neurother 6(4):39–49
Kahana MJ (2006) The cognitive correlates of human brain oscillations. J Neurosci 26(6):1669–1672
Kaiser J (2003) Induced gamma-band activity and human brain function. Neuroscientist 9(6):475–484
Keizer AW, Verschoor M, Verment RS, Hommel B (2010a) The effect of gamma enhancing neurofeedback on the control of feature bindings and intelligence measures. Int J Psychophysiol 75(1):25–32
Keizer AW, Verment RS, Hommel B (2010b) Enhancing cognitive control through neurofeedback: a role of gamma-band activity in managing episodic retrieval. NeuroImage 49(4):3404–3413
Kerson C, Collaborative Neurofeedback Group (2013) A proposed multisite double-blind randomized clinical trial of neurofeedback for ADHD: need, rationale, and strategy. J Atten Disord 17(5):420–436

Klusek J, Roberts JE, Losh M (2015) Cardiac autonomic regulation in autism and fragile X syndrome: a review. Psychol Bull 141:141–175

Kober SE, Witte M, Ninaus M, Neuper C, Wood G (2013) Learning to modulate one's own brain activity: the effect of spontaneous mental strategies. Front Hum Neurosci 7:695

Kober SE, Witte M, Neuper C, Wood G (2017) Evaluation of band, baseline, and cognitive specificity of sensorimotor rhythm- and gamma-based neurofeedback: specific or nonspecific? Int J Psychophysiol 120:1–13

Kouijzer M, de Moor J, Gerrits B, Congedo M, van Schie HT (2009a) Neurofeedback improves executive functioning in children with autism spectrum disorders. Res Autism Spectr Disord 3(1):145–162

Kouijzer MEJ, de Moor JMH, Gerrits BJL, Buitelaar JK, van Schie HT (2009b) Long-term effects of neurofeedback treatment in autism. Res Autism Spectr Disord 3:496–501

Kouijzer MEJ, Van Schie HT, De Moor JMH, Gerrits BJL, Buitelaar JK (2010) Neurofeedback treatment in autism. Preliminary findings in behavioral, cognitive, and neurophysiological functioning. Res Autism Spectr Disord 4:386–389

Kushki A, Drumm E, Mobarak MP, Tanel N, Dupius A, Chau T, Anagnostou E (2013) Investigating the autonomic nervous system response to anxiety in children with autism spectrum disorders. PLoS One 8(4):e59730

Kushki A, Brian J, Dupius A, Anagnostou E (2014) Functional autonomic nervous system profile in children with autism spectrum disorder. Mol Autism 5:39

Lansbergen MM, Arns M, van Dongen-Boomsma M, Spronk D, Buitelaar JK (2011) The increase in theta/beta ratio on resting-state EEG in boys with attention-deficit/hyperactivity disorder in mediated by slow alpha peak frequency. Prog Neuropsychopharmacol Biol Psychiatry 35:47–52

Le Couteur A, Lord C, Rutter M (2003) The autism diagnostic interview – revised (ADI-R). Western Psychological Services, Los Angeles, CA

Leitner Y (2014) The co-occurrence of autism and attention deficit hyperactivity disorder in children – what do we know? Front Hum Neurosci 8:268

Linden M, Gunkelman J (2013) QEEG-guided neurofeedback for autism: clinical observations and outcomes. In: Casanova MF, El-Baz AS, Suri JS (eds) Imaging the brain in autism. Springer, New York, pp 45–60

Llinas R, Ribary U (1993) Coherent 40-Hz oscillation characterizes dream state in humans. Proc Natl Acad Sci U S A 90(5):2078–2081

Lofthouse N, Arnold LE, Hurt E (2010) A comment on Sherlin, Arns, Lubar, and Sokhadze. J Neurother 14: 301-306

Lord K, Rutter M (2005) Autism diagnostic observation schedule, 2nd edn. WPS, Torrance, CA

Loring DW, Sheer DE, Largen JW (1985) Forty Hertz EEG activity in dementia of the Alzheimer type and multi-infarct dementia. Psychophysiology 22(1):116–121

Milne E, Scope A, Pascalis O, Buckley D, Makeig S (2009) Independent component analysis reveals atypical electroencephalographic activity during visual perception in individuals with autism. Biol Psychiatry 65(1):22–30

Ming X, Bain JM, Smith D, Brimacombe M, Gold von-Simson G, Axelrod FB (2011) Assessing autonomic dysfunction symptoms in children: a pilot study. J Child Neurol 26(4):420–427

Ming X, Patel R, Kang V, Chokroverty S, Julu PO (2016) Respiratory and autonomic dysfunction in children with autism spectrum disorders. Brain Dev 38(2):225–232

Monastra VJ, Linden M, Lubar JF, Vandeusen P (1999) Assessing attention deficit hyperactivity disorder via quantitative electroencephalography: an initial validation study. Neuropsychology 13:424–433

Monastra VJ, Lubar JF, Linden M, York N (2001) The development of a quantitative electroencephalographic scanning process for attention deficit-hyperactivity disorder: reliability and validity studies. Neuropsychology 15:136–144

Murias M, Webb SJ, Greenson J, Dawson G (2007) Resting state cortical connectivity reflected in EEG coherence in individuals with autism. Biol Psychiatry 62(3):270–273

Niv S (2013) Clinical efficacy and potential mechanisms of neurofeedback. Personal Individ Differ 54(6):676–686

Orekhova EV, Stroganova TA, Nygren G, Tsetlin M, Posikera I, Gillberg C, Elam M (2007) Excess of high frequency electroencephalogram oscillations in boys with autism. Biol Psychiatry 62(9):1022–1029
Pigott HE, Cannon R (2014) Neurofeedback is the best available first-line treatment for ADHD: what is the evidence for this claim? NeuroRegulation 1(1):4–23
Pigott HE, Cannon R, Trullinger M (2018) The fallacy of sham-controlled neurofeedback trials: a reply to Thibault and colleagues (2018). J Atten Disord. https://doi.org/10.1177/1087054718790802
Pineda JA, Juavinett A, Datko M (2012) Self-regulation of brain oscillations as a treatment for aberrant brain connections in children with autism. Med Hypotheses 79(6):790–798
Pineda JA, Friedrich EV, LaMarca K (2014a) Neurorehabilitation of social dysfunctions: a model-based neurofeedback approach for low and high-functioning autism. Front Neuroeng 7:29
Pineda JA, Carrasco K, Datko M, Pillen S, Schalles M (2014b) Neurofeedback training produces normalization in behavioral and electrophysiological measures of high-functioning autism. Philos Trans R Soc Lond Ser B Biol Sci 369(1644):20130183
Rees CA (2014) Lost among trees? The autonomic nervous system and paediatrics. Arch Dis Child 99(6):552–562
Rippon G, Brock J, Brown C, Boucher J (2007) Disordered connectivity in the autistic brain: challenges for the "new psychophysiology". Int J Psychophysiol 63(2):164–172
Rojas DC, Wilson LB (2014) γ-Band abnormalities as markers of autism spectrum disorders. Biomark Med 8(3):353–368
Ros T, Baars BJ, Lanius RA, Vuilleumier P (2014) Tuning pathological brain oscillations with neurofeedback: a systems neuroscience framework. Front Hum Neurosci 8:1008
Salari N, Büchel C, Rose M (2014) Neurofeedback training of gamma band oscillations improves perceptual processing. Exp Brain Res 232(10):3353–3361
Sedley W, Cunningham MO (2013) Do cortical gamma oscillations promote or suppress perception? An under-asked question with an over-assumed answer. Front Hum Neurosci 7:595
Sheer DE (1989) Focused arousal and the cognitive 40-Hz event-related potentials: differential diagnosis of Alzheimer's disease. Prog Clin Biol Res 317:79–94
Sherlin L, Arns M, Lubar J, Sokhadze E (2010) A position paper on neurofeedback for the treatment of ADHD. J Neurother 14(2):66–78
Sherlin L, Arns M, Lubar J, Heinrich H, Kerson C, Strehl U, Sterman MB (2011) Neurofeedback and basic learning theory: implications for research and practice. J Neurother 15(4):292–304
Sokhadze E (2012) Peak performance training using prefrontal EEG biofeedback. Biofeedback 40(1):7–15
Sokhadze E, El-Baz A, Baruth J, Mathai G, Sears L, Casanova MF (2009) Effects of a low-frequency repetitive transcranial magnetic stimulation (rTMS) on gamma frequency oscillations and event-related potentials during processing of illusory figures in autism. J Autism Dev Disord 39(4):619–634
Sokhadze EM, El-Baz AS, Tasman A, Sears LL, Wang Y, Lamina EV, Casanova MF (2014) Neuromodulation integrating rTMS and neurofeedback for the treatment of autism spectrum disorder: an exploratory study. Appl Psychophysiol Biofeedback 39(3–4):237–257
Sokhadze EM, Casanova MF, Casanova EL, Klusek J, Roberts J (2019) Autonomic nervous system dysfunctions in children with autism spectrum disorder. In: Sokhadze EM, Casanova MF (eds) Autism spectrum disorder: neuromodulation, neurofeedback and sensory integration approaches to research and treatment. FNNR, Murfreesboro, TN, pp 169–206
Sokhadze EM, Lamina EV, Casanova EL, Kelly DP, Opris I, Tasman A, Casanova MF (2018) Exploratory study of rTMS neuromodulation effects on electrocortical functional measures of performance in an oddball test and behavioral symptoms in autism. Front Syst Neurosci 12:20
Staufenbiel SM, Brouwer AM, Keizer AW, van Wouwe NC (2014) Effect of beta and gamma neurofeedback on memory and intelligence in the elderly. Biol Psychol 95:74–85
Strehl U (2014) What learning theories can teach us in designing neurofeedback treatments. Front Hum Neurosci 8:894

Stroganova TA, Orekhova EV, Prokofyev AO, Tsetlin MM, Gratchev VV, Morozov AA, Obukhov YV (2012) High-frequency oscillatory response to illusory contour in typically developing boys and boys with autism spectrum disorders. Cortex 48:701–717
Stroganova TA, Butorina AV, Sysoeva OV, Prokofyev AO, Nikolaeva AY, Tsetlin MM, Orekova EV (2015) Altered modulation of gamma oscillation frequency by speed of visual motion in children with autism spectrum disorders. J Neurodev Disord 7(1):21
Tallon-Baudry C, Bertrand O, Delpuech C, Pernier J (1996) Stimulus specificity of phase-locked and non-phase-locked 40 Hz visual responses in human. J Neurosci 16(13):4240–4249
Thayer JF (2015) A neurovisceral integration perspective. The 46th annual meeting of the Association for Applied Psychophysiology and Biofeedback. Austin, TX, March 14
Thayer JF, Lane RD (2000) A model of neurovisceral integration in emotion regulation and dysregulation. J Affect Disord 61(3):201–216
Thayer JF, Lane RD (2005) The importance of inhibition in dynamical systems models of emotion and neurobiology. Brain Behav Sci 28(2):218–219
Thayer JF, Lane RD (2009) Claude Bernard and the heart–brain connection: further elaboration of a model of neurovisceral integration. Neurosci Biobehav Rev 33(2):81–88
Thayer JF, Hansen AL, Saus-Rose E, Johnsen BH (2009) Heart rate variability, prefrontal neural function, and cognitive performance: the neurovisceral integration perspective on self-regulation, adaptation, and health. Ann Behav Med 37(2):141–153
Thayer JF, Ahs F, Fredrikson M, Sollers JJ, Wager TD (2012) A meta-analysis of heart rate variability and neuroimaging studies: implications for heart rate variability as a marker of stress and health. Neurosci Biobehav Rev 36(2):747–756
Thibault RT, Veissière S, Olson JA, Raz A (2018) Treating ADHD with suggestion: Neurofeedback and placebo therapeutics. J Atten Disord 22(8):707–711
Thompson M, Thompson L (2013) The rationale for using EEG biofeedback for clients with Asperger's syndrome. Appl Psychophysiol Biofeedback 35(1):39–61
Uzunova G, Pallanti S, Hollander E (2016) Excitatory/inhibitory imbalance in autism spectrum disorders: implications for interventions and therapeutics. World J Biol Psychiatry 17(3):174–186
Vollebregt MA, van Dongen-Boomsma M, Slaats-Willemse D, Buitelaar JK (2014) What future research should bring to help resolving the debate about the efficacy of EEG-neurofeedback in children with ADHD. Front Hum Neurosci 8:321
Von Stein A, Rappelsberger P, Sarnthein J, Petsche H (1999) Synchronization between temporal and parietal cortex during multimodal object processing in man. Cereb Cortex 9(2):137–150
Wang Y, Sokhadze EM, El-Baz AS, Li X, Sears L, Casanova MF, Tasman A (2016) Relative power of specific EEG bands and their ratios during neurofeedback training in children with autism spectrum disorder. Front Hum Neurosci 9:723
Wechsler D (1999) Wechsler abbreviated scale of intelligence. Harcourt Assessment, Inc., San Antonio, TX
Wechsler D (2004) Wechsler intelligence scale for children - fourth edition integrated (WISC-IV integrated). Harcourt, San Antonio, TX
Wilson TW, Rojas DC, Reite ML, Teale PD, Rogers SJ (2007) Children and adolescents with autism exhibit reduced MEG steady-state gamma responses. Biol Psychiatry 62(3):192–197
Zivoder I, Martic-Biocina S, Kosic AV, Bosak J (2015) Neurofeedback application in the treatment of autistic spectrum disorders (ASD). Psychiatr Danub 27(1):391–394
Zuberer A, Brandeis D, Drechsler R (2015) Are treatment effects of neurofeedback training in children with ADHD related to the successful regulation of brain activity? A review on the learning of regulation of brain activity and a contribution to the discussion on specificity. Front Hum Neurosci 9:135

Part IV
Futuristic Approaches to Augmentation

Augmentation Through Interconnection: Brain-Nets and Telemedicine

Mikhail A. Lebedev, Igor A. Shaderkin, Ilia V. Ryabkov, and Georgy S. Lebedev

1 Brain-Nets

Brain-nets are a futuristic development in neuroengineering that has evolved from BCIs. A traditional BCI records neural activity in a single brain, extracts information of interest from it, and converts it into signals that are then sent to an external device, such a prosthetic limb, an exoskeleton, or an electrical stimulator of muscles (Lebedev 2014). A class of BCIs, called sensory BCIs, have the capacity of delivering sensory information from an external device to the brain, for example, delivering touch sensations arising from the sensors of a virtual or prosthetic hand using intracortical microstimulation (ICMS) of the somatosensory cortex (O'Doherty et al. 2011; Flesher et al. 2016). BCIs that simultaneously extract information from the brain and deliver information back to the brain are called bidirectional BCIs (Fagg et al. 2009; Lebedev 2016; Rao 2019). Brain-nets can both decode brain activity and deliver feedback, and they differ from traditional BCIs in that they

M. A. Lebedev (✉)
Center for Neuroengineering, Duke University, Durham, NC, USA

Center for Bioelectric Interfaces of the Institute for Cognitive Neuroscience, National Research University Higher School of Economics, Moscow, Russia

Department of Information and Internet Technologies of Digital Health Institute, I.M. Sechenov First Moscow State Medical University, Moscow, Russia

I. A. Shaderkin · I. V. Ryabkov
Department of Information and Internet Technologies of Digital Health Institute, I.M. Sechenov First Moscow State Medical University, Moscow, Russia

G. S. Lebedev (✉)
Department of Information and Internet Technologies of Digital Health Institute, I.M. Sechenov First Moscow State Medical University, Moscow, Russia

Federal Research Institute for Health Organization and Informatics, Moscow, Russia

I. Opris et al. (eds.), *Modern Approaches to Augmentation of Brain Function*, Contemporary Clinical Neuroscience,
https://doi.org/10.1007/978-3-030-54564-2_16

incorporate multiple brains instead of one. Brain-nets can enact tasks where subjects cooperate (Ramakrishnan et al. 2015) and/or communicate using direct links between their brains (Pais-Vieira et al. 2015). To emphasize the communication aspect of brain-nets, the term "brain-to-brain interface" (BTBI) is often used in the literature (Pais-Vieira et al. 2015). In one study, a brain-net was called an organic computer because it performed computer-like operations and enabled memory storage (Pais-Vieira et al. 2015).

The field of brain-nets is currently at the very initial stage of its development, and the published demonstrations of brain-nets are relatively simple. Yet, philosophers have been already alerted by this development, which prompted publications on the ethical issues of using brain-nets and BTBIs (Trimper et al. 2014; Hildt 2015; Hongladarom 2015). Historically, multibrain BCIs were first demonstrated using noninvasive recordings of brain activity (Nijholt 2015). The pioneering work on a BTBI that was based on invasive cortical implants was conducted by Pais-Vieira et al. (2013) who experimented with the possibility of direct information exchange between the rat brains. Two rats were housed in separate boxes. In several experimental sessions, the animals were located many thousands of miles apart and their brains communicated via the internet. One rat, called encoder, provided the neural signal that was sent to the other rat, called decoder. The encoder rat was engaged in a behavioral task that required responding to somatosensory or visual cues either by making a choice by a nose-poke in one of two possible locations or by pressing one of two levers. The rats were implanted with microelectrode arrays placed in the somatosensory and motor cortical areas. Spikes of cortical neuronal populations recorded in the encoder rate were processed using a sigmoid transform function and converted into the frequency of electrical pulses delivered to the cortex implant of the decoder rat. Both the encoder and decoder rats were rewarded for correct performance. Prior to the start of the BTBI experiments, the decoder rat underwent training to perform the same behaviors tasks as the encoder rat, with an important difference that ICMS, instead of the natural sensory stimuli, cued the responses. Given this overtraining of the decoder rat to respond to ICMS, it was not particularly surprising that this rat continued responding correctly when ICMS represented task-related cortical activity of the encoder rat. Yet, the experiment incorporated a sophistication: the encoder rat received feedback from the decoder. Each time the decoder rat responded correctly the encoder rat received an additional reward. Owing to this additional reinforcement, the encoder rat adapted its behavioral responses to send "clearer" messages for the decoder rat.

Following this pioneering work on an invasive BTBI, Pais-Viera and his colleagues added more sophistications to their experiments (Pais-Vieira et al. 2015). In the new experiment, four animals performed several BTBI tasks. Information was injected into the brain of each rat using ICMS applied to the primary somatosensory cortex. The somatosensory area of the hemisphere contralateral to the stimulated one was used for neural recordings and information extraction. The brain-net composed of the brains of four rats was called an organic computer because several computational algorithms were enabled, including simultaneous responses of multiple animals to ICMS, discrimination of ICMS patterns, and keeping information

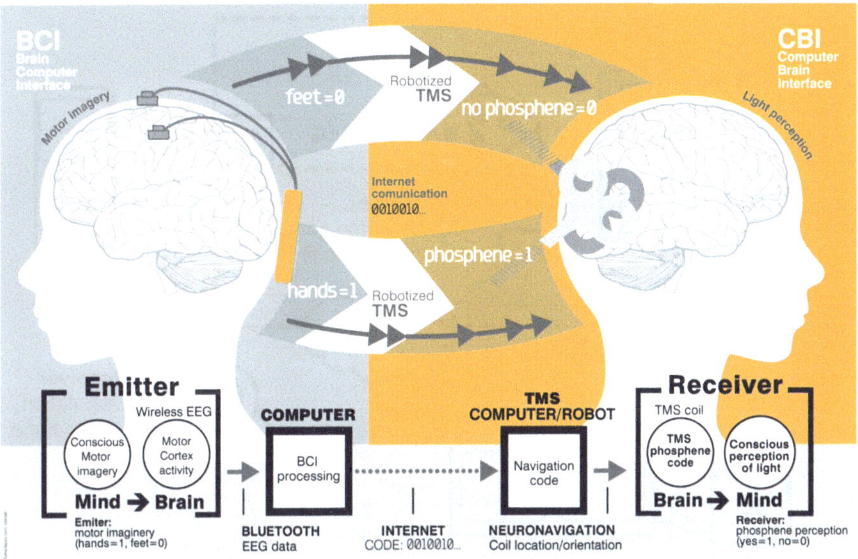

Fig. 1 A brain-to-brain interface connecting two human subjects. EEG activity was recorded in the emitter subject (shown on the left) while this subject performed a motor imagery task with two choices: 0 when imagining movements of the feet, and 1 when imagining movements of the hands. The receiver subject (on the right) received TMS in his visual cortex; a TMS pulse that evoked a phosphene corresponded to 1. The connection between the subjects was maintained by an internet link. (Reproduced from Grau et al. (2014), PLoS One 9(8): e105225)

in a memory buffer constructed of serially interconnected brains. While some of these tasks could be performed by a single rat, the brain-net composed of four animals outperformed a single rat.

The studies of Pais-Viera and his colleagues were highly impactful. Several research groups conducted experiments that reproduced these results in human subjects using noninvasive methods for recordings and stimulation. Grau and his colleagues connected the sensorimotor cortex of a human subject, called emitter, to the visual cortex of the other subject, called receiver (Grau et al. 2014) (Fig. 1). Cortical activity of the emitter was sampled using electroencephalography (EEG), and the receiver's cortex was activated by pulses of transcranial magnetic stimulation (TMS). The emitter performed a motor-imagery BCI task where he generated a binary message ("0" or "1") by imagined moving the feet or hands, respectively. The BCI detected the patterns of cortical activity associated with these types of imagery and decoded them into zeros or ones. This output was sent to the computer that controlled the TMS device. Depending on the message, a robotic manipulator positioned the TMS coil over the visual-cortex site whose stimulation evoked a perception of a flash of light (phosphene) or over the site where no phosphene was produced. This BTBI operated slowly (at rate of 2–3 bits per minute) and did not contain a feedback loop from the receiver to the emitter.

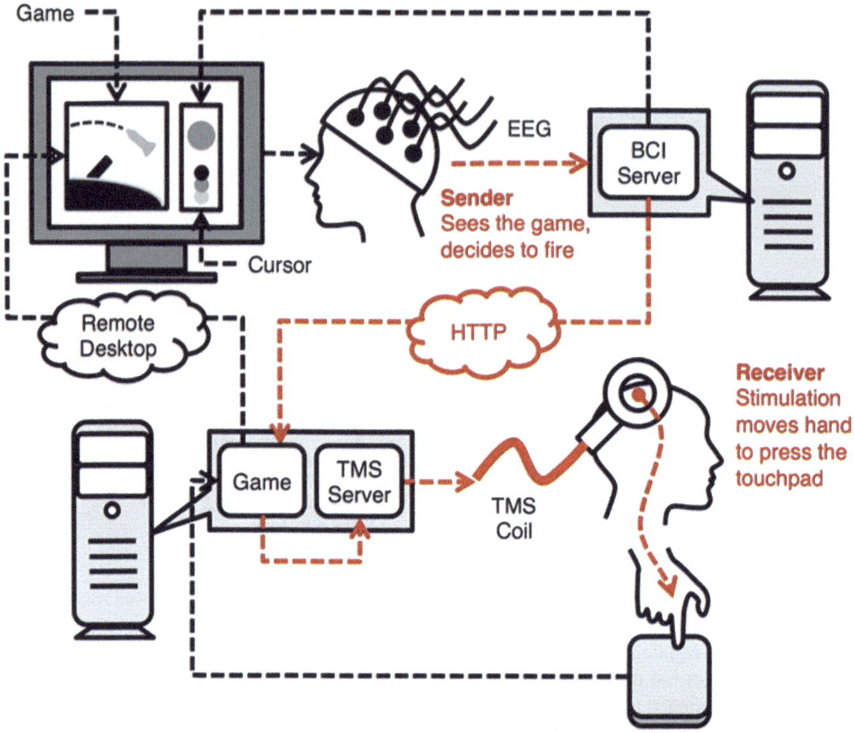

Fig. 2 Transmission of a motor command using a brain-to-brain interface. Two human subjects participated in these experiments: a sender and a receiver. The sender performed a motor-imagery BCI task where he imagined hand movements to issue a "fire" command. A BCI converted the sender's EEG decoded that command and sent it to a TMS device that stimulated the receiver's motor cortex. TMS pulses evoked finger movements that were detected by a touchpad. The touchpad signal in turn evoked an event in the BCI game played by the sender. (Reproduced from Rao et al. (2014), PLoS One 9(11): e111332)

Rao and his colleagues conducted a similar BTBI study with a difference that finger movements instead of visual perceptions were evoked with TMS (Rao et al. 2014) (Fig. 2). In this study, EEG of one subject was converted into a "fire" command that was delivered to the second subject's motor cortex as a TMS pulse. The TMS evoked a finger movement and touchpad click. Thus, in this experiment, the subject on the receiving end acted as a "slave," and the emitter was a "master" that commanded a movement of a different person. Overall, similar to the experiments of Pais-Vieira et al. (2013, 2015), the BTBI experiments in humans (Grau et al. 2014; Rao et al. 2014) had a relatively simple design, and even in more recent studies with an increased number of human participants (Stocco et al. 2015; Jiang et al. 2019), the design can be considered simple because the brain-to-brain connections were used for low-rate transfer of one-bit messages—something that could be easily achieved by traditional ways of communication like pressing a button. Thus, more meaningful usages of BTBI are something that should be expected in the future.

In addition to noninvasive BTBIs in humans, master-to-slave transmission has been demonstrated using monkey invasive implants (Shanechi et al. 2014). In this study, neuronal activity recorded in the premotor cortex of an awake monkey was decoded and sent to the spinal cord of the second, anesthetized monkey. The second monkey was called avatar. The avatar's hand was attached to a joystick. The master monkey controlled a BCI where the monkey intended to move a computer cursor toward a screen target but did not move its arm. Since neuronal activity recorded in premotor cortex was modulated depending on the direction of intended movement, movement direction could be decoded from that activity. Next, the decoded directional signal was converted into a pattern of electrical stimulation that was applied to the spinal cord of the avatar monkey and evoked joystick movement in the same direction. Thus, the anesthetized monkey reproduced the movement that the master monkey intended to perform.

BTBIs interconnecting different species have been demonstrated, as well. Yoo and his colleagues connected the brain of a human subject to the spinal cord of a rat; the rat was anesthetized (Yoo et al. 2013). The human controlled a BCI that was based on steady state visual evoked potentials (SSVEPs) and issued a "go" command. The command then triggered a transcranial focused ultrasound applied to the rat's motor cortex. This stimulation method induced a movement of the rat's tail. Li and his colleagues conducted a study where a human operated a BCI that governed locomotion of a Madagascar hissing cockroach in an S-shaped track (Li and Zhang 2016). The guidance was performed by electrical stimulation of the cockroach antennae.

In the study conducted by Folcher and his colleagues, the human brain was connected to cell culture (Folcher et al. 2014). This brain-to-culture interface controlled gene expression, as claimed by the authors. Indeed, the output of an EEG-based BCI, operated by a human, triggered an optogenetic paradigm where optical stimulation of designer cells controlled their expression.

In addition to intersubject communications, brain-nets can be used to enable cooperative tasks. Ramakrishnan and his colleagues reported several designs of brain-net where monkeys performed BCI tasks cooperatively (Ramakrishnan et al. 2015). Two or three monkeys participated in each experiment. The monkeys were chronically implanted with multielectrode arrays that sampled activity of large neuronal populations (several hundred neurons recorded simultaneously). The monkeys collectively controlled the movements of a virtual arm displayed on a computer screen by their cortical activity. Three brain-net types were tested. The first brain-net performed a shared-control, where cortical activities recorded in two monkey cortices were combined to move the virtual arm. Separate BCIs generated the arm coordinates from the cortical activity of each monkey. Next, the outputs of BCIs were combined, which resulted in an improvement of the performance because the noisy outputs of both BCIs canceled each other. In the second brain-net, two monkeys were engaged, as well. The performance improved because each monkey was assigned a separate, one-dimensional task. The first monkey was assigned control of the X-coordinate of the virtual arm, whereas the second monkey was assigned control of the Y-coordinate. With these easy tasks, two monkeys performed better than each of them on the two-dimensional task. In the third brain-net, three animals were engaged. Each monkey had a two-dimensional display where the position of the cursor was displayed, but all three monkeys together controlled the movements of the virtual arm in three dimensions. Since individual monkeys were unaware of the three-dimensional task, this brain-net can be considered to be a "super-brain," where only the net of interconnected brains could handle high-order tasks, not the single brains.

Several brain-nets for performing cooperative tasks were implemented in humans using noninvasive recordings (Nijholt 2015). Thus, Poli and his colleagues demonstrated a brain-net where two human subjects cooperatively controlled a spacecraft (Poli et al. 2013). Additionally, cooperative brain-nets have been demonstrated for movement planning (Wang and Jung 2011; Yuan et al. 2013) and decision-making (Eckstein et al. 2012; Yuan et al. 2013; Poli et al. 2014; Mirabella and Lebedev 2017).

Overall, these demonstrations of brain-nets can be considered as a step toward transition of science fiction into reality (Lebedev et al. 2018b). However, skeptics could object to the expectation that brain-nets would evolve into something practical. The major issue with the existing brain-nets is that they do not enable high-throughput transfer of information between the participating subjects. Thus, it will be critical for the advancement of this field to increase the amount of information transferred through the communication channels connecting individual brains. This can be achieved by improving neural decoding and stimulation methods, increasing the number of channels, and increasing the numbers of brain-net participants. Additionally, clinically relevant applications of brain-nets will have to emerge, for example, bioelectrical interfaces where both a patient and therapist participate.

2 Telemedicine for Neurological Disorders

Telemedicine bears relevance to brain-to-brain interfaces in that this is also a technology that connects people. The American Telemedicine Association defines telemedicine and telehealth as "the use of medical information exchanged from one site to another via electronic communications to improve the patient's health status" (Lustig 2012). Modern usages of telemedicine include electronic document transfer between the health-care provider, patients and medical organization, information technologies, and application of mathematical methods to medical data (Ekeland et al. 2010; Dorsey and Topol 2016). For a telemedicine system to be effective, it should enable a set of interactive tests and methods for patient diagnostics and monitoring. For example, such a system could incorporate augmented-reality and virtual-reality components that assist distant communication with patients. Additionally, participating personnel should be well trained to perform medical service enabled by telemedicine tools.

Telemedicine is applicable to treatment of neurological disorders. Here, we describe a telemedicine system that we are developing for autism spectral disorder (ASD). ASD is a significant global problem that affects 67 million people worldwide (~1% of the total population) (Elsabbagh et al. 2012; Kopetz and Endowed 2012; APA 2013). ASD is characterized by a person's inability to interact socially and stereotyped behaviors that lead to social maladaptation. Patients suffering from ASD exhibit phobias, somatic symptoms, and eating disorder (code 299.00/F84.0) (WHO 2004; APA 2013). The sooner ASD symptoms are noticed in a child and the sooner the child is engaged in treatment and rehabilitation, the better the prognosis for successful social adaptation. Thus, ASD is a disease that requires constant observation and complex training to achieve social adaptation. ASD typically progresses over time, with complications arising, including epilepsy, catatonia, mutism, catalepsy, and wax flexibility (Wing and Shah 2000; Fatemi and Clayton 2008). Overall, it is challenging to provide care ASD patients, and in this aspect, telemedicine can greatly help.

An information resource is an essential part of a telemedicine system for treating ASD. Such a resource contains a range of technologies and enables communications between the child suffering from ASD, his/her parents, and medical and social workers. At the Institute of digital medicine (ICM) of Sechenov University, we have developed an information resource for monitoring families that have children with ASD. Among the information technologies, this resource supports deep-learning neural network trained to recognize the behavior and facial expressions of a child, especially deviations in behavior. The system also supports virtual reality and augmented reality tools that facilitate the child social adaptation. These implementations are based on the Internet of things methods for management of mobile health devices. Physicians use this telemedicine technology for remote monitoring of the child's symptoms. The information resource essentially constitutes a telemedicine hospital. Similar to a regular hospital, telemedicine hospital has a set of rules, such as requirements for the delivery of remote medical care, rules for medical care contracts, procedures for collecting and handling of medical records, and documentation of information technologies and mathematical methods.

The telemedicine system for ASD strives to achieve several the following essential goals: (1) recognition and prevention of epileptic attack and catatonias, (2) monitoring and correction of behavior, (3) improving socialization and communication, and (4) early diagnostics of a child and monitoring his/her development. A portable EEG device is used for monitoring of epileptic seizures. Additionally, the child wears a wrist tracker that performs monitoring of physiological activities, such as onset and phases of sleep and night rising. The wrist tracker incorporates iBeacon indoor technology for determining the device location in the house. Additionally, the wrist tracker enables photoplethysmography, monitoring of heart rate, blood pressure, and skin moisture. Moreover, an infrared ear thermometer monitors body temperature. Child's movements and facial expressions are monitored with a video camera. The device management software handles mobile medical devices, collects of child's health data from the devices, visualizes data on the computers, and transfers data to the physician. Physicians or guardians monitor and supervise the child using an automated workplace (Fig. 3). The system's software handles Internet communications, processing of signals from multiple devices, visualization of patient health data, mathematical methods, communications with

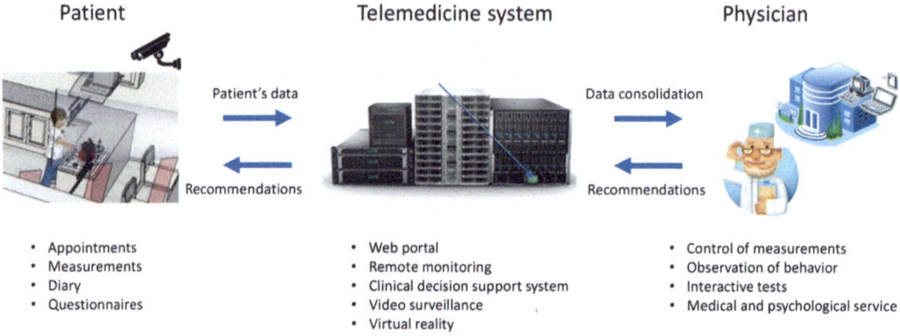

Fig. 3 Telemedicine system implemented at Sechenov University. The system includes patients' personal accounts (left), hardware and software (middle), and physicians' accounts (right). The system collects information from patients, processes it, and sends to the physician. The physician sends recommendations, which are delivered to the patient. During the video monitoring of a patient at home (left), an alarm is issued when dangerous actions occur, for example, patient approaches the stove

the clinical decision support system, video sessions with the patient, and electronic record keeping. The clinical decision support system processes patient data with mathematical methods and warns the physicians about the deviations of clinical indicators from normal ranges.

Mathematical methods are key for the successful operation of a telemedicine system. In our system, a dynamic video surveillance system employs learning methods based on deep neural networks (Lebedev et al. 2018a). These algorithms conduct early diagnostics of behavioral deviations. The surveillance system is placed at the patient's home. The surveillance software collects videos from the cameras and recognizes behaviors (waking, sleep, movements, falling, etc.) and facial expressions. Additionally, the video surveillance system detects the development of epileptic attacks. The system's knowledge database uses the examples collected from patients and actors. Interactive tests are used to diagnose age-specific ASD features (Schopler et al. 1980; Gotham et al. 2009, 2012; Baenskaya 2017; Sorokin and Davydova 2017; Lebedev et al. 2019). These interactive tests are designed to identify possible disabilities and provide recommendations to the physician. The system includes virtual toolsets for distance training of children with ASD (Bellani et al. 2011; Kandalaft et al. 2013). These virtual reality tools allow to model real-world situations in a controlled and safe environment (Gal et al. 2009; Alves et al. 2013). The Sechenov University system is currently being tested in three families with children with ASD. After the system is fully developed, it will be distributed in Europe and Russia under the program ERASMUS+ (European Commission 2018).

Several other publications have been published on the development of telemedicine systems similar to the one developed at the Sechenov University (Terry 2009; Boisvert et al. 2010; Knutsen et al. 2016; Sutherland et al. 2018). Sutherland and his colleagues recently reviewed the literature containing the search terms "autism" and "telehealth" (Sutherland et al. 2018). The search yielded 155 articles. Next, the search results were reduced to 14 articles that satisfied the following inclusion criteria: (1) inclusion an patient suffering from ASD; (2) the usage a telemedicine system for patient assessment; (3) a properly controlled experimental design; (4) quantification of outcomes; and (5) publication of a peer-reviewed article. The results reported in these articles support our suggestions regarding the composition of a telemedicine system for ASD. Most of these works used off-the-shelf equipment for telecommunication, such as tablets, laptop and desktop computers, and web cameras. Videoconferencing was enabled with such software as Adobe Connect and Skype. ASD diagnostics was performed using telecommunication tools (Reese et al. 2013, 2015; Schutte et al. 2015). Several intervention methods were tested, including iPICS (Meadan et al. 2016) and ImPACT (Ingersoll and Berger 2015; Pickard et al. 2016). Telemedicine approaches were employed for teaching (Ruble et al. 2013), functional communication training (Lindgren et al. 2016; Suess et al. 2016), and anxiety intervention (Hepburn et al. 2016). Hrates of participant satisfaction were reported, and numerous positive outcomes.

While telemedicine has been growing during the last decade, many issues with this novel type of medical devices still have to be researched because of the small samples of subjects in the previous studies. We suggest that the next steps in the development of telemedicine systems, including the systems for ASD, should consist of the development of appropriate mathematical algorithms, such as deep-learning methods, for recognition and analysis of disease symptoms. These mathematical methods and computer algorithms should be combined with the equipment for distant communications and monitoring. As telemedicine approaches are advancing and new implementations are being developed, there is also a growth in patient databases that can be used for further refinement of telemedicine methods. Of particular interest, is the addition of BCI technologies to telemedicine approaches, including the novel technology of brain-nets.

3 Future Merger of Telemedicine and Brain-Nets

Both brain-nets and telemedicine applications rely on interconnection between multiple subjects and the usage of advanced technologies to optimize information processing and exchange. In the case of brain-nets, information is taken from neural recordings and delivered to the brain using neurostimulation. In the case of telemedicine, traditional interaction between the doctor and patient is replaced by patient remote monitoring assisted by computer analysis, and Internet communications replace physical contact. Based on the comparison of these approaches, it is reasonable to suggest that at some point, they will merge. Indeed, many capabilities of BCIs (Lebedev 2014, 2016), such as decoding motor intentions, decisions, choices, and brain states, could be used in telemedicine applications to improve patient diagnostics and monitoring, provide means of communication (Birbaumer et al. 1999), and implement neurofeedback (Ossadtchi et al. 2017; Ros et al. 2020). Clearly, these clinically relevant approaches could be incorporated in telemedicine applications to bring benefits to patients. Brain-nets, i.e., BCIs composed of multiple brains, could be relevant for telemedicine, as well. Here, one could envision telemedicine therapy based on BCI-based social interactions (Obbink et al. 2011; Mattout 2012) and even BTBIs where the brains of physicians and patients are interconnected. Based on these considerations, we envision brain-nets as a foundation for telemedicine of the future.

Acknowledgments This work was supported by the Center for Bioelectric Interfaces of the Institute for Cognitive Neuroscience of the National Research University Higher School of Economics, RF Government grant, ag. No. 14.641.31.0003 and by the RFBR grant 18-07-00987.

References

Alves S, Marques A, Queirós C, Orvalho V (2013) LIFEisGAME Prototype: a serious game about emotions for children with autism spectrum disorders. PsychNology J 11(3):191–211

APA (2013) Diagnostic and statistical manual of mental disorders (DSM-5®). American Psychiatric Publishing, Arlington

Baenskaya E (2017) Early diagnostics and correction of ASD in the field of emotional-semantic approach. Autism Dev Disord 15(2):32–37

Bellani M, Fornasari L, Chittaro L, Brambilla P (2011) Virtual reality in autism: state of the art. Epidemiol Psychiatr Sci 20(3):235–238

Birbaumer N, Ghanayim N, Hinterberger T, Iversen I, Kotchoubey B, Kübler A, Perelmouter J, Taub E, Flor H (1999) A spelling device for the paralysed. Nature 398(6725):297–298

Boisvert M, Lang R, Andrianopoulos M, Boscardin ML (2010) Telepractice in the assessment and treatment of individuals with autism spectrum disorders: a systematic review. Dev Neurorehabil 13(6):423–432

Dorsey ER, Topol EJ (2016) State of telehealth. N Engl J Med 375(2):154–161

Eckstein MP, Das K, Pham BT, Peterson MF, Abbey CK, Sy JL, Giesbrecht B (2012) Neural decoding of collective wisdom with multi-brain computing. Neuroimage 59(1):94–108

Ekeland AG, Bowes A, Flottorp S (2010) Effectiveness of telemedicine: a systematic review of reviews. Int J Med Inform 79(11):736–771

Elsabbagh M, Divan G, Koh YJ, Kim YS, Kauchali S, Marcín C, Montiel-Nava C, Patel V, Paula CS, Wang C (2012) Global prevalence of autism and other pervasive developmental disorders. Autism Res 5(3):160–179

European Commission (2018) Erasmus+ Programme Guide. Version 1, Erasmus+ Brussels

Fagg AH, Hatsopoulos NG, London BM, Reimer J, Solla SA, Wang D, Miller LE (2009) Toward a biomimetic, bidirectional, brain machine interface. In: 2009 annual international conference of the IEEE Engineering in Medicine and Biology Society, IEEE

Fatemi SH, Clayton PJ (2008) The medical basis of psychiatry. Springer, New York

Flesher SN, Collinger JL, Foldes ST, Weiss JM, Downey JE, Tyler-Kabara EC, Bensmaia SJ, Schwartz AB, Boninger ML, Gaunt RA (2016) Intracortical microstimulation of human somatosensory cortex. Sci Transl Med 8(361):361ra141

Folcher M, Oesterle S, Zwicky K, Thekkottil T, Heymoz J, Hohmann M, Christen M, El-Baba MD, Buchmann P, Fussenegger M (2014) Mind-controlled transgene expression by a wireless-powered optogenetic designer cell implant. Nat Commun 5:5392

Gal E, Bauminger N, Goren-Bar D, Pianesi F, Stock O, Zancanaro M, Weiss PLT (2009) Enhancing social communication of children with high-functioning autism through a co-located interface. AI Soc 24(1):75

Gotham K, Pickles A, Lord C (2009) Standardizing ADOS scores for a measure of severity in autism spectrum disorders. J Autism Dev Disord 39(5):693–705

Gotham K, Pickles A, Lord C (2012) Trajectories of autism severity in children using standardized ADOS scores. Pediatrics 130(5):e1278–e1284

Grau C, Ginhoux R, Riera A, Nguyen TL, Chauvat H, Berg M, Amengual JL, Pascual-Leone A, Ruffini G (2014) Conscious brain-to-brain communication in humans using non-invasive technologies. PLoS One 9(8):e105225

Hepburn SL, Blakeley-Smith A, Wolff B, Reaven JA (2016) Telehealth delivery of cognitive-behavioral intervention to youth with autism spectrum disorder and anxiety: a pilot study. Autism 20(2):207–218

Hildt E (2015) What will this do to me and my brain? Ethical issues in brain-to-brain interfacing. Front Syst Neurosci 9:17

Hongladarom S (2015) Brain-brain integration in 2035: metaphysical and ethical implications. J Inf Commun Ethics Soc 13(3–4):205–217

Ingersoll B, Berger NI (2015) Parent engagement with a telehealth-based parent-mediated intervention program for children with autism spectrum disorders: predictors of program use and parent outcomes. J Med Internet Res 17(10):e227

Jiang L, Stocco A, Losey DM, Abernethy JA, Prat CS, Rao RP (2019) BrainNet: a multi-person brain-to-brain interface for direct collaboration between brains. Sci Rep 9(1):1–11

Kandalaft MR, Didehbani N, Krawczyk DC, Allen TT, Chapman SB (2013) Virtual reality social cognition training for young adults with high-functioning autism. J Autism Dev Disord 43(1):34–44

Knutsen J, Wolfe A, Burke BL, Hepburn S, Lindgren S, Coury D (2016) A systematic review of telemedicine in autism spectrum disorders. Rev J Autism Dev Disord 3(4):330–344

Kopetz PB, Endowed EDL (2012) Autism worldwide: prevalence, perceptions, acceptance, action. J Soc Sci 8(2):196

Lebedev M (2014) Brain-machine interfaces: an overview. Transl Neurosci 5(1):99–110

Lebedev M (2016) Augmentation of sensorimotor functions with neural prostheses. Opera Med Physiol 2(3-4):211–227

Lebedev G, Klimenko H, Kachkovskiy S, Konushin V, Ryabkov I, Gromov A (2018a) Application of artificial intelligence methods to recognize pathologies on medical images. Proced Comput Sci 126:1171–1177

Lebedev MA, Opris I, Casanova MF (2018b) Augmentation of brain function: facts, fiction and controversy. Front Syst Neurosci 12:45

Lebedev G, Klimenko H, Fartushniy E, Shaderkin I, Kozhin P, Galitskaya D (2019) Building a Telemedicine System for monitoring the health status and supporting the social adaptation of children with autism spectrum disorders. Intelligent Decision Technologies 2019. Springer, New York, pp 287–294

Li G, Zhang D (2016) Brain-computer interface controlled cyborg: establishing a functional information transfer pathway from human brain to cockroach brain. PLoS One 11(3):e0150667

Lindgren S, Wacker D, Suess A, Schieltz K, Pelzel K, Kopelman T, Lee J, Romani P, Waldron D (2016) Telehealth and autism: treating challenging behavior at lower cost. Pediatrics 137(Suppl 2):S167–S175

Lustig TA (2012) The role of telehealth in an evolving health care environment: workshop summary. National Academies Press, Washington, DC

Mattout J (2012) Brain-computer interfaces: a neuroscience paradigm of social interaction? A matter of perspective. Front Hum Neurosci 6:114

Meadan H, Snodgrass MR, Meyer LE, Fisher KW, Chung MY, Halle JW (2016) Internet-based parent-implemented intervention for young children with autism: a pilot study. J Early Interv 38(1):3–23

Mirabella G, Lebedev MA (2017) Interfacing to the brain's motor decisions. J Neurophysiol 117(3):1305–1319

Nijholt A (2015) Competing and collaborating brains: multi-brain computer interfacing. Brain-computer interfaces. Springer, New York, pp 313–335
O'Doherty JE, Lebedev MA, Ifft PJ, Zhuang KZ, Shokur S, Bleuler H, Nicolelis MA (2011) Active tactile exploration using a brain-machine-brain interface. Nature 479(7372):228–231
Obbink M, Gürkök H, Bos DP-O, Hakvoort G, Poel M, Nijholt A (2011) Social interaction in a cooperative brain-computer interface game. International conference on intelligent technologies for interactive entertainment. Springer, New York
Ossadtchi A, Shamaeva T, Okorokova E, Moiseeva V, Lebedev MA (2017) Neurofeedback learning modifies the incidence rate of alpha spindles, but not their duration and amplitude. Sci Rep 7(1):1–12
Pais-Vieira M, Lebedev M, Kunicki C, Wang J, Nicolelis MA (2013) A brain-to-brain interface for real-time sharing of sensorimotor information. Sci Rep 3:1319
Pais-Vieira M, Chiuffa G, Lebedev M, Yadav A, Nicolelis MA (2015) Building an organic computing device with multiple interconnected brains. Sci Rep 5:11869
Pickard KE, Wainer AL, Bailey KM, Ingersoll BR (2016) A mixed-method evaluation of the feasibility and acceptability of a telehealth-based parent-mediated intervention for children with autism spectrum disorder. Autism 20(7):845–855
Poli R, Cinel C, Matran-Fernandez A, Sepulveda F, Stoica A (2013) Towards cooperative brain-computer interfaces for space navigation. In: Proceedings of the 2013 international conference on intelligent user interfaces, ACM
Poli R, Valeriani D, Cinel C (2014) Collaborative brain-computer interface for aiding decision-making. PLoS One 9(7):e102693
Ramakrishnan A, Ifft PJ, Pais-Vieira M, Byun YW, Zhuang KZ, Lebedev MA, Nicolelis MA (2015) Computing arm movements with a monkey brainet. Sci Rep 5:10767
Rao RP (2019) Towards neural co-processors for the brain: combining decoding and encoding in brain–computer interfaces. Curr Opin Neurobiol 55:142–151
Rao RP, Stocco A, Bryan M, Sarma D, Youngquist TM, Wu J, Prat CS (2014) A direct brain-to-brain interface in humans. PLoS One 9(11):e111332
Reese RM, Jamison R, Wendland M, Fleming K, Braun MJ, Schuttler JO, Turek J (2013) Evaluating interactive videoconferencing for assessing symptoms of autism. Telemed e-Health 19(9):671–677
Reese RM, Braun MJ, Hoffmeier S, Stickle L, Rinner L, Smith C, Ellerbeck K, Jamison R, Wendland M, Jarrett L (2015) Preliminary evidence for the integrated systems using telemedicine. Telemed e-Health 21(7):581–587
Ros T, Enriquez-Geppert S, Zotev V, Young KD, Wood G, Whitfield-Gabrieli S, Wan F, Vuilleumier P, Vialatte F, Van De Ville D (2020) Consensus on the reporting and experimental design of clinical and cognitive-behavioural neurofeedback studies (CRED-nf checklist). Brain 143(6):1674–1685
Ruble LA, McGrew JH, Toland MD, Dalrymple NJ, Jung LA (2013) A randomized controlled trial of COMPASS web-based and face-to-face teacher coaching in autism. J Consult Clin Psychol 81(3):566
Schopler E, Reichler RJ, DeVellis RF, Daly K (1980) Toward objective classification of childhood autism: Childhood Autism Rating Scale (CARS). J Autism Dev Disord 10(1):91–103
Schutte JL, McCue MP, Parmanto B, McGonigle J, Handen B, Lewis A, Pulantara IW, Saptono A (2015) Usability and reliability of a remotely administered adult autism assessment, the autism diagnostic observation schedule (ADOS) module 4. Telemed e-Health 21(3):176–184
Shanechi MM, Hu RC, Williams ZM (2014) A cortical–spinal prosthesis for targeted limb movement in paralysed primate avatars. Nat Commun 5:3237
Sorokin A, Davydova EY (2017) Autism diagnostic evaluation schedule (ADOS-2) for evaluation of behavior and communication in toddlers with concern of autism spectrum disorder. Autism Dev Disord 15(2):38–44
Stocco A, Prat CS, Losey DM, Cronin JA, Wu J, Abernethy JA, Rao RP (2015) Playing 20 questions with the mind: collaborative problem solving by humans using a brain-to-brain interface. PLoS One 10(9):e0137303

Suess AN, Wacker DP, Schwartz JE, Lustig N, Detrick J (2016) Preliminary evidence on the use of telehealth in an outpatient behavior clinic. J Appl Behav Anal 49(3):686–692

Sutherland R, Trembath D, Roberts J (2018) Telehealth and autism: a systematic search and review of the literature. Int J Speech Lang Pathol 20(3):324–336

Terry M (2009) Telemedicine and autism: researchers and clinicians are just starting to consider telemedicine applications for the diagnosis and treatment of autism. Telemed e-Health 15(5):416–419

Trimper JB, Root Wolpe P, Rommelfanger KS (2014) When "I" becomes "We": ethical implications of emerging brain-to-brain interfacing technologies. Front Neuroeng 7:4

Wang Y, Jung T-P (2011) A collaborative brain-computer interface for improving human performance. PLoS One 6(5):e20422

WHO (2004) International statistical classification of diseases and related health problems. World Health Organization, Geneva

Wing L, Shah A (2000) Catatonia in autistic spectrum disorders. Br J Psychiatry 176(4):357–362

Yoo S-S, Kim H, Filandrianos E, Taghados SJ, Park S (2013) Non-invasive brain-to-brain interface (BBI): establishing functional links between two brains. PLoS One 8(4):e60410

Yuan P, Wang Y, Gao X, Jung T-P, Gao S (2013) A collaborative brain-computer interface for accelerating human decision making. International conference on universal access in human-computer interaction. Springer, New York

Cognitive Augmentation Via a Brain/Cloud Interface

A. Angelica, I. Opris, Mikhail A. Lebedev, and F. J. Boehm

Abbreviations

AIS	Axon initial segment
B/CI	Brain/cloud interface
BCI	Brain-computer interface
BMI	Brain-machine interface
BTBI	Brain-to-brain interface
EEG	Electroencephalography
fmri	Functional magnetic resonance imaging
FNIRS	Functional near-infrared spectroscopy.
TS	Transparent shadowing

1 Introduction

The next significant technological leap from today's internet/cloud/edge, which cumulatively operates as a decentralized global scale system to assist humanity with generation, processing, storage, and access to massive amounts of data (~2.5 quintillion bytes of data created each day) (Marr 2018), is a future development referred

A. Angelica (✉)
Kurzweil Technologies, Newton, MA, USA

I. Opris
Department of Neurosurgery, University of Miami, Miami, FL, USA

M. A. Lebedev
Center for Neuroengineering, Duke University, Durham, NC, USA

Center for Bioelectric Interfaces of the Institute for Cognitive Neuroscience, National Research University Higher School of Economics, Moscow, Russia

Department of Information and Internet Technologies of Digital Health Institute, I.M. Sechenov First Moscow State Medical University, Moscow, Russia

F. J. Boehm
Nanoapps Medical, Inc., Vancouver, BC, Canada

I. Opris et al. (eds.), *Modern Approaches to Augmentation of Brain Function*, Contemporary Clinical Neuroscience,
https://doi.org/10.1007/978-3-030-54564-2_17

to here as a "brain/cloud interface" ("B/CI"). This envisaged, highly sophisticated B/CI platform could significantly improve the speed of our access to information. "We'll have nanobots that… connect our neocortex to a synthetic neocortex in the cloud… Our thinking will be a …. biological and nonbiological hybrid" (Kurzweil 2014). A B/CI may be enabled by synergies between developments in nanotechnology, nanomedicine, and AI. If eventually adopted ubiquitously, this platform may emerge as the next evolutionary step for humankind, one where we seamlessly merge with our technologies. Such a potential B/CI system might be facilitated by employing "nanorobotic" devices or "nanobots"—"complex engineered objects fabricated by positioning matter with molecular control" (Freitas 1999b). "Medical nanodevices based on this concept may include large doses of independent, micron-sized individual nanorobots or alternatively may include macroscale nanoorgans (nanorobotic organs) assembled either as solid objects or built up from trillions of smaller artificial cells or docked nanorobots inside the body" (Freitas 2003).

Advanced hypothetical autonomous nanorobotic nanomedical devices or "neuralnanorobotics" (Martins et al. 2012, 2015, 2016) may encompass "endoneurorobots," "gliabots," and "synapobots," each of which would operate within particular domains of the brain and work in conjunction with specific cells to transmit and receive data to and from the cloud.

There is an unrelenting and accelerating impetus in medicine to continuously develop more compact, sophisticated, multifunctional, and economical devices and platforms. One of the primary aims behind this drive is related to the nature of human disease, which emerges from dysfunctional cascades that extend through molecular, organellar, and cellular domains. Due to their diminutive size, functionalized nanoparticles and (in the future) autonomous AI-imbued nanomedical devices may have the capacity to directly interact, influence, and correct these micro-/nanoscale physiological/biochemical aberrations. They may also be able to effectively prevent the onset of many conditions, with great beneficial significance in terms of addressing diseases of the brain (Kandel 2001; Kandel et al. 2000; Falk et al. 2016; Chaudhury et al. 2015; Fornito et al. 2015; Zigmond et al. 2014).

Ultimately, advanced nanomedicine may finally equip those in the medical profession with the ultrahigh resolution tools they require to resolve a myriad of diseases at the individual cellular/subcellular level (Freitas 1998, 1999a, b, 2002, 2003, 2005a, b, 2007, 2016; Morris 2001; Astier et al. 2005; Patel et al. 2006; Park et al. 2007; Boehm 2013; Popov et al. 2007; Martel et al. 2009; Mallouk and Sen 2009; Kostarelos 2010; Mavroides and Ferreira 2011). Envisioned neuralnanorobotic platforms may enable precision diagnostics and therapeutics, at single cell resolution (Freitas 2007), to effectively address the approximately 400 conditions that the human brain is subject to (NINDS 2017). Some of the most prominent of these conditions are Parkinson's and Alzheimer's (Freitas 2016), dementia, epilepsy, spinal cord disorders, and addiction.

Neuralnanorobotic devices may also serve as critical B/CI infrastructures to mediate robust and seamless connections between specific cognitive functions and cloud-based processing and data storage. This would initially entail directly monitoring some of the ~86×10^9 neurons and ~2×10^{14} synapses of the brain (Herculano-

Houzel 2009). As applied to the human brain, these devices may, for example, enable minimally invasive, real-time surveillance and recording of neuroelectric activity associated with individual neurons and synapses, as well as localized neuropeptide flows, and more. Further, these devices may facilitate detailed mapping of the surface features of neurons and other important brain microstructures, which may lead to highly detailed next-generation connectome maps (Martins et al. 2012, 2015, 2016; Sporns et al. 2005; Lu et al. 2009; Anderson et al. 2011; Kleinfeld et al. 2011; Seung, 2011).

In one scenario, once nanorobots traverse the vasculature, they would cross the blood-brain barrier (BBB) and self-migrate into the parenchyma and their targeted brain cells. "Endoneurobots" (neuron-residing neuralnanorobots that monitor/interface with neuronal action potentials) would automatically orient themselves at the neuronal axon initial segments, whereas gliabots would enter the glial cells, and synaptobots (neuralnanorobots that monitor/interact with synaptic gap traffic) would closely position themselves at each synapse. Approximately 6×10^{16} bits/s of electrical information (preprocessed at the synapses by nanorobots) would be wirelessly transferred through a total of 30 cm^3 integrated self-assembled nanorobotic fiber optics. These nanometric conduits with a 10^{18} bit/s transmission capacity may facilitate rapid transfer of data for processing in edge devices, which could then transmit at high speed to the cloud, enabling real-time monitoring and transfer of data from and to the brain.

Further to the medical applications described above, it is conceivable that neuralnanorobotic devices applied to a B/CI may facilitate a wide array of nonmedical applications and new paradigms in the augmentation of human brain function and cognition. This may be made possible by coupling the neocortex directly with the processing and storage capacities of cloud-based supercomputers via edge-based AI systems. Other future technologies such as molecular manufacturing (MM) (Domschke and Boehm 2014) may further extend the development of these devices.

A critical aspect of autonomous nanomedical devices will be biocompatibility, as some species will be expected to be resident within the human body for extended periods (Freitas 2003). So these devices may comprise diamondoid or sapphire, as these materials exhibit the highest reliability, resilience, and strength in vivo (Freitas 2010). Diamondoid researchers Robert A. Freitas, Jr. and Ralph Merkle have established an international "Nanofactory Collaboration" with the goal of developing a MM-based nanofactory. This nanofactory would have the capacity to mass fabricate sophisticated autonomous neuralnanorobots, spanning medical and nonmedical applications (Freitas 2009, 2010; Freitas and Merkle 2004, 2006).

2 Hypothetical Human Brain/Cloud Interface: Premise

Recent rapid technological advances such as digital communications, the internet/cloud/edge, nanotechnology, nanomedicine, AI, robotics, 3D printing, and blockchain suggest that the capacity to develop neuralnanorobotics within the next few decades may also be a reality. This achievement could result in the emergence of a set of new technologies that manifest as instantaneous, real-time, safe, and secure brain-computer interfaces (BCIs), brain-to-brain interfaces (BBIs), and highly sophisticated self-installing/uninstalling brain/cloud interfaces (B/CIs). Such systems could significantly advance human/machine communications and could have strong potential for significant human cognitive enhancement (Kurzweil 2014; Swan 2016).

2.1 *Quantifying the Human Brain*

The human brain (1400 g average weight ~1350 cm^3 volume, encased in an ~1700 cm^3 intracranial volume) is an extraordinarily complex and highly dynamic system for processing and storage of information. It is remarkably efficient in terms of computational capacity per volume. Rengachary and Ellenbogen (2005) revealed a basic quantitative survey of the brain's constituents, which encompass ~1350 cm^3 (~75%) brain cells, ~200 cm^3 (15%) blood, and up to ~150 cm^3 (10%) of cerebrospinal fluid. Furthermore, the brain's basic computational power approximately spans from 10^{13} to 10^{16} operations/s (Merkle 1989), whereas its power output is estimated to range from 15 to 25 W (Kandel and Schwartz 1985), with a power density of $1.1–1.8 \times 10^4$ W/m^3, operating at 37.3 °C (Freitas 1999b).

In the cellular domain, the estimated number of neurons (Fig. 1) within the average whole human brain ranges from $(86.06 \pm 8.2) \times 10^9$ to $(94.2 \pm 11.3) \times 10^9$, most of which reside within the cerebellum $(69.03 \pm 6.65) \times 10^9$ (~80.2%) and cerebral cortex $(16.34 \pm 2.17) \times 10^9$ (~19%), with the remaining neurons $(0.69 \pm 0.12) \times 10^9$ (~0.8%) distributed throughout the rest of the brain (Azevedo et al. 2009; Martins et al. 2012). (See Table 1 for further details on populations of brain constituents.)

The most ubiquitous subcellular cognitive information processing structures are the synapses, with an average whole brain population of $(\sim 2.42 \pm 0.29) \times 10^{14}$. These entities are critical for learning, memory storage and elimination (both long- and short-term), as well as temporal data processing. Remarkably, according to Smith et al. (2019), "In healthy individuals, the number of synapses formed and

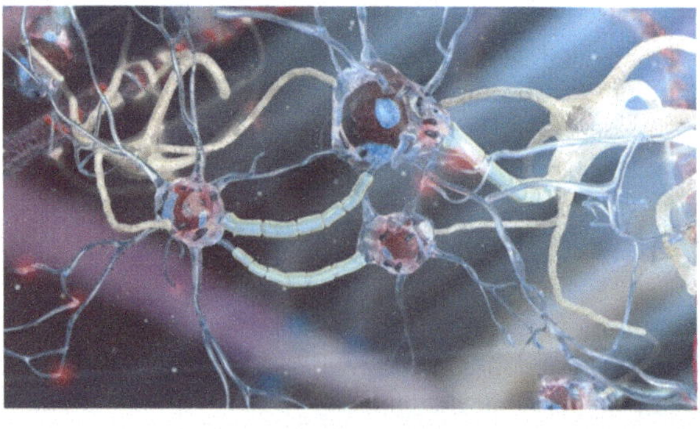

Fig. 1 Artistic portrayal of neurons (with blue processes) and glial (white) cells. (Image credit: Yuriy Svidinenko, Nanobotmodels Company)

Table 1 Enumeration of neurons and synapses in the human neocortex (Sandberg and Bostrom 2008; Tang et al. 2001; Karlsen and Pakkenberg 2011)

Neocortex region	Total neocortex volume (cm^3)	Number of synapses (10^{12})	Number of neurons (10^9)	Number of synapses per neuron (10^3)	Glial cell number (10^9)
Occipital	69	22.0	3–4.65	4.36	3
Parietal	149	41.5	4–6.61	6.33	4
Temporal	133	42.0	4–4.80	8.95	5
Frontal	239	58.9	6–7.89	7.54	7
Total	590	164.0	17–23.9	6.93	18

actively removed is estimated to be one trillion per day, presumably associated with constant learning and establishment of new memories" (Ashford 2015). In Alzheimer's patients, the loss of synapses is closely correlated with cognitive decline (Dekosky and Scheff 1990; Terry et al. 1991; Scheff and Price 2006).

Computer simulations have verified the role of synapses as processing units by demonstrating that a simulated computational network could be enhanced through the use of dynamic synapse-like models. Therefore, it is likely that biomimetic artificial synapses will be a prerequisite toward the emergence of computational brain analogs (Kuzum et al. 2012; Fuhrmann et al. 2002; Maass and Zador 1999). Recently, an ultralow-power synaptic analog for neural computing was developed, which exhibited the ability to configure 500 distinct states (Van de Burgt et al. 2017).

2.2 The Neocortex: Tapping Six Layers of Complexity

The primate neocortex has a well-organized neural architecture (Fig. 2a) encompassing the sensorimotor, cognitive, and emotional domains (Alexanderet al. 1986; Fuster and Bressler 2012). This cortical structure consists of minicolumnar and laminar arrangements of neurons, which are linked into an integrated network by afferent and efferent connections that are distributed across many regions of the brain and facilitate a multifunctional spectrum (Lorente de No 1938; Mountcastle 1997; Opris et al. 2011, 2012a, b, 2013; Shepherd and Grillner 2010).

Cortical minicolumns consist of chains of pyramidal neurons that are surrounded by a "curtain of inhibition", which is formed by double bouquet interneurons (Szentágothai and Arbib 1975). The neocortical mantle, on the other hand, has a laminar structure (Fig. 2b) composed of six layers that are grouped into supragranular, granular, and infragranular laminae, with granular lamina 4 (L4) receiving sensory input from the thalamus (Jones 2000; DeFelipe et al. 2012; Constantinople and Bruno 2013). Granular L4 segregates the functionality of cortical minicolumns into perceptual and executive/behavior/action components (Opris et al. 2012a, b; Douglas and Martin 1991).

In essence, the supragranular layers consist of small pyramidal neurons that generate a complex network of intracortical connections. Cajal was the first to empha-

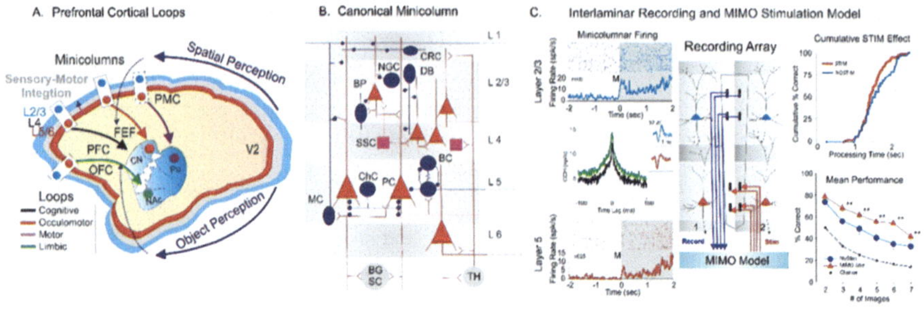

Fig. 2 InterLaminar Microcircuits across the Neocortex. (**a**) Primate brain showing the cortical mantle split into cortical layers and minicolumns. Minicolumns across neocortex work cooperatively to translate perception into complex action via thalamo-cortical loops (adapted with permission from Opris et al.) (Opris et al. 2015b). (**b**) Cortical minicolumns with pyramidal labeled in red and dark blue are distributed across the supra- and infragranular layers. Stellate cells in layer L4 are highlighted in pink (adapted with permission from Markram et al.) (Markram et al. 2004). (**c**) Interlaminar recording of pyramidal cells in the prefrontal cortex and the MIMO stimulation model. Rasters and peri-event histograms (in blue and red) depict the activity of supra-and infragranular layers. Cross-correlation showed that interlaminar firing increased following the presentation of targets, in comparison to pretarget epoch. Recording array with the MIMO model, for recording in layer L2/3 and stimulation in layer L5. Stimulation effect compared to the population tuning for MIMO stim (red) versus layer L5 prefrontal cortical activity (dark blue dotted line). The overall MIMO stimulation effect (red) was significantly greater than no-stim and chance levels (with permission from Opris et al.) (Opris et al. 2012a, b)

size the possibility that this network is primarily responsible for high level functions that spanned thought, learning, and memory (Cajal 1899; Lorente de No 1938). The supragranular layers also provide a major input to the infragranular layers of relatively large pyramidal neurons that generate the majority of output of cerebral cortex to other parts of the brain (Buxhoeveden and Casanova 2001). In fact, infragranular layers are essentially the "executive" portion of the cerebral cortex (Opris and Casanova 2014).

According to this three layer diagram of the functional module, infragranular layers execute "cognitive" computations, which are elaborated in the supragranular layers (Opris and Casanova 2014). Cortical minicolumnar arrangements in the frontal cortex may be regarded as modules that integrate the perceptual stimuli from various sensory modalities (visual, auditory, or somatosensory) with the supragranular layers and select behaviorally relevant signals in the infragranular layers (Opris et al. 2011, 2012a, b, 2013, 2014). Table 2 enumerates the neurons and synapses in the human male and female brains.

Evidence for neocortical columnar organization was initially obtained through single neuron recording studies of the somatic sensory cortexes of anesthetized cats and monkeys (Mountcastle 1957; Powell and Mountcastle 1959). Microelectrode penetrations made perpendicular to the pial surface encounter neurons with similar properties of place and modality within each cellular layer. Penetrations parallel to the pial surface and crossing the vertical axis of the cortex pass through blocks of tissue (300–500 mm in size) with identical properties. Dramatic transitions may be

Table 2 Neocortical measures (Stark et al. 2007a, 2007b; Pakkenberg and Gundersen 1997)

	Surface (cm^2)	Thickness (mm)	Volume (cm^3)	Neuron number density ($10^6/cm^3$)	Neurons (N, 10^9)
Female	1678–1680	2.61–2.74	440–458	43.1–43.8	19.3–19.7
Male	1883–1900	2.72–2.79	517–524	44.0–44.1	22.8–22.9
Humans	1820	2.69	489	44.0	21.5

observed when transitioning from one block, which possesses one set of properties, to an adjacent block that has different properties. The minicolumnar organization of the somatic sensory cortex for place has been confirmed in microelectrode mapping experiments with cats and monkeys (Favorov and Whitsel 1988a, b; Tommerdahl et al. 1993).

In the auditory cortex, Atencio and Schreiner revealed similar features with visual and somatosensory cortexes, namely, that the supragranular laminae process independent acoustic stimuli with a more-defined spatiotemporal receptive field, while deeper laminae process stimuli with a broader receptive field (Atencio and Schreiner 2010). In relation to the chemical sense modality, complementary sensory and associative microcircuitry have been reported in the three-layered primary olfactory cortex (Wiegand et al. 2011). For the motor cortex, a top-down laminar organization of the excitatory network (Weiler et al. 2008), which was fractionated into distinct sublayer-specific microcircuits of corticospinal and corticostriatal neurons, has been reported in slice studies (Anderson et al. 2010). A role of directional inhibition in directional accuracy was demonstrated by Mahan and Georgopoulos, in which the width of the directional tuning curve (modulated by local inhibitory mechanisms) determined the accuracy of directional commands (Mahan and Georgopoulos 2013).

Two of the most critical roles for cognitive microcircuits are executive control within the prefrontal cortex (Opris and Bruce 2005; Opris and Casanova 2014).and memory within the temporal cortex (Hirabayashi et al. 2013). The role of prefrontal cortical minicolumns in executive control was shown by contrasting the correlated firing of neurons (Fig. 2c, left panel) between supragranular and infragranular laminae, while a monkey selected correct versus error targets. The results revealed a reduction in correlated firing between cell pairs within single minicolumns on error trials with inappropriate target selection (Opris et al. 2011, 2012a, b).

This comprised a direct demonstration of task-specific, real-time columnar processing within the prefrontal cortex, which verified the role of this interlaminar microcircuit in the executive control of decision-making in the primate brain. Subsequently, it was shown that in the perception and executive selection phases of the task, cell firing within the prefrontal layers and caudate-putamen exhibited similar location preferences in spatial trials. Even stimulation-induced spatial preferences in a percentage of correct performance in spatial trials were similar to neural tuning, which indicated that interlaminar prefrontal microcircuits played causal roles in perception-to-action cycles (Opris et al. 2013). Prefrontal cortical microcircuits bridge perception with action across many sensory/motor modalities. A great deal remains to be learned in regard to the distinct and modality specific modularity aspects of perceptual and action microcircuits on both ventral and dorsal visual streams, auditory, somatosensory, and even chemical senses.

Memory neurons within the microcircuitry for object retrieval in the temporal cortex were dynamically modulated by object association memory (Hirabayashi et al. 2013; Wang 2006). Interestingly, the interlaminar signal flow during sensory (encoding) and memory processing was reversed, which indicated the differential recruiting of interlaminar microcircuits for sensory and mnemonic processing (Takeuchi et al. 2011). In the emotional domain, the acquisition of associative fear memories was contingent on the recruitment of disinhibitory microcircuits in the mouse auditory cortex (Letzkus et al. 2011). Fear-conditioning-associated disinhibition in the auditory cortex is driven by foot-shock-mediated cholinergic activation of layer 1 interneurons, which in turn generates the inhibition of laminae 2/3 parvalbumin-positive interneurons.

Recent progress in the understanding of cortical microcircuits is due to recent advancements in microelectrode technologies that provide multielectrode arrays (MEA) for the simultaneous recording of supra- and infragranular cortical laminae in adjacent minicolumns, which have imparted unprecedented insights into cortical microcircuit functionality (Opris et al. 2014; Vidu et al. 2014). MEAs are ideal for interlaminar recordings from neurons in the prefrontal laminae L2/3 and L5. Such recordings are analyzed using a nonlinear multi-input/multi-output (MIMO) microstimulator (Berger et al. 2011).

The MIMO model (Fig. 2c, right panel) confirmed that the individually recorded prefrontal minicolumns responded to entrained target selections in patterns that were critical for successful cognitive performance. This allowed for the substitution of task-related laminar L5 neuron firing patterns with electrical stimulation in the same recording regions, during columnar transmission from lamina L2/3, at the time of target selection. Such stimulation improved normal task performance, but more importantly, had the capacity to recover task performance when applied as a neuroprosthesis, following the pharmacological disruption of decision-making related to the same task (Hampson et al. 2012).

These recent findings have provided the first successful application of neuroprosthesis in the primate brain, which were specifically designed to repair or restore disrupted cognitive function (Berger et al. 2011; Hampson et al. 2012; Opris 2013; Opris and Ferrera 2014). This work is focused on the utilization of neuromodulation of cortical microcircuits to enable brain/cloud interfaces for multiple purposes, including the repair and augmentation of cognitive function in patients with neurological and psychiatric disorders.

The disruption of interlaminar microcircuits within cortical minicolumns is the signature of a broad spectrum of neurologic and psychiatric disorders, such as autism, schizophrenia, Alzheimer's, and drug addiction (Opris et al. 2015a; Opris and Casanova 2014). The targeting of this cortical microcircuitry through novel approaches will be key toward the successful development of methods and treatments for cognitive enhancement.

3 Precursor Technologies Toward a Brain/Cloud Interface

From a historical perspective, the first elucidation of the brain's electrical activities arrived in 1924 with the advent of electroencephalography (EEG) (Stone and Hughes 2013), which involved patients having to endure the insertion of silver wires into their scalps. Almost 50 years later (1973), the term "brain-computer interface" was used for the first time, with the insight that EEG signals might be utilized to convey data between the brain and a computer to facilitate communications (Vidal 1973).

At present, both experimental invasive and noninvasive brain-computer interface (BCI) and brain-to-brain interface (BBI) systems have been shown to be feasible. These advances have spurred intensive ongoing research in these areas, with the aim of developing therapies that might assist in restoring the movement of limbs in partially and completely paralyzed patients, as well as enable the control of prosthetic limbs for amputees (Birbaumer 2006). The first instance of direct human brain-to-brain communications took place in 2014 using a technique called "hyperinteraction" (Grau et al. 2014).

3.1 Functionalized Nanomaterials

Various classes of nanomaterials may have strong potential to serve as precursors and testbeds toward facilitating a robust, safe, and secure B/CI. For example, magnetoelectric nanoparticles have exhibited the capacity to improve the interface between electric fields generated by neural networks and external magnetic fields (Guduru et al. 2015). With the application of an appropriate direct-current magnetic field to the cranium, these nanoparticles may be able to pass through the blood-brain barrier (BBB) (D'Agata et al. 2017). Freitas (1999a, b) suggests that the ability to deliver "nanoparticles to the perineuronal environment is expected to provide a means to access and eventually stimulate selected populations of neurons" (Martins et al. 2019).

There will be formidable challenges associated with the precision delivery of any species of nanoparticles into the brain, and so innovative strategies will be required to resolve these issues. To illustrate, it was found that ~90% of intravenously injected nanoparticles were captured by various organs and tissues before reaching the brain (Calvo et al. 2001); however, as alluded to above, guiding nanoparticles to specific brain regions might be accomplished by employing external magnetic fields (Li et al. 2018). Further, it was observed that particular nanoparticle species can impair dopaminergic and serotonergic systems. Thus, comprehensive assessments of nanoparticle biodistribution, metabolism, risks of infection and inflammation, cytotoxicity, immunogenicity, and tumorigenicity will be indispensable toward assuring their safe and efficaciously beneficial application related to the brain (Cupaioli et al. 2014).

It has been proposed that carbon nanotubes might be employed to unidirectionally electrically stimulate deep brain targets to assist with alleviating Parkinson's and other CNS conditions (Srikanth and Kessler 2012). This is likely to be far more precise and less invasive than the use of deeply inserted microelectrodes (Taghva et al. 2013; Mayberg et al. 2005). Carbon nanotube or carbon nanofibers may be configured to serve as bidirectional data conduits to record the electrochemical activities of individual neurons.

Fluorescent carbon nanodots (Ø2.28 ± 0.42 nm) might be employed as an element of a B/CI, as they have been utilized to precisely target and image C6 glioma cells in the brains of mice. Advantages of these nanometric entities include the capacity to easily traverse the BBB, high biocompatibility, and customizable full color emission. So they may serve as useful tagging beacons to facilitate the navigation and emplacement of myriad B/CI components (Zheng et al. 2015). Another example of nanomaterials for optical imaging strategies is the use of quantum dot fullerenes, which have been employed in vivo to quantify cellular membrane potentials (Nag et al. 2017).

3.2 Neural Lace and Nanowires

A microscale "neural lace" has recently been developed, which may 1 day serve to integrate biological neural networks with computers and also enable multiplexed neural network recording in vivo. This technology would involve the injection of very flexible (submicron thick) three-dimensional mesh nanoelectronics into living brain tissues (~100 μm needle diameter) to enable neural interfaces to continually stimulate and monitor specific neurons and networks (Liu et al. 2015). It was demonstrated that these mesh nanoelectronics were stable for at least 1 year, with no indications of chronic immune response or glial scarring, which are typically the case with traditional implants (Dai et al. 2018). Further, through the use of field-effect transistors composed of silicon nanowires, plug and play I/O linkages were established (Schuhmann et al. 2017).

Nanoelectronics platforms using carbon nanotubes and silicon nanowires were shown to have the ability to detect/identify the secreted biomolecular chemicals and bioelectrical activities of neurons (Veliev 2016). Various configurations of nanowire arrays have been or are under investigation, which have the capacity to detect, stimulate, or inhibit nerve impulses as they traverse individual neurites (Patolsky et al. 2006; Zeck and Fromherz 2001; Freitas 1999a, b); cytosolically record the action potentials of neurons (Tian et al. 2010; Duan et al. 2011); simultaneously detect and stimulate nanoelectronic activity at various sites within neurons (Saha et al. 2008); and enable subcellular signal mapping (Timko et al. 2010).

3.3 Neural Dust

Another proposed neural interface technology would aim to deploy thousands of distinct free-floating 10–100 μm "neural dust" CMOS sensor nodes enabled with two-way communications, to allow for monitoring localized groups of neurons. These low-power microscale entities would be powered by external ultrasound (~500 μw of received power) and could be employed in backscatter communications. These neural dust nodes, with a 1 mm^2 interrogator, had the capacity to detect and transmit localized extracellular electrophysiological data. However, this technology has limited ability to efficiently access and interact with specific 3D neural whole human brain regions, such as the neocortex, to facilitate a formative B/CI (Seo et al. 2013).

3.4 Brain-Machine Interface (BMI)

Various Brain-Machine Interface (BMI) neural prosthetics (e.g., applied as cochlear implants, artificial retinas, motor cortex and hippocampus prosthetics, and more) composed of microchip sensor/electrode arrays are being investigated and implemented, which have the ability to interrogate multicellular signals (Berger et al. 2005; BrainGate 2009). One current BMI strategy interrogates neural activity to unidirectionally control an external device (Lebedev 2014) and another that transmits sensory feedback from a device to the brain (O'Doherty et al. 2011). Examples of technologies that enable noninvasive neural BMI interfaces encompass functional magnetic resonance imaging (fMRI), magnetoencephalography (MEG), and electroencephalography (EEG) (Miyawaki et al. 2008), as well as functional near-infrared spectroscopy (fNIRS) and photonic/optical strategies (Naseer and Hong, 2015).

There are a number of increasingly sophisticated prosthetics that are approaching parity with their biological counterparts in specific performance domains (e.g., synthetic hands with similar dexterity, precision, sensitivity, and strength to those of human hands). In a further advance, prosthetics imbued with neurophotonics (e.g., via high-speed bidirectional fiber-optic data transfer between the brain and prosthetic limbs) will establish high-resolution interfaces with peripheral nerves to enable users to feel pressure and temperature (Tabot et al. 2013).

In the cellular domain, the establishment of direct interfaces between solitary nerve cells and silicon microstructures is being investigated. Several researchers have reported on the spontaneous creation of neuro-silica interfaces using mammalian neurons to allow the direct excitation of nerve cells (Vassanelli and Fromherz 1997; Fromherz and Stett 1995; Schätzthauer and Fromherz 1998; Stett et al. 1997; Offenhausser 1996). Further advanced technologies may facilitate the development of brain/computer interfaces.

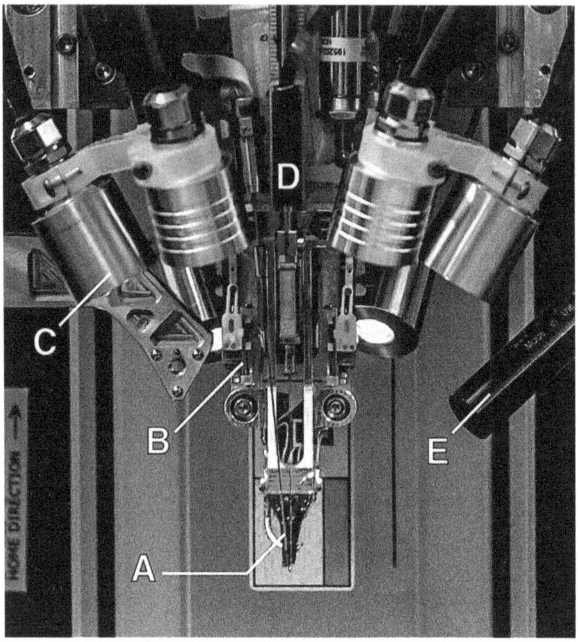

Fig. 3 The robotic electrode inserter. (*a*) Loaded needle pincher cartridge. (*b*) Low-force contact brain position sensor. (*c*) Light modules with multiple independent wavelengths. (*d*) Needle motor. (*e*) One of the four cameras focused on the needle during insertion. (*f*) Camera with wide angle view of surgical field (not shown). (*g*) Stereoscopic cameras (not shown). (Image credit: Elon Musk)

The most advanced BMI may be the recently developed Neuralink technology. "Brain-machine interfaces (BMIs) hold promise for the restoration of sensory and motor function and the treatment of neurological disorders, but clinical BMIs have not yet been widely adopted, in part because modest channel counts have limited their potential (Musk 2019). … In Neuralink's first steps toward a scalable high-bandwidth BMI system, the researchers have built arrays of small and flexible electrode 'threads,' with as many as 3072 electrodes per array distributed across 96 threads" (Fig. 3).

The Neuralink researchers have also "built a neurosurgical robot capable of inserting six threads (192 electrodes) per minute." Each thread can be individually inserted into the brain with micron precision for avoidance of surface vasculature and targeting specific brain regions.

"The electrode array is packaged into a small implantable device that contains custom chips for low-power onboard amplification and digitization: the package for 3072 channels occupies less than $(23 \times 18.5 \times 2)$ mm^3. A single USB-C cable provides full-bandwidth data streaming from the device, recording from all channels simultaneously."

This system has achieved "a spiking yield of up to 70% in chronically implanted electrodes. Neuralink's approach to BMI has unprecedented packaging density and scalability in a clinically relevant package."

3.5 Brainets

A "brainet" (brain network) system has enabled the brains of two rats with permanently implanted microelectrodes in the primary somatosensory cortex to directly communicate, even though they were located in different continents (Pais-Vieira et al. 2013).

Neuronal signals of multiple interfaced rat brains were recorded and processed to facilitate internet-transmitted data exchange and the performance of cooperative tasks (Pais-Vieira et al. 2015; Ramakrishnan et al. 2015). In another experimental brainet system, researchers investigated three different control systems that allowed multiple implanted monkeys to shared BMI-enabled control of a virtual "arm" (Ramakrishnan et al. 2015). This might be considered evidence for a nascent "superbrain" —the cooperative efforts of multiple individuals enabled higher-level functions that were not possible by solitary individuals.

In other research projects, a number of collaborative BMI tasks were investigated with humans, which involved the cooperative navigation of a spacecraft (Poli et al. 2013); decision-making (Eckstein et al. 2012; Yuan et al. 2013; Poli et al. 2014); and the design of movements (Wang and Jung 2011). "Clinically relevant brainets that connect patients with therapists, or healthy to unhealthy individuals, would be a particularly interesting application" (Martins et al. 2019).

3.6 Photometric Recording of Neural Dynamics

Burton et al. (2020) (Gutruf Lab, Dept. of Biomedical Engineering, University of Arizona) have developed a novel wireless, battery-free (energy harvesting), and extremely miniaturized subdermally implantable photometric system (Fig. 4), which has the capacity to continually record fine neuronal dynamics (e.g., calcium transients (0.0625–32 μM)). This platform has the advantages of facilitating real-time neural dynamics monitoring and clarifying the connectivity of neural circuits in completely unconstrained subjects. This low power-consuming (10.37 mW) system obtains power wirelessly through magnetic resonant coupling (13.56 MHz) "between a primary antenna that encircles the experimental arena and a millimeter-scale receiving antenna on the implant" and has strong potential for extensive utility, spanning a myriad of areas of neuroscience research.

3.7 Limitations of Current Techniques

The brief survey of technologies listed above illustrates the steady progress that is being made toward potentially interfacing the human brain with the cloud/edge in the coming decades. Specific criteria have been proposed for bona fide robust, safe,

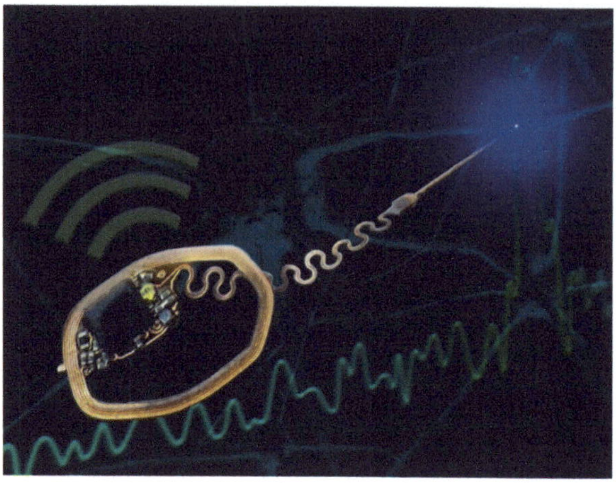

Fig. 4 Artistic illustration of wireless subdermally implantable photometric probe for continuous neural interrogation. (Image credit: Gutruf Lab, Biomedical Engineering, The University of Arizona)

and secure B/CI. "Current technological trajectories appear to be converging toward the creation of systems that will have the capacity to empower a human B/CI. However, since the human brain possesses cellular (neuron) and subcellular (synapse) processing elements, any technology that is capable of establishing a long-term and nondestructive, real-time human interface with the cloud must embody the following capabilities: (1) ultrahigh resolution mobility, (2) autonomous or semiautonomous activity, (3) nonintrusive (ideally, physiologically imperceptible) ingress/egress from the human body, and (4) supplying sufficient and robust information transfer bandwidth for interfacing with external supercomputing systems. Current techniques, whether in present day or extrapolated future forms, appear to be unscalable and incapable of fulfilling all of the temporal or spatial resolution requirements necessary for a properly comprehensive fully functional human B/CI" (Martins et al. 2019).

4 Neuralnanorobotic Species: Toward Enabling a Brain/Cloud Interface (B/CI)

Within the next several decades, we may witness the emergence of neuralnanorobotics via synergies between nanotechnology, nanomedicine, and AI, which may have the capacity to facilitate minimally invasive, nondestructive coupling of the human neocortex with a synthetic neocortex that resides in the cloud/edge, referred to here as a Brain/Cloud Interface (B/CI). Neuralnanorobotics, which proposes the application of highly advanced autonomous nanomedical robots to the human brain, was initially conceived by Robert Freitas Jr., who articulated how they might enable real-time in vivo monitoring of neural traffic, in conjunction with translation of neuronal/synaptic messages (Freitas 1999b, 2003; Martins et al. 2016).

A hypothetical B/CI would be based on neuralnanorobotics enabled by three classes of autonomous medical nanorobots, designated as endoneurorobots, gliabots, and synapobots, which would self-migrate and interface with neocortical neurons, glial cells, and synapses. "Real-time monitoring of the whole human brain (by placing neuralnanorobots within each neuron and nearby synaptic connections to record/transmit data from localized neuron and synapse spiking) will very likely provide redundant data that may be employed in the development of validation protocols" (Martins et al. 2019).

The achievement of a safe and reliable high performance B/CI will require stable, intimate connectivity between neuralnanorobotic entities and the ~86×10^9 neurons and ~2×10^{14} synapses of the human brain at an appropriate repetition rate (400–800 Hz) (Contreras 2004; Wilson 1999). A critical prerequisite prior to the installation of a B/CI would be the acquisition/storage of high-detailed structural and functional connectome-associated data for a specific individual brain, that is, highly specific mapping of the spatial coordinates/orientations and dendritic and axon configurations of different classes of neurons, inclusive of their typical electrophysiological spiking patterns (e.g., regular spiking, bursting, and fast spiking) (Seung 2011), as well as glial cells. These data would be uploaded to the various neuralnanorobots to facilitate programing. Dedicated autonomous nanomedical mapping devices, such as a conceptual Vascular Cartographic Scanning Nanodevice (VCSN), might be deployed for this task (Domschke and Boehm 2014).

Envisaged neuralnanorobotics-enabled B/CI technologies may have multiple advantages over their macroscale neural interface counterparts, including design at a useful scale (~2 μm) for optimization of neuronal interactivity, superior mobility, and minimal invasiveness/damage to brain tissues (self-implanting). Further, they may have the ability to be massively distributed within the brain, resolving a myriad of challenges that are currently faced in the development of safe and highly efficient B/CI (Martins et al. 2019). One design to facilitate a robust, safe, secure, and completely reversible B/CI based on these criteria is described below.

4.1 Endoneurobots

Endoneurobots are envisioned as autonomous neuron-resident neuralnanorobots that would interface with all ~86×10^9 neurons of the human brain at the axon initial segment (AIS) to directly monitor and bidirectionally interface with electrically processed action potential-based information. A 10 μm^3 volume of endoneurobots (Fig. 5) might be transdermally injected and subsequently proceed through the bloodstream, traverse the BBB (Freitas 2016), perhaps via a process akin to diapedesis (Boehm 2013), enter the brain parenchyma, and navigate the neuropil. There, they would have access to the concentrated microvasculature of the human brain, encasing ~100 billion capillaries, with a cumulative surface area of ~20 m^2 and an overall length of ~400 miles. The spacing of capillaries in the brain is ~40 μm on

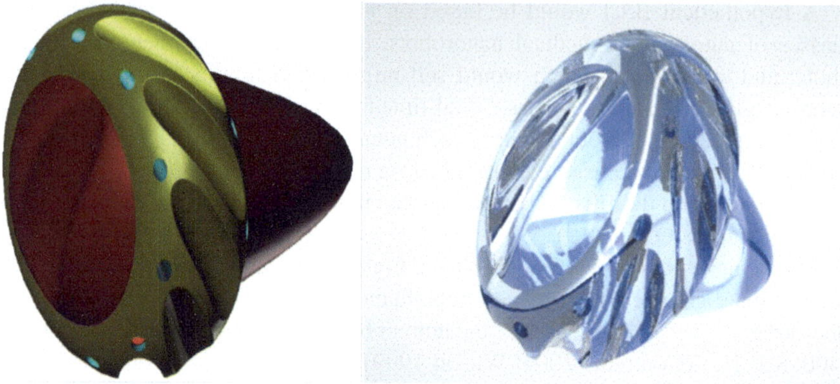

Fig. 5 Artistic illustration of endoneurobot (L) with diamondoid portrait (R). Grooves and orifices may enable propulsion within the neurons. Extendable tendrils might project from several of these orifices to facilitate solid tethering and slight orientation adjustments. (Image credits: ((L) Frank Boehm—Nanoapps Medical, Inc. and (R) Yuriy Svidinenko—Nanobotmodels Company). (These conceptual illustrations do not represent the actual neuralnanorobot design of the endoneurobots)

average; thus, every neuron of the brain is at a maximum the equivalent of 2–3 neurons distant from a microcapillary (Pardridge 2011).

Next, they would ingress into the neuron cell soma and intracellularly orient themselves within the Axon Initial Segment (AIS) (Martins et al. 2016). Once appropriately emplaced, the endoneurobots would monitor the action potentials and structural modifications derived from action potential-based functional data. Data acquired by the synapobots would be transmitted to their associated endoneurobots, which could be supported by gliabots if required. Subsequently, these data would be transmitted in vivo through a previously self-assembled high-speed auxiliary nanofiber-optic network to primary data transmission nodes for transfer to dedicated cloud-based supercomputers for postprocessing. The auxiliary nanofiber-optic system would serve to appreciably reduce the required onboard data storage for both the endoneurobots and synaptobots.

4.2 Gliabots

Gliabots would comprise autonomous glia-residing neuralnanorobots that would serve to monitor the pertinent activities of brain resident glial cells, with the further role as a facilitative infrastructure for endoneurobots and synaptobots and the overall B/CI system. Similar to endoneurobots, a 10 μm^3 volume of gliabots (Fig. 6) would ingress the brain through the capillaries and then their assigned glial cells, to optimally orient themselves at intracellular glial sites.

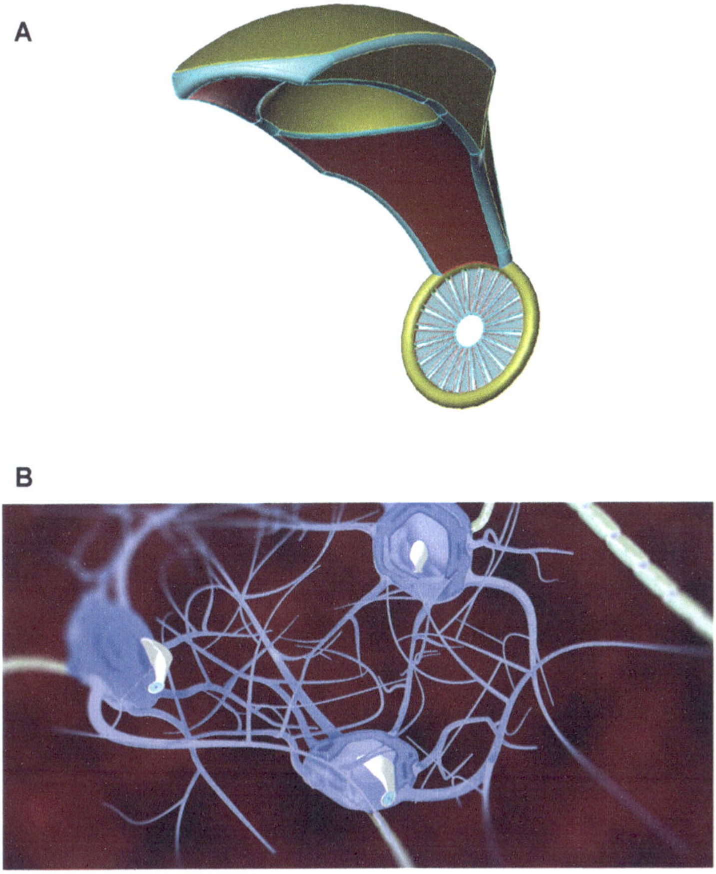

Fig. 6 Artistic illustration of gliabots that would self-migrate to glial cells and intracellularly orient themselves in the most suitable intraglial regions to conduct facilitative B/CI operations. (Image credits: (**a**) Frank Boehm—Nanoapps Medical, Inc.). (**b**) Julia Walker, Department of Chemical Engineering, Monash University. (These conceptual illustrations do not represent the actual neuralnanorobot design of the gliabots)

4.3 Synaptobots

Synaptobots, being the smallest (0.5 μm^3) of the three neuralnanorobotic species, would be tasked with monitoring the synapses, which are the most critical subcellular structures of the human brain. Since synapses (with 5–25% electrical and 75–95% chemical functionality) (DeFelipe and Fariñas 1992) are the core information processing elements of the brain's neural network, they play indispensable roles in learning and memory (Black et al. 1990; Bliss and Collingridge 1993; Holtmaat and Svoboda 2009; Liu et al. 2012); long- and short-term memory retention and erasure (Kandel 2001; Lee et al. 2008); and temporal data conversion (Fuhrmann et al. 2002). Thus, they are essential enabling components of the human

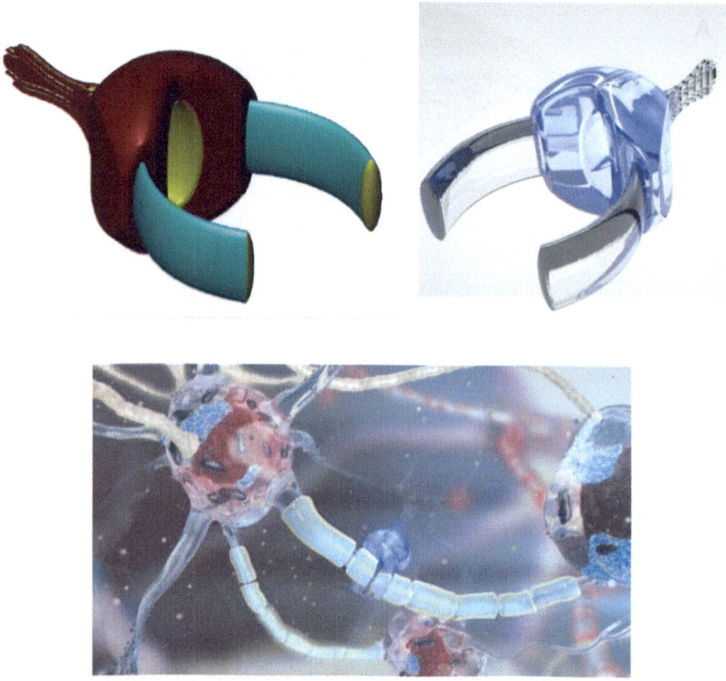

Fig. 7 Artistic illustrations of synaptobot (left) with diamondoid portrayal (right) and engaged in axon calibration (below). Oscillating piezo "fins" operating in conjunction with a medial ovoid orifice may facilitate flow-through propulsion. In one functional mode, hypersensitive extendible "cuff" nanosensors might circumscribe synaptic gaps to interrogate neurotransmitter traffic. (Image credits: (left) Frank Boehm, Nanoapps Medical, Inc. and (right & below) Yuriy Svidinenko, Nanobotmodels Company). (These conceptual illustrations do not represent the actual neural-nanorobot design of the synaptobots)

brain, in facilitating signal transduction and plasticity (Rollenhagen and Lübke 2006; Rollenhagen et al. 2007). The high resolution interrogation of synapses will likely be invaluable toward the establishment of a robust, stable, and fully operational real-time B/CI.

Synaptobots (Fig. 7) would serve as autonomous neuron-resident neuralnanorobots that employ multiple pliable nanostalk-mounted nanosensors to couple with each of the brain's ~2×10^{14} synapses to directly interrogate and interact with synaptically handled and stored information in the presynaptic or postsynaptic structures. As described by Martins et al. (2019), in contrast to endoneurobots and gliabots:

> Synaptobots would be delivered via the brain microvasculature to avoid long-distance navigation within the brain parenchyma. Auxiliary transport nanorobots having a volume of ~20 μm^3 (~$3.2 \times 2.5 \times 2.5$ μm) might each convey cargos of 24 synaptobots (total of ~12 μm^3) through the circulatory system and into the neuron soma. "The full complement of synaptobots would be transported by a fleet of ~1 trillion auxiliary transport nanorobots, which perform ~10 round trips to complete the insertion of all synaptobots," toward the

implementation of the neuralnanorobotic system prior to the activation of the B/CI system. Individual neurons, on average, would obtain ~117 such shipments, for an average overall distribution of 2800 synaptobots ($\approx 2.42 \times 10^{14}$ synapses/86×10^{9} neurons), which would assign one nanorobot per synapse.

Since the synaptically processed action potential-based information at these sites is considered to be elemental information (Shepherd 2003; Abbott and Regehr 2004; Fuhrmann et al. 2002), synaptobots would be tasked with identifying essentially all of the action potentials processed at the synapses, as well as their waveforms. They would then communicate "synaptically processed spikes into the data handling system. Consequently, neuralnanorobots would assist with the prediction of neurotransmitter bursts that traverse each synaptic gap. All these data will be continually processed at submillisecond resolution, enabling a virtually real-time data stream between the human brain and the cloud" (Martins et al. 2019). It may be the case that the interrogation of neuron- and synapse-prepared action potential-based electrical brain activity by itself (without the requirement of chemically based data) may be adequate to enable a robust human B/CI system. For instance, a recent investigation revealed that quantum dots had the capacity to serve as voltage-sensitive nanosensors for real-time observation of neuronal cell membrane potential (Nag et al. 2017).

Protocols initiated by endoneurobots, charged with communicating synaptic monitoring deficits, would ensure that synaptobots are regularly repositioned or replenished as necessary, due to nanorobot damage, neuronal death, or synapse elimination/formation (~1 trillion per day). Approximately ~1 trillion auxiliary transport nanorobots may be sufficient to handle the dynamic adjustments in the physical repositioning or deployment of synaptobots as is necessary. "The auxiliary transport nanorobots (~2.5 μm) will adhere to a similar transit protocol for crossing the BBB and traversing the neuropil as the endoneurobots and gliabots, which are of comparable size (~2.2 μm)" (Martins et al. 2019).

The combined data transmission support provided by the endoneurobots, gliabots, and nanofiber-optic system would significantly reduce the onboard synaptobot data storage requirements. The integrated synaptobot nanocomputer is anticipated to be an ~0.01 μm^3 CPU device with a capacity of ~100 megaflops (Freitas 2009). The overall computational volume required by synaptobots may be 0.11 μm^3 to satisfy redundancy requirements. This allocated volume would be equivalent to other nanorobot models with proportionate levels of mission design complexity (Freitas 2005a).

4.4 *Wireless Neuralnanorobot Transmitters*

Subsequent iterations of an initial high-speed nanofiber-optic network might integrate wireless transmitters (Fig. 8), which could be self-embedding at the periphery of the human brain or within the skull. These transmitters may be configured as a homogeneously distributed network with the capacity to wirelessly enable an interface with neurons, axons, and synapses, receiving and transmitting data from and to the cloud via an external I/O communication device.

5 Human Brain/Cloud Interface Applications

Beyond the many beneficial therapeutic medical applications of a neuralnanorobotically enabled B/CI for the diagnosis and treatment of all manner of cognitive impairments and diseases will be many nonmedical applications. Some of these are briefly described below.

5.1 Enhanced Education and Intelligence

Neuralnanorobotics residing within the human brain, interfaced with cloud/edge computing and manifest as a B/CI, may enable us to dramatically exceed our presently limited biological cognitive abilities, in terms of learning in a world of exponentially expanding knowledge. A mature B/CI might facilitate instantaneous access to the full extent of cumulative human knowledge, where the ultimate learning experience might involve the direct conveyance of knowledge to the human brain. Complex skills like playing the piano or performing a delicate brain operation might be "injected" into the brain, which may drastically reduce the time and practice typically required to gain expertise in these areas.

The rapid injection of facts and knowledge, however, "may not necessarily translate to cognition, understanding, meta-analysis, or meta thought, where imagination

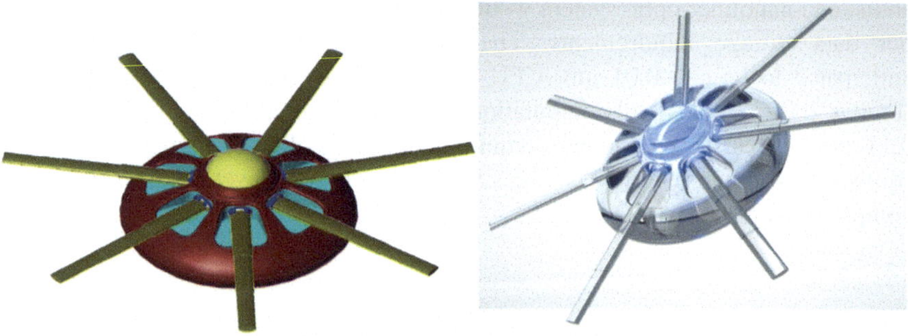

Fig. 8 Artistic illustrations of wireless nanoscale transmitter (left) and portrayed in diamondoid (right), which could interconnect to establish a homogeneously distributed mesh network, following self-migration at the periphery of the brain and on or within the skull. (Image credits: (left) Frank Boehm—Nanoapps Medical, Inc.; (right) Yuriy Svidinenko—Nanobotmodels Company). These conceptual illustrations do not represent the actual neuralnanorobot design of the wireless nanoscale transmitter

and creativity follow on" (Martins et al. 2019). With that said, all specific manual skills are in fact imprinted across various brain domains; thus, it follows that further access to the hippocampus and cerebellum for memory injection would also likely be necessary, in addition to the cerebellum and basal ganglia for the injection of complex motor tasks.

An initial proof-of-principle "instant learning" demonstration has been achieved by employing decoded functional magnetic resonance imaging. Patterns of human visual cortex brain activity were induced to align with a known earlier target state to enhance the completion of visual tasks (Shibata et al. 2011). New skills were also induced through the use of transcranial magnetic stimulation, where a strong pulsed magnetic field was applied to the skull using an external magnetic coil precisely positioned over the head. The energizing of a "virtual lesion" of small domains of the brain was observed to either degrade or augment skills, where ~40% of the subjects exhibited impressive new skills, such as drawing (Mottaghy et al. 1999).

Further augmentation of education may include virtual, fully immersive discussions with historical figures or other influential personalities to gain deep insights, virtual travel, or experiential classes on any topic (real or conceptual) occurring in the past, present, or future. This educational enhancement capacity may be facilitated in part by a Transparent Shadowing (TS) application, as described below.

In terms of human intelligence, neuralnanorobotically facilitated B/CI systems may enable significantly augmented mental acuity by optimizing/increasing the connective density of the intrinsic neural domain architectures of the brain. This might robustly enhance overall cognition, memory capacities, pattern recognition, inference, and associative abilities, which may cumulatively be associated with increased intelligence.

5.2 *Reduction of AI-Driven Existential Risk*

There are increasingly serious concerns that humans may become redundant in the near future due to exponentially advancing narrow AI and emergent general AI (Bostrom 2002, 2013; Bostrom and Cir 2008; Whitby and Oliver 2000; Schneider 2009; Yudkowsky 2008; Joy 2007). B/CI technologies may assist with allaying these fears to a certain degree. However, this will involve nothing less than the next evolutionary step for humankind, as it will entail our intimately melding with our technologies to become biosynthetic hybrids (Dewey 2015). This would allow us to keep pace with certain aspects of AI; however, whether this would be an ultimate solution remains to be seen.

5.3 *Virtual/Augmented Reality*

With the advent of mature neuralnanorobotics applied as a B/CI, full-immersion virtual reality, replete with ultrahigh tactile and sensory resolution, might be indiscernible from reality. This may relegate most types of physical travels for business or pleasure to obsolescence. As Martins et al. (2019) observe,

> Office buildings might be replaced by virtual-reality environments, wherein conferences could be attended virtually, replacing today's VoIP conference calls and Internet-based video conference calls with highly realistic fully immersive virtual reality conferences in virtual reality spaces. Immersive virtual reality will likely enable long-distance communications in engaging ways within environments that are indistinguishable from reality. The economic and environmental benefits of significantly reducing travel requirements will be enormous. For example, Cisco has reported savings of millions of dollars through the use of highly realistic telepresence systems.

Since artificially induced B/CI signals will likely have equivalence to actual physiological sensory inputs, relevant outputs from the brain to limbs and eyes, etc., may be neuralnanorobotically suppressed to negate their movement. All brain output signals might be suppressed by neuralnanorobots, and in their place, virtual limbs and sensory organs would respond appropriately while accommodating the surrounding virtual world. Neuralnanorobotics may also be applied to enable high-resolution, real-time augmented reality via superimposition of data related to the immediate environment on the retinas of the user, which might take the form of explanatory imagery or text. Further, auditory information provided by a virtual guide or avatar or instantaneous translation may likely be a commonplace application.

5.4 *"Transparent Shadowing"*

Neuralnanorobotically driven B/CI technologies operating in conjunction with cloud-based supercomputers may provide opportunities for users to literally engage in real-time episodic experiences (with ultrahigh fully immersive/sensory resolution) from the lives of any other concurring participants on the planet through what is referred to as "transparent shadowing" (TS). This ability might serve to enhance human cooperation, empathy, respect, and understanding to an unprecedented degree. This may ultimately lead, once adopted by a sufficiently large demographic on a global scale, to the eventual minimization and eradication of armed conflict (Domschke and Boehm 2014). As Kurzweil observes, "We will be able to change our appearance and effectively become other people" (Kurzweil 2005). An overview of the TS is articulated by Martins et al. (2019) below:

> *With neuralnanorobotically enabled B/CI, individuals might engage in the TS of voluntary or remunerated "spatial hosts." Under strict protocols, accredited spatial hosts would agree to allow single or multiple attendees (conceivably numbering in the millions) to literally experience portions of their life experiences over a predetermined timeline/schedule.*

These transparent shadowing sessions might be akin to today's seminars or lecture series, where the knowledge or specific skills of the host would be experientially imparted to the "attendees." However, these sessions would be exponentially more intimate, with seamless experiential resolution. The full sensorial realm (e.g., physical presence, tactile sensations, olfactory, visual, tastes, and auditory) would be experienced by the attendees as if they inhabited the body of the spatial host. Although they would perceive the vocal instructions of the host, to temporally experience exactly what the spatial host is experiencing, for the sake of personal privacy attendees might, by default, be completely blocked from any access to the thoughts, emotions, or self-speak of their spatial hosts.

Experiencing episodes of the lives of those in other cultures and ethnic groups could promote cross-cultural understanding and tolerance, improving prospects for the reduction of hatred and racism. For example, perhaps those of majority ethnic groups might be more sensitized with the issue of racism against those of minority ethnic groups, once they "experience" it for themselves through TS sessions. Similarly, minorities who experience a majority host might come to realize that many actions perceived by them as purposeful racism were entirely unintentional. Cross-gender experiences might impact real-life relationships between genders, due to increased empathy and understanding might be possible that an eventual shift in gender attitudes could lead to decreased gender-related and domestic violence.

6 Conclusion

The exponential pace of digitized human knowledge increasingly relies on cloud/edge computing to facilitate its processing and storage. It is becoming clear that the biologically constrained cognitive capacity of the human brain is quickly being outpaced by the generation of new knowledge and the advent of AI. So, it appears reasonable that if humanity is to keep abreast of its technologies and meaningfully interact with them, it may soon be facing the sobering choice between either becoming redundant or fusing with the technologies.

Neuralnanorobotics may facilitate the development of a robust, safe, and secure real-time interface between the human neocortex and cloud-based supercomputers, referred to as a brain/cloud interface (B/CI). This technology, which may be enabled through synergies between nanotechnology, nanomedicine, and AI, may allow us to instantaneously access the full breadth of human knowledge and engage in fully immersive learning, travel, and entertainment. Further, it might enable a specialized application referred to as "transparent shadowing" (TS), where we might experience episodes of other willing individuals' lives, anywhere on the planet in ultra-high fully immersive resolution. These insightful experiences might significantly enhance human understanding, empathy, and tolerance to unprecedented levels toward the ultimately positive benefit of humanity.

References

Abbott LF, Regehr WG (2004) Synaptic computation. Nature 431:796–803

Alexander GE, DeLong M, Strick PL (1986) Parallel organization of functionally segregated circuits linking basal ganglia and cortex. Ann Rev Neurosci 9:357–381

Anderson CT, Sheets PL, Kiritani T, Shepherd GM (2010) Sublayer-specific microcircuits of corticospinal and corticostriatal neurons in motor cortex. Nat Neurosci 13(6):739–744

Anderson JR, Jones BW, Watt CB, Shaw MV, Yang JH, Demill D et al (2011) Exploring the retinal connectome. Mol Vis 17:355–379

Ashford JW (2015) Treatment of Alzheimer's disease: the legacy of the cholinergic hypothesis, neuroplasticity, and future directions. J Alzheimers Dis 47:149–156

Astier Y, Bayley H, Howorka S (2005) Protein components for nanodevices. Curr Opin Chem Biol 9:576–584

Atencio CA, Schreiner CE (2010) Columnar connectivity and laminar processing in cat primary auditory cortex. PLoS One 5:e9521

Azevedo FA, Carvalho LR, Grinberg LT, Farfel JM, Ferretti RE, Leite RE et al (2009) Equal numbers of neuronal and nonneuronal cells make the human brain an isometrically scaled-up primate brain. J Comp Neurol 513:532–541

Berger TW, Ahuja A, Courellis SH, Deadwyler SA, Erinjippurath G, Gerhardt GA et al (2005) Restoring lost cognitive function. IEEE Eng Med Biol Mag 4:30–44

Berger TW, Hampson RE, Song D, Goonawardena A, Marmarelis VZ, Deadwyler SA (2011) A cortical neural prosthesis for restoring and enhancing memory. J Neural Eng 8(4):046017

Birbaumer N (2006) Breaking the silence: brain-computer interfaces (BCI) for communication and motor control. Psychophysiology 43:517–532

Black JE, Isaacs KR, Anderson BJ, Alcantara AA, Greenough WT (1990) Learning causes synaptogenesis, whereas motor activity causes angiogenesis, in cerebellar cortex of adult rats. Proc Natl Acad Sci U S A 87:5568–5572

Bliss TV, Collingridge GL (1993) A synaptic model of memory: long-term potentiation in the hippocampus. Nature 361:31–39

Boehm FJ (2013) Nanomedical device and systems design—challenges, possibilities, visions. CRC Press, Boca Raton

Bostrom N (2002) Existential risks: analyzing human extinction scenarios and related hazards. J Evol Technol 9:31–33

Bostrom N (2013) Existential risks reduction as global priority. Glob Policy 4:15–31

Bostrom N, Cir MM (2008) Global catastrophic risks, 1st edn. Oxford University Press, Oxford

BrainGate Wired for Thought (2009). http://www.braingate.com/

Burton A, Obaid SN, Vázquez-Guardado A, Schmit MB, Stuart T, Cai L, Chen Z, Kandela I et al (2020) Wireless, battery-free subdermally implantable photometry systems for chronic recording of neural dynamics. Proc Natl Acad Sci U S A 117(6):2835–2845

Buxhoeveden DP, Casanova MF (2001) The minicolumn hypothesis in neuroscience. Brain 125:935–951

Cajal SR (1899) Comparative study of the sensory areas of the human cortex. Harvard University, Cambridge

Calvo P, Gouritin B, Chacun H, Desmaële D, D'Angelo J, Noel JP et al (2001) Long-circulating pegylated polycyanoacrylate nanoparticles as new drug carrier for brain delivery. Pharm Res 18:1157–1166

Chaudhury D, Liu H, Han MH (2015) Neuronal correlates of depression. Cell Mol Life Sci 72:4825–4848

Constantinople CM, Bruno RM (2013) Deep cortical layers are activated directly by thalamus. Science 340(6140):1591–1594

Contreras D (2004). Electrophysiological classes of neocortical neurons. Neural Netw 17:633–646. https://doi.org/10.1016/j.neunet.2004.04.003

Cupaioli FA, Zucca FA, Boraschi D, Zecca L (2014) Engineered nanoparticles, how brain friendly is this new guest? Prog Neurobiol 119–120:20–38

D'Agata F, Ruffinatti FA, Boschi S, Stura I, Rainero I, Abollino O, Cavalli R, Guiot C (2017) Magnetic nanoparticles in the central nervous system: targeting principles, applications and safety issues. Molecules 23(1):pii:E9
Dai X, Hong G, Gao T, Lieber CM (2018) Mesh nanoelectronics: seamless integration of electronics with tissues. Acc Chem Res 51:309–318
DeFelipe J, Fariñas I (1992) The pyramidal neuron of the cerebral cortex: morphological and chemical characteristics of the synaptic inputs. Prog Neurobiol 39:563–607
DeFelipe J, Markram H, Rockland KS (2012) The neocortical column. Front Neuroanat 6:22
Dekosky ST, Scheff SW (1990) Synapse loss in frontal cortex biopsies in Alzheimer's disease: correlation with cognitive severity. Ann Neurol 27:457–464
Dewey D (2015) Long-term strategies for ending existential risk from fast takeoff. In: Müller V (ed) Risks of artificial intelligence. CRC Press, Boca Raton
Domschke A, Boehm FJ (2014) Quandary—are molecularly manufactured burgers imbued with the life force? (Fqxi essay contest, in response to the question, how should humanity steer the future?). https://ieet.org/index.php/IEET2/more/boehm20160115
Douglas RJ, Martin KA (1991) A functional microcircuit for cat visual cortex. J Physiol 440:735–769
Duan X, Gao R, Xie P, Cohen-Karni T, Qing Q, Choe HS et al (2011) Intracellular recordings of action potentials by an extracellular nanoscale field-effect transistor. Nat Nanotechnol 7:174–179
Eckstein MP, Das K, Pham BT, Peterson MF, Abbey CK, Sy JL et al (2012) Neural decoding of collective wisdom with multi-brain computing. Neuroimage 59:94–108
Falk A, Heine VM, Harwood AJ, Sullivan PF, Peitz M, Brüstle O et al (2016) Modeling psychiatric disorders: from genomic findings to cellular phenotypes. Mol Psychiatry 21:1167–1179
Favorov O, Whitsel BL (1988a) Spatial organization of the peripheral input to area 1 cell columns I. The detection of 'segregates'. Brain Res 472(1):25–42
Favorov O, Whitsel BL (1988b) Spatial organization of the peripheral input to area 1 cell columns II. The forelimb representation achieved by a mosaic of segregates. Brain Res 472(1):43–56
Fornito A, Zalesky A, Breakspea M (2015) The connectomics of brain disorders. Nat Rev Neurosci 16:159–172
Freitas RA Jr (1998) Exploratory design in medical nanotechnology: a mechanical artificial red cell. Artif Cells Blood Substit Immobil Biotechnol 26:411–430
Freitas RA Jr (1999a) Is diamond biocompatible with living cells? IMM report 12. Institute for Molecular Manufacturing, Palo Alto
Freitas RA Jr (1999b) Nanomedicine, volume I: Basic capabilities. Georgetown, Landes Bioscience
Freitas RA Jr (2002) Is sapphire biocompatible with living cells? IMM report 35. Institute for Molecular Manufacturing, Palo Alto
Freitas RA Jr (2003) Nanomedicine, volume IIA: Biocompatibility. Georgetown, Landes Bioscience
Freitas RA Jr (2005a) Current status of nanomedicine and medical nanorobotics (Invited Survey). J Comput Theor Nanosci 2:1–25
Freitas RA Jr (2005b) Microbivores: artificial mechanical phagocytes using digest and discharge protocol. J Evol Technol 14:1–52
Freitas RA Jr (2007) The ideal gene delivery vector: chromallocytes, cell repair nanorobots for chromosome replacement therapy. J Evol Technol 16:1–97
Freitas RA Jr (2009) Chapter 15. Computational tasks in medical nanorobotics. In: Eshaghian-Wilner MM (ed) Bio-inspired and nano-scale integrated computing. Wiley, New York, pp 391–428
Freitas RA Jr (2010) Chapter 23. Comprehensive nanorobotic control of human morbidity and aging. In: Fahy GM, West MD, Coles LS, Harris SB (eds) The future of aging: pathways to human life extension. Springer, New York, pp 685–805
Freitas RA Jr (2016) The Alzheimer protocols: a nanorobotic cure for Alzheimer's disease and related neurodegenerative conditions. http://www.imm.org/Reports/rep048.pdf

Freitas RA Jr, Merkle RC (2004) Kinematic self-replicating machines. Georgetown, Landes Bioscience

Freitas RA Jr, Merkle RC (2006) Nanofactory collaboration. http://www.molecularassembler.com/Nanofactory

Fromherz P, Stett A (1995) Silicon-neuron junction: capacitive stimulation of an individual neuron on a silicon chip. Phys Rev Lett 75:1670–1673

Fuhrmann G, Segev I, Markram H, Tsodyks M (2002) Coding of temporal information by activity-dependent synapses. J Neurophysiol 87:140–148

Fuster JM, Bressler SL (2012) Cognit activation: a mechanism enabling temporal integration in working memory. Trends Cogn Sci 16(4):207–218

Grau C, Ginhoux R, Riera A, Nguyen TL, Chauvat H, Berg M et al (2014) Conscious brain-to-brain communication in humans using non-invasive technologies. PLoS One 9:e105225

Guduru R, Liang P, Hong J, Rodzinski A, Hadjikhani A, Horstmyer J et al (2015) Magnetoelectric 'spin' on stimulating the brain. Nanomedicine 10:2051–2061

Hampson RE, Gerhardt GA, Marmarelis V, Song D, Opris I, Santos L, Berger TW, Deadwyler SA (2012) Facilitation and restoration of cognitive function in primate prefrontal cortex by a neuroprosthesis that utilizes minicolumn-specific neural firing. J Neural Eng 9(5):056012

Herculano-Houzel S (2009) The human brain in numbers: a linearly scaled-up primate brain. Front Hum Neurosci 3:31

Hirabayashi T, Takeuchi D, Tamura K, Miyashita Y (2013) Microcircuits for hierarchical elaboration of object coding across primate temporal areas. Science 341(6142):191–195

Holtmaat A, Svoboda K (2009) Experience-dependent structural synaptic plasticity in the mammalian brain. Nat Rev Neurosci 10:647–658

Jones EG (2000) Microcolumns in the cerebral cortex. Proc Natl Acad Sci U S A 97:5019–5021

Joy B (2007) Why the future doesn't need us: how 21st century technologies threaten to make humans an endangered species. New York, Random House Audio

Kandel ER (2001) The molecular biology of memory storage: a dialogue between genes and synapses. Science 294:1030–1038

Kandel ER, Schwartz JH (1985) Principles of neural science, 2nd edn. Elsevier, Amsterdam

Kandel ER, Schwartz J, Jessell T (2000) Principles of neural science, 4th edn. McGraw Hill, New York

Karlsen AS, Pakkenberg B (2011) Total numbers of neurons and glial cells in cortex and basal ganglia of aged brains with Down syndrome-a stereological study. Cereb Cortex 21:2519–2524

Kleinfeld D, Bharioke A, Blinder P, Bock DD, Briggman KL et al (2011) Large-scale automated histology in the pursuit of connectomes. J Neurosci 31:16125–16138

Kostarelos K (2010) Nanorobots for medicine: how close are we? Nanomedicine 5:341–342

Kurzweil R (2005) The Singularity Is Near: When Humans Transcend Biology. New York, NY: Viking Press

Kurzweil R (2014) Get ready for hybrid thinking. Ted Talk. https://www.ted.com/talks/ray_kurzweil_get_ready_for_hybrid_thinking/transcript

Kuzum D, Jeyasingh RG, Lee B, Wong HS (2012) Nanoelectronic programmable synapses based on phase change materials for brain-inspired computing. Nano Lett 12:2179–2186

Lebedev MA (2014) Brain-machine interfaces: an overview. Transl Neurosci 5:99–110

Lee SH, Choi JH, Lee N, Lee HR, Kim JI, Yu NK et al (2008) Synaptic protein degradation underlies destabilization of retrieved fear memory. Science 319:1253–1125

Letzkus JJ, Wolff SB, Meyer EM, Tovote P, Courtin J, Herry C, Lüthi A (2011) A disinhibitory microcircuit for associative fear learning in the auditory cortex. Nature 480(7377):331–335

Li J, Li X, Luo T, Wang R, Liu C, Chen S et al (2018) Development of a magnetic microrobot for carrying and delivering targeted cells. Sci Rob 3:eaat8829

Liu X, Ramirez S, Pang PT, Puryear CB, Govindarajan A, Deisseroth K et al (2012) Optogenetic stimulation of a hippocampal engram activates fear memory recall. Nature 484:381–385

Liu J, Fu TM, Cheng Z, Hong G, Zhou T, Jin L et al (2015) Syringe-injectable electronics. Nat Nanotechnol 10:629–636

Lorente de No R (1938) Analysis of the activity of the chains of internuncial neurons. J Neurophysiol 1:207–244
Lu J, Tapia JC, White OL, Lichtman JW (2009) The interscutularis muscle connectome. PLoS Biol 7:e32
Maass W, Zador AM (1999) Dynamic stochastic synapses as computational units. Neural Comput 11:903–917
Mahan MY, Georgopoulos AP (2013) Motor directional tuning across brain areas: directional resonance and the role of inhibition for directional accuracy. Front Neural Circuits 7:92
Mallouk TE, Sen A (2009) Powering nanorobots. Sci Am 300:72–77
Markram H, Toledo-Rodriguez M, Wang Y, Gupta A, Silberberg G, Wu C (2004) Interneurons of the neocortical inhibitory system. Nat Rev Neurosci 5(10):793–807
Marr B (2018) How much data do we create every day? The mind-blowing stats everyone should read. Forbes. Accessed 8 Feb 2020
Martel S, Mohammadi M, Felfoul O, Lu Z, Pouponneau P (2009) Flagellated magnetotactic bacteria as controlled, M.R.I-trackable propulsion and steering systems for medical nanorobots operating in the human microvasculature. Int J Robot Res 28:571–582
Martins NRB, Erlhagen W, Freitas RA Jr (2012) Non-destructive whole-brain monitoring using nanorobots: neural electrical data rate requirements. Int J Mach Conscious 4:109–140
Martins NRB, Erlhagen W, Freitas RA Jr (2015) Action potential monitoring using neuronanorobotics: neuroelectric nanosensors. Int J Nanomater Nanostruct 1:20–41
Martins NRB, Erlhagen W, Freitas RA Jr (2016) Human connectome mapping and monitoring using neuronanorobotics. J Evol Technol 26:1–24
Martins NRB, Angelica A, Chakravarthy K, Svidinenko Y, Boehm FJ, Opris I, Lebedev MA, Swan M, Garan SA, Rosenfeld JV, Hogg T, Freitas RA Jr (2019) Human brain/cloud interface. Front Neurosci 13:112
Mavroides D, Ferreira A (eds) (2011) Nanorobotics: current approaches and techniques. Springer, New York
Mayberg HS, Lozano AM, Voon V, Mcneely HE, Seminowicz D, Hamani C et al (2005) Deep brain stimulation for treatment-resistant depression. Neuron 45:651–660
Merkle R (1989) Energy limits to the computational power of the human brain. Palo Alto, Foresight Institute
Miyawaki Y, Uchida H, Yamashita O, Sato MA, Morito Y, Tanabe HC et al (2008) Visual image reconstruction from human brain activity using a combination of multiscale local image decoders. Neuron 60:915–929
Morris K (2001) Macrodoctor, come meet the nanodoctors. Lancet 357:778
Mottaghy FM, Hungs M, Brügmann M, Sparing R, Boroojerdi B, Foltys H et al (1999) Facilitation of picture naming after repetitive transcranial magnetic stimulation. Neurology 53:1806–1812
Mountcastle VB (1957) Modality and topographic properties of single neurons of cat's somatic sensory cortex. J Neurophysiol 20(4):408–434
Mountcastle VB (1997) The columnar organization of the neocortex. Brain 120(4):701–722
Musk E (2019) An integrated brain-machine interface platform with thousands of channels. bioRxiv
Nag K, Stewart MH, Deschamps JR, Susumu K, Oh E, Tsytsarev V et al (2017) Quantum dot-peptide-fullerene bioconjugates for visualization of in vitro and in vivo cellular membrane potential. ACS Nano 11:5598–5613
Naseer N, Hong KS (2015) fNIRS-based brain-computer interfaces: a review. Front Hum Neurosci 9:3
National Institute of Neurological Disorders and Stroke (NINDS) (2017). https://www.ninds.nih.gov/Disorders/All-Disorders. Accessed 8 Feb 2020
O'Doherty JE, Lebedev MA, Ifft PJ, Zhuang KZ, Shokur S, Bleuler H et al (2011) Active tactile exploration using a brain-machine-brain interface. Nature 479:228–231
Offenhausser A (1996) Neuron-silicon junction: electrical recordings from neural cells cultured on modified microelectronic device surfaces, In: Proceedings of the 18th annual international

conference of the IEEE on engineering in medicine and biology society, bridging disciplines for biomedicine, Piscataway, NJ

Opris I (2013) Inter-laminar microcircuits across neocortex: repair and augmentation. Front Syst Neurosci 19(7):80

Opris I, Bruce CJ (2005) Neural circuitry of judgment and decision mechanisms. Brain Res Rev 48:509–526

Opris I, Casanova MF (2014) Prefrontal cortical minicolumn: from executive control to disrupted cognitive processing. Brain 137(Pt 7):1863–1875

Opris I, Ferrera VP (2014) Modifying cognition and behavior with electrical microstimulation: implications for cognitive prostheses. Neurosci Biobehav Rev 47:321–335

Opris I, Hampso RE, Stanford TR, Gerhardt GA, Deadwyler SA (2011) Neural activity in frontal cortical cell layers: evidence for columnar sensorimotor processing. J Cogn Neurosci 23:1507–1521

Opris I, Hampson RE, Gerhardt GA, Berger TW, Deadwyler SA (2012a) Columnar processing in primate pFC: evidence for executive control microcircuits. J Cogn Neurosci 24(12):2334–2347

Opris I, Fuqua JL, Huettl PF, Gerhardt GA, Berger TW, Hampson RE, Deadwyler SA (2012b) Closing the loop in primate prefrontal cortex: inter-laminar processing. Front Neurosci 6:88

Opris I, Santos L, Gerhardt GA, Song D, Berger TW, Hampson RE, Deadwyler SA (2013) Prefrontal cortical microcircuits bind perception to executive control. Sci Rep 3:2285

Opris I, Fuqua JL, Gerhardt GA, Hampson RE, Deadwyler SA (2014) Prefrontal cortical recordings with biomorphic MEAs reveal complex columnar-laminar microcircuits for BCI/BMI implementation. J Neurosci Methods 244:104–113

Opris I, Gerhardt GA, Hampson RE, Deadwyler SA (2015a) Disruption of columnar and laminar cognitive processing in primate prefrontal cortex following cocaine exposure. Front Syst Neurosci 9:79

Opris I, Popa IL, Casanova MF (2015b) Prefrontal cortical microcircuits for executive control of behavior. Ch 10. In: Casanova MF, Opris I (eds) Recent advances in the modular organization of the cortex. Springer, New York

Pais-Vieira M, Lebedev M, Kunicki C, Wang J, Nicolelis MA (2013) A brain-to-brain interface for real-time sharing of sensorimotor information. Sci Rep 3:1319

Pais-Vieira M, Chiuffa G, Lebedev M, Yadav A, Nicolelis MA (2015) Building an organic computing device with multiple interconnected brains. Sci Rep 5:11869

Pakkenberg B, Gundersen HJ (1997) Neocortical neuron number in humans: effect of sex and age. J Comp Neurol 384:312–320. https://doi.org/10.1002/(SICI)1096-9861(19970728)384:2<312::AID-CNE10>3.0.CO;2-K

Pardridge WM (2011) Drug transport in brain via the cerebrospinal fluid. Fluids Barriers CNS 8:7

Park HH, Jamison AC, Lee TR (2007) Rise of the nanomachine: the evolution of a revolution in medicine. Nanomedicine 2:425–439

Patel GM, Patel GC, Patel RB, Patel JK, Patel M (2006) Nanorobot: a versatile tool in nanomedicine. J Drug Target 14:63–67

Patolsky F, Timko BP, Yu G, Fang Y, Greytak AB, Zheng G et al (2006) Detection, stimulation, and inhibition of neuronal signals with high-density nanowire transistor arrays. Science 313:1100–1104

Poli R, Cinel C, Matran-Fernandez A, Sepulveda F, Stoica A (2013) Towards cooperative brain-computer interfaces for space navigation. In: Proceedings of the international conference on intelligent user interfaces, Santa Monica, CA, p 149–160

Poli R, Valeriani D, Cinel C (2014) Collaborative brain-computer interface for aiding decision-making. PLoS One 9:e102693

Popov AM, Lozovik YE, Fiorito S, Yahia L (2007) Biocompatibility and applications of carbon nanotubes in medical nanorobots. Int J Nanomedicine 2:361–372

Powell TP, Mountcastle VB (1959) Some aspects of the functional organization of the cortex of the postcentral gyrus of the monkey: a correlation of findings obtained in a single unit analysis with cytoarchitecture. Bull Johns Hopkins Hosp 105:133–162

Ramakrishnan A, Ifft PJ, Pais-Vieira M, Byun YW, Zhuang KZ, Lebedev MA et al (2015) Computing arm movements with a monkey brainet. Sci Rep 5:10767

Rengachary SS, Ellenbogen RG (eds) (2005) Principles of neurosurgery. Elsevier, Edinburgh

Rollenhagen A, Lübke JH (2006) The morphology of excitatory central synapses: from structure to function. Cell Tissue Res 326:221–237

Rollenhagen A, Sätzler K, Rodríguez EP, Jonas P, Frotscher M, Lübke JH (2007) Structural determinants of transmission at large hippocampal mossy fiber synapses. J Neurosci 27:10434–10444

Saha S, O'Malley DM, Menon L (2008) Vertically arranged gold nanowires: an interface for live neuronal recordings. NSTI, Boston, pp 1–5

Sandberg A, Bostrom N (2008) Whole Brain Emulation: A Roadmap. Technical Report #2008-3. Oxford: Oxford University

Schätzthauer R, Fromherz P (1998) Neuron-silicon junction with voltage-gated ionic currents. Eur J Neurosci 10:1956–1962

Scheff SW, Price DA (2006) Alzheimer's disease-related alterations in synaptic density: neocortex and hippocampus. J Alzheimers Dis 9(3 Suppl):101–115

Schneider S (2009) Science fiction and philosophy: from time travel to superintelligence, 1st edn. Malden, Wiley-Blackwell

Schuhmann TG Jr, Yao J, Hong G, Fu TM, Lieber CM (2017) Syringe-injectable electronics with a plug-and-play input/output interface. Nano Lett 17:5836–5842

Seo D, Carmena JM, Rabaey JM, Alon E, Maharbiz MM (2013) Neural dust: an ultrasonic, low power solution for chronic brain-machine interfaces. http://arxiv.org/pdf/1307.2196v1.pdf

Seung HS (2011) Neuroscience: towards functional connectomics. Nature 471:170–172

Shepherd GM (2003) The synaptic organization of the brain. Oxford University Press, Oxford

Shepherd G, Grillner S (2010) Handbook of brain microcircuits. Oxford University Press, Oxford

Shibata K, Watanabe T, Sasaki Y, Kawato M (2011) Perceptual learning incepted by decoded fmri neurofeedback without stimulus presentation. Science 334:1413–1415

Smith CJ, Ashford JW, Perfetti TA (2019) Putative survival advantages in young apolipoprotein ε4 carriers are associated with increased neural stress. J Alzheimers Dis 68(3):885–923

Sporns O, Tononi G, Kötter R (2005) The human connectome: a structural description of the human brain. PLoS Comput Biol 1:e42

Srikanth M, Kessler JA (2012) Nanotechnology-novel therapeutics for CNS disorders. Nat Rev Neurol 8:307–318

Stark AK, Petersen AO, Gardi J, Gundersen HJ, Pakkenberg B (2007a) Spatial distribution of human neocortical neurons and glial cells according to sex and age measured by the saucer method. J Neurosci Methods 164:19–26. https://doi.org/10.1016/j.jneumeth.2007.03.019

Stark AK, Toft MH, Pakkenberg H, Fabricius K, Eriksen N, Pelvig DP et al. (2007b).The effect of age and gender on the volume and size distribution of neocortical neurons. Neuroscience 150:121–130

Stett A, Müller B, Fromherz P (1997) Two-way silicon-neuron interface by electrical induction. Physical Review E 55:1779–1782

Stone JL, Hughes JR (2013) Early history of electroencephalography and establishment of the American Clinical Neurophysiology Society. J Clin Neurophysiol 30:28–44

Swan M (2016) The future of brain-computer interfaces: blockchaining your way into a cloud-mind. J Evol Technol 26:60–81

Szentágothai J, Arbib MA (1975) Conceptual models of neural organization. MIT Press, Cambridge

Tabot GA, Dammann JF, Berg JA, Tenore FV, Boback JL, Vogelstein RJ et al (2013) Restoring the sense of touch with a prosthetic hand through a brain interface. Proc Natl Acad Sci U SA 110:18279–18284

Taghva AS, Malone DA, Rezai AR (2013) Deep brain stimulation for treatment-resistant depression. World Neurosurg 80:S27.e17–S27.e24

Takeuchi D, Hirabayashi T, Tamura K, Miyashita Y (2011) Reversal of interlaminar signal between sensory and memory processing in monkey temporal cortex. Science 331:1443–1447

Tang Y, Nyengaard JR, De Groot DM, Gundersen HJ (2001) Total regional and global number of synapses in the human brain neocortex. Synapse 41:258–273. https://doi.org/10.1002/syn.1083)

Terry RD, Masliah E, Salmon DP, Butters N, Deteresa R, Hill R et al (1991) Physical basis of cognitive alterations in Alzheimer's disease: synapse loss is the major correlate of cognitive impairment. Ann Neurol 30:572–580

Tian B, Cohen-Karni T, Qing Q, Duan X, Xie P, Lieber CM (2010) Three-dimensional, flexible nanoscale field-effect transistors as localized bioprobes. Science 329:830–834

Timko BP, Cohen-Karni T, Qing Q, Tian B, Lieber CM (2010) Design and implementation of functional nanoelectronic interfaces with biomolecules, cells, and tissue using nanowire device arrays. IEEE Trans Nanotechnol 9:269–280

Tommerdahl M, Favorov O, Whitsel BL, Nakhle B, Gonchar YA (1993) Minicolumnar activation patterns in cat and monkey SI cortex. Cereb Cortex 3(5):399–411

Van de Burgt Y, Lubberman E, Fuller EJ, Keene ST, Faria GC, Agarwal S et al (2017) A non-volatile organic electrochemical device as a low-voltage artificial synapse for neuromorphic computing. Nat Mater 16:414–418

Vassanelli S, Fromherz P (1997) Neurons from rat brain coupled to transistors. Appl Phys A 65:85–88

Veliev F (2016) Interfacing neurons with nanoelectronics: from silicon nanowires to carbon devices materials. Universiteì Grenoble Alpes, Grenoble

Vidal JJ (1973) Toward direct brain-computer communication. Annu Rev Biophys Bioeng 2:157–180

Vidu R, Rahman M, Mahmoudi M, Enachescu M, Poteca TD, Opris I (2014) Nanostructures: a platform for brain repair and augmentation. Front Syst Neurosci 8:91

Wang XJ (2006) Toward a prefrontal microcircuit model for cognitive deficits in schizophrenia. Pharmacopsychiatry 39(Suppl 1):S80–S87

Wang Y, Jung TP (2011) A collaborative brain-computer interface for improving human performance. PLoS One 6:e20422

Weiler N, Wood L, Yu J, Solla SA, Shepherd GM (2008) Top-down laminar organization of the excitatory network in motor cortex. Nat Neurosci 11:360–366

Whitby B, Oliver K (2000) How to avoid a robot takeover: political and ethical choices in the design and introduction of intelligent artifacts. Presented presented at the AISB-00 symposium on artificial intelligence, ethics an (quasi-) human rights, Birmingham

Wiegand HF, Beed P, Bendels MH, Leibold C, Schmitz D, Johenning FW (2011) Complementary sensory and associative microcircuitry in primary olfactory cortex. J Neurosci 31(34):12149–12158

Wilson HR (1999) Simplified dynamics of human and mammalian neocortical neurons. J Theor Biol 200:375–388

Yuan P, Wang Y, Gao X, Jung TP, Gao S (2013) A collaborative brain-computer interface for accelerating human decision making. In: Stephanidis C, Antona M (eds) Proceedings of the 7th international conference on universal access in human-computer interaction: design methods, tools, and interaction techniques for eInclusion, UAHCI 2013. Springer, Berlin, p 672–681

Yudkowsky E (2008) Artificial intelligence as a positive and negative factor in global risk. In: Bostrom N, Cirkovic M (eds) Global catastrophic risks. Oxford University Press, Oxford

Zeck G, Fromherz P (2001) Noninvasive neuroelectronic interfacing with synaptically connected snail neurons immobilized on a semiconductor chip. Proc Natl Acad Sci U S A 98:10457–10462

Zheng M, Ruan S, Liu S, Sun T, Qu D, Zhao H et al (2015) Self-targeting fluorescent carbon dots for diagnosis of brain cancer cells. ACS Nano 9:11455–11461. https://doi.org/10.1021/acsnano.5b05575

Zigmond MJ, Coyle JT, Rowland LP (2014) Neurobiology of brain disorders: biological basis of neurological and psychiatric disorders. Academic Press, Cambridge

Augmentation of Neuromarketing by Neural Technology

Qingguo Ma, Jia Jin, Tao Liu, and Xiaoyi Wang

1 Introduction

1.1 Great Advances in Neurotechnology Have Promoted the Development of Economics and Management, Including Marketing

In recent decades, there have been great advances in neural technology. Specifically, the huge progress of the noninvasive neural technology, which can detect human brain activity, makes the research platform for social science (such as sociology, economics, and management) greatly expanded. It also makes people have new

Q. Ma (✉)
School of Management, Zhejiang University, Hangzhou, China

Neuromanagement Lab, Zhejiang University, Hangzhou, China

School of Management, Zhejiang University of Technology, Hangzhou, China

Business School, Ningbo University, Ningbo, China

Academy of Neuroeconomics and Neuromanagement, Ningbo University, Ningbo, China
e-mail: maqingguo3669@zju.edu.cn

J. Jin
Business School, Ningbo University, Ningbo, China

Academy of Neuroeconomics and Neuromanagement, Ningbo University, Ningbo, China

T. Liu
School of Management, Zhejiang University, Hangzhou, China

X. Wang
School of Management, Zhejiang University, Hangzhou, China

Neuromanagement Lab, Zhejiang University, Hangzhou, China

I. Opris et al. (eds.), *Modern Approaches to Augmentation of Brain Function*, Contemporary Clinical Neuroscience,
https://doi.org/10.1007/978-3-030-54564-2_18

understanding of social science based on the neuroscience, i.e., a new perspective to understand the social science.

In the history of scientific development, people always try their best to use new technologies to explore the unknown world, that is, to expand the application boundary of new technologies as far as possible, so as to expand human scientific knowledge and promote the progress of human civilization. Among them, the emergence and development of interdisciplinary subjects is one of the most important forms of pushing scientific progress.

Neuromarketing is a typical branch of these interdisciplinary subjects. This is the reason we focus on augmentation of neuromarketing by neural technology.

1.2 Marketing and Its Traditional Research Methods

The central objective of marketing is to help match products with people and then provide values to them. It is important to understand consumer preferences and their choice process (Ariely and Berns 2010). Consumer behavior and decision-making need to be directed by the brain, and so it is necessary to uncover how the brain works to respond to different marketing promotions.

The traditional approaches to marketing research mainly include surveys and other types of self-reports such as focus groups. These traditional questionnaire survey approaches used to evaluate consumer preferences, attitudes, and purchase intentions may have deviations and lead to incorrect results (Ariely and Berns 2010). On the one hand, consumers may refuse to reveal their truth willingness for some reason; on the other hand, consumers may also cause unconscious deviations due to inaccurate expressions. Neural technology can objectively reflect the neural mechanism in the course of consumer behavior by directly detecting changes in consumers' brains. In essence, most of the current empirical research studies on neuromarketing use the objectivity of neuroscience and neurotechnology to verify the existing marketing theory. Brain scanning tools can be a useful supplement to consumer surveys.

The traditional questionnaire market survey is based on the assumption that people's real needs are self-knowledge. No matter how objective the questionnaire's targets are, the final result is still judged by the brain. Neuroscience provides a reliable method for revealing consumers' true decision-making process. This feature is currently used by many business consulting organizations for market research and advertising effectiveness testing.

Self-reports rely on ability and willingness of individuals to accurately report their attitudes and/or prior behaviors (Petty et al. 1983). Even more, when it comes to the factors that may elicit nearly automatic visceral, 'gut,' or emotional reactions, individuals asked to subjectively report may not be able to articulate how or what they are feeling or why they like or dislike certain stimulus (Schwarz and Clore 2003; Gerger et al. 2018), raising the need for more 'objective' measures of direct and implicit processes of human perception and decision-making (Griffin and Hauser 1993; Ariely and Berns 2010; Waldman et al. 2017).

1.3 New Technology of the Augmentation of Neuromarketing: Neural Technology

A particularly powerful solution to the more objective and actionable assessment of human-environment interactions may be found through brain imaging (Plassmann et al. 2015; Karmarkar and Yoon 2016). This approach, or the wider, emerging field of neuromarketing, involves considering consumer perceptions, responses, or behavior as they are related to activity or processes in the brain, and has attracted increasing attention as a potential complement to traditional behavioral measures (Ariely and Berns 2010; Yoon et al. 2012). As Plassmann et al. (2015) proposed, neuromarketing could provide access to more objective, direct and implicit processes of human perception, and decision-making and thereby help to better understand consumer behaviors that are difficult to access using traditional behavioral measures (Plassmann et al. 2015). It is thus believed that neuromarketing may turn black box of consumer's mind into an aquarium, providing additional, often less subjective, and in-the-moment information about consumption decisions and interactive experiences (Shiv et al. 2005; Yoon et al. 2012; Karmarkar and Yoon 2016).

In the 15 years from 2004 to 2019, neuromarketing has accumulated a large number of academic research results (McClure et al. 2004; Lee et al. 2007; Eser et al. 2011; Shiv and Yoon 2012; Agarwal and Dutta 2015; Plassmann et al. 2015). Marketing scholars have borrowed various neuroscience and neurotechnology methods and approaches to study the effects of marketing stimuli on consumers' brains. These neuroscience methods cover functional magnetic resonance imaging (fMRI), event-related potentials (ERPs), functional near-infrared spectroscopy (fNIRS) brain imaging, transcranial electrical stimulation (TES), transcranial magnetic stimulation (TMS), and other technologies.

These neural techniques can be categorized into three classes. The first class is the techniques based on the directly measuring electrical activity associated with neuronal firing, such as ERP (Gazzaniga et al. 2006). The second class indirectly measures the neuronal activity, which is supported by increased local blood flow and metabolic activity (Bunge and Kahn 2009) such as fMRI and fNIRS. The third class is a kind of technique that influences or modulates the activity of the human brain (Lewis et al. 2016), such as TES and TMS.

We will introduce three neural techniques, ERP, fMRI, and fNIRS, which are used widely.

1.4 The Major Fields of Neuromarketing Augmented by Neural Technology

Almost all aspects of neuromarketing can be augmented by neural technology, for example, the brand perception and the factors influenced the brand perception (such as the facial attractiveness, the continuous wining, and the fame of online product description), the marketing strategy of brand extension, the perception of advertisement and the advertising design, the perception of price and the factors that impacted the price perception, the fake rating of the online products and the strategy of cash coupons, the social factors that influenced the purchasing willingness (such as social risks), the luxury purchase and the prosocial behavior, the factors of the human sensory which influenced the purchasing decision, and so on.

1.5 How Does the Neurotechnology Augment Neuromarketing

The main ways in which neurotechnology augments neuromarketing are as follows:

- Neurotechnology can help reveal the implicit processes of psychology, which are typically difficult to be assessed using traditional approaches in marketing studies such as self-report. Specifically, the issues leading moral consequences are to be subject to self-deception and social desirability biases (Weaver and Prelec 2013) or situations and contexts in which the decision maker is unaware of or unable to articulate why he or she exhibits a specific behavior (Plassmann et al. 2015). Investigation of the neural processes can provide a window into the consumers' implicit motivations and serve as a complement to or explanation of self-reported results (Ahlert et al. 2006; Yoon et al. 2006).
- Neural technology can be used to find the real subtle factors influencing the purchase decision and their interaction among the factors, for example, how the interaction between color saturation and sound frequency affects consumers' perception of product (Wang et al. 2019).
- Neurotechnology can be used to study the cognitive neural basis of individual personalized consumption, for example, consumers' implicit subjective neural responses to experience good designs (Ma et al. 2018b).
- Neurotechnology can be employed to check whether some previous marketing theories are correct or not; in other words, if these theories can be supported by neuroscience, they should be correct. For example, Hilke Plassmann et al. (2008) using fMRI examined that increasing the price of a wine increased flavor pleasantness of the participants, as well as blood oxygen level-dependent activity in medial orbitofrontal cortex, which was widely thought as the neural reflection of experienced pleasantness (Plassmann et al. 2008). This study confirmed the inductive effect of price on consumption and proved that the basic assumptions in economics about consumption experienced pleasantness only depending on the intrinsic properties of the product and on the state of the individual are incomplete (Kahneman et al. 1997).
- In addition, neurotechnology can be used to discover new factors that influence consumers' brand perception, product perception, price perception, etc. from the perspective of neuroscience. For example, the facial attractiveness of the brand referrers affects the perception of the brand, the continuous wining affects the perception of product price, etc. In a word, neurotechnology can be used to augment neuromarketing.

1.6 The Important Challenges and the Trends in Neuromarketing

1.6.1 Three Important Challenges in Neuromarketing

The development of neuromarketing faces many challenges. Here are a few that deserve our first attention:

1. The feeling in real scenarios
   There is a natural gap between the laboratory environment using neural technology and the real marketing environment (for example, fMRI is a semiclosed high-noise environment, which is very different from the environment experienced by consumers in stores), which has long been questioned (James 2004) and has not been solved well. To find a better solution is the challenge that neuromarketing must face with.
2. Whether the behavioral data are consistent with the neural data
   If the behavioral data are consistent with the neural data, both can confirm each other. In some studies, however, the behavioral data are inconsistent with the neural data. What is the cause? Is it an experiment design problem, an experiment manipulation problem, or an inevitable problem under a certain experimental paradigm (including tasks)? It is a real challenge.
3. The Reverse Inference Issue
   Reverse Inference refers to the process of inferring mental states from neural data. Hilke Plassmann et al. (2015) gave a clear illustration for it (Plassmann et al. 2015).

   (a) In the current study, task A activated brain region Z,
   (b) Previous studies found that cognitive process C activated brain region Z,
   (c) Thus, the activity in area Z in the current study demonstrates that cognitive process C occurred in task A.

   Reverse Inference is often used in the study of neuromarketing augmented by neural technology. But in many cases, the inference is not correct. Because cognitive process E may also activate brain region Z, you cannot simply say that activation of brain region Z does not reflect cognitive process E.
   How to solve this problem is the third important challenge for us.

1.6.2 An important Trends in Neuromarketing: Hyperscanning

Marketing is a social activity, and consumer behavior is inherently social in nature. Therefore, a new trend of scanning multiple brains simultaneously in a more natural social context, in terms of hyperscanning, has emerged in recent years (for reviews, see Hasson et al. 2012; Hasson and Honey 2012; Pozharliev et al. 2017), making it possible to examine consumer behaviors in real-world marketing environments (Barnett and Cerf 2017a, b). It will open new frontiers in neuromarketing.

2 The Neural Techniques to Augment Neuromarketing

Prior to introduction of neuromarketing topics, we briefly introduce three main brain-imaging techniques that are potentially important for the neuromarketing research (Table 1). The first one is the technique of ERP, which is based on electroencephalography (EEG), the second one is the hemodynamic technique and based on magnetics, i.e., functional magnetic resonance imaging (fMRI), and the third one is also the hemodynamic technique but based the Optics, i.e., functional near-infrared spectroscopy (fNIRS).

2.1 *Principles of ERP, fMRI, and fNIRS*

Considering marketing activities as environmental stimuli, perception of certain marketing stimulus activates neurons in related brain regions, inducing neuroelectric activity, for cognitive or emotional processing. The neural firing will in turn increases metabolic activity of neurons, and the required energy is then supplied from local blood supply. Accordingly, the cerebral blood flow (CBF) to that region will also increase cerebral blood volume (CBV), oxygen metabolic rate ($CMRO_2$), and oxygenated blood (oxy-Hb) (Attwell and Iadecola 2002). Because the degree of increase in CBF is larger than that of the increase in CBV and $CMRO_2$, this leads to a decrease in deoxy-Hb as more oxygenated blood flows into the area (see Buxton 2012 for details). Therefore, increase in oxy-Hb with a decrease in deoxy-Hb in the brain tissue provides an indicator of neural activity in a specific region (Hoshi 2003).

EEG directly measures the gross electrical activity of the surface of the brain (Carter and Shieh 2015). ERP technique is formed by a series of data processing, such as superposition and artifacts removal, based on EEG that is related to an event. ERPs are distinct, stereotyped waveforms in an EEG report that correspond to a specific sensory, cognitive, or motor event. It can be used to ascertain certain particular states of consciousness with a temporal resolution of milliseconds (Carter and Shieh 2015).

fMRI is one of the tools to study the neural basis of cognition, which measures hemodynamic activations. Specifically, the fMRI uses magnetic properties to mea-

Table 1 Comparison of three brain-imaging techniques of fMRI, EEG/ERP, and fNIRS

Items	fMRI	EEG/ERP	fNIRS
Spatial resolution	+	–	~
Temporal resolution	–	+	~
Constraints on body movement	–	~	+
Continuous, long-time measurement	–	~	+
Application cost	–	+	+

Note: '+,' '~,' and '–' represent good, moderate, and poor, respectively

sure blood oxygen level-dependent (BOLD) response resulting from local concentration changes in paramagnetic deoxy-Hb (Ogawa et al. 1990). The changes of BOLD in an area are potentially correlated with a given task or action (Gratton et al. 2008).

The fNIRS is also a noninvasive brain-imaging technology. Ranging from 650 to 950 nm, NIR is relatively transparent to biological tissues, with its main absorbers being oxygenated and the deoxygenated hemoglobin itself. Emitted NIR light thus penetrates through the scalp and the skull to a maximum depth of approximately 20 mm, spreading in a banana shape, and is then detected by a light receiver measuring relative changes in light intensity (Gratton and Fabiani 2010). When fNIRS probes are placed—most often according to the 10–20 international system for EEG recording—cerebrocranial correlation to generalized brain models is considered to vary within 10 mm, allowing resolution at the level of the gyrus (Okamoto et al. 2004).

2.2 *The Comparison of fMRI, EEG/ERP, and fNIRS*

Compared to neural technologies such as fMRI, fNIRS, and TES, EEG/ERP records brain neuron potential changes corresponding to individual cognitive processes in real time. First, the time resolution of EEG/ERP reaches milliseconds, far higher with any other brain scanning technologies, it can provide a more subtle insight into the occurrence of consumer perception; Second, the experimental environment can adopt a design that is closer to the real world, which makes EEG/ERP more widely used to measure daily consumer behavior; Third, compared to the high cost and maintenance costs of other brain scanning equipment, EEG/ERP costs are very low. Over the past decade, scholars have used EEG/ERP to start with specific marketing issues, focusing on product design (Handy et al. 2010; Wang et al. 2012a), and brand extension (Ma et al. 2007, 2008, 2010; Jin et al. 2015; Fudali-Czyż et al. 2016), advertising (Varan et al. 2015; Barnett and Cerf 2017a, b), and other fields.

The fNIRS is a technique between EEG/ERP and fMRI. Compared to fMRI, fNIRS has a higher temporal resolution (up to tens of milliseconds) and allows for monitoring during much more ecologically valid (upright, hands-free) activities. Compared to EEG/ERP, fNIRS has higher spatial resolution, better electromagnetic and motion artifact performance, and the ability to more directly connect activity to specific brain regions without need for postprocessing (Scholkmann et al. 2014; Liu et al. 2019). Although limited to only cortical regions, fNIRS does allow for specific targeting of higher-order cognitive functional areas concerning attentional control, decision-making, social cognition, etc. (Leff et al. 2011).

Previous studies comparing fNIRS and fMRI across multiple paradigms (see (Cui et al. 2011) for review) suggest that fNIRS signals can be conceptually compared to the BOLD response, making fNIRS capable of similar brain mapping, with results comparable to lab-based fMRI findings. At the same time, literature using simultaneous fNIRS and EEG/ERP has shown a generally linear relationship between the hemodynamic response and the neuronal activity as measured through event-related electromagnetic potentials (Gratton et al. 2001; Obrig et al. 2002; Koch et al. 2008).

Of course, different neuroscience technologies can be used in combination with each other. For example, ERP and fMRI are combined with each other, and the two can complement each other, so that the research results can reach a better level in both temporal and spatial resolution. We can not only understand the continuous change process that marketing stimuli bring to consumer psychology but also understand which region of the individual brain this psychology can activate. For example, Clark et al. also collected data such as visual fixation, heart rate, electro-encephalogram, and galvanic skin to evaluate mobile users 'attention and participation in advertising content (Clark et al. 2018). Venkatraman et al. even used up to six experimental methods (traditional questionnaire survey, implicit association test, eye tracking technology, biometrics technology, ERP, and fMRI) to measure consumers' reaction to TV advertising (Venkatraman et al. 2015).

3 The Major Areas of Marketing Augmented by Neural Technology

The enhancement of neurotechnology to marketing involves almost all fields of marketing, mainly including the following topics.

3.1 Brand Perception

Brand perception, namely, consumers' thoughts, feelings, and behaviors toward brands, is one of the fundamental issues in consumer research. Neural technology is used to probe into consumers' perception of brand. Yoon et al. (2006) using fMRI examined whether semantic judgments about products and persons are processed similarly or not (Yoon et al. 2006). The results revealed that perception of persons yielded higher activation in the medial prefrontal cortex (mPFC), indicating self-related processing. In contrast, perception of brands showed increased activation in the left inferior prefrontal cortex (PFC), an area known to be involved in object processing. This pioneering study challenges the brand personality theory. Although processing of brands is different from that of humans, brands may also activate mPFC inducing a brand self-matching process.

Wang et al. (2012b) using ERP technology further revealed that strong associations with famous brands might trigger enhanced amplification of N400. Since the design of their experiment did not require participants to assess the appropriateness of brand extension, the N400 they recorded should reflect the psychological process of brand processing. Brand priming evoked participants' association with typical products associated with the brand and retrieval of attributes in long-term memory. The product name activates an unconscious process of comparison between the brand and the product. In this process, participants view the brand as a spiritual category and categorize the product as a member of it. If the attributes of the product are different from the category of the brand, there will be a huge cognitive response, leading to the increase of N400 amplitude. These findings may help to understand unconscious brand processing, which is crucial in consumer psychology (Wang et al. 2012b).

3.2 *Facial Attractiveness as an Important Factor to Influence the Perception of Brand*

In 2015, Ma et al. published a paper at the Frontiers of Neuroscience on the undermining effect of facial attractiveness on the perception of fairness, which has a high social impact (Ma et al. 2015). Its Altmetric social impact index is 145, ranking in the top 1%. After this study, Ma et al. (2017) further studied on the female facial attractiveness effects on brand perception (Ma et al. 2017). They found that the brand recommended by beautiful women would be able to get more male consumer preferences. In the experiment, they used beautiful or unattractive women to recommend fake brands of digital products. The ERP components of male consumers' brain were recorded. It was found that when facing the brands recommended by attractive women, the components of N1, P2, and later positive potentials (LPP) of male consumers were significantly higher (negative-polarity) than the components elicited by the brands recommended by unattractive women (Fig. 1). This finding suggested that a woman's attractiveness might change a man's perception of brand.

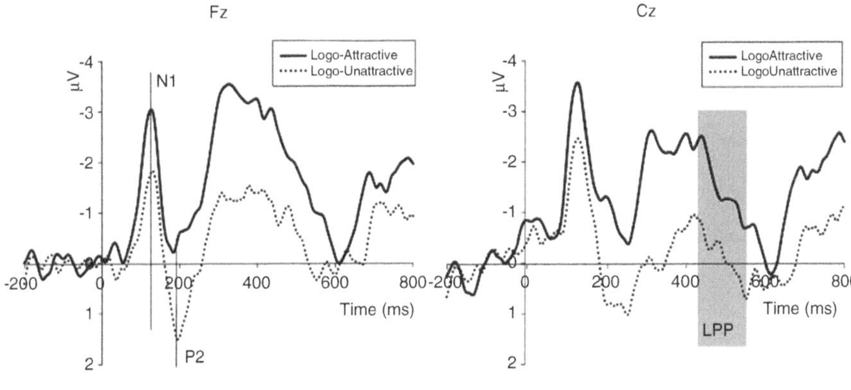

Fig. 1 Waveforms and brain topographic maps of N1, P2, and LPP of logo onset at Fz and Cz

3.3 Victory or Defeat Affects the Brand Perception

Yu et al. (2018) explored whether and how victory and defeat, as two critical factors in competition outcomes, would affect consumers' preference of unfamiliar brands by using EEG (Yu et al. 2018). Specifically, participants' status of victory or defeat was induced by a pseudo-online game, followed by a main task of brand preference rating.

Behavioral data showed that individuals had a stronger preference for unfamiliar brands in victory condition than in defeat condition, even if the brand was completely unrelated to the competition; this indicated a transfer of valence. At the brain level, as shown in Fig. 2, three emotion-related event-related potential components, N1, P2, and LPP, were elicited more negatively in victory than in defeat condition, which suggested that victory and defeat experiences would arouse opposite emotions and then further induce different impacts on the preference for brands in the early processing phase. It is inferred that the initial perception of a new brand undergoes the following stages. A positive or negative emotion is first evoked by an experience of victory or defeat; then, attentional distribution is biased for the subsequent stimuli (even stimuli that are completely unrelated to the competition) during the maintenance period and finally affects the decision-making process. This result is also in line with the ABC model of attitudes (Breckler 1984), both decision-related effect and decision-unrelated incidental effect have informational value that increases the

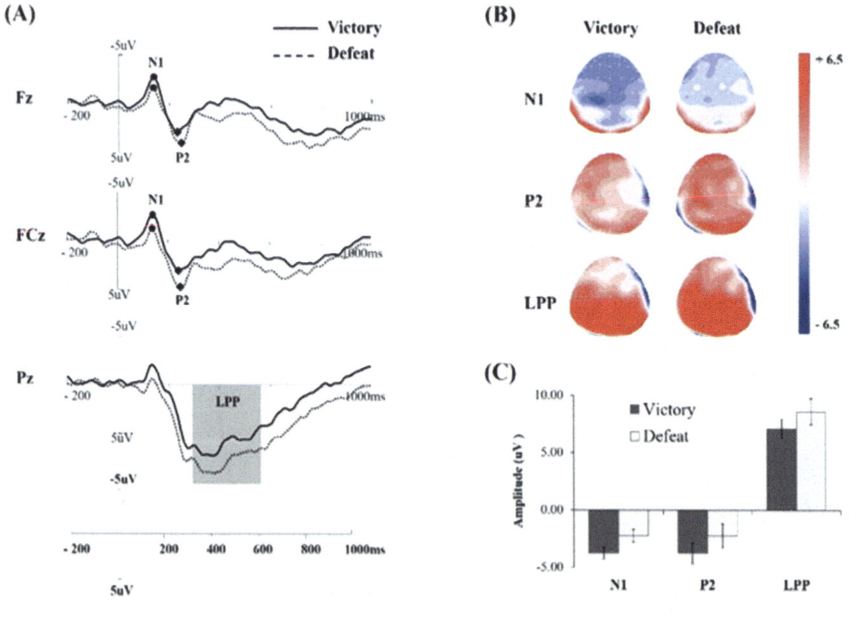

Fig. 2 Waveforms and brain topographic maps of N1, P2, and LPP at Fz, FCz, and Pz; (**A**) is the waveform, (**B**) is the brain topographic, and (**C**) is the mean voltage

cognitive availability of effect-congruent information and impacts the individual's judgment. Therefore, playing a game before presenting brand information might help promote the brand by inducing a good impression of the brand in consumers.

3.4 Fame of Description Affects Brand Perception

Jin et al. (2017) employed EEG to investigate the neural correlates of attribute framing effect of online product description, which is one of the most important channels that online consumers know the product (Jin et al. 2017). The attribute framing effect refers to the phenomenon in which consumers show inconsistent preferences or choices when identical attribute information of products is differentially described (positively or negatively) (Tversky and Kahneman 1974, 1981). It has been well studied by numerous scholars; however, the associated underlying neural mechanisms with a critical temporal resolution have not been revealed. The ERPs were extracted from EEG signal to directly examine the role of attribute framing in information processing and decision-making in online shopping.

The behavioral results showed that participants demonstrated a higher purchase intention with a shorter reaction time under a positive framing condition compared to the negative framing condition. The results of ERPs indicated that compared to positive framing messages, negative framing messages attracted more attention resources at the early stage of rapid automatic processing (larger P2 amplitude) and resulted in greater cognitive conflict and decision difficulty (larger P2-N2 complex). Moreover, compared to negative messages, positive framing messages allowed consumers to perceive a better future performance of products and to classify these products as a categorization of higher evaluation (larger LPP amplitude) at the late cognitive processing stage of evaluation. We supposed that positive framing messages were more desirable to participants' expectation compared to negative framing messages; thus, they made purchase decisions more easily in positive frames compared to negative frames. Furthermore, negative framing messages attracted more attentional resources and elicited higher decision conflict than positive attribute framing messages. The significant positive correlation between LPP amplitudes and the purchase intentions of the participants across conditions also supported the above speculation. Due to the higher evaluation, participants had a higher purchase intention with a shorter reaction time in positive frames compared to negative frames. These results provide evidence for a better understanding of how different attribute framing messages are processed and ultimately lead to the framing effect.

3.5 Studies on Brand Extension

A successful brand extension means that the consumers already accepted the similarity of some features between the original brand and the extension product. Many explicit techniques, including focus groups, questionnaires (e.g., Barone et al.

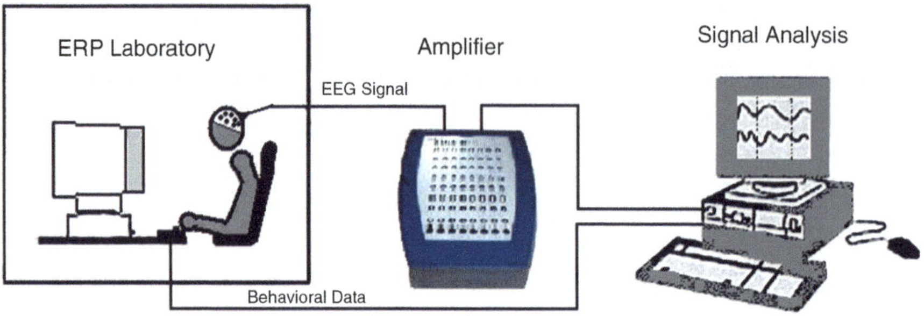

Fig. 3 The ERP experiment that collected behavioral data and ERP data

2000), and category similarity judgments (e.g., Zhang and Sood 2002), are often applied to test consumers' perception of brand extensions. Many scholars pointed out that participants cannot make the decisions in laboratory as the same as in the real market (Graeff 2002). When participants responded to questions about fictitious brand extensions, they had little awareness about their reaction in a real market context and had to consciously conform to the view of majority. For these reasons, since 2006, Ma and Wang had published a series of papers measuring the evaluation of brand extensions by using ERP (Ma et al. 2007, 2008, 2010; Jin et al. 2015).

A typical study on the brand extension is as follows. The stimuli consisted of 225 pairs of "brand name-product name," i.e., 15 (brand names) × 15 (product names), in which S1 was one of the original beverage brands and S2 was one of the extension products in three categories: beverage (close extension), snack (moderate extension), and household appliance (far extension). EEG was continuously recorded using a Neuroscan Synamp2 Amplifier with an electrode cap with 64 Ag/AgCl electrodes mounted according to the extended international 10–20 system and referenced to linked mastoids (Fig. 3).

The amplitude of N2 in far extension was significantly larger than that in the moderate extension, which is, in turn, larger than close extension. It showed that frontocentral N2 might serve as an endogenous indicator of perceived fit in evaluating brand extension. The amplitude of d-N2 in difference wave between far and close extension in the frontocentral area was significantly larger than that in difference wave between the moderate and close extension, showing that the comparison between the far and close extension produced stronger conflict compared to the comparison between the moderate and close extension. The topographic plots (top down view: nose up) of maximal amplitudes of N2 (from 250 to 450 ms) in three extension types were distributed on the frontal scalp areas. As is consistent with the ERP components of N2 and d-N2, the behavioral data from the experiment also predicted the conflict process in evaluating moderate, especially far, brand extensions. The affirmative rate of answer correlated negatively with the amplitude of N2 in three extension conditions: the larger the N2 amplitude was, the smaller the affirmative rate.

3.6 The Perception of Advertisement and the Advertising Design

A collection of studies have presented compelling evidence for specific neural responses to advertising (Ambler et al. 2000; Rossiter et al. 2001; Yoon et al. 2006; Mostafa 2013) or media in general (Pelowski et al. 2017), documenting impacts on a number of brain regions from reward to basic perception to more complex cognitive processing, associations, and social or introspective connections, which, in turn, may show differing activations in cases where ads are more or less preferred. For instance, high-frequency EEG components of participants while watching movie trailers have been correlated with both individual preference (beta activity) and population preference (gamma activity) of the upcoming movies (Boksem and Smidts 2015). In addition, Venkatraman et al. (2015) have demonstrated that the ventral striatum activations, measured by fMRI and related to reward and motivation, explain the most variance in advertising preference beyond baseline traditional measures (Venkatraman et al. 2015).

Following the suggestion of multibrain frame, Liu and colleagues (2018) have examined advertising effectiveness using fNIRS coupled with a new approach based on graph theory and network modeling in three studies (Liu and Liu 2018). In Study 1, 20 female participants were presented with 20 video advertisements (10 high-scored and 10 low-scored selected via a pilot study). After viewing each advertisement, participants made a rating on both liking and willing to pay (WTP) for the product/service on a 7-point scale. A postquestionnaire was then given at the end of the study, in which participants needed to rate their understanding level with detailed descriptions. In addition, whether or not participants remembered each advertisement was also assessed. The fNIRS results revealed higher mPFC activation when viewing the low-scored than the high-scored advertisements, and the mPFC activation positively correlated with their understanding score on the advertisements.

Using a network modeling method, the authors constructed a multibrain network, considering each participant's brain as a node, and then calculated interpersonal neural synchronization across all pairs of nodes forming network edges. Finally, they counted the total number of edges as an index of network density. The results showed higher neural network density in participants when viewing high-scored than the low-scored advertisements. Importantly, the density of the multibrain network in the right inferior frontal gyrus (IFG) positively predicted participants' attitude (mean value of liking and WTP scores) toward the advertisements.

In Study 2, the authors further confirmed that the neural network density in the right IFG can not only predict participants' attitude toward the chorus/refrain sections of 30 music songs obtained from '2014 Hot 100' published by Billboard.com but also predict population-level ratings on the songs taken from Douban.com. Studies 1 and 2 consistently demonstrated that the multibrain network density in the right IFG may represent cognitive/emotional resonance among groups of consumers due to embodied experiences. In Study 3, 12 commercial movie trailers were presented to participants as a reverse example since movie trailers describe no full story, but consist of discrete clips of scenarios and thus cannot be effectively experienced. As expected, Liu and colleagues revealed no significant relation between the multibrain network density in the right IFG and both the sample-level scores and the population-level box office. In contrast, participants' right IFG activations negatively predicted the attitude scores and the box office.

Taken together, on one hand, this research established a new multibrain network model, focusing on the neural network density in the right IFG, to evaluate experiential media effectiveness objectively. On the other hand, the multibrain network model may help to construct the advertising, which can convey precise information to a target consumer group to arouse resonance in their emotions and understandings, achieving goal of precision marketing.

3.7 The Perception of Price Affected by Winning or Losing

Price played an important role in most purchases. The consumers' perception of price strongly influences their purchasing decisions. Ma et al. (2018a) creatively integrated a competitive game, a two-player "Finger Play" (FP) game, without monetary gains or losses as an emotional priming paradigm to investigate how emotions influence subjects' price perception (Ma et al. 2018a). The rules of FP are that "scissors > paper > stone > scissors".

Knutson et al. (2007) deemed that consumers mostly avoided excessive price products in addition to those deeply preferred ones (Knutson et al. 2007). In the study by Ma et al. (2018a), subjects were assumed to choose cheaper hard disks as all the product quality was the same. Interestingly, subjects' choices were impacted to a great extent by different arousal emotions that were induced by continuous winning (CW) or continuous losing (CL) in the FP game (Ma et al. 2018a).

Behavioral data show that when the subjects experienced a continuous winning outcome, they added more hard drives to the cart than they experienced a continuous losing outcome. From a new point of view, the study by Ma, Zhang, and Wang verifies the important role of emotion in consumer behavior and marketing.

The P2 and LPP components of ERP also showed consistent results: When the price was high, P2 and LPP were larger (positive) in the condition of continuous winning and the same quality hard disk-labeled high price (CWHP) when compared with the condition of continuous losing and the same quality hard disk-labeled low price (CLLP). This means that more positive emotion was evoked in CWHP and then continuously affected the price perception, and in turn, the subjects added the proportion of more products to the shopping cart in CWHP.

3.8 Personalized Design of Products

Ma Q et al. (2018a, Ma Y et al. 2018b) using a revised go/no-go paradigm and ERP technique investigated consumers' implicit neural responses to experience good designs since product design is an important source of utility of experience goods

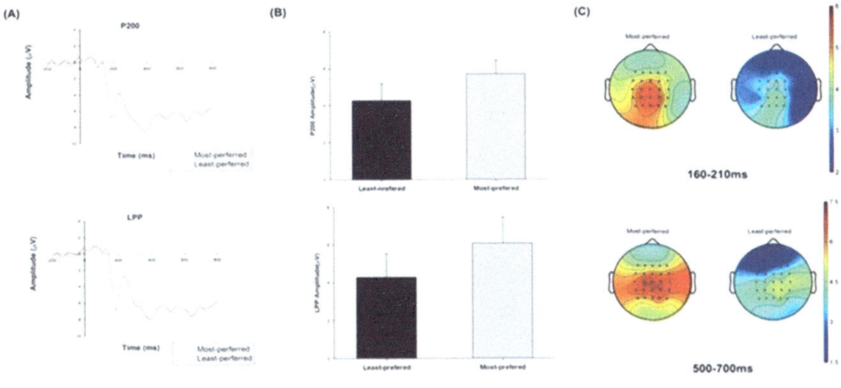

Fig. 4 Waveforms and brain topographic maps of P200 and LPP, (**A**) is the waveform, (**B**) is the mean voltage, and (**C**) is the brain topographic

(Ma et al. 2018b). To explore the personal and subjective nature of product design evaluation, the personalized product designs rather than the same stimulus for all the participants were employed in their experiment since personalized stimuli may reflect individuals' preferences for product designs more accurately.

The brain level data, as shown in Fig. 4, indicated that the P200 and LPP amplitude elicited by the most preferred experience good designs were significantly larger than those elicited by least preferred designs, and the two ERP components were positively correlated with the personalized rating scores. P200 and the LPP can represent the early attention arousal and the late emotional cognition assessment, respectively. This is consistent with previous studies, which suggested that experience good designs are related to symbolic and esthetical dimensions of product design, and related to emotional experience and arousal (Homburg et al. 2015). The results indicated that ERP signals may provide important information regarding consumer preferences for experience good designs and can shed light on why consumers like customized products.

3.9 *The Fake Rating and the Strategy of Cash Coupons*

Online ratings impose significant effects on the behaviors of potential customers. Thus, online merchants try to adopt strategies that affect this rating behavior, and most of these strategies are connected to money, such as the strategies of returning cash coupons if a consumer gives a five-star rating (RI strategy, an acronym for "returning" and "if") or returning cash coupons directly with no additional requirements (RN strategy, an acronym for "returning" and "no"). Wang et al. (2018) using ERP technique explored whether a certain strategy (RN or RI) was more likely to give rise to fake rating behaviors (Wang et al. 2018).

In their experiment, the first stimulus (S1) was the picture of a product with four Chinese characters that reflected the product quality (slightly defective vs. seriously

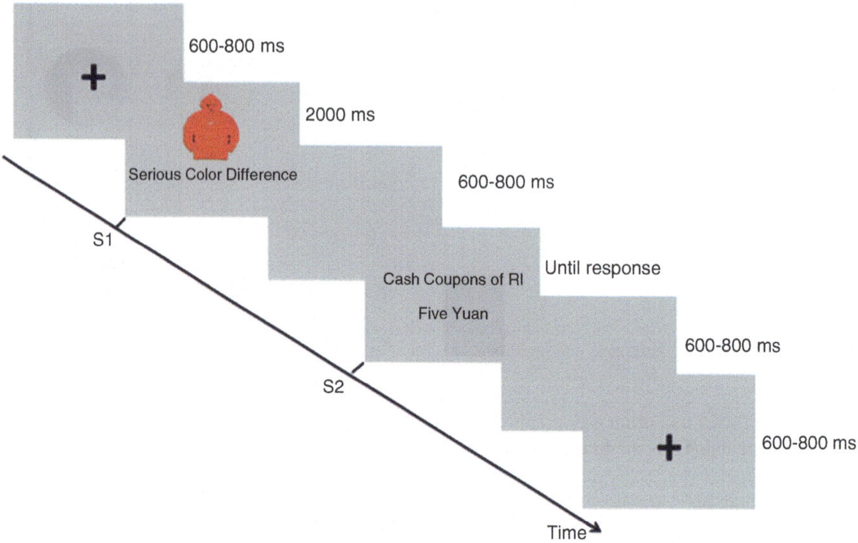

Fig. 5 Experimental task

defective vs. nondefective), and the second stimulus (S2) displayed the coupon strategy (RN or RI) (shown in Fig. 5).

Behavioral results revealed that customers were more willing to give a five-star rating with the RI strategy compared to the RN strategy. A significant effect was found for the product quality factor; slightly defective products corresponded to longer RTs than nondefective and seriously defective products, with no difference in RT between the latter two. Previous studies have found that the task difficulty affects RT, with more difficult tasks requiring more time to process.

ERP results showed that a main effect of product quality on the amplitude of the N1 was significant. The slightly defective products elicited higher N1 amplitudes than the seriously defective products and nondefective products, respectively. No significant difference in N1 amplitude between the seriously defective products and nondefective products was found. The findings indicated that the participants perceived difficultly the slightly defective products at an early stage. The RI strategy elicited a less negative N2 component than RN, indicating that the participants detected less conflict in giving a five-star rating in response to the RI strategy than to RN. The LPP amplitude evoked by RI was larger than that by RN, showing that the RI strategy had a greater incentive effect for the subjects to resolve the conflict than the RN strategy.

The study revealed the neural process of fake online comments, which were evoked by monetary rewards. Network platforms or government agencies should strengthen the supervision of false evaluation to reduce the harm to consumers.

3.10 Social Risks Affect Consumers' Willingness to Buy

Shang et al. (2017) investigated the neural correlates of consumers' evaluation toward marketing-related stimuli (Shang et al. 2017). They explored how social risk influences consumers' purchase intention. The social risk refers to the potential disapproval from significant others (especially family or friends), and it is crucial in dissuading consumers from making decisions to purchase. The electrophysiological process was detected by using ERPs when subjects evaluated their purchase intention with social risk factors. The behavioral data showed that the social risk condition inhibited people's purchase intention compared to the control condition. This might be related to a mechanism of normative conformity motivation, which suggests that people desire to obtain social approval from others or to behave correctly in a social sense, and they are therefore likely to change their decision to match the response of others. At the brain level, larger anterior N2 amplitude was induced by the social risk condition in contrast to the control condition. It is suggested that this anterior N2 may reflect the cognitive control or conflict monitoring. The participant may have to manage the conflict between an internal desire to purchase the item and the discordant information obtained from the social risk sentence, which would pressure the participant to not purchase the item in accordance with social norms. This study depicts the cognitive process by using the electrophysiological index, which has high temporal resolution. Its findings clarify the neural basis of social risk perception during purchase decisions.

3.11 Luxury Purchases Undermining the Prosocial Behavior: The Cognitive Neural Mechanism Studies

It is well accepted that human beings are social by nature. Understanding of factors influencing human prosocial behavior and the underlying mechanism is one important topic in marketing research. Some scholars have criticized luxury brands for undermining interpersonal harmony (Jiang et al. 2014) and revealed that luxury brands may function as strong emotional stimuli affecting consumers' attitudes and behaviors (Nia and Zaichkowsky 2000), especially in social contexts (Pozharliev et al. 2015). However, the underlying cognitive mechanisms are as yet unclear.

To address this issue, Li et al. (2018) have conducted three experiments focusing on effects of luxury brands on human prosocial behavior using both behavioral and neural techniques (Li et al. 2018). In Study 1, 94 female undergraduates were randomly subdivided into 2 groups, in which 10 luxury or nonluxury branded bags were presented. After viewing each bag, participants rated four questions at artistry level, emotional response, and degrees of liking and willing to pay (WTP) on a 7-point scale, followed by a question checking their luxury perception of the bag brands. A donation advertising that proposes a reading project for poverty-stricken children was then displayed, and participants were instructed to rate degree of their donation willingness on a 7-point scale and to report their monetary donation in consideration of their monthly living expenses. The *t*-tests revealed lower donation attitude (mean value of donation willingness and monetary donation scores) in the luxury group compared to the nonluxury group (referred to as negative luxury effect), which was moderated by participants' emotional response.

To examine the neural mechanism underlying the negative luxury effect, in Study 2, the authors simultaneously measure pairs of participants' brain activations in the frontotemporoparietal regions using a functional near-infrared spectroscopy (fNIRS) based hyperscanning technique. Twenty-two pairs of participants, recruited from MBA courses and the company of China Eastern Airlines, were instructed to performed a cooperation-competition key-press task modified from the study by Cui et al. (2012) with luxury or nonluxury branded goods as task background, forming four conditions: cooperation with luxury brand, cooperation without luxury brand, competition with luxury brand, and competition without luxury brand. A donation poster was displayed to participant pairs who were then asked to make a monetary-donation decision (up to 999 RMB maximum) after all interactive tasks.

Participants in the luxury condition showed worse cooperative performance but better competitive performance than in the nonluxury condition, confirming negative luxury effect. Although no monetary difference was found between two conditions, the luxury condition indeed induced longer reaction time to make the donation decision compared to the nonluxury condition. Focusing on the fNIRS signals, when performing the cooperative task, exposure to luxury relative to nonluxury brands led to higher activations in the right dlPFC and the right IFG. Additionally, in the luxury condition, correlation results revealed that increased dlPFC activation in competition yielded decreased cooperative performance, while the increased dlPFC activation in cooperation promoted competitive performance.

More importantly, participants' IFG activation while cooperating negatively correlated with their subsequent donation willingness in the luxury condition. By contrast, in the nonluxury condition, a positive correlation was found between the dlPFC activation in cooperation and their donation willingness and the monetary donation as well. These results suggest that the luxury brands may distract participants' attention from the cooperative task-inducing social comparison between the dyads, which in turn impaired their prosocial behavior.

Study 3 further excluded a possibility that any attractive and desired stimuli could create the same negative effect on human prosocial behavior, by showing decreased donation performance even in the luxury-without-artistry group compared to nonluxury group. This study demonstrates the cognitive mechanisms underlying negative luxury effect on human prosocial behavior and provides a new paradigm to examine neural basis of consumer behavior in daily social activities as well. Practically, these results also suggest that when promoting luxury brands, marketers need to pay much attention to the specific social contexts to make sure that the brands are aligned with social contexts.

3.12 How the Sound Frequency and Color Saturation Affect Consumers' Perception of Product Size

Technological progress and demand upgrading have greatly evolved consumers' shopping habits, from the pursuit of affordable price and convenience to the enjoyment of holistic shopping experience. However, almost all retailers choose the sensory stimulation in-store (e.g., background music) according to their own preferences, ignoring how each subtle dimension of the stimulus affects each stage of the consumer decision-making process. Therefore, how to enhance consumers' holistic shopping experience through scientific screening and manipulation of various types of stimuli is a common problem faced by retail enterprises in this new retail era.

There were competing explanatory mechanisms in decision-making-related studies based on sensory experience. However, they mainly focused on how a single sense affects a certain stage of the consumer decision-making process (e.g., evaluation). Therefore, it is necessary to explore the explanatory mechanism of consumer decision-making based on multisensory interaction. Interestingly, the widely recognized Cross-modal correspondence theory (CMC) from psychology can provide guidance for the above theoretical needs.

Wang et al. (2019) have conducted three studies to integrate these theories, methods, and techniques from management science, decision science, psychology, and cognitive neuroscience, to systematically delineate how ambient sound in a retail environment interacts with product color to affect the entire consumer decision-making process (i.e., perception-preference-purchase) from the novel perspective of CMC (Wang et al. 2019).

Study 1 conducted a number of behavioral experiments and found that the higher the color saturation of the product, or the lower the sound frequency in the environment, the larger the perceived size of the product, and vice versa. Interestingly, only when the color of the product is in high saturation, can the sound frequency significantly influence the perceived size of the product. When the product color is in low saturation, the effect of sound frequency on perceived size disappears.

Study 2 adopted the ERP technique to explore the underlying mechanism behind the CMC phenomenon in study 1 and further tested the downstream effect brought by "sound frequency-color saturation interaction" on product preferences. They found that high (vs. low) saturation elicited more salient P300 amplitudes (i.e. emotional arousal), and this ERP component was only affected by sound frequency under high saturation condition. They further found that sound frequency is first moderated by color saturation, and it was then mediated by arousal to affect the perceived product size. In addition, when subjects' usage goal calls for bigger product size, the combination of "low frequency sound-high color saturation" will lead to higher product preferences.

Study 3 further replicated the laboratory results (studies 1 and 2) in a field study and explored whether the "sound frequency-color saturation" interaction would further affect product purchase. Result illustrated that when consumers' usage goal calls for smaller size, unit sales of products under low saturation were significantly higher than those under high saturation. However, music frequency had null effect on unit sales under low saturation condition. High (vs. low) frequency music significantly increased unit sales when the products were in high saturation.

The theoretical and practical contributions of their studies are threefold:

1. They applied CMC concept from psychology into the field of consumer decision-making and revealed its downstream effects on multiple stages of consumer decision-making process, enriching the explanatory mechanism of multisensory integration on decision-making. Specifically, they found empirical evidence that the sound frequency effect on perceived product size is moderated by the product's own color saturation. This effect appears robust even if there is no connection between the sound and product and is only active when the product is of high color saturation.
2. They investigated interplay between color saturation (an understudied but very important dimension of color) and sound frequency and further examined its downstream effect on product preference and purchase to complement extant management research exploring color effects on consumer decision-making.
3. They found an easy and costless manipulation for retailers to achieve favorable sale outcome by extrapolating the laboratory results into a dynamic retail environment.

4 Three Important Challenges in Neuromarketing

The development of neuromarketing faces many challenges. The following three challenges are important that we must address firstly.

- Challenge #1: The feeling in real scenarios
  The key factor for a successful neuromarketing experiment is to mobilize the desired psychological processes of the subjects. However, there is a natural gap between the laboratory environment and the real marketing environment (for example, fMRI is a semiclosed high-noise environment, which is very different from the environment experienced by consumers in stores), which has long been questioned (James 2004) and has not been solved well. To find a better solution is the challenge that neuromarketing must face with.
  Hommel et al. (2012) cleverly used fMRI noise to study the effect of noise on cognitive control. Skouras et al. (2013) also used fMRI noise to study the neural process of the interaction between noise and affective. Of course, these are only two very special examples. For the gap between the laboratory environment and the real marketing environment, how to skillfully solve or narrow the gap one by one (to mobilize the psychological process required by the researchers) is still a huge challenge.
  It should be noted that the use of virtual reality (VR) technology to enhance the subjects' feeling in real scenarios such as the consumption field is an important trend, which should be paid attention.

- Challenge #2: The neural data are often inconsistent with behavioral data
  If the behavioral data are consistent with the neural data, both can confirm each other. In some studies, however, the behavioral data are inconsistent with the neural data. What is the cause? Is it an experiment design problem, an experiment manipulation problem, or an inevitable problem under a certain experimental paradigm (including tasks)? It is a real challenge.
  In general, most researchers expect the neural data to be consistent with the behavioral data. However, one of the common phenomena in studies is the inconsistency between the neural data and the behavioral data. For example, in choosing option A or D, consumers often quickly decide to press the A button when they think that option A has clear advantage compared to option D. This mental process is instantaneous in the brain. The representation in the neural data is that the peak latency of an ERP component (such as P2) in condition A is usually shorter than that in condition D, but the reaction time (RT) for condition A is often not significantly different from condition D; even on the contrary, the RT for option A is longer than that for D.
  Many psychologists believe that the neural data and behavioral data must be consistent; otherwise, there may be problems in the experiment and the process needs to be checked again. On the contrary, many marketing scholars believe that this is the significance of the application of neural technology and that the data of neural activity reflect the psychological processes. Ma, Q has speculated that this difference may be caused by the difference in the execution speed of different subjects from the choosing decision to the finger touched the key. Anyway, this is a question that needs to be answered experimentally.
- Challenge #3: The Reverse Inference Issue
  Reverse Inference refers to the process of inferring mental states from neural data. Hilke Plassmann et al. (2015) gave a clear illustration (see Sect. 1.6 of this chapter). In many cases, reverse inference resulted in mistakes. Neural activity Z, in general, is not evoked uniquely by cognition process C since cognitive process E may also evoke Z, and in general, Z is only a part of neural activities to finish task A (Poldrack 2006).
  Researchers are trying to solve or improve the problem. One of these efforts is to use the big data processing methods such as support vector machine to predict (with high probability) cognitive processes based on a large amount of neuroscience data (Wei et al. 2018).

5 Hyperscanning: An Important Trend in Neuromarketing

Marketing is a social activity, and consumer behavior is inherently social in nature. Therefore, a new trend of scanning multiple brains simultaneously in a more natural social context, in terms of hyperscanning, has emerged in recent years (for reviews, see Hasson et al. 2012; Hasson and Honey 2012; Pozharliev et al. 2017), making it possible to examine consumer behaviors in real-world marketing environments (Barnett and Cerf 2017a, b). It will open up new frontiers in neuromarketing.

Marketing research strives to explain how consumers process information and interact with the consumption environments in natural contexts. However, isolating and manipulating independent variables in controlled laboratory environments hinder utility of consumer neuroscience (Hasson and Honey 2012). Therefore, another important trend in consumer neuroscience research is the shift from lab paradigms to daily-life scenarios such as consumption scenarios.

5.1 *Shift from Single-Brain to MultiBrain Frame*

Most neuromarketing studies to date have mainly focused on single individuals' brain activity in well-controlled experimental settings. As Cozolino (2006) noted, the human brain is "an organ of adaptation that builds its structures through interactions with others" (p. 6). "The individual neuron or single human brain does not exist in nature. Without mutually stimulating interactions, people and neurons wither and die" (p. 11) (Cozolino 2006). Thus, the single-brain frame used in previous studies may omit the most unique aspects of social processing (for a review, see Hasson et al. 2012). In particular, marketing is a social activity, and consumer behavior is inherently social in nature. Therefore, a growing number of researchers have called for new investigation of social interaction and its relationship to social cognitive abilities in more ecologically valid ways (Schilbach et al. 2013). Recent technical advances in hyperscanning allow researchers to scan multiple brains simultaneously in a more natural social context, making it possible to examine consumer behaviors in real-world marketing environments (Barnett and Cerf 2017a, b). This is an inevitable future trend in the field of neuromarketing.

5.2 *Neural Coupling Across Customers*

Previous hyperscanning studies on social interactions have demonstrated that interacting members tend to show interpersonal neural synchronization (INS), which is similarity of brain activations among them, indicating shared mental representations and mutual understanding of emotions and intentions, which may serve as a basic underlying mechanism supporting human social behaviors (Hasson et al. 2012; Liu and Pelowski 2014; Hu et al. 2017; Ahn et al. 2018). As far as marketing is concerned, INS is the neural basis of consumers' common consumption tendency, which is also largely related to the group behavior of consumption.

In the domain of neuromarketing, Hasson et al. (2004) firstly examined neural relations to five subjects using fMRI when they viewed half an hour of a popular movie and reported neural synchronization not only in primary and secondary visual and auditory areas but also in association cortices, especially during the moments showing emotionally arousing scenes and regionally selective components (Hasson et al. 2004). The results indicate a tendency of individual brains to "tick collectively" during natural movie vision. In the same vein, Lankinen et al. (2014) also revealed such similar patterns of brain activities among spectators of a movie.

These studies suggest that multibrain frame examining interpersonal neural relations could depict shared mental representations across different customers effectively and in turn may open a window for studying social features of consumer perception and purchasing behavior in more natural marketing scenarios, such as shopping with friends or recommending a product to others.

5.3 Graph Theory-Based Approach Depicting MultiBrain Interactions

Besides the INS mentioned above, we also propose a means of assessing neural relations among a group of customers. This employs network science and graph theory, a branch of mathematics that uses the properties of graphs or networks involving nodes (or specific variables such as individuals) and edges (or positive or negative relation with certain strength) to represent system elements and their interrelations (Bullmore and Sporns 2009; Fornito et al. 2016). The theory of networks has been applied to examine various social phenomena such as kinship structure, social mobility, class structure, and social network (for review, see Sporns 2018; Karwowski et al. 2019). Recently, they have also been employed to map shared patterns of reactions to media and their interconnections to music or art (Cotter et al. 2018).

At the neural level, graph theory has further opened a new window into the study of complex neural network organizations and multibrain network modeling methods (Duan et al. 2015; Bassett and Sporns 2017). This allows us to consider each individuals' brain as a network node and to use an index of neural synchronization over a specific cortical area between pairs of brains in the entire network, beyond a certain set threshold, to construct network edges between the nodes. By looking at the resulting network strength, density, and global/local efficiency (Van Wijk et al. 2010; Keown et al. 2017), this may provide measures of cognitive and emotional resonance across entire customer group and of efficiency of information exchange among the group, thus providing an opportunity to examine neural mechanisms underlying group consumption behaviors.

5.4 A Successful Example of Hyperscanning

In the field of neuromarketing, Barnett and Cerf (2017a, b) measured neural activity from multiple moviegoers as they watched movie trailers in a commercial theater using EEG-based hyperscanning (Barnett and Cerf 2017a, b). One hundred and twenty-two participants watched trailers and movies at a commercial theater and reported a survey following the movie viewing. Fifty-eight of the participants additionally received neural and physiological recordings throughout the trailers and movies. Six months later, 36 of the original 58 participants responded to the same survey examining their memory on the trailers and the liking and WTP ratings.

The authors computed the interpersonal neural synchronization as moment-to-moment synchronization in EEG data across participants experiencing the same audio-visual stimuli. Specifically, for a given pair of participants and a given electrode site, they calculated the Pearson correlation for each timestep and then averaged across all 32 electrode sites and all participant pairs viewing the same trailer to arrive at a single value of neural similarity at each timestep. The neural similarity positively predicted the movie trailer recall, and no correlations were found between the free recall and behavioral ratings of liking and WTP. In addition, the neural similarity also showed a positive relation with the average weekly ticket sales of the advertised film although free recall of trailers predicted weekly ticket sales as well, but less strongly than the neural-sale link. This study suggests that advertisements that generate elevated neural similarity across consumers are more memorable and persuasive, and the neural similarity was more predictive of population-level sales than subjective measurements.

References

Agarwal S, Dutta T (2015) Neuromarketing and consumer neuroscience: current understanding and the way forward. Decision 42(4):457–462

Ahlert D et al (2006) A window to the consumer's mind: application of functional brain imaging techniques to advertising research. International advertising and communication. Springer, New York, pp 163–178

Ahn S et al (2018) Interbrain phase synchronization during turn-taking verbal interaction—a hyperscanning study using simultaneous EEG/MEG. Hum Brain Mapp 39(1):171–188

Ambler T et al (2000) Brands on the brain: neuro-images of advertising. Bus Strategy Rev 11(3):17–30

Ariely D, Berns GS (2010) Neuromarketing: the hope and hype of neuroimaging in business. Nat Rev Neurosci 11(4):284–292

Attwell D, Iadecola C (2002) The neural basis of functional brain imaging signals. Trends Neurosci 25(12):621–625

Barnett SB, Cerf M (2017a) Few and far between: identifying measures of advertising visuals that correlate with neural engagement and sales. In: Gneezy A, Griskevicius V, Williams P (eds) NA—advances in consumer research, vol 45. Association for Consumer Research, Duluth

Barnett SB, Cerf M (2017b) A ticket for your thoughts: method for predicting content recall and sales using neural similarity of moviegoers. J Consum Res 44(1):160–181

Barone MJ et al (2000) The influence of positive mood on brand extension evaluations. J Consum Res 26(4):386–400

Bassett DS, Sporns O (2017) Network neuroscience. Nat Neurosci 20(3):353

Boksem MA, Smidts A (2015) Brain responses to movie trailers predict individual preferences for movies and their population-wide commercial success. J Mark Res 52(4):482–492

Breckler SJ (1984) Empirical validation of affect, behavior, and cognition as distinct components of attitude. J Pers Soc Psychol 47(6):1191–1205

Bullmore E, Sporns O (2009) Complex brain networks: graph theoretical analysis of structural and functional systems. Nat Rev Neurosci 10(3):186–198
Bunge SA, Kahn I (2009) Cognition: an overview of neuroimaging techniques. Encyclopedia of Neuroscience 1063–1067
Buxton RB (2012) Dynamic models of BOLD contrast. Neuroimage 62(2):953–961
Carter M, Shieh JC (2015) Guide to research techniques in neuroscience. Academic Press, New York
Clark KR et al (2018) How advertisers can keep mobile users engaged and reduce video-ad blocking: best practices for video-ad placement and delivery based on consumer neuroscience measures. J Advert Res 58(3):311–325
Cotter KN et al (2018) What does feeling like crying when listening to music feel like? Psychol Aesthet Creat Arts 12(2):216
Cozolino L (2006) The neuroscience of human relationships: attachment and the developing social brain. WW Norton&Company, Inc, New York
Cui X et al (2011) A quantitative comparison of NIRS and fMRI across multiple cognitive tasks. Neuroimage 54(4):2808–2821
Cui X et al (2012). NIRS-based hyperscanning reveals increased interpersonal coherence in superior frontal cortex during cooperation. Neuroimage, 59(3):2430–2437
Duan L et al (2015) Cluster imaging of multi-brain networks (CIMBN): a general framework for hyperscanning and modeling a group of interacting brains. Front Neurosci 9:267
Eser Z et al (2011) Perceptions of marketing academics, neurologists, and marketing professionals about neuromarketing. J Mark Manag 27(7–8):854–868
Fornito A et al (2016) Fundamentals of brain network analysis. Academic Press, New York
Fudali-Czyż A et al (2016) Controlled categorisation processing in brand extension evaluation by Indo-European language speakers. An ERP study. Neurosci Lett 628:30–34
Gazzaniga MS et al (2006) Cognitive neuroscience. The biology of the mind. Norton, New York
Gerger G et al (2018) Empathy, Einfühlung, and aesthetic experience: the effect of emotion contagion on appreciation of representational and abstract art using fEMG and SCR. Cogn Process 19(2):147–165
Graeff TR (2002) Uninformed response bias in telephone surveys. J Bus Res 55(3):251–259
Gratton G, Fabiani M (2010) Fast optical imaging of human brain function. Front Hum Neurosci 4:52
Gratton G et al (2001) Comparison of neuronal and hemodynamic measures of the brain response to visual stimulation: an optical imaging study. Hum Brain Mapp 13(1):13–25
Gratton G et al (2008) Time course of executive processes: Data from the event-related optical signal. In: Bunge SA, Wallis JD (eds) Perspectives on rule-guided behavior. Oxford University Press, Oxford, pp 197–223
Griffin A, Hauser JR (1993) The voice of the customer. Mark Sci 12(1):1–27
Handy TC et al (2010) ERP evidence for rapid hedonic evaluation of logos. J Cogn Neurosci 22(1):124–138
Hasson U, Honey CJ (2012) Future trends in neuroimaging: neural processes as expressed within real-life contexts. Neuroimage 62(2):1272–1278
Hasson U et al (2004) Intersubject synchronization of cortical activity during natural vision. Science 303(5664):1634–1640
Hasson U et al (2012) Brain-to-brain coupling: a mechanism for creating and sharing a social world. Trends Cogn Sci 16(2):114–121
Homburg C et al (2015) New product design: concept, measurement, and consequences. J Mark 79(3):41–56
Hommel B et al (2012) The effect of fMRI (noise) on cognitive control. J Exp Psychol Hum Percept Perform 38(2):290
Hoshi Y (2003) Functional near-infrared optical imaging: utility and limitations in human brain mapping. Psychophysiology 40(4):511–520

Hu Y et al (2017) Brain-to-brain synchronization across two persons predicts mutual prosociality. Soc Cogn Affect Neurosci 12(12):1835–1844
James S (2004) Neuromarketing is no brainwave if you just think about it. Precis Mark 24(9):12–13
Jiang M et al (2014) The devil wears P rada: advertisements of luxury brands evoke feelings of social exclusion. Asian J Soc Psychol 17(4):245–254
Jin J et al (2015) Extending or creating a new brand: evidence from a study on event-related potentials. Neuroreport 26(10):572
Jin J et al (2017) How consumers are affected by product descriptions in online shopping: event-related potentials evidence of the attribute framing effect. Neurosci Res 125:21–28
Kahneman D et al (1997) Back to Bentham? Explorations of experienced utility. Quart J Econ 112(2):375–406
Karmarkar UR, Yoon C (2016) Consumer neuroscience: advances in understanding consumer psychology. Curr Opin Psychol 10:160–165
Karwowski W et al (2019) Application of graph theory for identifying connectivity patterns in human brain networks: a systematic review. Front Neurosci 13:585
Keown CL et al (2017) Network organization is globally atypical in autism: a graph theory study of intrinsic functional connectivity. Biol Psychiatry Cogn Neurosci Neuroimaging 2(1):66–75
Knutson B et al (2007) Neural predictors of purchases. Neuron 53(1):147–156
Koch SP et al (2008) Individual alpha-frequency correlates with amplitude of visual evoked potential and hemodynamic response. Neuroimage 41(2):233–242
Lankinen K et al (2014) Intersubject consistency of cortical MEG signals during movie viewing. Neuroimage 92:217–224
Lee N et al (2007) What is 'neuromarketing'? A discussion and agenda for future research. Int J Psychophysiol 63(2):199–204
Leff DR et al (2011) Assessment of the cerebral cortex during motor task behaviours in adults: a systematic review of functional near infrared spectroscopy (fNIRS) studies. Neuroimage 54(4):2922–2936
Lewis PM et al (2016) Brain neuromodulation techniques: a review. Neuroscientist 22(4):406–421
Li J, et al (2018) Exposure to luxury brands reduces interpersonal neural synchronization across cooperative dyads: an fNIRS-based hyperscanning study. In: 2018 interdisciplinary symposium of decision neuroscience, Ann Arbor, Michigan, USA
Liu T, Liu X (2018) Multi-brain network across consumers'right inferior frontal gyrus predicts their attitudes toward advertising. In: 2018 global marketing conference at Tokyo
Liu T, Pelowski M (2014) A new research trend in social neuroscience: towards an interactive-brain neuroscience. PsyCh J 3(3):177–188
Liu T et al (2019) Assessing autism at its social and developmental roots: a review of autism spectrum disorder studies using functional near-infrared spectroscopy. Neuroimage 185:955–967
Ma Q et al (2007) Event-related potential N270 correlates of brand extension. Neuroreport 18(10):1031
Ma Q et al (2008) P300 and categorization in brand extension. Neurosci Lett 431(1):57
Ma Q et al (2010) The influence of negative emotion on brand extension as reflected by the change of N2: a preliminary study. Neurosci Lett 485(3):237–240
Ma Q et al (2015) The undermining effect of facial attractiveness on brain responses to fairness in the Ultimatum Game: an ERP study. Front Neurosci 9:77
Ma Q et al (2017) Neural process of the preference cross-category transfer effect: evidence from an event-related potential study. Sci Rep 7(1):3177
Ma Q et al (2018a) "You Win, You Buy"—how continuous win effect influence consumers' price perception: an ERP study. Front Neurosci 12:691
Ma Y et al (2018b) How is the neural response to the design of experience goods related to personalized preference? An implicit view. Front Neurosci 12:760
McClure SM et al (2004) Neural correlates of behavioral preference for culturally familiar drinks. Neuron 44(2):379–387

Mostafa MM (2013) The persistence of memory: an fMRI investigation of the brain processing of surrealistic imagery in advertising. J Mark Commun 19(5):341–359
Nia A, Zaichkowsky JL (2000) Do counterfeits devalue the ownership of luxury brands? J Prod Brand Manag 9(7):485–497
Obrig H et al (2002) Habituation of the visually evoked potential and its vascular response: implications for neurovascular coupling in the healthy adult. Neuroimage 17(1):1–18
Ogawa S et al (1990) Brain magnetic resonance imaging with contrast dependent on blood oxygenation. Proc Natl Acad Sci 87(24):9868–9872
Okamoto M et al (2004) Three-dimensional probabilistic anatomical cranio-cerebral correlation via the international 10–20 system oriented for transcranial functional brain mapping. Neuroimage 21(1):99–111
Pelowski M et al (2017) Move me, astonish me… delight my eyes and brain: the Vienna integrated model of top-down and bottom-up processes in art perception (VIMAP) and corresponding affective, evaluative, and neurophysiological correlates. Phys Life Rev 21:80–125
Petty RE et al (1983) Central and peripheral routes to advertising effectiveness: the moderating role of involvement. J Consum Res 10(2):135–146
Plassmann H et al (2008) Marketing actions can modulate neural representations of experienced pleasantness. Proc Natl Acad Sci 105(3):1050–1054
Plassmann H et al (2015) Consumer neuroscience: applications, challenges, and possible solutions. J Mark Res 52(4):427–435
Poldrack RA (2006) Can cognitive processes be inferred from neuroimaging data? Trends Cogn Sci 10(2):59–63
Pozharliev R et al (2015) Merely being with you increases my attention to luxury products: using EEG to understand consumers' emotional experience with luxury branded products. J Mark Res 52(4):546–558
Pozharliev R et al (2017) Social consumer neuroscience: neurophysiological measures of advertising effectiveness in a social context. J Advert 46(3):351–362
Rossiter JR et al (2001) Brain-imaging detection of visual scene encoding in long-term memory for TV commercials. J Advert Res 41(2):13–21
Schilbach L et al (2013) Toward a second-person neuroscience 1. Behav Brain Sci 36(4):393–414
Scholkmann F et al (2014) A review on continuous wave functional near-infrared spectroscopy and imaging instrumentation and methodology. Neuroimage 85:6–27
Schwarz N, Clore GL (2003) Mood as information: 20 years later. Psychol Inquiry 14(3–4):296–303
Shang Q et al (2017) My friends have a word for it: event-related potentials evidence of how social risk inhibits purchase intention. Neurosci Lett 643:70–75
Shiv B, Yoon C (2012) Integrating neurophysiological and psychological approaches: towards an advancement of brand insights. J Consum Psychol 22:3–6
Shiv B et al (2005) Placebo effects of marketing actions: consumers may get what they pay for. J Mark Res 42(4):383–393
Skouras S et al (2013) fMRI scanner noise interaction with affective neural processes. Plos One 8(11)
Sporns O (2018) Graph theory methods: applications in brain networks. Dialogues Clin Neurosci 20(2):111
Tversky A, Kahneman D (1974) Judgment under uncertainty: heuristics and biases. Science 185(4157):1124–1131
Tversky A, Kahneman D (1981) The framing of decisions and the psychology of choice. Science 211(4481):453–458
Van Wijk BC et al (2010) Comparing brain networks of different size and connectivity density using graph theory. Plos One 5(10):e13701
Varan D et al (2015) How reliable are neuromarketers' measures of advertising effectiveness?: data from ongoing research holds no common truth among vendors. J Advert Res 55(2):176–191

Venkatraman V et al (2015) Predicting advertising success beyond traditional measures: new insights from neurophysiological methods and market response modeling. J Mark Res 52(4):436–452

Waldman DA et al (2017) Neuroscience in organizational behavior. Annu Rev Organ Psych Organ Behav 4:425–444

Wang X et al (2012a) Event-related potential P2 correlates of implicit aesthetic experience. Neuroreport 23(14):862–866

Wang X et al (2012b) N400 as an index of uncontrolled categorization processing in brand extension. Neurosci Lett 525(1):76–81

Wang C et al (2018) The effects of money on fake rating behavior in e-commerce: electrophysiological time course evidence from consumers. Front Neurosci 12:156

Wang L et al (2019) The cross-modal interaction between sound frequency and color saturation on consumer's product size perception, preference, and purchase. Psychol Mark 37(7):876–899

Weaver R, Prelec D (2013) Creating truth-telling incentives with the Bayesian truth serum. J Mark Res 50(3):289–302

Wei Z et al (2018) Using support vector machine on EEG for advertisement impact assessment. Front Neurosci 12:76

Yoon C et al (2006) A functional magnetic resonance imaging study of neural dissociations between brand and person judgments. J Consum Res 33(1):31–40

Yoon C et al (2012) Decision neuroscience and consumer decision making. Mark Lett 23(2):473–485

Yu W et al (2018) Things become appealing when I win: NEURAL evidence of the influence of competition outcomes on brand preference. Front Neurosci 12:779

Zhang S, Sood S (2002) "Deep" and "surface" cues: brand extension evaluations by children and adults. J Consum Res 29(1):129–141

Augmentation of Nutrition by Nanotechnology

Cosmin Sonea, Mircea Lupusoru, and Ioan Opris

1 Nanomaterials and Nanofoods

Nanotechnology is a recent technology that has great potential for use in a broad array of Food and Drug Administration (FDA)-regulated products, including foods, medical products, and cosmetics. So, nanomaterials have different chemical, biological, and physical properties and can be used in medical products, foods, or cosmetics. Nanomaterials are measured in nanometers—equal to about one-billionth of a meter—so small that they cannot be seen with a regular light microscope. These nanomaterials can have different physical, chemical, or biological properties than their conventionally scaled-up counterpart materials that are used in many products regulated by the FDA. The very changes in biological, chemical, and other properties that can make nanotechnology applications so exciting may also merit examination to determine any effects on product safety, effectiveness, or other attributes.

2 Nanofoods

The domain of nanotechnology has emerged into a huge field of activity, from simple items to extremely complex ones (cleaning agents, edible items, medicine, science). Something that interests us is nanotechnology in the food industry (production

C. Sonea (✉)
University of Agronomical Studies and Veterinary Medicine, Bucharest, Romania
e-mail: cosmin_sn@yahoo.com

M. Lupusoru
Carol Davila University of Medicine and Pharmacy, Bucharest, Romania

I. Opris
Department of Neurosurgery, University of Miami, Miami, FL, USA

I. Opris et al. (eds.), *Modern Approaches to Augmentation of Brain Function*,
Contemporary Clinical Neuroscience,
https://doi.org/10.1007/978-3-030-54564-2_19

efficiency and food safety—waste reduction), which is closely correlated to electronics, data storage, and advancement of integrated devices. Moreover, this type of technology has really improved, besides food safety, some aspects such as packaging, processing, and nutrition.

Nanofood describes that type of food that, at least once, got in touch with nanotechnology techniques or tools. Recently, nanotechnology is focused on the detection of food pathogens using very sensitive nanosensors. Even if there are some evident advantages to the use of nanotechnology (health-promoting additives, longer shelf life, new flavors, smart food packaging), there can also be some disadvantages not been thoroughly studied by far, such as toxicity.

The most used nanostructures in the food industry are engineered nanomaterials, nanoemulsions, nanoliposomes, and nanofibers. Thus, the nanomaterials can be divided into three categories: inorganic (nanoparticles of gold, silver, iron, calcium, magnesium, silicon, selenium), surface functionalized nanomaterials (nanoclays), and organic nanoparticles (liposomes, proteins, polymers). The first two categories are generally used as food additives, food packaging, and storage, while the third category is used in products to enhance their uptake.

Nanoemulsions have been synthesized in order to reduce the contamination on different surfaces (packages, equipment, fruits, vegetables, chicken skin). Nanoliposomes are synthesized from natural ingredients (soy, egg, milk). For example, the first nanoliposomes were used in the composition of cheese. Inorganic nanoparticles (zinc oxide, silver, copper, titanium oxide) are used for improving the physical performance, durability, barrier properties, and biodegradation of foods. For example, plastic beer bottles contain nanoclays that help keeping oxygen outside of the bottle and carbon dioxide inside.

Nanosensors can detect viruses, pathogenic microorganisms, detrimental chemicals, and physical contaminants in food and gases, as well as pathogens and toxins in packaged foods. In food processing, nanoparticles can be used in many ways such as improving the nutritional quality of food, flavor, color, but in addition, they also protect functional ingredients (antimicrobials and vitamins). Utilization of nanomaterials in food packaging can protect consumers from infestation with microorganisms that leads to spoilage. In the field of agriculture, nanotechnology has had an extremely powerful impact, a lot of concepts being nowadays developed (nanofertilizers, nanomaterials in machinery and tools), and it supports the so-called smart agriculture. A very important aspect is represented by the potential benefits of nanotechnology for agriculture, food, fisheries, and aquaculture. In all these domains, a necessary requirement is a balance between natural elements (soil, water, environment) and the occupational health of workers.

Using nanotechnology, mainly in food industry, might have toxic effects because food tends to be modified at the atomic level. In a study, performed on rats, the scientists observed acute pulmonary inflammation and stress response. Some nanoparticles create oxidative stress which can generate free radicals, leading to DNA mutation, cancer, and possible death.

3 Nutrition

The main applications of nanotechnology in nutrition relate to food safety, waste reduction, packaging, and processing. Nanotechnology can also be of use in specific applications such as modifying taste, color, texture of food and detecting food pathogens and microorganisms. For example, thymoquinone has been well studied, a particle that has a lot of beneficial effects (e.g., it is hepatoprotective of central nervous system and controls inflammations).

The main applications of nanotechnology may assist with obtaining accurate spatial information about the location of a nutrient or bioactive food component in tissues, cells, or cellular components. Some unexpected and extremely useful things became possible, for example, the ultrasensitive detection of nutrients and metabolites and the understanding of nutrient and biomolecular interactions in specific tissues. In theory, such new technologies have the potential to improve nutritional assessment and measure the bioavailability. They may help to identify and characterize molecular targets of nutrient activity and biomarkers of effect and, therefore, may inform on a "personalized/customized" nutrition. Until now, the specific applications of nanotechnology in food and nutrition include modification of taste, color, and texture of foods. Moreover, detection of food pathogens and spoilage microorganisms and enhancing the nutrition quality of foods have improved a lot. Nanotechnology in food can be used as novel vehicles for nutrient delivery and can also serve as a tool to enable further elucidation of nutrient metabolism and physiology (Surai et al. 2017; Ezzat et al. 2017). For example, one food nanotechnology application involves creating coatings for foods and food packaging, serving as barriers to bacteria or to additional nutrients (Zhang et al. 2017). Nutritional products claiming to use nanotechnology are currently available in the market. It is important to recognize that the potential toxicity of nutrients can be affected by a change in particle size (Abdolahi et al. 2017).

4 Applications of Nanomaterials and Nanodevices in Food Industry

There are a lot of applications for nanotechnology in the food industry (food safety, quality control, additives, supplements, flavors, food storage, food formulation, food packaging). For example, nanotechnology enhances thermal and mechanical properties and safety, regarding food packaging (Fig. 1).

Nanosensors are also able to alert consumers when food is expired. Approximative 9% from the available nanomaterials are used in agriculture and food industry. Their main purpose is to improve nutritional values.

The following nanomaterials are used in food industry: nanoparticles (Ag, ZnO, Mg, SiO_2), nanosieves (specific nanoparticles), nanocapsules (bioactive compounds), nanoemulsions (tweens or spans, arabica gum or modified starch, soy,

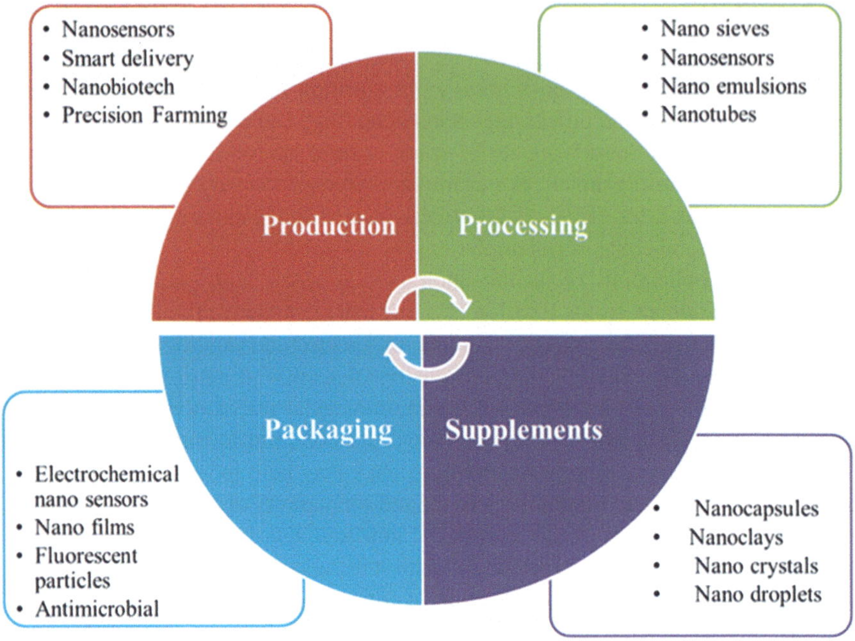

Fig. 1 An overview of applications of nanomaterials in the food industry

caseinate), nanospheres (starch nanospheres), nanosensors (aptasensors), nanocochleates (coiled nanoparticles), nanocomposite (Fe-Cr/Al_2O_3 and Ni/Al_2O_3), and nanomicelles (aquanova, novasol).

All of these nanomaterials have various applications in the food industry, such as food protection, stability, deterioration control, encapsulation, processing, packaging; liquid carrier; local and controlled release of nutrients; colorant; and synthetic adhesives. They also have antimicrobial and antioxidant characteristics, and they can detect microorganisms and remove pathogens or contaminants. In addition, nanomaterials own certain properties. For example, they increase efficacy and water solubility and enhance nutritional value of food and shelf life of food. There are four categories of nanomaterials but, of course, some of them can be catalogued into multiple classes. Nanosensors, smart delivery nanobiotech, and precision farming belong to PRODUCTION category; nanosieves, nanosensors, nanoemulsions, and nanotubes fall under the PROCESSING category; electrochemical nanosensors, nanofilms, fluorescent particles, and antimicrobial belong to the PACKIGING; and one and the last category, SUPPLEMENTS, includes nanocapsules, nanoclays, nanocrystals, and nanodroplets.

Although this field has been developed a lot lately (all the devices have become more sensitive and smaller in size), the nanotechnology still has a great potential to exploit. In this regard, new applications are explored in various areas of the food industry. Among these new applications is the detection of toxins: mycotoxin, ochratoxin, botulinum neurotoxin, and aflatoxin using different methods, especially immunological ones (immunoassay and enzyme-linked immunosorbent assay). In this case, scientists use nanoparticles (e.g., magnetic and superparamagnetic, cerium dioxide, zinc oxide, gold, iron oxide nanoparticles), ionic liquids, nanotubes, and quartz nanopipettes. Another application is the detection of microbes such as *Salmonella enterica, Escherichia coli, Salmonella typhimurium, Streptococcus aureus,* and *Listeria monocytogenes* by using microbiological methods. Detection of pesticides and chemicals uses spectroscopy, chromatography, or voltammetry in order to identify chloramphenicol, sudan I, cadmium ions, bisphenol, carbamate pesticides, sulfites, and melamine. The last application consists of detection of unstable key food ingredients: ethanol, folate, hydrogen peroxide, glucose, fructose, sucrose, glutamic acid, succinic acid, ascorbic acid, tryptophan, and acetaminophen using voltammetry or spectroscopy.

5 Augmentation of Nutrition

We consider the improvement of quality, taste, perception, and packaging of foods as augmentation of nutrition. Here, we look at the following nanoaugmentations:

1. Curcumin: Sivasami and Hemalatha (2018) demonstrated the therapeutic potential of curcumin for inflammatory, neoplastic, and preneoplastic diseases using nanotechnology.
2. Electric taste augmentation: Ranasinghe et al. (2017) designed and implemented two utensils (a pair of chopsticks and a soup bowl) for generating electric taste sensations. Both the utensils apply electrical taste sensations to users' tongues when used as part of dining interactions. The experimental findings show a significant increase in perceived saltiness and sourness when consuming unsalted mashed potato and a significant increase in perceived sourness and bitterness, when consuming diluted miso soup.
3. Thymoquinone: El-Far et al. (2018) provided protective roles of thymoquinone nanoformulations: potential nanonutraceuticals in human diseases. The effects of thymoquinone are discussed below.

6 Nanonutrients

How can nanotechnology augment the nutritive function? Obviously, by nanonutrients.

Some food ingredients that work on the nanoscale as nanonutrients have different properties as they improve texture, consistency, taste, etc. Various types of food nanotechnology have been involved in improving shelf life (Pradhan et al. 2015). Today, nanocarriers are used in the same way as delivery structures for different

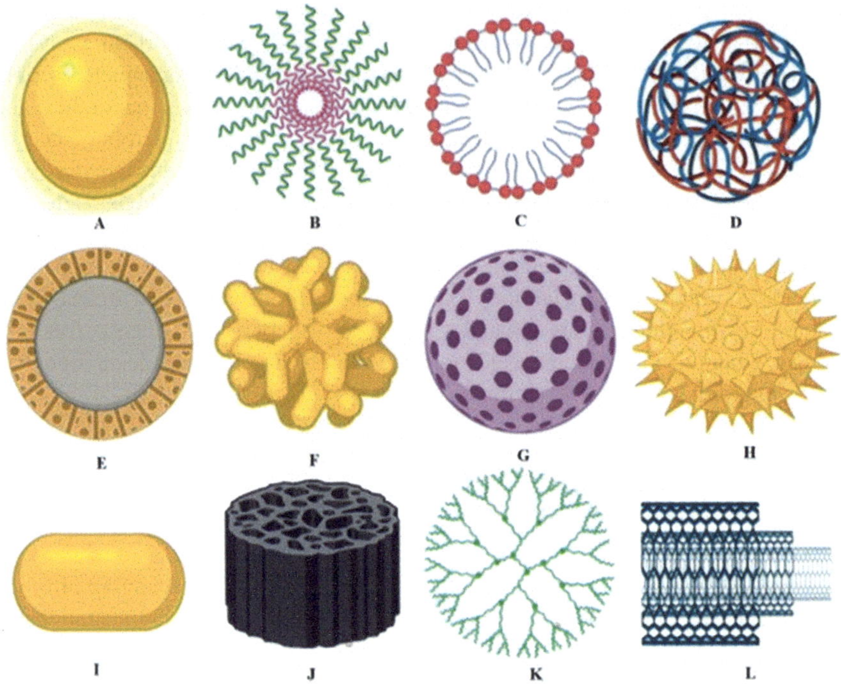

Fig. 2 Graphical representation of different types of nanomaterials used in the food industry. (**a**) Metallic nanoparticle, (**b**) polymeric micelle, (**c**) liposome nanoparticle, (**d**) polymeric nanoparticle, (**e**) solid core mesoporous nanoparticle, (**f**) branched gold nanoparticle, (**g**) mesoporous nanoparticle, (**h**) surface functionalized nanoparticle, (**i**) nanorod, (**j**) porous silica nanoparticle, (**k**) dendrimer, (**l**) carbon nanotubes

flavors of food in food yields without manipulating the morphology (Fig. 2). Particle sizes might openly influence the supply of bioactive complex to numerous sites inside the body; for example, it was observed that the submicron particles can only work on the nanoscale. This nanoscopic scale refers to structures with a length scale (1–100 nm) applicable to nanotechnology (Dube et al. 2010). A perfect delivery method has the following properties: (1) the ability to bring energetic material specifically to the target position; (2) the accessibility of the target period along with a definite rate; (3) efficiency at keeping active compounds at appropriate levels for a long time (storage state).

Nanotechnology has an important role in the development of emulsions, encapsulations, simple solutions, the association of colloids biopolymer matrices, and well-organized delivery organizations with the above-mentioned abilities (Jafarizadeh-Malmiri et al. 2019).

Nanoparticles have enhanced capabilities such as the release efficiency and encapsulation properties, when compared with traditional encapsulation methods. Nanoencapsulation covers tastes and the odor control connections of potent ingredients by means of the food matrix, release of active compounds, confirmation of the availability of the target time, and protection from different sources of contamination such as heat, moisture, and biological and chemical degradation (Drusch and Mannino 2009) during storage, processing, and utilization. These nanoparticles also show compatibility when compared with other compounds in the system (Weiss et al. 2006). Furthermore, these delivery classifications retain the capability to enter into tissues owing to their small size and thus permit the well-organized distribution of active complexes to mark entry sites in the body (Lamprecht et al. 2004). In order to own a better-quality bioavailability and to offer protection of active components of food, various artificial and ordinary polymer-based encapsulation has been used. Moreover, the significance of nanotechnology such as in food processing can be estimated by examining its role in the augmentation of food items (Sastry et al. 2013).

7 Nanotechnology and Its Impact on Food and Nutrition

Some domains in which nanotechnology will play an important role are the agribusiness sector, sustainable agriculture, or food industry (Kah et al. 2019).

Protein nanoaggregates have an interesting future because they can improve functional properties of foods. For example, soy-protein isolate nanoaggregates were used to prepare oil-in-water nanoemulsions with canola oil (that was stable over 21 days and 4 °C), with some pH adjustments (Yildiz et al. 2017).

Nanotechnology is trying to provide safe food, waste reduction, and, of course, consumer health. But another important purpose is to reduce toxicity of nanoparticles (Sonkaria et al. 2012). In terms of augmentation, for food industry and nutrition nanoemulsions, solid lipid nanoparticles, micelles, and liposomes are included (Nakajima et al. 2015). Nanoemulsions have been synthesized in order to reduce contamination of different surfaces (packaging equipment, fruits, vegetables, chicken skin). Nanosensors can detect viruses, pathogenic microorganisms, detrimental chemicals, and physical contaminants in food and gases, pathogens, and toxins in packaged foods. In food processing, nanoparticles can be used in many ways such as improving the nutritional quality of food, flavor, and color, but they also protect functional ingredients like antimicrobials and vitamins (Thiruvengadam et al. 2018).

Nanotechnology has many vital biological applications and can be used in human food industry and also for diets of poultry, broilers, layers, turkeys, or quails. From all their benefits, we can mention that nanoparticles provide antimicrobial activities, increase the number of gut's good microbes as well as their products, e.g., short-chain fatty acids (Hassan et al. 2019). If selenium is supplied in proper form, it has huge bioavailability for fish. Thus, the quantity of selenium that we can add in fish food is important because the border between what is beneficial and what is toxic is very narrow (Khan et al. 2017). Curcumin used on non-alcoholic fatty liver disease has antioxidant and anti-inflammatory effects. Nanocurcumin on overweight patients improved indices of glucose, lipids, and liver transaminases and that of fatty liver (Jazayeri-Tehrani et al. 2019).

8 Nutrition of Plants

Nanotechnology has many applications in agriculture; it is used to solve some problems such as environmental pollution with pesticides and other substances. On this line of thinking, nanosensors can detect pesticides at minute levels (Ghormade et al. 2011). Moreover, the nanostructures can release fertilizer slower and nanophosphate more slowly in order to protect soil composition, fertility, and integrity of environment. There are some methods that can be used for this purpose: germination, specific activity of enzymes, carbohydrates, protein, photosynthetic pigments, root nodule number, and microbial population (Mala et al. 2017).

9 Animal Nutrition

Nanoparticles are widely used in biomedical applications like silica nanoparticles which have tropism on the gastrointestinal tract of animals (Pieszka et al. 2019). Currently, the use of tin (Sn) in animal nutrition, with a focus on gastrointestinal tract function, is of great interest. Regarding nanomaterials, even if chemical methods are more beneficial, the biological ones are more protective of the environment (Swain et al. 2015). In animal feeding, the nanominerals are synthesized by chemical methods and have significant growth-promoting, immunomodulatory, antibacterial effects than the conventional counterparts. In ovo feeding has a lot of benefits for chicken embryo development in modern broilers: it reduces post-hatch mortality and skeletal disorders and increases muscle growth and breast meat yield. On the 20th day of incubation, it was found that an application of ATP to chicken embryos increases the expression of fibroblast growth factor 2 (FGF2), vascular endothelial growth factor (VEGF), and Na(+)/K(+) transporting ATPase (Sawosz et al. 2013).

10 Neurotransmitters Produced During Nutrition

One of the most important flavonoids that can be isolated from some edible fruit is naringenin. It has anti-inflammatory, antioxidant, and insulin-like effects as well as different types of effects on sex hormone metabolism and lipid (Zobeiri et al. 2018). Nonalcoholic fatty liver disease (NAFLD) is the most common chronic liver disease mainly caused by high-fat diets and sudden feed changes, vitamin and energy deficiency, and inflammatory processes. Fatty liver leads to cirrhosis, hepatocellular carcinoma, as well as liver failure. Lifestyle modifications such as weight loss and a healthy diet have an inverse correlation with the risk of fatty liver.

11 Effects on Foods and Nutrition

Food grounding has impact on physiological and clinical outcomes, having a strategy against the global epidemic of noncommunicable, degenerative, and inflammatory-related diseases. But probably the most important fact is to explain the potential of food grounding to clinicians as a simple strategy for prevention, therapy, and improving patient outcomes (Sinatra et al. 2017).

Ranasinghe et al. (2017) have applied electrical taste sensations/flavor modulation to users' tongues when used as part of dining interactions. Ionizing sources have applicability because they inactivate microorganisms, but they are also used in food and drug industries. Both thermal and radiation resistance concepts were reviewed, and some typical values of radiation resistance are given for sensitive vegetative bacterial cells, yeasts, and molds and for resistant bacterial spores and viruses (Mudgett 1988). This study was performed on pigs, and the main purpose was to discover a connection between the brain and food signals. So, it was found that sweet and bitter have both oral and visceral levels and have an impact on weight gain, obesity, and brain metabolism (Val-Laillet 2019).

12 Protective Roles of Thymoquinone Nanoformulations

Thymoquinone, TQ, is considered an excellent nutraceutical product that has many properties such as anti-inflammatory, antioxidant, anticancer, antimicrobial, immunomodulatory, antidiabetic, and antihistaminic. TQ is one of the most important constituents of *Nigella sativa*. TQ should have the following characteristics: (1) it should be distributed to the target tissue or organ, (2) it needs to reach a therapeutic concentration, (3) it should be stored a well-determined interval of time, and (4) it should protect and reduce renal clearance. The nutraceutical product therapeutic effects have been highlighted by different methods. Thus, the single-emulsion method shows the enhanced anticancer activity, whereas the double-emulsion method indicates the inhibition of tumor growth. Another method which consists of nanoprecipitate expresses reduced tumor and increased lifespan. The ionic gelation method shows an increase in the brain's targeting efficiency, and the film rehydration method indicates more potent antiproliferative activity. Last, but not least, the cold wet-milling method expresses an enhancement of oral bioavailability.

Regarding the anticancer activity, it was demonstrated that a combination between doxorubicin and TQ in F2 gel (fully acetylated poly-*N*-acetyl glucosamine nanofiber) can reduce the tumor volume due to B-cell lymphoma 2 (Bcl2) downregulation and P53 upregulation. Moreover, another combination between poly(D,L-lactide-*co*-glycolide) nanoparticles encapsulating paclitaxel and TQ, using the single emulsion solvent evaporation method, can induce apoptosis in breast cancer cells. Some of the disadvantages of TQ (low aqueous solubility, thermal properties, photosensitivity) had been reduced by synthesized TQ encapsulated nanoparticles. These structures were more effective on breast cancer cells (the mechanism is also apoptosis) and had less toxicity to the normal cells. Talking about antidiabetic effects, a nanoformulation (nano-TQ and metformin) was administrated to diabetic rats for 21 days in a row. It showed a sustained release profile, a decrease in the levels of blood glucose that overcame hyperglycemia.

TQ is an efficient particle that protects the Central Nervous System against different diseases. In the same time, TQ-loaded solid lipid nanoparticles on the brain increased the TQ delivery to the brain tissues (this fact could be objectified by monitoring the level of 5-hydroxytryptamine, dopamine, norepinephrine in the brain). In addition, if TQ-rich fraction is administrated in rats, it is noticed that a diminution in the rats' memory deficit is observed.

Nanoformulations that contain TQ lipospheres are antipsoriatic drugs for topical use and control inflammation. The seeds of *Nigella sativa* are used as hepatoprotective medicinal herb (mainly due to the antioxidant potential). TQ-loaded solid lipid nanoparticles protect the liver against cirrhosis. Nano-TQs can be used in cosmetics because they have some benefits such as anti-aging effect and moisturizing due to their antioxidant and anti-apoptosis powers. Nanotechnology can provide various ways of improving formulations of TQs in order to use them in different cases (e.g., drug delivery).

13 Potential Nanonutraceuticals in Human Diseases

One example introduces applications of nanomedicine to enhance the biological activities of nano-TQ to control different diseases in several studies as a preliminary investigation for human disease treatment with nano-TQ. Nano-TQ effectively augments the anticancer roles of doxorubicin by upregulation of P53 and downregulation of BCl_2 and potentiates paclitaxel's apoptosis in MCF-7 breast cancer cells. Moreover, nano-TQ protects against diabetes, inflammation, CNS diseases, and hepatotoxicity, mainly by enhancement of antioxidant status of organs. Nano-TQ may be considered as a promising nutraceutical for human health.

As an ideal nanonutraceutical, TQ would be delivered to the target tissue or organ, where free TQ could then reach a therapeutic concentration and be maintained for a required time. Nanomaterials can easily penetrate the biological membranes and provide sustained release of TQ to different body parts (Srinivas et al. 2010), and therefore, nanoformulated TQ would enhance its bioavailability. Enhancement of drug and food constituents' delivery and therapeutic effects have been achieved by a wide assortment of bottom-up encapsulation methods, such as single emulsions, double emulsions, nanoprecipitation, or the ionic gelation method (Li et al. 2017), biopolymer side chain conjugations (Weiss et al. 2006), and top-down methods such as the cold wet-milling method (Nihei et al. 2016) (Table 1, Fig. 3). In addition, nanoparticles (NPs) have been extensively used for drug delivery enhancement such as carbon, ceramic, and chitosan NPs.

Table 1 TQ nanoformulations and characteristics

Method	Materials, stabilizer	Size (nm)	Therapeutic effect	Reference
Single emulsion	PLGA	200–300	Enhanced anticancer activity	Soni et al. (2015)
	Compritol ATO 888 Gelucire	~200	Enhancement of oral bioavailability	Enck et al. (2016)
Double emulsion	PVA	185	Inhibited tumor growth	El-Ashmawy et al. (2017)
Nanoprecipitate	PVP, PEG200, PEG4000, P123	20–40	Reduced tumor and increased lifespan	Bhattacharya et al. (2015)
	Gum rosin, oleic acid, PVA, polysorbate 80	50–90	Decreased blood glucose level and glycated hemoglobin	Rani et al. (2018)
Ionic gelation	TPP	150	Higher brain targeting efficiency	Alam et al. (2012)
Film rehydration	Liposomes	100	More potent antiproliferative activity	Odeh et al. (2012)
Cold wet-milling	HPC-SSL	143	Enhancement of oral bioavailability	Nihei et al. (2016)

PLGA poly(D,L-lactide-*co*-glycolide), *PVA* poly(vinyl alcohol), *PVP* polyvinylpyrrolidone, *PEG* poly(ethylene glycol), *P123* poly(ethylene glycol)-b-poly(propylene glycol)-b-poly(ethylene glycol), *TPP* sodium triphosphate, *HPC-SSL* hydroxypropyl cellulose grade SSL

Some studies have been done to enhance the bioavailability of TQ, especially by the oral route, including micelle NPs, chitosan NPs, and liposomes (Alam et al. 2012; Ganea et al. 2010). Oral liquid formulations of TQ enhanced the solubility and bioavailability of TQ and protected it from photodegradation (Dube et al. 2010). Nano-TQ has more photostability and a sixfold increase of oral bioavailability than free TQ solution (Kah et al. 2019). In addition, Tubesha et al. (2013) indicated that TQ nanoemulsion is stable for 6 months. We conducted this review to shed light on the effect of TQ nanoformulations on the efficacy of TQ in control of human diseases. Finally, the future directions for improving the pharmacokinetics of TQ and perspectives on the clinical translation of nano-TQ are discussed.

14 Brain–Gut Connectome

The enteric nervous system includes a huge number of neurons, in the range of 100 million neurons, more than in the spinal cord, reason for which it is often called the "second brain." It consists of fibers of neurons "embedded" in the walls of the long tube of alimentary canal or the gut. This ensemble of neurons in the enteric nervous system "enables us" to "feel" the contents of the inner world of

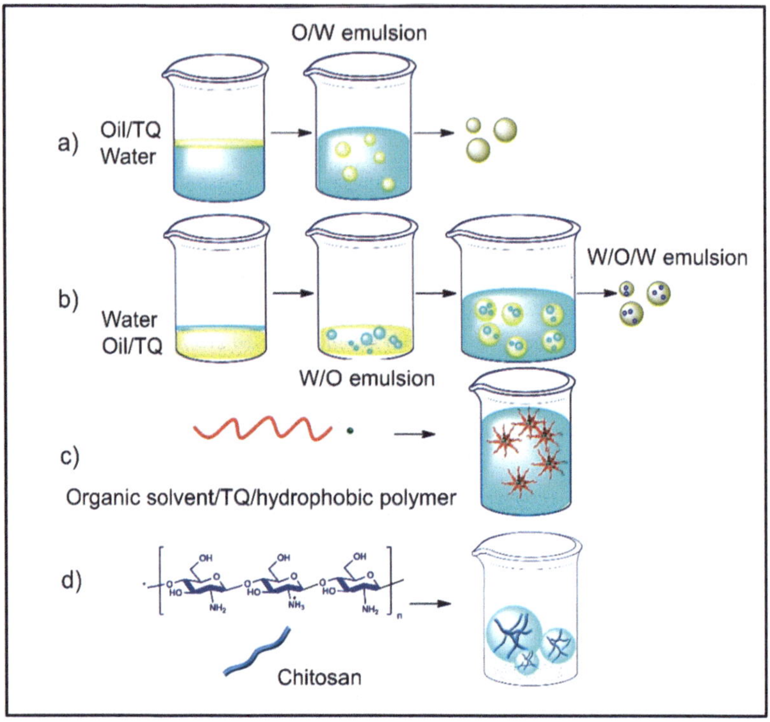

Fig. 3 Examples of nanonutrients. Thymoquinone nanoformulations prepared through (**a**) single emulsion; (**b**) double emulsion; (**c**) nanoprecipitation; (**d**) ionic gelation methods. (With permission from El-Far, Nutrients, 2018)

our gut, as well as in the elaborate daily routine of digestion. Performing tasks like (1) breaking down food, (2) absorbing nutrients, and (3) removing of waste requires chemical processing, mechanical mixing, and rhythmic muscle contractions that move everything down the line. The most interesting characteristic of the enteric system is that of being equipped with reflexes and senses, it can control the complex behavior of the gut, while being totally independent of the brain. "The brain in the head doesn't need to get its hands dirty with the messy business of digestion, which is delegated to the brain in the gut", Gershon says (citation taken from Hadhazy 2010). The complexity of the enteric system cannot be interpreted through this process alone. The interesting part is that about 90% of the fibers in the primary visceral nerve, the vagus nerve, carry information from the gut to the brain and not the other way around (Fig. 4).

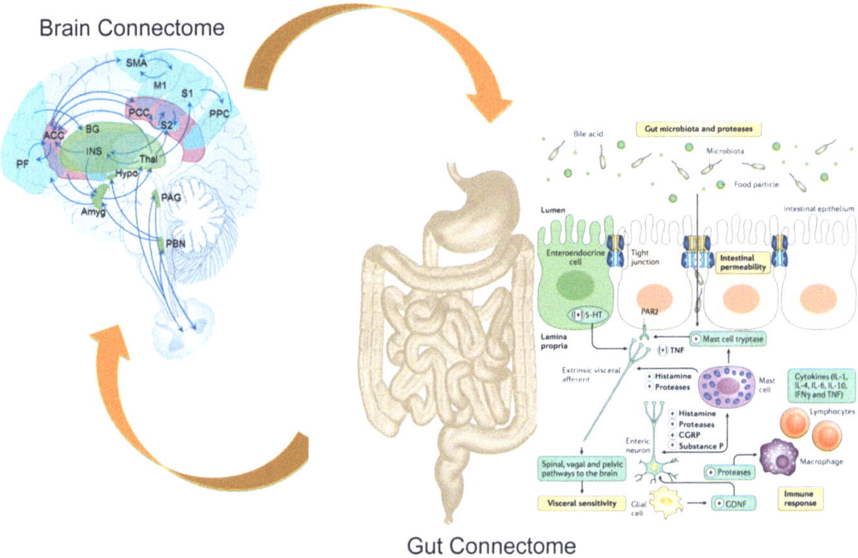

Fig. 4 A proposed integrative model for disorders of gut–brain interactions. Replacing the conventional focus on individual brain regions and cell types in the gut, this integrative model posits reciprocal interactions between brain networks (brain connectome) and networks made up of multiple cells in the gut, including the gut microbiota (gut connectome). Gut-to-brain communication is mediated by neural, endocrine, and inflammatory pathways, while brain-to-gut communication relies mainly on autonomic nervous system output to the gut. (Modified with permission from Enck et al. 2016)

15 Phosphate Binders

Besides low-phosphorous diet and dialysis, the management of hyperphosphatemia by phosphate binders are considered as the cornerstones in ESRD (end-stage renal disease) patients (Biruete et al. 2020). Phosphate binders prevent the dietary Pi absorption within the GIT (gastrointestinal tract) by exchanging an active cation (e.g., carbonate, acetate, oxyhydroxide, and citrate) with anion phosphate to make a nonabsorbable compound that is excreted in the feces. Long-term ingestion of these compounds may alter the gut microbial composition by lowering the intestinal phosphate burden through the formation of phosphate complexes (see Fig. 5). There are different classes of phosphate binders to be reviewed in the following section.

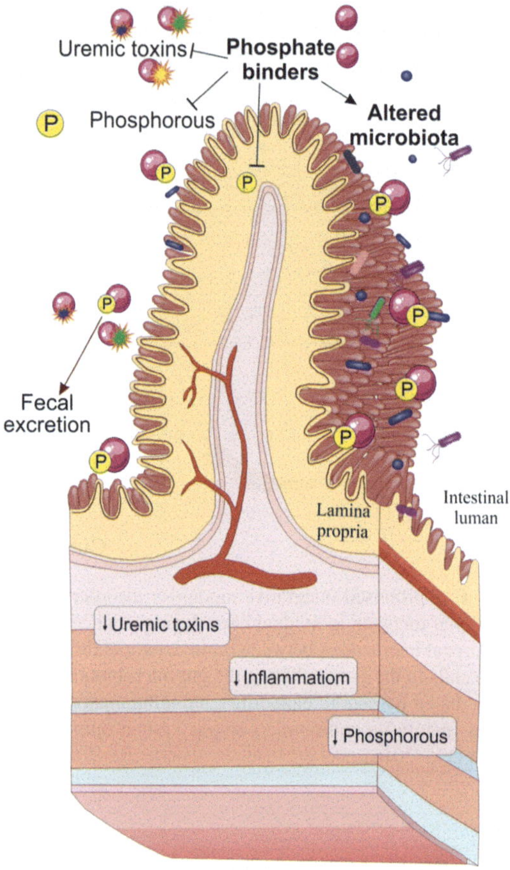

Fig. 5 The effects of phosphate binders on intestinal phosphate and microbiota. Phosphate binders are prescribed to the chronic kidney disease (CKD) patients in order to lower hyper-phosphatemia. These medications aim to lower serum Pi (phosphorous) within intestine by forming nonabsorbable compounds which removed by fecal excretion to lessen the adverse outcomes of elevated Pi levels. Consequently, low levels of Pi may alter the gut microbial communities. Besides, these drugs decrease the inflammation and accumulation of uremictoxins (indoles, amines, and phenols). (With permission from Enck et al. 2016)

16 Conclusion

Nanotechnology has the potential to provide us healthier, safer, and better-tasting foods, as well as improved food packaging. The hesitation of the food industry and public fears in some countries about tampering with nature may be holding back the introduction of nanofoods.

Nanotechnology is a multidisciplinary science that investigates matter at the molecular and atomic levels. It has nowadays lots of applications in different fields, but the most studied are medicine and food industry (including animal nutrition).

When we talk about nanotechnology within the food industry, it is important to highlight that there is a wide variety in which it is used; from food safety to some aspects such as packaging, processing, and nutrition. There are many nanotechnology techniques and tools that can improve food industry. Thus, nanoemulsions are used to reduce the contamination on different surfaces, nanoliposomes can be one of the components of cheese, nanoclays are used in plastic beer bottles in order to keep oxygen outside, nanosensors can detect chemical and physical contaminants in food, and there are many more such examples. All in all, the food industry had a huge advantage by using nanotechnology: food protection, stability, deterioration control, encapsulation; liquid carrier; colorant; and synthetic adhesives. Nanomaterials have antimicrobial and antioxidant characteristics, can remove pathogens or contaminants, and can enhance nutritional value of food and shelf life of food.

From the medical point of view, nanotechnology is used for human disease treatment. It has been shown that nano-thymoquinone protects against breast cancer, diabetes, inflammation, hepatotoxicity, or diseases of central nervous system.

In addition, we should correlate nanotechnology to physiological functions of the organism. For example, the enteric nervous system has a huge role in breaking down food, absorbing nutrients, and removing what is not necessary. Another use is represented by the phosphate binders that can manage the hyperphosphatemia (besides other methods such as low-phosphorous diet or dialysis).

Acknowledgment To the always helpful people from Springer Publishing Company.

References

Abdolahi M, Tafakhori A, Togha M, Okhovat AA, Siassi F, Eshraghian MR, Sedighiyan M, Djalali M, Mohammadzadeh Honarvar N, Djalali M (2017) The synergistic effects of ω-3 fatty acids and nano-curcumin supplementation on tumor necrosis factor (TNF)-α gene expression and serum level in migraine patients. Immunogenetics 69(6):371–378. https://doi.org/10.1007/s00251-017-0992-8

Alam S, Khan ZI, Mustafa G, Kumar M, Islam F, Bhatnagar A, Ahmad FJ (2012) Development and evaluation of thymoquinone-encapsulated chitosan nanoparticles for nose-to-brain targeting: a pharmacoscintigraphic study. Int J Nanomed 7:5705–5718

Bhattacharya S, Ahir M, Patra P, Mukherjee S, Ghosh S, Mazumdar M, Chattopadhyay S, Das T, Chattopadhyay D, Adhikary A (2015) PEGylated-thymoquinone-nanoparticle mediated retardation of breast cancer cell migration by deregulation of cytoskeletal actin polymerization through miR-34a. Biomaterials 51:91–107

Biruete A, Hill Gallant KM, Lindemann SR, Wiese GN, Chen NX, Moe SM (2020) Phosphate Binders and Nonphosphate Effects in the Gastrointestinal Tract. J Ren Nutr 30(1):4–10. https://doi.org/10.1053/j.jrn.2019.01.004

Drusch S, Mannino S (2009) Patent-based review on industrial approaches for the microencapsulation of oils rich in polyunsaturated fatty acids. Trends Food Sci Technol 20:237–244

Dube A, Ng K, Nicolazzo JA, Larson I (2010) Effective use of reducing agents and nanoparticle encapsulation in stabilizing catechins in alkaline solution. Food Chem 122:662–667

El-Ashmawy NE, Khedr EG, Ebeid E-ZM, Salem ML, Zidan A-AA, Mosalam EM (2017) Enhanced anticancer effect and reduced toxicity of doxorubicin in combination with thymoquinone released from poly-N-acetyl glucosamine nanomatrix in mice bearing solid Ehrlish carcinoma. Eur J Pharm Sci 109:525–532

El-Far AH, Al Jaouni SK, Li W, Mousa SA (2018) Protective roles of thymoquinone nanoformulations: potential nanonutraceuticals in human diseases. Nutrients 10(10):1369

Enck P, Aziz Q, Barbara G et al (2016) Irritable bowel syndrome. Nat Rev Dis Primers 2:16014. https://doi.org/10.1038/nrdp.2016

Ezzat A, Abdelhamid AO, El Awady MK, Abd El Azeem AS, Mohammed DM (2017) The biochemical effects of nano tamoxifen and some bioactive components in experimental breast cancer. Biomed Pharmacother 95:571–576. https://doi.org/10.1016/j.biopha.2017.08.099
Ganea GM, Fakayode SO, Losso JN, van Nostrum CF, Sabliov CM, Warner IM (2010) Delivery of phytochemical thymoquinone using molecular micelle modified poly(D,L-lactide-co-glycolide) (PLGA) nanoparticles. Nanotechnology 21:285104
Ghormade V, Deshpande MV, Paknikar KM (2011) Perspectives for nano-biotechnology enabled protection and nutrition of plants. Biotechnol Adv 29(6):792–803. https://doi.org/10.1016/j.biotechadv.2011.06.007
Hassan S, Hassan FU, Rehman MS (2019) Nano-particles of trace minerals in poultry nutrition: potential applications and future prospects. Biol Trace Elem Res 195(2):591–612. https://doi.org/10.1007/s12011-019-01862-9
Jafarizadeh-Malmiri H, Sayyar Z, Anarjan N, Berenjian A (2019) Nano-sensors in food nanobiotechnology. In: Nanobiotechnology in food: concepts, applications and perspectives. Springer, Cham, pp 81–94
Jazayeri-Tehrani SA, Rezayat SM, Mansouri S, Qorbani M, Alavian SM, Daneshi-Maskooni M, Hosseinzadeh-Attar MJ (2019) Nano-curcumin improves glucose indices, lipids, inflammation, and nesfatin in overweight and obese patients with non-alcoholic fatty liver disease (NAFLD): a double-blind randomized placebo-controlled clinical trial. Nutr Metab (Lond) 16:8. https://doi.org/10.1186/s12986-019-0331-1. eCollection 2019.
Kah M, Tufenkji N, White JC (2019) Nano-enabled strategies to enhance crop nutrition and protection. Nat Nanotechnol 14(6):532–540. https://doi.org/10.1038/s41565-019-0439-5
Khan KU, Zuberi A, Fernandes JBK, Ullah I, Sarwar H (2017) An overview of the ongoing insights in selenium research and its role in fish nutrition and fish health. Fish Physiol Biochem 43(6):1689–1705. https://doi.org/10.1007/s10695-017-0402-z
Lamprecht A, Saumet J-L, Roux J, Benoit J-P (2004) Lipid nanocarriers as drug delivery system for ibuprofen in pain treatment. Int J Pharm 278:407–414
Li W, Yalcin M, Lin Q, Ardawi MM, Mousa SA (2017) Self-assembly of green tea catechin derivatives in nanoparticles for oral lycopene delivery. J Control Release 248:117–124
Mala R, Selvaraj RCA, Sundaram VB, Rajan RBSS, Gurusamy UM (2017) Evaluation of nano structured slow release fertilizer on the soil fertility, yield and nutritional profile of Vigna radiata. Recent Pat Nanotechnol 11(1):50–62. https://doi.org/10.2174/18722105106661607270 93554
Mudgett R (1988) Electromagnetic energy and food processing. J Microw Power Electromagn Energy 23(4):225–230
Nakajima M, Wang Z, Chaudhry Q, Park HJ, Juneja LR (2015) Nano-science-engineering-technology applications to food and nutrition. J Nutr Sci Vitaminol (Tokyo) 61(Suppl):S180–S182. https://doi.org/10.3177/jnsv.61.S180
Nihei T, Suzuki H, Aoki A, Yuminoki K, Hashimoto N, Sato H, Seto Y, Onoue S (2016) Development of a novel nanoparticle formulation of thymoquinone with a cold wet-milling system and its pharmacokinetic analysis. Int J Pharm 511:455–461. [CrossRef] [PubMed]
Odeh F, Ismail SI, Abu-Dahab R, Mahmoud IS, Al Bawab A (2012) Thymoquinone in liposomes: a study of loading efficiency and biological activity towards breast cancer. Drug Deliv 19:371–377
Pieszka M, Bederska-Łojewska D, Szczurek P, Pieszka M (2019) The membrane interactions of nano-silica and its potential application in animal nutrition. Animals (Basel) 9(12). pii: E1041. https://doi.org/10.3390/ani9121041
Pradhan N, Singh S, Ojha N, Shrivastava A, Barla A, Rai V, Bose S (2015) Facets of nanotechnology as seen in food processing, packaging, and preservation industry. Biomed Res Int 2015:365672
Ranasinghe N, Tolley D, Nguyen TNT, Yan L, Chew B, Do EY (2017) Augmented flavours: modulation of flavour experiences through electric taste augmentation. Altern Ther Health Med 23(5):8–16

Rani R, Dahiya S, Dhingra D, Dilbaghi N, Kim K-H, Kumar S (2018) Improvement of antihyperglycemic activity of nano-thymoquinone in rat model of type-2 diabetes. Chem Biol Interact 295:119–132
Sastry RK, Anshul S, Rao N (2013) Nanotechnology in food processing sector–an assessment of emerging trends. J Food Sci Technol 50:831–841
Sawosz F, Pineda L, Hotowy A, Jaworski S, Prasek M, Sawosz E, Chwalibog A (2013) Nano-nutrition of chicken embryos. The effect of silver nanoparticles and ATP on expression of chosen genes involved in myogenesis. Arch Anim Nutr 67(5):347–355. https://doi.org/10.1080/1745039X.2013.830520
Sinatra ST, Oschman JL, Chevalier G, Sinatra D (2017) Electric nutrition: the surprising health and healing benefits of biological grounding (earthing). Altern Ther Health Med 23(5):8–16
Sivasami P, Hemalatha T (2018) Augmentation of therapeutic potential of curcumin using nanotechnology: current perspectives. Artif Cells 46(7):1–12
Soni P, Kaur J, Tikoo K (2015) Dual drug-loaded paclitaxel–thymoquinone nanoparticles for effective breast cancer therapy. J Nanopart Res 17:18
Sonkaria S, Ahn SH, Khare V (2012) Nanotechnology and its impact on food and nutrition: a review. Recent Pat Food Nutr Agric 4(1):8–18
Srinivas PR, Philbert M, Vu TQ, Huang Q, Kokini JL, Saltos E, Chen H, Peterson CM, Friedl KE, McDade-Ngutter C et al (2010) Nanotechnology research: applications in nutritional sciences. J Nutr 140:119–124
Surai PF, Kochish II, Velichko OA (2017) Nano-Se assimilation and action in poultry and other monogastric animals: is gut microbiota an answer? Nanoscale Res Lett 12(1):612. https://doi.org/10.1186/s11671-017-2383-3
Swain PS, Rajendran D, Rao SB, Dominic G (2015 Jul) Preparation and effects of nano mineral particle feeding in livestock: a review. Vet World 8(7):888–891. https://doi.org/10.14202/vetworld.2015.888-891
Thiruvengadam M, Rajakumar G, Chung I-M (2018) Nanotechnology: current uses and future applications in the food industry. 3 Biotech 8:74
Tubesha Z, Abu Bakar Z, Ismail M (2013) Characterization and stability evaluation of thymoquinone nanoemulsions prepared by high-pressure homogenization. J Nanomater 2013:453290
Val-Laillet D (2019) Review: Impact of food, gut-brain signals and metabolic status on brain activity in the pig model: 10 years of nutrition research using in vivo brain imaging. Animal 13(11):2699–2713. https://doi.org/10.1017/S1751731119001745
Weiss J, Takhistov P, McClements DJ (2006) Functional materials in food nanotechnology. J Food Sci 71:R107–R116
Yildiz G, Andrade J, Engeseth NE, Feng H (2017) Functionalizing soy protein nano-aggregates with pH-shifting and mano-thermo-sonication. J Colloid Interface Sci 505:836–846. https://doi.org/10.1016/j.jcis.2017.06.088
Zhang J, Zu Y, Dhanasekara CS, Li J, Wu D, Fan Z, Wang S (2017 Jan) Detection and treatment of atherosclerosis using nanoparticles. Wiley Interdiscip Rev Nanomed Nanobiotechnol 9(1). https://doi.org/10.1002/wnan.1412
Zobeiri M, Belwal T, Parvizi F, Naseri R, Farzaei MH, Nabavi SF, Sureda A, Nabavi SM (2018) Naringenin and its nano-formulations for fatty liver: cellular modes of action and clinical perspective. Curr Pharm Biotechnol 19(3):196–205. https://doi.org/10.2174/1389201019666180514170122

Neural Spintronics: Noninvasive Augmentation of Brain Functions

Stewart E. Barnes, Ioan Opris, Brian R. Noga, Sunxiang Huang, and Fulin Zuo

1 Introduction

Electronics corresponds to the use of the electron's charge −e to make functional devices. Spintronics uses, in addition, the spin, that is, the quantized intrinsic angular $\pm\hbar/2$, where $\hbar$ is Plank's constant, to add new functionalities (Wolf et al. 2006; Bhatti et al. 2017; Umesh and Mittal 2019). This electronic spin is responsible for the magnetism of Mn, Co, Fe, or Ni, and their alloys and compounds and spintronics invariably use such materials to realize new devices. The origin of spintronics may be traced back to the initial experiments on magnetic tunnel junctions by Julliere in the 1970s (Julliere 1975). The use of semiconductors for spintronics started with the theoretical proposal of a spin field-effect-transistor by Datta and Das in 1990 (Datta and Das 1990) and of the electric dipole spin resonance by Rashba in 1960 (Rashba 1960; Torrejon et al. 2017).

Reflecting the spin-torque transfer (STT) effect (Ralph and Stiles 2008), the magnetization of thin magnetic wires can be manipulated as is done in magnetic random-access memory (MRAM) (Bhatti et al. 2017). When excited by the same STT effect, a small magnetic particle will undergo a precessional motion known as ferromagnetic resonance (FMR). In turn, the precessional frequency can be modified by a small applied magnetic field. Neural spintronics is the name given to the

S. E. Barnes (✉) · S. Huang · F. Zuo
Physics Department, University of Miami, Coral Gables, FL, USA
e-mail: sbarnes@miami.edu

I. Opris
Department of Neurosurgery, University of Miami, Miami, FL, USA

B. R. Noga
Department of Neurological Surgery, University of Miami, Miller School of Medicine, Miami, FL, USA

I. Opris et al. (eds.), *Modern Approaches to Augmentation of Brain Function*, Contemporary Clinical Neuroscience,
https://doi.org/10.1007/978-3-030-54564-2_20

marriage between spintronics and neurology. From the physics point of view, a neuron is considered as an integral part of a spintronics circuit. The coupling with the extracranial circuits is via the magnetic field. From the medical perspective, transcranial magnetic stimulation (TMS) is performed using a programmable magnet system resulting in a new level of precision and high-resolution magnetoencephalography (MEG) is realized by using the FMR of a spin-torque nano-oscillators (STNO) (Collet et al. 2016) to measure neural magnetic field.

The Hodgkin–Huxley model is a well-known mathematical approach that describes how action potentials in neurons are initiated and propagated (Wang et al. 2017; Hodgkin and Rushton 1946). An axon is reduced to a nonlinear electronic circuit containing batteries, capacitors, and conductances that reflect the various ionic channels. There are many extensions to this model. In the medical context, the aphorism "all models are wrong, but some are useful," is often used. While this might apply to the Hodgkin–Huxley model, despite its wide acceptance, it is not relevant to the laws of physics such as those of electromagnetism relevant here. The simulation and verification, for example, of the magnetic field produced by a magnetic wire, or some solenoid, make no sense since the underlying law of Nature has been established for some centuries. Even the more recent theory of spintronics, based as it is on quantum theory, is tested in its own context. There is no progress made by "simulate and verify" this theory in the context of each application to neural spintronics.

2 Transcranial Magnetic Stimulation

Transcranial magnetic stimulation (TMS) is a noninvasive tool widely used to better understand the neurobiology of cognitive function, behavior, and emotional processing (Wagner et al. 2007; Tang et al. 2017; Roth and Basser 1990; Sokhadze et al. 2014). It is routinely used in a clinical setting to treat several disorders. A representation of a TMS device using a standard butterfly coil in which a time-varying magnetic field is used to induce an electric current within the brain. A very large current is required to produce a field of 1–2 T at the center of the coils and limits the time the current flows to ~10^{-3} to 10^{-4} s. This can be rather short, and inflexible, compared to the response of the neural system that rather occurs as spikes of 1 ms duration on a timescale of 10–100 ms. Typical coils have diameter 4–9 cm, with 10–20 turns. The relatively large dimensions limit the spatial resolution. The inductance L ~15 to 150 μH and resistance R of the coils determine the time constant. A 100 Hz repetition rate of the pulses is possible. Given that 500 J is required per discharge and a 100 Hz repetition rate, this corresponds to an enormous average power of 50 kW. Given the absence of apparent adverse effects on the subjects, it is evident that most of the energy is wasted by inducing dissipative current in the coils and not a magnetic field in the brain.

Based on the Hodgkin–Huxley model (Wang et al. 2017; Hodgkin and Rushton 1946), the energy supplied to a neuron by adenosine triphosphate (ATP) during one action potential is $\approx 2.5 \times 10^{-7}$ J. There are about 20 billion neurons in the cerebral cortex, but in the primary motor cortical area with dimensions of about 5 mm, there are about 30,000 spiking neurons that might be activated by TMS. If all of these are activated once a second, the energy requirement is less that 10^{-2} W illustrating the inefficiency of traditional machines.

TMS with butterfly coils works has already been used for significant medical applications. However, from a physicist's point of view, how this works and why it gives rather good spatial resolution is far from clear. Of particular importance is the threshold electric field E_{th} necessary to fire individually, or collectively, a neuron or more complex networks. An electric field of $E = 10$ V/m would correspond well to the direct stimulation of individual neurons but is an order of magnitude smaller than estimated from extracranial TMS experiments using coils. A typical 7 cm dimensioned buttery coil is estimated to produce 200 V/m at the surface of the brain. However, the orientation of the magnetic field $\vec{B}$ is parallel and the electrical field $\vec{E}$ therefore perpendicular to the nerve axis whereas it is the field along the axis that counts. The useful effective field is perhaps closer to 10 V/m mentioned above. More serious is if E or rather dE/dz, i.e., the gradient in the electrical field, is the relevant parameter. That E is not the appropriate parameter is suggested by the orders of magnitude difference between the voltage gradients required for tDCS (transient DC stimulation) ~0.5 1 V/m and TMS stimulations. A butterfly coil is estimated to produce a dE/dx ~3000 V/m^2 at the surface of the brain but again not correctly aligned. These numbers reflect the fact that for buttery coils, the positive and negative regions of the E field are separated by a little less than 1 cm.

3 Neuronal Cable Theory

Neuronal cable theory is a simplification of that of Hodgkin–Huxley and corresponds to the partial differential equation:

$$\tau \frac{\partial u}{\partial t} + f(u) = \lambda^2 \frac{\partial^2 u}{\partial x^2}$$

where in the simplest case $f(u) = u$ and where $u = V - V_0$ is the voltage across the cell membrane minus its equilibrium value, τ the time, and λ the length constant. The time to return to equilibrium τ is of the order of milliseconds while the length constant varies but a value of 100 μm seems typical. Here, x is the distance measured along the axon, so $E_x = -\frac{\partial V}{\partial x}$ is the electric field along that direction and $\frac{\partial^2 V}{\partial x^2} = -\frac{\partial E_x}{\partial x}$ is the (negative of) the spatial derivative of the component of the electric field along the axon. It is to be noted that a bend in the axon $\frac{\partial E_x}{\partial x}$ can be large even though $\vec{E}$ is constant.

In this model, the intra- and extracellular currents are equal and opposite. The current circulates with a length scale λ due to the internal magnetic field generated by the compact intracellular current. It is emphasized that this cancellation of intra- and extracellular currents cannot occur without an internal magnetic field. This length is therefore the key parameter that also determines the coupling to an external such field. The "speed" of the action potential is $v \approx \frac{2\lambda}{\tau}$ so $\lambda \approx \frac{1}{2} v\tau$ and given

$\tau \sim 10^{-3}$ s and $\upsilon = 20$ m/s are typical in the central nervous system (CNS) $\lambda \approx 10^{-2}$ m (or 1 cm). However, λ can be almost ten times larger for the fastest neurons or ten times less for small un-myelinated axons.

An external field produces, at a given point, a circulating electron field given by Faraday's law of electromagnetic induction $\vec{\nabla} \times \vec{E} = -\frac{\partial \vec{B}}{\partial t}$, the time derivative of the magnetic field. With conventional TMS coils $\frac{\partial \vec{B}}{\partial t}$, is essentially parallel to $\vec{B}$ but this need not be the case.

Assuming an action potential is initiated just beyond the axon hillock, and that the axon is perpendicular to the surface of the brain, a neuron is sensitive to a $\frac{\partial \vec{B}}{\partial t}$ parallel to the surface of the brain over an area λ^2 whereas for the usual TMS coils $\frac{\partial \vec{B}}{\partial t}$ is perpendicular. Given the appropriate orientation, there is an EMF $\mathcal{E}$ or order $\lambda^2 \frac{\partial \vec{B}}{\partial t}$ that adds to $u \to u + \mathcal{E}$. If an additional voltage of 10 meV is needed to initiate an action potential, then a $\frac{\partial \vec{B}}{\partial t}$ of about 10^2 T/s is needed. For conventional TMS, the pulse lasts about 1 ms and a minimum field B of about 0.1 T is required, but again with $\frac{\partial \vec{B}}{\partial t}$ locally perpendicular to the axon. For an integrate-and-fire neuron, many shorter monophase pulses with a smaller field could also initiate an action potential.

4 Programmable Permanent Magnets

These objectives can be realized, and the electrical energy required for TMS might be reduced, using a programmable magnet in which the magnetization of permanent magnets is switched using the STT effect. The idea is based upon the race track memory (Parkin and Yang 2015) illustrated in Fig. 1a. Shown in Fig. 1b is that a thin wire of Ni (or the Ni0.8Fe0.2 alloy Permalloy) amounts to a bar magnet which in turn produces the magnetic field of two magnetic monopoles. Even though they do not exist in Nature, the field due to a system of permanent magnets can be imagined as coming from a distribution of monopoles. In Fig. 1c is shown the situation in which there is a domain wall near the center of the magnetic wire. If the wall is moved all the way to the top, there is a North pole at the top, whereas if it is pushed to the bottom, this pole is at the bottom. The direction of the magnetization of a permanent magnet can be reversed by simply moving a domain wall. Such a motion can be induced by briefly passing a current through the wire (Barnes and Maekawa 2007) or by applying an external magnetic field for a short time.

It cannot be emphasized too strongly that the magnetization is *not* produced by the current in the Ni wires itself, and by applying a magnetic field just below the

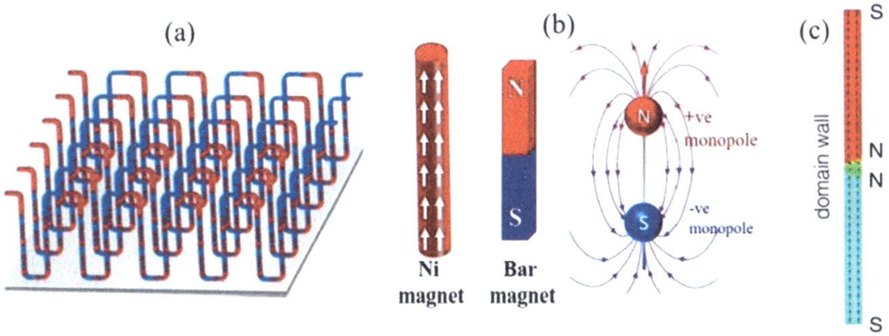

Fig. 1 Ilustrating (**a**) a racetrack memory, (**b**) a magnetic nanowire compared with a bar magnet and a pair of magnetic monopoles, and (**c**) a magnetic nanowire with a domain wall (DW) at its center. (With permission from Torrejon et al. 2017)

critical value for the Walker breakdown, this current can be made quite small. In contrast, for TMS with butterfly coils, the field *is* produced by a large current, usually, in copper wires. Such coils made of Ni wire would be more resistive and dissipate much more energy, although strong controlled dissipation is in fact needed in order to get a large $\frac{\partial \vec{B}}{\partial t}$ with such a conventional system. It is not proposed to make conventional coils of Ni.

The race-track system of Fig. 1a can be programmed to simulate any magnetic system by passing small currents to switch the direction of the permanent magnet wires. In Fig. 2b, the programmable magnets mimics, at no energy cost, two oppositely orientated reducing a field similar to that of the conventional coils, of Fig. 2a, when a large dissipative current is flowing. For programmable magnets, the magnitude of the magnetic fields remains a constant, but the simulated magnets rotate producing an electric field along an axon directed towards the interior. The rotation of these virtual magnets avoids changing the intrinsic energy of permanent magnets wires, thereby reducing dissipation.

Programmable permanent magnets can focus the TMS in a manner not possible with conventional coils. The switching of a wire requires a μs or less, and many such evens can occur on the timescale of a ms required to evoke an action potential. With the rotating magnet scheme, the stimulation at all but a small region will be biphasic on a sub-ms scale and the stimulation will be reduced to almost zero in integrate-and-fire neurons. The trade-off of such focusing is a reduction in the magnitude stimulating $\mathcal{E}$. It is estimated that mm resolution should be possible.

5 Nano-TMS Application to Neuromodulation

The nano-TMS devices described herein can have overwhelming advantages over existing TMS devices. The nano-TMS devices are programmable with TMS-bits on the order of a few μm^2, allowing the devices to have focal stimulation with unprecedentedly high spatial resolution, depth control as well as multisite stimulation. The fast switching of TMS-bits can also generate large dB/dt (hence large E

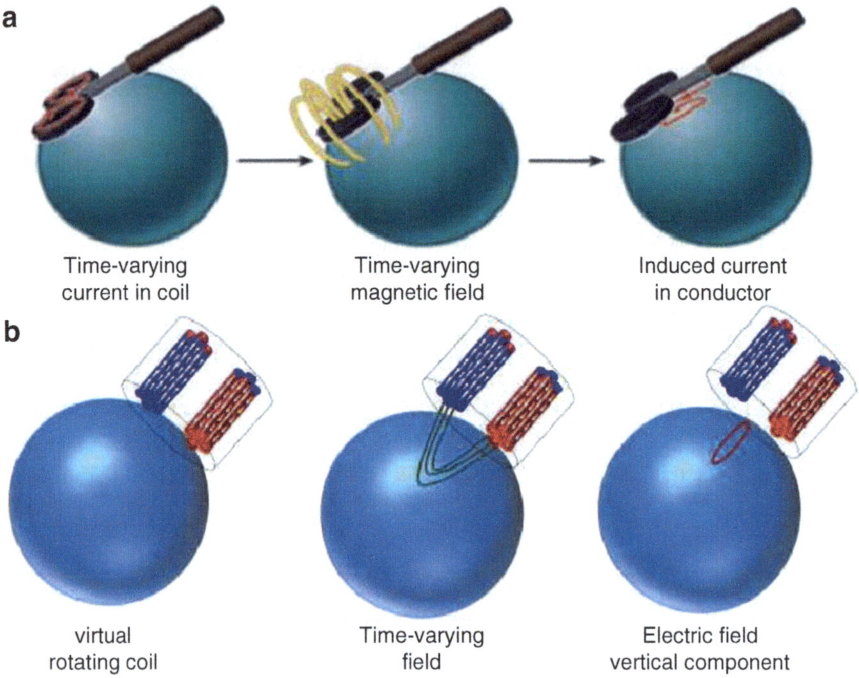

Fig. 2 (**a**) Transcranial magnetic stimulation (TMS). Graphical representation of (**a**) a TMS device using a butterfly coil and (**b**) a nano-TMS device

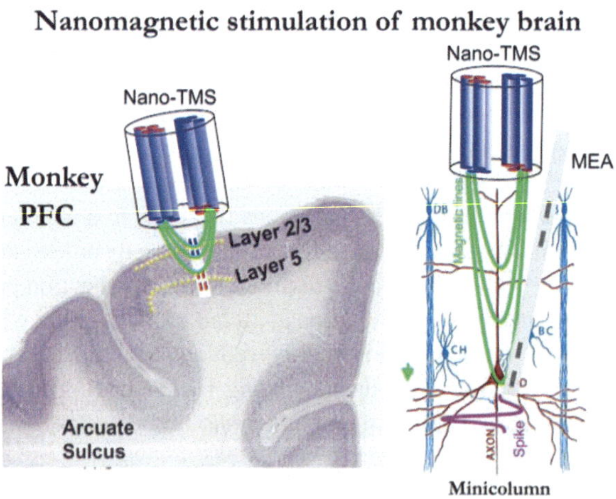

Fig. 3 Illustration of noninvasive nano-TMS stimulation over the premotor cortex of a nonhuman primate (left), and (right) while recording neuron firing with a multielectrode array (MEA)

field) in a short time, allowing the devices to have exceptionally high temporal precision. In addition, flexible nano-TMS devices have many practical applications, e.g., nanomagnetic stimulation of the frontal cortex in monkeys (Fig. 3)

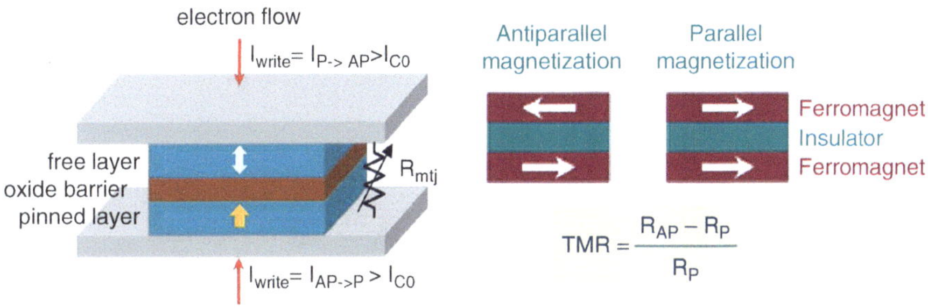

Fig. 4 Left: A magnetic tunnel junction (MTJ) diagram depicting the flow of current and the resistance of the MTJ. Right: Graphical depiction of a MTJ with parallel and antiparallel magnetization. The tunneling magnetoresistance is given by the ratio of the difference between R_{ap} and R_p, divided by R_p. R_{ap} is the electrical resistance in the antiparallel state, whereas R_p is the resistance in the parallel state. While such traditional junctions are the work-horses of our hard-drives, they are limited to a pT sensitivity and of limited use for MEG, but see below

(Fetterhoff et al. 2015; Tischler et al. 2011; Opris 2013; Opris et al. 2014; Perez and Cohen 2009; Sokhadze et al. 2014; Casanova et al. 2015).

6 Magnetic Tunnel Junctions

Spintronics elements based upon giant magnetic resistance (MR) can be implemented. Such elements can measure the magnetic field of small neuron populations down to individual neuron (Egelhoff et al. 2009). A magnetic tunnel junction (MTJ) consists of two ferromagnets separated by a thin insulator [see Fig. 4]. If the insulating layer is thin (nanometers), the electrons can "tunnel" from one ferromagnet into the other. Since this process is forbidden in classical physics, the tunnel magnetoresistance is a strictly quantum mechanical phenomenon.

Magnetic tunnel junctions are manufactured in thin-film technology.

7 Spintronics Approach Involving MTJ-STNOs

Of importance for MTJ-STNOs is the extremely narrow line width of the FMR signal with appropriate combinations of field and current. Shown in Fig. 5 are data for a MTJ from Houssameddine et al. (2008). The current passes through the MTJ to and from a fixed polarization layer, and it is the DC current that drives the FMR of a free layer. The detected signal arises through the tunneling magnetoresistance (TMR) effect, and an important aim of the work in this reference is to maximize the output power. The STT effect increases the coherence time of the magnetic tunnel junction by up to 10^3. By using an optimal composition (Schoen et al. 2016) of 25% Co, for the CoFeB free layer in this junction, the coherence time might be increased

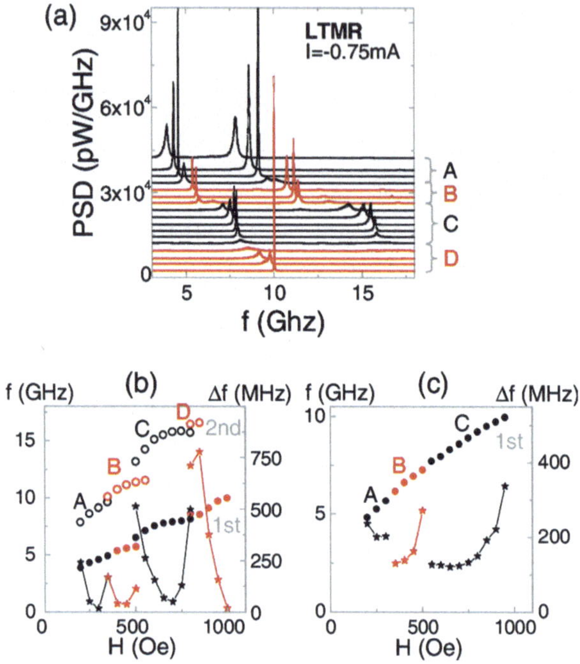

Fig. 5 Magnetic tunnel junction (MTJ); (**a**) Spectra for a sample with a low tunnel magnetoresistance (LTMR) at constant current $I = 0.75$ mA and for fields of 0.21 kOe in steps of 50 Oe from top to bottom, offset for clarity. Peak frequency vs. bias field at $I = 0.75$ mA dots for (**b**) the LTMR sample first and second harmonics and (**c**) the high tunnel magnetoresistance sample first harmonics. The linewidths (stars) of the first harmonics are superposed in (**b**, **c**). (Adapted from Houssameddine et al. 2008)

by about another factor of 10. Also, of importance is the existence of several discrete branches ABCD of the FMR spectrum. There are barriers between these modes that are large compared with the thermal energy of ~30 meV. The intrinsic line width (probably due to thermal jumps between modes) has been examined (Houssameddine et al. 2009).

8 High-Resolution Magnetoencephalography (MEG)

The magnetic field produced by the brain is extremely small (~10^{-14} T or 10^{-2} pT), and its measurement requires special techniques and extraordinary screening of stray magnetic fields. Commercial devices for performing MEG use SQUIDs (superconducting quantum interference device) and require cooling to a few degrees above absolute zero and cost about $2 million. In contrast, an atomic magnetometer can achieve a similar sensitivity at somewhat above room temperature. Both technologies involve cm sized magnetometers, which represents a limitation on resolution. The spintronic technology used to measure very small magnetic fields produced by magnetic hard disks has also been considered in the biological context. This is based upon magnetic tunnel junctions but is limited to ~ 10^{-9} T and not very useful for extracranial MEG (Iivanainen et al. 2017; Cohen 1968, 1972; Hamalainen et al. 1993; Xia et al. 2006; Boto et al. 2016) but has limited usefulness as discussed below.

A simple compass needle can measure changes in fields of ~10^{-5} T, and it has been suggested that the ultimate magnetometer is essentially a floating magnetic needle perpendicular to the field to be measured and can detect maybe ~10^{-22} T. Apart from what is described here, there is not a practical device based upon these ideas.

Both superconductivity and magnetism correspond to a spontaneous broken symmetry, and in both cases, there is a quantum mechanical order parameter with a phase that can be used to exhibit sensitive interference effects. A SQUID performs interference in space whereas the spintronic MEG device, described here, uses interference in time. The idea is to compare the time given by a clock that is sensitive to a magnetic field with one that is not. Yttrium Iron Garnet (YIG) is a crystal with an extremely sharp FMR, and YIG oscillators are routinely use in frequency synthesizers. In a synthesizer, a phase-locked loop (PLL) is used to lock the FMR frequency to a simple multiple of that of an accurate, usually quartz, oscillator. Of importance here is that the line width of the FMR is determined not by the large intrinsic Q-factor of the YIG oscillators but rather the much larger value of the quartz oscillator. In order to measure the very small precession change of the YIG oscillator, a rotating frame is used. Imagine a turntable that is rotating with a period that T_0/N, where N is an integer and T_0 the period of the quartz oscillator. Standing one this platform, the precession of the direction of the magnetization $\vec{M}$ of the YIG magnet is observed. If, as in a frequency synthesizer, the YIG period $T = T_0/N$ than the direction of the "compass needle" $\vec{M}$ appears stationary. Adding a field ~10^{-14} T, due to the brain, will cause this needle to process with a period of about 100 s. If the duration of the action potential responsible for the magnetic field is 1 ms, then the angle subtended by the needle to its initial position in the rotating frame is $2\pi \times 10^{-5}$ rad but easily measurable with suitable electronics. The YIG oscillator used in commercial frequency synthesizers contains a magnetic sphere off about 1 mm in diameter. The spintronic device, a spin-torque oscillator (STO), is much smaller comprising a pillar of height 30 nm and diameter of only about 200 nm (0.2 μm) and is susceptible to larger scale integration techniques imply designs with millions, or more, of such elements is easy to envisage. Only a single reference quartz or atomic clock is needed.

Much discussed is the MEG inverse problem. The determination of the current source given the observed signal. True current sources cannot exist. A source or sink of current density $\vec{J}_T$ corresponds to a finite value of the divergence $\vec{\nabla} \cdot \vec{J}_T$. Since currents obey Ampère's law $\vec{\nabla} \times \vec{B} = \mu_0 \left(\vec{J} + \vec{J}_d \right)$, and since it is a mathematical identity that $\vec{\nabla} \cdot \vec{\nabla} \times \vec{B} = 0$, it follows $\vec{\nabla} \cdot \left(\vec{J} + \vec{J}_d \right) = 0$ and there are no sources/sinks of currents. Here, the total current density is $J_T = \left(\vec{J} + \vec{J}_d \right)$ where $\vec{J}$ is the current due to the transport of ions and/or electrons while $\vec{J}_d$ is the displacement current that is invariably ignored in the MEG context. In electroencephalography (EEG), measured is the potential difference between different points on the scalp. That such a time-dependent signal, reflecting ultimately the effect of action potentials, exists, implies there exists a significant time-dependent electric field $\vec{E}$ to the exterior of the brain. Since $\vec{J}_d = \epsilon_0 \mu_0 \frac{\partial \vec{E}}{\partial t}$, there is a net current J_T flowing through the surface of the brain. The transport current directed toward the surface is converted to a displacement current by the accumulation of charge near the surface. The argument that displacement currents are negligible because the frequency is too low is incorrect. Such currents are essential for not only MEG and EEG but also to propagation of the action potential along the distributed capacity of an axon.

So, while there are no current sources, the initiation of an action potential constitutes a source of a current loop that expands as action potential propagates. MEG should be envisaged as a means of measuring the neural currents associated with action potential initiation of these current loops. Commercial contactless current meters use a closed magnetic circuit. It is magnetic flux that flows in a magnetic circuit, and this flux is proportional to the electrical current that flows through the loop. Ferrite with a relative magnetic permeability μ/μ_0 as large as 10,000 would increase the field to 10^{-10} T and reduce the required coherence time to 10^{-2} s. The use of many such magnetic circuit loops permits the currents directed to the surface to be mapped as a function of time. This boundary value problem has a unique solution for the source of the action potential. A considerable advantage of such STO-MEG instrumentation is that it is sensitive only to currents, including the displacement currents, that flow through the magntic circuit loop. Stray magnetic fields have no effect. Displacement currents due to time-varying magnetic fields are very small and easily screened.

Figure 6 shows measurements of alpha waves in the MEG signal using a regular MTJ-TMR sensor. Figure 6a depicts the experimental setup for the MEG measurements. The TMR sensor was placed 60 mm to the left of the center of the occipital area on a male subject (aged 30 years) in the prone position, and the transverse component of the magnetic field was measured. Electroencephalography (EEG) was simultaneously performed using two Ag = AgCl electrodes placed on the midline occipital and forehead areas. Under voice instructions from outside the shielded room, the subject alternatively closed and opened his eyes for periods of about 10 s. Prominent peaks of EEG alpha waves at around 10 Hz during the eye-closure period were used as the trigger to average the simultaneously measured MEG signals. Figure 6b, c show the result of averaging the MEG and EEG signals at 3750 times per second. Despite a phase shift, a MEG signal with almost the same frequency as the EEG alpha wave was detected. The amplitude of the magnetic field was approximately 2pT peak-to-peak, which is consistent with the literatutre value. Averaging 3750 times reduced the noise of the TMR sensor so that measurement was possible with an adequate signal-to-noise ratio for the 2pT peak-to-peak signal. Since the frequency of the alpha wave fluctuates, the signal attenuates as time elapses from the center of the trigger. The attenuation of the alpha wave due to this fluctuation can be seen in both the EEG and MEG measurements.

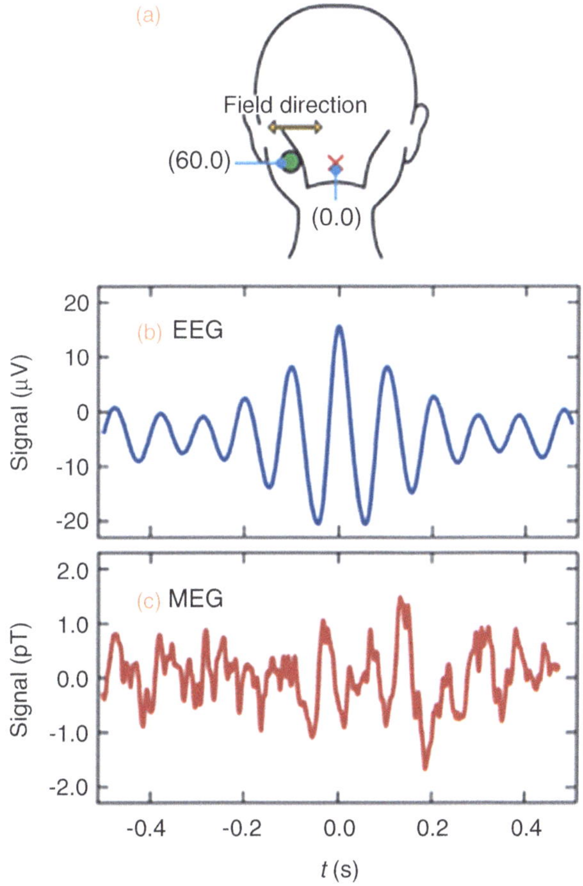

Fig. 6 (**a**) Experimental setup for MEG measurement. (**b**) Reference EEG signal and (**c**) MEG signal using TMR sensor with averaging 3750 times. A signal of almost the same frequency as the EEG alpha wave was seen in MEG, and the MEG signal was attenuated further from the center of the trigger. With permission from Fujiwara et al. (2018)

Overall, this new nanotechnology can be used in clinical neuroscience (Perez and Cohen 2009; Sokhadze et al. 2014; Chae and Kim 2017) and preclinical research, including in vivo animal experiments (Tang et al. 2017) and in culture (Moretti et al. 2018). In preclinical science, it will provide the possibility to interact with the neural tissue without damaging the cells. Single unit extracellular recordings of neurons require the penetration of the electrode in the tissue which can damage cells around the electrode (Santos et al. 2012). In contrast, the nano-MEG will record the activity of cells without such damage. Similarly, the nano-TMS will activate the localized ensemble of cells within cortical minicolumns. In clinical studies involving autism spectrum disorder (Perez and Cohen 2009; Sokhadze et al. 2014), the application of TMS has significantly improved the behavioral reactions of patients. This may be explained by the improved lateral inhibition within the cortical minicolumns, following repetitive TMS sessions.

Neural Spintronics is emerging as a new branch of Neuroscience. Figure 7 depicts the increasing number of articles published in Spintronics [red] in the last

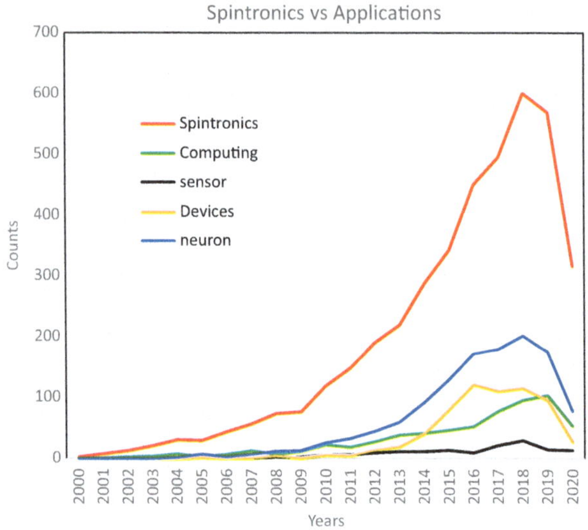

Fig. 7 Neural Spintronics an emerging augmentation approach. Number of publications versus years. Please note that the last year has only 5 months up to date under consideration

twenty years compared to the increasing number dealing with neural applications: devices, computing, memory, neuron, and memories that altogether augment neural technology.

9 Conclusion

Nano-TMS and MEG devices can be effectively used in a broad range of research fields and technological applications where programmable focusing magnetic field and/or detection of weak field are required. This may have a groundbreaking effect in decoding the human brain and also therapeutics over a multitude of brain disorders.

References

Barnes SE, Maekawa S (2007) Generalization of Faraday's law to include nonconservative spin forces. Phys Rev Lett 98:246601

Bhatti S et al (2017) Spintronics based random access memory: a review. Mater Today 20(9):530–548. https://doi.org/10.1016/j.mattod.2017.07.007

Boto E, Bowtell R, Krüger P, Fromhold TM, Morris PG, Meyer SS, Barnes GR, Brookes MJ (2016) On the potential of a new generation of magnetometers for MEG: a beamformer simulation study. PLoS One 11(8):e0157655. https://doi.org/10.1371/journal.pone.0157655

Casanova MF, Sokhadze E, Opris I, Wang Y, Li X (2015) Autism spectrum disorders: linking neuropathological findings to treatment with transcranial magnetic stimulation. Acta Paediatr 104(4):346–355; 21

Chae K-S, Kim Y-H (2017) Potential impact of geomagnetic field in transcranial magnetic stimulation for the treatment of neurodegenerative diseases. Front Hum Neurosci 11:478. https://doi.org/10.3389/fnhum.2017.00478

Cohen D (1968) Magnetoencephalography: evidence of magnetic fields produced by alpha-rhythm currents. Science 161:784

Cohen D (1972) Magnetoencephalography: detection of the brain's electrical activity with a superconducting magnetometer. Science 175:664–666

Collet M et al (2016) Generation of coherent spin-wave modes in yttrium iron garnet microdiscs by spin-orbit torque. Nat Commun 7:10377. https://doi.org/10.1038/ncomms10377

Datta S, Das B (1990) Electronic analog of the electrooptic modulator. Appl Phys Lett 56(7):665–667. https://doi.org/10.1063/1.102730. Bibcode:1990ApPhL..56..665D

Egelhoff WF Jr, Ponga PWT, Ungurisb J, McMichaelb RD, Nowakc ER, Edelsteind AS, Burnetted JE, Fischerd GA (2009) Critical challenges for picoTesla magnetic-tunnel-junction sensors. Sens Actuators A 155:217–225

Fetterhoff D, Kraft RA, Sandler RA, Opris I, Sexton CA, Marmarelis VZ, Hampson RE, Deadwyler SA (2015) Distinguishing cognitive state with multifractal complexity of hippocampal interspike interval sequences. Front Syst Neurosci 9:130. https://doi.org/10.3389/fnsys.2015.00130. eCollection 2015

Fujiwara K, Oogane M, Kanno A, Imada M, Jono J, Terauchi T, Okuno T, Aritomi Y, Morikawa M, Tsuchida M, Nakasato N, Ando Y (2018) Magnetocardiography and magnetoencephalography measurements at room temperature using tunnel magneto-resistance sensors. Appl Phys Express 11 (2):023001. https://doi.org/10.7567/APEX.11.023001

Hamalainen M, Hari R, Ilmoniemi RJ, Knuutila J, Lounasmaa OV (1993) Magnetoencephalography theory, instrumentation, and applications to noninvasive studies of the working human brain. Rev Mod Phys 65:413

Hodgkin AL, Rushton WAH (1946) The electrical constants of a crustacean nerve fibre. Proc Roy Soc B-133:444–479

Houssameddine D, Florez SH, Katine JA, Michel J-P, Ebels U, Mauri D, Ozatay O, Delaet B, Viala B, Folks L, Terris BD, Cyrille M-C (2008) Spin transfer induced coherent microwave emission with large power from nanoscale MgO tunnel junctions. Appl Phys Lett 93:022505. https://doi.org/10.1063/1.2956418

Houssameddine D, Ebels U, Dieny B, Garello K, Michel J-P, Delaet B, Viala B, Cyrille M-C, Katine JA, Mauri D (2009) Temporal coherence of MgO based magnetic tunnel junction spin torque oscillators. Phys Rev Lett 102:257202

Iivanainen J, Stenroos M, Parkkonen L (2017) Measuring MEG closer to the brain: performance of on-scalp sensor arrays. Neuroimage 147:542–553. https://doi.org/10.1016/j.neuroimage.2016.12.048

Julliere M (1975) Tunneling between ferromagnetic films. Phys Lett A 54(3):225–226. https://doi.org/10.1016/0375-9601(75)90174-7. Bibcode:1975PhLA...54..225J

Moretti D, DiFrancesco ML, Sharma PP, Dante S, Albisetti E, Monticelli M, Bertacco R, Petti D, Baldelli P, Benfenati F (2018) Biocompatibility of a magnetic tunnel junction sensor array for the detection of neuronal signals in culture. Front Neurosci 12:909. https://doi.org/10.3389/fnins.2018.00909

Opris I (2013) Inter-laminar microcircuits across neocortex: repair and augmentation. Front Syst Neurosci 7:80. https://doi.org/10.3389/fnsys.2013.00080

Opris I, Fuqua JL, Gerhardt GA, Hampson RE, Deadwyler SA (2014) Prefrontal cortical recordings with biomorphic MEAs reveal complex columnar-laminar microcircuits for BCI/BMI implementation. J Neurosci Methods 244:104–113.pii: S0165-0270(14)00197-6

Parkin S, Yang S-H (2015) Memory on the racetrack. Nat Nanotechnol 10(3):195–198. https://doi.org/10.1038/nnano.2015.41

Perez MA, Cohen LG (2009) The corticospinal system and transcranial magnetic stimulation in stroke. Top Stroke Rehabil 16(4):254–269. https://doi.org/10.1310/tsr1604-254

Ralph DC, Stiles MD (2008) Spin transfer torques. J Magn Magn Mater 320(7):1190–1216. https://doi.org/10.1016/j.jmmm.2007.12.019. ISSN 0304-8853

Rashba EI (1960) Cyclotron and combined resonances in a perpendicular field. Sov Phys Solid State 2:1109–1122

Roth BJ, Basser PJ (1990) A model of the stimulation of a nerve fiber by electromagnetic induction. IEEE Trans Biomed Eng 37:588

Santos L, Opris I, Fuqua J, Hampson RE, Deadwyler SA (2012) A novel tetrode microdrive for simultaneous multi-neuron recording from different regions of primate brain. J Neurosci Methods 205(2):368–374. https://doi.org/10.1016/j.jneumeth.2012.01.006

Schoen MAW, Thonig D, Schneider ML, Silva TJ, Nembach HT, Eriksson O, Karis O, Shaw JM (2016) Ultra-low magnetic damping of a metallic ferromagnet. Nat Phys 12:839

Sokhadze EM, El-Baz AS, Sears LL, Opris I, Casanova MF (2014) rTMS neuromodulation improves electrocortical functional measures of information processing and behavioral responses in autism. Front Syst Neurosci 8:134. https://doi.org/10.3389/fnsys.2014.00134

Tang A, Thickbroom G, Rodger J (2017) Repetitive transcranial magnetic stimulation of the brain: mechanisms from animal and experimental models. Neuroscientist 23(1):82–94. https://doi.org/10.1177/1073858415618897

Tischler H, Wolfus S, Friedman A, Perel E, Pashut T, Lavidor M, Korngreen A, Yeshurun Y, Bar-Gad I (2011) Mini-coil for magnetic stimulation in the behaving primate. J Neurosci Methods 194:242–251

Torrejon J, Riou M, Araujo FA, Tsunegi S, Khalsa G, Querlioz D, Bortolotti P, Cros V, Yakushiji K, Fukushima A, Kubota H, Yuasa S, Stiles MD, Grollier J (2017) Neuromorphic computing with nanoscale spintronic oscillators. Nature 547(7664):428–431. https://doi.org/10.1038/nature23011

Umesh S, Mittal S (2019) A survey of spintronic architectures for processing-in-memory and neural networks. J Syst Archit 97:349–372. https://doi.org/10.1016/j.sysarc.2018.11.005

Wagner T, Valero-Cabre A, Pascual-Leone A (2007) Noninvasive human brain stimulation. Annu Rev Biomed Eng 9:527–565

Wang Y et al (2017) Neural energy supply-consumption properties based on Hodgkin-Huxley model. Neural Plast 2017:6207141. https://doi.org/10.1155/2017/6207141

Wolf SA, Chtchelkanova AY, Treger DM (2006) Spintronics—a retrospective and perspective. IBM J Res Dev 50:101. https://doi.org/10.1147/rd.501.0101

Xia H, Ben-Amar Baranga A, Hoffman D, Romalis MV (2006) Magnetoencephalography with an atomic magnetometer. Appl Phys Lett 89(21):211104

Part V
Augmenting Behavior

Does the Power to Suppress an Action Make Us 'Free'?

Giovanni Mirabella

In shaping our lives, are we the cues, the cue-holders, or the billiard balls? Are we players, or are we played?
Bauman (2008)

The Polish philosopher Zygmut Bauman raises a question which frames the debate that, for centuries, has been crucial to theologists and philosophers, and in the last 50 years, to neuroscientists as well. From the moment when one wakes up to when one goes to sleep, one performs hundreds of actions, some as simple as picking up a pen to write, others more complex as deciding what to say during an imminent business meeting. Even though in the majority of the circumstances, one puts little or no attention in what one does, one still has the impression of having control over every single action. Unconsciously, one reckons that actions are determined by one's own conscious will, namely free will. This concept functions as the basis not only for the idea of self-control and moral responsibility but also of every legal system that shapes social interactions in modern societies (Lavazza and Inglese 2015). For instance, in the penal codes of several Western countries, mental capacity is pivotal when questioning punishability. In other words, a person could be considered guilty only if one was fully cognizant of the action undertaken.

However, the idea that conscious intention to perform an action is actually its cause, which is in strong contrast with determinism for which every event is completely determined by previously existing causes. Under this view, performing an action represents a need and not a choice. Determinism finds its roots in Newton's physics laws for which a system's initial conditions are sufficient to undoubtedly predict the future behavior of a system as such. For instance, knowing precisely both the initial conditions of movement and characteristics of the planets of our

G. Mirabella (✉)
IRCCS Neuromed, Pozzilli, IS, Italy

Department of Clinical and Experimental Sciences. Section: Neuroscience, Brescia University, Brescia, Italy
e-mail: Giovanni.mirabella@unibs.it

I. Opris et al. (eds.), *Modern Approaches to Augmentation of Brain Function*, Contemporary Clinical Neuroscience,
https://doi.org/10.1007/978-3-030-54564-2_21

Solar System makes it possible to accurately predict their orbit around the Sun. Since our brains are made of the same matter as planets are made of, they are thus subject to the same rules. In conclusion, determinism asserts that our future actions are as certain as the orbits of the planets. According to philosopher Daniel Dennet (1984) "At best one has the illusion of control. One is in fact entirely controlled by external factors, locked into a life story that was written at the dawn of creation, like a puppet that was destined to play Punch and Judy even before its wooden face was carved and painted".

1 The Irruption of Neurosciences in the Free Will Debate

At the beginning of the 80s, an experiment carried out by Benjamin Libet and colleagues marks the official breach of neuroscience in the debate on the intentionality of actions (Libet et al. 1983). In this study, healthy subjects were asked to perform simple voluntary actions, i.e., pressing a button when they are intended to (Fig. 1). The participants were also asked to watch a clock hand, and after having pressed the button, report the exact moment in which they first became aware of the will to move. While participants performed the task, scalp electroencephalography (EEG) was recorded. As predicted, the electrodes located on the motor cortex of the frontal lobe registered a physiological correlation between the neural activity, which precedes voluntary movements, the so-called readiness potential (RP; Kornhuber and Deecke 1965). RP significantly increased over the baseline level almost a second and a half before the action was performed. Surprisingly, participants affirmed that they felt the intention to act just about 200 ms before actually doing it. In other words, the intention of moving seemed to have appeared several hundred milliseconds after the increase in neural activity connected to its motion. Clearly, the fact that the conscious will to move is preceded by the motor activity challenged with the classic idea of free will.

As easily foreseeable, this experimental procedure has been widely criticized. The main critique concerns the self-reported measure of the moment in which the conscious experience of intentionality emerged. This measure has been considered invalid and inaccurate as the participants need to focus simultaneously on both the clock hand and the action they are about to perform. A further objection has been made in regards to the RP role in generating intentionality. Alexander et al. (2016) showed that RP is registered even when the subjects move whilst being under hypnosis—a context in which one does not expect people to perceive any intentionality of the movement. This could suggest that RP might not be neurally linked to the conscious will to move. Nevertheless, it has also been shown that patients affected by primary-complex motor stereotypes, i.e., a disorder characterized by involuntary, complex, repetitive, and apparently purposeless movements, are not preceded by RP (Houdayer et al. 2014). Such evidence would suggest that the absence of RP does not allow patients to feel the awareness of the movements.

Despite criticisms, the result of Libet and colleagues has been replicated several times. For example, Soon et al. (2008) used a model that elaborates the cerebral activity

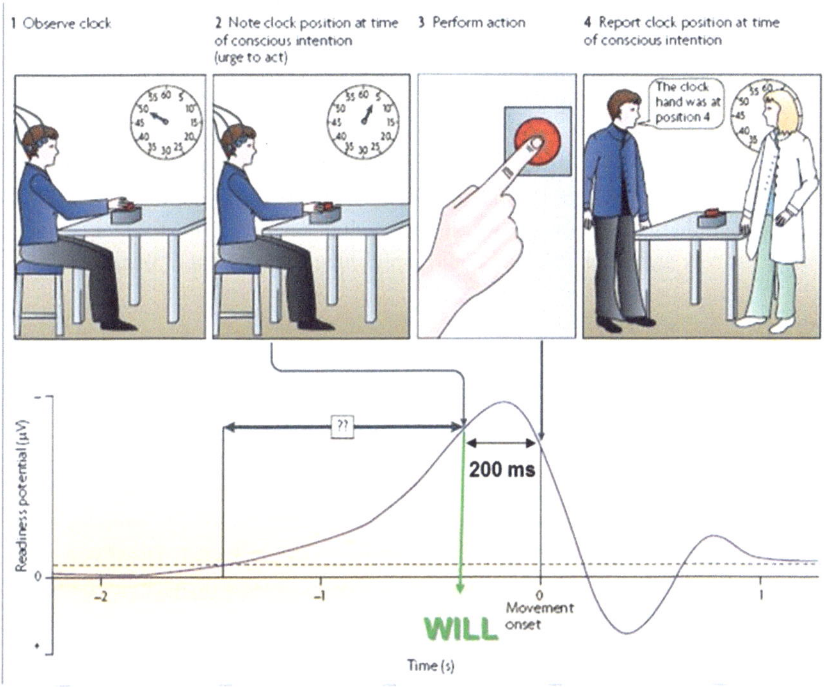

Fig. 1 The time of will. In the above panels, the phases of the experiment are reported. Participants were requested to fixate the clock hand that completed one full revolution in 2.56 s and at the same time to press a key at will. After the movement, participants reported the position of the clock hand at the time when they felt the urge to move. In the panel below, the time course of the readiness potential that aligned to movement onset is represented. As can be observed, while the readiness potential overcomes the baseline activity (dashed-black line) about 1.5 s before the movement onset, the participants reported the feeling of willing to move just about 200 ms before the movement onset (freely modified from Haggard 2008)

by measuring it with the functional magnetic resonance imaging (fMRI) during a decision-making task. Results showed that the activation of the frontopolar and posterior cingulate cortex could precede the awareness of intentionality by 8 s. More recently, Fried et al. (2011) repeated Libet's experiment, while recording single-unit activity from the supplementary motor area (SMA), pre-SMA, and the anterior cingulate cortex in drug-resistant epileptic patients.[1] They discovered that the neurons start to modify their firings roughly about 1000 ms before the subjects become conscious of their wiliness of executing the action. Furthermore, by using a mathematical method, they

[1] This rare opportunity is allowed because of the clinical approach to refractory epilepsy. Patients undergo surgery, during which the region of the brain whose activity generates epileptic crises, i.e., the epileptic focus, is removed. To locate precisely the focus, and to avoid the removal of parts of the primary motor cortex, primary sensory and eloquent areas, brain activity is recorded with intraparenchymal electrodes (or in other instances with subdural electrodes). This functional mapping is fundamental to inform subsequent surgery. Obviously, both the position and the number of implanted electrodes are chosen solely based on clinical criteria.

have also shown that the spiking activity of a small population of neurons allows predicting the moment in which the movement would take place about 700 ms before the persons report the conscious intention to move. These results suggest that the emergence of volition occurs well before the subject becomes aware of his intention to act. Could this be the case? Is the subjective experience of being able to freely choose between different courses of an action anything else than an illusion? Surely, whatever it is, the free will must emerge from some physical processing occurring in the brain. Otherwise, it could not be a concept within the reach of science. In the following, I will try to suggest why and when free will could represent a key element of our mental life.

2 What Does the Brain Push to Prepare for a Movement If Not the Free Will?

Even though it is appropriate to be very cautious when interpreting the results mentioned above, one could say that these seem to indicate that our brains' activity starts well before the moment in which we consciously decide to act. However, what could force the brain to prepare for a movement? A reasonable answer can be found in the theory of "affordance" (Gibson 1979). According to this theory, when perceiving an object, one immediately and automatically selects the actions that could potentially be performed concerning that specific object. For instance, this happens even when one sees a fork for the first time, one knows how to use it thanks to the fact that specific characteristics are the same for all the types of forks (Fig. 2). Following this idea, Grafton et al. (1997) have demonstrated that the mere observation of images of manipulable objects generates an increase in activity in the motor areas that control the body parts involved in the actions that one would have potentially carried out to interact with the depicted object. This happened even if the subject had no intention to move at all, thus indicating that the environmental stimuli can spark potential actions. These types of activations are rapid and unconscious, i.e., they are definitely not part of our conscious experience. Clearly, we do not perform all potential actions triggered by objects in the surrounding environment, otherwise stepping in a kitchen would be a nightmare. Thus, the overwhelming part of those actions is filtered out. Healthy people perform just those actions that are coupled with an internal state congruent with the primed action; e.g., the sight of a cup of coffee will prime the action if and only if an individual feels the need for caffeine. Accordingly, action implementation might be seen not as an active process but rather as an automatic consequence of an evaluative procedure aimed at determining whether or not the individual's current needs are satisfied (Mirabella 2014).

Interestingly two rare neuropsychological disorders, the alien limb syndrome and utilization behavior are characterized by a reduced capability of controlling automatically triggered actions. In both cases, patients cannot resist performing stimulus-driven motor responses even when they do not need those objects (Humphreys and Riddoch 2000). Patients with the alien hand syndrome lose the control of the upper limb contralateral to a focal brain lesion of the medial frontal cortex, often involving SMA (Biran and Chatterjee 2004). Patients suffering from utilization behavior compulsively act on objects placed within their reach that, for

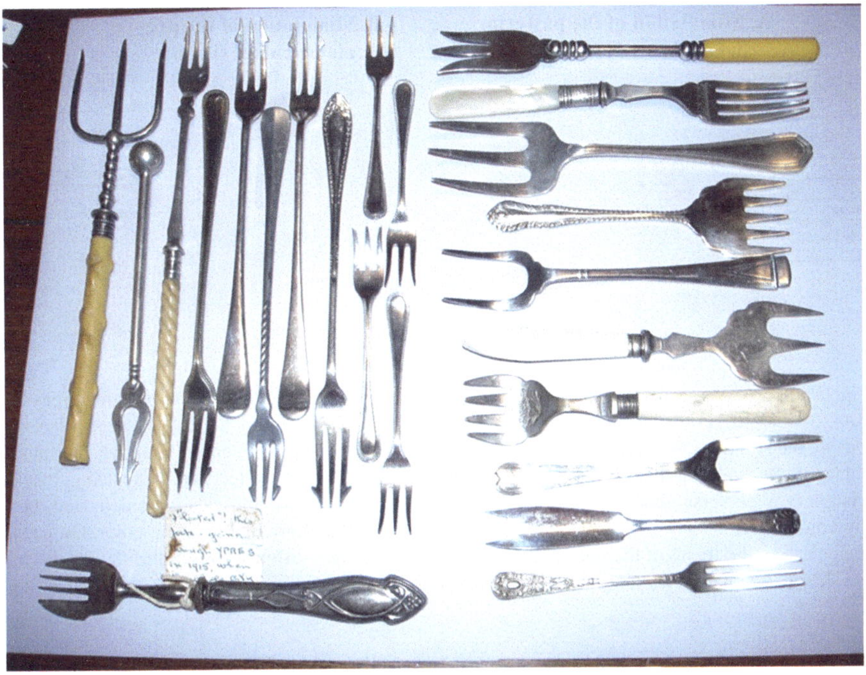

Fig. 2 As far as the theory of affordance is concerned, the characteristics that belong to different types of forks are intrinsically linked to all the potential actions that allow the interaction with that object. This means that a person who interacts with a new type of fork already knows how to use it

some reason, they consider salient. This syndrome arises after the bilateral damage of the medial frontal areas (SMA, pre-SMA, and cingulate cortex, Boccardi et al. 2002). Possibly, in both syndromes, the lesion affects the circuitry underlying the process, coupling stimulus-driven activations with our internal needs with the outcome that many unwanted actions have to be performed (Mirabella 2014).

3 Electric Stimulation of the Posterior Parietal Cortex and of Pre-supplementary Motor Cortex Evoke the Feeling of Intentionality

To complete the state-of-the-art, two surprising findings have to be mentioned. First, Desmurget et al. (2009) demonstrated that the electric stimulation of the posterior parietal cortex in patients undergoing tumor surgery,[2] can induce the conscious

---

[2]To circumscribe as much as possible, the regions that need to be removed, avoiding to damage parts of the brain underlying fundamental functions (e.g., speech and movement), patients with tumors undergo pre-surgical implantation of subdural electrodes for functional mapping. This is aimed to guarantee patients to maintain a good quality of life. By using those electrodes, one can administer low-intensity electric shocks in relatively circumscribed regions.

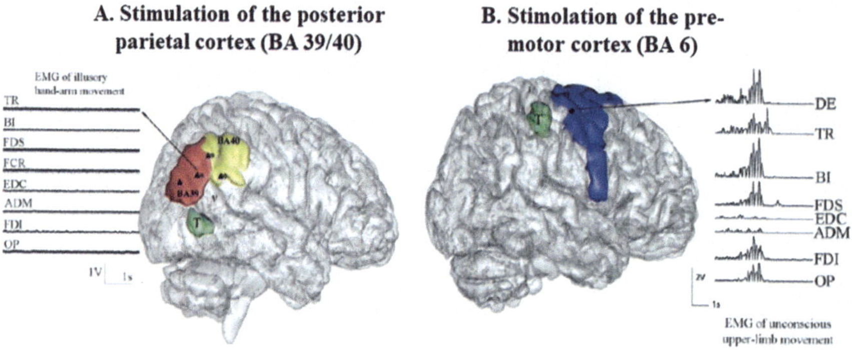

Fig. 3 Effects of electric stimulation of (**a**) the posterior parietal cortex (Broadman's areas—BA—39 and 40, respectively in red and yellow) and (**b**) of the premotor cortex (BA 6, in purple). During the stimulation of the parietal cortex, the patients reported the intention to move, but muscular activity is absent (the spots where the stimulation induced these effects are indicated in black triangles). Vice versa, during the stimulation of BA 6, the patients' muscles were activated, but they did not report any willingness to act. The tumor (T) is represented in green color. On each image's side, the traces of the muscular activity (EMG) of eight different muscles of the hand/arm are shown (adaptation from Desmurget et al. 2009)

experience of movement, even when muscular activity is absent (Fig. 3a). Importantly, at low-stimulation intensity, patients reported the intention of moving, but at higher stimulation intensities, some patients paradoxically stated to have performed a movement, even though they did not. By contrast, at low intensities, the stimulation of the premotor cortex evoked an evident muscular activity in the contralateral limb (Fig. 3b). At high intensities, an actual movement was induced. Nevertheless, in both instances, patients stated that they never felt the conscious intention of moving.

The posterior parietal cortex is not the only site that, when stimulated, produces the experience of the urge to move. Fried et al. (1991) demonstrated that also a low-intensity stimulation of the pre-SMA also induces the feeling of intentionality. However, in contrast with parietal stimulation, frontal stimulation at higher intensities caused actual movement. Given that the stimulation at the two sites produces qualitatively different effects, it has been argued that the two areas house distinct components of the experience of voluntary action (Haggard 2009). Pre-SMA would implement the conscious correlate of preparing motor plans, whereas the sense of volition in the parietal cortex would be needed for computing sensory predictions of the consequences of motor commands. At their turn, sensory predictions would help the perception of the sense of agency, i.e. the sense of authorship over one's own voluntary movements (Wegner 2003). In fact, it has been shown that damage to the parietal cortex impairs the awareness of voluntary actions (Sirigu et al. 2004).

4 Is Free Will a Mere Illusion? The Hypothesis of 'Free Won't'

All in all, the reviewed scientific evidence suggests that the motor system plans movements and subsequently, other brain regions read this activity to produce the subjective experience of willingness to move (Hallett 2007). Therefore, are we victims of unconscious processes that we have no power on? This idea conflicts with the common experience that, at least to some extent, we can consciously exert control over our actions. Furthermore, if this is the case, then would it still make sense to have a legal system based on the principle of self-control and moral responsibility, i.e., on the idea that we can freely choose what to do if this is nothing but some illusion? Almost ironically, Libet himself provided a way out (Libet 1985). He was no determinist and thought that people have free will. He reasoned that as awareness of intention precedes movements onset by a few hundreds of milliseconds, there is still time to consciously withhold the upcoming action, i.e., to exert a veto power. However, Libet did not find any identifiable neural correlate of this process. Even though this hypothesis is attractive, a veto being always dictated by the subject's conscious will is absolutely improbable. If this was the case, then entering a kitchen would require an unsustainable level of cognitive control. If we were to make a conscious effort to suppress all actions triggered by the affordances of objects such as mugs, forks, or pot handles, our computational resources would run out. Everyday experience teaches us that entering a kitchen is not so challenging. This is because our decisional processes are strongly linked to the valuation of the available resources (Mirabella 2014; Rangel et al. 2008). Every animal selects the actions that, on the basis of subjective values, allow it to maximize its biological fitness, i.e., to maximize the probability to survive and reproduce. Even though the way the nervous system assigns differential value to available behavioral options is not yet fully understood, it is evident that most of those processes are automatic, rooted in the cerebral circuits shaped by the evolution and especially in mammals, by learning. This concept holds not only for controlling actions triggered by affordances but more generally for all goal-directed actions. Both in animals and humans, well-learned goal-directed movements automatically link the incoming sensory information to motor actions, while the awareness of these processes remains minimal (Scott 2016).

However, despite the ability to predict the results of an action, in the world animals live in, events cannot be fully predicted. Therefore, all decisions intrinsically bear a certain degree of uncertainty. To minimize risks, the value of actions is continuously evaluated, from their initial selection to the instant potentially preceding their execution. If a change in the environmental conditions or the subject's internal mental state turns the action into something inconvenient, then the actions are suppressed (Mirabella 2014). Hence, the inhibitory control represents a crucial executive function, underlying behavioral flexibility. Inhibition is not a unitary construct, but a multifaceted one. In fact, a distinction must be made between motor inhibition and interference inhibition (Bari and Robbins 2013). Motor inhibition refers to the

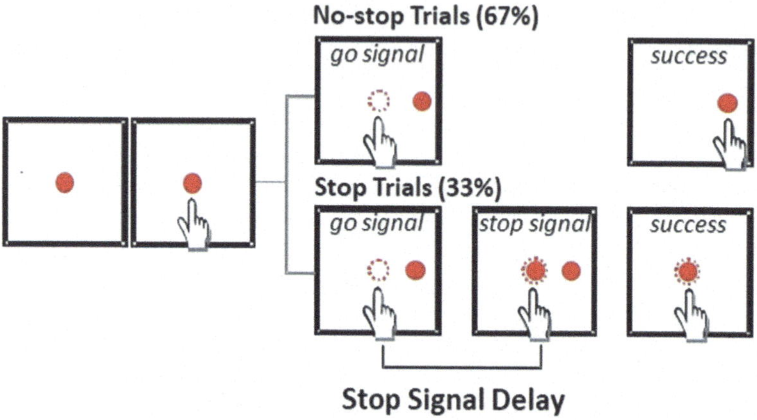

Fig. 4 Arm-reaching version of the stop-signal task. Subjects are seated in front of a touch-sensitive screen, and they could comfortably reach the stimuli projected on the screen. At the start of all types of trials, subjects were required to reach and hold a central stimulus. After a variable time (500–800 ms), the central stimulus disappears, and a new one appears on its left or right ('go-signal'). Only in no-stop trials, which are the more frequent, i.e., 67% of all trials, subjects have to reach and hold the peripheral target. By contrast, in stop trials after the delivery of the go-signal, but before the onset of the movement, the central stimulus reappears. In these trials (33% of all trials), subjects need to inhibit the already planned movement and keep holding the stop signal for about 300–400 ms. The time interval between the presentation of the go-signal and the stop signal is named stop-signal-delay. The dotted circle (blind to the subjects) indicates the size of the tolerance window for the touches (freely adapted from Mirabella et al. 2006)

ability to stop a potential or ongoing action; interference inhibition refers to the ability to resolve conflicts due to irrelevant but incompatible stimulus features that might lead to an erroneous response. Accordingly, different paradigms have been developed to measure these types of inhibitory control. On the one hand, motor response inhibition is measured via the go/no-go (Donders 1969) or the stop-signal tasks (Logan et al. 1984). On the other hand, interference inhibition is measured via the Simon (Simon and Rudell 1967), Eriksen flanker (Eriksen and Eriksen 1974), and Stroop (Stroop 1935) tasks. Among all those different paradigms, the one closer to the experimental setting of Libet is the stop-signal task, as it requires subjects to withhold an already initiated response. This paradigm, (Fig. 4), consists of a random mix of trials in which speeded responses are needed when a go-signal is presented (no-stop trials) with trials in which movements have to be withheld because, randomly and infrequently, an imperative stop-signal is presented after the delivery of the go-signal (stop trials).

Exploiting this task, the neural underpinnings of inhibitory control have been discovered in humans (e.g., Aron et al. 2003; Li et al. 2008; Mattia et al. 2012; Mirabella et al. 2012; Richard Ridderinkhof et al. 2011), monkeys (e.g., Chen et al. 2010; Hanes and Schall 1996; Mattia et al. 2013; Mirabella et al. 2011), and also rats (i.e. Eagle et al. 2008; Eagle and Robbins 2003; Mallet et al. 2016). These results are completely compatible with the fact that almost every animal is able to modify its behavior depending on the context in which it is embedded and therefore,

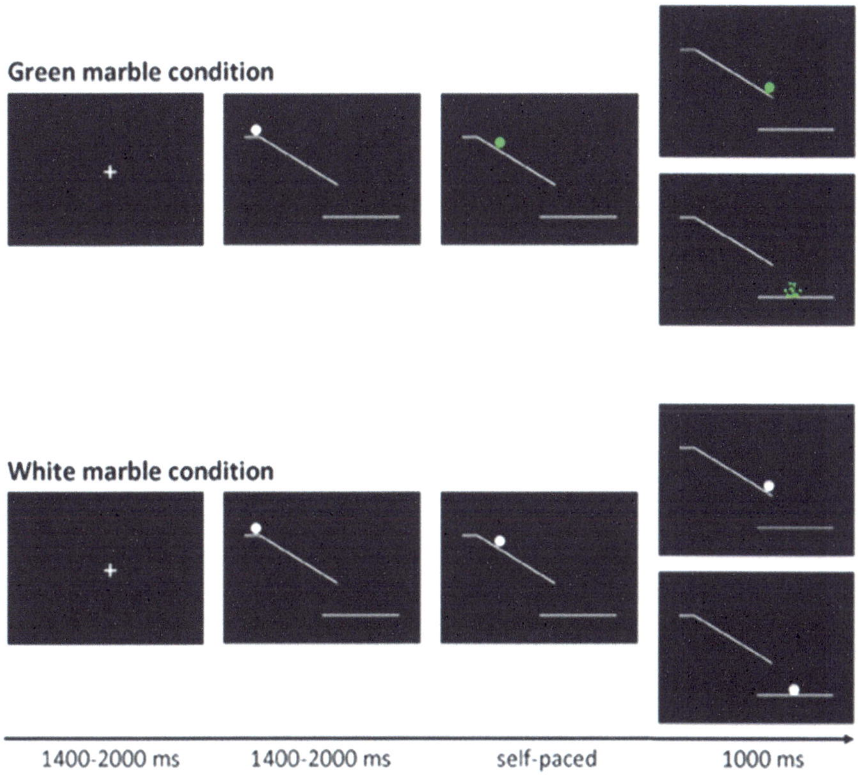

Fig. 5 Marble task. At the beginning of each trial, subjects have to fixate a white cross (1400–2000 ms), thereafter, a white marble appeared on the top of a ramp. After a variable delay (1400–2000 ms), the marble started to roll down the ramp, and in one condition change the color to green whereas in the other it remains white. In the former case, participants were instructed to stop the marble from crashing via a key-press. In the latter case, participants were instructed to choose between responding and inhibiting at their will (from Schel et al. 2014)

able to cancel invaluable actions. However, we do not attribute animals the ability to exercise free will. Hence, the mere ability to suppress undesired incipient actions cannot explain this capacity per se. Clearly, the conscious deliberation to move or not to move, as Libet intended it, derives from an internal mental process free from environmental stimuli. However, in all these experiments, the stop-signal was always an external stimulus (e.g., a visual or an acoustic signal). To overcome this critical limitation, recently, an effort was made to create new experimental paradigms to allow the study of internally driven inhibition, i.e., the so-called intentional inhibition (Filevich et al. 2012). Probably, the most successful attempt is the marble task (Fig. 5). This paradigm, as the stop-signal task, consists of a random mix of two types of trials. In one type of trial, a rotating marble changes color from white to green, and subjects are explicitly requested to stop it from rolling down the ramp by pressing a key. In the other type of trial, the marble remains white, and subjects can choose between responding or inhibiting the key-press, allowing the marble to be stopped or to crash at their will, respectively. However, overall, subjects are requested to inhibit intentionally the key-press about 50% of times. Interestingly, fMRI on healthy subjects revealed that the neural network subserving intentional inhibition largely overlaps with that of stimulus-driven inhibition, even though internally driven inhibitory control is characterized by the recruitment of some more brain regions located in the medial prefrontal cortex (Schel et al. 2014).

Although of great interest, these results provide just a partial frame of the neuronal mechanisms subserving our veto power. Many questions still remain unanswered. The marble task still partially relies on the external stimuli and on a number of instructions that might limit the expression of free choices of individuals, (e.g., subjects can suppress the movement at will, but overall they are requested to 'freely' inhibit just in half of trials). Thus, the extent to which this paradigm could capture the context of human volition is disputable. In addition, the experimental paradigms that have been used are extremely simple in comparison with real-life situations. Humans have the power to inhibit their survival needs whenever there is a valuable abstract reason (Montague 2006). For instance, we are the only living beings that can start hunger strikes for political ideas. How to test experimentally these complex mental phenomena is yet unknown.

Nevertheless, the uncovering of such mechanisms could be of extreme importance for understanding the origin of disorders characterized by the inability to control impulses typical of many psychiatric and neurological conditions, such as Parkinson's disease, obsessive-compulsive disorders, attention deficit-hyperactivity disorder, and many types of addiction.

5 What Free Will Is For? Why Does Free Will Could Have Sparked Human Evolution?

Even though the neural mechanisms of awareness and volition are only partially within reach of neuroscience, it is still possible to speculate about the function that free will might have in our mental life.

As previously mentioned, in the majority of the cases, the fate of our actions is shaped by automatic or semi-automatic processes that require minimal or null conscious participation. If that were not the case, then our processing capacities would not be enough to carry out simple tasks such as entering a store, preparing some food, or visiting art collections. Most probably, the need for making a fully conscious decision emerges when confronted with controversial choices in which the outcomes could have significant consequences on the person's future life. For instance, let us imagine a situation where an employee is told that he/she is not getting the expected promotion. The employee would feel anger towards the boss, yet, very likely, he/she will be capable of suppressing any violent reaction. This happens because, on the basis of every possible future scenario, the brain elaborated that a potential fight between the employer and the employee would only lead to the employee being fired and reported to the police. Supposedly, in the mental simulation of future scenarios, the representation of one's self requires conscious participation of the individual in such a way that he/she can have a vivid experience of the sensations associated with each potential action. In this case, the subject would experience the feeling that he/she would have potentially felt if he/she had hit the boss. In situations as such, inhibitory control can have a crucial role in suppressing choices which our instinct that would have led us to follow. The constructions of possible scenarios in which a conscious self moves and the ability to suppress actions triggered by emotions, such as anger or rancor, can stand at the basis of our free will, and thus, at the core of the wise choice not to hit the boss. According to Martin Seligman and his colleagues (Seligman et al. 2016), the ability to evaluate the pros and cons of potential action is what led to the human race's success. In fact, there is no animal able to foresee the repercussions of its actions on a long-term basis. No other animal studies in college, hoping for a brilliant career. For this reason, men and men only have been able to build such complex societies that are not based on family relationships, as for bees and ants.

6 Conclusion

Although, there are still no clear and univocal answers, the bottom-up approach of neuroscience to the problem of free will seems to be more promising than that of philosophy, which is deductive or top-down by nature. Obviously, even the empirical method has its limits. Firstly, the experience of intentionality exists only in the agents who experience it. This means that the experimenter needs to rely on a verbal report of a subjective experience, which could be biased or extremely variable between subjects. Secondly, the idea of using an experiment to prove whether a human being can exercise its free will implies a context in which free choices can be made as in real life. However, due to the intrinsic limits of the experimental activity, what is measured in laboratories does not reproduce what typically happens in the external world. Despite these limitations, neuroscience is getting closer to the comprehension of nervous mechanisms that are prone to complex mental phenomena as the ones that give birth to the conscious control of movements, at the very least, a component of our free will. Overall, the experimental evidence gathered so far suggests that, except specific medical conditions, we are free of choosing how to act as much as we are responsible for what we do. Besides, this same evidence provides some hints about the neural basis of the genesis of our internal 'veto power.'

References

Alexander P, Schlegel A, Sinnott-Armstrong W, Roskies AL, Wheatley T, Tse PU (2016) Readiness potentials driven by non-motoric processes. Conscious Cogn 39:38–47

Aron AR, Fletcher PC, Bullmore ET, Sahakian BJ, Robbins TW (2003) Stop-signal inhibition disrupted by damage to right inferior frontal gyrus in humans. Nat Neurosci 6:115–116

Bari A, Robbins TW (2013) Inhibition and impulsivity: behavioral and neural basis of response control. Prog Neurobiol 108:44–79
Bauman Z (2008) The art of life. Polity, Cambridge
Biran I, Chatterjee A (2004) Alien hand syndrome. Arch Neurol 61:292–294
Boccardi E, Della Sala S, Motto C, Spinnler H (2002) Utilisation behaviour consequent to bilateral SMA softening. Cortex 38:289–308
Chen X, Scangos KW, Stuphorn V (2010) Supplementary motor area exerts proactive and reactive control of arm movements. J Neurosci 30:14657–14675
Dennett DC (1984) Elbow room: the varieties of free will worth wanting. MIT Press, Cambridge
Desmurget M, Reilly KT, Richard N, Szathmari A, Mottolese C, Sirigu A (2009) Movement intention after parietal cortex stimulation in humans. Science 324:811–813
Donders FC (1969) On the speed of mental processes. Acta Psychol (Amst) 30:412–431
Eagle DM, Robbins TW (2003) Lesions of the medial prefrontal cortex or nucleus accumbens core do not impair inhibitory control in rats performing a stop-signal reaction time task. Behav Brain Res 146:131–144
Eagle DM, Baunez C, Hutcheson DM, Lehmann O, Shah AP, Robbins TW (2008) Stop-signal reaction-time task performance: role of prefrontal cortex and subthalamic nucleus. Cereb Cortex 18:178–188
Eriksen BA, Eriksen CW (1974) Effects of noise letters upon the identification of a target letter in a nonsearch task. Percept Psychophys 16:143–149
Filevich E, Kuhn S, Haggard P (2012) Intentional inhibition in human action: the power of 'no'. Neurosci Biobehav Rev 36:1107–1118
Fried I, Katz A, McCarthy G, Sass KJ, Williamson P, Spencer SS, Spencer DD (1991) Functional organization of human supplementary motor cortex studied by electrical stimulation. J Neurosci 11:3656–3666
Fried I, Mukamel R, Kreiman G (2011) Internally generated preactivation of single neurons in human medial frontal cortex predicts volition. Neuron 69:548–562
Gibson JJ (1979) The ecological approach to visual perception. Houghton, Mifflin and Company, Boston
Grafton ST, Fadiga L, Arbib MA, Rizzolatti G (1997) Premotor cortex activation during observation and naming of familiar tools. Neuroimage 6:231–236
Haggard P (2008) Human volition: towards a neuroscience of will. Nat Rev Neurosci 9:934–946
Haggard P (2009) Neuroscience. The sources of human volition. Science 324:731–733
Hallett M (2007) Volitional control of movement: the physiology of free will. Clin Neurophysiol 118:1179–1192
Hanes DP, Schall JD (1996) Neural control of voluntary movement initiation. Science 274:427–430
Houdayer E, Walthall J, Belluscio BA, Vorbach S, Singer HS, Hallett M (2014) Absent movement-related cortical potentials in children with primary motor stereotypies. Mov Disord 29:1134–1140
Humphreys GW, Riddoch MJ (2000) One more cup of coffee for the road: object-action assemblies, response blocking and response capture after frontal lobe damage. Exp Brain Res 133:81–93
Kornhuber HH, Deecke L (1965) Changes in the brain potential in voluntary movements and passive movements in man: readiness potential and reafferent potentials. Pflugers Arch Gesamte Physiol Menschen Tiere 284:1–17
Lavazza A, Inglese S (2015) Operationalizing and measuring (a kind of) free will (and responsibility). Towards a new framework for psychology, ethics and law. Riv Int di Filos e Psicol 6:37–35
Li CS, Yan P, Sinha R, Lee TW (2008) Subcortical processes of motor response inhibition during a stop signal task. Neuroimage 41:1352–1363
Libet B (1985) Unconscious cerebral initiative and the role of conscious will in voluntary action. Behav Brain Sci 8:529–539
Libet B, Gleason CA, Wright EW, Pearl DK (1983) Time of conscious intention to act in relation to onset of cerebral activity (readiness-potential). The unconscious initiation of a freely voluntary act. Brain 106(Pt 3):623–642

Logan GD, Cowan WB, Davis KA (1984) On the ability to inhibit simple and choice reaction time responses: a model and a method. J Exp Psychol Hum Percept Perform 10:276–291
Mallet N, Schmidt R, Leventhal D, Chen F, Amer N, Boraud T, Berke JD (2016) Arkypallidal cells send a stop signal to striatum. Neuron 89:308–316
Mattia M, Spadacenta S, Pavone L, Quarato P, Esposito V, Sparano A, Sebastiano F, Di Gennaro G, Morace R, Cantore G et al (2012) Stop-event-related potentials from intracranial electrodes reveal a key role of premotor and motor cortices in stopping ongoing movements. Front Neuroeng 5:12
Mattia M, Pani P, Mirabella G, Costa S, Del Giudice P, Ferraina S (2013) Heterogeneous attractor cell assemblies for motor planning in premotor cortex. J Neurosci 33:11155–11168
Mirabella G (2014) Should I stay or should I go? Conceptual underpinnings of goal-directed actions. Front Syst Neurosci 8:206
Mirabella G, Pani P, Pare M, Ferraina S (2006) Inhibitory control of reaching movements in humans. Exp Brain Res 174:240–255
Mirabella G, Pani P, Ferraina S (2011) Neural correlates of cognitive control of reaching movements in the dorsal premotor cortex of rhesus monkeys. J Neurophysiol 106:1454–1466
Mirabella G, Iaconelli S, Romanelli P, Modugno N, Lena F, Manfredi M, Cantore G (2012) Deep brain stimulation of subthalamic nuclei affects arm response inhibition in Parkinson's patients. Cereb Cortex 22:1124–1132
Montague R (2006) Why choose this book?: how we make decisions. Dutton Adult, New York
Rangel A, Camerer C, Montague PR (2008) A framework for studying the neurobiology of value-based decision making. Nat Rev Neurosci 9:545–556
Richard Ridderinkhof K, Forstmann BU, Wylie SA, Burle B, van den Wildenberg WPM (2011) Neurocognitive mechanisms of action control: resisting the call of the Sirens. Wiley Interdiscip Rev Cogn Sci 2:174–192
Schel MA, Kuhn S, Brass M, Haggard P, Ridderinkhof KR, Crone EA (2014) Neural correlates of intentional and stimulus-driven inhibition: a comparison. Front Hum Neurosci 8:27
Scott SH (2016) A functional taxonomy of bottom-up sensory feedback processing for motor actions. Trends Neurosci 39:512–526
Seligman MEP, Railton P, Baumeister RF, Sripada C (2016) Homo prospectus. Oxford University Press, Oxford
Simon JR, Rudell AP (1967) Auditory S-R compatibility: the effect of an irrelevant cue on information processing. J Appl Psychol 51:300–304
Sirigu A, Daprati E, Ciancia S, Giraux P, Nighoghossian N, Posada A, Haggard P (2004) Altered awareness of voluntary action after damage to the parietal cortex. Nat Neurosci 7:80–84
Soon CS, Brass M, Heinze HJ, Haynes JD (2008) Unconscious determinants of free decisions in the human brain. Nat Neurosci 11:543–545
Stroop JR (1935) Studies of interference in serial verbal reactions. J Exp Psychol 18:643–666
Wegner DM (2003) The illusion of conscious will. MIT Press, Cambridge

Deep Brain Stimulation for Parkinson's Disease: Clinical Efficacy and Future Directions for Enhancing Motor Function

Corneliu C. Luca, Joacir Graciolli Cordeiro, Iahn Cajigas, and Jonathan Jagid

1 Introduction

Deep brain stimulation is the most effective treatment available for the management of advanced Parkinson's disease. The technique was introduced in the field of movement disorders in the 80s when successful application of high-frequency stimulation to the thalamus produced a remarkable improvement in tremor severity in people with disabling essential tremor. The ability to modulate neural activity using electrical stimulation has led to the approval of DBS for a variety of clinical applications in neurological and psychiatric conditions. The effects of DBS in Parkinson's disease are for the most part immediate, reversible, and titratable, without permanently damaging neural tissue. Deep brain stimulation (DBS) of the subthalamic nucleus (STN) or the internal globus pallidus (GPi) has been shown to improve the quality of life and motor function in patients with advanced Parkinson's disease (PD). A multitude of clinical data supports the use of DBS for the management of motor symptoms of Parkinson's disease and motor complications of therapy and shows the superiority of DBS over medical management alone.

Advancements in the field of imaging, operative techniques, and device technology as well as its safety profile have allowed for increased acceptance and use of DBS. Additionally, a trend towards earlier use in the disease process has occurred. However, despite these facts, the procedure remains underutilized and it is estimated that less than 5% of patients are referred for DBS. The need for cranial surgery, cost, and the fact that axial symptoms do not respond as well to DBS have

C. C. Luca (✉)
Department of Neurology, University of Miami Miller School of Medicine, Miami, FL, USA
e-mail: cluca@med.miami.edu

J. G. Cordeiro · I. Cajigas · J. Jagid
Department of Neurosurgery, University of Miami Miller School of Medicine, Miami, FL, USA

I. Opris et al. (eds.), *Modern Approaches to Augmentation of Brain Function*, Contemporary Clinical Neuroscience,
https://doi.org/10.1007/978-3-030-54564-2_22

hindered widespread adoption of DBS. Emerging technologies using adaptive DBS, field steering, and different wave forms of stimulation will potentially improve our ability to restore physiological brain function and likely increase the adoption of this technique. It is critical for the DBS field to continue to innovate technologies that allow for more accurate stimulation in the desired target as well as deliver stimulation that is temporally patterned to the individual neural circuits in order to obtain an optimal clinical benefit.

2 Deep Brain Stimulation Efficacy in Parkinson's Disease

DBS is a well-established procedure for the treatment of movement disorders such as Parkinson's disease, essential tremors, and dystonia. The mechanism of action of DBS is complex and only partially understood. The initial hypothesis of the direct inhibition of neural activity by high-frequency stimulation has been supported by numerous experiments that used somatic recordings from neurons close to the recording electrode. The direct excitation of neural activity has also been shown in postsynaptic recordings and supported by the biophysics of axonal responses to electrical stimulation leading to the formulation of a model where stimulation can directly activate axons and modulate upstream/downstream activity and at the same time inhibits neural activity at the site of stimulation (McIntyre et al. 2004). The consensus today is that DBS can modulate excessive oscillatory electrical activity and revert the network to a physiological pattern (Eusebio et al. 2010). Many signs and symptoms of movement disorders are secondary to abnormal synchronization within and between structures in motor networks, and DBS can profoundly alter these synchronized oscillations (Swann et al. 2018). Emerging evidence suggests that in Parkinson's disease there is a significant increase in pathological synchronous activity in the basal ganglia structures and that high-frequency DBS is able to suppress this excessive activity (Blumenfeld and Bronte-Stewart 2015).

2.1 Deep Brain Stimulation for Motor Symptoms in Parkinson's Disease

Multiple randomized clinical trials have shown that DBS is effective in the treatment of motor fluctuations and improves the quality of life. To achieve optimal results in careful patient selection, individualized anatomical target localization, and proper evaluation of stimulation parameters are necessary. The targets used for DBS in Parkinson's disease are the subthalamic nucleus (STN) and globus pallidus pars interna (GPi). Both targets are associated with a significant improvement in motor symptoms (bradykinesia, rigidity, and tremors), reduction in OFF time, and improvement in the quality of life. The choice between targets depends on a patient's profile and goals of the therapy. Overall, the STN target seems to be the preferred target and the fact is that it is associated with a higher likelihood of levodopa reduction, while GPi is used when dystonia, mood disorders, or mild cognitive impairment are the predominant features.

Multiple randomized clinical trials have shown a significant improvement in motor symptoms, reduction in the total daily OFF time, and concomitant increase in ON time in patients with advanced PD post-DBS (see Table 1). One of the first large studies to compare DBS vs. best medical therapy conducted by Deuschl showed the superiority of STN DBS when compared with best medical therapy (BMT) (Deuschl et al. 2006). At 6 months, the quality of life as assessed by PDQ-39 and the unified Parkinson's disease rating scale (UPDRS) motor scores were significantly better in the STN group when compared with the best medical therapy group (Deuschl et al.

Table 1 Summary of the main controlled clinical trials for deep brain stimulation in Parkinson's disease

Study	No of patients	Design	Age	UPDRS III	Outcomes
Deuschl et al. (2006)	156	STN DBS + best medical therapy (BMT) vs. BMT	60.5	48.0	– 4.4 h increase in ON time – 41% improvement in the UPDRS score in the DBS group vs. no change in BMT
VA study Weaver et al. (2009)	255	STN or GPi DBS+ BMT vs. BMT	62.4	43.0	3–1% improvement in the DBS stim group vs. no change in the BMT group – 4 h improvement in ON time
PD surg trial Williams et al. (2010)	366	STN DBS + BMT vs. BMT	59	47.6	– 16.8 points improvement in the UPDRS score in the DBS group vs. 0.4 in BMT – 4-h increase in ON time
Okun et al. (2012)	136	STN DBS vs. 3-month delayed DBS activation	60.6	40.8	– 39% improvement in the DBS group vs. 4% improvement in the group with delayed activation
NSTAPS Odekerken et al. (2013)	128	STN DBS vs. GPi DBS	59.1	44.0	– Similar functional health – 11.4 points improvement in UPDRS in GPi and 20.3 points in STN
EARLYSTIM Schuepbach et al. (2013)	251	STN DBS + best medical therapy vs. best medical therapy	52.9	33.2	– 53% improvement in the stim group vs. 4% in the BMT group – Improved the quality of life in STN DBS
INTREPID study Vitek et al. (2019)	160	Randomized (3:1), controlled trial DBS vs. sham stimulation	59.9	43.4	– 49.2% improvement in UPDRS in the stim group versus the baseline – 3.7 h increase in ON time

UPDRS unified Parkinson's disease rating scale, *BMT* best medical therapy

2006). Other randomized trials have shown a similar improvement in the quality of life after bilateral STN DBS (Williams et al. 2010; Okun et al. 2012; Schuepbach et al. 2013).

In the VA study, both STN and GPi were assessed and compared with BMT: ON time without troubling dyskinesia in the DBS group improved by a mean of 4.6 h per day versus 0 h per day for the best medical therapy group. The total daily levodopa dose required to control the participants' symptoms was reduced significantly, and the quality of life improved (Weaver et al. 2009). To compare directly the effects of STN vs. GPI on motor function, a randomized trial enrolled 128 patients, who were assigned either GPi DBS or STN DBS (1:1). The primary outcome, functional health measured by Academic Medical Center Linear Disability Scale did not show a difference between the two targets. However, there was a larger improvement in the UPDRS motor scores and a higher reduction in the levodopa dose in patients undergoing DBS in the STN group (Odekerken et al. 2013).

The most recent study evaluating DBS effects compared 292 patients, who were randomized to bilateral STN DBS versus sham DBS (Vitek et al. 2019). The primary end point at 12 weeks showed an increase in the total ON time in the active group with about 3.7 h when compared with the baseline. The overall improvement in UPDRS was 49% in the DBS group compared with sham stimulation, similar to other clinical trials (Vitek et al. 2020).

2.2 Deep Brain Stimulation for Non-motor Symptoms in Parkinson's Disease

While the motor symptoms are the defining feature of PD, there is a growing awareness that the presence of non-motor features such as cognitive and neuropsychiatric symptoms, autonomic dysfunction, and sleep disorders have a significant toll on the quality of life. Emerging evidence suggests that deep brain stimulation has beneficial effects on a variety of non-motor symptoms.

The goal of DBS is to improve motor function without a significant risk to behavioral and neurocognitive functions. Multiple studies have looked at long term cognitive and behavioral outcomes and suggested that while both STN and GPi DBS are safe, they may produce subtle cognitive changes. The evaluation of cognitive decline, as it relates to DBS, is confounded by the progression of the underlying disease itself; therefore, it is not surprising that cognitive outcomes are hard to evaluate (Cernera et al. 2019). STN DBS, for example, is associated with a mild decline in the psychomotor speed, memory, attention and executive function, and a moderate decrease in semantic and phonemic fluencies (Combs et al. 2015). Interestingly, GPi DBS resulted in fewer neurocognitive deficits when compared with STN DBS, a feature that should be taken into consideration in patients with mild cognitive impairment during the target selection.

With regards to its effect on depression, both GPi DBS and STN DBS are associated with lower levels of depressive symptoms postsurgery, changes that have been associated with an increase in the quality of life and patient satisfaction (Ramirez-Zamora and Ostrem 2018) . With regards to sleep disorders, DBS improves sleep quality as quantified by objective and subjective sleep parameters in patients with PD. Polysomnography (PSG) has demonstrated that sleep latency and wake after sleep onset improve after both STN and GPi DBS (Zuzuarregui and Ostrem 2020). Subjective sleep measures such as nocturnal motor symptoms and some non-motor symptoms (nocturia) improved with an overall increase in the total sleep time of about 1 h.

2.3 *Deep Brain Stimulation for the Axial Symptoms in Parkinson's Disease*

One of the biggest unmet needs of DBS for Parkinson's disease is the inability to improve the axial motor symptoms such as gait, postural instability, and dysphagia. While DBS is effective in the treatment of freezing of gait (FOG) in patients with early motor complications, the long-term effect on FOG is not as robust and as the disease progresses the benefit on FOG wanes. Improvement of gait and FOG is sustained at up to 4 years after STN DBS in the meds OFF, stim ON condition (Schlenstedt et al. 2017). In an attempt to address these shortcomings, different targets and stimulation paradigms have been tested in people with PD. Several studies have reported that pedunculopontine nucleus (PPN) DBS has the potential effect to alleviate gait and motor symptoms in PD; however, the studies are small and methodology is heterogeneous, perhaps explaining the inconsistent results. A recent meta-analysis suggests that bilateral low frequency PPN DBS may improve FOG and falls acutely in people with PD (Wang et al. 2017). The mechanism is unknown but is believed to be that PPN DBS at a low frequency may modulate the physiological alpha activity related with attention allocation and therefore, enhance the ability to optimally use resources within the motor system and improve FOG (Thevathasan and Moro 2019). One ongoing clinical trial is currently exploring stimulation dorsal to the PPN region (cuneiform nucleus), an area evidenced by optogenetic studies to contain the principal neurons eliciting locomotion (NCT04218526).

Another approach to levodopa-resistant FOG was the simultaneous stimulation of STN/substantia nigra pars reticulata (SNr). Experimental evidence suggests that low-frequency stimulation of SNr could be useful in regulating gait disorders refractory to medical treatment in Parkinson's disease (PD). A recent study showed that the combined SNr/STN stimulation is associated with an improvement in gait in people with levodopa-resistant PD FOG (Valldeoriola et al. 2019).

While PPN and SNr have the potential to become the target of choice for treating FOG, current evidence does not support its use in clinical practice. Multicenter, randomized, blind, controlled trials are needed to determine the optimal frequency of stimulation, the optimal lead location, and the ideal combination of nuclei to be stimulated. Advances in DBS technology, including directional stimulation and imaging, will hopefully lead to a better understanding of the midbrain locomotor circuitry and its pathophysiologic state in movement disorders ultimately allowing for better strategies to restore brain function.

3 Advances in DBS Technology: Robotics, Imaging, and Directional Stimulation

Although DBS has been in the clinical practice for movement disorders for more than 20 years, the adoption of technological advances in biotechnology has been slow. Recently, the field has witnessed a surge in innovative technologies that allowed a more precise mapping of neuroanatomic targets using robotic techniques and improved imaging modalities. Together with the use of microelectrode recordings and the ability to deliver stimulation directionally, these new developments have provided the means by which DBS can modulate brain networks with precision and provides optimal therapy without side effects.

3.1 Robotic Advances for Stereotactic Surgery

Regardless of disease and targeted structure, the placement of DBS leads is done using stereotactic methods via a small burr-hole. The procedure is classically performed using a stereotactic frame rigidly affixed to the patient's skull. However, there are many frameless techniques that have been developed as well for patient comfort. All of these techniques require some form of rigid registration primarily via bone fiducials placed either prior to the surgical procedure or on the day of the surgical procedure. Another approach, robotic driven stereotaxy (Fig. 1) has gained popularity since the first description of DBS made by Benabid in 1987 (Benabid et al. 1987). This approach combines the use of any rigid frame fixation and bone fiducials with the advantage of potential reductions in workflow and human error.

There are two commonly used robotic systems: NeuroMate (Renishaw Mayfield SA, Nyon, Switzerland) and the ROSA Brain system (Zimmer Biomet, IN, US). Robotic assistance has been described in a variety of procedures including brain biopsy, stereo-electroencephalography (SEEG), laser interstitial thermal therapy (LiTT), endoscopy, responsive neural stimulation (RNS), and lastly, DBS (Candela et al. 2018; Faraji et al. 2020; Yasin et al. 2019). In the case of SEEG, robotic stereotaxy has gained significant attention due to the ability to substantially reduce surgical time when placing a large number of intracranial electrodes. Likewise, it also bears the potential advantage of minimizing human error compared to traditional frame-based surgery as well as decreasing operative time for bilateral

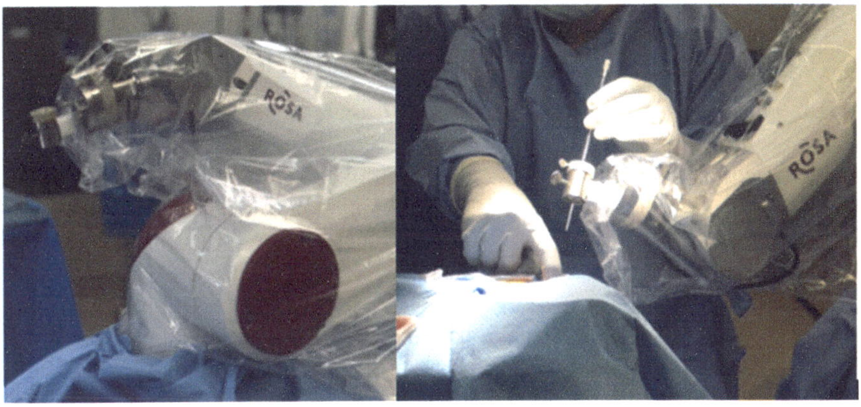

Fig. 1 Pictures from the robotic surgical assistant (ROSA™) from the University of Miami Hospital. On the left, the robotic arm with the adaptor used for lead implant. On the right, ROSA is being used to guide to the entry point for permanent lead insertion

implants. Data suggest that this technique does not negatively impact precision, safety, or the ability to perform awake neurophysiology (Faraji et al. 2020). Neudorfer et al. reported on a series of 80 consecutive DBS patients, 40 frame-based and 40 robotic-assisted procedures. In this series, the robotic-assisted subset of cases showed superior lead placement accuracy. In addition, a statistically significant reduction of 69 min in operative time was reported (Neudorfer et al. 2018). The use of robotic guidance combined with a stereotactic frame is also feasible, safe, and may have advantages. A recent series reviewed the results of 20 consecutive patients, who underwent frame-based ROSA-assisted DBS. In this study, the researchers took advantage of the Leksell frame fixation screws by using them as rigid bone fiducials for registration; thus, negating the need for additional bone fiducial placement, a potential advantage for patient comfort. The reported overall radial error for lead placement was 1.14 ± 0.11 mm. Of note, the steep learning curve illustrated by lower radial errors in the last 10 subjects (0.86 ± 0.09 mm) was compared to the first 10 subjects (1.46 ± 0.19 mm) (Faraji et al. 2020). It remains, nevertheless, unclear if any of the technical advantages translate to an improvement in clinical outcome.

3.2 *Imaging Advances for Surgical Planning*

In order to achieve precise electrode placement, an accurate stereotactic system coupled to advanced planning software is desirable. Optimal imaging and precise co-registration of pre- and intraoperative images are key for targeting. Most centers perform a preoperative high-resolution MRI, which is used to plan the electrode trajectory. Optimal trajectories avoid sulci, blood vessels, ventricles and terminate at the intended subpart of the targeted structure (e.g. dorsolateral STN). More recently, the use of diffusion tensor imaging (DTI) for fiber tractography (FT) has been investigated. The visualization of particular tracts may assist in targeting MRI invisible anatomy such as the ventral intermediate thalamic nucleus (Vim) for tremor control. Morishita et al. reported on seven successful cases of Vim-DBS in which targeting was made with a combination of fast gray matter acquisition T1 inversion recovery (FGATIR) and DTI-FT. Parcellation of the thalamic nuclei was done using the FGATIR sequence and Vim identification was confirmed by the localization of the dentato-rubro-thalamic tract (DRTt) on DTI (Morishita et al. 2019).

Other significant advances in neuroimaging include high-resolution MRI such as 3 and 7 T strength, the latter has been used primarily for research currently. Three Tesla imaging has the advantage of delineating certain deep nuclear structures often used for DBS. Over the last 15 years, MR sequences have been refined allowing for the direct visualization and more consistent and precise lead placement. One such advance was the introduction of fast gray matter acquisition inversion recovery sequences (FGATIR) for the direct visualization of subnuclei of the globus pallidus, improving the accuracy of GPi lead placement (Sudhyadhom et al. 2009). Unfortunately, GPi laminar borders are not always precisely visible on FGATIR. An alternative approach utilizing the putamen (PUT) as a surrogate anatomical marker to target GPi was described allowing consistent and precise patient-specific GPi targeting (Thompson et al. 2017). Three Tesla T2-weighted MRI provides adequate STN identification on most occasions. However, given the average age of DBS subjects at the time of the implant with the decreased white matter density, the suboptimal visualization of the structure may occur (Ranjan et al. 2018). Recently, a method that incorporates a database of 7-Tesla (T) MRIs of PD patients together with machine-learning (ML) has been developed to improve its identification on routine MRI. Shamir et al. reported that this 7T-ML method was highly consistent with MER data, offering reliable and accurate STN targeting (Boutet et al. 2019).

In addition, a better understanding of brain interconnectivity has led to the discovery of different targets. Coenen et al. demonstrated the utility of identifying the dentato-rubro-thalamic tract (DRT) in a patient with prior bilateral Vim implants. After the patient became refractory to Vim DBS, the patient underwent a DTI-FT-assisted DRT DBS revision surgery. As a result, the patient experienced a sustained marked improvement for over 6 months (Coenen et al. 2017). In the last few years, DTI studies have also been conducted in patients with Parkinson's disease going STN DBS. DTI-FT reconstruction was able to show the projection of neurons connecting STN to the (pre)motor cortex. This study reinforced the hyper-direct pathway as a potential therapeutic target in PD (Vassal et al. 2020). Different stereotactic planning software is available, and some include specific tools that are able to integrate DTI into stereotactic planning. (Fig. 2).

For placement accuracy verification, postoperative imaging is routinely performed. In our department, a postoperative head CT without contrast is done immediately after the implant. CT is merged with the preoperative plain MRI to verify lead location since the usual targets are only visible on the latter sequence (Cordeiro

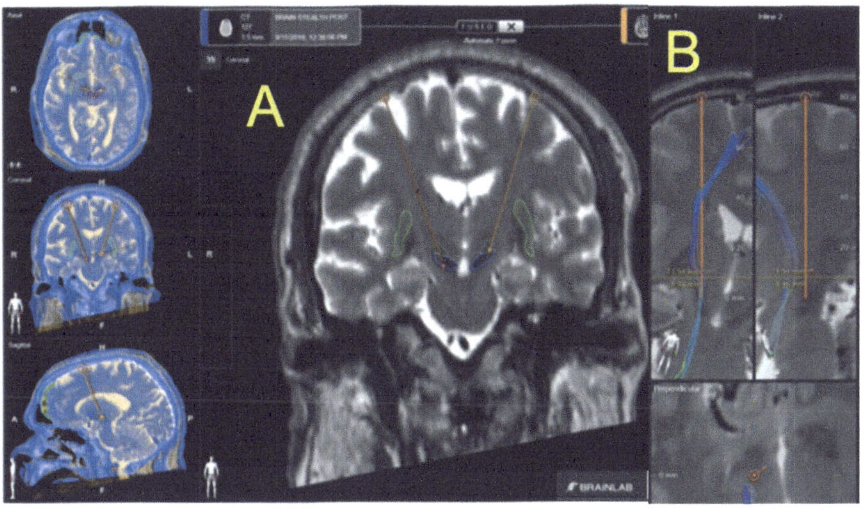

Fig. 2 (**a**) Illustration of a preoperative stereotactic plan to target STN on both sides. Note that the trajectory avoids sulci, vessels, and ventricles. Both trajectories cross STN, which were automatically segmented. (**b**) This illustration depicts the integration of DTI Fiber Tractography to the preoperative stereotactic plan. (Courtesy from Brainlab)

et al. 2020). In fact, the lead artifact visible on CT is used for this purpose. There is, however, a concern for brain shift on the postoperative CT due to the presence of pneumocephalus. Recently developed planning software (e.g. Brainlab, Munich, Germany) is equipped with tools that may be able to improve verification imaging via novel image fusion techniques. Most stereotactic planning software uses a rigid image fusion algorithm, which may result in spatial error caused by pneumocephalus. Recently developed fusion protocols use a tool to overcome this problem, called distortion correction fusion. This tool was initially utilized to correct inherent MRI distortion during image acquisition with the goal of improving preoperative imaging homogeneity. Potentially, this fusion technique may allow for improved intra- and postoperative placement verification. This distortion correction protocol would minimize pneumocephalus-induced error (Fig. 3).

Advances in imaging methods, such as high-field MRI and use of tractography to map neuroanatomical structures, will help refine and optimize strategies to deliver personalized-DBS therapy (Hell et al. 2019).

3.3 Role of Intraoperative Neurophysiology

The use of microelectrode recordings (MER) allows for real-time physiological verification of the appropriate target as well as intraoperative refinement of ultimate lead location.

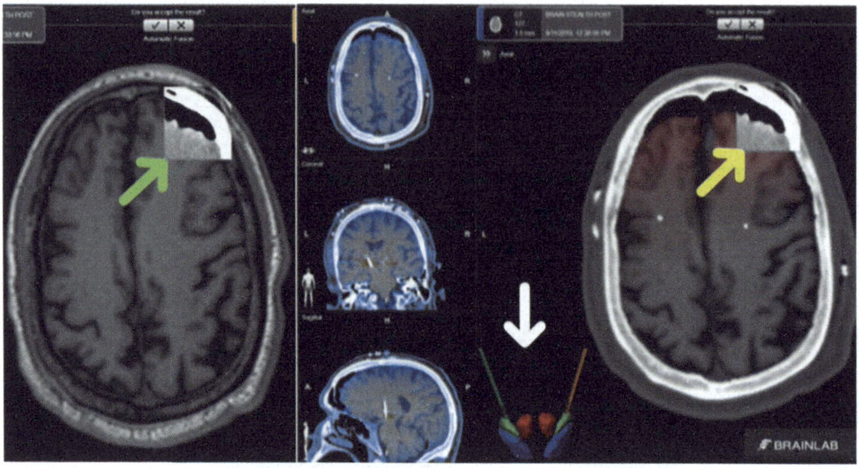

Fig. 3 LEFT—Illustration of a rigid coregistration of T1-weighted preoperative MRI to the postoperative CT. Note the presence of pneumocephalus (green arrow). RIGHT—This illustration depicts the elastic merge tool that deformed the preoperative MRI sequence to fit the postoperative image, correcting for the presence of the pneumocephalus (yellow arrow). WHITE ARROW—Postoperative reconstruction depicting the real DBS lead position in STN (green). This tool is able to automatically identify the electrode orientation (arrows), which is valuable when directional DBS is desired. (Courtesy from Brainlab)

MER is a valuable tool when targeting deep nuclei for patients with movement disorders. It is commonly performed using a hybrid micro-macro electrode for recording and stimulation purposes. During MER, the neuronal firing is monitored by the surgical team as the electrode passes through the critical brain anatomy along the planned trajectory. Frequency content analysis provides valuable data to further define the morphology of the target. As an example, the PD patients exhibit a high-power content in the β band (13–35 Hz) in the MER signal, mainly in the dorsolateral region of STN (Tepper et al. 2017). The increased oscillatory activity in the beta band is considered as a biomarker of bradykinesia and there is merging evidence that a high-frequency DBS is associated with a suppression of this firing and improved motor outcomes (Eusebio et al. 2010).

Intraoperative neurophysiological recordings and stimulation help to confirm proper electrode placement and efficacy at the time of implantation by visibly improving symptoms including rigidity, tremor, and bradykinesia. When the recording pattern indicates proper positioning, a macrocontact is used to deliver electric current (i.e. macrostimulation). This macrostimulation paradigm serves to verify the presence of a meaningful therapeutic window by determining efficacy in the motor symptom control as well as amplitude threshold for stimulation-induced side effects. Electrophysiologic testing is an invaluable tool to minimize targeting errors resulting from MRI distortion, image fusion error, and intraoperative brain shift of various causes.

Intraoperative MER is classically performed with the patient awake in order to map motor-evoked potentials and asses intraoperative stimulation effects. More recently, studies evaluated the feasibility of MER under general anesthesia (GA). The equivalent intraoperative yield of MER under GA when compared to local anesthesia (LA) was reported for PD (Tsai et al. 2019), Tourette syndrome (Bos et al. 2019), and Dystonia (Venkatraghavan et al. 2016) in both GPi and STN targeting. The possibility of performing MER in asleep patients is currently considered to be a valid alternative for those patients who cannot tolerate an awake procedure. In fact, the literature shows an ever-growing number of cases being performed asleep without intraoperative MER. A recent systematic review included 14 cohort studies involving 1523 PD patients comparing DBS done in LA and GA. The meta-analysis results showed that there were no significant differences between the GA and LA groups in motor score improvement and postoperative medication requirements. Additionally, there was no significant difference in the incidence of adverse events including postoperative speech impairment and intracranial hemorrhage. As expected, pneumocephalus was significantly lower in the GA group. In a subgroup analysis, there was no significant difference in clinical efficacy between MER and non-MER groups. The review suggested that MER might not be necessary for successful DBS implantation (Liu et al. 2019). A larger review included 145 studies ($n = 2563$) with DBS for PD, 16 performed under GA and the rest in LA. There was a significantly lower number of DBS lead passes, intracerebral hemorrhages, and infections with GA. There were no significant differences in clinical motor outcomes and medication reduction between both techniques. The awake DBS cohort had a significantly greater decrease in treatment-related side effects as measured by the UPDRS off medication score (78.4% awake vs. 59.7% asleep, $p = 0.022$) supporting asleep DBS as an acceptable alternative to the awake procedure (Ho et al. 2018). The finding that asleep DBS was associated with a narrower therapeutic window is not surprising given that intraoperative macrostimulation cannot be performed to ensure a robust therapeutic window.

3.4 Directional DBS

Deep brain stimulation for neurologic diseases continues to expand with rapid adaptation and major technological advances. It is a fully reversible, surgical technique that allows the surgeon to target different structures without destroying the target. For these reasons, as well as consistent therapeutic benefit, new targets, innovative stimulation paradigms, and novel devices are being developed. One example is the advent of DBS leads with the ability to steer current, potentially mitigating surgical problems such as suboptimal lead placement and narrow therapeutic windows. The first DBS leads were composed of four cylindrical contacts that delivered stimulation in an omnidirectional fashion. The new electrodes are composed of two cylindrical and six segmented contacts that allow steering of current titrated to the patient's needs (i.e. directional stimulation).

Currently, there are two FDA-approved systems that are able to deliver directional stimulation. Both have similar electrode configurations but differ in the way the current is delivered. One system uses a single current source (SCS) and the other multiple independent current sources (MICS). In the SCS system, the current is divided into the active contacts. Hence, the current will be divided according to the bioimpedance in the tissues surrounding the active contacts. This increases the likelihood of distorting the volume of activated tissue (VTA). On the other hand, the MICS system will deliver the current in a more controlled fashion. In the SCS system, the segmented contact can be active or inactive, whereas in the MICS system the current can be fractionalized horizontally and vertically in increments of 1° (Schüpbach et al. 2017). In the case of suboptimal lead placement, steering the current towards the intended target could improve efficacy and prevent the need for further surgical intervention for lead refinement. There is emerging evidence suggesting that directional DBS could enlarge the therapeutic window by means of steering the stimulation field away from structures responsible for side effects. In one pilot study, Steigerwald et al. investigated the effects of directional DBS in PD and showed that axially asymmetric stimulations can increase the therapeutic window and decrease the threshold for side effects as compared to ring-mode DBS (Steigerwald et al. 2016). PROGRESS, a recent large multicenter, randomized trial of DBS comparing omnidirectional versus directional stimulation (Merola et al. 2020) has shown the superiority of directional stimulation: the mean therapeutic window was 35% higher with directional stimulation (3.00 ± 1.39 vs. 2.22 ± 1.27 mA), while therapeutic current strength was reduced by 30% using the optimal directional configuration (Schnitzler et al. 2019)..

Just as lead technology has advanced, so too has pulse generator technology. Newer systems are able to stimulate using shorter pulse width and so a higher amplitude of stimulation can be delivered. There are reports in which this strategy yielded more efficacious treatment (Soh et al. 2019). Steigerwald et al. reported on a series of 15 PD patients with STN DBS using a 30 μs pulse width with the same efficacy but less side effects and less energy expenditure when compared to the usual paradigm of 60 μs (Steigerwald et al. 2018). Reich et al. explained that based on measured chronaxies and model data, the use of pulse widths lower than 60 μs in STN DBS would better focus the stimulation effect on smaller diameter axons adjacent to the electrode and therefore, cause less excitation of the pyramidal tract fibers (Reich et al. 2015). Implantable pulse generators (IPG) of newer generation offer more versatile stimulation paradigms; however, this may come at some expenses (Israeli-Korn et al. 2019). Schlichting et al. reported on a series of 281 IPGs from a single company and showed a shorter battery life (de Schlichting et al. 2019).

Simulation software is another area that may help to refine the advantages of current steerability over omnidirectional technology. Several tools are now available: Lead DBS, SureTune, and Guide XT. They provide a visual representation of the volume of tissue activated (VTA) and its relationship to surrounding critical circuitry and the implanted lead (Fig. 4). The field is adjusted according to DBS parameters and its centroid can be used to compare VTAs across different patients in order to better locate specific "sweet spots" and optimal DBS paradigms for symptom control. It is important to note that all VTA models available for programming have limitations since they do not account for individual patient tissue impedance and anisotropy (Howell and McIntyre 2017).

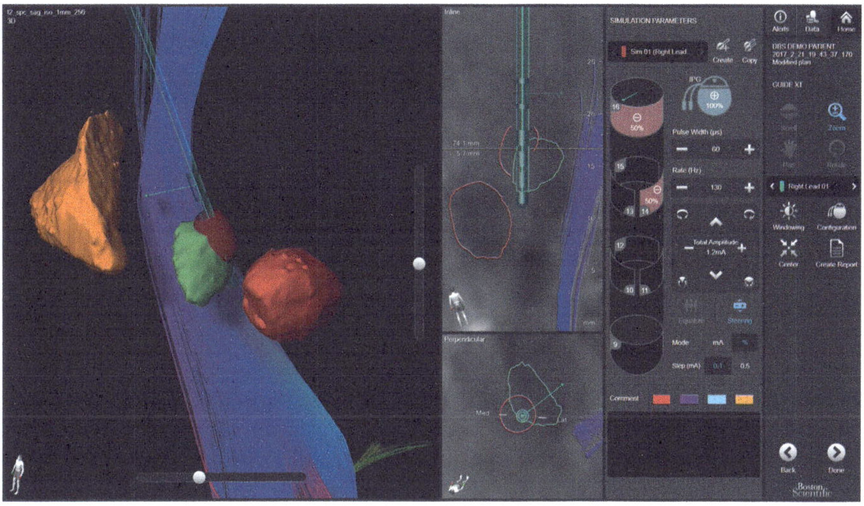

Fig. 4 Software ilustrating the STN (green) internal capsule(blue) and volume of tissue activation(dark red). Programing platform for directional DBS system allows custom field shaping and simultanous visualization of activated structures (courtesy to Boston Scientific)

Stimulation parameters and targeting optimization both play a critical role in treatment success and therefore, improvement in the quality of life. As an example, one can look at the effect of interleaving stimulation in which the simultaneous but desynchronized stimulation of the subthalamic nucleus (STN) and zona incerta (ZI) in PD patients improved tremor control, among other symptoms (França et al. 2019). Coordinated reset stimulation has been proposed as a new stimulation paradigm that can desynchronize neuronal populations using patterns of short pulses (Wang et al. 2016). This will allow to target pathological oscillatory network activity with both desynchronizing effects and associated therapeutic at the same time. More recently, conventional cathodic stimulation was compared to anodic simulation. The physiologic effect of anodic current is different since it first induces an axonal hyperpolarization before triggering ortho- and antidromic depolarization. Despite the higher energy expenditure of anodic stimulation, it has been shown that it may provide a better reduction of refractory off-period PD motor symptoms when compared to cathodic stimulation (Kirsch et al. 2018).

4 Closed-Loop DBS: Role of Local Field Potentials to Modulate Brain Function

4.1 Rationale

Despite the marked improvements that are seen with standard open-loop (OL) DBS in PD, there are several limitations with current strategies, including the need to decrease device power consumption (thereby increasing battery life) and to minimize stimulation-induced side effects in patients who continue on levodopa, that have driven the development of closed-loop (CL) stimulation paradigms. Current OL DBS strategies provide continuous stimulation at the same settings established during clinical visits by the treating neurologist and do not respond to the temporal variations in symptoms that patients experience during their daily medication use ("ON"/"OFF" periods) or to slower symptom variations due to disease progression or tachyphylaxis to medications. Additionally, it has been suggested that adjusting the temporal pattern of stimulation could spare residual functional physiological neural processing. For example, the physiological bursts of beta activity in the subthalamic nucleus related to decision conflict are suppressed by the continuous DBS stimulation and this is thought to lead to motor impulsivity that is seen in some patients after STN DBS (Herz et al. 2018). The idea behind CL DBS is that if somehow the patient's clinical state could be monitored or predicted, the DBS stimulation settings could be adjusted in such a way to provide symptomatic relief (increasing stimulation as symptoms worsen or decreasing stimulation if stimulation-induced side effects are being elicited). Various classes of signals have emerged as potential biomarkers including pathological neural activity (Swann et al. 2018; Kühn et al. 2006; Kuhn et al. 2004; Little et al. 2013, 2016a; Rosa et al. 2015) and peripheral measurements (Cagnan et al. 2015, 2016). Importantly, biomarkers need not be directly related to the underlying disease mechanism, but should co-vary with disease symptom severity and response to therapeutic interventions (Cagnan et al. 2019).

4.2 Potential Biomarkers for Closed-Loop DBS

The instantaneous power of the rhythmic neural activity within the beta band (13–30 Hz) can be tracked from the local field potential at the site of stimulation (Little et al. 2013, 2016b; Little and Brown 2012). Increasing evidence suggests that the beta frequency band can be consistently decoded from LFP recorded from the stimulating electrodes within STN and that the power in this frequency band correlates with the current state of motor impairment (rigidity-bradykinesia) (Little and Brown 2012). LFP analysis at the level of STN has also demonstrated the presence of increased signal power at the tremor frequency, which was well-localized within STN and coherent with peripheral EMG (Reck et al. 2009). Little et al. demonstrated the potential utility of the increased beta power as a marker of motor impairment by using a CL-DBS algorithm, where stimulation was administered until the beta power amplitude fell below a pre-specified threshold (Little et al. 2013). Using this CL-DBS paradigm, they were able to show a 29% (unblinded) and 27% (blinded) improvements in motor scores compared to a continuous stimulation strategy (OL DBS). Moreover, these improvements were achieved with a 56% reduction in the stimulation time compared to OL DBS, which corresponded to a significantly decreased total amount of electrical energy delivered (132 ± 21 μW vs. 270 ± 37 μW, $p < 0.0001$). Therefore, better clinical improvements were achieved with less than half the total electrical energy delivered. Unfortunately, beta band changes are strongly modulated by normal voluntary movements, which may complicate their use as signatures of motor impairment (Kuhn et al. 2004).

Biomarkers that correlate with clinical symptoms can be found in the motor cortex as well as the basal ganglia. Swann et al. characterized a narrowband gamma oscillation (60–90 Hz), detectable in the human motor cortex, which occurred following medication-induced dyskinesia and also dyskinesias occurring during DBS (Swann et al. 2016). More recently, they sought to test the utility of this gamma-band signal related to dyskinesia to decrease the stimulation voltage, when gamma oscillatory activity was high (indicating dyskinesia) and increase stimulation voltage, when it was low (Swann et al. 2018). During their short-term clinical testing on two patients that had been implanted with a DBS system that allowed brain sensing of this gamma-band signal from the motor cortex to control stimulation amplitude (Activa PC + S), they demonstrated a substantial 38–45% reduction in energy usage while maintaining therapeutic efficacy. Interestingly, it has been reported that patients with PD demonstrate exaggerated coupling between the beta-phase (13–30 Hz) and gamma-amplitude in M1 compared to patients with craniocervical dystonia and humans without movement disorders and that this excessive coupling may be reduced by therapeutic STN stimulation (de Hemptinne et al. 2013) Fig. 5.

Potential biomarkers may also be obtained with the use of wearable inertial sensors or electromyography (EMG) that capture clinical phenomena such as tremor or movement initiation. Herron et al. recently employed the use of an inertial measuring unit (consisting of a combined accelerometer and gyroscope) to detect arm movement and/or EMG to control when stimulation was applied to a patient with essential tremor (Herron et al. 2017). However, the use of external sensors has currently been limited by increased system complexity and additional power consumption that arises from the wireless communication between the implanted neurostimulator and the sensor. One potential drawback of external sensors is that despite capturing the clinical symptoms traditionally associated with PD (e.g. tremor and bradykinesia) directly, these symptoms are already present. As such, any CL-DBS strategies based on these external measures will not be predictive of the symptoms and will thus lag in performance compared to CL-DBS strategies based on internal signals that may precede symptom onset. Lastly, as of yet, no biomarker has been found that is related to all clinical features of PD and therefore, even with CL-DBS strategies, there will likely need to be flexibile to select an appropriate biomarker on a per-patient basis that captures the clinical features that are being targeted for treatment.

4.3 Current Applications

While there is much interest in the development of CL-DBS strategies, there are no commercial devices that allow these adaptive strategies to be deployed. Several recent studies have been able to deploy CL DBS within completely implanted sys-

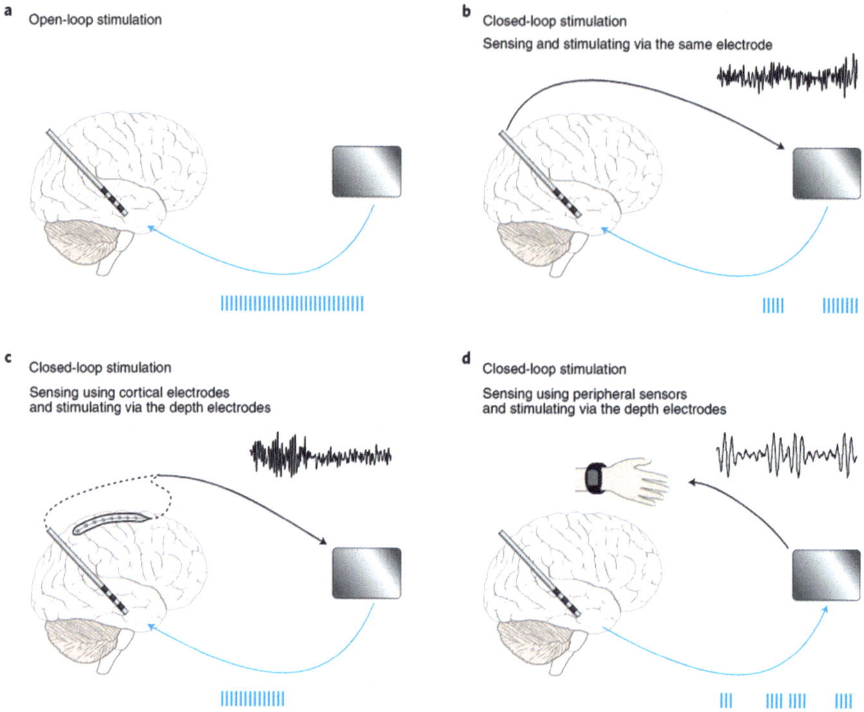

Fig. 5 A comparison of different stimulation strategies. (**a**) Stimulation timing and parameters are not automatically adjusted according to a disease biomarker, although the clinician will fine-tune stimulation during follow-up visits (usually twice a year). (**b**) Local field potentials sensed using depth electrodes are continuously used to automatically determine stimulation timing or intensity. Stimulation is delivered via the same depth electrodes. (**c**) Cortical signals sensed using an electrocorticography array are continuously used to automatically determine stimulation timing or intensity. Stimulation is delivered across the depth electrodes, creating a spatial separation between sensing and stimulation sites. (**d**) Peripheral signals obtained from noninvasive measurement devices, such as accelerometers and/or electromyography, are used to automatically determine stimulation timing or intensity. As in (**c**), this allows a separation between sensing and stimulation sites and therefore, minimizes the impact of stimulation artifacts. The gray box represents a computing device and could be an implantable pulse generator, a computer, or cloud-based computing. The computing device is used to process signals and extract features such as the intensity of neural activity in a certain frequency band or phase-amplitude coupling to control stimulation timing and parameters. (**Figure from** Cagnan et al. 2019 ** ***NEED TO GET PERMISSION OR RE-MAKE*****).https://s100.copyright.com/CustomerAdmin/PI.jsp?ref=4fd13048-8416-43ec-b021-b816849fdcf4

tems within the context of limited research studies using the Activa PC + S device (Swann et al. 2016, 2018; de Hemptinne et al. 2013; Herron et al. 2017). There is hope that subsequent generations of DBS hardware incorporate the ability to sense LFPs at the site of stimulation or allow the implantation of cortical electrodes that could be used to implement CL-DBS strategies in a clinical setting. Early this year, the first device that is able to record FFP became available for clinical practice in

Europe. There has already been some early success with closed-loop neuromodulation strategies for other pathologies, such as for the treatment of pain with spinal cord stimulation (Russo et al. 2018; Schade et al. 2011, 2012) and responsive neurostimulation for intractable epilepsy (Sun and Morrell 2014).

5 Conclusion

Deep brain stimulation for Parkinson's Disease has stood the test of time. Multiple studies have shown a robust, consistent improvement in motor symptoms of the disease, and significant improvement in the quality of life. The inherent advantages of this therapy make it very appealing, including reversibility, titratability, immediacy, consistency, and beneficial safety profile. Given these qualities, the field of deep brain stimulation will continue to expand rapidly beyond movement disorders. This chapter has provided a review of current innovations as it relates to imaging, planning, surgical techniques, stimulation paradigms, as well as future directions. From its original description by Luis Alim Benabid, deep brain stimulation has become a standard of care in the surgical treatment of movement disorders. DBS has evolved from a simple reversible lesion to a complex platform that modulates with increasing spatial and temporal resolution in the pathological brain networks. Advancements in imaging methods, such as high-field MRI, use of tractography to map neuroanatomical structures, and use of high-resolution electrodes will help refine the spatial accuracy of DBS. In the future, the adaptive closed-loop stimulation will allow real-time adjustment of stimulation based on the feedback from brain recordings or external wearable sensors and implementation of automated programming algorithms that will lead to optimal therapy.

References

Benabid AL, Cinquin P, Lavalle S, Le Bas JF, Demongeot J, de Rougemont J (1987) Computer-driven robot for stereotactic surgery connected to CT scan and magnetic resonance imaging. Technological design and preliminary results. Appl Neurophysiol 50(1–6):153–154

Blumenfeld Z, Bronte-Stewart H (2015) High frequency deep brain stimulation and neural rhythms in Parkinson's disease. Neuropsychol Rev 25(4):384–397

Bos MJ, Alzate Sanchez AM, Smeets AYJM, Bancone R, Ackermans L, Absalom AR et al (2019) Effect of anesthesia on microelectrode recordings during deep brain stimulation surgery in tourette syndrome patients. Stereotact Funct Neurosurg 97(4):225–231

Boutet A, Hancu I, Saha U, Crawley A, Xu DS, Ranjan M et al (2019) 3-Tesla MRI of deep brain stimulation patients: safety assessment of coils and pulse sequences. J Neurosurg 132(2):586–594

Cagnan H, Brown P, Bourget D, Denison T (2015) Inertial-based control system concepts for the treatment of movement disorders. Int Solid State Sens Actuators Microsyst Conf 18:70–73

Cagnan H, Pedrosa D, Little S, Pogosyan A, Cheeran B, Aziz T et al (2016) Stimulating at the right time: phase-specific deep brain stimulation. Brain J Neurol 140(Pt 1):132–145

Cagnan H, Denison T, McIntyre C, Brown P (2019) Emerging technologies for improved deep brain stimulation. Nat Biotechnol 37(9):1024–1033

Candela S, Vanegas MI, Darling A, Ortigoza-Escobar JD, Alamar M, Muchart J et al (2018) Frameless robot-assisted pallidal deep brain stimulation surgery in pediatric patients with movement disorders: precision and short-term clinical results. J Neurosurg Pediatr 22(4):416–425

Cernera S, Okun MS, Gunduz A (2019) A review of cognitive outcomes across movement disorder patients undergoing deep brain stimulation. Front Neurol 10:419

Coenen VA, Varkuti B, Parpaley Y, Skodda S, Prokop T, Urbach H et al (2017) Postoperative neuroimaging analysis of DRT deep brain stimulation revision surgery for complicated essential tremor. Acta Neurochir 159(5):779–787

Combs HL, Folley BS, Berry DT, Segerstrom SC, Han DY, Anderson-Mooney AJ et al (2015) Cognition and depression following deep brain stimulation of the subthalamic nucleus and globus pallidus pars internus in Parkinson's disease: a meta-analysis. Neuropsychol Rev 25(4):439–454

Cordeiro JG, Diaz A, Davis JK, Di Luca DG, Farooq G, Luca CC et al (2020) Safety of noncontrast imaging–guided deep brain stimulation electrode placement in Parkinson disease. World Neurosurg 134:e1008–e1e14

de Hemptinne C, Ryapolova-Webb ES, Air EL, Garcia PA, Miller KJ, Ojemann JG et al (2013) Exaggerated phase-amplitude coupling in the primary motor cortex in Parkinson disease. Proc Natl Acad Sci U S A 110(12):4780–4785

de Schlichting E, Coll G, Zaldivar-Jolissaint JF, Coste J, Marques AR, Mulliez A et al (2019) Pulse generator battery life in deep brain stimulation: out with the old... in with the less durable? Acta Neurochir 161(10):2043–2046

Deuschl G, Schade-Brittinger C, Krack P, Volkmann J, Schafer H, Botzel K et al (2006) A randomized trial of deep-brain stimulation for Parkinson's disease. N Engl J Med 355(9):896–908

Eusebio A, Thevathasan W, Gaynor LD, Pogosyan A, Bye E, Foltynie T et al (2010) Deep brain stimulation can suppress pathological synchronisation in parkinsonian patients. J Neurol Neurosurg Psychiatry 82(5):569–573

Faraji AH, Kokkinos V, Sweat JC, Crammond DJ, Richardson RM. Robotic-Assisted Stereotaxy for Deep Brain Stimulation Lead Implantation in Awake Patients. Oper Neurosurg (Hagerstown). 2020 Sep 15;19(4):444-452. https://doi.org/10.1093/ons/opaa029. PMID: 32147722

França C, Barbosa ER, Iglesio R, Teixeira MJ, Cury RG (2019) Interleaving stimulation in Parkinson disease: interesting to whom? World Neurosurg 130:e786–e793

Hell F, Palleis C, Mehrkens JH, Koeglsperger T, Bötzel K (2019) Deep brain stimulation programming 2.0: future perspectives for target identification and adaptive closed loop stimulation. Front Neurol 10:314

Herron JA, Thompson MC, Brown T, Chizeck HJ, Ojemann JG, Ko AL (2017) Chronic electrocorticography for sensing movement intention and closed-loop deep brain stimulation with wearable sensors in an essential tremor patient. J Neurosurg 127(3):580–587

Herz DM, Little S, Pedrosa DJ, Tinkhauser G, Cheeran B, Foltynie T et al (2018) Mechanisms underlying decision-making as revealed by deep-brain stimulation in patients with Parkinson's disease. Curr Biol 28(8):1169–78.e6

Ho AL, Ali R, Connolly ID, Henderson JM, Dhall R, Stein SC et al (2018) Awake versus asleep deep brain stimulation for Parkinson's disease: a critical comparison and meta-analysis. J Neurol Neurosurg Psychiatry 89:687–691

Howell B, McIntyre CC (2017) Role of soft-tissue heterogeneity in computational models of deep brain stimulation. Brain Stimul 10(1):46–50

Israeli-Korn SD, Fay-Karmon T, Tessler S, Yahalom G, Benizri S, Strauss H et al (2019) Decreasing battery life in subthalamic deep brain stimulation for Parkinson's disease with repeated replacements: just a matter of energy delivered? Brain Stimul 12(4):845–850

Kirsch AD, Hassin-Baer S, Matthies C, Volkmann J, Steigerwald F (2018) Anodic versus cathodic neurostimulation of the subthalamic nucleus: a randomized-controlled study of acute clinical effects. Parkinsonism Relat Disord 55:61–67

Kuhn AA, Williams D, Kupsch A, Limousin P, Hariz M, Schneider GH et al (2004) Event-related beta desynchronization in human subthalamic nucleus correlates with motor performance. Brain 127(Pt 4):735–746

Kühn AA, Kupsch A, Schneider G-H, Brown P (2006) Reduction in subthalamic 8-35 Hz oscillatory activity correlates with clinical improvement in Parkinson's disease. Eur J Neurosci 23(7):1956–1960

Little S, Brown P (2012) What brain signals are suitable for feedback control of deep brain stimulation in Parkinson's disease? Ann N Y Acad Sci 1265(1):9–24

Little S, Pogosyan A, Neal S, Zavala B, Zrinzo L, Hariz M et al (2013) Adaptive deep brain stimulation in advanced Parkinson disease. Ann Neurol 74(3):449–457

Little S, Beudel M, Zrinzo L, Foltynie T, Limousin P, Hariz M et al (2016a) Bilateral adaptive deep brain stimulation is effective in Parkinson's disease. J Neurol Neurosurg Psychiatry 87(7):717

Little S, Tripoliti E, Beudel M, Pogosyan A, Cagnan H, Herz D et al (2016b) Adaptive deep brain stimulation for Parkinson's disease demonstrates reduced speech side effects compared to conventional stimulation in the acute setting. J Neurol Neurosurg Psychiatry 87(12):1388–1389

Liu Z, He S, Li L (2019) General anesthesia versus local anesthesia for deep brain stimulation in Parkinson's disease: a meta-analysis. Stereotact Funct Neurosurg 97(5–6):381–390

McIntyre CC, Savasta M, Walter BL, Vitek JL (2004) How does deep brain stimulation work? Present understanding and future questions. J Clin Neurophysiol 21(1):40–50

Merola A, Romagnolo A, Krishna V, Pallavaram S, Carcieri S, Goetz S et al (2020) Current directions in deep brain stimulation for Parkinson's disease—directing current to maximize clinical benefit. Neurol Ther 9(1):25–41

Morishita T, Higuchi M-A, Kobayashi H, Abe H, Higashi T, Inoue T (2019) A retrospective evaluation of thalamic targeting for tremor deep brain stimulation using high-resolution anatomical imaging with supplementary fiber tractography. J Neurol Sci 398:148–156

Neudorfer C, Hunsche S, Hellmich M, El Majdoub F, Maarouf M (2018) Comparative study of robot-assisted versus conventional frame-based deep brain stimulation stereotactic neurosurgery. Stereotact Funct Neurosurg 96(5):327–334

Odekerken VJ, van Laar T, Staal MJ, Mosch A, Hoffmann CF, Nijssen PC et al (2013) Subthalamic nucleus versus globus pallidus bilateral deep brain stimulation for advanced Parkinson's disease (NSTAPS study): a randomised controlled trial. Lancet Neurol 12(1):37–44

Okun MS, Gallo BV, Mandybur G, Jagid J, Foote KD, Revilla FJ et al (2012) Subthalamic deep brain stimulation with a constant-current device in Parkinson's disease: an open-label randomised controlled trial. Lancet Neurol 11(2):140–149

Ramirez-Zamora A, Ostrem JL (2018) Globus pallidus interna or subthalamic nucleus deep brain stimulation for Parkinson disease: a review. JAMA Neurol 75(3):367–372

Ranjan M, Boutet A, Xu DS, Lozano CS, Kumar R, Fasano A et al (2018) Subthalamic nucleus visualization on routine clinical preoperative MRI scans: a retrospective study of clinical and image characteristics predicting its visualization. Stereotact Funct Neurosurg 96(2):120–126

Reck C, Florin E, Wojtecki L, Krause H, Groiss S, Voges J et al (2009) Characterisation of tremor-associated local field potentials in the subthalamic nucleus in Parkinson's disease. Eur J Neurosci 29(3):599–612

Reich MM, Steigerwald F, Sawalhe AD, Reese R, Gunalan K, Johannes S et al (2015) Short pulse width widens the therapeutic window of subthalamic neurostimulation. Ann Clin Transl Neurol 2(4):427–432

Rosa M, Arlotti M, Ardolino G, Cogiamanian F, Marceglia S, Di Fonzo A et al (2015) Adaptive deep brain stimulation in a freely moving Parkinsonian patient. Mov Disord 30(7):1003–1005

Russo M, Cousins MJ, Brooker C, Taylor N, Boesel T, Sullivan R et al (2018) Effective relief of pain and associated symptoms with closed-loop spinal cord stimulation system: preliminary results of the avalon study. Neuromodulation 21(1):38–47

Schade CM, Schultz DM, Tamayo N, Iyer S, Panken E (2011) Automatic adaptation of neurostimulation therapy in response to changes in patient position: results of the Posture Responsive Spinal Cord Stimulation (PRS) Research Study. Pain Physician 14(5):407–417

Schlenstedt C, Shalash A, Muthuraman M, Falk D, Witt K, Deuschl G (2017) Effect of high-frequency subthalamic neurostimulation on gait and freezing of gait in Parkinson's disease: a systematic review and meta-analysis. Eur J Neurol 24(1):18–26

Schnitzler AS, Mir PM, Brodsky MB, Verhagen LV, Groppa SG, Cheeran BC, Karst EK, Defresne FD, Vesper JV (2019) Directional versus omnidirectional deep brain stimulation for Parkinson's disease: results of a prospective, blinded, multi-center, single-arm crossover study [abstract]. Mov Disord 34(Suppl 2)

Schuepbach WM, Rau J, Knudsen K, Volkmann J, Krack P, Timmermann L et al (2013) Neurostimulation for Parkinson's disease with early motor complications. N Engl J Med 368(7):610–622

Schultz DM, Webster L, Kosek P, Dar U, Tan Y, Sun M (2012) Sensor-driven position-adaptive spinal cord stimulation for chronic pain. Pain Physician 15(1):1–12

Schüpbach WMM, Chabardes S, Matthies C, Pollo C, Steigerwald F, Timmermann L et al (2017) Directional leads for deep brain stimulation: opportunities and challenges. Mov Disord 32(10):1371–1375

Soh D, Lozano AM, Fasano A (2019) Hybrid deep brain stimulation system to manage stimulation-induced side effects in essential tremor patients. Parkinsonism Relat Disord 58:85–86

Steigerwald F, Muller L, Johannes S, Matthies C, Volkmann J (2016) Directional deep brain stimulation of the subthalamic nucleus: a pilot study using a novel neurostimulation device. Mov Disord 31(8):1240–1243

Steigerwald F, Timmermann L, Kühn A, Schnitzler A, Reich MM, Kirsch AD et al (2018) Pulse duration settings in subthalamic stimulation for Parkinson's disease. Mov Disord 33(1):165–169

Sudhyadhom A, Haq IU, Foote KD, Okun MS, Bova FJ (2009) A high resolution and high contrast MRI for differentiation of subcortical structures for DBS targeting: the Fast Gray Matter Acquisition T1 Inversion Recovery (FGATIR). Neuroimage 47(Suppl 2):T44–T52

Sun FT, Morrell MJ (2014) Closed-loop neurostimulation: the clinical experience. Neurotherapeutics 11(3):553–563

Swann NC, de Hemptinne C, Miocinovic S, Qasim S, Wang SS, Ziman N et al (2016) Gamma oscillations in the hyperkinetic state detected with chronic human brain recordings in Parkinson's disease. J Neurosci 36(24):6445–6458

Swann NC, Hemptinne C, Thompson MC, Miocinovic S, Miller AM, Gilron R et al (2018) Adaptive deep brain stimulation for Parkinson's disease using motor cortex sensing. J Neural Eng 15(4):046006

Tepper Á, Henrich MC, Schiaffino L, Rosado Muñoz A, Gutiérrez A, Guerrero MJ (2017) Selection of the optimal algorithm for real-time estimation of beta band power during DBS surgeries in patients with Parkinson's disease. Comput Intell Neurosci 2017:1512504

Thevathasan W, Moro E (2019) What is the therapeutic mechanism of pedunculopontine nucleus stimulation in Parkinson's disease? Neurobiol Dis 128:67–74

Thompson JA, Yin D, Ojemann SG, Abosch A (2017) Use of the putamen as a surrogate anatomical marker for the internal segment of the globus pallidus in deep brain stimulation surgery. Stereotact Funct Neurosurg 95(4):229–235

Tsai S-T, Chen T-Y, Lin S-H, Chen S-Y (2019) Five-year clinical outcomes of local versus general anesthesia deep brain stimulation for Parkinson's disease. Parkinsons Dis 2019:5676345

Valldeoriola F, Munoz E, Rumia J, Roldan P, Camara A, Compta Y et al (2019) Simultaneous low-frequency deep brain stimulation of the substantia nigra pars reticulata and high-frequency stimulation of the subthalamic nucleus to treat levodopa unresponsive freezing of gait in Parkinson's disease: a pilot study. Parkinsonism Relat Disord 60:153–157

Vassal F, Dilly D, Boutet C, Bertholon F, Charier D, Pommier B (2020) White matter tracts involved by deep brain stimulation of the subthalamic nucleus in Parkinson's disease: a connectivity study based on preoperative diffusion tensor imaging tractography. Br J Neurosurg 34(2):187–195

Venkatraghavan L, Rakhman E, Krishna V, Sammartino F, Manninen P, Hutchison W (2016) The effect of general anesthesia on the microelectrode recordings from pallidal neurons in patients with dystonia. J Neurosurg Anesthesiol 28(3):256–261

Vitek J, Jain R, Starr P (2019) Two year outcomes: a prospective, double blinded, multicenter randomized controlled trial evaluating deep brain stimulation with a new multiple source, constant current rechargeable system in Parkinson's disease (INTREPID) (P1.8-026). Neurology 92(15 Suppl):P1.8-026

Vitek J, Jain R, Starr P (2020) Three-year follow-up of a prospective, double-blinded, multi-center randomized controlled trial evaluating deep brain stimulation with multiple source, constant-current rechargeable system for treatment of Parkinson's disease (INTREPID) (1365). Neurology 94(15 Suppl):1365

Wang J, Nebeck S, Muralidharan A, Johnson MD, Vitek JL, Baker KB (2016) Coordinated reset deep brain stimulation of subthalamic nucleus produces long-lasting, dose-dependent motor improvements in the 1-methyl-4-phenyl-1,2,3,6-tetrahydropyridine non-human primate model of parkinsonism. Brain Stimul 9(4):609–617

Wang JW, Zhang YQ, Zhang XH, Wang YP, Li JP, Li YJ (2017) Deep brain stimulation of pedunculopontine nucleus for postural instability and gait disorder after Parkinson disease: a meta-analysis of individual patient data. World Neurosurg 102:72–78

Weaver FM, Follett K, Stern M, Hur K, Harris C, Marks WJ Jr et al (2009) Bilateral deep brain stimulation vs best medical therapy for patients with advanced Parkinson disease: a randomized controlled trial. JAMA 301(1):63–73

Williams A, Gill S, Varma T, Jenkinson C, Quinn N, Mitchell R et al (2010) Deep brain stimulation plus best medical therapy versus best medical therapy alone for advanced Parkinson's disease (PD SURG trial): a randomised, open-label trial. Lancet Neurol 9(6):581–591

Yasin H, Hoff H-J, Blümcke I, Simon M (2019) Experience with 102 frameless stereotactic biopsies using the neuromate robotic device. World Neurosurg 123:e450–e456

Zuzuarregui JRP, Ostrem JL (2020) The impact of deep brain stimulation on sleep in Parkinson's disease: an update. J Parkinsons Dis 10(2):393–404

Neuromodulation for Gait Disorders

Stephano J. Chang, Ioan Opris, James D. Guest, and Brian R. Noga

1 Introduction

Despite ever more impressive advances in transportation technologies, bipedal gait remains our most essential and versatile means of mobilizing within and navigating local environments and is instrumental to our daily functioning. While few of us give a second thought about our ability to walk and balance in an upright position, gait disorders greatly affect quality of life, and are associated with falls and numerous medical comorbidities (Tinetti et al. 1988). Neurological gait disorders represent a

The original version of this chapter was revised. The correction to this chapter is available at https://doi.org/10.1007/978-3-030-54564-2_31

S. J. Chang
Neuroscience Graduate Program, University of Miami Miller School of Medicine, Miami, FL, USA

The Miami Project to Cure Paralysis, University of Miami Miller School of Medicine, Miami, FL, USA

Division of Neurosurgery, Department of Surgery, University of British Columbia, Vancouver, BC, Canada

I. Opris
Department of Neurosurgery, University of Miami, Miami, FL, USA

J. D. Guest · B. R. Noga (✉)
Neuroscience Graduate Program, University of Miami Miller School of Medicine, Miami, FL, USA

The Miami Project to Cure Paralysis, University of Miami Miller School of Medicine, Miami, FL, USA

Department of Neurological Surgery, University of Miami Miller School of Medicine, Miami, FL, USA
e-mail: bnoga@med.miami.edu

I. Opris et al. (eds.), *Modern Approaches to Augmentation of Brain Function*, Contemporary Clinical Neuroscience,
https://doi.org/10.1007/978-3-030-54564-2_23

diverse group of etiologies, ranging from neurodegenerative diseases such as Parkinson's disease (PD), to traumatic injuries such as spinal cord injury (SCI), and reflect the widespread involvement of the nervous system in the control of gait. Neuromodulatory techniques, such as deep brain stimulation (DBS), which have demonstrated striking results in the treatment of movement disorders, represent a promising category in the treatment of neurological gait disorders. While the initial successes of DBS were empirical in nature, greater understanding of the neural circuits controlling locomotion, and recognition of where they fail in gait disorders, will allow us to develop new strategies to compensate for pathologies and augment gait function. This chapter will review the current literature regarding the neural control of locomotion, common neurological gait disorders, and the research behind the neuromodulation strategies being investigated for the augmentation of gait function.

2 Key Aspects of Gait: A brief Overview

Bipedal gait is a complex motor function that can be broken down conceptually both mechanically and by its component modules: initiation, termination, acceleration/deceleration, turning, and balance. Mechanically, gait can be described in terms of a step cycle with a stance phase that begins with a heel strike and ends with a toe-off into a swing phase before the next stance phase (Fig. 1). Initiation of gait begins with an anticipatory postural adjustment (APA) of the body's center of mass (CoM) over the leg beginning in stance so that the swinging leg can be lifted while maintaining sagittal balance (Santos et al. 2010). Termination of gait can be an active or passive process, and can take place abruptly mid step cycle, or with a natural completion of the cycle to a neutral standing position. Gait acceleration and deceleration are determined by parameters such as push-off force, stride length, and cadence (stepping frequency), with walking and running differentiated by whether at least 1 ft always remains in contact with the ground during the step cycle. Turning usually follows a sequential pattern of head, trunk, and pelvis rotation, and in bipedal gait, is primarily facilitated by combinations of internal and external hip rotations in the stance and swing legs to pivot slightly toward the turning direction over the course of each step (Hong et al. 2009).

Finally, dynamic balance during gait entails the active maintenance of the CoM over a moving base of support (MacKinnon 2018). Center of pressure (CoP) is defined as the centroid of all external forces acting on the plantar surface of the foot to keep the CoM over the base of support, and CoP measures have been identified as a reliable proxy for postural and balance control (Lugade and Kaufman 2014). In particular, CoP velocity has been proposed as the most sensitive measure of detecting changes in balance ability due to aging and neurological diseases (Masani et al. 2007; Rocchi et al. 2006). Balance involves numerous mechanisms, such as the aforementioned APAs, crossed synchrony between leg movements and arm swings, and several postural reflexes that correct perturbations to the CoM, including cutaneous reflexes in the feet and hands to detect and avoid obstacles (Zehr et al. 1997), and vestibular, proprioceptive, and ocular righting reflexes (Takakusaki 2017). While it is remarkable to consider that many if not all these processes can be automatic and take place with little to no thought during gait, it is important to note

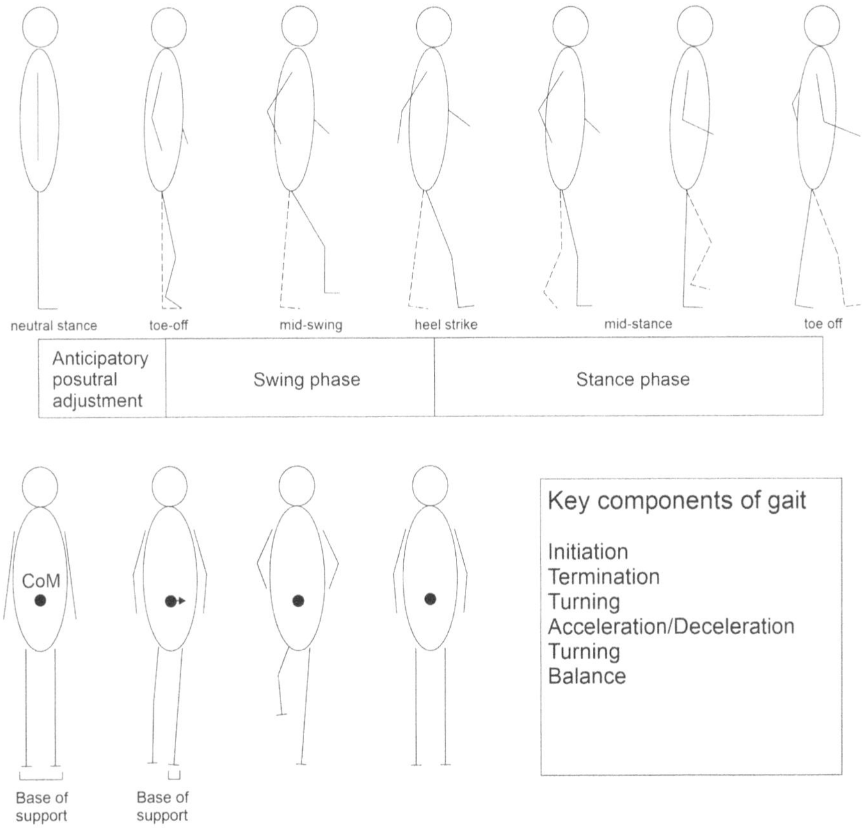

Fig. 1 The gait cycle and key aspects of gait. Illustrated are lateral (above) and anteroposterior (below) views of the gait cycle, broken down into swing and stance phases. An anticipatory postural adjustment (APA) is required to shift the center of mass (CoM) and maintain sagittal balance over the base of support prior to taking the first step. Gait is a complex motor function that consists of several different modules, such as initiation, termination, acceleration/deceleration, turning, and balance

that there also is an executive and cognitive aspect to gait. This is especially evident during more complex tasks such as turning, passing through narrow corridors, and when navigating obstacles (Ho et al. 2019; Chen et al. 1996), and further evidenced by impairment of gait parameters during dual-tasking (Lee 2017), as well as by the fact that we can voluntarily modulate our gait parameters executively.

3 Neural Control of Locomotion

Much of the scientific groundwork in our understanding of the neural control of vertebrate locomotion stems from early experiments in cats, using electrical or pharmacological stimulation along with various transections along the neuraxis to infer the contributions of different neural structures to locomotor control (Shik and Orlovsky 1976). Besides a few key species differences that will be discussed below, the relevance of these pioneering studies to our understanding of human locomotion is supported by multiple electrophysiological and functional imaging studies, which suggest that the basic features of quadrupedal locomotion, including spinal and supraspinal mechanisms, remain conserved in human bipedal gait (Jahn et al. 2008a; Dietz et al. 2001; Wannier et al. 2001; Debaere et al. 2001; Dietz and Michel 2009).

3.1 Spinal Mechanisms Controlling Gait: Central Pattern Generators

Experiments in cats with low thoracic spinal transections and deafferentation of the hind limbs have revealed that fictive locomotion, or alternating neural activity of hind limb flexors and extensors, can be elicited even in the absence of descending or afferent inputs to the spinal cord (Grillner and Zangger 1979; Brown and Sherrington 1911). This gave credence to the concept of central pattern generators (CPGs) residing in the spinal cord—neuronal circuits that could intrinsically produce the stereotyped rhythmic and patterned motor output for gait, as well as flying and swimming in other species (Steuer and Guertin 2019).

Functionally, CPG circuits are composed of rhythm-generating circuits, which produce the intrinsic drive and rhythm for locomotion, and pattern-generating circuits, which then distribute the generated rhythms to motoneurons to produce flexor-extensor and left-right alternations or synchronies (Kiehn 2016). In mice, molecular genetics and electrophysiological approaches have identified several distinct classes of spinal interneurons thought to comprise the locomotor CPGs, including the 'core' V0, V1, V2, and V3 interneurons (Goulding 2009). For example, genetic ablation experiments suggest that V0 commissural interneurons crossing the midline are crucial for generating left-right alternation (Kiehn 2016). Descending motor pathways, such as the corticospinal tract (CST) and the reticulospinal tract (RST), project to CPG interneurons and motoneurons, to allow for supraspinal control of locomotion (Guertin 2013).

Locomotor CPGs are also modulated by afferent input from the limbs. In fact, spinal cord transected kittens with intact hind limb afferents can be trained to walk on a moving treadmill, indicating that the afferent input from the hind limbs is sufficient to regulate its gait to match the speed of the treadmill (Shik and Orlovsky 1976). This locally autoregulating feature of CPGs allows for environmentally adaptive locomotor function and may serve to simplify the supraspinal control of locomotion (Fig. 2).

Despite these advances in our understanding of the organization of CPGs in animals, the evidence for CPGs in humans is indirect and remains under some debate (Minassian et al. 2017). The limited recovery of gait in patients with spinal cord injury (SCI) is argued by some to suggest a greater importance of supraspinal mech-

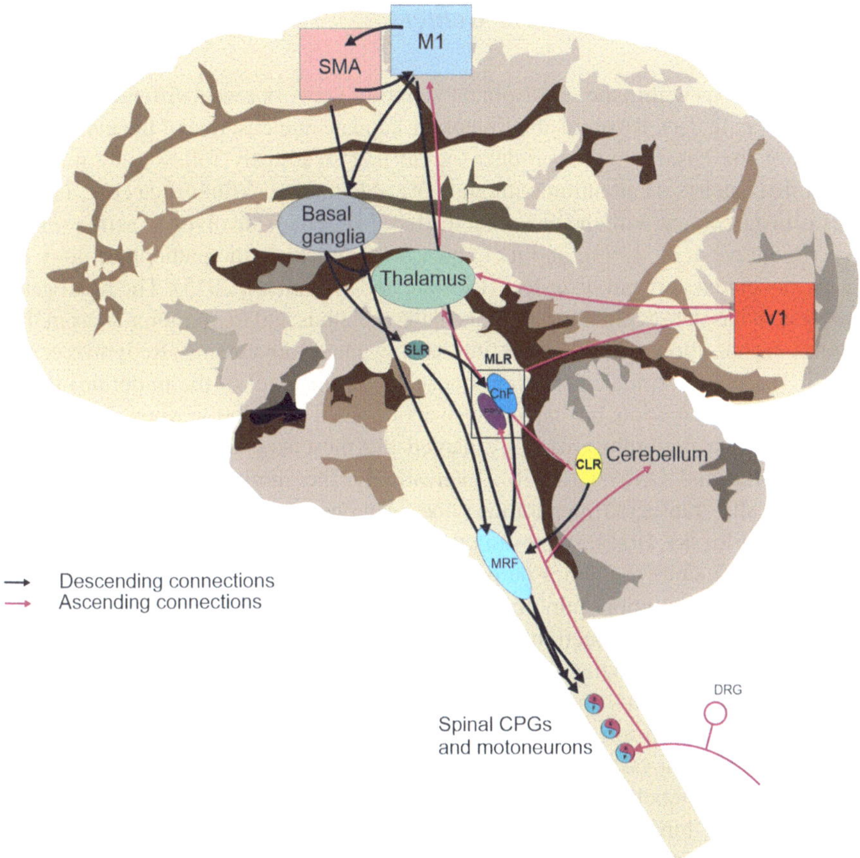

Fig. 2 Schematic of the neural control of locomotion. This drawing shows the cortical, subcortical, brainstem, and spinal cord regions involved in the neural control of locomotion, with select important connections between regions illustrated. *SMA* supplementary motor area, *M1* primary sensorimotor cortex, *V1* primary visual cortex, *SLR* subthalamic locomotor region, *MLR* mesencephalic locomotor region, *CnF* cuneiform nucleus, *PPN* pedunculopontine nucleus, *CLR* cerebellar locomotor region, *MRF* medullary reticular formation, *CPG* central pattern generator, *DRG* dorsal root ganglion

anisms over spinal CPGs in human locomotion (Duysens and Van de Crommert 1998), although observations of involuntary, rhythmic locomotor-like leg movements in patients with SCI, either spontaneously or with electrical spinal cord stimulation, are also cited as evidence for the existence of CPGs in humans (Calancie 2006; Dietz et al. 1995; Bussel et al. 1988). Newborn infants also exhibit involuntary stepping through a reflex that is triggered when they are held upright with their feet placed on the floor (Forssberg 1985). This stepping reflex is proposed to reflect the activity of spinal locomotor CPGs prior to the maturation of descending pathways (Forssberg 1985; Ryczko and Dubuc 2013).

3.2 Supraspinal Locomotor Centers

Different lesions within the supraspinal locomotor network reveal different locomotor behaviors in cats. Bilateral ablation of the caudate nucleus, one of the main input centers of the basal ganglia, produces a cat that steadfastly follows any moving object that catches its attention without being able to stop (Villablanca et al. 1976). This is in agreement with the view that the basal ganglia is involved in action selection (Redgrave et al. 1999), and its known involvement in the pathophysiology of many movement and gait disorders (DeLong and Wichmann 2015). The basal ganglia system has dense inhibitory outputs to the thalamus and to premotor areas in the brainstem that allow it to tonically inhibit unwanted movements (Redgrave et al. 1999). The mesencephalic cat (brain transected at the level of the midbrain) lacks spontaneous locomotion, but can produce coordinated locomotion when electrically stimulated in a midbrain area aptly named the mesencephalic locomotor region (MLR) (Shik and Orlovsky 1976). Interestingly, the subthalamic cat (a slightly more rostrally transected preparation) does demonstrate spontaneous locomotion (Shik and Orlovsky 1976), suggesting that while the brainstem and spinal cord are sufficient to generate the motor output patterns for gait, more rostral structures are required for the initiation, inhibition, and disinhibition of walking behaviors. The region between these transections can also elicit locomotion with electrical stimulation and has been termed the subthalamic locomotor region (SLR). It corresponds to the lateral hypothalamus and zona incerta (ZI) area, which has connections to the MLR, but which can also elicit locomotion when the MLR is ablated (Shik and Orlovsky 1976; Orlovsky 1969).

Finally, the cerebellum, long known to contribute to the control of balance and gait, contains a third site where electrical stimulation elicits locomotion (Table 1) (Thomas Thach and Bastian 2004). This site is known as the cerebellar locomotor region (CLR), and is located near the fastigial nucleus and hook bundle of Russel in the midline cerebellar white matter (Mori et al. 1999). Combined stimulation of the CLR and MLR has an additive effect on locomotion, such that stimulating each site at individually subthreshold strengths can elicit regular locomotion, and stimulating

Table 1 Supraspinal sites initiating locomotion in mammals

Stimulation site	Putative anatomical targets	Animals studied in	Key references
Mesencephalic locomotor region	PPN	Cat, rat	Garcia-Rill et al. (1987)
	CnF	Cat, rat, salamander, stingray, macaque, pig	Shik et al. (1966), Takakusaki et al. (2016); Opris et al. (2019)
	Medioventral to PPN	Rat	Sherman et al. (2015)
Subthalamic locomotor region	Lateral hypothalamus, zona incerta, H field of Forel	Cat	Orlovsky (1969)
Cerebellar locomotor region	Hook bundle of Russell	Cat	Mori et al. (1999)

both at suprathreshold levels has a much stronger locomotor response (Mori et al. 1999). Ablation of either of these sites bilaterally does not block the locomotor effect of the other site, suggesting that these centers activate parallel pathways to the medullary reticular formation (MRF) to initiate locomotion. In fact, the reticulospinal neurons within the MRF are known to be innervated by multiple regions beyond the MLR, SLR, and CLR (Mori et al. 1999; Steeves and Jordan 1984; Sinnamon and Karen 1987), including periaqueductal gray (PAG) (Dampney et al. 2013), the superior colliculus (Furigo et al. 2010), vestibular nuclei, proprioceptive inputs from the limbs (Miller et al. 2017), as well as the motor cortex (Matsuyama et al. 2004), which will be discussed further below. Thus, locomotion may be initiated or modulated through multiple different independent pathways that converge onto the MRF (Fig. 2) (Opris et al. 2019).

Optogenetics studies in mice have revealed important details about neuronal populations in the MRF that mediate locomotion, including why electrical stimulation of the MRF generally fails to initiate locomotion despite being a well-known locomotor center (Capelli et al. 2017). Capelli et al. showed that locomotor functions in the medulla are segregated by neurotransmitter phenotype, and identified glutamatergic reticulospinal neurons within the lateral paragigantocellular (LPGi) nucleus that are necessary and sufficient for mediating MLR-evoked locomotor function to interneurons and motor neurons in the spinal cord (Capelli et al. 2017). Activation of these neurons elicited high-speed locomotion, while ablation significantly reduced the speed of MLR-evoked locomotion (Capelli et al. 2017). Capelli et al. also identified several groups of glycinergic and GABAergic neuronal subpopulations within the reticular nuclei that could stop locomotion (Capelli et al. 2017). Furthermore, the authors determined that while glutamatergic LPGi neurons primarily targeted laminae VII and VIII in the spinal cord (where pattern- and rhythm-generating interneurons reside), glycinergic subpopulations made significant connections directly onto motor neurons, acting as a final gate for locomotor output (Capelli et al. 2017). Although this close intermingling of inhibitory and excitatory populations in the medulla may hinder attempts to evoke locomotor activity electrically (Capelli et al. 2017), stimulation of medullary cells either electrically or pharmacologically in the cat can produce locomotion (Noga et al. 1988).

Bouvier et al. previously also identified a group of brainstem V2a neurons capable of halting locomotion; however, this parallel pathway was found to be glutamatergic, and primarily inhibited locomotion at the rhythm-generating layer of the locomotor CPGs through projections to lamina VII, VIII, and X (Bouvier et al. 2015). Interestingly, the latency for terminating locomotion through photoactivation of these neurons (140 ms) was long enough to allow the completion of the ongoing step, while activation of glycinergic LPGi neurons by Capelli et al. (2017) stopped locomotion mid-step, reflecting the distinct mechanisms of these pathways. Together, these studies demonstrate that multiple parallel descending reticulospinal pathways exist by which locomotion can be suppressed, and are more anatomically dispersed and mechanistically diverse than those promoting locomotion (Capelli et al. 2017).

The MLR is the most studied of the supraspinal locomotor centers, and has been identified as a phylogenetically conserved control center in multiple vertebrate species (Ryczko and Dubuc 2013), with electrophysiological and functional imaging evidence establishing its existence in humans (Jahn et al. 2008b; Piallat et al. 2009). It is located in the upper brainstem, where it regulates locomotion both through descending reticulospinal and monoaminergic circuits to spinal CPGs (Steeves and Jordan 1980, 1984; Noga et al. 2003, 2017), as well as through ascending connections to higher brain centers (Martinez-Gonzalez et al. 2011). These latter connections are hypothesized to couple attentional, arousal, and cortical states to various locomotor states (Lee et al. 2014).

Despite these insights from a large number of studies on the MLR, there is significant controversy in the literature as to the optimal location within the MLR for electrical stimulation to promote locomotion. Much of the animal literature, including the original description of the MLR in cats and electrical mapping studies, support a location near the dorsally located cuneiform nucleus (CnF) (Fig. 3) (Shik et al. 1966; Opris et al. 2019; Takakusaki et al. 2016); others advocate the importance of cholinergic neurons in the more ventrally located pedunculopontine nucleus (PPN) (Garcia-Rill et al. 1987), despite its mixed electrical mapping results (Takakusaki et al. 2016). A third putative site just medial to the PPN has also been

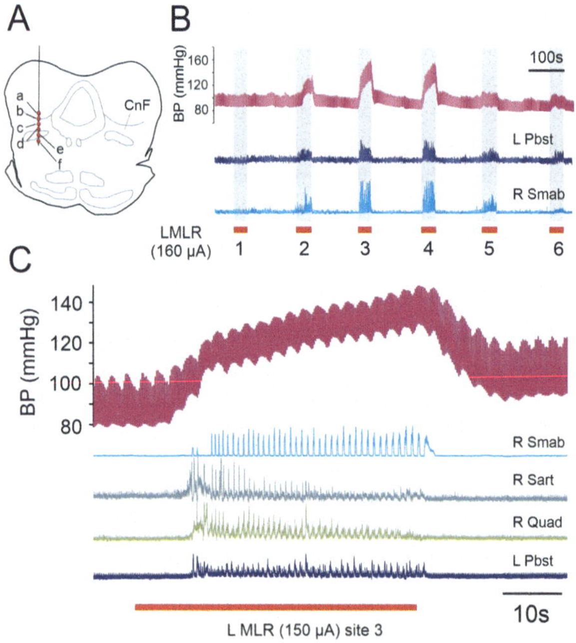

Fig. 3 Electrical mapping of the MLR and surrounding region in the cat. (Adapted from Opris et al. 2019, with permission). (**a**) Stimulation of the CnF (site c) and Sub-cuneiform (site d) produces the largest locomotor responses and associated pressor responses (from arterial line), shown in (**b**) (sites 3 and 4, respectively). (**c**) Detailed plot of blood pressure and EMG responses to stimulation of the CnF (b: site 3). *R* right, *L* left, *Pbst* posterior biceps/semitendinosus, *Smab* semimembranosus/anterior biceps, *Sart* sartorius, *Quad* quadriceps

proposed (Sherman et al. 2015). Interestingly, all clinical studies investigating this area for neuromodulation to alleviate gait disorders have targeted the PPN, and this potential mistargeting has been proposed as a reason for the variable outcomes to date (Thevathasan et al. 2018). In favor of this hypothesis, recent optogenetics studies in mice have shown that glutamatergic neurons near the CnF have the strongest effect on promoting locomotion, while activation of cholinergic neurons in the PPN failed to initiate locomotion (Josset et al. 2018; Caggiano et al. 2018). Furthermore, a recent anatomo-clinical correlation study involving DBS of the PPN area in Parkinson's patients with freezing of gait showed the best gait outcomes were associated with more dorsally located active electrode locations (Goetz et al. 2019).

3.3 Cortical Control of Locomotion

Functional neuroimaging studies such as single-photon emission computed tomography (SPECT), positron emission tomography (PET), and near-infrared spectroscopic topography (NIRS) demonstrate the importance of cortical control in locomotion, showing increased activity of the medial primary sensorimotor cortex, premotor cortex (PMC), supplementary motor area (SMA), and visual cortex in healthy human subjects during running and walking (Hanakawa et al. 1999a, b; Fukuyama et al. 1997; Tashiro et al. 2001; Miyai et al. 2001). To distinguish between cortical activity involved in the generation of locomotion as opposed to sensory processing, Christensen et al. performed similar PET scan experiments in subjects during active and passive bicycling (Christensen et al. 2000). By subtracting these datasets, the authors showed that while much of the observed cortical activity was likely due to sensory feedback from the moving limbs, there was significant additional activity in the leg region of the primary motor cortex during active cycling, suggesting that neurons in this area are involved in generating rhythmic locomotor-like leg activity (Christensen et al. 2000). Furthermore, functional MRI (fMRI) and PET during imaginary gait or bicycling movements increased activity of the SMA when compared to rest, suggesting that the SMA plays a role in the planning of these movements (Miyai et al. 2001; Christensen et al. 2000). Asking participants to imagine walking along a curved path also activated the visual cortex and parahippocampal and fusiform gyri on fMRI—regions considered to be involved in visuospatial navigation (Jahn et al. 2004).

Of the descending anatomical pathways connecting these cortical regions to the brainstem and spinal cord, the corticospinal tract (CST) is considered to be the most important in mediating voluntary movements in humans, especially skilled movements involving distal muscles (Welniarz et al. 2017; Davidoff 1990). Thus, it is generally accepted and taught in clinical neurology that CST lesions, such as stroke or spinal cord injury, result in significant and long-lasting, spastic deficits of voluntary motor control, including gait (Eidelberg 1981). It may be of surprise to some clinicians then, that motor cortex and CST lesions in cats and rodents result in minimal motor or locomotor deficits (Magoun and Ranson 1938; Eidelberg and Yu 1981;

Bieler et al. 2018). In cats, these lesions generally result in a mildly decreased flexion of the limbs during locomotion, with circumduction during the swing phase, and which resolves after a few weeks (Eidelberg and Yu 1981). While some of this disparity can be ascribed to species differences in the functional roles of the CST and other descending pathways, such as the cortico-reticulospinal tract (CRT) (Welniarz et al. 2017; Nudo and Masterton 1988), it may come as even further surprise that surgical sectioning of the corticospinal tract at the level of the cerebral peduncles in patients with hemiballismus and tremor is reported to produce only a transient hemiplegia without spasticity, mostly in the hands, and with nearly complete recovery of gait a few weeks postoperatively (Bucy and Keplinger 1960; Bucy et al. 1964; Walker 1949). This discrepancy is often explained with the following: (1) CST lesions with significant hemiplegia may also involve other nearby or intermingled fiber tracts of functional importance; and (2) other descending pathways may either contribute to, or can compensate for lost CST function, especially with regards to gait (Davidoff 1990; Bucy 1957). As previously mentioned, the reticulospinal tract (RST) has been identified as having a phylogenetically preserved function in locomotion (Nudo and Masterton 1988), and its role in human locomotion and in the rehabilitation of gait is under active investigation (Matsuyama et al. 2004; Jang et al. 2013; Ahn et al. 2006; Jang and Lee 2017, 2019; Yeo et al. 2013).

Taken together, the abovementioned brain regions form a locomotor network that underlies the neural control of gait (Weiss et al. 2020). This hierarchical and modular organization of control, where a supraspinal network receives and processes various sensorimotor, cognitive, and limbic inputs to control a locally autoregulating (through afferent input and spinal reflexes) network of spinal CPGs, allows for the expedient control of complex motor functions (Fig. 2). And while damage to specific regions can impair locomotor function, network derangements can also result in significant deficits. Finally, there appear to be some parallel and redundant components to the system, such that damage to one or more pathways may not completely abolish locomotion with opportunities for compensation for by other pathways.

4 Neurological Gait Disorders

Gait disturbances present in numerous neurological disorders, including Parkinson's disease, spinal cord injuries, and even aging. They are especially common in elderly populations, with neurological gait disorders having a prevalence of more than 20% in those over 60 years of age (Mahlknecht et al. 2013). These impairments lead to immobility and falls, as well as their associated comorbidities, and contribute to social isolation, reduced quality of life, and loss of independence (Mahlknecht et al. 2013; Sudarsky 2001).

Table 2 Gait problems in Parkinson's disease

General	Initiation	Turning	Balance
Decreased stride length	Freezing of gait	Freezing of gait	Postural instability
Decreased arm swing	Poor anticipatory postural adjustments	En bloc turning	Camptocormia (forward flexion of trunk)
Increased step-to-step variability	Increased movement preparation time	Slow turning with multiple steps	Freezing and festination

4.1 *Parkinson's Disease and Freezing of Gait*

Patients with Parkinson's disease are prone to gait difficulties due to the cardinal motor features of bradykinesia, rigidity, tremor, and in later stages of the disease, postural instability (Table 2) (Postuma et al. 2015). Thus, Parkinsonian gait is often described as hypokinetic and rigid, with reduced stride length, reduced arm swing amplitude, and a flexed and stiffened trunk posture (Wu et al. 2015). Many patients exhibit a festinating gait, in which they resort to small rapid steps to try to keep the CoM of their forward-flexed trunk over their feet (Grabli et al. 2012). Turns are slow and rigid, involve numerous small steps, and are high risk for falls (Hong and Earhart 2010). Patients lose the usual sequential pattern of head, trunk, and pelvis rotation, and instead keep these segments fixed to demonstrate "en bloc" turning (Hong et al. 2009). Although most of these aspects of gait improve with dopaminergic therapy, some aspects, such as postural stability and freezing of gait, may actually worsen (Galna et al. 2015). Many of the motor and non-motor manifestations of PD are quantified using the Unified Parkinson Disease Rating Scale (UPDRS), which rates patients based on items in multiple domains that pertain to the primary signs and symptoms of the disease (Goetz et al. 2007).

Freezing of gait (FOG) is considered the most debilitating and poorly understood of these gait deficits, and is seen in a subset of patients with Parkinson's disease and Parkinson-plus syndromes such as progressive supranuclear palsy (PSP) (Nonnekes et al. 2015). It is described as the transient and episodic inability, or significant reduction in ability, to progress forward with one's feet, despite intending to walk (Giladi and Nieuwboer 2008). Patients express feeling as if their feet are "stuck" to the ground, and these incapacitating episodes can significantly decrease patients' quality of life and also contribute to falls and their associated morbidities (Bloem et al. 2004; Perez-Lloret et al. 2014; Canning et al. 2014). It is triggered in situations that highlight the cognitive and sensory influences on gait: when attempting to initiate stepping or turning, when navigating narrow corridors and obstacles, as well as when the patient is distracted or under stress (Table 3) (Giladi et al. 1992; Lamberti et al. 1997). Conversely, these episodes often improve or resolve with the help of auditory or visual cues, such as a metronome set to a desired step cadence, or lines marked on the floor set to a desired stride length (Table 3) (Nonnekes et al. 2015). One remarkable example often attributed to sensorimotor cueing is the ability of patients with severe FOG to use a bicycle with relative ease (Snijders et al. 2011).

Table 3 Factors alleviating and exacerbating freezing of gait

Alleviating factors	Exacerbating factors
Visual cues (lines marked on the floor)	Attempting to initiate gait or turning
Auditory cues (metronome)	Approaching an object
Sensorimotor cues (cycling, tapping patient)	Passing through a narrow hallway or doorway
	Cognitive dual-tasking (serial sevens) and other distractions
	Anxiety and other emotional states

FOG is associated with advanced Parkinson's disease (Macht et al. 2007), though it can present early (Nonnekes et al. 2015). While some controversy remains as to whether severe gait difficulties represent a distinct spectrum of the disease or merely a more advanced stage of the disease (Lee et al. 2019), many clinicians distinguish between tremor dominant (TD) and postural instability and gait difficulty dominant (PIGD) subtypes of Parkinson's disease, both in terms of their clinical prognoses as well as their management (Stebbins et al. 2013). Treating FOG is complicated by its heterogeneity in response to dopaminergic therapy (Nonnekes et al. 2015): while some patients' FOG improve with levodopa treatment, others have freezing that is refractory to levodopa and conventional deep brain stimulation therapies (Thevathasan et al. 2018; Follett and Torres-Russotto 2012; Weaver et al. 2009; St George et al. 2010). Furthermore, a small proportion of patients' FOG even appears to be made worse or induced by treatment with levodopa or other dopamine agonists (Espay et al. 2012).

The pathophysiology of FOG is not completely understood, though the fact that it can be refractory to dopaminergic therapy suggests that it involves both dopaminergic and non-dopaminergic factors (Santens 2018). Early histopathological observations that the PPN in PD and PSP patients showed cholinergic neuronal degeneration (Hirsch et al. 1987; Jellinger 1988), coupled with the idea that the PPN may play a role in locomotion as part of the MLR, gave rise to the hypothesis that cholinergic neurons in the PPN may represent this non-dopaminergic deficit (Aziz et al. 1998). However, the cholinergic PPN hypothesis remains in serious question today, as newer studies have revealed that: noncholinergic neurons in both the PPN and CnF also undergo significant degeneration in PD and PSP (Sebille et al. 2019; Pienaar et al. 2013); optogenetic activation of cholinergic neurons in the PPN of mice fails to initiate locomotion (Josset et al. 2018; Caggiano et al. 2018); and that cholinergic therapies and PPN DBS have largely failed to improve FOG in PD patients (Thevathasan et al. 2018; Nonnekes et al. 2015).

Another hypothesis to explain gait dysfunctions in PD is that automatic motor processes become impaired in PD (Wu et al. 2015). That is, motor programs that normally require little to no attentional control, such as arm swings, repetitive hand movements, writing, facial expressions, and gait are impaired in PD (Wu et al. 2015). This is attributed to basal ganglia dysfunction resulting from neurodegenerative processes, but also to impairments in network connectivity between multiple cortical (premotor cortex, SMA, primary motor cortex) and subcortical regions, including pathological beta oscillation synchronies between these regions (Wu et al. 2015; Stein and Bar-Gad 2013; Pozzi et al. 2019; Ehgoetz Martens et al. 2018). In addition to accounting for many of the signs and symptoms of PD, this idea explains why external cues can alleviate FOG, and why cognitive and emotional loads can exacerbate it (Wu et al. 2015).

4.2 Spinal Cord Injury

Traumatic spinal cord injury (SCI) causes paralysis and multiple other functional losses by disrupting the communication between supraspinal and spinal circuits, leading to a reduced quality of life, dependency, and medical complications (Fuhrer et al. 1993; Budh and Osteraker 2007). Estimates of prevalence vary, but it is projected that some 300,000 individuals are affected by SCI in the United States alone, with the majority of these being young male adults (Lasfargues et al. 1995). Neurologic recovery is poor, as the initial insult results in irreversible neurologic injury, given the limited ability of CNS axons to regenerate, and is followed by a pathophysiological cascade of inflammation, ischemia, excitotoxicity, and free radical formation that results in secondary injury (Kwon et al. 2004; Belegu et al. 2007). Current management of acute SCI involves supportive care and prevention of hypoxia and hypotension (Ryken et al. 2013; Vale et al. 1997), with some evidence supporting early surgical decompression (Fehlings et al. 2012; Wilson et al. 2020; Sewell et al. 2018); however, no definite treatments exist to aid with axonal regeneration or to prevent secondary injury after SCI. In chronic injury, spasticity and excessive co-contractions of muscles can further contribute to gait dysfunction (Nadeau et al. 2011).

Anatomically, there is frequently sparing of a subset of fiber tracts—even in cases of functionally complete paralysis (Bunge et al. 1993; Harkema et al. 2011; Rejc et al. 2017a; Angeli et al. 2014). In particular, the reticulospinal tract (RST) has been identified as both having a phylogenetically preserved function in locomotion (Nudo and Masterton 1988), and being predisposed to survive after SCI in rodents and humans alike due to its distributed nature within the spinal cord (Asboth et al. 2018; Ballermann and Fouad 2006; Nathan et al. 1996). As lumbar level spinal CPGs and motoneurons involved in walking may be intact with higher level injuries, strategies that augment or maximize the function of spared tracts may restore useful gait and mobility in SCI patients and provide opportunities for activity-based rehabilitation. In fact, recent experimental results demonstrating that epidural spinal cord stimulation (SCS) can restore voluntary leg movements, standing, and even stepping in patients with chronic motor-complete SCI have brought new promise in SCI research (Rejc et al. 2015, 2017a, b; Angeli et al. 2014).

4.3 Age-Related Gait Changes

Several changes to gait occur with age, including decreased gait speed (Bohannon and Williams 2011), increased postural sway (Maki and McIlroy 1996), as well as increased fear of falling and actual risk of falls (Pirker and Katzenschlager 2017). While some of these changes are likely due to musculoskeletal declines over time, neurological changes take place as well. A comparison of blood oxygen level-dependent (BOLD) signals in fMRI between younger (<40 years old) and older (>60 years old) healthy subjects demonstrated a relative increase in the activation of sensory cortices during imagined gait (Zwergal et al. 2012). This was most prominent in the vestibular cortices, followed by motion-sensitive visual areas, and somatosensory areas, and was noted to be less prominent during imagined running (Zwergal et al. 2012). The authors hypothesized that this could reflect a shift from automatic (subcortical) to attentional (cortical) control of gait with age, as a compensatory mechanism for peripheral sensory decline (Zwergal et al. 2012). This agrees with studies showing that elderly people who are less capable of multitasking during gait are also more at risk for falls (Lundin-Olsson et al. 1997; Ansai et al. 2017), and implies a cognitive role for gait that increases in importance with age. As the prevalence of gait disorders increases with age, substantial increases in the number of people suffering from immobility and falls can be expected due to anticipated demographic shifts in our population. Greater understanding of age-related neurological changes contributing to these gait deficits could help develop neuromodulatory strategies to improve function in the elderly.

5 Invasive Neuromodulation Strategies for Gait

Invasive strategies involve surgeries or procedures to more directly access parts of the nervous system and require device implantation to yield its effect. While these strategies have the advantage of placing electrodes closer to target neural structures for potentially greater specificity of electrical effects and fewer side effects, this must be balanced against the risks of the procedures and device implantation.

5.1 Deep Brain Stimulation of the STN and GPi

Deep brain stimulation of the subthalamic nucleus (STN) and the globus pallidus interna (GPi) are established treatments for advanced Parkinson's disease (PD), with significant improvements of bradykinesia, rigidity, and tremor (Obeso et al. 2001; Pahwa et al. 2006). However, the reported effects of these conventional DBS targets on balance and gait have been more controversial (St George et al. 2014; Fasano et al. 2015). As the GPi is the less commonly used target, there are fewer studies, especially long-term studies, to establish conclusions from (Mei et al. 2020). Most of these studies, including a meta-analysis, suggest a minor improvement in gait, although smaller than that seen with levodopa (Defebvre et al. 2002; Bakker et al. 2004).

In line with its known effects on bradykinesia, STN DBS is reported to improve gait speed (Roper et al. 2016; Collomb-Clerc and Welter 2015). Furthermore, compared to GPi DBS, STN DBS generally allows for a greater reduction of patients' levodopa dosages, reducing dyskinesias that could affect gait and balance (Peng et al. 2018; Curtze et al. 2015). However, a meta-analysis investigating the effect of STN DBS on gait (UPDRS III item 29) and freezing of gait (UPDRS II item 14) concluded that while there was an overall positive effect of STN DBS on gait and FOG in the medication-OFF condition in both short- and long-term assessments, it did not improve gait in the medication-ON condition (Schlenstedt et al. 2017). Furthermore, multiple studies have documented the deterioration of gait with the onset of STN DBS (Celiker et al. 2019), resulting in increased risk of falls during ambulation (Farris and Giroux 2016), and inhibiting the beneficial effects of levodopa on gait (Fleury et al. 2016). Precise location of the active contacts likely plays an important role, as several studies have correlated specific active contact positions around the STN with positive clinical outcomes, including the posterodorsal STN (Conrad et al. 2018; Mostofi et al. 2019), and the caudal zona incerta (Mostofi et al. 2019). Conversely, studies have identified positions in the anterodorsal STN that correlated to either worsening FOG (Tommasi et al. 2007), or inhibition of levodopa's positive effect on gait (Fleury et al. 2016). Fleury et al. suggested that this effect may be due to stimulation of GABAergic pallido-PPN fibers that pass along the anterior zona incerta, reinforcing abnormal neuronal information responsible for gait disturbances (Fleury et al. 2016; Tommasi et al. 2007).

Stimulation frequency is another important parameter in clinical effects, with worsening of bradykinesia with STN DBS at 5 Hz, and improvement in rigidity and bradykinesia beginning at 33 and 50 Hz, respectively (Moro et al. 2002). While optimal improvement of these signs in the 130–185 Hz range has resulted in the nearly universal adoption of high-frequency stimulation for STN DBS, findings of beneficial effects on gait in other studies in the 60–80 Hz range suggest that there may be a trade-off between axial and appendicular effects based on stimulation frequency (Khoo et al. 2014; Moreau et al. 2008; Zibetti et al. 2016). Further studies clearly delineating the postural and gait effects of electrode position and stimulation parameters in STN DBS are required to clarify this issue, with low-frequency stimulation remaining as a potential alternative for the time being in select PD patients with predominant postural and gait disturbances (Santens 2018).

The strongest predictor of whether STN DBS could improve gait was levodopa-responsiveness, supporting the general view that PD patients with levodopa-refractory gait disorders are poor candidates for STN DBS (Schlenstedt et al. 2017; Odin et al. 2015). These observations have led clinicians to postulate that gait impairments may have both dopaminergic and non-dopaminergic components, with levodopa and STN DBS primarily addressing the former (Schlenstedt et al. 2017). A practical approach to FOG should then first determine if the patient's signs and symptoms are levodopa-responsive, levodopa-resistant, or levodopa-induced (Nonnekes et al. 2015). If levodopa-responsive, and the patient is otherwise a good candidate, STN DBS may work well to improve gait disturbances, especially gait speed and other parameters related to bradykinesia. Rare patients with levodopa-induced FOG may also benefit from STN DBS, especially low-frequency DBS, if reducing levodopa medication and dyskinesias are the primary goals (Nonnekes et al. 2015; Ferraye et al. 2008); other studies suggest that GPi DBS may be an option in this setting to avoid some of the gait and balance deficits seen with STN DBS (Celiker et al. 2019; St George et al. 2012). The general consensus for levodopa-resistant FOG is that DBS of conventional targets such as STN and GPi is contraindicated (Nonnekes et al. 2015; Schlenstedt et al. 2017; Odin et al. 2015).

5.2 *Deep Brain Stimulation of the Mesencephalic Locomotor Region*

Over the past 15 years, several centers have reported on DBS of the pedunculopontine nucleus (PPN), a putative anatomical component of the mesencephalic locomotor region (MLR), for postural instability and gait dysfunction in Parkinson's disease (Thevathasan et al. 2018), based on successful animal studies (Nandi et al. 2002, 2008; Jenkinson et al. 2004). Despite initially promising case reports (Mazzone et al. 2005; Plaha and Gill 2005), the efficacy of this therapy has since been called into dispute through the results of double-blinded studies and meta-analyses (Moro et al. 2010; Wang et al. 2016, 2017; Golestanirad et al. 2016).

Several possible issues have been put forth to explain this lack of efficacy, including species differences in MLR function, differences between bipedal gait and quadrupedal gait, and that stimulating the MLR may have inadequate effectiveness in recovering function in the setting of a diseased basal ganglia and degenerated neurons (Alam et al. 2011; Albin et al. 2018; Benarroch 2013). However, in favor of the view that DBS of this region can meaningfully improve gait dysfunction, a new anatomo-clinical correlation study showed that while there was significant variability in cohort FOG outcomes, categorizing the group by these outcomes revealed a majority "good responder" cluster that had a significant reduction in % time spent in FOG with DBS ON compared to OFF (34.1 ± 14% vs. 2.7 ± 2.6%) (Goetz et al. 2019). Furthermore, all good responders had active electrode contacts in the dorsal part of the PPN area bordering the cuneiform nucleus (CnF), with the two best responders, who had complete resolution of FOG, having their electrodes dorsal to the PPN, in the CnF (Fig. 4) (Goetz et al. 2019). This is in agreement with the classical electrophysiological experiments in cats and optogenetics studies in mice, which show that stimulation of glutamatergic neurons in the CnF have the clearest role in promoting and initiating locomotion (Shik et al. 1966; Opris et al. 2019; Josset et al. 2018; Caggiano et al. 2018; Takakusaki et al. 2003). This importance of electrode targeting on efficacy is reiterated by computer modeling studies of DBS in the region, which demonstrate that targeting errors as little as 1 mm in this area significantly decrease target activation selectivity (Zitella et al. 2013).

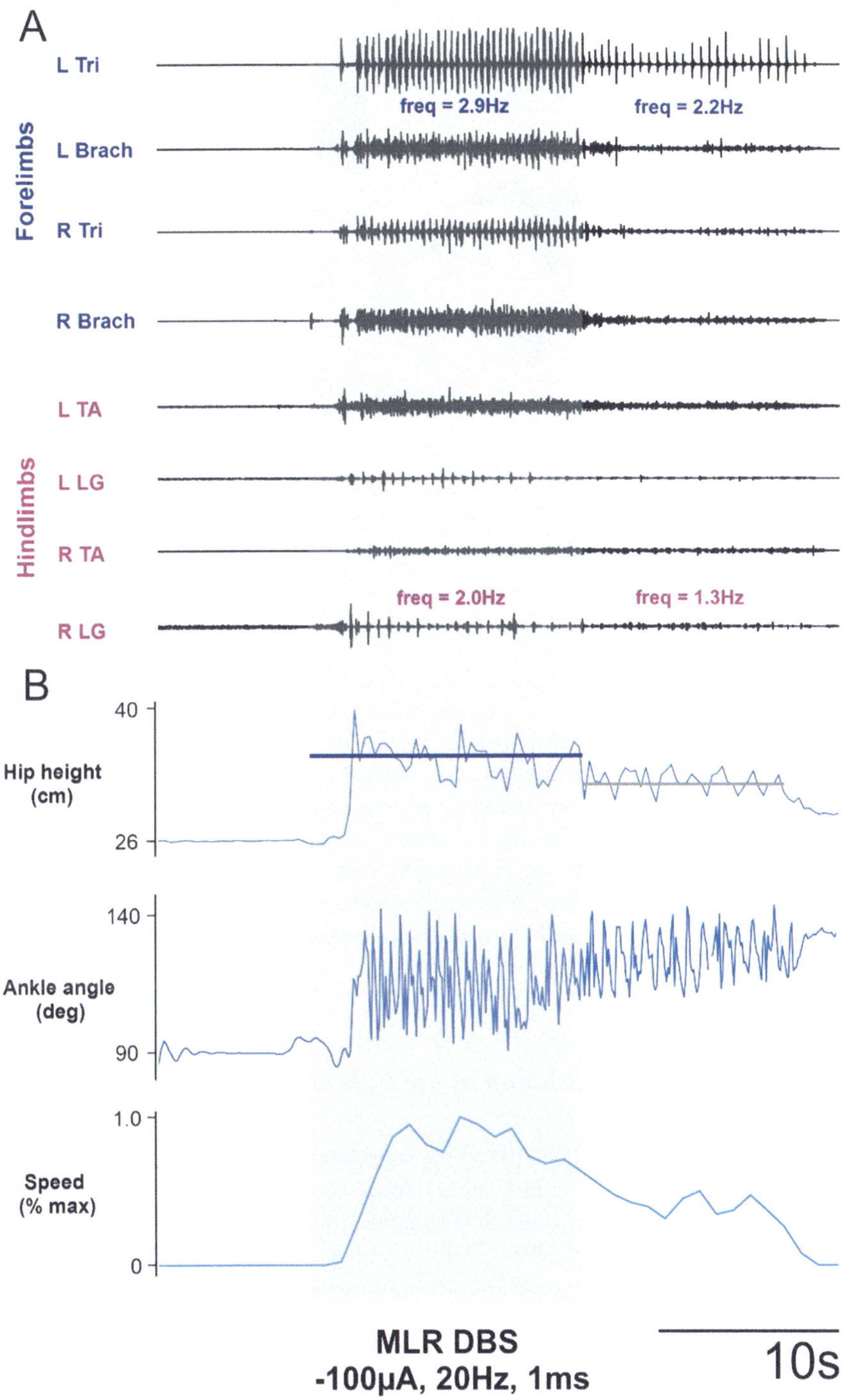

Fig. 4 MLR DBS improves partially weight-supported gait in a large animal (porcine) model of severe thoracic contusive SCI. (**a**) Limb EMG just before, during (grey), and after MLR-evoked locomotion. Frequency of EMG bursts in representative forelimbs and hind limbs is shown during and after stimulation. (**b**) Hip height, left ankle joint excursion, and treadmill speed are shown during locomotion. *R* right, *L* left, *Tri* triceps, *Brach* brachialis, *TA* tibialis anterior, *LG* lateral gastrocnemius

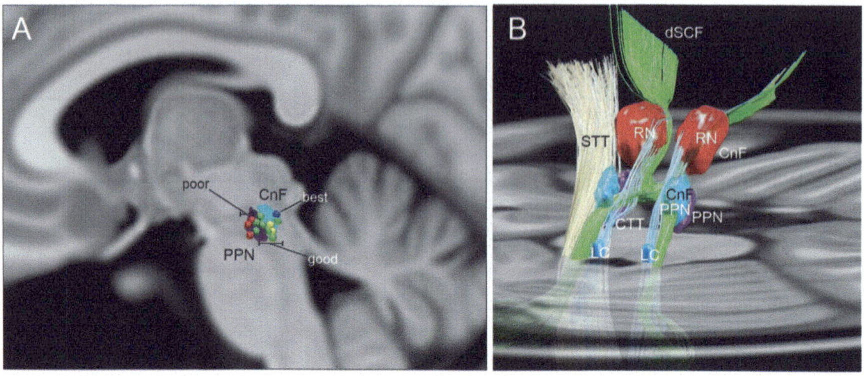

Fig. 5 Three-dimensional reconstruction of the human MLR and anatomo-clinical correlation of DBS contacts used for MLR DBS. (**a**) Sagittal projection of the CnF and PPN with overlay of active contacts from PPN DBS patients with poor (red), good (green), best (blue), and unevaluated (yellow) gait outcomes from Goetz et al. (2019). (**b**) 3D view with right ML and STT removed, projected onto a transverse slice of the brain at the pons level. *CnF* cuneiform nucleus, *CTT* central tegmental tract, *dSCF* decussating superior cerebellar fibers, *LC* locus coeruleus, *ML* medial lemniscus, *PPN* pedunculopontine nucleus, *RN* red nucleus, *STT* spinothalamic tract. Adapted from Fig. 2 of Chang et al. (2020) with permission

Several potential mechanisms for the effects of MLR DBS on gait have been posited: through ascending stimulation on arousal; via afferent-mediated interruption of cortical/subcortical pathological oscillations; through the modulation of other structures such as the subthalamic nucleus and substantia nigra; and through descending stimulation of spinal locomotor circuits (Garcia-Rill et al. 2019). Ongoing animal work in the pig model of SCI (Fig. 5), and future clinical studies targeting the CnF for DBS for gait disorders, including a recently approved clinical trial at our own institution, will hopefully help to clarify the potential clinical utility of this important locomotor center.

5.3 Deep Brain Stimulation of the Substantia Nigra

A few studies have reported on DBS of the substantia nigra pars reticulata (SNr) for the management of refractory FOG in PD, based on its reciprocal interconnections to the MLR and it being the other major output center of the basal ganglia system beside the GPi (Weiss et al. 2011, 2013; Chastan et al. 2009; Valldeoriola et al. 2019). From a pragmatic standpoint, the ventromedially adjacent location of the SNr to the STN has allowed for investigation of SNr DBS and combined STN/SNr DBS using a conventional quadripolar electrode that passes through both (Pötter-Nerger and Volkmann 2013).

Supporting a dichotomy between axial and appendicular disturbances within PD, Chastan et al. found that SNr DBS only improved axial subscores (items 27–30 of UPDRS III: rising from chair, posture, postural stability, gait), while STN DBS improved both axial and appendicular scores (Chastan et al. 2009). Furthermore, biomechanical gait analysis showed that SNr DBS was able to improve vertical (axial) control of CoM just prior to heel strike, but did not improve stride length or gait speed, while STN DBS was able to improve both (Chastan et al. 2009). Alternatively, a more recent study suggested a distinction between spatial (stride length) and temporal (swing time variability) features of gait, with STN DBS improving both, while SNr DBS only improved temporal gait parameters (Scholten et al. 2017).

This distinction between STN and SNr DBS effects is further investigated in studies of combined STN/SNr DBS, with one double-blinded randomized control study employing high-frequency DBS at both sites (125 Hz) (Weiss et al. 2013), while another pilot study used a lower frequency (63 Hz) at the SNr (Valldeoriola et al. 2019). The double-blinded study found no difference between combined SNr/STN DBS and STN DBS in its primary outcome, a composite score of axial items from UPDRS II and III (Weiss et al. 2013), or in most of its secondary outcome measures of FOG and balance at 3 weeks follow-up (Weiss et al. 2013). The study did report a significant improvement with combined stimulation in one secondary outcome measure for FOG, the FOG Assessment Course Score (Weiss et al. 2013); however, it should be noted that parametric analysis was used despite this score being known to produce non-normal distributions (Ziegler et al. 2010).

More recently, a pilot study of 6 patients used a lower frequency (63 Hz) to stimulate the SNr based on literature that low-frequency STN DBS had superior gait outcomes (Valldeoriola et al. 2019). Although no statistical comparison between combined SNr/STN DBS and STN DBS was performed in this study, four of the six patients were reported to improve most in their FOG with combined stimulation, and preferred to combined stimulation over STN DBS or SNr DBS for 3 years of follow-up (Valldeoriola et al. 2019). While promising, the rationale for the choice of stimulation frequency has been questioned (Weiss et al. 2019), given recent intraoperative neurophysiological data demonstrating that the SNr is silenced by stimulation at 50 Hz and above (Milosevic et al. 2017).

Overall, the evidence for SNr DBS and combined SNr/STN DBS remains preliminary, and more rigorous studies with more convincing outcomes are required to establish the SNr as a useful target for gait. Fortunately, the proximity of the SNr to the STN presents an attractive practical opportunity for groups interested in doing so, especially in patients who have previously undergone STN DBS and later develop refractory FOG. At least one such case report found significant improvement in FOG with simple reprogramming of STN DBS to include the SNr (Brosius et al. 2015).

5.4 *Motor Cortex Stimulation*

Motor cortex stimulation (MCS) is a neuromodulation technique historically used to treat refractory pain, but which has also been investigated in movement disorders (Slotty et al. 2015). Initial reports of motor effects were described incidentally in patients receiving MCS for pain relief in conditions such as stroke (Katayama et al. 1998) and neuropathic pain (Nguyen et al. 1998). Based on observations that intraoperative stimulation of the primary motor cortex could relieve tremor and rigidity in a PD patient (Woolsey et al. 1979), Canavero et al. investigated the motor effects of unilateral, low-frequency (30 Hz) epidural MCS in three advanced PD patients with contraindications for DBS surgery, and found significant improvements in tremor, bradykinesia, rigidity, and gait (Canavero et al. 2002). He also implanted bilateral epidural MCS in a multiple system atrophy (MSA) patient who had undergone unsuccessful STN DBS, and reported significant but temporary improvements in gait (Canavero et al. 2003). This was supported by experiments in a MPTP-treated Parkinsonian baboon model, where high-frequency (130 Hz), but not low-frequency (10 Hz) unilateral epidural MCS was able to significantly improve bradykinesia and gait initiation (Drouot et al. 2004). In this study, MCS was shown to reduce synchronized oscillatory activity between the GPi and STN, and increase regional glucose consumption in the ipsilateral mesial premotor cortex/SMA of these animals on PET scan, suggesting that MCS could exert its motor benefits by disrupting pathological cortico-striatal oscillations (Drouot et al. 2004). However, the observation that motor benefits were delayed by several minutes after the onset of unilateral MCS and seen bilaterally, suggested a mechanism distinct from that involved in STN DBS (Drouot et al. 2004).

Touted as a potentially less invasive and more cost-effective alternative to DBS for Parkinsonism, this was followed by several case reports and small series from other groups with varied results, ranging from: no motor benefit (Kleiner-Fisman et al. 2003; Gutierrez et al. 2009); only subjective improvements in gait and axial symptoms (Cilia et al. 2007); temporary motor benefits lasting less than 12 months (Arle et al. 2008); temporary improvement in gait lasting 5 months (Fasano et al. 2008); to significant motor and gait improvement (Benvenuti et al. 2006; Pagni et al. 2005). Moro et al. reported the first double-blinded evaluation of MCS for PD and essential tremor (ET), using subdural rather than epidural electrodes, to avoid the electrical variability attributable to changes in CSF thickness (Chang et al. 2015), and to directly visualize the motor cortex. This study found significant improvement of tremor in ET patients, but no motor or gait benefits in PD patients (Moro et al. 2011).

As with many other neuromodulatory techniques, analysis of efficacy is complicated by methodological heterogeneity in a limited number of studies. Overall, the lack of consistent or persistent motor and gait benefits demonstrated in these studies, and the lack of mechanistic understanding of the circuits and pathways that may be involved in exerting motor benefits, suggests that additional preclinical studies are needed before this modality is pursued further clinically.

5.5 Spinal Cord Stimulation

Spinal cord stimulation (SCS) first began as a treatment for chronic, intractable pain in 1967 (Shealy et al. 1967), based on Melzack and Wall's gate theory of pain (Melzack and Wall 1965). It was eventually noted to also improve motor signs in some of these populations, such as spasticity (Cook 1976). More recently, in addition to demonstrating benefits for cough and bladder control (Kowalski et al. 2013; Gad et al. 2016), epidural SCS has been shown to increase the excitability of locomotor circuits and restore voluntary limb movements and even gait in multiple preclinical (Asboth et al. 2018; Ichiyama et al. 2005; Hachmann et al. 2013; Wenger et al. 2014) and clinical SCI studies (Harkema et al. 2011; Rejc et al. 2015, 2017a; Angeli et al. 2014; Hofstoetter et al. 2015; Wagner et al. 2018). Many of these recoveries occurred in chronic, motor-complete SCI patients, revealing that some neural function may be preserved in patients despite the absence of motor function, and that this subthreshold function can be harnessed to restore volitional control of neurons below the level of injury. Computer simulations and experimental evidence suggest that SCS facilitates sensorimotor circuits by activating large diameter muscle afferents (e.g., Group Ia, Ib, and II axons) at the dorsal roots, increasing the excitability of the spinal locomotor networks they feed into, and enhancing their receptivity to residual descending signals (Fig. 6) (Angeli et al. 2014; Hofstoetter

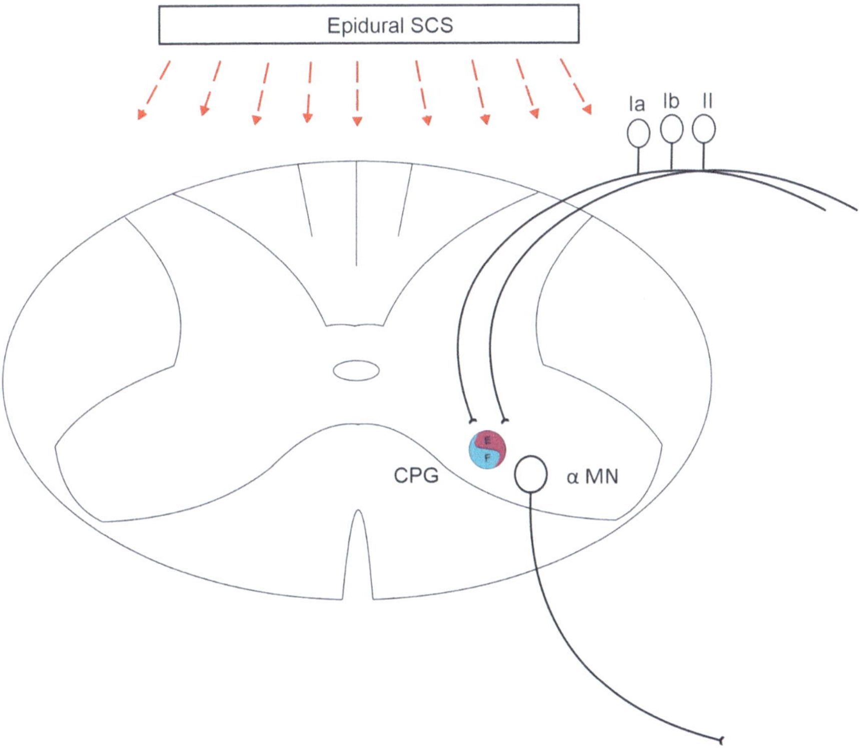

Fig. 6 Schematic of epidural spinal cord stimulation mechanism of action. Large diameter, myelinated fibers such as Ia, Ib, and II afferents are preferentially activated by epidural spinal cord stimulation (SCS). These afferents send ascending information to supraspinal structures, and feed into local spinal central pattern generator (CPG) networks and eventually, alpha motoneurons (MN)

et al. 2015; Capogrosso et al. 2013; Rattay et al. 2000; Gerasimenko et al. 2006; Danner et al. 2015).

While these improvements have not been sufficient to enable safe, independent gait in the community setting, animal studies using more sophisticated closed-loop stimulation protocols indicate that further progress with SCS is possible (Wenger et al. 2014; Capogrosso et al. 2016, 2018; Moraud et al. 2016). Furthermore, a recent study in rodents has identified descending signal transmission along spared glutamatergic reticulospinal tract (RST) neurons originating from the ventrocaudal medulla as necessary for SCS to re-enable locomotion after SCI (Asboth et al. 2018). Inactivation of these RST neurons abolished the ability of these injured animals to walk with SCS, while chronic intensive training with SCS resulting in significant cortico-reticulospinal sprouting allowed the injured rats to walk even without SCS (Asboth et al. 2018). In decerebrate cats, electrical or pharmacological activation of RST neurons in this area can initiate locomotion (Noga et al. 1988), which can be blocked by spinal intrathecal injection of glutamatergic antagonists (Douglas et al. 1993). In intact rodents, optogenetic activation of these RST neurons evokes locomotion, while optogenetic inactivation abolishes high-speed locomotion (Capelli et al. 2017). These mechanistic insights suggest that: (1) the effect of SCS is largely dependent upon, and works to maximize receptiveness to, descending glutamatergic RST signal transmission; (2) SCS effects may be enhanced by increasing glutamatergic RST activity; and (3) neuromodulation-based rehabilitation may bring about activity-dependent cortico-reticulospinal circuit adaptations resulting in permanent functional gains.

Epidural SCS has also been considered as a potential treatment for FOG and postural instability in patients with Parkinsonism (de Andrade et al. 2016; Fonoff et al. 2019), based on positive studies in mice and primate models of PD (Fuentes et al. 2009; Santana Maxwell et al. 2014). While it is too early to draw definitive conclusions, several of these preliminary clinical series and case reports have demonstrated promising results for FOG (Rohani et al. 2017; Pinto de Souza et al. 2017; Hassan et al. 2013; Fenelon et al. 2012; Samotus et al. 2018), with the largest study showing persistent improvements in UPDRS motor scores, postural stability motor subscores, and Timed 10 m walk tests at 12 months post-surgery (Agari and Date 2012). Another more recent study demonstrated that high-frequency (300 Hz), but not lower frequency (60 Hz) SCS stimulation, reduces the duration of anticipatory postural adjustments (APAs) thought to trigger FOG in PD patients (de Lima-Pardini et al. 2018). Insight into the mechanisms of action underlying these beneficial effects is provided in primate experiments, where SCS has been shown to increase activity in the primary motor cortex and reduce pathological oscillations between the cortex and the basal ganglia (Fuentes et al. 2009). Other proposed mechanisms include the ascending modulation of other supraspinal locomotor centers such as the supplementary motor area, a cortical region proposed to contribute to the generation of APAs (Jacobs et al. 2009), and local modulation of spinal locomotor networks (Rohani et al. 2017; de Lima-Pardini et al. 2018). Larger prospective studies with longer follow-up will be needed to confirm these results and clarify optimal spinal levels, electrode configurations, and stimulation parameters (Samotus et al. 2018).

6 Noninvasive Neuromodulatory Strategies for Gait

The success of invasive neuromodulatory techniques such as DBS and SCS, and the shift toward modulation of entire brain networks rather than brain regions, has led investigators to explore alternative, less-invasive methods of neurostimulation. Repetitive transcranial magnetic stimulation (rTMS) and transcranial direct current stimulation (tDCS) are the most common techniques, and have been studied for a wide number of neurological conditions, including depression and other psychiatric disorders, pain, PD, SCI, stroke, and multiple sclerosis (MS) (To et al. 2018). Both techniques influence neuronal excitability, although using electromagnetic induction in the case of rTMS, and a low level of continuously applied electrical current for tDCS (To et al. 2018; Valero-Cabre et al. 2017). Importantly, as opposed to rTMS, the electrical currents delivered by tDCS are not strong enough to fire an action potential (To et al. 2018).

Many clinical studies have investigated the motor effects (UPDRS-III score) of rTMS in PD, and several meta-analyses consistently demonstrate a positive moderate effect size (Elahi et al. 2009; Fregni et al. 2005; Zanjani et al. 2015; Chou et al. 2015). However, few studies have investigated the effect of repetitive TMS on FOG, and most studies do not report the breakdown of the UPDRS-III scores to allow for subgroup meta-analysis of specific motor deficits such as gait. Moreover, studies differ in their stimulation targets and stimulation protocols, with common targets including the primary motor cortex, the supplementary motor area (SMA), and the dorsolateral prefrontal cortex (Chou et al. 2015). Heterogeneity in parameters such as stimulation frequency is important given that high-frequency stimulation (≥5 Hz) is known to enhance motor cortex excitability (Pascual-Leone et al. 1994), while low-frequency stimulation (≤1 Hz) has been shown to depress cortical excitability (Chen et al. 1997). Two more recent studies, including one double-blinded and sham-controlled randomized control study, found that high-frequency (10 Hz) repetitive TMS over the SMA significantly improved FOG Questionnaire (FOGQ) scores in PD patients (Mi et al. 2019; Kim et al. 2018). A third study using high-frequency repetitive TMS over the leg area of the primary motor cortex also found significant improvement in FOGQ scores (Kim et al. 2015). Additional high-quality studies are still required to replicate these findings, and support the rationale that these cortical regions may be underactive in PD patients with FOG, and that high-frequency TMS may alleviate FOG and gait deficits by increasing the excitability of these regions (Mi et al. 2019). High-frequency rTMS over the leg motor cortex has also been investigated for gait effects in patients with incomplete SCI (Belci et al. 2004). A prospective and double-blinded randomized control study found significant improvement in lower extremity motor scores, spasticity, and various gait test scores (Benito et al. 2012). This reiterated previous findings in MS patients, where repetitive magnetic stimulation over either the motor cortex or the spinal cord was found to ameliorate leg spasticity (Nielsen and Sinkjaer 1997; Centonze et al. 2007).

TDCS has the benefit of being cheaper and more portable than TMS, and is established as an effective neuromodulatory therapy for depression, addiction, and chronic pain (Lefaucheur et al. 2017). Given studies showing that tDCS of the motor cortex is able to increase corticospinal plasticity (Frazer et al. 2016), and alter gait in healthy individuals (Koganemaru et al. 2018), multiple studies have proposed the integration of tDCS into gait rehabilitation training. A systematic review of such studies concluded that there was insufficient evidence to support a role for tDCS in enhancing gait rehabilitation training, citing an inadequate number of studies demonstrating improvements beyond sham treatment (de Paz et al. 2019). Notably, the review included a diversity of pathologies and methodologies, and of the three studies in stroke patients using a minimum of 10 min of stimulation, all showed improvement in at least one parameter of gait (de Paz et al. 2019).

Overall, these studies indicate that rTMS and tDCS can noninvasively modulate corticospinal and potentially other corticofugal pathways that may be of relevance in the treatment and rehabilitation of gait disorders, with rTMS having a more potent neuromodulatory effect. Future studies may help identify optimal therapeutic locations and stimulation parameters.

7 Conclusion and Future Directions

Gait is a quintessential and complex motor function that relies on the proper functioning of multiple bodily and neural systems, the latter of which have been summarized here. Studies attempting to investigate the therapeutic effects of neuromodulatory techniques in gait disorders are complicated both by the heterogeneity of underlying pathophysiologies, as well as the multifaceted nature of measuring gait and improvements in gait. These realizations have led to the careful classification of gait pathologies and the development of useful tools with which to evaluate the efficacy of interventions. As a result, many high-quality studies and quantitative reviews are emerging in the field of gait disorders.

Overall, neuromodulation strategies hold significant clinical promise in the treatment of gait disorders. Future advances in our understanding of the neural basis of locomotor control and the circuit effects of neuromodulatory techniques will continue to refine our selection of network targets and the application of these techniques. Furthermore, the ongoing development of new and less-invasive methods of neuromodulation with increased spatial resolution and target specificity, many of which are discussed in other chapters of this book, will undoubtedly promote further adoption and expansion of the field. Initial studies of combinations of techniques and targets have already started and may contribute to our understanding of synergistic network interactions, other ecological principles of neuromodulation, and hopefully an overarching hypothesis on mechanisms of action.

References

Agari T, Date I (2012) Spinal cord stimulation for the treatment of abnormal posture and gait disorder in patients with Parkinson's disease. Neurol Med Chir 52(7):470–474

Ahn YH, Ahn SH, Kim H, Hong JH, Jang SH (2006) Can stroke patients walk after complete lateral corticospinal tract injury of the affected hemisphere? Neuroreport 17(10):987–990

Alam M, Schwabe K, Krauss JK (2011) The pedunculopontine nucleus area: critical evaluation of interspecies differences relevant for its use as a target for deep brain stimulation. Brain 134(Pt 1):11–23

Albin RL, Surmeier DJ, Tubert C, Sarter M, Müller MLTM, Bohnen NI et al (2018) Targeting the pedunculopontine nucleus in Parkinson's disease: time to go back to the drawing board. Mov Disord 33(12):1871–1875

Angeli CA, Edgerton VR, Gerasimenko YP, Harkema SJ (2014) Altering spinal cord excitability enables voluntary movements after chronic complete paralysis in humans. Brain 137(Pt 5):1394–1409

Ansai JH, Andrade LP, Rossi PG, Takahashi ACM, Vale FAC, Rebelatto JR (2017) Gait, dual task and history of falls in elderly with preserved cognition, mild cognitive impairment, and mild Alzheimer's disease. Braz J Phys Ther 21(2):144–151

Arle JE, Apetauerova D, Zani J, Deletis DV, Penney DL, Hoit D et al (2008) Motor cortex stimulation in patients with Parkinson disease: 12-month follow-up in 4 patients. J Neurosurg 109(1):133–139

Asboth L, Friedli L, Beauparlant J, Martinez-Gonzalez C, Anil S, Rey E et al (2018) Cortico-reticulo-spinal circuit reorganization enables functional recovery after severe spinal cord contusion. Nat Neurosci 21(4):576–588

Aziz TZ, Davies L, Stein J, France S (1998) The role of descending basal ganglia connections to the brain stem in parkinsonian akinesia. Br J Neurosurg 12(3):245–249

Bakker M, Esselink RA, Munneke M, Limousin-Dowsey P, Speelman HD, Bloem BR (2004) Effects of stereotactic neurosurgery on postural instability and gait in Parkinson's disease. Mov Disord 19(9):1092–1099

Ballermann M, Fouad K (2006) Spontaneous locomotor recovery in spinal cord injured rats is accompanied by anatomical plasticity of reticulospinal fibers. Eur J Neurosci 23(8):1988–1996

Belci M, Catley M, Husain M, Frankel HL, Davey NJ (2004) Magnetic brain stimulation can improve clinical outcome in incomplete spinal cord injured patients. Spinal Cord 42(7):417–419

Belegu V, Oudega M, Gary DS, McDonald JW (2007) Restoring function after spinal cord injury: promoting spontaneous regeneration with stem cells and activity-based therapies. Neurosurg Clin N Am 18(1):143–168, xi

Benarroch EE (2013) Pedunculopontine nucleus: functional organization and clinical implications. Neurology 80(12):1148–1155

Benito J, Kumru H, Murillo N, Costa U, Medina J, Tormos JM et al (2012) Motor and gait improvement in patients with incomplete spinal cord injury induced by high-frequency repetitive transcranial magnetic stimulation. Top Spinal Cord Inj Rehabil 18(2):106–112

Benvenuti E, Cecchi F, Colombini A, Gori G (2006) Extradural motor cortex stimulation as a method to treat advanced Parkinson's disease: new perspectives in geriatric medicine. Aging Clin Exp Res 18(4):347–348

Bieler L, Grassner L, Zaunmair P, Kreutzer C, Lampe L, Trinka E et al (2018) Motor deficits following dorsal corticospinal tract transection in rats: voluntary versus skilled locomotion readouts. Heliyon 4(2):e00540

Bloem BR, Hausdorff JM, Visser JE, Giladi N (2004) Falls and freezing of gait in Parkinson's disease: a review of two interconnected, episodic phenomena. Mov Disord 19(8):871–884

Bohannon RW, Williams AA (2011) Normal walking speed: a descriptive meta-analysis. Physiotherapy 97(3):182–189

Bouvier J, Caggiano V, Leiras R, Caldeira V, Bellardita C, Balueva K, Fuchs A, Kiehn O (2015) Descending command neurons in the brain-stem that halt locomotion. Cell 163:1191–1203

Brosius SN, Gonzalez CL, Shuresh J, Walker HC (2015) Reversible improvement in severe freezing of gait from Parkinson's disease with unilateral interleaved subthalamic brain stimulation. Parkinsonism Relat Disord 21(12):1469–1470

Brown TG, Sherrington CS (1911) The intrinsic factors in the act of progression in the mammal. Proc R Soc Lond Ser B 84(572):308–319

Bucy PC (1957) Is there a pyramidal tract. Brain 80(3):376–392

Bucy PC, Keplinger JE (1960) Section of the cerebral peduncles. Trans Am Neurol Assoc 85:65–66

Bucy PC, Keplinger JE, Siqueira EB (1964) Destruction of the "pyramidal tract" in man. J Neurosurg 21:285–298

Budh CN, Osteraker AL (2007) Life satisfaction in individuals with a spinal cord injury and pain. Clin Rehabil 21(1):89–96
Bunge RP, Puckett WR, Becerra JL, Marcillo A, Quencer RM (1993) Observations on the pathology of human spinal cord injury. A review and classification of 22 new cases with details from a case of chronic cord compression with extensive focal demyelination. Adv Neurol 59:75–89
Bussel B, Roby-Brami A, Azouvi P, Biraben A, Yakovleff A, Held JP (1988) Myoclonus in a patient with spinal cord transection. Possible involvement of the spinal stepping generator. Brain 111(Pt 5):1235–1245
Caggiano V, Leiras R, Goni-Erro H, Masini D, Bellardita C, Bouvier J et al (2018) Midbrain circuits that set locomotor speed and gait selection. Nature 553(7689):455–460
Calancie B (2006) Spinal myoclonus after spinal cord injury. J Spinal Cord Med 29(4):413–424
Canavero S, Paolotti R, Bonicalzi V, Castellano G, Greco-Crasto S, Rizzo L et al (2002) Extradural motor cortex stimulation for advanced Parkinson disease. Report of two cases. J Neurosurg 97(5):1208–1211
Canavero S, Bonicalzi V, Paolotti R, Castellano G, Greco-Crasto S, Rizzo L et al (2003) Therapeutic extradural cortical stimulation for movement disorders: a review. Neurol Res 25(2):118–122
Canning CG, Paul SS, Nieuwboer A (2014) Prevention of falls in Parkinson's disease: a review of fall risk factors and the role of physical interventions. Neurodegen Dis Manag 4(3):203–221
Capelli P, Pivetta C, Soledad Esposito M, Arber S (2017) Locomotor speed control circuits in the caudal brainstem. Nature 551(7680):373–377
Capogrosso M, Wenger N, Raspopovic S, Musienko P, Beauparlant J, Bassi Luciani L et al (2013) A computational model for epidural electrical stimulation of spinal sensorimotor circuits. J Neurosci 33(49):19326–19340
Capogrosso M, Milekovic T, Borton D, Wagner F, Moraud EM, Mignardot JB et al (2016) A brain-spine interface alleviating gait deficits after spinal cord injury in primates. Nature 539(7628):284–288
Capogrosso M, Wagner FB, Gandar J, Moraud EM, Wenger N, Milekovic T et al (2018) Configuration of electrical spinal cord stimulation through real-time processing of gait kinematics. Nat Protoc 13(9):2031–2061
Celiker O, Demir G, Kocaoglu M, Altug F, Acar F (2019) Comparison of subthalamic nucleus vs. globus pallidus interna deep brain stimulation in terms of gait and balance; a two year follow-up study. Turk Neurosurg 29(3):355–361
Centonze D, Koch G, Versace V, Mori F, Rossi S, Brusa L et al (2007) Repetitive transcranial magnetic stimulation of the motor cortex ameliorates spasticity in multiple sclerosis. Neurology 68(13):1045–1050
Chang S, Ranjan M, Slotty PJ, Honey CR (2015) The influence of positioning and muscle activity on motor threshold during motor cortex stimulation programming. Stereotact Funct Neurosurg 93(2):122–126
Chang SJ, Iahn Cajigas, Ioan Opris, James D. Guest, Brian R. Noga, (2020) Dissecting Brainstem Locomotor Circuits: Converging Evidence for Cuneiform Nucleus Stimulation. Frontiers in Systems Neuroscience 14
Chastan N, Westby GW, Yelnik J, Bardinet E, Do MC, Agid Y et al (2009) Effects of nigral stimulation on locomotion and postural stability in patients with Parkinson's disease. Brain 132(Pt 1):172–184
Chen HC, Schultz AB, Ashton-Miller JA, Giordani B, Alexander NB, Guire KE (1996) Stepping over obstacles: dividing attention impairs performance of old more than young adults. J Gerontol A Biol Sci Med Sci 51(3):M116–M122
Chen R, Classen J, Gerloff C, Celnik P, Wassermann EM, Hallett M et al (1997) Depression of motor cortex excitability by low-frequency transcranial magnetic stimulation. Neurology 48(5):1398–1403
Chou Y-h, Hickey PT, Sundman M, Song AW, Chen N-k (2015) Effects of repetitive transcranial magnetic stimulation on motor symptoms in Parkinson disease: a systematic review and meta-analysis. JAMA Neurol 72(4):432–440
Christensen LO, Johannsen P, Sinkjaer T, Petersen N, Pyndt HS, Nielsen JB (2000) Cerebral activation during bicycle movements in man. Exp Brain Res 135(1):66–72

Cilia R, Landi A, Vergani F, Sganzerla E, Pezzoli G, Antonini A (2007) Extradural motor cortex stimulation in Parkinson's disease. Mov Disord 22(1):111–114
Collomb-Clerc A, Welter ML (2015) Effects of deep brain stimulation on balance and gait in patients with Parkinson's disease: a systematic neurophysiological review. Neurophysiol Clin 45(4–5):371–388
Conrad EC, Mossner JM, Chou KL, Patil PG (2018) Atlas-independent, electrophysiological mapping of the optimal locus of subthalamic deep brain stimulation for the motor symptoms of Parkinson disease. Stereotact Funct Neurosurg 96(2):91–99
Cook AW (1976) Electrical stimulation in multiple sclerosis. Hosp Pract 11(4):51–58
Curtze C, Nutt JG, Carlson-Kuhta P, Mancini M, Horak FB (2015) Levodopa is a double-edged sword for balance and gait in people with Parkinson's disease. Mov Disord 30(10):1361–1370
Dampney RA, Furlong TM, Horiuchi J, Iigaya K (2013) Role of dorsolateral periaqueductal grey in the coordinated regulation of cardiovascular and respiratory function. Auton Neurosci 175(1–2):17–25
Danner SM, Hofstoetter US, Freundl B, Binder H, Mayr W, Rattay F et al (2015) Human spinal locomotor control is based on flexibly organized burst generators. Brain 138(Pt 3):577–588
Davidoff RA (1990) The pyramidal tract. Neurology 40(2):332–339
de Andrade EM, Ghilardi MG, Cury RG, Barbosa ER, Fuentes R, Teixeira MJ et al (2016) Spinal cord stimulation for Parkinson's disease: a systematic review. Neurosurg Rev 39(1):27–35
de Lima-Pardini AC, Coelho DB, Souza CP, Souza CO, Ghilardi M, Garcia T et al (2018) Effects of spinal cord stimulation on postural control in Parkinson's disease patients with freezing of gait. Elife 7:e37727
de Paz RH, Serrano-Muñoz D, Pérez-Nombela S, Bravo-Esteban E, Avendaño-Coy J, Gómez-Soriano J (2019) Combining transcranial direct-current stimulation with gait training in patients with neurological disorders: a systematic review. J Neuroeng Rehabil 16(1):114
Debaere F, Swinnen SP, Beatse E, Sunaert S, Van Hecke P, Duysens J (2001) Brain areas involved in interlimb coordination: a distributed network. Neuroimage 14(5):947–958
Defebvre LJ, Krystkowiak P, Blatt JL, Duhamel A, Bourriez JL, Perina M et al (2002) Influence of pallidal stimulation and levodopa on gait and preparatory postural adjustments in Parkinson's disease. Mov Disord 17(1):76–83
DeLong MR, Wichmann T (2015) Basal ganglia circuits as targets for neuromodulation in Parkinson disease. JAMA Neurol 72(11):1354–1360
Dietz V, Michel J (2009) Human bipeds use quadrupedal coordination during locomotion. Ann N Y Acad Sci 1164:97–103
Dietz V, Colombo G, Jensen L, Baumgartner L (1995) Locomotor capacity of spinal cord in paraplegic patients. Ann Neurol 37(5):574–582
Dietz V, Fouad K, Bastiaanse CM (2001) Neuronal coordination of arm and leg movements during human locomotion. Eur J Neurosci 14(11):1906–1914
Douglas JR, Noga BR, Dai X, Jordan LM (1993) The effects of intrathecal administration of excitatory amino acid agonists and antagonists on the initiation of locomotion in the adult cat. J Neurosci 13(3):990–1000
Drouot X, Oshino S, Jarraya B, Besret L, Kishima H, Remy P et al (2004) Functional recovery in a primate model of Parkinson's disease following motor cortex stimulation. Neuron 44(5):769–778
Duysens J, Van de Crommert HW (1998) Neural control of locomotion; the central pattern generator from cats to humans. Gait Posture 7(2):131–141
Ehgoetz Martens KA, Hall JM, Georgiades MJ, Gilat M, Walton CC, Matar E et al (2018) The functional network signature of heterogeneity in freezing of gait. Brain 141(4):1145–1160
Eidelberg E (1981) Consequences of spinal cord lesions upon motor function, with special reference to locomotor activity. Prog Neurobiol 17(3):185–202
Eidelberg E, Yu J (1981) Effects of corticospinal lesions upon treadmill locomotion by cats. Exp Brain Res 43(1):101–103
Elahi B, Elahi B, Chen R (2009) Effect of transcranial magnetic stimulation on Parkinson motor function—systematic review of controlled clinical trials. Mov Disord 24(3):357–363

Espay AJ, Fasano A, van Nuenen BFL, Payne MM, Snijders AH, Bloem BR (2012) "On" state freezing of gait in Parkinson disease. Neurology 78(7):454

Farris SM, Giroux ML (2016) Rapid assessment of gait and speech after subthalamic deep brain stimulation. Surg Neurol Int 7(Suppl 19):S545–S550

Fasano A, Piano C, De Simone C, Cioni B, Di Giuda D, Zinno M et al (2008) High frequency extradural motor cortex stimulation transiently improves axial symptoms in a patient with Parkinson's disease. Mov Disord 23(13):1916–1919

Fasano A, Aquino CC, Krauss JK, Honey CR, Bloem BR (2015) Axial disability and deep brain stimulation in patients with Parkinson disease. Nat Rev Neurol 11(2):98–110

Fehlings MG, Vaccaro A, Wilson JR, Singh A, Cadotte DW, Harrop JS et al (2012) Early versus delayed decompression for traumatic cervical spinal cord injury: results of the Surgical Timing in Acute Spinal Cord Injury Study (STASCIS). PloS One 7(2):e32037

Fenelon G, Goujon C, Gurruchaga JM, Cesaro P, Jarraya B, Palfi S et al (2012) Spinal cord stimulation for chronic pain improved motor function in a patient with Parkinson's disease. Parkinsonism Relat Disord 18(2):213–214

Ferraye MU, Debu B, Pollak P (2008) Deep brain stimulation effect on freezing of gait. Mov Disord 23(Suppl 2):S489–S494

Fleury V, Pollak P, Gere J, Tommasi G, Romito L, Combescure C et al (2016) Subthalamic stimulation may inhibit the beneficial effects of levodopa on akinesia and gait. Mov Disord 31(9):1389–1397

Follett KA, Torres-Russotto D (2012) Deep brain stimulation of globus pallidus interna, subthalamic nucleus, and pedunculopontine nucleus for Parkinson's disease: which target? Parkinsonism Relat Disord 18:S165–S1S7

Fonoff ET, de Lima-Pardini AC, Coelho DB, Monaco BA, Machado B, Pinto de Souza C et al (2019) Spinal cord stimulation for freezing of gait: from bench to bedside. Front Neurol 10:905

Forssberg H (1985) Ontogeny of human locomotor control. I. Infant stepping, supported locomotion and transition to independent locomotion. Exp Brain Res 57(3):480–493

Frazer A, Williams J, Spittles M, Rantalainen T, Kidgell D (2016) Anodal transcranial direct current stimulation of the motor cortex increases cortical voluntary activation and neural plasticity. Muscle Nerve 54(5):903–913

Fregni F, Simon DK, Wu A, Pascual-Leone A (2005) Non-invasive brain stimulation for Parkinson's disease: a systematic review and meta-analysis of the literature. J Neurol Neurosurg Psychiatry 76(12):1614–1623

Fuentes R, Petersson P, Siesser WB, Caron MG, Nicolelis MAL (2009) Spinal cord stimulation restores locomotion in animal models of Parkinson's disease. Science 323(5921):1578

Fuhrer MJ, Garber SL, Rintala DH, Clearman R, Hart KA (1993) Pressure ulcers in community-resident persons with spinal cord injury: prevalence and risk factors. Arch Phys Med Rehabil 74(11):1172–1177

Fukuyama H, Ouchi Y, Matsuzaki S, Nagahama Y, Yamauchi H, Ogawa M et al (1997) Brain functional activity during gait in normal subjects: a SPECT study. Neurosci Lett 228(3):183–186

Furigo IC, de Oliveira WF, de Oliveira AR, Comoli E, Baldo MV, Mota-Ortiz SR et al (2010) The role of the superior colliculus in predatory hunting. Neuroscience 165(1):1–15

Gad PN, Roy RR, Zhong H, Gerasimenko YP, Taccola G, Edgerton VR (2016) Neuromodulation of the neural circuits controlling the lower urinary tract. Exp Neurol 285(Pt B):182–189

Galna B, Lord S, Burn DJ, Rochester L (2015) Progression of gait dysfunction in incident Parkinson's disease: impact of medication and phenotype. Mov Disord 30(3):359–367

Garcia-Rill E, Houser CR, Skinner RD, Smith W, Woodward DJ (1987) Locomotion-inducing sites in the vicinity of the pedunculopontine nucleus. Brain Res Bull 18(6):731–738

Garcia-Rill E, Saper CB, Rye DB, Kofler M, Nonnekes J, Lozano A et al (2019) Focus on the pedunculopontine nucleus. Consensus review from the May 2018 brainstem society meeting in Washington, DC, USA. Clin Neurophysiol 130(6):925–940

Gerasimenko YP, Lavrov IA, Courtine G, Ichiyama RM, Dy CJ, Zhong H et al (2006) Spinal cord reflexes induced by epidural spinal cord stimulation in normal awake rats. J Neurosci Methods 157(2):253–263

Giladi N, Nieuwboer A (2008) Understanding and treating freezing of gait in parkinsonism, proposed working definition, and setting the stage. Mov Disord 23(Suppl 2):S423–S425
Giladi N, McMahon D, Przedborski S, Flaster E, Guillory S, Kostic V et al (1992) Motor blocks in Parkinson's disease. Neurology 42(2):333–339
Goetz CG, Fahn S, Martinez-Martin P, Poewe W, Sampaio C, Stebbins GT et al (2007) Movement Disorder Society-sponsored revision of the Unified Parkinson's Disease Rating Scale (MDS-UPDRS): process, format, and clinimetric testing plan. Mov Disord 22(1):41–47
Goetz L, Bhattacharjee M, Ferraye MU, Fraix V, Maineri C, Nosko D et al (2019) Deep brain stimulation of the pedunculopontine nucleus area in Parkinson disease: MRI-based anatomoclinical correlations and optimal target. Neurosurgery 84(2):506–518
Golestanirad L, Elahi B, Graham SJ, Das S, Wald LL (2016) Efficacy and safety of pedunculopontine nuclei (PPN) deep brain stimulation in the treatment of gait disorders: a meta-analysis of clinical studies. Can J Neurol Sci 43(1):120–126
Goulding M (2009) Circuits controlling vertebrate locomotion: moving in a new direction. Nat Rev Neurosci 10(7):507–518
Grabli D, Karachi C, Welter M-L, Lau B, Hirsch EC, Vidailhet M et al (2012) Normal and pathological gait: what we learn from Parkinson's disease. J Neurol Neurosurg Psychiatry 83(10):979–985
Grillner S, Zangger P (1979) On the central generation of locomotion in the low spinal cat. Exp Brain Res 34(2):241–261
Guertin P (2013) Central pattern generator for locomotion: anatomical, physiological, and pathophysiological considerations. Front Neurol 3:183
Gutierrez JC, Seijo FJ, Alvarez Vega MA, Fernandez Gonzalez F, Lozano Aragoneses B, Blazquez M (2009) Therapeutic extradural cortical stimulation for Parkinson's disease: report of six cases and review of the literature. Clin Neurol Neurosurg 111(8):703–707
Hachmann JT, Jeong JH, Grahn PJ, Mallory GW, Evertz LQ, Bieber AJ et al (2013) Large animal model for development of functional restoration paradigms using epidural and intraspinal stimulation. PLoS One 8(12):e81443
Hanakawa T, Fukuyama H, Katsumi Y (1999a) Enhanced lateral premotor activity during paradoxical gait in Parkinson's disease. Ann Neurol 45:329–336
Hanakawa T, Katsumi Y, Fukuyama H (1999b) Mechanisms underlying gait disturbance in Parkinson's disease: a single photon emission computed tomography study. Brain 122:1271–1282
Harkema S, Gerasimenko Y, Hodes J, Burdick J, Angeli C, Chen Y et al (2011) Effect of epidural stimulation of the lumbosacral spinal cord on voluntary movement, standing, and assisted stepping after motor complete paraplegia: a case study. Lancet (London, England) 377(9781):1938–1947
Hassan S, Amer S, Alwaki A, Elborno A (2013) A patient with Parkinson's disease benefits from spinal cord stimulation. J Clin Neurosci 20(8):1155–1156
Hirsch EC, Graybiel AM, Duyckaerts C, Javoy-Agid F (1987) Neuronal loss in the pedunculopontine tegmental nucleus in Parkinson disease and in progressive supranuclear palsy. Proc Natl Acad Sci U S A 84(16):5976–5980
Ho S, Mohtadi A, Daud K, Leonards U, Handy TC (2019) Using smartphone accelerometry to assess the relationship between cognitive load and gait dynamics during outdoor walking. Sci Rep 9(1):3119
Hofstoetter US, Danner SM, Freundl B, Binder H, Mayr W, Rattay F et al (2015) Periodic modulation of repetitively elicited monosynaptic reflexes of the human lumbosacral spinal cord. J Neurophysiol 114(1):400–410
Hong M, Earhart GM (2010) Effects of medication on turning deficits in individuals with Parkinson's disease. J Neurol Phys Ther 34(1):11–16
Hong M, Perlmutter JS, Earhart GM (2009) A kinematic and electromyographic analysis of turning in people with Parkinson disease. Neurorehabil Neural Repair 23(2):166–176

Ichiyama RM, Gerasimenko YP, Zhong H, Roy RR, Edgerton VR (2005) Hindlimb stepping movements in complete spinal rats induced by epidural spinal cord stimulation. Neurosci Lett 383(3):339–344

Jacobs JV, Lou JS, Kraakevik JA, Horak FB (2009) The supplementary motor area contributes to the timing of the anticipatory postural adjustment during step initiation in participants with and without Parkinson's disease. Neuroscience 164(2):877–885

Jahn K, Deutschlander A, Stephan T, Strupp M, Wiesmann M, Brandt T (2004) Brain activation patterns during imagined stance and locomotion in functional magnetic resonance imaging. Neuroimage 22(4):1722–1731

Jahn K, Deutschlander A, Stephan T, Kalla R, Hufner K, Wagner J et al (2008a) Supraspinal locomotor control in quadrupeds and humans. Prog Brain Res 171:353–362

Jahn K, Deutschlander A, Stephan T, Kalla R, Wiesmann M, Strupp M et al (2008b) Imaging human supraspinal locomotor centers in brainstem and cerebellum. Neuroimage 39(2):786–792

Jang SH, Lee HD (2017) Gait recovery by activation of the unaffected corticoreticulospinal tract in a stroke patient: a case report. Medicine (Baltimore) 96(50):e9123

Jang SH, Lee HD (2019) Late recovery of walking ability in a person with chronic stroke after an individualized rehabilitation program. Ann Phys Rehabil Med 62(5):386–388

Jang SH, Chang CH, Lee J, Kim CS, Seo JP, Yeo SS (2013) Functional role of the corticoreticular pathway in chronic stroke patients. Stroke 44(4):1099–1104

Jellinger K (1988) The pedunculopontine nucleus in Parkinson's disease, progressive supranuclear palsy and Alzheimer's disease. J Neurol Neurosurg Psychiatry 51(4):540–543

Jenkinson N, Nandi D, Miall RC, Stein JF, Aziz TZ (2004) Pedunculopontine nucleus stimulation improves akinesia in a Parkinsonian monkey. Neuroreport 15(17):2621–2624

Josset N, Roussel M, Lemieux M, Lafrance-Zoubga D, Rastqar A, Bretzner F (2018) Distinct contributions of mesencephalic locomotor region nuclei to locomotor control in the freely behaving mouse. Curr Biol 28(6):884–901.e3

Katayama Y, Fukaya C, Yamamoto T (1998) Poststroke pain control by chronic motor cortex stimulation: neurological characteristics predicting a favorable response. J Neurosurg 89(4):585–591

Khoo HM, Kishima H, Hosomi K, Maruo T, Tani N, Oshino S et al (2014) Low-frequency subthalamic nucleus stimulation in Parkinson's disease: a randomized clinical trial. Mov Disord 29(2):270–274

Kiehn O (2016) Decoding the organization of spinal circuits that control locomotion. Nat Rev Neurosci 17(4):224–238

Kim MS, Chang WH, Cho JW, Youn J, Kim YK, Kim SW et al (2015) Efficacy of cumulative high-frequency rTMS on freezing of gait in Parkinson's disease. Restor Neurol Neurosci 33:521–530

Kim SJ, Paeng SH, Kang SY (2018) Stimulation in supplementary motor area versus motor cortex for freezing of gait in Parkinson's disease. J Clin Neurol (Seoul, Korea) 14(3):320–326

Kleiner-Fisman G, Fisman DN, Kahn FI, Sime E, Lozano AM, Lang AE (2003) Motor cortical stimulation for parkinsonism in multiple system atrophy. Arch Neurol 60(11):1554–1558

Koganemaru S, Mikami Y, Maezawa H, Matsuhashi M, Ikeda S, Ikoma K et al (2018) Anodal transcranial patterned stimulation of the motor cortex during gait can induce activity-dependent corticospinal plasticity to alter human gait. PLoS One 13(12):e0208691

Kowalski KE, Hsieh YH, Dick TE, DiMarco AF (2013) Diaphragm activation via high frequency spinal cord stimulation in a rodent model of spinal cord injury. Exp Neurol 247:689–693

Kwon BK, Tetzlaff W, Grauer JN, Beiner J, Vaccaro AR (2004) Pathophysiology and pharmacologic treatment of acute spinal cord injury. Spine J 4(4):451–464

Lamberti P, Armenise S, Castaldo V, de Mari M, Iliceto G, Tronci P et al (1997) Freezing gait in Parkinson's disease. Eur Neurol 38(4):297–301

Lasfargues JE, Custis D, Morrone F, Carswell J, Nguyen T (1995) A model for estimating spinal cord injury prevalence in the United States. Paraplegia 33(2):62–68

Lee K (2017) Effects of single and dual tasks during walking on spatiotemporal gait parameters of community-dwelling older. J Phys Ther Sci 29(10):1874–1877

Lee AM, Hoy JL, Bonci A, Wilbrecht L, Stryker MP, Niell CM (2014) Identification of a brainstem circuit regulating visual cortical state in parallel with locomotion. Neuron 83(2):455–466
Lee JW, Song YS, Kim H, Ku BD, Lee WW (2019) Alteration of tremor dominant and postural instability gait difficulty subtypes during the progression of Parkinson's disease: analysis of the PPMI cohort. Front Neurol 10:471
Lefaucheur J-P, Antal A, Ayache SS, Benninger DH, Brunelin J, Cogiamanian F et al (2017) Evidence-based guidelines on the therapeutic use of transcranial direct current stimulation (tDCS). Clin Neurophysiol 128(1):56–92
Lugade V, Kaufman K (2014) Dynamic stability margin using a marker based system and Tekscan: a comparison of four gait conditions. Gait Posture 40(1):252–254
Lundin-Olsson L, Nyberg L, Gustafson Y (1997) "Stops walking when talking" as a predictor of falls in elderly people. Lancet (London, England) 349(9052):617
Macht M, Kaussner Y, Möller JC, Stiasny-Kolster K, Eggert KM, Krüger H-P et al (2007) Predictors of freezing in Parkinson's disease: a survey of 6,620 patients. Mov Disord 22(7):953–956
MacKinnon CD (2018) Sensorimotor anatomy of gait, balance, and falls. Handb Clin Neurol 159:3–26
Magoun HW, Ranson SW (1938) The behavior of cats following bilateral removal of the rostral portion of the cerebral hemispheres. J Neurophysiol 1(1):39–44
Mahlknecht P, Kiechl S, Bloem BR, Willeit J, Scherfler C, Gasperi A et al (2013) Prevalence and burden of gait disorders in elderly men and women aged 60-97 years: a population-based study. PLoS One 8(7):e69627
Maki BE, McIlroy WE (1996) Postural control in the older adult. Clin Geriatr Med 12(4):635–658
Martinez-Gonzalez C, Bolam JP, Mena-Segovia J (2011) Topographical organization of the pedunculopontine nucleus. Front Neuroanat 5:22
Masani K, Vette AH, Kouzaki M, Kanehisa H, Fukunaga T, Popovic MR (2007) Larger center of pressure minus center of gravity in the elderly induces larger body acceleration during quiet standing. Neurosci Lett 422(3):202–206
Matsuyama K, Mori F, Nakajima K, Drew T, Aoki M, Mori S (2004) Locomotor role of the corticoreticular-reticulospinal-spinal interneuronal system. Prog Brain Res 143:239–249
Mazzone P, Lozano A, Stanzione P, Galati S, Scarnati E, Peppe A et al (2005) Implantation of human pedunculopontine nucleus: a safe and clinically relevant target in Parkinson's disease. Neuroreport 16(17):1877–1881
Mei S, Eisinger RS, Hu W, Tsuboi T, Foote KD, Hass CJ et al (2020) Three-year gait and axial outcomes of bilateral STN and GPi Parkinson's disease deep brain stimulation. Front Hum Neurosci 14:1
Melzack R, Wall PD (1965) Pain mechanisms: a new theory. Science 150(3699):971–979
Mi TM, Garg S, Ba F, Liu AP, Wu T, Gao LL et al (2019) High-frequency rTMS over the supplementary motor area improves freezing of gait in Parkinson's disease: a randomized controlled trial. Parkinsonism Relat Disord 68:85–90
Miller DM, DeMayo WM, Bourdages GH, Wittman SR, Yates BJ, McCall AA (2017) Neurons in the pontomedullary reticular formation receive converging inputs from the hindlimb and labyrinth. Exp Brain Res 235(4):1195–1207
Milosevic L, Kalia SK, Hodaie M, Lozano AM, Fasano A, Popovic MR et al (2017) Neuronal inhibition and synaptic plasticity of basal ganglia neurons in Parkinson's disease. Brain 141(1):177–190
Minassian K, Hofstoetter US, Dzeladini F, Guertin PA, Ijspeert A (2017) The human central pattern generator for locomotion: does it exist and contribute to walking? Neuroscientist 23(6):649–663
Miyai I, Tanabe HC, Sase I, Eda H, Oda I, Konishi I et al (2001) Cortical mapping of gait in humans: a near-infrared spectroscopic topography study. Neuroimage 14(5):1186–1192
Moraud EM, Capogrosso M, Formento E, Wenger N, DiGiovanna J, Courtine G et al (2016) Mechanisms underlying the neuromodulation of spinal circuits for correcting gait and balance deficits after spinal cord injury. Neuron 89(4):814–828

Moreau C, Defebvre L, Destee A, Bleuse S, Clement F, Blatt JL et al (2008) STN-DBS frequency effects on freezing of gait in advanced Parkinson disease. Neurology 71(2):80–84

Mori S, Matsui T, Kuze B, Asanome M, Nakajima K, Matsuyama K (1999) Stimulation of a restricted region in the midline cerebellar white matter evokes coordinated quadrupedal locomotion in the decerebrate cat. J Neurophysiol 82(1):290–300

Moro E, Esselink RJ, Xie J, Hommel M, Benabid AL, Pollak P (2002) The impact on Parkinson's disease of electrical parameter settings in STN stimulation. Neurology 59(5):706–713

Moro E, Hamani C, Poon YY, Al-Khairallah T, Dostrovsky JO, Hutchison WD et al (2010) Unilateral pedunculopontine stimulation improves falls in Parkinson's disease. Brain 133(Pt 1):215–224

Moro E, Schwalb JM, Piboolnurak P, Poon Y-YW, Hamani C, Hung SW et al (2011) Unilateral subdural motor cortex stimulation improves essential tremor but not Parkinson's disease. Brain 134(7):2096–2105

Mostofi A, Evans JM, Partington-Smith L, Yu K, Chen C, Silverdale MA (2019) Outcomes from deep brain stimulation targeting subthalamic nucleus and caudal zona incerta for Parkinson's disease. NPJ Parkinsons Dis 5:17

Nadeau S, Duclos C, Bouyer L, Richards CL (2011) Chapter 11—Guiding task-oriented gait training after stroke or spinal cord injury by means of a biomechanical gait analysis. In: Green A, Chapman CE, Kalaska JF, Lepore F (eds) Progress in brain research, vol 192. Elsevier, Amsterdam, pp 161–180

Nandi D, Liu X, Winter JL, Aziz TZ, Stein JF (2002) Deep brain stimulation of the pedunculopontine region in the normal non-human primate. J Clin Neurosci 9(2):170–174

Nandi D, Jenkinson N, Stein J, Aziz T (2008) The pedunculopontine nucleus in Parkinson's disease: primate studies. Br J Neurosurg 22(Suppl 1):S4–S8

Nathan PW, Smith M, Deacon P (1996) Vestibulospinal, reticulospinal and descending propriospinal nerve fibres in man. Brain 119(Pt 6):1809–1833

Nguyen JP, Pollin B, Feve A, Geny C, Cesaro P (1998) Improvement of action tremor by chronic cortical stimulation. Mov Disord 13(1):84–88

Nielsen JF, Sinkjaer T (1997) Long-lasting depression of soleus motoneurons excitability following repetitive magnetic stimuli of the spinal cord in multiple sclerosis patients. Mult Scler 3(1):18–30

Noga BR, Kettler J, Jordan LM (1988) Locomotion produced in mesencephalic cats by injections of putative transmitter substances and antagonists into the medial reticular formation and the pontomedullary locomotor strip. J Neurosci 8(6):2074–2086

Noga BR, Kriellaars DJ, Brownstone RM, Jordan LM (2003) Mechanism for activation of locomotor centers in the spinal cord by stimulation of the mesencephalic locomotor region. J Neurophysiol 90(3):1464–1478

Noga BR, Turkson RP, Xie S, Taberner A, Pinzon A, Hentall ID (2017) Monoamine release in the cat lumbar spinal cord during fictive locomotion evoked by the mesencephalic locomotor region. Front Neural Circuits 11:59

Nonnekes J, Snijders AH, Nutt JG, Deuschl G, Giladi N, Bloem BR (2015) Freezing of gait: a practical approach to management. Lancet Neurol 14(7):768–778

Nudo RJ, Masterton RB (1988) Descending pathways to the spinal cord: a comparative study of 22 mammals. J Comp Neurol 277(1):53–79

Obeso JA, Olanow CW, Rodriguez-Oroz MC, Krack P, Kumar R, Lang AE (2001) Deep-brain stimulation of the subthalamic nucleus or the pars interna of the globus pallidus in Parkinson's disease. N Engl J Med 345(13):956–963

Odin P, Ray Chaudhuri K, Slevin JT, Volkmann J, Dietrichs E, Martinez-Martin P et al (2015) Collective physician perspectives on non-oral medication approaches for the management of clinically relevant unresolved issues in Parkinson's disease: consensus from an international survey and discussion program. Parkinsonism Relat Disord 21(10):1133–1144

Opris I, Dai X, Johnson DMG, Sanchez FJ, Villamil LM, Xie S et al (2019) Activation of brainstem neurons during mesencephalic locomotor region-evoked locomotion in the cat. Front Syst Neurosci 13:69

Orlovsky GN (1969) Spontaneous and induced locomotion of the thalamic cat. Biophysics 15:1154–1162. (Translated from Russian)

Pagni CA, Altibrandi MG, Bentivoglio A, Caruso G, Cioni B, Fiorella C et al (2005) Extradural motor cortex stimulation (EMCS) for Parkinson's disease. History and first results by the study group of the Italian Neurosurgical Society. Acta Neurochir Suppl 93:113–119

Pahwa R, Factor SA, Lyons KE, Ondo WG, Gronseth G, Bronte-Stewart H et al (2006) Practice parameter: treatment of Parkinson disease with motor fluctuations and dyskinesia (an evidence-based review): report of the Quality Standards Subcommittee of the American Academy of Neurology. Neurology 66(7):983–995

Pascual-Leone A, Valls-Sole J, Wassermann EM, Hallett M (1994) Responses to rapid-rate transcranial magnetic stimulation of the human motor cortex. Brain 117(Pt 4):847–858

Peng L, Fu J, Ming Y, Zeng S, He H, Chen L (2018) The long-term efficacy of STN vs GPi deep brain stimulation for Parkinson disease: a meta-analysis. Medicine (Baltimore) 97(35):e12153

Perez-Lloret S, Negre-Pages L, Damier P, Delval A, Derkinderen P, Destee A et al (2014) Prevalence, determinants, and effect on quality of life of freezing of gait in Parkinson disease. JAMA Neurol 71(7):884–890

Piallat B, Chabardes S, Torres N, Fraix V, Goetz L, Seigneuret E et al (2009) Gait is associated with an increase in tonic firing of the sub-cuneiform nucleus neurons. Neuroscience 158(4):1201–1205

Pienaar IS, Elson JL, Racca C, Nelson G, Turnbull DM, Morris CM (2013) Mitochondrial abnormality associates with type-specific neuronal loss and cell morphology changes in the pedunculopontine nucleus in Parkinson disease. Am J Pathol 183(6):1826–1840

Pinto de Souza C, Hamani C, Oliveira Souza C, Lopez Contreras WO, dos Santos Ghilardi MG, Cury RG et al (2017) Spinal cord stimulation improves gait in patients with Parkinson's disease previously treated with deep brain stimulation. Mov Disord 32(2):278–282

Pirker W, Katzenschlager R (2017) Gait disorders in adults and the elderly: a clinical guide. Wien Klin Wochenschr 129(3–4):81–95

Plaha P, Gill SS (2005) Bilateral deep brain stimulation of the pedunculopontine nucleus for Parkinson's disease. Neuroreport 16(17):1883–1887

Postuma RB, Berg D, Stern M, Poewe W, Olanow CW, Oertel W et al (2015) MDS clinical diagnostic criteria for Parkinson's disease. Mov Disord 30(12):1591–1601

Pötter-Nerger M, Volkmann J (2013) Deep brain stimulation for gait and postural symptoms in Parkinson's disease. Mov Disord 28(11):1609–1615

Pozzi NG, Canessa A, Palmisano C, Brumberg J, Steigerwald F, Reich MM et al (2019) Freezing of gait in Parkinson's disease reflects a sudden derangement of locomotor network dynamics. Brain 142(7):2037–2050

Rattay F, Minassian K, Dimitrijevic MR (2000) Epidural electrical stimulation of posterior structures of the human lumbosacral cord: 2. quantitative analysis by computer modeling. Spinal Cord 38(8):473–489

Redgrave P, Prescott TJ, Gurney K (1999) The basal ganglia: a vertebrate solution to the selection problem? Neuroscience 89(4):1009–1023

Rejc E, Angeli C, Harkema S (2015) Effects of lumbosacral spinal cord epidural stimulation for standing after chronic complete paralysis in humans. PLoS One 10(7):e0133998

Rejc E, Angeli CA, Bryant N, Harkema SJ (2017a) Effects of stand and step training with epidural stimulation on motor function for standing in chronic complete paraplegics. J Neurotrauma 34(9):1787–1802

Rejc E, Angeli CA, Atkinson D, Harkema SJ (2017b) Motor recovery after activity-based training with spinal cord epidural stimulation in a chronic motor complete paraplegic. Sci Rep 7(1):13476

Rocchi L, Chiari L, Cappello A, Horak FB (2006) Identification of distinct characteristics of postural sway in Parkinson's disease: a feature selection procedure based on principal component analysis. Neurosci Lett 394(2):140–145

Rohani M, Kalsi-Ryan S, Lozano AM, Fasano A (2017) Spinal cord stimulation in primary progressive freezing of gait. Mov Disord 32(9):1336–1337

Roper JA, Kang N, Ben J, Cauraugh JH, Okun MS, Hass CJ (2016) Deep brain stimulation improves gait velocity in Parkinson's disease: a systematic review and meta-analysis. J Neurol 263(6):1195–1203

Ryczko D, Dubuc R (2013) The multifunctional mesencephalic locomotor region. Curr Pharm Des 19(24):4448–4470

Ryken TC, Hurlbert RJ, Hadley MN, Aarabi B, Dhall SS, Gelb DE et al (2013) The acute cardiopulmonary management of patients with cervical spinal cord injuries. Neurosurgery 72(Suppl 2):84–92

Samotus O, Parrent A, Jog M (2018) Spinal cord stimulation therapy for gait dysfunction in advanced Parkinson's disease patients. Mov Disord 33(5):783–792

Santana Maxwell B, Halje P, Simplício H, Richter U, Freire Marco Aurelio M, Petersson P et al (2014) Spinal cord stimulation alleviates motor deficits in a primate model of Parkinson disease. Neuron 84(4):716–722

Santens P (2018) Neuromodulatory procedures for gait disorders in Parkinson's disease. Acta Neurol Belg 118(1):13–19

Santos MJ, Kanekar N, Aruin AS (2010) The role of anticipatory postural adjustments in compensatory control of posture: 2. Biomechanical analysis. J Electromyogr Kinesiol 20(3):398–405

Schlenstedt C, Shalash A, Muthuraman M, Falk D, Witt K, Deuschl G (2017) Effect of high-frequency subthalamic neurostimulation on gait and freezing of gait in Parkinson's disease: a systematic review and meta-analysis. Eur J Neurol 24(1):18–26

Scholten M, Klemt J, Heilbronn M, Plewnia C, Bloem BR, Bunjes F et al (2017) Effects of subthalamic and nigral stimulation on gait kinematics in Parkinson's disease. Front Neurol 8:543

Sebille SB, Rolland AS, Faillot M, Perez-Garcia F, Colomb-Clerc A, Lau B et al (2019) Normal and pathological neuronal distribution of the human mesencephalic locomotor region. Mov Disord 34(2):218–227

Sewell MD, Vachhani K, Alrawi A, Williams R (2018) Results of early and late surgical decompression and stabilization for acute traumatic cervical spinal cord injury in patients with concomitant chest injuries. World Neurosurg 118:e161–e1e5

Shealy CN, Mortimer JT, Reswick JB (1967) Electrical inhibition of pain by stimulation of the dorsal columns: preliminary clinical report. Anesth Analg 46(4):489–491

Sherman D, Fuller PM, Marcus J, Yu J, Zhang P, Chamberlin NL et al (2015) Anatomical location of the mesencephalic locomotor region and its possible role in locomotion, posture, cataplexy, and parkinsonism. Front Neurol 6:140

Shik ML, Orlovsky GN (1976) Neurophysiology of locomotor automatism. Physiol Rev 56(3):465–501

Shik ML, Severin FV, Orlovskii GN (1966) [Control of walking and running by means of electric stimulation of the midbrain]. Biofizika 11(4):659–666

Sinnamon HM, Karen SC (1987) Locomotion elicited by lateral hypothalamic stimulation in the anesthetized rat does not require the dorsal midbrain. Brain Res 402(1):78–86

Slotty PJ, Chang S, Honey CR (2015) Motor threshold: a possible guide to optimizing stimulation parameters for motor cortex stimulation. Neuromodulation 18(7):566–571; discussion 71–3

Snijders AH, Toni I, Ružička E, Bloem BR (2011) Bicycling breaks the ice for freezers of gait. Mov Disord 26(3):367–371

St George RJ, Nutt JG, Burchiel KJ, Horak FB (2010) A meta-regression of the long-term effects of deep brain stimulation on balance and gait in PD. Neurology 75(14):1292–1299

St George RJ, Carlson-Kuhta P, Burchiel KJ, Hogarth P, Frank N, Horak FB (2012) The effects of subthalamic and pallidal deep brain stimulation on postural responses in patients with Parkinson disease. J Neurosurg 116(6):1347–1356

St George RJ, Carlson-Kuhta P, Nutt JG, Hogarth P, Burchiel KJ, Horak FB (2014) The effect of deep brain stimulation randomized by site on balance in Parkinson's disease. Mov Disord 29(7):949–953

Stebbins GT, Goetz CG, Burn DJ, Jankovic J, Khoo TK, Tilley BC (2013) How to identify tremor dominant and postural instability/gait difficulty groups with the movement disorder society

unified Parkinson's disease rating scale: comparison with the unified Parkinson's disease rating scale. Mov Disord 28(5):668–670
Steeves JD, Jordan LM (1980) Localization of a descending pathway in the spinal cord which is necessary for controlled treadmill locomotion. Neurosci Lett 20(3):283–288
Steeves JD, Jordan LM (1984) Autoradiographic demonstration of the projections from the mesencephalic locomotor region. Brain Res 307(1–2):263–276
Stein E, Bar-Gad I (2013) Beta oscillations in the cortico-basal ganglia loop during parkinsonism. Exp Neurol 245:52–59
Steuer I, Guertin PA (2019) Central pattern generators in the brainstem and spinal cord: an overview of basic principles, similarities and differences. Rev Neurosci 30(2):107–164
Sudarsky L (2001) Gait disorders: prevalence, morbidity, and etiology. Adv Neurol 87:111–117
Takakusaki K (2017) Functional neuroanatomy for posture and gait control. J Mov Disord 10(1):1–17
Takakusaki K, Habaguchi T, Ohtinata-Sugimoto J, Saitoh K, Sakamoto T (2003) Basal ganglia efferents to the brainstem centers controlling postural muscle tone and locomotion: a new concept for understanding motor disorders in basal ganglia dysfunction. Neuroscience 119(1):293–308
Takakusaki K, Chiba R, Nozu T, Okumura T (2016) Brainstem control of locomotion and muscle tone with special reference to the role of the mesopontine tegmentum and medullary reticulospinal systems. J Neural Transm (Vienna, Austria : 1996) 123(7):695–729
Tashiro M, Itoh M, Fujimoto T, Fujiwara T, Ota H, Kubota K et al (2001) 18F-FDG PET mapping of regional brain activity in runners. J Sports Med Phys Fitness 41(1):11–17
Thevathasan W, Debu B, Aziz T, Bloem BR, Blahak C, Butson C et al (2018) Pedunculopontine nucleus deep brain stimulation in Parkinson's disease: a clinical review. Mov Disord 33(1):10–20
Thomas Thach W, Bastian AJ (2004) Role of the cerebellum in the control and adaptation of gait in health and disease. Prog Brain Res 143:353–366
Tinetti ME, Speechley M, Ginter SF (1988) Risk factors for falls among elderly persons living in the community. N Engl J Med 319(26):1701–1707
To WT, De Ridder D, Hart J Jr, Vanneste S (2018) Changing brain networks through non-invasive neuromodulation. Front Hum Neurosci 12:128
Tommasi G, Lopiano L, Zibetti M, Cinquepalmi A, Fronda C, Bergamasco B et al (2007) Freezing and hypokinesia of gait induced by stimulation of the subthalamic region. J Neurol Sci 258(1):99–103
Vale FL, Burns J, Jackson AB, Hadley MN (1997) Combined medical and surgical treatment after acute spinal cord injury: results of a prospective pilot study to assess the merits of aggressive medical resuscitation and blood pressure management. J Neurosurg 87(2):239–246
Valero-Cabre A, Amengual JL, Stengel C, Pascual-Leone A, Coubard OA (2017) Transcranial magnetic stimulation in basic and clinical neuroscience: a comprehensive review of fundamental principles and novel insights. Neurosci Biobehav Rev 83:381–404
Valldeoriola F, Munoz E, Rumia J, Roldan P, Camara A, Compta Y et al (2019) Simultaneous low-frequency deep brain stimulation of the substantia nigra pars reticulata and high-frequency stimulation of the subthalamic nucleus to treat levodopa unresponsive freezing of gait in Parkinson's disease: a pilot study. Parkinsonism Relat Disord 60:153–157
Villablanca JR, Marcus RJ, Olmstead CE (1976) Effects of caudate nuclei or frontal cortical ablations in cats. I. Neurology and gross behavior. Exp Neurol 52(3):389–420
Wagner FB, Mignardot J-B, Le Goff-Mignardot CG, Demesmaeker R, Komi S, Capogrosso M et al (2018) Targeted neurotechnology restores walking in humans with spinal cord injury. Nature 563(7729):65–71
Walker AE (1949) Cerebral pedunculotomy for the relief of involuntary movements; hemiballismus. Acta Psychiatr Neurol 24(3–4):723–729
Wang H, Gao H, Jiao T, Luo Z (2016) A meta-analysis of the pedunculopontine nucleus deep-brain stimulation effects on Parkinson's disease. Neuroreport 27(18):1336–1344

Wang JW, Zhang YQ, Zhang XH, Wang YP, Li JP, Li YJ (2017) Deep brain stimulation of pedunculopontine nucleus for postural instability and gait disorder after parkinson disease: a meta-analysis of individual patient data. World Neurosurg 102:72–78

Wannier T, Bastiaanse C, Colombo G, Dietz V (2001) Arm to leg coordination in humans during walking, creeping and swimming activities. Exp Brain Res 141(3):375–379

Weaver FM, Follett K, Stern M, Hur K, Harris C, Marks WJ Jr et al (2009) Bilateral deep brain stimulation vs best medical therapy for patients with advanced Parkinson disease: a randomized controlled trial. JAMA 301(1):63–73

Weiss D, Wachter T, Meisner C, Fritz M, Gharabaghi A, Plewnia C et al (2011) Combined STN/SNr-DBS for the treatment of refractory gait disturbances in Parkinson's disease: study protocol for a randomized controlled trial. Trials 12:222

Weiss D, Walach M, Meisner C, Fritz M, Scholten M, Breit S et al (2013) Nigral stimulation for resistant axial motor impairment in Parkinson's disease? A randomized controlled trial. Brain 136(Pt 7):2098–2108

Weiss D, Milosevic L, Gharabaghi A (2019) Deep brain stimulation of the substantia nigra for freezing of gait in Parkinson's disease: is it about stimulation frequency? Parkinsonism Relat Disord 63:229–230

Weiss D, Schoellmann A, Fox MD, Bohnen NI, Factor SA, Nieuwboer A et al (2020) Freezing of gait: understanding the complexity of an enigmatic phenomenon. Brain 143(1):14–30

Welniarz Q, Dusart I, Roze E (2017) The corticospinal tract: evolution, development, and human disorders. Dev Neurobiol 77(7):810–829

Wenger N, Moraud EM, Raspopovic S, Bonizzato M, DiGiovanna J, Musienko P et al (2014) Closed-loop neuromodulation of spinal sensorimotor circuits controls refined locomotion after complete spinal cord injury. Sci Transl Med 6(255):255ra133

Wilson JR, Witiw CD, Badhiwala J, Kwon BK, Fehlings MG, Harrop JS (2020) Early surgery for traumatic spinal cord injury: where are we now? Global Spine J 10(1 Suppl):84s–91s

Woolsey CN, Erickson TC, Gilson WE (1979) Localization in somatic sensory and motor areas of human cerebral cortex as determined by direct recording of evoked potentials and electrical stimulation. J Neurosurg 51(4):476–506

Wu T, Hallett M, Chan P (2015) Motor automaticity in Parkinson's disease. Neurobiol Dis 82:226–234

Yeo SS, Kim SH, Jang SH (2013) Proximal weakness due to injury of the corticoreticular pathway in a patient with traumatic brain injury. NeuroRehabilitation 32(3):665–669

Zanjani A, Zakzanis KK, Daskalakis ZJ, Chen R (2015) Repetitive transcranial magnetic stimulation of the primary motor cortex in the treatment of motor signs in Parkinson's disease: a quantitative review of the literature. Mov Disord 30(6):750–758

Zehr EP, Komiyama T, Stein RB (1997) Cutaneous reflexes during human gait: electromyographic and kinematic responses to electrical stimulation. J Neurophysiol 77(6):3311–3325

Zibetti M, Moro E, Krishna V, Sammartino F, Picillo M, Munhoz RP et al (2016) Low-frequency subthalamic stimulation in Parkinson's disease: long-term outcome and predictors. Brain Stimul 9(5):774–779

Ziegler K, Schroeteler F, Ceballos-Baumann AO, Fietzek UM (2010) A new rating instrument to assess festination and freezing gait in Parkinsonian patients. Mov Disord 25(8):1012–1018

Zitella LM, Mohsenian K, Pahwa M, Gloeckner C, Johnson MD (2013) Computational modeling of pedunculopontine nucleus deep brain stimulation. J Neural Eng 10(4):045005

Zwergal A, Linn J, Xiong G, Brandt T, Strupp M, Jahn K (2012) Aging of human supraspinal locomotor and postural control in fMRI. Neurobiol Aging 33(6):1073–1084

Augmentation and Rehabilitation with Active Orthotic Devices

Gelu Onose, Maria Veronica Morcov, Corina Sporea, Andrada Mirea, and Vlad Ciobanu

1 Introduction: Background

Despite the great research efforts, especially in the last decades, although there are several reasons for enthusiasm (Kordower and Tuszynski 2008) regarding the possibility to heal and/or regenerate damaged parts of the central nervous system (CNS), unfortunately the expectations have not yet been materialized.

Thus, regarding cerebral pathology "…there are no neuroprotective agents clinically available to counteract damage or promote repair after brain trauma ..." (Watts et al. 2015) and respectively, spinal cord injuries have "…limited therapeutic opportunities" (Zhou et al. 2014)—practically "... for which there currently is no cure ..." (Fan et al. 2018).

Overall, "neurological disorders"—encompassing predominantly vascular, degenerative/inflammatory, traumatic, infectious, tumoral conditions, and those that

G. Onose (✉)
Department of Physical and Rehabilitation Medicine, Carol Davila University of Medicine and Pharmacy, Bucharest, Romania

Teaching Emergency Hospital Bagdasar Arseni, Bucharest, Romania

M. V. Morcov · C. Sporea
National Teaching Center for Neuro-Psychomotor Rehabilitation in Children (NTCNRC) N. Robanescu, Bucharest, Romania

A. Mirea
Department of Physical and Rehabilitation Medicine, Carol Davila University of Medicine and Pharmacy, Bucharest, Romania

National Teaching Center for Neuro-Psychomotor Rehabilitation in Children (NTCNRC) N. Robanescu, Bucharest, Romania

V. Ciobanu
Computer Science Department, Politehnica University of Bucharest, Bucharest, Romania

I. Opris et al. (eds.), *Modern Approaches to Augmentation of Brain Function*, Contemporary Clinical Neuroscience,
https://doi.org/10.1007/978-3-030-54564-2_24

may affect CNS components: the brain and the spinal cord—are the "leading cause group of disability" (Feigin et al. 2017) because of their often devastating and long-life consequences, stroke being the "largest contributor to this burden globally" (Feigin et al. 2017).

It is necessary to ensure the highest possible quality of life (QOL)—according to the severity of impairment—and an enhanced motivation for training (within more complex procedural structure patterns, such as noninvasive advanced active orthotic device-assisted interventions, alongside/together with—on a case-by-case basis—classical related methods applied just by therapists) for people with neurological disabilities, too. Considering the important favorable role, in general, of exercise (of course, depending on the patient's general health state—mainly cardiovascular and respiratory—and cognitive capabilities, these enable him/her to cooperate during the use of advanced active orthotic devices (mainly but not exclusively mechatronic/robotic exoskeletons).

Taking into account, for instance: "… results indicate that the intervention, which involves massed and sustained practice of functional arm movements, also produces a massive use-dependent cortical reorganization that may provide the basis for the long-term persistence of the treatment effect for the 6 months studied…" (Liepert et al. 2000) such an assertion as the following arises accordingly: "There is substantial evidence of the impact of therapeutic exercise on cortical reorganization after stroke, with associated improvements in motor control and functional use of the affected limbs" (Stein 2019).

Consequently, important advances in modern technology have made possible consistent, non-pharmacological, therapeutic-rehabilitative, and/or compensatory/assistive device-mediated interventions, some of them very sophisticated, like for instance, of mechatronic/robotic, and/or sensor-based computer-aided "serious games" (Michmizos and Krebs 2012), and/or augmented/virtual reality, and/or neuromuscular/functional electrical stimulation, types. Many of them also entail/target augmentation of brain functions and thus, will be approached in this overview. Self-autonomy—encompassing critical components like mobility and accessibility—is one of the key dimensions of QOL. We thus emphasize the complex, including assistive, capabilities, of mobile mechatronic/robotic exoskeletons, that are continuously evolving in their constructive and related assistive-rehabilitative performances. They encouragingly progress toward replacing the (mainly but not exclusively) severe neurodisabled persons' classical means of circulation—the wheelchair (which itself still must be subject to optimizations until then)—and/or, respectively, contribute to substantial improvement of the upper extremity functionality and hence, self-care possibilities.

It should be mentioned that such an approach paradigm, realistically considering the current possibilities to properly compensate for the CNS damages and their (usually) severe and permanent disabling consequences, falls in line with the International Association of Neurorestoratology (IANR)'s statement (point/article no. 7): "Neurorestoratology recognizes the importance of small functional gains that have significant effects on quality of life. Neurorestoratology is interested in the mechanisms of spontaneous activity and enhancing this recovery" (Young et al. 2015).

Of course, the quest to overcome the enormous hurdle and challenge represented by the neurological condition, especially of the CNS, must and will continue, and in this respect the synthetic formulation of Kordower and Tuszynski ought to be cited: "…the 'holy grail' remains neuroprotection and regeneration of remaining systems and augmentation of existing pathways" (Kordower and Tuszynski 2008).

With respect to the topic of the book that includes this chapter, we begin by very briefly summarizing several defining considerations regarding the possibilities to enhance brain functions through apparatus interventions, starting from several conceptual outlines.

Thus, nowadays it is considered that the nervous system, and especially the brain as part of the CNS, bases its functioning on the morphological complex constituted by the tissue ensemble represented by neurons and their cell support populations and, respectively, the corresponding vascular structures, and subsequently, by a permanent control, running up to genic level, of four fundamental neurobiological processes: neurotrophicity, neuroprotection, neuroplasticity, and neurogenesis—considered the basic items of the so-called endogenous defense activity (EDA) (Muresanu et al. 2012).

Both the specific physiological performances and the reactivity in pathological situations (including for the last three EDA components) (Muresanu et al. 2012) are further conditioned/modulated by the activity of the complex entities and array of the neural circuits and connectivity network known as "the brain connectome" (Soiza-Reilly et al. 2015) within the subtle and complicated phenomena of synchronization/desynchronization on dynamic large scale (Bressler and Menon 2010; DeGracia 2010).

Brain irrigation improvement may be also considered a modality, although an indirect one, for the augmentation of brain functions. Acting on the previously mentioned cerebral morpho-functional components, there are currently, along with other medical approaches, a series of non-pharmacological, device-assisted interventional possibilities, which are recorded in the literature with references to having such augmentative effects, as indicated above.

From this perspective, modern research and instruments have been able—at least partially–to objectify the influence of sensory inputs and/or motion, being especially active, on brain activity and local nutritive blood flow.

Accordingly, it should be specified that functional Magnetic Resonance Imaging (fMRI) acts to measure the blood oxygen level-dependent (BOLD) contrast signal—specifically "through the deoxy-hemoglobin concentration [rHb]" (Parker et al. 2013) in venous blood, considering its paramagnetic effects "... based on the assumption that increased neural activity requires increased metabolism and therefore increased blood flow to the active region" (Parker et al. 2013). Thus, fMRI is able to give a "feedback of the hemodynamic brain activity" which also seems to be closely interrelated with the electric activity measured by EEG (Weiskopf et al. 2004).

Additionally, activation of brain function was indirectly assessed, by measuring oxyhemoglobin (OxyHb) through functional Near-InfraRed Spectroscopy (fNIRS) in the cerebral blood flow during passive and, respectively, active ("motivative"—performed with the "Pata" device) exercises. Thereby, including recently published, it has been "… found that the motivative exercise on an ankle activated the cerebral function more than the passive exercise [$P < 0.05$] and the motivative exercise and resistive self-exercise had no significant difference" (Takizawa et al. 2018).

For the moment, fNIRS is an insufficient standardized neuroimaging method, that consists in the assay of "…the concentration of oxygenated hemoglobin (oxyHb) and deoxygenated hemoglobin (deoxyHb—or rHb/HbR: o.n.), non-invasively, in the human brain" (Herold et al. 2018), conceptually relying on the same idea presented above, that there is a connection "between cerebral oxygenation/hemodynamics, physical activity, and cognition" (Herold et al. 2018).

More recently there is also magnetoencephalographic confirmation of sensorimotor stimulations' augmenting influence on brain-related functions (and thus substantiation for their use in "motor rehabilitation"). "Magnetoencephalography signals were recorded with a bandpass of 0.1–330 Hz": index finger movements (fast abductions), performed as a reaction to skin-delivered electrical stimulation (ES), produced higher activation of the "…primary motor (M1) and sensory (S1) cortices…" than "self-paced (SP) voluntary movements without external command". This represents a very recent confirmation of the—including physiologically based—possibility that skin-delivered electrical stimulations generate motor responses related to cortical activation, and thus "…support the application of cutaneous stimulation to assist motor rehabilitation" (Tarkka and Hautasaari 2019).

Further, we propose a synthetic overview, with systematization attempts, of the main approaches advanced active orthotic device-facilitated, with related rehabilitative and/or assistive, interventions.

2 Advanced Active Orthotic Devices Focusing on Mechatronic/Robotic Exoskeletons and Related Assistive/Therapeutic-Rehabilitative Types of Interventions

Taxonomically, the current advanced exoskeletons may be classified by their main indications and technologically conceived possibilities of use.

By the first systematic criterion, there are medical purposes and nonmedical ones: military purposes (to magnify soldiers' overall physical possibilities and consequent combat skills) (https://bleex.me.berkeley.edu/research/exoskeleton/bleex/; https://bleex.me.berkeley.edu/research/exoskeleton/exohiker/; https://bleex.me.berkeley.edu/research/exoskeleton/hulc/), mixed military and medical necessity situations [such as to improve the capabilities of specific personnel that acts to rescue individuals in war or civil disaster situations (https://www.cyberdyne.jp/

english/products/supporting.html)], work patterns involving the need to manipulate heavy patients and/or loads (https://www.cyberdyne.jp/english/products/supporting.html) and also, with various uses such as improving/making more comfortable the orthostatic posture; this latter purpose is an important objective for many people with different health conditions and linked stance difficulties and, respectively, gait-related limitation capability (especially in an aging population, but which, at the same time, has a growing appetite for being active, including with leisure/touristic aims) (https://asimo.honda.com/innovations/default.aspx?ID=body-weight-support-assist).

Synthetically, Neugebauer asserts four related fields: Military, Industry, Medical, and First Responders (Neugebauer 2017).

Regarding the medical applications of mechatronic/robotic exoskeletons, they may be divided, by and large, topographically, into devices designated to assist-rehabilitate neuromotor impairments in upper limbs or in lower limbs (including with locomotor conditions) (Onose et al. 2016, 2018). Further, either those for the upper limbs or the lower ones, may be designated to assist/train either a single-joint "target hitting task" or more than one; in connection with this, several basic related properties are defined, namely the range of motion (ROM) for every targeted joint, and the specific degrees of freedom (DOFs) the device can mobilize. Considering the biomechanical particularities of the assisted joint(s) either "single-DOF" or respectively, "multi-DOF movements" may be performed (Chang et al. 2019a).

In addition, such exoskeletons may be divided by their technologically conceived possibilities of use/interventions in: stationary and mobile (portable and/or wearable) devices.

It should be mentioned that, regarding "Mechatronic Wearable Exoskeletons for Bionic Bipedal Standing and Walking" and, respectively, "Mobile Mechatronic/Robotic Orthotic Devices to Assist–Rehabilitate Neuromotor Impairments in the Upper Limb", the main author and coauthors—some of them contributors to this chapter, too—have achieved recent consistent reviews including with items of technical/constructional kind—(Onose et al. 2016, 2018) with the abovementioned titles. So, this work mainly aims to emphasize current complementary-related data and thus minimize redundancy with the details already presented in there.

However, before going on to this chapter, a point that has to be reaffirmed is that "... currently there is still no such thing as an optimal, fully functional assistive-rehabilitative device (in the common sense of the term)" (Onose et al. 2018). This refers to many aspects: safety concerns regarding both, the man-machine interaction (including, for the ones dedicated to assist-rehabilitate severe motor deficits in the lower limbs: risk of falling), technical working ("appropriate autonomous power duty", durability and maintenance demands), availability for being set up, either by caregivers and/or by the customer him/her self, affordability (cost-effectiveness). For the wearable exoskeletons, it is also important to have: "...enough miniaturization and cosmetics..." (ideally aiming at reaching "... underwear dimensions") and "...very low/practically imperceptible noise when in service..." (Onose et al. 2018), thus giving the user inner and external emotional and mentality coping/assent, as well.

Stationary mechatronic/robotic exoskeletons are designed to be used, basically, in clinical settings, whereas some of the portable, and especially the wearable ones, can be of service to outpatients, mainly at home/indoors, and hopefully they will be sufficiently developed to be really effective in the community, too.

The main beneficial effects of contemporary mechatronic/robotic exoskeletons rely on two basic types of actions: the assistive and the training/rehabilitative, components.

Therefore, their structure must be as strong (especially for those used in lower limb pathologies) but as light as possible. At the same time, they should have: the best available partial gravity offload (Onose et al. 2016; Gandolla et al. 2017; Mazzoleni et al. 2017; Grimm et al. 2016) ["pelvis segment" (Stewart et al. 2017; Tu et al. 2017)—to prevent "... hairline fracture of the talus ..." (Benson et al. 2016)], protection (padding)—against mechanical factors/friction, heat and/or moisture, and/or skin-electrode dis-adherence—facilities (including a system to compensate the deficient venous lymphatic circulation in the lower limbs, mainly for complete paraplegics (Onose et al. 2016)—which, although necessary, are not achieved in current exoskeletons, to our knowledge). They also should allow at least functional ranges of motion (ROM) in the joints they work upon, and very importantly: with avoiding the danger to harm them by motorized movements, especially when acting on completely paralyzed territories.

As emphasized above, the stationary mechatronic/robotic exoskeletons provide therapeutic-rehabilitative interventions, the associated assistive dimension being possible with advanced mobile (portable and especially wearable) devices, such advanced apparatus also being able to function (to a limited extent) in a real overground environment, including in the community.

As for the training/rehabilitative type of action, such complex devices must provide, from high motion sensoristics, system control signals—including for assessing baseline and dynamic motility evolution and/or for calibrating the interventions they offer, according to the rehabilitation goals and the patient's current movement capabilities/progress—to real-time purposeful and balanced actuated movements, feedback-based [including by biofeedback electromyography—EMG), made possible by added virtual reality (VR) facilities (Mazzoleni et al. 2017; Grimm et al. 2016; Gilliaux et al. 2015; Buongiorno et al. 2018; Frisoli et al. 2016; Li et al. 2008; Kiguchi et al. 2008; Kawase et al. 2017; Mironov et al. 2017; Stein 2009; Dowling et al. 2014; Huang et al. 2012; Kim and Rosen 2015; Song et al. 2014; Thielbar et al. 2016; Wei et al. 2013a; Hesse et al. 2008, 2012)]. This may also be enforced by functional electrical stimulation (FES) (Li et al. 2008; Chen et al. 2013), thus using/training possibly residual neuromuscular capabilities, and/or by noninvasively collected cerebral voluntary motor commands within brain-computer-interfaces/brain-machine-interfaces (BCI/BMI), for supplementing the inputs to the exoskeletons' actuators through wireless transmitted brain motricity orders from the user, deciphered in digital signals from EEG (Grimm et al. 2016; Barsotti et al. 2015; Chéron et al. 2012; Contreras-Vidal et al. 2018; Kwak et al. 2015; Lebedev 2014; Nicolelis and Lebedev 2009; Onose et al. 2012; Presacco et al. 2012) [also: "Walk Again" (https://walk-again-project.org/#/en) and "Mindwalker", projects (https://www.utwente.nl/en/et/be/research/projects/MINDWALKER/)].

The main rehabilitative capabilities and consequent kinds of interventions thus possibly provided are:

- Fatigueless, "reproducible motor learning experience" (Charles et al. 2005), with constant accuracy, either passive or active movements—including with resistive components (and so-called "challenge-based" training for/of motor (neuro)plasticity, respectively, "assistance-as-needed" within a dialectical paradigm: "assistance is simply less challenge, and challenge is less assistance" (Marchal-Crespo and Reinkensmeyer 2009),
- (Accordingly) the treatment/training they can perform consists of numberless repetitive (rather distressful tedious/tiresome for a human caregiver) application of coordinated movements—prone to help the client in "motor learning" (Charles et al. 2005) and to achieve task-oriented goals (Grimm et al. 2016).
- They offer the possibility to avoid overloading the therapists in clinical and residential settings, as well (considering also the current and future overall "shortage" of "…professionals handy to deliver domiciliary physiotherapy/rehabilitation and nursing…") (Onose et al. 2018; Maciejasz et al. 2014; Norouzi-Gheidari et al. 2012; Noveanu et al. 2013).
- In addition, adequate—including on longer training sessions—isokinetic movements may be delivered (Mavroidis et al. 2005).

Taxonomically, the following related summarization is worth mentioning:

- Facilitator "assistive" (for)/functional movements aid,
- "Challenge-based" training for/of motor (neuro) "plasticity"
- "Normal tasks" and/or "haptic" simulation—including with the use of VR therapeutic "environment".
- ("Non-contact") "coaching"—more commonly used as an adjunct to task-oriented acquiring engrams (Marchal-Crespo and Reinkensmeyer 2009).

Subsequently, there is a very recent related synthetic design: "Fatigue free Novel Resistive; Variable, Can facilitate neuroplasticity, Functional and Treatment Outcome Assessment" (Chang et al. 2019a).

Regarding mechatronic/robotic devices currently used to assist/rehabilitate severe impairments in upper limbs' functionality, a well-known, by structure and methodology topographically targeted, classification, appropriately applicable, refers to: "…end-effector devices, exoskeletons and soft robots" (Chang et al. 2019a).

Aside from all the abovementioned technical and usage considerations and, respectively, the related details (Onose et al. 2018) concerning this kind of devices, two more specific aspects should also be emphasized, because of the extremely complex and complicated tasks necessary to be achieved by the upper limbs.

These are: retraining of mind/"eye, head and hand" coordination (Pelz et al. 2001; Carnahan 1992) and, respectively, "transparency" (Kim and Rosen 2015; Proietti et al. 2016) precision feedback (including with ensured "feedforward compensation of the gravity and friction, for example") (Proietti et al. 2016) for task-oriented fulfilled practices provided—making thereby the trained/assistive movements more natural—based on advanced overall technological performances of the device, characterized by the capability of the idle state 'choice' if action is not necessary: due to an artificial-based high level of abstractness (Proietti et al. 2016).

Consequently, regarding the mechatronic/robotic advanced active orthotic devices designated to rehabilitate upper limbs disabilities, in order to train dexterity for re-achievement of activities of daily living (ADL): "wearable robots…" are "… also used in (neuro)rehabilitation applications with the aim of enhancing the recovery process and minimizing functional disability, with consequent earlier reintegration in activities of daily living" (Buongiorno et al. 2018). Additionally, computer technology (sensor-based including VR and/or serious games "facilities") are nowadays both applied and considered, at least partially, useful.

VR—synonymous in certain works with the term of virtual environment (VE) (Ma and Zheng 2011; Montana et al. 2019) consists of "a computer technology that simulates real-life learning while providing augmented feedback and increased frequency, duration, and intensity of practiced tasks" (Sisto et al. 2002).

Basically, the spectrum of so-called VR/VE approaches is tightly connected to the devices used and the outcomes provided concerning the user's interaction within related procedures. Accordingly, an unanimously accepted classification of this field is difficult to find and often this is intended to be systematized in "manifold" manner (Weidig et al. 2014). However, although a clear border cannot be drawn between the components of this field—that encompasses Augmented Reality (AR), too (see further)—a taxonomic criterion within the VR/VE/AR field could be considered the capability of computing technology to bring the user in contact with real-world projections. This may enhance digital perception—even by using supplementary stimuli such as of the hearing, tactile, and/or olfactory kinds—but without limiting perceptions from the physical ambient (i.e. AR), intermingling, to this purpose, "synthetic elements like 3D objects, multimedia context or text information" (Martín-Gutiérrez et al. 2017). It should be specified that AR also uses specs, specially manufactured, that overlap virtual 3D items of the real environment, thus augmenting the reality displayed through a "transparent glass": prone for interaction with the user (a specific key related term) (Martín-Gutiérrez et al. 2017).

Aside from this paradigm, there is the one of a "whole simulated reality" (Martín-Gutiérrez et al. 2017), alternative to the natural surrounding, which is able, in variable degrees, to separate the involved person(s) from the actual surroundings, through immersion. This latter pathway, i.e. VR/VE, is based on advanced informatic technology entailing adequate hardware and software (3D), also including dedicated accessories such as head-mounted displays (HMD) (Buhrman 2018), specific gloves, video cameras, joysticks, (Ma and Zheng 2011; Weber et al. 2019), "VR helmets or dedicated glasses" (Martín-Gutiérrez et al. 2017), etc. Characterized by immersion of different degrees, VR/VE is able to provide "transportation" (see below), also including individual levels of psychologic merging within a virtual ambient, and/or even "telepresence", a "sense of being physically present with virtual object(s) at the remote teleoperator site" (Sheridan 1992)—enabling for high delicacy remotely manipulating physical objects. VR/VE thereby has, as key term: immersivity.

It should be added that transportation may be characterized as "…the extent to which a group of participants and objects leave behind their local space and enter into some new remote space in order to meet with others…"; this may also be compared to a converse situation: the respective participants "…remain in their local space and the remote participants and objects are brought to them" (Benford et al. 1998). In this respect, a necessary specification points a difference between the apparatus referring term of immersion and the features of the psycho-emotional one of presence or "co-presence" (Sheridan 1992; Benford et al. 1998).

Considering the profoundness of the immersion, there are three main immersive VR/VE categories described:

- "Fully" (Weber et al. 2019) immersive (Ma and Zheng 2011; Weber et al. 2019; Bamodu and Ye 2013) ("visual, auditory and tactile sensory aspects… delivered… through visual display units and speakers within a HMD unit, data gloves or body suits") (Ma and Zheng 2011);
- Semi-immersive (Ma and Zheng 2011; Weber et al. 2019; Bamodu and Ye 2013) using "projection systems" (Ma and Zheng 2011) able to achieve partial feeling of "presence" (Ma and Zheng 2011) in the artificial space, without complete detachment from reality;
- Non-immersive—the individual keeps awareness on the true contiguity interacting with a 3D graphic system through classical computer items (Ma and Zheng 2011; Bamodu and Ye 2013).

The neurobiological basis of the VR/VE effects is the stimulation of neuroplasticity resulting in a so-called "reinforcement learning" (De Luca et al. 2019).

Learning processes also characterize neurophysiologic activity, VR/VE being mentioned in different bibliographic resources; these include nonpathological situations of learning, such as attempts to complement by specific simulations: driving, flight, military, trainings, entertainment (Buhrman 2018)/gaming or, respectively, medical training—for instance, in laparoscopic surgery (Ma and Zheng 2011). They also refer to pathological circumstances: adaptive to new, abnormal morpho-functional conditions (Muresanu et al. 2012), thus contributing also to brain function augmentation.

At this point, the related category of "serious games" must be emphasized—"… nurturing autonomy, entertainment, and gratification of gaming, while serving a nobler purpose" (Michmizos and Krebs 2012). They may "…contribute to increase motivation in rehabilitation sessions, which is the major problem in therapy sessions, caused by the repetitive nature of exercises" (Rego et al. 2010).

Therefore, VR/VE/AR—with medical use since more than two decades ago (Adamovich et al. 2009)—enhancing neuroplasticity phenomena, including by "the reactivation of brain neurotransmitter capacities" (Montana et al. 2019), is nowadays frequently mentioned in the literature for its indications in neurorehabilitation.

More specifically, there are reports, although without unanimous recognition for all, of positive effects of VR/VE as types of interventions in cognitive rehabilitation—regarding “memory, attention, executive function” (De Luca et al. 2019), spatial recognition (Ma and Zheng 2011), respectively, in dementias (Montana et al. 2019), phobias (Ma and Zheng 2011), and other neuropsychological impairments—depression, anxiety, posttraumatic stress disorder (Montana et al. 2019), inappropriate eating behaviors and related reactions (Clus et al. 2018)—in conditions such as brain injuries, traumatic (Montana et al. 2019) or stroke, and even for pain (Ma and Zheng 2011)—especially, highly immersive VR systems” (Hoffman et al. 2015).

Furthermore, certain works consider that VR/VE might have a contribution for motor rehabilitation in strokes [possibly with better outcomes if including mirroring technologies/interventions and even, as sensory types of inputs, haptic stimulation/feedback (Ma and Zheng 2011; Montana et al. 2019; Weber et al. 2019; De Luca et al. 2019); the most appropriate timing for such kind of approach, i.e. early (Weber et al. 2019) or still in the chronic stage (Burdea 2003) has yet to be clarified], cerebral palsy (Ma and Zheng 2011), and also musculo-skeletal (orthopedic) pathology (Burdea 2003).

The implementation of serious/therapy games can provide “amplification of rewarding stimuli, e.g. by linking feedback about motor performance to a reward …” that, for instance, might “stimulate the dopaminergic system of stroke patients…” prone to stimulate neuroplasticity and “…recovery of motor functions” (Widmer 2018).

In this respect, different modalities have been proposed to develop games by projecting therapy sessions to “…trigger physical and cognitive behavioral…” models necessary for neurological treatment and recovery (Janssen et al. 2017).

VR/VE interventions appear to be quite safe—except, especially for “simulation sickness” (Buhrman 2018)/“simulator sickness” (Weber et al. 2019)/“motion sickness” (Buhrman 2018; Weber et al. 2019)/“cybersickness” (Montana et al. 2019; Weber et al. 2019) and as regards immersive VR/VE, mainly consisting of “dizziness, general discomfort, nausea, headache and in the worst cases even vomiting” (Buhrman 2018).

Several stationary mechatronic robotic advanced (most of them orthotic) devices, frequently encountered in bibliographic resources, that have served to train/rehabilitate disabilities in the upper limbs, are listed alphabetically below. They deliver their motility functions/interventions, using: electric [ALEx (Pirondini et al. 2014), Amadeo (Maciejasz et al. 2014), ArmeoPower (Maciejasz et al. 2014; Gijbels et al. 2011), Armotion (Chang et al. 2019b; Mazzoleni et al. 2018), InMotion (Maciejasz et al. 2014), respectively, mechanic and/or pneumatic (supinator extender—SUE—added to ArmeoSpring) (Maciejasz et al. 2014)] actuators.

As regards the more frequently used mobile (portable/wearable) devices, we can mention (alphabetically): Gloreha (Milia et al. 2019), Myomo (Stein 2009; https://exoskeletonreport.com/2016/09/myopro-the-assistive-arm-exoskeleton-by-myomo-featured-in-solidworks/; https://myomo.com/what-is-a-myopro-orthosis/), Ness H200 (Weber et al. 2010; Ring and Rosenthal 2005) (Figs. 1 and 2).

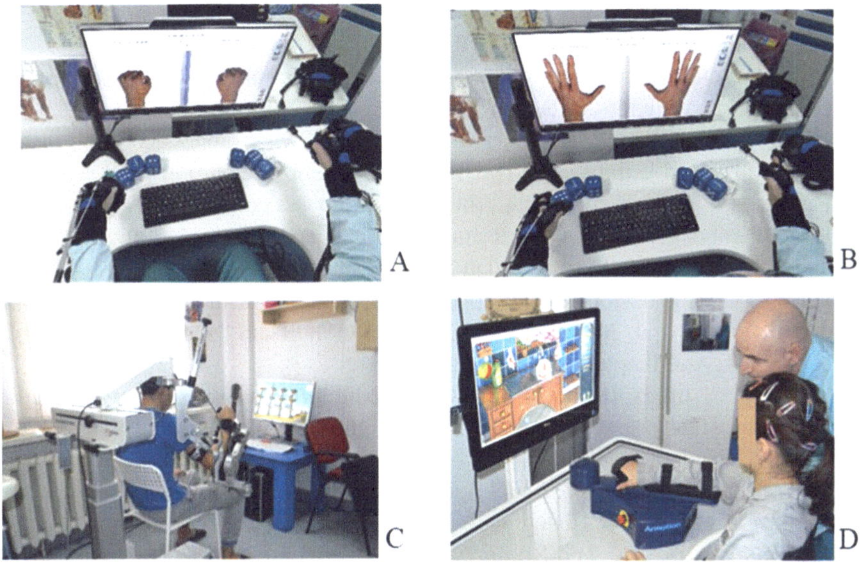

Fig. 1 Photos from the NTCNRC casuistry. Stationary mechatronic/robotic advanced (most of them orthotic, even exoskeletons) active devices—(**a**, **b**) Gloreha; (**c**) Armeo Spring; (**d**) Armotion

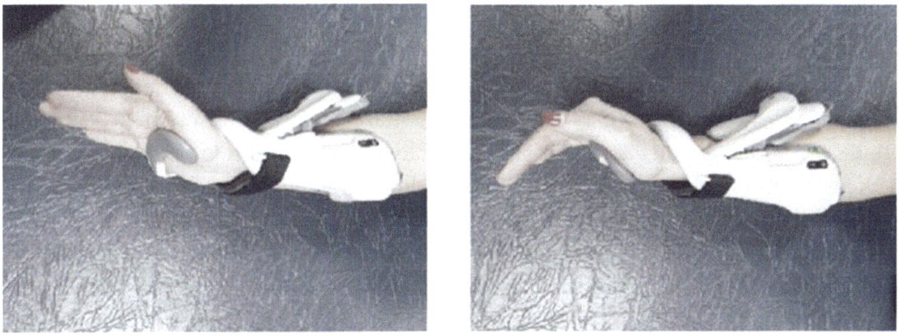

Fig. 2 Photos from the NTCNRC casuistry. Mobile (portable/wearable) advanced active orthotic device: NESS H200

The main upper limb disabling conditions we found in our previous systematic literature review (plus spare found related works) (Onose et al. 2018), to be reported as possible indications for mobile mechatronic/robotic device-assisted interventions, are (alphabetically):

- Arthrogryposis Multiplex Congenita (Gilliaux et al. 2015; Maciejasz et al. 2014; Haumont et al. 2011; Rahman et al. 2006; López et al. 2014).
- Brachial plexus injury (BPI) (Gilliaux et al. 2015; Maciejasz et al. 2014; https://myomo.com/what-is-a-myopro-orthosis/; Haumont et al. 2011; Rahman et al. 2006; López et al. 2014).

- Cerebral Palsy (Gilliaux et al. 2015; Maciejasz et al. 2014; Haumont et al. 2011; Rahman et al. 2006; López et al. 2014).
- Multiple Sclerosis (Gilliaux et al. 2015; Maciejasz et al. 2014; https://myomo.com/what-is-a-myopro-orthosis/; Haumont et al. 2011; Rahman et al. 2006; López et al. 2014; Robinson et al. 2019; http://www.bioness.com/Products/H200_for_Hand_Paralysis/Is_The_H200_Right_For_Me.php).
- Parkinson's disease/tremor (http://www.bioness.com/Products/H200_for_Hand_Paralysis/Is_The_H200_Right_For_Me.php; Rocon et al. 2007; Nimawat and Jailiya 2015; Shull and Damian 2015; Freer et al. 2017).
- Peripheral nerve lesions in the upper limb, including with carpal tunnel syndrome (Frisoli et al. 2016; Noveanu et al. 2013; Shull and Damian 2015; Giberti et al. 2014; Andrikopoulos et al. 2015).
- Spinal Cord Injury (Robinson et al. 2019; http://www.bioness.com/Products/H200_for_Hand_Paralysis/Is_The_H200_Right_For_Me.php; Jung et al. 2019).
- Spinal Muscular Atrophy (Gilliaux et al. 2015; Maciejasz et al. 2014; Haumont et al. 2011; Rahman et al. 2006; López et al. 2014).
- Stroke (Tu et al. 2017; Huang et al. 2012; Kim et al. 2012; Wei et al. 2013a, b; Xiao et al. 2014; Kim and Rosen 2015; https://myomo.com/what-is-a-myopro-orthosis/; Robinson et al. 2019; http://www.bioness.com/Products/H200_for_Hand_Paralysis/Is_The_H200_Right_For_Me.php; Nycz et al. 2015; Guo et al. 2016; Song et al. 2012).
- Traumatic brain injury (Robinson et al. 2019; http://www.bioness.com/Products/H200_for_Hand_Paralysis/Is_The_H200_Right_For_Me.php; Giberti et al. 2014).

Also, in the supplementary material (https://www.frontiersin.org/articles/10.3389/fnins.2018.00577/full#supplementary-material) afferent to (Onose et al. 2018) in Appendix 2, there is a tabular synthesis of mobile prototype-related devices, most encountered in the literature, chronologically, since the beginning of this millennium.

Additionally, in the last few decades "all-in-one" device types have appeared, with different degrees of complexity, being able to deliver electrical stimulation (ES) through generators embedded within portable/wearable devices (some of them being, at the same time, active orthoses or even advanced exoskeletons—as pointed out previously). Therefore, synthetically, a few connected, main items will be herein further, presented.

In regard to the theme of this chapter, ES—at least regarding stroke—"... may modulate a recovery factor …" referring also to "… mechanisms of brain repair, neurogenesis, ... rewiring of electrical brain circuits, growth, ... remodulation of collateral blood flow, functional dynamics of intra-brain networking and its changes. ... these recovery factors are exemplary of brain plasticity" (Danielsson 2009).

Hence, the noninvasively (transcutaneously administered) concept and interventions of ES (Sheffer et al. 2010) type—or quite synonymously: "transcutaneous electrical nerve stimulation (TENS)", but it should be specified that TENS is also the naming of a well-known electrotherapeutic analgesic method, i.e. "In the clinical context, it is most commonly assumed to refer to the use of electrical stimulation with the specific intention of providing symptomatic pain relief" (Watson 2007)—entails basically two proactive dimensions: motor learning—"neuromuscular electrical stimulation (NMES)" (Sheffer et al. 2010)—and functional (electrical) stimulation (FES) (Sheffer et al. 2010).

More specifically, ES used in NeuroRehabilitation is "broadly classified"—by its rationale/outcomes targeted—"as therapeutic or functional"(Sheffer et al. 2010). According to this point of view, ES encompasses "neuromuscular electrical stimulation (NMES)" (Sheffer et al. 2010), which is predominantly therapeutic ("motor re-learning") and/or prophylactic (Sheffer et al. 2010), and, respectively, a basically assistive dimension: functional electrical stimulation (FES), which "… substitutes for or replaces lost (or severely impaired—o. n.) neuromuscular function and is considered to be an ongoing or permanent intervention" (Knutson et al. 2010). This latter feature includes a more modern and useful applicative development by adding/incorporating, with a resulting neuroprosthesis (as exemplified above: NESS H200). Actually, this conceptual and constructive advancement may enable for such neuroprosthetic FES systems to be "worn by the user", giving them therapeutic/prophylactic properties and, respectively—very importantly—capabilities of purposeful functional compensation, kind (Sheffer et al. 2010; Knutson et al. 2020).

NMES for motor relearning: it has—as already mentioned—mainly therapeutic-rehabilitative and/or prophylactic goals, as it does not provide functional relevant muscle contractions (possibly just partially relevant, i.e. to strengthen an—for the moment insufficiently intense and/or controlled—active movement); it "… incorporates novel tasks and includes cognitive investment" (Sheffer et al. 2010).

Such interventions can also be electromyography (EMG)-mediated: "EMG-mediated NMES couples cognitive intent and NMES-mediated muscle contraction." (Sheffer et al. 2010), possibly with self-evaluated outcomes by EMG bio-feedback, too—an example is the Myomo exoskeleton (Stein 2009; https://exoskeletonreport.com/2016/09/myopro-the-assistive-arm-exoskeleton-by-myomo-featured-in-solidworks/; https://myomo.com/what-is-a-myopro-orthosis/).

Some commonly used stationary mechatronic robotic exoskeletons for lower limbs motor deficits' approaches are Lokomat (https://www.hocoma.com/solutions/lokomat/) and G-EO (Palermo et al. 2017; Louie et al. 2015; Louie and Eng 2016; Hesse et al. 2010; https://exoskeletonreport.com/product/g-eo-system/), but the literature also mentions such devices as (alphabetically): ALEX (Hesse et al. 2010; Banala et al. 2008), AutoAmbulator (Hesse et al. 2010; https://www.medgadget.com/2006/07/autoambulator_r.html), LOPES (Hesse et al. 2010; Veneman et al. 2007), Optimal—GPro (http://www.motorika.com/optimal-g-pro/), Pelvic Assist Manipulator (PAM) (Reinkensmeyer et al. 2014), Pneumatically Operated Gait Orthosis (POGO) (Reinkensmeyer et al. 2014), ReoAmbulator (http://motorika.com/reoambulator/), and University of Auckland system (Chen et al. 2013).

Most of these stationary mechatronic/robotic advanced rehabilitative devices (some of them exoskeletons) acting in lower limb pathologies entail, as an intervention provided principle, the usage of a body weight support (BWS) (Chen et al. 2013) or partial body weight support (PBWS), structure ("… a modified parachute harness …") (Hesse 2007). The robotically controlled gait is often trained on a

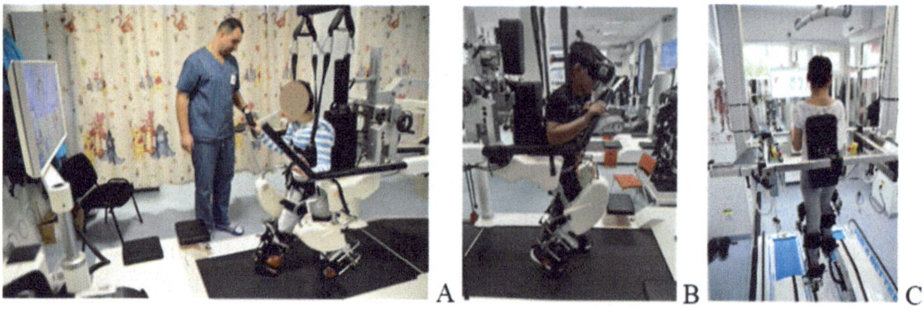

Fig. 3 Photos from the NTCNRC casuistry. Stationary mechatronic/robotic advanced rehabilitative exoskeleton/device system—(a, b) Lokomat; (c) G-EO

treadmill, barring G-EO (https://exoskeletonreport.com/product/g-eo-system/). The main related difference in between concerning the walk training pattern: although both have a BWS/PBWS and a fixed framing system, the first paradigm uses a treadmill, and the second one, the patients' feet placed on actuated supports passively reproducing gait sequences from bottom to top, prone for stairs climbing. Thus, all these kinds of systems address, somehow complementarily, main aspects of stance and gait rehabilitation (Hesse et al. 2010, 2012) (Fig. 3).

Alternatively to the abovementioned "classic type" of BWS/PBWS, a complex pneumatic-related type of system can be exemplified, such as, for instance, AlterG (https://www.alterg.com/anti-gravity-treadmills)—appropriate for use, in principle, if there is a good stability of the user's torso.

Intermediate between stationary robotic exoskeletons—having embedded either a treadmill (for instance Lokomat - https://www.hocoma.com/solutions/lokomat/) or actuatedspecificlegssupports(G-EO)(https://exoskeletonreport.com/product/g-eo-system/)—and wearable ones, there are several portable devices of this kind (not all of them properly exoskeletons), some of them equipped with FES facilities (Li et al. 2008), designated to train gait recovery on common indoor overgrounds (WalkTrainer) (Bouri et al. 2006), (Andago -https://www.hocoma.com/solutions/andago/).

For the mobile—portable/wearable mechatronic/robotic exoskeletons, the balance (Onose et al. 2016; Chen et al. 2013; Palermo et al. 2017; Louie et al. 2015; Louie and Eng 2016; Kannape and Lenggenhager 2016; Kolakowsky-Hayner et al. 2013; Lajeunesse et al. 2016; Veneman 2014; Yoshimoto et al. 2015; http://balance-fp7.eu/project_structure.php; http://balance-fp7.eu/objectives.php) subject matter is the main issue encountered by such devices, designated to assist/rehabilitate severe neuromotor impairments in the lower limbs. This, at the same time, makes the main difference to the stationary ones, because they are not fixed systems (with the client safely contained inside) but thereby have the possibility to be used overground—assistive action/function—and thus prone to customer's autonomy, including with the possible use in the community.

The most frequently mobile (portable/wearable)-related powered exoskeletons reviewed in the literature are the following (alphabetically): Ekso Bionics/ eLEGS™ (https://peteredwards2012.wordpress.com/ekso-bionics-elegs/), ExoAtlet (http://

www.exoatlet.com/), Exo-H2 (https://www.technaid.com/products/robotic-exo-skeleton-exo-exoesqueleto/), and Exo H3 (https://www.technaid.com/products/robotic-exoskeleton-exo-exoesqueleto-h3/)—some continuing endeavors aiming to improve BCI or human-machine interface (HMI) adding (Contreras-Vidal et al. 2018; https://www.technaid.com/products/robotic-exoskeleton-exo-exoesqueleto/; https://www.technaid.com/products/robotic-exoskeleton-exo-exoesqueleto-h3/; Abdelkarim and Brahim 2019)—Hybrid Assistive Leg/Limb (HAL) (https://www.cyberdyne.jp/english/products/LowerLimb_medical.html), Indego (http://www.indego.com/indego/en/home), Isocentric Reciprocating Gait Orthosis (IRGO) (Arazpour et al. 2015; Ko 2019), Keeogo (https://keeogo.com/success-stories/testimonials; https://www.cadth.ca/sites/default/files/pdf/htis/july-2015/RA0784%20Powered%20Wearable%20Walking%20Assistive%20Devices%20Final.pdf), Kickstart (https://www.cadth.ca/sites/default/files/pdf/htis/july-2015/RA0784%20Powered%20Wearable%20Walking%20Assistive%20Devices%20Final.pdf; https://www.cadencebiomedical.com), Mina (http://robots.ihmc.us/x1-mina-exo-skeleton), ReWalk (https://www.cadth.ca/sites/default/files/pdf/htis/july-2015/RA0784%20Powered%20Wearable%20Walking%20Assistive%20Devices%20Final.pdf), Rex (https://www.rexbionics.com), Stride Management Assist (https://www.cadth.ca/sites/default/files/pdf/htis/july-2015/RA0784%20Powered%20Wearable%20Walking%20Assistive%20Devices%20Final.pdf; Nagarajan and Goswami 2015; https://www.honda.com/mobility/walking-assist-technology), Wearable Power-Assist Locomotor (WPAL) (Kagawa and Uno 2009; Yatsuya et al. 2018; Hirano et al. 2015; https://www.youtube.com/watch?v=76sF3rsHMYo) (Fig. 4).

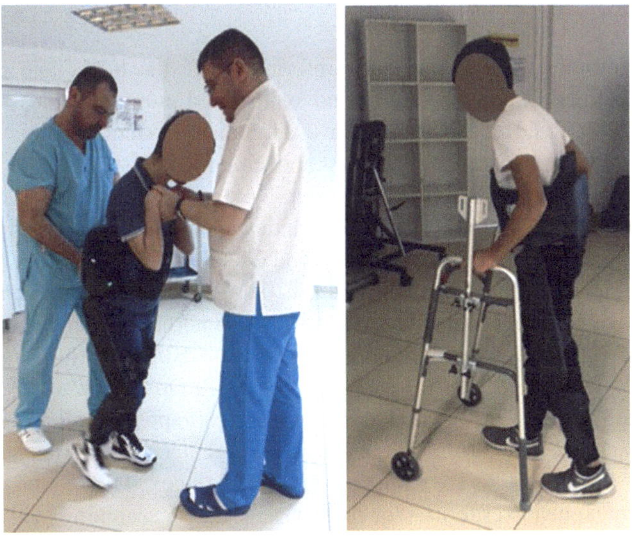

Fig. 4 Photos from the NTCNRC casuistry. Mobile (portable/wearable) mechatronic/robotic exoskeleton—Indego

The main lower limb pathology domains indicated for the use of mechatronic/robotic exoskeletons-based interventions are (alphabetically):

- Cerebral palsy (http://motorika.com/reoambulator/; Zeilig et al. 2012).
- Hip and/knee osteoarthritis (https://keeogo.com/success-stories/testimonials).
- Myelomeningocele (Zeilig et al. 2012).
- Multiple Sclerosis (Robinson et al. 2019; https://keeogo.com/success-stories/testimonials).
- Neuropathic pain (Zeilig et al. 2012).
- Orthopedic conditions that result in gait abnormality (http://motorika.com/reoambulator/).
- Parkinson's Diseases (http://motorika.com/reoambulator/; https://keeogo.com/success-stories/testimonials).
- Polyneuropathy (Zeilig et al. 2012).
- Post-surgery rehabilitation (http://motorika.com/reoambulator/; https://keeogo.com/walking-aids-for-leg-weakness-or-knee-problems?keyword=exoskeleton&gclid=EAIaIQobChMIlZzxjoLs5gIVy5AYCh0tfgCdEAAYASAAEgKIFPD_BwE).
- Spinal cord injury-complete and incomplete (Robinson et al. 2019; http://motorika.com/reoambulator/; https://keeogo.com/success-stories/testimonials).
- Stroke (Robinson et al. 2019; Hesse 2007; https://keeogo.com/success-stories/testimonials; Zeilig et al. 2012).
- Traumatic brain injury (Robinson et al. 2019; Zeilig et al. 2012).

3 Recent Related Systematic Literature Review

In order to provide the most updated information in the field—and taking into account the exhaustive data searched and synthetized in our previous two related works (Onose et al. 2016, 2018)—we have also done a systematic literature review covering the period elapsed since the first abovementioned issue.

Thereby, we have achieved a focus step-by-step classification—see Fig. 5—according to the stages of the well-known and accepted method named "Preferred Reporting Items for Systematic Reviews and Meta-Analyses (PRISMA)" (http://prisma-statement.org/documents/PRISMA%202009%20flow%20diagram.pdf). It was initiated by interrogation of several prestigious international medical databases: NCBI/PubMed (https://www.ncbi.nlm.nih.gov), NCBI/PMC (https://www.ncbi.nlm.nih.gov), Elsevier (https://www.elsevier.com/), PEDro (http://search.pedro.org.au/search), ISI Web of Knowledge/Science (https://apps.webofknowledge.com; https://www.e-nformation.ro/resurse/bnr_thomson-reuters) (in order to identify which of the found articles are published in journals indexed there, thus supporting our inclusion/exclusion criteria—see further) and use for searching the specific key words combinations/syntaxes ("lower limb mechatronic exoskeleton", "systematic

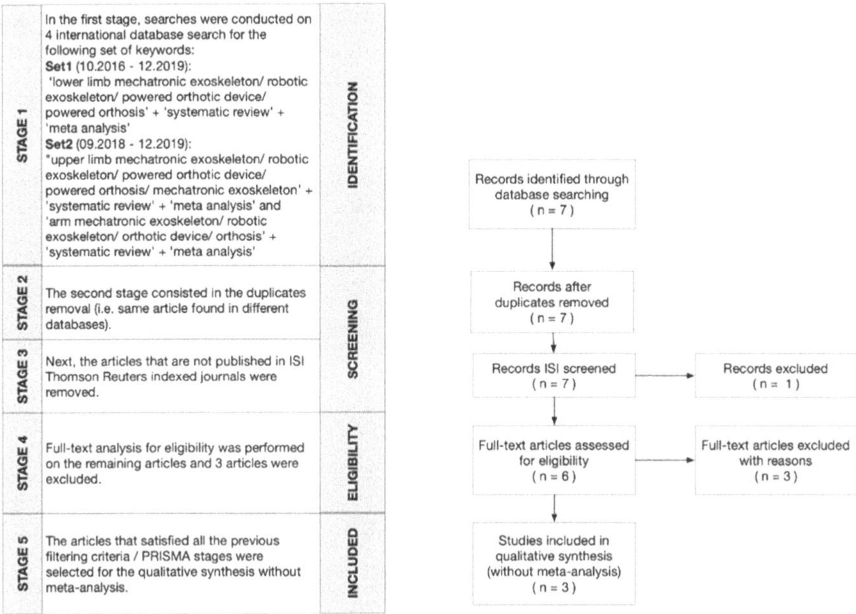

Fig. 5 PRISMA kind flow diagram customized for our articles' screening

review", "meta analysis"; "lower limb powered orthosis", "systematic review", "meta analysis"; "lower limb robotic exoskeleton", "systematic review", "meta analysis"; "lower limb powered orthotic device", "systematic review", "meta analysis"; "upper limb mechatronic exoskeleton", "systematic review", "meta analysis"; "upper limb robotic exoskeleton", "systematic review", "meta analysis"; "upper limb powered orthotic device", "systematic review", "meta analysis"; "upper limb powered orthosis", "systematic review", "meta analysis"; "arm mechatronic exoskeleton", "systematic review", "meta analysis"; "arm robotic exoskeleton", "systematic review", "meta analysis"; "arm powered orthotic device", "systematic review", "meta analysis"; "arm powered orthosis", "systematic review", "meta analysis").

Open-access works have been searched for, all of them being written in English and issued in ISI Thomson Reuters indexed publications. In order to obtain updated synthetic-related information, which had to be as exhaustively based as possible, the focus was on systematic reviews and/or meta-analyses. The coverage of the related publishing time frame was: October 2016–December 2019.

Nevertheless, although we have rigorously applied the abovementioned methodology for the selection of the works we have searched for, certain related data might have remained uncovered. On the other hand, not all the articles identified by our query, contained information necessary to be included within our afferent references (Tables 1 and 2).

Table 1 Keywords used for our most recent related systematic literature review

Keywords	Period	Elsevier	PubMed	PMC	PEDro	Total
Lower limb mechatronic exoskeleton', 'systematic review', 'meta analysis'	10.2016 – 12.2019	0	0	0	0	**0**
'Lower limb powered orthosis', 'systematic review', 'meta analysis'		2	0	0	0	**2**
'Lower limb robotic exoskeleton', 'systematic review', 'meta analysis'		0	0	5	0	**5**
'Lower limb powered orthotic device', 'systematic review', 'meta analysis'		0	0	0	0	**0**
Upper limb mechatronic exoskeleton', 'systematic review', 'meta analysis'	09.2018 – 12.2019	0	0	0	0	**0**
'Upper limb robotic exoskeleton', 'systematic review', 'meta analysis'		0	0	0	0	**0**
'Upper limb powered orthotic device', 'systematic review', 'meta analysis'		0	0	0	0	**0**
'Upper limb powered orthosis', 'systematic review', 'meta analysis'		0	0	0	0	**0**
'Arm mechatronic exoskeleton', 'systematic review', 'meta -analysis'		0	0	0	0	**0**
'Arm robotic exoskeleton', 'systematic review', 'meta analysis'		0	0	0	0	**0**
'Arm powered orthotic device', 'systematic review', 'meta -analysis'		0	0	0	0	**0**
'Arm powered orthosis', 'systematic review', 'meta analysis'		0	0	0	0	**0**
Total		**2**	**0**	**5**	**0**	**7**

Table 2 Works identified in the databases NCBI/PubMed, NCBI/PMC, Elsevier, PEDro, ISI Web of Knowledge/Science, according to the paradigm based on the keyword combinations/syntaxes, presented above

Article	Link
Robinson et al., Psychosocial Health Interventions by Social Robots: Systematic Review of Randomized Controlled Trials, J Med Internet Res. 2019 May; 21(5): e13203. Published online 2019 May 10. https://doi.org/10.2196/13203	link
Clark et al., Evaluating the use of robotic and virtual reality rehabilitation technologies to improve function in stroke survivors: A narrative review, J Rehabil Assist Technol Eng. 2019 Jan-Dec; 6: 2055668319863557. Published online 2019 Nov 13. https://doi.org/10.1177/2055668319863557	link
Onose et al., Mobile Mechatronic/Robotic Orthotic Devices to Assist/Rehabilitate Neuromotor Impairments in the Upper Limb: A Systematic and Synthetic Review, Front Neurosci. 2018; 12: 577. Published online 2018 Sep 5. https://doi.org/10.3389/fnins.2018.00577	link

4 Discussion and Conclusions

Considering the benefits of the abovementioned active orthotic devices and related assistive/therapeutic-rehabilitative types of interventions, synthetically overviewed in this chapter, there is an overall rather favorable opinion in the literature regarding the use of mechatronic/robotic/orthotic devices/exoskeletons in the rehabilitation of different—especially severe—neuro-/loco-motor impairments: "… we predict that this technology will become instrumental in treating a wide range of orthopedic and neurologic disorders across the continuum of care, from acute care hospitals to home and outpatient settings" (Fasoli et al. 2004), and: "Robotic technologies … encourage motor re-learning with the goal of reducing impairment" (Clark et al. 2019). It should be mentioned, too, in this respect, that there is an inherently consequent psychological investment in this newer high technology aiming to support therapeutic exercises.

Yet, this opinion is not unanimous, certain data considering there are better results with the classical approach: "Therapist-assisted LT (locomotor training—o. n.) facilitates greater improvements in walking ability in ambulatory stroke survivors as compared to a similar dosage of robotic-assisted LT" (Hornby et al. 2008)/"Locomotor training, including the use of body-weight support in stepping on a treadmill, was not shown to be superior to progressive exercise at home managed by a physical therapist" (Duncan et al. 2011).

The occurrence of such kinds of wearable apparatus, able to provide efficient assistive capabilities, toward improving users' functional autonomy and QOL, is still a desired perspective.

Additionally, lately it has been considered that the combination between robotic-assisted therapies "… and virtual reality technologies in a rehabilitation programme may further improve … outcomes…". On the other hand, claims of the unimportant "… benefit in ADLs with VR technologies as compared to dose-controlled conventional therapy …" and, respectively, of "… no significant difference for upper-limb function, gait speed or balance" (Clark et al. 2019), are asserted.

At the same time, high costs (especially for the most advanced such devices), cost-effectiveness and reimbursement by health insurance systems, are still current issues and sources of limitations—hopefully to be, at least partially, overcome as technology is progressing—in their practical, especially everyday use, assistive availability (Onose et al. 2018; Chang et al. 2019b).

Acknowledgments We would like to thank Ioana Andone, Valeriu Avramescu, Vladimir Cârdei, Marian Vladimir Constantinescu, Ştefan T. Crăciunoiu, Cristina Daia, Mikhail A. Lebedev, Ioan Opriş, Marian-Daniel Mirea, Constantin Munteanu, Nirvana Popescu, and Aura Spînu for their contributions to two important articles for the achievement of which we have all fulfilled substantial endeavors that resulted in consistent enhancement of the knowledge in the field of mechatronic/robotic exoskeletons/orthotic devices, and hence, we were able to make in this chapter a related synthetic step forward to a broader approach having included such solid bases.

This work has the Ethics Commission's approvals of the TEHBA in Bucharest, Romania (with No. 42291/06.01.2020) and, respectively, of the NTCNRC (with No. 10650/16.12.2019) in Bucharest, Romania.

Author Contributions All authors listed have made a substantial, direct, and intellectual contribution to the work, and approved it for publication.

Conflict of Interest Statement No conflicts of interest: the research was conducted in the absence of any commercial or financial relationships that could be construed as a potential conflict of interest.

References

Abdelkarim E, Brahim SM (2019) Development and control of a low cost Exoskeleton system with an interactive HMI designed for paraplegic children. Int J Sci Eng Res 10(1):4

Adamovich SV, Fluet GG, Tunik E, Merians AS (2009) Sensorimotor training in virtual reality: a review. NeuroRehabilitation 25(1):29–44. https://doi.org/10.3233/NRE-2009-0497

Andrikopoulos G, Nikolakopoulos G, Manesis S (2015) Motion control of a novel robotic wrist exoskeleton via pneumatic muscle actuators. In: 2015 IEEE 20th conference on emerging technologies & factory automation (ETFA). IEEE, pp 1–8. https://doi.org/10.1109/ETFA.2015.7301464

Arazpour M, Hutchins SW, Ahmadi Bani M (2015) The efficacy of powered orthoses on walking in persons with paraplegia. Prosthetics Orthot Int 39(2):90–99. https://doi.org/10.1177/0309364613520031

Bamodu O, Ye XM (2013) Virtual reality and virtual reality system components. In: Advanced materials research, vol 765. Trans Tech Publications Ltd, pp 1169–1172. https://doi.org/10.2991/icsem.2013.192

Banala SK, Kim SH, Agrawal SK, Scholz JP (2008) Robot assisted gait training with active leg exoskeleton (ALEX). IEEE Trans Neural Syst Rehabil Eng 17(1):2–8. https://doi.org/10.1109/BIOROB.2008.4762885

Barsotti M, Leonardis D, Loconsole C, Solazzi M, Sotgiu E, Procopio C et al (2015) A full upper limb robotic exoskeleton for reaching and grasping rehabilitation triggered by MI-BCI. In: 2015 IEEE international conference on rehabilitation robotics (ICORR). IEEE, pp 49–54. https://doi.org/10.1109/ICORR.2015.7281174

Benford S, Greenhalgh C, Reynard G, Brown C, Koleva B (1998) Understanding and constructing shared spaces with mixed-reality boundaries. ACM Trans Comput Hum Interact 5(3):185–223. https://doi.org/10.1145/292834.292836

Benson I, Hart K, Tussler D, van Middendorp JJ (2016) Lower-limb exoskeletons for individuals with chronic spinal cord injury: findings from a feasibility study. Clin Rehabil 30(1):73–84. https://doi.org/10.1177/0269215515575166

Bouri M, Stauffer Y, Schmitt C, Allemand Y, Gnemmi S, Clavel R et al (2006) The WalkTrainer: a robotic system for walking rehabilitation. In: 2006 IEEE international conference on robotics and biomimetics. IEEE, pp 1616–1621. https://doi.org/10.1109/ROBIO.2006.340186

Bressler SL, Menon V (2010) Large-scale brain networks in cognition: emerging methods and principles. Trends Cogn Sci 14(6):277–290. https://doi.org/10.1016/j.tics.2010.04.004

Buhrman J, Åkesson L (2018) DriVR A driving school simulator in VR. Master Thesis published by: Department of Design Sciences, Faculty of Engineering LTH, Lund University. Lund, Sweden

Buongiorno D, Barsotti M, Barone F, Bevilacqua V, Frisoli A (2018) A linear approach to optimize an EMG-driven neuromusculoskeletal model for movement intention detection in myo-control: a case study on shoulder and elbow joints. Front Neurorobot 12:74. https://doi.org/10.3389/fnbot.2018.00074

Burdea G (2003) Review paper—virtual rehabilitation-benefits and challenges. Yearb Med Inform 1:170–176

Carnahan H (1992) Chapter 8: eye, head and hand coordination during manual aiming. In: Advances in psychology, pp 179–196. https://doi.org/10.1016/S0166-4115(08)62015-4

Chang S-HJ, Sullivan JL, Kadivar Z, O'Malley M, Francisco GE (2019a) Rehabilitation robotics. In: DeLisa's physical medicine and rehabilitation: principles and practice. Lippincott Williams & Wilkins, Philadelphia, pp 1394–1407

Chang JL, Saul M, Volpe BT (2019b) Practical review of robotics in the treatment of chronic impairment after acquired brain injury. In: Acquired brain injury. Springer, Cham, pp 71–88. https://doi.org/10.1007/978-3-030-16613-7_5

Charles SK, Krebs HI, Volpe BT, Lynch D, Hogan N (2005) Wrist rehabilitation following stroke: initial clinical results. In: 9th international conference on rehabilitation robotics, 2005. ICORR 2005. IEEE, pp 13–16. https://doi.org/10.1109/ICORR.2005.1501040

Chen G, Chan CK, Guo Z, Yu H (2013) A review of lower extremity assistive robotic exoskeletons in rehabilitation therapy. Crit Rev Biomed Eng 41(4–5):343–363. https://doi.org/10.1615/CritRevBiomedEng.2014010453

Chéron G, Duvinage M, De Saedeleer C, Castermans T, Bengoetxea A, Petieau M et al (2012) From spinal central pattern generators to cortical network: integrated BCI for walking rehabilitation. Neural Plast 2012:375148. https://doi.org/10.1155/2012/375148

Clark WE, Sivan M, O'Connor RJ (2019) Evaluating the use of robotic and virtual reality rehabilitation technologies to improve function in stroke survivors: a narrative review. J Rehabil Assist Technol Eng 6:2055668319863557. https://doi.org/10.1177/2055668319863557

Clus D, Larsen ME, Lemey C, Berrouiguet S (2018) The use of virtual reality in patients with eating disorders: systematic review. J Med Internet Res 20(4):e157. https://doi.org/10.2196/jmir.7898

Contreras-Vidal JL, Bortole M, Zhu F, Nathan K, Venkatakrishnan A, Francisco GE et al (2018) Neural decoding of robot-assisted gait during rehabilitation after stroke. Am J Phys Med Rehabil 97(8):541–550. https://doi.org/10.1097/PHM.0000000000000914

Danielsson I (2009) Method for improving functional recovery after stroke by electrical stimulation of a cranial nerve. https://patents.google.com/patent/US9079031

De Luca R, Maggio MG, Maresca G, Latella D, Cannavò A, Sciarrone F et al (2019) Improving cognitive function after traumatic brain injury: a clinical trial on the potential use of the semi-immersive virtual reality. Behav Neurol 2019:9268179. https://doi.org/10.1155/2019/9268179

DeGracia DJ (2010) Towards a dynamical network view of brain ischemia and reperfusion. Part III: therapeutic implications. J Exp Stroke Transl Med 3(1):90. https://doi.org/10.6030/1939-067X-3.1.90

Dowling AV, Barzilay O, Lombrozo Y, Wolf A (2014) An adaptive home-use robotic rehabilitation system for the upper body. IEEE J Transl Eng Health Med 2:1–10

Duncan PW, Sullivan KJ, Behrman AL, Azen SP, Wu SS, Nadeau SE et al (2011) Body-weight–supported treadmill rehabilitation after stroke. N Engl J Med 364(21):2026–2036. https://doi.org/10.1056/NEJMoa1010790

Fan B, Wei Z, Yao X, Shi G, Cheng X, Zhou X et al (2018) Microenvironment imbalance of spinal cord injury. Cell Transplant 27(6):853–866. https://doi.org/10.1177/0963689718755778

Fasoli SE, Krebs HI, Hogan N (2004) Robotic technology and stroke rehabilitation: translating research into practice. Top Stroke Rehabil 11(4):11–19. https://doi.org/10.1310/G8XB-VM23-1TK7-PWQU

Feigin VL, Abajobir AA, Abate KH, Abd-Allah F, Abdulle AM, Abera SF et al (2017) Global, regional, and national burden of neurological disorders during 1990–2015: a systematic analysis for the Global Burden of Disease Study 2015. Lancet Neurol 16(11):877–897. https://doi.org/10.1016/S1474-4422(17)30299-5

Freer DR, Liu J, Yang GZ (2017) Optimization of EMG movement recognition for use in an upper limb wearable robot. In: 2017 IEEE 14th international conference on wearable and implantable body sensor networks (BSN). IEEE, pp 202–205. https://doi.org/10.1109/BSN.2017.7936041

Frisoli A, Solazzi M, Loconsole C, Barsotti M (2016) New generation emerging technologies for neurorehabilitation and motor assistance. Acta Myol 35(3):141

Gandolla M, Costa A, Aquilante L, Gfoehler M, Puchinger M, Braghin F, Pedrocchi A (2017) BRIDGE—behavioural reaching interfaces during daily antigravity activities through upper limb exoskeleton: preliminary results. In: 2017 international conference on rehabilitation robotics (ICORR). IEEE, pp 1007–1012. https://doi.org/10.1109/ICORR.2017.8009381

Giberti H, Bertoni V, Coppola G (2014) Conceptual design and feasibility study of a novel upper-limb exoskeleton. In: 2014 IEEE/ASME 10th international conference on mechatronic and embedded systems and applications (MESA). IEEE, pp 1–6. https://doi.org/10.1109/MESA.2014.6935548

Gijbels D, Lamers I, Kerkhofs L, Alders G, Knippenberg E, Feys P (2011) The Armeo Spring as training tool to improve upper limb functionality in multiple sclerosis: a pilot study. J Neuroeng Rehabil 8(1):5. https://doi.org/10.1186/1743-0003-8-5

Gilliaux M, Renders A, Dispa D, Holvoet D, Sapin J, Dehez B et al (2015) Upper limb robot-assisted therapy in cerebral palsy: a single-blind randomized controlled trial. Neurorehabil Neural Repair 29(2):183–192. https://doi.org/10.1177/1545968314541172

Grimm F, Walter A, Spüler M, Naros G, Rosenstiel W, Gharabaghi A (2016) Hybrid neuroprosthesis for the upper limb: combining brain-controlled neuromuscular stimulation with a multi-joint arm exoskeleton. Front Neurosci 10:367. https://doi.org/10.3389/fnins.2016.00367

Guo S, Gao J, Guo J, Zhang W, Hu Y (2016) Design of the structural optimization for the upper limb rehabilitation robot. In: 2016 IEEE international conference on mechatronics and automation. IEEE, pp 1185–1190. https://doi.org/10.1109/ICMA.2016.7558730

Haumont T, Rahman T, Sample W, King MM, Church C, Henley J, Jayakumar S (2011) Wilmington robotic exoskeleton: a novel device to maintain arm improvement in muscular disease. J Pediatr Orthop 31(5):e44–e49. https://doi.org/10.1097/BPO.0b013e31821f50b5

Herold F, Wiegel P, Scholkmann F, Müller NG (2018) Applications of functional near-infrared spectroscopy (fNIRS) neuroimaging in exercise–cognition science: a systematic, methodology-focused review. J Clin Med 7(12):466. https://doi.org/10.3390/jcm7120466

Hesse S (2007) Treadmill training with partial body weight support after stroke: a review. NeuroRehabilitation 22:1–11. https://doi.org/10.3233/NRE-2008-23106. IOS Press. File: nre394.tex; BOKCTP/wyy p 1

Hesse S, Kuhlmann H, Wilk J, Tomelleri C, Kirker SG (2008) A new electromechanical trainer for sensorimotor rehabilitation of paralysed fingers: a case series in chronic and acute stroke patients. J Neuroeng Rehabil 5(1):21. https://doi.org/10.1186/1743-0003-5-21

Hesse S, Waldner A, Tomelleri C (2010) Innovative gait robot for the repetitive practice of floor walking and stair climbing up and down in stroke patients. J Neuroeng Rehabil 7(1):30. https://doi.org/10.1186/1743-0003-7-30

Hesse S, Tomelleri C, Bardeleben A, Werner C, Waldner A (2012) Robot-assisted practice of gait and stair climbing in nonambulatory stroke patients. J Rehabil Res Dev 49(4):613–622. https://doi.org/10.1682/JRRD.2011.08.0142

Hirano S, Saitoh E, Tanabe S, Katoh M, Shimizu Y, Yatsuya K et al (2015) Comparison between gait-assisting robot (WPAL) and bilateral knee-ankle-foot orthoses with a medial single hip joint in gait reconstruction for patients with paraplegia. Jpn J Compr Rehabil Sci 6:21–26. https://doi.org/10.11336/jjcrs.6.21

Hoffman HG, Interface H, Chambers GT, Arceneaux LL, Russell WJ, Seibel EJ et al (2015) Virtual reality as an adjunctive non-pharmacologic analgesic for acute burn pain during medical procedures. Ann Behav Med 41(2):183–191. https://doi.org/10.1007/s12160-010-9248-7

Hornby TG, Campbell DD, Kahn JH, Demott T, Moore JL, Roth HR (2008) Enhanced gait-related improvements after therapist-versus robotic-assisted locomotor training in subjects with chronic stroke: a randomized controlled study. Stroke 39(6):1786–1792. https://doi.org/10.1161/STROKEAHA.107.504779

http://balance-fp7.eu/objectives.php

http://balance-fp7.eu/project_structure.php

http://motorika.com/reoambulator/

http://prisma-statement.org/documents/PRISMA%202009%20flow%20diagram.pdf

http://robots.ihmc.us/x1-mina-exoskeleton
http://search.pedro.org.au/search
http://www.bioness.com/Products/H200_for_Hand_Paralysis/Is_The_H200_Right_For_Me.php
http://www.exoatlet.com/
http://www.indego.com/indego/en/home
http://www.motorika.com/optimal-g-pro/
https://apps.webofknowledge.com
https://asimo.honda.com/innovations/default.aspx?ID=body-weight-support-assist
https://bleex.me.berkeley.edu/research/exoskeleton/bleex/
https://bleex.me.berkeley.edu/research/exoskeleton/exohiker/
https://bleex.me.berkeley.edu/research/exoskeleton/hulc/
https://exoskeletonreport.com/2016/09/myopro-the-assistive-arm-exoskeleton-by-myomo-featured-in-solidworks/
https://exoskeletonreport.com/product/g-eo-system/
https://keeogo.com/success-stories/testimonials
https://keeogo.com/walking-aids-for-leg-weakness-or-knee-problems?keyword=exoskeleton&gclid=EAIaIQobChMIlZzxjoLs5gIVy5AYCh0tfgCdEAAYASAAEgKIFPD_BwE
https://myomo.com/what-is-a-myopro-orthosis/
https://peteredwards2012.wordpress.com/ekso-bionics-elegs/
https://walk-again-project.org/#/en
https://www.alterg.com/anti-gravity-treadmills
https://www.cadencebiomedical.com
https://www.cadth.ca/sites/default/files/pdf/htis/july-2015/RA0784%20Powered%20Wearable%20Walking%20Assistive%20Devices%20Final.pdf
https://www.cyberdyne.jp/english/products/LowerLimb_medical.html
https://www.cyberdyne.jp/english/products/supporting.html
https://www.elsevier.com/
https://www.e-nformation.ro/resurse/bnr_thomson-reuters, https://www.e-nformation.ro/resurse/bnr_thomson-reuters
https://www.frontiersin.org/articles/10.3389/fnins.2018.00577/full#supplementary-material
https://www.hocoma.com/solutions/lokomat/
https://www.honda.com/mobility/walking-assist-technology
https://www.medgadget.com/2006/07/autoambulator_r.html
https://www.ncbi.nlm.nih.gov
https://www.rexbionics.com
https://www.technaid.com/products/robotic-exoskeleton-exo-exoesqueleto/
https://www.technaid.com/products/robotic-exoskeleton-exo-exoesqueleto-h3/
https://www.utwente.nl/en/et/be/research/projects/MINDWALKER/
https://www.youtube.com/watch?v=76sF3rsHMYo
Huang S, Luo C, Ye S, Liu F, Xie B, Wang C et al (2012) Motor impairment evaluation for upper limb in stroke patients on the basis of a microsensor. Int J Rehabil Res 35(2):161–169. https://doi.org/10.1097/MRR.0b013e328353053a
Janssen J, Verschuren O, Renger WJ, Ermers J, Ketelaar M, van Ee R (2017) Gamification in physical therapy: more than using games. Pediatr Phys Ther 29(1):95–99. https://doi.org/10.1097/PEP.0000000000000326
Jung JH, Lee HJ, Cho DY, Lim JE, Lee BS, Kwon SH et al (2019) Effects of combined upper limb robotic therapy in patients with tetraplegic spinal cord injury. Ann Rehabil Med 43(4):445
Kagawa T, Uno Y (2009) Gait pattern generation for a power-assist device of paraplegic gait. In: RO-MAN 2009—the 18th IEEE international symposium on robot and human interactive communication. IEEE, pp 633–638. https://doi.org/10.1109/ROMAN.2009.5326348
Kannape OA, Lenggenhager B (2016) Engineered embodiment: comment on "the embodiment of assistive devices-from wheelchair to exoskeleton" by M. Pazzaglia and M. Molinari. Phys Life Rev 16:181–183. https://doi.org/10.1016/j.plrev.2016.01.011

Kawase T, Sakurada T, Koike Y, Kansaku K (2017) A hybrid BMI-based exoskeleton for paresis: EMG control for assisting arm movements. J Neural Eng 14(1):016015. https://doi.org/10.1088/1741-2552/aa525f

Kiguchi K, Rahman MH, Sasaki M, Teramoto K (2008) Development of a 3DOF mobile exoskeleton robot for human upper-limb motion assist. Robot Auton Syst 56(8):678–691. https://doi.org/10.1016/j.robot.2007.11.007

Kim H, Rosen J (2015) Predicting redundancy of a 7 dof upper limb exoskeleton toward improved transparency between human and robot. J Intell Robot Syst 80(1):99–119. https://doi.org/10.1007/s10846-015-0212-4

Kim H, Miller LM, Fedulow I, Simkins M, Abrams GM, Byl N, Rosen J (2012) Kinematic data analysis for post-stroke patients following bilateral versus unilateral rehabilitation with an upper limb wearable robotic system. IEEE Trans Neural Syst Rehabil Eng 21(2):153–164. https://doi.org/10.1109/TNSRE.2012.2207462

Knutson JS, Sheffer LR, Chae J (2010) Functional neuromuscular electrical stimulation (Chapter 72). In: Frontera WR, DeLisa JA (eds) DeLisa's physical medicine & rehabilitation principles and practice. Wolters Kluwer Health/Lippincott Williams & Wilkins, Philadelphia, pp 1977–1996

Knutson JS, Brose SW, Plow EB et al (2020) Electrical stimulation (therapeutic and functional). In: Frontera WR, DeLisa JA (eds) DeLisa's physical medicine and rehabilitation: principles and practice. Wolters Kluwer Health/Lippincott Williams & Wilkins, Philadelphia, pp 1246–1263

Ko HY (2019) Management and rehabilitation of spinal cord injuries. Springer, Singapore

Kolakowsky-Hayner SA, Crew J, Moran S, Shah A (2013) Safety and feasibility of using the EksoTM bionic exoskeleton to aid ambulation after spinal cord injury. J Spine 4(3). https://doi.org/10.4172/2165-7939.s4-003

Kordower JH, Tuszynski MH (2008) CNS regeneration: basic science and clinical advances. Elsevier. https://www.sciencedirect.com/science/book/9780123739940

Kwak NS, Müller KR, Lee SW (2015) A lower limb exoskeleton control system based on steady state visual evoked potentials. J Neural Eng 12(5):056009. https://doi.org/10.1088/1741-2560/12/5/056009

Lajeunesse V, Vincent C, Routhier F, Careau E, Michaud F (2016) Exoskeletons' design and usefulness evidence according to a systematic review of lower limb exoskeletons used for functional mobility by people with spinal cord injury. Disabil Rehabil Assist Technol 11(7):535–547. https://doi.org/10.3109/17483107.2015.1080766

Lebedev M (2014) Brain-machine interfaces: an overview. Transl Neurosci 5(1):99–110. https://doi.org/10.2478/s13380-014-0212-z

Li R, Hu XL, Tong KY (2008) Combined electromyography (EMG)-driven system with functional electrical stimulation (FES) for poststroke rehabilitation. In: 2008 2nd IEEE RAS & EMBS international conference on biomedical robotics and biomechatronics. IEEE, pp 642–646 https://doi.org/10.1109/BIOROB.2008.4762821

Liepert J, Bauder H, Miltner WH, Taub E, Weiller C (2000) Treatment-induced cortical reorganization after stroke in humans. Stroke 31(6):1210–1216. https://doi.org/10.1161/01.STR.31.6.1210

López NM, de Diego N, Hernández R, Pérez E, Ensinck G, Valentinuzzi ME (2014) Customized device for pediatric upper limb rehabilitation in obstetric brachial palsy. Am J Phys Med Rehabil 93(3):263–266. https://doi.org/10.1097/PHM.0b013e3182a51c95

Louie DR, Eng JJ (2016) Powered robotic exoskeletons in post-stroke rehabilitation of gait: a scoping review. J Neuroeng Rehabil 13(1):53. https://doi.org/10.1186/s12984-016-0162-5

Louie DR, Eng JJ, Lam T (2015) Gait speed using powered robotic exoskeletons after spinal cord injury: a systematic review and correlational study. J Neuroeng Rehabil 12(1):82. https://doi.org/10.1186/s12984-015-0074-9

Ma M, Zheng H (2011) Virtual reality and serious games in healthcare. In: Advanced computational intelligence paradigms in healthcare 6. Virtual reality in psychotherapy, rehabilitation, and assessment. Springer, Berlin, pp 169–192

Maciejasz P, Eschweiler J, Gerlach-Hahn K, Jansen-Troy A, Leonhardt S (2014) A survey on robotic devices for upper limb rehabilitation. J Neuroeng Rehabil 11(1):3

Marchal-Crespo L, Reinkensmeyer DJ (2009) Review of control strategies for robotic movement training after neurologic injury. J Neuroeng Rehabil 6(1):20. https://doi.org/10.1186/1743-0003-6-20

Martín-Gutiérrez J, Mora CE, Añorbe-Díaz B, González-Marrero A (2017) Virtual technologies trends in education. EURASIA J Math Sci Technol Educ 13(2):469–486. https://doi.org/10.12973/eurasia.2017.00626a

Mavroidis C, Nikitczuk J, Weinberg B, Arango R, Danaher G, Jensen K et al (2005) Smart portable rehabilitation devices. In: International design engineering technical conferences and computers and information in engineering conference, vol 47446, pp 501–510. https://doi.org/10.1186/1743-Received

Mazzoleni S, Duret C, Grosmaire AG, Battini E (2017) Combining upper limb robotic rehabilitation with other therapeutic approaches after stroke: current status, rationale, and challenges. BioMed Res Int 2017. https://doi.org/10.1155/2017/8905637

Mazzoleni S, Battini E, Crecchi R, Dario P, Posteraro F (2018) Upper limb robot-assisted therapy in subacute and chronic stroke patients using an innovative end-effector haptic device: a pilot study. NeuroRehabilitation 42(1):43–52. https://doi.org/10.3233/NRE-172166

Michmizos KP, Krebs HI (2012) Serious games for the pediatric anklebot. In: 2012 4th IEEE RAS & EMBS international conference on biomedical robotics and biomechatronics (BioRob). IEEE, pp 1710–1714. https://doi.org/10.1109/BioRob.2012.6290706

Milia P, Peccini MC, De Salvo F, Sfaldaroli A, Grelli C, Lucchesi G et al (2019) Rehabilitation with robotic glove (Gloreha) in poststroke patients. Digit Med 5(2):62. https://doi.org/10.4103/digm.digm_3_19

Mironov VI, Kastalskiy I, Lobov S, Kazantsev VB (2017) A biofeedback control system of the exoskeleton trainer for lower limbs motor function recovery. In: NEUROTECHNIX. pp 54–59. https://doi.org/10.5220/0006584700540059

Montana JI, Tuena C, Serino S, Cipresso P, Riva G (2019) Neurorehabilitation of spatial memory using virtual environments: a systematic review. J Clin Med 8(10):1516. https://doi.org/10.3390/jcm8101516

Muresanu DF, Buzoianu A, Florian SI, von Wild T (2012) Towards a roadmap in brain protection and recovery. J Cell Mol Med 16(12):2861–2871. https://doi.org/10.1111/j.1582-4934.2012.01605.x

Nagarajan U, Goswami A (2015) Improved mobility with a neutral, motion-amplifying controller for an experimental exoskeleton. SAE Int J Passeng Cars Mech Syst 8(2):606–613. https://doi.org/10.4271/2015-01-1400

Neugebauer J (2017) Robotics/exoskeleton/augmentation definitions and taxonomy. pp 1–16. https://www.nist.gov/system/files/documents/2017/02/06/2a._nist_taxonomy_and_definitions_v2.pdf

Nicolelis MA, Lebedev MA (2009) Principles of neural ensemble physiology underlying the operation of brain–machine interfaces. Nat Rev Neurosci 10(7):530–540. https://doi.org/10.1038/nrn2653

Nimawat D, Jailiya PRS (2015) Requirement of wearable robots in current scenario. Eur J Adv Eng Technol 2(2):19–23

Norouzi-Gheidari N, Archambault PS, Fung J (2012) Effects of robot-assisted therapy on stroke rehabilitation in upper limbs: systematic review and meta-analysis of the literature. J Rehabil Res Dev 49(4):479–496. https://doi.org/10.1682/JRRD.2010.10.0210

Noveanu S, Chetran B, Tatar O, Raducanu G, Mândru D (2013) Structural synthesis of the upper limb modular wearable exercisers. In: 2013 17th international conference on system theory, control and computing (ICSTCC). IEEE, pp 693–698

Nycz CJ, Delph MA, Fischer GS (2015) Modeling and design of a tendon actuated soft robotic exoskeleton for hemiparetic upper limb rehabilitation. In: 2015 37th annual international con-

ference of the IEEE engineering in medicine and biology society (EMBC). IEEE, pp 3889–3892. https://doi.org/10.1109/EMBC.2015.7319243
Onose G, Grozea C, Anghelescu A, Daia C, Sinescu CJ, Ciurea AV et al (2012) On the feasibility of using motor imagery EEG-based brain–computer interface in chronic tetraplegics for assistive robotic arm control: a clinical test and long-term post-trial follow-up. Spinal Cord 50(8):599–608. https://doi.org/10.1038/sc.2012.14
Onose G, Cârdei V, Crăciunoiu ŞT, Avramescu V, Opriş I, Lebedev MA, Constantinescu MV (2016) Mechatronic wearable exoskeletons for bionic bipedal standing and walking: a new synthetic approach. Front Neurosci 10:343. https://doi.org/10.3389/fnins.2016.00343
Onose G, Popescu N, Munteanu C, Ciobanu V, Sporea C, Mirea MD et al (2018) Mobile mechatronic/robotic orthotic devices to assist–rehabilitate neuromotor impairments in the upper limb: a systematic and synthetic review. Front Neurosci 12:577. https://doi.org/10.3389/fnins.2018.00577
Palermo AE, Maher JL, Baunsgaard CB, Nash MS (2017) Clinician-focused overview of bionic exoskeleton use after spinal cord injury. Top Spinal Cord Inj Rehabil 23(3):234–244. https://doi.org/10.1310/sci2303-234
Parker J, Sherwood M, Kane J (2013) A real-time functional magnetic resonance imaging (fMRI) neurofeedback system. IFAC Proc 46(15):341–348. https://doi.org/10.3182/20130811-5-US-2037.00078
Pelz J, Hayhoe M, Loeber R (2001) The coordination of eye, head, and hand movements in a natural task. Exp Brain Res 139(3):266–277. https://doi.org/10.1007/s002210100745
Pirondini E, Coscia M, Marcheschi S, Roas G, Salsedo F, Frisoli A et al (2014) Evaluation of a new exoskeleton for upper limb post-stroke neuro-rehabilitation: preliminary results. In: Replace, repair, restore, relieve–bridging clinical and engineering solutions in neurorehabilitation. Springer, Cham, pp 637–645. https://doi.org/10.1007/978-3-319-08072-7
Presacco A, Forrester LW, Contreras-Vidal JL (2012) Decoding intra-limb and inter-limb kinematics during treadmill walking from scalp electroencephalographic (EEG) signals. IEEE Trans Neural Syst Rehabil Eng 20(2):212–219
Proietti T, Crocher V, Roby-Brami A, & Jarrasse N (2016) Upper-limb robotic exoskeletons for neurorehabilitation: a review on control strategies. IEEE reviews in biomedical engineering,9:4–14
Rahman T, Sample W, Jayakumar S, King MM (2006) Passive exoskeletons for assisting limb movement. J Rehabil Res Dev 43(5):583. https://doi.org/10.1682/JRRD.2005.04.0070
Rego P, Moreira PM, Reis LP (2010) Serious games for rehabilitation: a survey and a classification towards a taxonomy. In: 5th Iberian conference on information systems and technologies. IEEE, pp 1–6
Reinkensmeyer DJ, Aoyagi D, Emken JL, Galvez JA, Ichinose W, Kerdanyan G et al (2014) Tools for understanding and optimizing robotic gait training. J Rehabil Res Dev 43(5):657–670. https://doi.org/10.1682/JRRD.2005.04.0073
Ring H, Rosenthal N (2005) Controlled study of neuroprosthetic functional electrical stimulation in sub-acute post-stroke rehabilitation. J Rehabil Med 37(1):32–36. https://doi.org/10.1080/16501970410035387
Robinson NL, Cottier TV, Kavanagh DJ (2019) Psychosocial health interventions by social robots: systematic review of randomized controlled trials. J Med Internet Res 21(5):e13203. https://doi.org/10.2196/13203
Rocon E, Belda-Lois JM, Ruiz AF, Manto M, Moreno JC, Pons JL (2007) Design and validation of a rehabilitation robotic exoskeleton for tremor assessment and suppression. IEEE Trans Neural Syst Rehabil Eng 15(3):367–378. https://doi.org/10.1109/TNSRE.2007.903917
Sheffer LR, Knutson JS, Chae J (2010) Therapeutic electrical stimulation in neurorehabilitation (Chapter 71). In: Frontera WR, DeLisa JA (eds) DeLisa's physical medicine and rehabilitation: principles and practice. Wolters Kluwer Health/Lippincott Williams & Wilkins, Philadelphia, pp 1953–1975
Sheridan TB (1992) Musings on telepresence and virtual presence. Presence 1(1):120–126. https://doi.org/10.1162/pres.1992.1.1.120

Shull PB, Damian DD (2015) Haptic wearables as sensory replacement, sensory augmentation and trainer—a review. J Neuroeng Rehabil 12(1):59. https://doi.org/10.1186/s12984-015-0055-z
Sisto SA, Forrest GF, Glendinning D (2002) Virtual reality applications for motor rehabilitation after stroke. Top Stroke Rehabil 8(4):11–23
Soiza-Reilly M, Saggau P, Arenkiel BR (2015) Neural circuits revealed. Front Neural Circuits 9:35. https://doi.org/10.3389/fncir.2015.00035
Song Z, Guo S, Xiao N, Gao B, Shi L (2012) Implementation of human-machine synchronization control for active rehabilitation using an inertia sensor. Sensors 12(12):16046–16059. https://doi.org/10.3390/s121216046
Song Z, Guo S, Pang M, Zhang S, Xiao N, Gao B, Shi L (2014) Implementation of resistance training using an upper-limb exoskeleton rehabilitation device for elbow joint. J Med Biol Eng 34(2):188–196. https://doi.org/10.5405/jmbe.1337
Stein J (2009) e100 NeuroRobotic system. Expert Rev Med Devices 6(1):15–19. https://doi.org/10.1586/17434440.6.1.15
Stein J (2019) Stroke. In: Frontera W, Silver J, Rizzo T (eds) Essentials of physical medicine and rehabilitation: musculoskeletal disorders, pain, and rehabilitation. Elsevier, Philadelphia, pp 931–936
Stewart AM, Pretty CG, Adams M, Chen X (2017) Review of upper limb hybrid exoskeletons. IFAC-Papers Online 50(1):15169–15178. https://doi.org/10.1016/j.ifacol.2017.08.2266
Takizawa S, Wada R, Tachibana T et al (2018) Implementation and result of the takizawa method to the outpatient rehabilitation facility and evaluation of the brain activity by fNIRS. J Clin Diagn Treat 1:26–31
Tarkka IM, Hautasaari P (2019) Motor action execution in reaction-time movements: magnetoencephalographic study. Am J Phys Med Rehabil 98(9):771–776. https://doi.org/10.1097/PHM.0000000000001187
Thielbar KO, Triandafilou KM, Fischer HC, O'Toole JM, Corrigan ML, Ochoa JM et al (2016) Benefits of using a voice and EMG-driven actuated glove to support occupational therapy for stroke survivors. IEEE Trans Neural Syst Rehabil Eng 25(3):297–305. https://doi.org/10.1109/TNSRE.2016.2569070
Tu X, Han H, Huang J, Li J, Su C, Jiang X, He J (2017) Upper limb rehabilitation robot powered by PAMs cooperates with FES arrays to realize reach-to-grasp trainings. J Healthc Eng 2017:1282934. https://doi.org/10.1155/2017/1282934
Veneman JF (2014) Exoskeletons supporting postural balance–the balance project. In: Replace, repair, restore, relieve–bridging clinical and engineering solutions in neurorehabilitation. Springer, Cham, pp 203–208. https://doi.org/10.1007/978-3-319-08072-7_38
Veneman JF, Kruidhof R, Hekman EE, Ekkelenkamp R, Van Asseldonk EH, Van Der Kooij H (2007) Design and evaluation of the LOPES exoskeleton robot for interactive gait rehabilitation. IEEE Trans Neural Syst Rehabil Eng 15(3):379–386. https://doi.org/10.1109/TNSRE.2007.903919
Watson T (2007) Transcutaneous Electrical Nerve Stimulation (TENS). http://www.electrotherapy.org/modality/transcutaneous-electrical-nerve-stimulation-tens
Watts LT, Long JA, Manga VH, Huang S, Shen Q, Duong TQ (2015) Normobaric oxygen worsens outcome after a moderate traumatic brain injury. J Cereb Blood Flow Metab 35(7):1137–1144. https://doi.org/10.1038/jcbfm.2015.18
Weber DJ, Skidmore ER, Niyonkuru C, Chang CL, Huber LM, Munin MC (2010) Cyclic functional electrical stimulation does not enhance gains in hand grasp function when used as an adjunct to onabotulinumtoxinA and task practice therapy: a single-blind, randomized controlled pilot study. Arch Phys Med Rehabil 91(5):679–686. https://doi.org/10.1016/j.apmr.2010.01.010
Weber LM, Nilsen DM, Gillen G, Yoon J, Stein J (2019) Immersive virtual reality mirror therapy for upper limb recovery after stroke: a pilot study. Am J Phys Med Rehabil 98(9):783–788. https://doi.org/10.1097/PHM.0000000000001190
Wei W, Guo S, Zhang F, Guo J, Ji Y, Wang Y (2013a) A novel upper limb rehabilitation system with hand exoskeleton mechanism. In: 2013 IEEE international conference on mechatronics and automation. IEEE, pp 285–290. https://doi.org/10.1109/ICMA.2013.6617932

Wei W, Guo S, Zhang W, Guo J, Wang Y (2013b) A novel VR-based upper limb rehabilitation robot system. In: 2013 ICME international conference on complex medical engineering. IEEE, pp 302–306. https://doi.org/10.1109/ICCME.2013.6548259

Weidig C, Mestre DR, Israel JH, Noel F, Perrot V, Aurich JC (2014) Classification of VR interaction techniques, based on user intention. Eurographics Digital Library. https://doi.org/10.2312/eurovr.20141339

Weiskopf N, Mathiak K, Bock SW, Scharnowski F, Veit R, Grodd W et al (2004) Principles of a brain-computer interface (BCI) based on real-time functional magnetic resonance imaging (fMRI). IEEE Trans Biomed Eng 51(6):966–970. https://doi.org/10.1109/TBME.2004.827063

Widmer M (2018) Research collection, (November). https://doi.org/10.3929/ethz-a-010870008

Xiao ZG, Elnady AM, Webb J, Menon C (2014) Towards a brain computer interface driven exoskeleton for upper extremity rehabilitation. In: 5th IEEE RAS/EMBS international conference on biomedical robotics and biomechatronics. IEEE, pp 432–437. https://doi.org/10.1109/biorob.2014.6913815

Yatsuya K, Hirano S, Saitoh E, Tanabe S, Tanaka H, Eguchi M et al (2018) Comparison of energy efficiency between wearable power-assist locomotor (WPAL) and two types of knee-ankle-foot orthoses with a medial single hip joint (MSH-KAFO). J Spinal Cord Med 41(1):48–54. https://doi.org/10.1080/10790268.2016.1226701

Yoshimoto T, Shimizu I, Hiroi Y, Kawaki M, Sato D, Nagasawa M (2015) Feasibility and efficacy of high-speed gait training with a voluntary driven exoskeleton robot for gait and balance dysfunction in patients with chronic stroke: nonrandomized pilot study with concurrent control. Int J Rehabil Res 38(4):338–343. https://doi.org/10.1097/MRR.0000000000000132

Young W, AlZoubi ZM, Saberi H, Sharma A, Muresanu D, Feng S, Chen L, Huang H (2015) Beijing declaration of international association of neurorestoratology. Journal of Neurorestoratology 3: 121–122

Zeilig G, Weingarden H, Zwecker M, Dudkiewicz I, Bloch A, Esquenazi A (2012) Safety and tolerance of the ReWalk™ exoskeleton suit for ambulation by people with complete spinal cord injury: a pilot study. J Spinal Cord Med 35(2):96–101. https://doi.org/10.1179/2045772312Y.0000000003

Zhou X, He X, Ren Y (2014) Function of microglia and macrophages in secondary damage after spinal cord injury. Neural Regen Res 9(20):1787. https://doi.org/10.4103/1673-5374.143423

Part VI
Augmenting Cognition and Emotion

Effects of rTMS on Behavioral and Electrocortical Measures of Error Monitoring and Correction Function in Children with Autism Spectrum Disorder

Estate M. Sokhadze, Ioan Opris, Lonnie Sears, Ayman S. El-Baz, Allan Tasman, and Manuel F. Casanova

1 Introduction

Autism spectrum disorder (ASD) is pervasive developmental disorder characterized by the early onset of impairments in social and communication skills along with restricted and repetitive interests and activities (APA 2013). There is growing evidence that executive function deficits may contribute to these core symptoms (Hill 2004). Executive functioning skills are usually referred to as the prefrontal functions that facilitate the ability to maintain problem-solving, flexible set-shifting, and forward planning to implement a goal-directed behavior (Ozonoff 1997; Ozonoff et al. 1991). One important executive function known to be compromised in ASD is

E. M. Sokhadze · M. F. Casanova (✉)
Department of Psychiatry and Behavioral Sciences, University of Louisville, Louisville, KY, USA

Department of Biomedical Sciences, University of South Carolina School of Medicine, Greenville, SC, USA
e-mail: manuel.casanova@louisville.edu

I. Opris
Department of Neurosurgery, University of Miami, Miami, FL, USA

L. Sears
Department of Pediatrics, University of Louisville, Louisville, KY, USA

A. S. El-Baz
Department of Bioengineering, Speed School of Engineering, University of Louisville, Louisville, KY, USA
e-mail: ayman.elbaz@louisville.edu

A. Tasman
Department of Psychiatry and Behavioral Sciences, University of Louisville, Louisville, KY, USA

I. Opris et al. (eds.), *Modern Approaches to Augmentation of Brain Function*, Contemporary Clinical Neuroscience,
https://doi.org/10.1007/978-3-030-54564-2_25

the ability to select a contextually appropriate response among several competing ones and simultaneously inhibit contextually inappropriate responses to avoid committing an error. Another executive deficit observed during performance on speeded reaction time tasks in autism is manifested in an abnormality related to response error monitoring and post-error response correction.

Current theory and research suggest that deficits in response monitoring may contribute to social-emotional and social-cognitive impairments in autism (Henderson et al. 2006). Executive deficit hypotheses of autism emphasize that many of the everyday behaviors of autistic individuals, such as, perseverative responding, repetitive behaviors, poor imitation skills, and joint attention impairments, may involve an inability to consistently and accurately monitor ongoing behaviors (Mundy 2003). Therefore, impairments specific to self-monitoring function have been already outlined in earlier models of autism (Russell 1997; Russell and Jarrold 1998). Several reports (Bogte et al. 2007; Chang 2017; Chmielewski and Beste 2015; Henderson et al. 2006; Hüpen et al. 2016; South et al. 2010; Thakkar et al. 2008; Vlamings et al. 2008) indicate that children and adult patients with ASD show reduced error processing and deficient behavioral correction after an error is committed. There are at the same time several reports in the literature that did not support these findings. For instance, in a study of Kim et al. (2018) on kindergarten children with ASD in a Go/NoGo task showed significantly greater amplitudes of ERN compared to matched controls, suggesting heightened response monitoring.

Performance on behavioral tasks is monitored by a brain system that is responsive to errors (Falkenstein et al. 2000; Gehring et al. 1993; Luu et al. 2000, 2003). The evidence both from functional magnetic resonance imaging (fMRI), electroencephalographic (EEG), and event-related potential (ERP) studies outlines that error monitoring is a function of the medial frontal cortex (MFC), including the supplementary eye fields, the rostral cingulate motor area, and the dorsal anterior cingulate cortex (ACC) (reviewed in Ridderinkhof et al. 2004). The error sensitivity can be readily examined by measuring response-locked event-related potential components associated with brain responses to errors. Two specific components relevant in this context are the error-related negativity (ERN, more rarely referred to as Ne) and the error-related positivity (Pe). The ERN is a response-locked negative ERP deflection, emerging between 0 and 150 ms after the onset of the incorrect behavioral response—a commission error. Usually, this ERN is followed by a positive wave referred to as the Pe potential. Although there is a discussion about the exact meaning of the Pe (Overbeek et al. 2005), most studies indicate that the Pe is related to the conscious recognition of the error (Nieuwenhuis et al. 2001) or the attribution of motivational significance to committed error (Falkenstein et al. 2000). This suggests that the ERN reflects an initial automatic brain response as a result of an error and the Pe possibly reflects the conscious reflections on the error and comprehension (Overbeek et al. 2005). The magnitude of the ERN is associated with behavioral evidence of self-monitoring (i.e., self-correction and post-error slowing responses), and therefore, is interpreted as a biomarker of error processing (van Veen and Carter 2002). Dipole modeling has localized ERN sources to the caudal ACC, while Pe to the more rostral ACC division (Bush et al. 2000; Gehring and Knight 2000; Herrmann et al. 2004; van Veen and Carter 2002; West 2003). The ERN and Pe are generally accepted as neural indices of response-monitoring processes in psychophysiological research and clinical neurophysiology.

One of the important research questions is whether this error-related frontal activity is associated with a premorbid trait reflecting initial deficiency of behavioral control and regulation, and whether this deficit can be generated as a result of neuropathological states that are known to be associated with behavioral control deficits, with pervasive developmental disorders being one of such states. Several clinical research studies have demonstrated an excessive error processing in patients with obsessive–compulsive disorders (OCD) (Johannes et al. 2001), anxiety disorders (Markela-Lerenc et al. 2004), and Tourette syndrome (Gehring et al. 2000). On the contrary, reduced error-processing manifestations were reported in borderline personality disorder (de Bruijn et al. 2006) and schizophrenia (Mathalon et al. 2002). In psychiatric studies, a decreased ERN is typically related to increased severity of psychomotor poverty symptoms (Bates et al. 2004). Furthermore, error processing has also been found to be reduced in nonclinical traits such as high impulsivity (Groen et al. 2008; Ruchsow et al. 2005). South et al. (2010) used a modified Flanker task to test the hypothesis that high-functioning individuals with ASD would show decreased amplitude ERN in 24 individuals with ASD and 21 age- and IQ-matched typically developing control participants. Behaviorally, individuals with ASD committed more errors than controls, but groups did not significantly differ on reaction time (RT), although there was a difference in post-error slowing. The authors found that ERN amplitude was significantly attenuated in individuals with ASD relative to controls. At the same time, groups did not differ in Pe amplitude. These findings were explained as a reflection of ASD patients' lower sensitivity to behavioral errors and reduced behavior correction ability.

Neuroanatomically and functionally, the anterior cingulate cortex (ACC) provides an interface between frontal action selection processes, limbic emotion or motivation processes, and motor output regulation (Coles et al. 2001; Holroyd and Coles 2002; Taylor et al. 2007). The integral role of the ACC in self-monitoring and guiding attention in goal-directed actions suggests that it may be an important focus for autism research. The disturbances in attention regulation and behavioral rigidity may result in social-orienting deficits and a chronic disruption of social information processing and social learning that together contribute to the social-cognitive and emotional deficits observed in autistic children (Dawson et al. 1998; Mundy 1995; Mundy and Neal 2001). Several neuroimaging studies (Barnea-Goraly et al. 2004; Hall et al. 2003) suggest that anomalous functioning of the ACC may distinguish between individuals with autism and controls. Haznedar et al. (2000) observed that a sample of children with autism displayed hypometabolism in the right ACC relative to controls, while an Asperger disorder subsample displayed left ACC hypometabolism relative to controls. A number of studies suggest ACC abnormalities in ASD, which might underlie response monitoring and social impairments exhibited by children and adolescents with ASD. Santesso et al. (2011) examined error and correct response monitoring using event-related potentials (ERN, Pe) and LORETA source localization in high-functioning adults with ASD and controls. Adults with ASD showed reduced ERN and Pe amplitudes and reduced rostral ACC activation compared with controls. These findings suggest that reduced ACC activity may reflect a putative brain mechanism involved in the origins and maintenance of social impairments and raise the possibility of the presence of stable brain-behavior relation impairment across development in some individuals with ASD.

There have been also several ERN-based empirical demonstrations of connections between ACC functions and autism. Children with the high-functioning autism displayed longer ERN latencies, but did not differ in the amplitude of the ERN relative to children in the control group in the Eriksen flanker task in a study of Henderson et al. (2006). There is other evidence of abnormal response monitoring in autism, in particular reduced error self-correction (Russell and Jarrold 1998) and reduced post-error slowing, a compensatory mechanism to improve performance on the subsequent trial (Bogte et al. 2007). Since the evaluation of ongoing behavior and its consequences is necessary to determine whether or not current behavior adjustment strategies should be maintained, abnormal response monitoring and deficient adaptive correction may contribute to behavioral inflexibility observed in ASD.

2 Studies of Our Group on Reaction Time, Accuracy, and ERN/Pe in Children with ASD

Novelty Task In our first prior study (Sokhadze et al. 2010) focused on error monitoring processing, we used a three-stimuli oddball task to examine the possibility that children with ASD exhibit a deficiency in the processing of error, reflected by a reduction and delays in the ERN and Pe response-locked brain potentials. The mean age of 14 participants enrolled in the ASD group was 13.0 ± 2.5 years, while the mean age of the Control (CNT) group ($N = 14$) was 14.1 ± 3.9 years. The test represents a traditional visual three-stimuli oddball task. Stimuli letters “X”, “O”, and novel distracters (“v,” “^,” “>,” and “<” signs) were presented on the screen after fixation mark “+”. One of the stimuli (“O”) was presented on 50% of the trials (frequent standard); the novel stimuli stimulus (e.g., “>”) was presented on 25% of the trials (rare distracter), whereas the third (“X”) is presented on the remaining 25% of the trials and represents the target. Subjects are instructed to press a key when they see the target letter on the screen.

Behavioral response measures were mean reaction time (RT) and response accuracy (percent of correct hits). Response-locked ERP-dependent measures recorded using dense-array EGI NetStation (Electrical Geodesics Inc., Eugene, OR) were amplitude and latency of ERN and Pe across the region-of-interest (ROI) channel groups at the midline fronto-central area. Our results showed that ASD patients had high rate of errors in the visual oddball task with novel distracters. Difference in commission error rate was significant (21.7 ± 29.2% in ASD vs. 4.8 ± 6.1% in CNT,

$F = 4.41$, $p = 0.046$); mean post-error RT was faster in ASD as compared to CNT group (420 ± 94 ms in ASD vs. 519 ± 99 ms in CNT, $F = 7.21$, $p = 0.012$). Difference between mean RT in correct trials and post-error trials (i.e., mean post-error RT-minus-correct trial RT) was negative in autism, but positive in controls, and this between-group difference was significant ($F = 5.22$, $p = 0.031$). In particular, in typically developing subjects, it has been observed that after an error has been committed, subjects show slower RT and decreased error rates. These normative changes have been interpreted as revealing changes in the speed–accuracy strategy of the subject possibly due to error-induced control processes and concomitant corrective adjustments. The patients with ASD showed opposite response: faster post-error RT instead of slowing down. Amplitude of the ERN across the fronto-central ROI in ASD group as compared to controls was significantly less negative (−0.29 ± 6.68 μV in ASD vs. −5.50 ± 5.76 μV in CNT, $F = 4.88$, $p = 0.036$). Amplitude of the Pe was not different between groups. Dipole source localization analysis allowed placing a dipole (PCA loading 93.6%) for the ERN in a more caudal division of the ingulated cortex (ACC), while for the Pe a dipole with 76.9% loading in a more rostral division of the ACC (Sokhadze et al. 2010). The results were replicated in our later study of our group (Sokhadze et al. 2017) using the same oddball paradigm with a larger sample size (ASD, $N = 32$; CNT, $N = 24$).

In both of these studies, we found lower ERN amplitude and prolonged latency of ERN in ASD as compared to typical controls. The reduced ERN along with a lack of post-error RT slowing in autism was interpreted as an insensitivity to detect and monitor response errors and reduced ability of execute corrective actions (Sokhadze et al. 2010, 2017). Results were indicative of reduced error awareness and a failure in adjustment in ASD when dealing with situations where erroneous responses may occur.

Visual Odd Ball Task with Illusory Kanizsa Figures In a subsequent study using visual odd ball task with illusory figures (Sokhadze et al. 2012a), participants were children with ASD ($N = 16$), ADHD ($N = 16$), and typical controls ($N = 16$). The task involved the recognition of a specific illusory shape (Kanizsa figures, Kanizsa 1976), in this case a square or triangle, created by three or four inducer disks. There were no between-group differences in RT to rare target stimuli, but both ASD and ADHD committed more errors, specifically the ASD group had statistically higher commission error rate than controls. A difference in total error rate in the ASD and ADHD groups vs. the CNT group was significant, $F = 4.63$, $p = 0.015$. A Post hoc Tukey test showed differences between ASD and control groups: 19.7 ± 20.4% in ASD vs. 2.6 ± 4.5% in CNT, $p = 0.011$. The percentage of commission errors was significantly lower in the typical children (1.9 ± 3.7% in CNT vs. 16.8 ± 19.7% in ASD). Post hoc analysis showed an ASD vs. CNT significant difference, −14.9%, $p = 0.018$. Mean post-error RT was different across groups ($F = 9.20$, $p < 0.001$); in particular, CNT and ADHD groups showed an increase of mean RT following committed errors (24.4 ± 65.2 ms in CNT and 7.6 ± 40.4 ms in ADHD), while the ASD group showed a decrease of post-error RTs (−52.6 ± 48.7 ms). This difference was confirmed by post hoc analysis both for ASD vs. CNT (−77.0 ± 18.9 ms, $p = 0.001$).

Amplitude of the ERN across a 5 frontal and fronto-central ROI showed significant group differences (−3.79 ± 6.83 μV in ADHD, 0.44 ± 7.29 μV in ASD, and −5.81 ± 5.12 μV in CNT, $F = 7.62$, $p = 0.002$). Post hoc analysis showed that the ERN amplitude in the ASD group as compared to controls was significantly less negative (−6.25 ms, $p = 0.001$). Amplitude of Pe between groups was not significantly different ($p = 0.36$, n.s.). Latency of the ERN showed between-group differences ($F = 3.44$, $p = 0.042$). Latency was prolonged in the ASD group (126 ± 32 ms in ASD vs. 89 ± 51 ms in CNT vs. 107 ± 28 ms in ADHD, $F = 4.97$, $p = 0.035$). Post hoc Tukey HSD test confirmed significant differences between ASD and CNT groups ($p = 0.032$). Latency of Pe across midline frontal and fronto-central channels did not yield any significant differences. To summarize the findings, the ASD group had more errors committed and post-error RT in ASD group was exhibited in a post-error speeding rather than corrective RT slowing typical for the controls. The ASD group also demonstrated an attenuated and prolonged ERN as compared to ADHD and controls.

In a more recent study (Sokhadze et al. 2019) using the same oddball task with Kanizsa figures, participants were age-matched children ($N = 18$ per group) with ASD, ADHD (13.2 ± 3.5 years old), comorbid ASD + ADHD diagnosis (13.4 ± 2.9 years old), and neurotypical controls (CNT, 14.2 ± 3.9 years old). Analysis of data revealed that there were no between-group differences in RT to target stimuli, but both ASD and ASD + ADHD groups committed more errors, specifically the ASD groups had higher omission error rate than neurotypical children. Post-error RT in ASD and ASD + ADHD groups was featured by a post-error response speeding rather than corrective RT slowing typical for the controls. Histogram of post-error RT difference compared to correct responses in four groups of children (ASD, ADHD, ASD + ADHD, Typical) is depicted n Fig. 1.

Accuracy of response was different between groups, in particular total error percentage showed significant differences ($F = 3.78$, $p = 0.015$). A post hoc Tukey test yielded significant difference between ASD and CNT groups (17.9 ± 14.3% in ASD vs. 2.4 ± 4.6% in CNT, $p = 0.009$). Post hoc analysis showed both ASD and ASD + ADD vs. CNT difference (5.5 ± 4.9% in ASD, 4.3 ± 5.3% in ASD + ADD vs. 0.4 ± 0.9% in typical children with $p < 0.01$ in both comparisons). The most pronounced group differences were found in the normative post-error RT slowing measure ($F_{3,71} = 16.45$, $p < 0.001$). Differences in mean post-error reaction time changes clearly separated groups with ASD from the typical children and ADHD groups, as both ASD and ASD + ADD groups showed post-error speeding (−46.1 ± 47.4 ms in ASD, −52.1 ± 51.7 ms in ASD + ADD), while CNT and ADHD groups showed normative slowing of RT following committed errors (49.1 ± 45.9 ms in CNT, 11.9 ± 14.2 ms in ADHD). The ASD diagnosis (combined ASD and ASD + ADHD) factored most significantly in affecting post-error RT change (−49.1 ± 48.3 ms in combined ASD vs. 49.2 ± 45.9 ms in CNT, $p < 0.001$). The ASD and ASD + ADHD group also demonstrated an attenuated error-related negativity (ERN) as compared to ADHD and controls.

Amplitude of the ERN measured at the midline fronto-central ROI (Fz-FCz) had significant between-group differences ($F = 3.15$, $p = 0.031$). The group differences

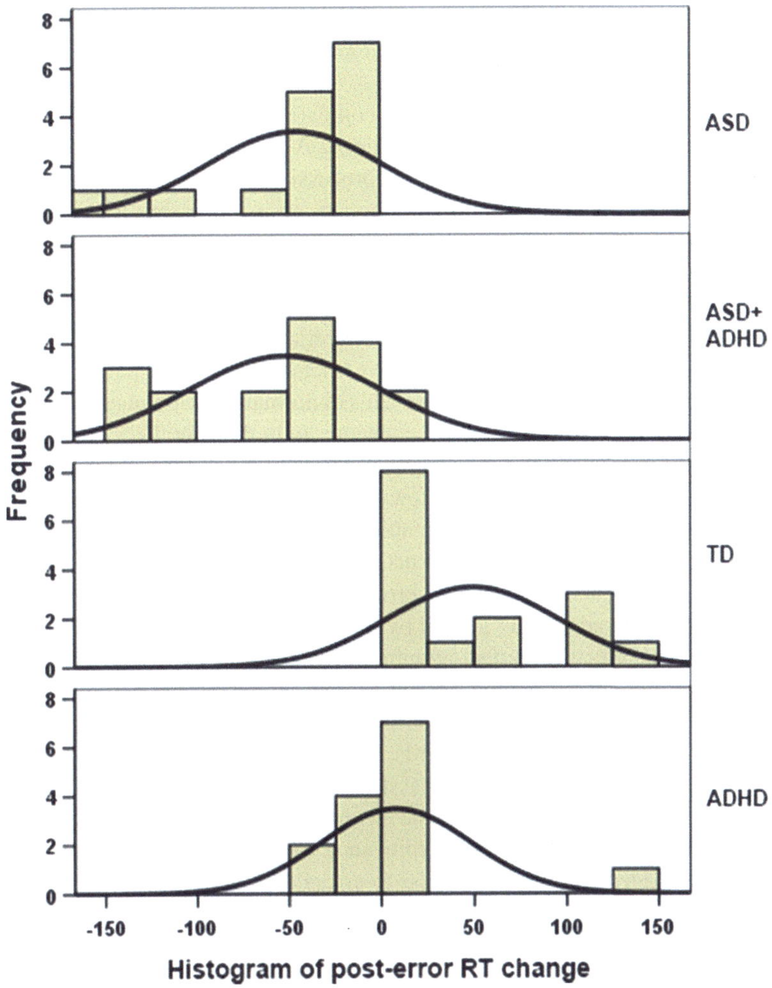

Fig. 1 Histogram of distribution of individual post-error reaction time (RT) in children with autism, children with ASD + ADHD, typically developing (TD) controls, and children with ADHD. Both ADHD and control groups demonstrate slower (positive) post-error RTs compared to correct response RTs. The ASD and ASD + ADHD groups show speeding of post-error RTs with a negative peak of distribution curve. The ADHD shows positive peak of the curve, though still less expressed post-error RT slowing as compared to controls (Sokhadze et al. 2019)

of the ERN amplitude were better pronounced across more expanded ROI that included 5 frontal and fronto-central sites ($F = 3.51$, $p = 0.02$). At this region, the differences were expressed as lower negative amplitudes of ERN in ASD and ASD + ADD groups as compared to typically developing children (difference respectively −5.32 μV and −5.15 μV, both $p < 0.05$). The ASD diagnosed combined group (ASD and ASD + ADHD) was statistically significantly different from the CNT group by ERN amplitude (by 5.43 μV, $p = 0.005$), while combined ADHD group (ADHD and ASD + ADHD) was not different from the group of control peers ($p = 0.487$, n. s.), thus pointing at the more important contribution of ASD factor on attenuated ERN amplitude.

In summary, the findings in our studies comparing error rate and ERN between ASD and other control groups (typical subjects, ADHD, ASD + ADHD) revealed that autism is associated with reduced error processing and impaired behavioral correction after an error is made. Because adequate error processing is necessary for optimal behavioral performance, it is plausible that these deficits contribute to the maintenance of the preservative behaviors typical for autism. Our studies (Sokhadze et al. 2010, 2017, 2019) reported abnormal response monitoring and correction functions observed in behavioral and electrocortical indices of the ACC activity (e.g., ERN) in ASD that might be related to restricted, repetitive behavior style typical for this neurodevelopmental disorder. This abnormal function may result from a compromised functional and structural connectivity in the neural circuitry subserving response monitoring and error correction. These findings suggest that functional abnormalities of the ACC reflected in lower amplitude and delayed ERN measures may compromise response monitoring and contribute to behavioral repetition and stereotype behaviors in ASD. Impairments in an ability to correctly and timely evaluate committed error and to learn from errors may lead to behavior that is rigid and repetitive rather than adaptively guided by action outcomes.

The ERN is an electroencephalographic measure associated with the commission of errors, thought to be independent of conscious perception (Franken et al. 2007), while the Pe is thought to reflect the motivational or emotional significance of occurred error or, in another words, the conscious evaluation of the error (Overbeek et al. 2005). The findings that the ERN is altered in autism may suggest that ASD patients are less sensitive to the errors they committed, less aware of their errors, and probably attribute less significance to the occurrence of errors. The inflexible and inadequate responsiveness to errors may underlie one of the typical characteristics of autism spectrum disorders, namely, the persistence of stereotype behaviors. It cannot be ruled out that the ERN and Pe findings are influenced by deficits in earlier perceptual processes in patients with autism, reflected in stimulus-locked early ERPs, or attentional and working memory processes, reflected in the stimulus-locked late ERP amplitudes. The ERN and Pe along with behavioral performance measures can be used as the functional outcome measures to assess effectiveness of behavioral intervention (e.g., social skills training) or neuromodulation (e.g., TMS) in children with ASD, and thus, may have useful practical applications.

3 Application of Transcranial Magnetic Stimulation (rTMS) to Treat ASD

In the last three decades, transcranial magnetic stimulation (TMS) has been used increasingly as an experimental investigation tool to explore the mechanisms and consequences of cortical plasticity in the human cortex (George et al. 2003; Wassermann and Lisanby 2001). Repetitive transcranial magnetic stimulation (rTMS) offers a noninvasive method for altering excitability of the brain. This method uses an electromagnet placed on the scalp that generates magnetic field pulses of very short duration (100–300 ms) approximately 1.5–2.2 T in strength. Magnetic fields pass largely undistorted through the scalp and skull (Enticott et al. 2014; George et al. 1999; Pascual-Leone et al. 2002; Wagner et al. 2009). By convention, rTMS in 0.3–1 Hz frequency range is referred to as "slow," whereas "fast" rTMS refers to stimulation greater than 1 Hz. Hoffman and Cavus (2002) in their review of slow rTMS studies proposed long-term depression and long-term depotentiation as potential models for understanding the mechanism of slow rTMS. Neocortical long-term depression and changes in the cortical excitability induced by slow rTMS appear to accumulate in an additive fashion as the number of stimulations is increased. Studies of both slow rTMS and long-term depression suggest additive efficacy when higher numbers of spaced, daily stimulations are administered. The reversal, or depotentiation, of previously enhanced synaptic transmission due to long- term potentiation may be the most relevant model for slow rTMS when used as a therapeutic tool.

Effects of rTMS are not limited to the stimulated target cortex, but give rise to functional changes in anatomically and functionally interconnected cortical areas. The lasting effects of rTMS offer new possibilities to study dynamic aspects of the pathophysiology of a variety of diseases and may have therapeutic potential in some psychiatric disorders, specifically ASD. TMS is considered as a safe, effective, and promising treatment in people of all ages who suffer from depression (George et al. 1999, 2003; Gershon et al. 2003; Loo and Mitchell 2005), obsessive-compulsive disorder (Alonso et al. 2001; Greenberg 2007), post-traumatic stress disorder (PTSD) (McCann et al. 1998; Rosenberg et al. 2002), and schizophrenia (Mathalon et al. 2002). Several reviews concluded that rTMS can be generally considered safe for use in pediatric populations, as no significant adverse effects or seizures have been reported (Croarkin et al. 2011; Garvey and Mall 2008; Quintana 2005).

We proposed that TMS may improve executive functioning in children with autism (Sokhadze et al. 2009). TMS provides a noninvasive method of induction of focal currents in the brain as well as transient modulation of the function of the targeted cortex areas. Despite the fact that TMS is now widely used as a diagnostic and therapeutic tool in adults, in children, its application to date has been limited, even though TMS offers unique opportunities to gain insights into the neurophysiology of a child's brain. We believe that rTMS may play an important role in the study and possibly in the therapy of autism. The overall aim of one of our initial exploratory studies of rTMS effects (Sokhadze et al. 2012b) was to investigate behavioral responses and ERP indices of performance monitoring in children with autism enrolled either in rTMS or wait-list group. The goal of the study was to investigate whether RT, error rate, post-error RT change, ERN, and Pe will show positive changes following 12-week long slow frequency rTMS over dorsolateral prefrontal cortex (DLPFC) in high-functioning children with ASD diagnosis. We hypothesized that 12 sessions of 1.0 Hz rTMS bilaterally applied over the DLPFC will result in improvements reflected in both behavioral performance measures and such error-related potentials as ERN and Pe.

RTMS was delivered using a Magstim Rapid (Model 220) instrument (Magstim Corporation, Sheffield, England) with a 70-mm wing span figure-eight coil. Motor threshold (MT) was determined for each hemisphere in all individuals by gradually increasing the output of the machine by 5% until a 50 μV deflection or a visible twitch in the First Dorsal Interosseous (FDI) muscle was identified in 2 out of 3 trials of stimulation over the cortical area controlling the contralateral FDI. Electromyographic (EMG) responses were monitored on a continuous base with a C-2 J&J Engineering Inc. physiological monitor (Poulsbo, WA). Motor-evoked potentials were recorded from the hand contralateral to stimulation using the C2 J&J system with USE-3 Physiodata software applications. The TMS treatment course was administered once per week for 12 weeks (a total of twelve 1.0 Hz rTMS treatments); the first six treatments were over the left DLPFC, while the remaining six were over the right DLPFC. The site for stimulation was found by placing the coil 5 cm anterior, and in a parasagittal plane, to the site of maximal FDI stimulation (Fig. 2). The figure-eight coil, with a 70-mm wing diameter, was kept flat over the scalp. Subjects were wearing a swimming cap to outline the TMS coil position and aid in its placement for each session. Stimulation was done at 1 Hz and 90% MT, with a total of 150 pulses/day (fifteen 10 s trains with a 20–30 s interval between the trains). The mean age of 20 participants enrolled in the rTMS treatment group (TMS group) was 13.5 ± 2.5 years, while the mean age of 20 participants assigned to

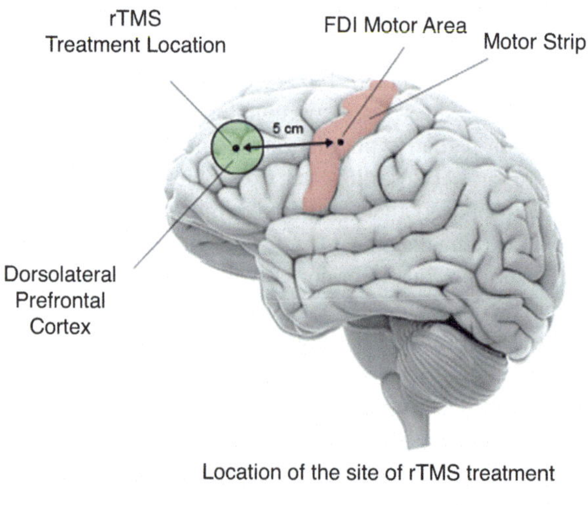

Fig. 2 Positioning of the TMS coil followed recommendations that take into consideration anatomical landmarks (Mir-Moghtadaei et al. 2015; Pommier et al. 2017) and could be approximately described as the scalp region used for F3 and F4 EEG electrode placements in the 10–20 International System (Sokhadze et al. 2018)

wait-list group (WTL group) was 14.1 ± 2.4 years. Both groups performed visual oddball task with illusory figures described above.

Between-group difference in the omission errors rate was significant, $F = 5.01$, $p = 0.031$. Paired samples test showed significant decrease of omission errors rate post-TMS (−33.0 ± 52.7%, $t = 2.26$, df = 19, $p = 0.034$). TMS group showed post-error slowing; post-error RT change was positive (37.33 ± 49.47 ms, $t = 2.87$, df = 19, $p = 0.009$). ANOVA showed that differences between TMS and WTL on this measure were also highly significant, $F = 7.35$, $p = 0.009$.

Comparison of ERN changes on repeated test (post-TMS, post-waiting period) revealed significant between-group differences on both amplitude and latency measures. Changes of the amplitude of ERN were more negative in the TMS group as compared to the WTL group (−4.74 ± 7.83 vs. 0.74 ± 7.53 μV, $F = 4.20$, $p = 0.049$), while the latency of ERN shortened in the TMS as compared to slight increase in the WTL group (−44.0 ± 43.7 ms in TMS vs. 3.92 ± 46.4 ms in WTL, $p = 0.003$). Post hoc examination using one-way ANOVA with pre- post TMS factor showed that the amplitude of ERN in TMS group decreased significantly, $F = 7.84$, $p = 0.008$, while the latency became shorter, $F = 8.81$, $p = 0.005$. The ERN changes following waiting period were highly insignificant (amplitude: $p = 0.326$, n.s.; latency: $p = 0.609$, n.s.). Changes in amplitude and latency characteristics of Pe wave were not significant (Fig. 3).

Our results show that low frequency rTMS over the DLPFC in children with autism resulted in enhanced behavioral performance in a visual attention task expressed in a lower error rate and in increased amplitude and shortened latency of the ERN, an electrocortical potential known to be reflecting error detection process.

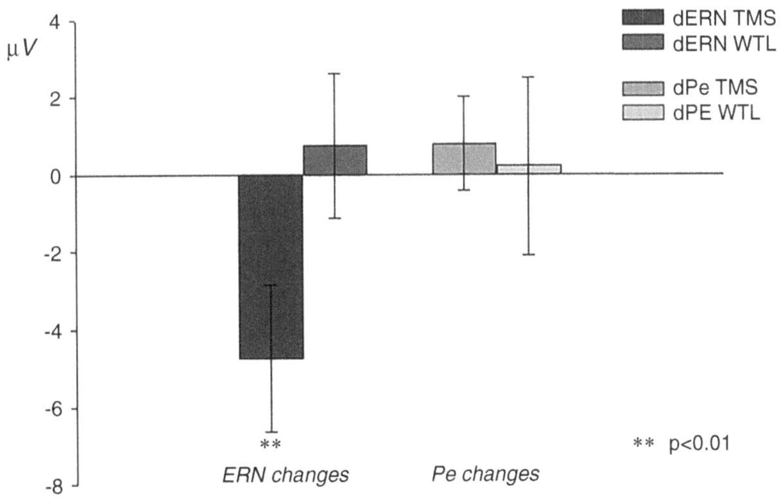

Fig. 3 Changes of ERN (dERN) and Pe (dPe) amplitudes from baseline levels in the TMS and WTL groups at posttreatment test. Bars are means with standard errors ($N = 20$ per group). Only post-TMS changes of the ERN were statistically significant (** $p < 0.05$)

It is important to emphasize that children in the TMS treatment group, as compared to the wait-list group, showed post-error slowing of reaction time, which is considered as a normative marker of a post-error corrective function. This is an interesting finding considering that autistic children consistently demonstrated post-error speeding of motor responses in our previous study where children with ASD were compared with typically developing children in a visual oddball task with novel distracters (Sokhadze et al. 2010, 2017). In most recent study, we used 18 sessions of rTMS of the same power and schedule by adding six more rTMS weekly treatments evenly distributed over the left and right DLPFC. We enrolled 54 autistic patients, 44 males and 10 females, with a mean age of 14.5 ± 2.9 years. Twenty-seven of them were assigned to active 1.0 Hz TMS treatment (TMS group), while 27 were assigned to the waiting-list group (WTL group). Mean age of subjects in the TMS group was 14.8 ± 3.2 years and 14.1 ± 2.6 years in the waiting-list group.

Commission and omission errors analysis yielded a significant between-group difference in the commission error percentage, $F(1, 52) = 4.32$, $p = 0.042$. T-test showed significant decrease of commission error rate in the TMS group (mean decrease −6.38 ± 2.54%, $t(26) = 2.50$, $p = 0.019$). Total error rate (% errors) change also showed decrease only in TMS group (−7.47 ± 2.82%, $t(26) = 2.64$, $p = 0.013$).

Main effect of *Time* (Pre-, Post-TMS) on normative post-error RT slowing was highly significant ($F = 15{,}14$, $p = 0.001$, $\eta^2) = 0.134$). Repeated measure ANOVA of post-error RT slowing revealed that TMS and WTL group differences on post-error RT changes were also statistically significant, i.e., *Time* × *Group* interaction, $F = 8.05$, $p = 0.006$, $\eta^2 = 0.134$. The TMS group showed post-error RT increase with significant positive change in post-error RT. This change was computed as post-TMS post-error RT change minus pre-treatment post-error RT change (49.9 ± 55.4 ms, $t(26) = 4.57$, $p < 0.001$). At the baseline test both in WTL and TMS groups, post-error RT was negative (mean post-error speeding was −23.1 ± 34.7 ms and not different between groups at pre-treatment stage), while in the TMS group post-error RT became positive (i.e., showed normative slowing), whereas it remained negative in the WTL group (Figs. 4 and 5).

TMS and WTL groups showed significant differences in ERN amplitude ($F = 6.20$, $p = 0.016$) and latency ($F = 5.82$, $p = 0.023$). Amplitude of ERN during commission errors across 5 fronto-central sites showed marginal *Time* × *Group* interaction ($F = 4.05$, $p = 0.05$), and paired-sample *t*-test showed significant increase of ERN negativity in the TMS group (2.97 ± 3.21 μV, $t(26) = 2.40$, $p = 0.023$). Analysis of ERN latency ANOVA yielded statistically significant *Time* × *Group* effect, ($F = 4.24$, $p = 0.041$, $\eta^2 = 0.099$). T-test of the ERN latency changes in the TMS group showed significant decrease (−28.1 ± 13.8 ms, $t(24) = 2.41$, $p = 0.023$). Amplitude and latency of Pe wave in both groups were not significantly changed posttreatment. Therefore, our results show significant changes in behavioral responses (accuracy, post-error RT slowing) and both early and later-stage ERP indices of task-relevant signal processing as a result of 18 sessions of low frequency rTMS treatment course in children with ASD.

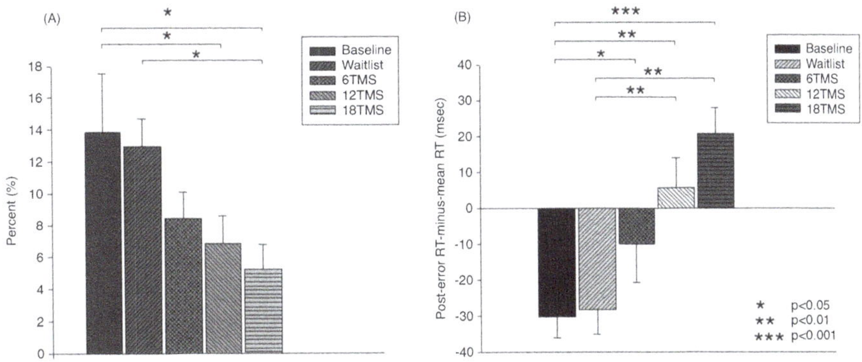

Fig. 4 (**a**) Total error rate (in %) in oddball test at baseline, post-wait period, and posttreatment in 6 TMS, 12 TMS, and 18 TMS groups of children with ASD. Accuracy gradually improved in all TMS groups. Most significant difference was between the 18 TMS group as compared to baseline and wait-list ($p < 0.05$). Accuracy difference between three TMS groups was not significant. (**b**) Post-error reaction time (RT), calculated as first RT post-error minus mean RT, in visual oddball test at baseline and post-treatment in wait-list, 6 TMS, 12 TMS, and 18 TMS groups. Most significant differences were noted between 18 TMS and baseline ($p < 0.001$), 18 TMS and wait-list ($p < 0.001$), as well as between 12 TMS and baseline ($p = 0.004$). Both 12 TMS and 18 TMS groups showed normative post-error slowing (Sokhadze et al. 2018)

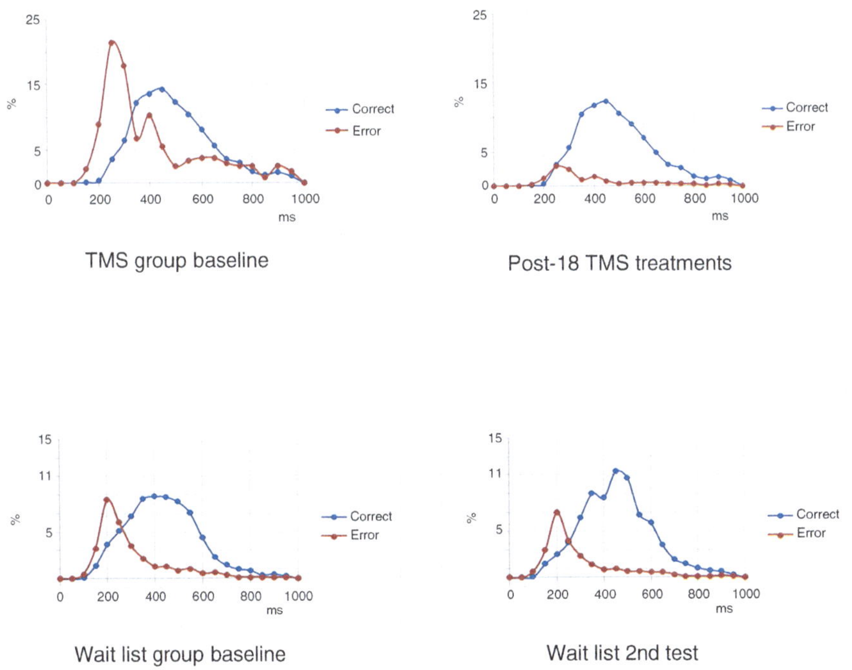

Fig. 5 Distribution of RT in correct (blue) and commission errors (blue) in the active rTMS group (N = 20, 18 sessions of 1.0 Hz rTMS sessions) and in the waiting-list (WTL) group (N = 20, 3 months waiting period). TMS group showed less errors and longer RTs during commission errors posttreatment

4 Discussion

Executive function deficits were always in the center of attention in autism research (Bishop 1993; Hill 2004; Ozonoff 1997; Ozonoff et al. 1991). Executive function of behavioral performance monitoring comprises error detection and response conflict monitoring, functions that can be measured using response-locked event-related potentials such as ERN and Pe (Arbel and Donchin 2009, 2011; Gehring et al. 1993; Mars et al. 2005; Carter et al. 1998; van Veen and Carter 2002). The ERN is a well-studied component whose parameters were investigated under different experimental task conditions, and its ties to error processing have been well-established (Carter et al. 1998; Falkenstein et al. 2000; Gehring and Knight 2000; van Veen and Carter 2002). There are an increased number of research studies examining ERN during commission errors in children (Davies et al. 2004). It is established that executive functions normally improve with age (Huizinga et al. 2006) along with demonstration that the ACC, which is now associated with executive performance monitoring, undergoes important maturation changes from childhood into adolescence, and then into adulthood (Arbel and Donchin 2009, 2011). Furthermore, the studying error-processing maturation can be used to understand mechanisms of various neurodevelopmental disorders, such as ADHD and ASD, which feature impairments in executing control (Liotti et al. 2005; Sokhadze et al. 2010; Vlamings et al. 2008; Zhang et al. 2009). The ERN abnormalities are interpreted as reflecting early error-processing impairments. A number of studies have investigated the functional relationship between the ERN and the fronto-central stimulus-locked N200; while some suggest that they represent distinct neurophysiological processes (Ridderinkhof et al. 2004), others suggest they represent different time points of the same process of response conflict monitoring (Yeung and Cohen 2006).

One of the most important findings of the series of our studies was replication of the increase of ERN amplitude and shortened latency post-TMS using 12 and 18 sessions of rTMS (Sokhadze et al. 2012b, 2014a, b). In accord with our first study focused on TMS effects on ERN/Pe indices (Sokhadze et al. 2012b), the Pe component did not change post-TMS. This component has a different topography and is expressed as a positivity elicited after the ERN (Falkenstein et al. 2000; Nieuwenhuis et al. 2001; Overbeek et al. 2005). One more critical methodological issue to be considered in absence of significant TMS effects on Pe in autism might be related to the number of commission errors as this measure depends on the actual number of committed errors (Franken et al. 2007). It is feasible to suggest that the magnitude of the Pe was affected by the reduced number of commission errors in the active TMS group. Our other investigations of ERN/Pe complex in autism (Sokhadze et al. 2010, 2017) also did show Pe differences between ASD and typical children on the similar visual oddball task. There is a possibility of a dissociation of ERN and Pe effects since generation of Pe wave might be affected by the absence of feedback about the accuracy of the motor response, resulting in lower awareness of committed errors. Our future studies will extend the number of trials in the test and will consider more traditional forced choice speeded reaction time tasks such as, Eriksens flanker task (Eriksen and Eriksen 1974) or classic Go/NoGo test to have more commission errors and more reliable Pe wave.

Within the context of ASD, rTMS may have unique potential applications as a treatment modality. It has been suggested that a wide range of deficits in autism might be understood by disrupted information integration in the brain, and more specifically, high local connectivity at the expense of deficiencies in long-range connectivity (Just et al. 2004; Rippon et al. 2007) and an increase in the ratio of cortical excitation to cortical inhibition (Casanova et al. 2003; Rubenstein and Merzenich 2003). Locally overconnected neural networks may explain the superior ability of autistic children in isolated tasks (e.g., visual discrimination), while, at the same time, deficiencies in long-range connectivity may explain other features of the disorder (e.g., behavioral stereotypy, lack of social reciprocity). Higher-than-normal cortical noise and an increase in the ratio of cortical excitation to inhibition may explain the strong aversive reactions to auditory, tactile, and visual stimuli frequently recorded in autistic individuals as well as a higher incidence of epilepsy (Gillberg and Billstedt 2000).

One possible explanation for higher-than-normal cortical noise and abnormal neural connectivity in ASD is the recent finding of minicolumnar abnormalities. Minicolumns are considered the basic anatomical and physiological unit of the cerebral cortex (Mountcastle 2003) and contain pyramidal cells that extend the cortical width surrounded by a neuropil space consisting of several species of GABAergic, inhibitory interneurons (i.e., double-bouquet, basket, and chandelier cells) (Casanova 2007; Casanova et al. 2002). Double-bouquet cells in the peripheral neuropil space of minicolumns provide a "vertical stream of negative inhibition" (Mountcastle 2003) surrounding the minicolumnar core. Our preliminary studies indicate that minicolumns are reduced in size and increased in number in the autistic brain, especially the prefrontal cortex (Casanova et al. 2002, 2006). More specifically, minicolumns in the brains of autistic patients are narrower and contain less peripheral, neuropil space (Casanova 2005; Casanova et al. 2006). The lack of a 'buffer zone' normally afforded by lateral inhibition and appropriate neuropil space may adversely affect the functional distinctiveness of minicolumnar activation and could result in isolated islands of coordinated excitatory activity (i.e., possible seizure foci); this autonomous cortical activity may hinder the binding of associated cortical areas, arguably promoting focus on particulars as opposed to general features. In addition, the effect of loss of surround inhibition may result in an increase in the ratio of cortical excitation to inhibition and signal/sensory amplification, which may impair executive functioning in patients with ASD. In terms of error monitoring, dysfunctions of DLPFC and ACC connectivity may result in poor processing of response errors and diminished ability to adjust behavioral outcome during performance on a speeded reaction time task.

In conclusion, the series of our studies (Casanova et al. 2012, 2015; Sokhadze et al. 2009, 2012b, 2014a, b, 2018) showed that treatment with "slow" rTMS improved ERP indices of attention to targets, reduced overreactivity to nontargets, significantly reduced motor response errors to target stimuli, and enhanced response-locked potentials reflective of error monitoring and correction (e.g., ERN to commission errors, post-error RT slowing, etc.). We also found significant reductions in both repetitive and stereotypic behaviors, reduced repetitive behaviors, hyperactivity, and irritability scores according to social and behavioral clinical evaluations post-TMS (Sokhadze et al. 2014a, b, 2018). We consider that it is possible to conclude that neuromodulation using low frequency, inhibitory rTMS improved executive functioning and behavior in autism. Our studies provide further support to the statement that TMS can be regarded as a prospective treatment targeting core symptoms of ASD such as executive function deficits.

It cannot be ruled out that the present ERN and Pe findings are influenced by deficits in earlier perceptual processes in patients with autism, reflected in stimulus-locked early ERPs, or attentional and working memory processes, reflected in the stimulus-locked late ERP amplitudes. Though we did not observe a significant effect of group on the frontal N200 amplitude (Sokhadze et al. 2009), we found significantly delayed latency of N200 to novel distracters in a similar three-category oddball task, suggesting that early processes that take place before responding may also be affected in autism. It has been suggested (Yeung and Cohen 2006; Yeung et al. 2004) that both the response-locked ERN and the stimulus-locked frontal N200 might reflect similar processes, i.e., response conflict detection and monitoring, and have similar neural correlates, i.e., the ACC.

On the behavioral response level, we found no group differences in RT, but only group differences between the percentages of commission (and not omission) errors during performance on visual oddball tasks. After an error, ASD patients did not show accuracy improvement through post-error RT slowing as typical controls did. Normally, performance on these trials is improved as a result of a change in speed–accuracy strategy, which reflects an executive control functioning (Burle et al. 2002). The worsened post-error performance of ASD children suggests the presence of an executive control deficiency. The impairment of adaptive error-correction behavior may have important consequences in daily life as optimal error-correction is necessary for adequate behavioral responses. As demonstrated in previous studies (Ridderinkhof et al. 2004), the posterior medial frontal cortex, more specifically the rostral ACC division, is the main brain area responsible for error processing, suggesting that ASD patients have reduced posterior medial frontal cortex functioning. This area is involved when there is a need for adjustments to achieve goals (Ridderinkhof et al. 2004). The current finding that children with ASD have an impaired ability to improve their response accuracy by slowing down the response speed on post-error trials corresponds with this notion. However, it is necessary to take into account that observed significant group differences between ASD and typical controls are manifested not only in the behavioral performance measures on reaction time tasks (RT, error rate) and associated response-monitoring indices (both to erroneous and correct), but also in terms of amplitude and latency characteristics of ERP components preceding motor response (frontal and parietal P100, N100, P200, N200) and those reflecting context update and closure (e.g., P300, N450) in visual oddball task (Sokhadze et al. 2009, 2012b, 2014a, b, 2018). The sum of the group differences across these behavioral and stimulus- and response-averaged ERP indices of the ASD patients' performance is that it reflects global deficits in attentional processes, more specifically deficits in effective differentiation of target and novel distracter stimuli. This latter interpretation is supported by the significant differences between the ASD patients and typically developing controls in terms of the stimulus-locked ERP amplitudes and latencies and the correlation between subjects' behavioral performance measures and specific ERP components magnitude.

Post-error adaptive correction of responses might be explained by some recent neurobiological findings. There are reports about an excessive preservation of short-distance connections (i.e., local over-connectivity) and relatively poor long-distance connections (i.e., distant under-connectivity) in the neocortex of individuals with autism (Casanova 2005, 2006; Just et al. 2004; Williams and Casanova 2010). These cortical connectivity abnormalities may explain why persons with autism tend to focus on details rather than perceiving the whole Gestalt. This overfocusing on details may imply an excessively laborious and ineffective way of handling each trial in the cognitive test and lower availability of resources after an error when the effort is needed to react appropriately. This may result in an insufficient activation of the ACC (Bogte et al. 2007), and thus, error detection and post-error reaction may be hampered (Minshew et al. 2005). Structural and functional deficiencies of the ACC may contribute to the atypical development of joint attention and social cognition in autism (Mundy 2003). Our interpretation of the results of this study is consistent with many aspects of theory and research which suggest that ACC-mediated response monitoring may contribute to the social-emotional and social-cognitive development in autism (Mundy 2003). However, while emphasizing the possible role of ACC-related self-monitoring deficits in autism, Mundy (2003) also noted that according to Devinsky and Luciano (1993) these ACC impairment-related behavioral deficits emerge only when they are combined with disturbances in other related functional neural networks, e.g., dorsolateral prefrontal cortex.

There are several limitations in series of our exploratory studies that should be mentioned. We could not rule out medication effects in studying the neurobiology of autism, since there is an obvious need to consider the effects of medication on both behavioral symptoms and neural functioning in EEG and behavioral studies of children with autism. Approximately half of the children in our ASD samples were medicated at the time of study and there was a variability in the type of medication children were taking. Therefore, it was not possible to analyze the associations between specific classes of medications and the ERN/Pe and behavioral measures in our studies. Another limitation or methodological issue to consider in this study was the large proportion of participants in both the ASD and control children samples who had relatively low numbers of commission error trials. The number of errors was introducing some additional variability in the amplitude and latency characteristics of individual subjects' data. This is a critical issue in all error monitoring researches which has to be considered given the large variance of number of error trials on which the ERN and Pe analyses are based. Finally, the dipole source localization in our studies was performed only on grand average ERN/Pe waveforms and this prevented us from being able to make any statistical analysis to make any definitive statements about the individual differences in the ERN/Pe dipole localizations. The BESA-based dipole source localization was aimed mostly for demonstration purposes taking into account the extensive literature on the likely ACC source localization of ERN and Pe dipoles (Holroyd and Coles 2002).

In summary, the present findings reveal that autism is associated with reduced error processing and impaired behavioral correction after an error is made. Because adequate error processing is necessary for optimal behavioral performance, it is plausible that these deficits contribute to the maintenance of the preservative behaviors typical for autism. Our studies report abnormal response monitoring and correction functions observed in behavioral and electrocortical indices of the ACC in children with ASD which might be related to restricted, repetitive behavior style typical for this neurodevelopmental disorder. This abnormal function may result from a compromised functional and structural connectivity in the neural circuitry subserving response monitoring and error correction. These findings suggest that functional abnormalities of the ACC reflected in lower amplitude and delayed ERN measure may compromise response monitoring and contribute to behavioral repetition in ASD. Impairments in an ability to correctly and timely evaluate committed error and to learn from errors may lead to behavior that is rigid and repetitive rather than adaptively guided by action outcomes. Elucidating the neurobiological basis and clinical significance of response monitoring and correction deficits in ASD represents a promising direction for further quantitative EEG-based research. The ERN and Pe along with behavioral performance measures can be used as the functional outcome measures to assess effectiveness of behavioral intervention (e.g., social skills training) or neurotherapy (e.g., TMS) in children with autism spectrum disorders, and thus, may have important practical implications.

Overall, our preliminary results (Baruth et al. 2011; Sokhadze et al. 2009, 2012b, 2014a, b, 2018) show promising results for TMS as a treatment modality targeting core symptom of ASD. Current finding of post-TMS improvement in a certain executive function such as error monitoring adds new angle to the understanding of neuropathological mechanisms of ASD symptoms. We consider that it is possible to conclude that neuromodulation using low frequency, inhibitory rTMS improved executive functioning and behavior in autism. This study provides further support to the statement that TMS can be regarded as a prospective treatment targeting core symptoms of ASD such as executive function deficits.

Acknowledgments This work was supported by the National Institutes of Health grant R01 MH086784 to Manuel F. Casanova.

References

Alonso P, Pujol J, Cardoner N, Benlloch L, Deus J, Menchón JM et al (2001) Right prefrontal repetitive transcranial magnetic stimulation in obsessive-compulsive disorder: a double-blind, placebo-controlled study. Am J Psychiatry 158(7):1143–1145

American Psychiatric Association (APA) (2013) Diagnostic and statistical manual of mental disorders (DSM-V), 5th edn. American Psychiatric Association, Washington, DC

Arbel Y, Donchin E (2009) Parsing the componential structure of post-error ERPs: a principal component analysis of ERPs following errors. Psychophysiology 46(6):1179–1189

Arbel Y, Donchin E (2011) When a child errs: the ERN and the Pe complex. Psychophysiology 48(1):55–63

Barnea-Goraly N, Kwon H, Menon V, Eliez S, Lotspeich L, Reiss A (2004) White matter structure in autism: preliminary evidence form diffusion tensor imaging. Biol Psychiatry 55(3):323–326

Baruth J, Williams E, Sokhadze E, El-Baz A, Sears L, Casanova MF (2011) Repetitive transcranial stimulation (rTMS) improves electroencephalographic and behavioral outcome measures in autism spectrum disorders (ASD). Autism Sci Digest 1:52–57

Bates AT, Liddle PF, Kiehl KA, Ngan ET (2004) State dependent changes in error monitoring in schizophrenia. J Psychiatr Res 38(3):347–356

Bishop DV (1993) Annotation: autism, executive functions and theory of mind: a neuropsychological perspective. J Child Psychol Psychiatry Allied Discip 34(3):279–293

Bogte H, Flamma B, van der Meere J, van Engeland H (2007) Post-error adaptation in adults with high functioning autism. Neuropsychologia 45(8):1707–1714

Burle C, Possamaï CA, Vidal F, Bonnet M, Hasbroucq T (2002) Executive control in the Simon effect: an electromyographic and distributional analysis. Psychol Res 66(4):324–336

Bush G, Luu P, Posner M (2000) Cognitive and emotional influences in the anterior cingulate cortex. Trends Cognit Sci 4(6):215–222

Carter CS, Braver TS, Barch DM, Botvinick MM, Noll D, Cohen JD (1998) Anterior cingulate cortex, error detection, and the online monitoring of performance. Science 280(5364):747–749

Casanova MF (2005) Minicolumnar pathology in autism. In: Casanova MF (ed) Recent developments in autism research. Nova Biomedical Books, New York, pp 133–144

Casanova MF (2006) Neuropathological and genetic findings in autism: the significance of a putative minicolumnopathy. Neuroscientist 12(5):435–441

Casanova MF (2007) Schizophrenia as a deficit in the modulation of cortical minicolumns by monoaminergic systems. Int Rev Psychiatry 19(4):361–372

Casanova MF, Buxhoeveden DP, Brown C (2002) Clinical and macroscopic correlates of minicolumnar pathology in autism. J Child Neurol 17(9):692–695

Casanova MF, Buxhoeveden D, Gomez J (2003) Disruption in the inhibitory architecture of the cell minicolumn: implications for autism. Neuroscientist 9:496–507

Casanova MF, van Kooten I, Switala AE, van England H, Heinsen H, Steinbuch HWM et al (2006) Abnormalities of cortical minicolumnar organization in the prefrontal lobes of autistic patients. Clin Neurosci Res 6(3–4):127–133

Casanova MF, Baruth JM, El-Baz AS, Tasman A, Sears L, Sokhadze EM (2012) Repetitive TMS (rTMS) modulates ERP indices of attention in autism. Transl Neurosci 3(2):170–180

Casanova MF, Sokhadze E, Opris I, Wang Y, Li X (2015) Autism spectrum disorders: linking neuropathological findings to treatment with transcranial magnetic stimulation. Acta Pediatr 104(4):346–355

Chang WP (2017) Electrophysiology of error processing in individuals with autism spectrum disorder. In: Casanova MF, El-Baz A, Suri JS (eds) Autism imaging and devices. CRC Press, Boca Raton, pp 437–456

Chmielewski WX, Beste C (2015) Action control processes in autism spectrum disorder—insights from a neurobiological and neuroanatomical perspective. Progr Neurobiol 124:49–83

Coles MG, Scheffers MK, Holroyd CB (2001) Why is there an ERN/Ne on correct trials. Response representations, stimulus-related components, and the theory of error-processing. Biol Psychol 56(3):173–189

Croarkin PE, Wall CA, Lee J (2011) Applications of transcranial magnetic stimulation (TMS) in child and adolescent psychiatry. Int Rev Psychiatry 23(5):445–453

Davies PL, Segalowitz SJ, Gavin WJ (2004) Development of response-monitoring ERPs in 7- to 25-year-olds. Dev Neuropsychol 25(3):355–376

Dawson G, Meltzoff AN, Osterling J, Rinaldi J, Brown E (1998) Children with autism fail to orient to naturally occurring social stimuli. J Autism Dev Disord *28*(6):479–485

de Bruijn ER, Grootens KP, Verkes RJ, Buchholz V, Hummelen JW, Hulstijn W (2006) Neural correlates of impulsive responding in borderline personality disorder: ERP evidence for reduced action monitoring. J Psychiatr Res 40(5):428–437

Devinsky O, Luciano D (1993) The contributions of cingulate cortex to human behavior. In: Gabriel M, Vogt BA (eds) Neurobiology of cingulate cortex and limbic thalamus: a comprehensive handbook. Birkhauser, Cambridge, MA, pp 527–556

Enticott PG, Fitzgibbon BM, Kennedy HA, Arnold SL, Elliot D, Peachey A et al (2014) A double-blind, randomized trial of deep repetitive transcranial magnetic stimulation (rTMS) for autism spectrum disorder. Brain Stimul 7(2):206–211

Eriksen BA, Eriksen CW (1974) Effects of noise letters on the identification of a target letter in a nonsearch task. Percept Psychophys 16:143–149

Falkenstein M, Hoormann J, Christ S, Hohnsbein J (2000) ERP components on reaction errors and their functional significance: a tutorial. Biol Psychol 51(2–3):87–107

Franken IH, van Strien JW, Franzek EJ, van de Wetering BJ (2007) Error-processing deficits in patients with cocaine dependence. Biol Psychol 75(1):45–51

Garvey MA, Mall V (2008) Transcranial magnetic stimulation in children. Clin Neurophysiol 119(5):973–984

Gehring WJ, Knight RT (2000) Prefrontal-cingulate interactions in action monitoring. Nat Neurosci 3(5):516–520

Gehring WJ, Goss B, Coles MGH, Meyer DE, Donchin E (1993) A neural system for error detection and compensation. Psychol Sci 4(6):385–390

Gehring WJ, Himle J, Nilsenson L (2000) Action monitoring dysfunction in obsessive-compulsive disorder. Psychol Sci 11(1):1–6

George MS, Lisanby SH, Sackeim HA (1999) Transcranial magnetic stimulation: applications in neuropsychiatry. Arch Gen Psychiatry 56(4):300–331

George MS, Nahas J, Kozol FA, Li X, Yamanaka K, Mishory A, Bohning DE (2003) Mechanisms and the current state of transcranial magnetic stimulation. CNS Spectr 8(7):496–514

Gershon AA, Dannon PN, Grunhaus L (2003) Transcranial magnetic stimulation in the treatment of depression. Am J Psychiatry 160(5):835–845

Gillberg C, Billstedt E (2000) Autism and Asperger syndrome: coexistence with other clinical disorders. Acta Psychiatr Scand 102(5):321–330

Greenberg BD (2007) Transcranial magnetic stimulation in anxiety disorders. In: George MS, Belmaker RH (eds) Transcranial magnetic stimulation in clinical psychiatry. American Psychiatric Publishing, Inc., Washington, DC, pp 165–178

Groen Y, Wijers AA, Mulder LJM, Waggeveld B, Minderaa RB, Althaus M (2008) Error and feedback processing in children with ADHD and children with autistic spectrum disorder: an EEG event-related potential study. Clin Neurophysiol 119(11):2476–2493

Hall G, Szectchman H, Hahmias C (2003) Enhanced salience and emotion recognition in autism: a PET study. Am J Psychiatry 160(8):1439–1441

Haznedar M, Buchsbaum M, Wei T, Hof P, Cartwright C, Bienstock C, Hollander E (2000) Limbic circuitry in patients with autism spectrum disorders studied with positron emission tomography and magnetic resonance imaging. Am J Psychiatry 157(12):1994–2001

Henderson H, Schwartz C, Mundy P, Burnette C, Sutton S, Zahka N, Pradella A (2006) Response monitoring, the error-related negativity, and differences in social behavior in autism. Brain Cognit 61(1):96–109
Herrmann MJ, Römmler J, Ehlis AC, Heindrich A, Fallgatter AJ (2004) Source localization (LORETA) of the error-related-negativity (ERN/Ne) and positivity (Pe). Cognit Brain Res 20(2):294–299
Hill EL (2004) Evaluating the theory of executive dysfunction in autism. Dev Rev 24:189–233
Hoffman RE, Cavus I (2002) Slow transcranial magnetic stimulation, long-term depotentiation, and brain hyperexcitability disorders. Am J Psychiatry 159(7):1093–1102
Holroyd CB, Coles MGH (2002) The neural basis of human error processing: reinforcement learning, dopamine, and the error related negativity. Psychol Rev 109(4):679–709
Huizinga M, Dolan CV, van der Molen MW (2006) Age-related change in executive function: developmental trends and a latent variable analysis. Neuropsychologia 44(11):2017–2036
Hüpen P, Groen Y, Gaastra G, Tucha L, Tucha O (2016) Performance monitoring in autism spectrum disorders: a systematic literature review of event-related potential studies. Int J Psychophysiol 102:33–46
Johannes S, Wieringa BM, Nager W, Rada D, Dengler R, Emrich HM et al (2001) Discrepant target detection and action monitoring in obsessive–compulsive disorder. Psychiatry Res 108(2):101–110
Just MA, Cherkassky VL, Keller TA, Minshew NJ (2004) Cortical activation and synchronization during sentence comprehension in high-functioning autism: evidence of underconnectivity. Brain 127(8):1811–1821
Kanizsa G (1976) Subjective contours. Sci Am 235:48–52
Kim SH, Grammer J, Benrey N, Morrison F, Lord C (2018) Stimulus processing and error monitoring in more-able kindergarteners with autism spectrum disorder: a short review and a preliminary event-related potentials study. Eur J Neurosci 47(6):556–567
Liotti M, Pliszka S, Perez R, Kothmann D, Woldorff M (2005) Abnormal brain activity related to performance monitoring and error detection in children with ADHD. Cortex 42(3):1–12
Loo C, Mitchell PB (2005) A review of the efficacy of transcranial magnetic stimulation (TMS) treatment for depression, and current and future strategies to optimize efficacy. J Aff Disord 88(3):255–267
Luu P, Flaisch T, Tucker DM (2000) Medial frontal cortex in action monitoring. J Neurosci 20(1):464–469
Luu P, Tucker DM, Derryberry D, Reed M, Poulsen C (2003) Electrophysiological responses to errors and feedback in the process of action regulation. Psychol Sci 14(1):47–53
Markela-Lerenc J, Ille N, Kaiser S, Fiedler P, Mundt C, Weisbrod M (2004) Prefrontal-cingulate activation during executive control: which comes first? Cognit Brain Res 18(3):278–287
Mars RB, Coles MG, Grol MJ, Holroyd CB, Nieuwenhuis S, Hulstijn W, Toni I (2005) Neural dynamics of error processing in medial frontal cortex. Neuroimage 28(4):1007–1013
Mathalon DH, Fedor M, Faustman WO, Gray M, Askari N, Ford JM (2002) Response-monitoring dysfunction in schizophrenia: an event-related brain potential study. J Abnormal Psychol 111(1):22–41
McCann UD, Kimbrell TA, Morgan CM, Geraci M, Benson BE, Wassermann EM et al (1998) Repetitive transcranial magnetic stimulation for posttraumatic stress disorder. Arch Gen Psychiatry 55(3):276–279
Minshew NJ, Sweeney JA, Bauman ML, Webb SJ (2005) Neurological aspects of autism. In: Volkmar FR, Paul R, Klin A, Cohen D (eds) Handbook of autism and pervasive developmental disorders, 3rd edn. Wiley, New York, pp 473–514
Mir-Moghtadaei A, Caballero R, Fried P, Fox MD, Lee K, Giacobbe P et al (2015) Concordance between BeamF3 and MRI-neuronavigated target sites for repetitive transcranial magnetic stimulation of the left dorsolateral prefrontal cortex. Brain Stimul 8(5):965–973
Mountcastle VB (2003) Introduction. computation in cortical columns. Cereb Cortex 13(1):2–4

Mundy P (1995) Joint attention and social-emotional approach behavior in children with autism. Dev Psychopathol 7(1):63–82

Mundy P (2003) The neural basis of social impairments in autism: the role of the dorsal medial-frontal cortex and anterior cingulate system. J Child Psychol Psychiatry Allied Discip 44(6):793–809

Mundy P, Neal R (2001) Neural plasticity, joint attention and a transactional social-orienting model of autism. Int Rev Res Ment Retard 23:139–168

Nieuwenhuis S, Ridderinkhof KR, Blom J, Band GP, Kok A (2001) Error-related brain potentials are differentially related to awareness of response errors: evidence from an antisaccade task. Psychophysiology 38(5):752–760

Overbeek TJM, Nieuwenhuis S, Ridderinkhof KR (2005) Dissociable components of error processing: on the functional significance of the Pe vis-à-vis the ERN/Ne. J Psychophysiol 19:319–329

Ozonoff S (1997) Casual mechanisms of autism: unifying perspectives from an information-processing framework. In: Cohen DJ, Volkmar FR (eds) Handbook of autism and pervasive developmental disorders. Wiley, New York, pp 868–879

Ozonoff S, Pennington B, Rogers SJ (1991) Executive function deficits in high-functioning autistic individuals: relationship to theory of mind. J Child Psychol Psychiatry Allied Discip 32(7):1081–1105

Pascual-Leone A, Davey NJ, Rothwell J, Wasserman EM, Puri BK (2002) Handbook of transcranial magnetic stimulation. Oxford University Press, New York

Pommier B, Vassal F, Boutet C, Jeannin S, Peyron R, Faillenot I (2017) Easy methods to make the neuronavigated targeting of DLPFC accurate and routinely accessible for rTMS. Neurophysiol Clin 7(1):35–46

Quintana H (2005) Transcranial magnetic stimulation in persons younger than the age of 18. J ECT 21(2):88–95

Ridderinkhof KR, Ullsperger M, Crone EA, Nieuwenhuis S (2004) The role of the medial frontal cortex in cognitive control. Science 306(5695):443–447

Rippon G, Brock J, Brown C, Boucher J (2007) Disordered connectivity in the autistic brain: challenges for the 'new psychophysiology'. Int J Psychophysiol 63(2):164–172

Rosenberg PB, Mehndiratta RB, Mehndiratta YP, Wamer A, Rosse RB, Balish M (2002) Repetitive magnetic stimulation treatment of comorbid posttraumatic stress disorder and major depression. J Neuropsychiatry Clin Neurosci 14(3):270–276

Rubenstein JL, Merzenich MM (2003) Model of autism: increased ratio of excitation/inhibition in key neural systems. Genes Brain Behav 2(5):255–267

Ruchsow M, Spitzer M, Groen G, Grothe J, Kiefer M (2005) Error processing and impulsiveness in normals: evidence from event-related potentials. Cognit Brain Res 24(2):317–325

Russell J (1997) How executive disorders can bring about an inadequate theory of mind. In: Russell J (ed) Autism as an executive disorder. Oxford University Press, Oxford, UK, pp 256–304

Russell J, Jarrold C (1998) Error-correction problems in autism: evidence for a monitoring impairment? J Autism Dev Disord 28(3):177–188

Santesso DL, Drmic IE, Jetha MK, Bryson SE, Goldberg JO, Hall GB et al (2011) An event-related source localization study of response monitoring and social impairments in autism spectrum disorder. Psychophysiology 48(2):241–251

Sokhadze E, El-Baz A, Baruth J, Mathai G, Sears L, Casanova MF (2009) Effects of a low-frequency repetitive transcranial magnetic stimulation (rTMS) on induced gamma frequency oscillations and event-related potentials during processing of illusory figures in autism spectrum disorders. J Autism Dev Disord 39(4):619–634

Sokhadze E, Baruth J, El-Baz A, Horrell T, Sokhadze G, Carroll T et al (2010) Impaired error monitoring and correction function in autism. J Neurother 14(2):79–95

Sokhadze EM, Baruth JM, Sears L, Sokhadze GE, El-Baz AS, Williams E et al (2012a) Event-related potential study of attention regulation during illusory figure categorization task in ADHD, autism spectrum disorder, and typical children. J Neurother 16(1):12–31

Sokhadze EM, Baruth JM, Sears L, Sokhadze GE, El-Baz AS, Casanova MF (2012b) Prefrontal neuromodulation using rTMS improves error monitoring and correction functions in autism. Appl Psychophysiol Biofeedback 37(2):91–102
Sokhadze EM, El-Baz AS, Tasman A, Sears LL, Wang Y, Lamina EV, Casanova MF (2014a) Neuromodulation integrating rTMS and neurofeedback for the treatment of autism spectrum disorder: an exploratory study. Appl Psychophysiol Biofeedback 39(3–4):237–257
Sokhadze EM, El-Baz AS, Sears LL, Opris I, Casanova MF (2014b) rTMS neuromodulation improves electrocortical functional measures of information processing and behavioral responses in autism. Front Syst Neurosci 8:134
Sokhadze EM, Lamina EV, Casanova EL, Kelly DP, Opris I, Khachidze I, Casanova MF (2017) Atypical processing of novel distracters in a visual oddball task in autism spectrum disorder. Behav Sci 7(4):e79
Sokhadze EM, Lamina EV, Casanova EL, Kelly DP, Opris I, Tasman A, Casanova MF (2018) Exploratory study of rTMS neuromodulation effects on electrocortical functional measures of performance in an oddball test and behavioral symptoms in autism. Front Syst Neurosci 12:20
Sokhadze EM, Sears L, Tasman A, Casanova EL, Casanova MF (2019) Comparative event-related potential study of performance in visual oddball task in children with autism spectrum disorder, ADHD, comorbid autism and ADHD, and neurotypical children. NeuroRegulation 6(3):134–152
South M, Larson MJ, Krauskopf E, Clawson A (2010) Error processing in high-functioning autism spectrum disorders. Biol Psychol 85(2):242–251
Taylor SF, Stern ER, Gehring WJ (2007) Neural systems for error monitoring: recent findings and theoretical perspectives. Neuroscientist 13(2):160–172
Thakkar KN, Polli FE, Joseph RM, Tuch DS, Hadjikhani N, Barton JJ, Manoach DS (2008) Response monitoring, repetitive behaviour and anterior cingulate abnormalities in autism spectrum disorders (ASD). Brain 131(9):2464–2478
van Veen V, Carter CS (2002) The timing of action-monitoring process in the anterior cingulate cortex. J Cognit Neurosci 14(4):593–602
Vlamings PH, Jonkman LM, Hoeksma MR, van Engeland H, Kemner C (2008) Reduced error monitoring in children with autism spectrum disorder: an ERP study. Eur J Neurosci 28:399–406
Wagner T, Rushmore J, Eden U, Valero-Cabre A (2009) Biophysical foundations underlying TMS: setting the stage for an effective use of neurostimulation in the cognitive neurosciences. Cortex 45(9):1025–1034
Wassermann EM, Lisanby SH (2001) Therapeutic application of repetitive transcranial magnetic stimulation: a review. Clin Neurophysiol 112(8):1367–1377
West R (2003) Neural correlates of cognitive control and conflict detection in the Stroop and digit-location tasks. Neuropsychologia 41(8):1122–1135
Williams EL, Casanova MF (2010) Autism and dyslexia: a spectrum of cognitive styles as defined by minicolumnar morphometry. Med Hypotheses 74(1):59–62
Yeung N, Cohen JD (2006) The impact of cognitive deficits on conflict monitoring. Predictable dissociations between the error-related negativity and N2. Psychol Sci 17(2):164–171
Yeung N, Cohen JD, Botvinick MM (2004) The neural basis of error detection: conflict monitoring and the error-related negativity. Psychol Rev 111(4):931–959
Zhang JS, Wang Y, Cai RG, Yan CH (2009) The brain regulation mechanism of error monitoring in impulsive children with ADHD—an analysis of error related potentials. Neurosci Lett 460(1):11–15

Affective Virtual Reality Gaming for Autism

Yi (Joy) Li, Estate M. Sokhadze, Hao (Irene) Luo, Ayman S. El-Baz, and Adel S. Elmaghraby

1 Introduction

In recent years, portable virtual reality is an emerging technology with hardware improvement that provides solutions to family entertainment and more serious, scientific applications. Even if the research related to virtual reality can be traced back to decades, the applications mostly were limited in training usage covered by laboratory, military or medical industries, due to the fact that equipment required to host virtual reality scenes were cumbersome and scenarios were insufficient. Now with much lighter headsets and a more seamless immersive environment, researches could turn to more fun directions and be concerned with fewer environment requirements such as space limit or power supply.

Y. (Joy) Li
Kennesaw State University, Kennesaw, GA, USA

E. M. Sokhadze
University of South Carolina School of Medicine, Columbia, SC, USA

H. (Irene) Luo
Renmin University of China, Beijing, China

A. S. El-Baz
Department of Bioengineering, Speed School of Engineering, University of Louisville, Louisville, KY, USA
e-mail: ayman.elbaz@louisville.edu

A. S. Elmaghraby (✉)
University of Louisville, Louisville, KY, USA
e-mail: adel@louisville.edu

I. Opris et al. (eds.), *Modern Approaches to Augmentation of Brain Function*, Contemporary Clinical Neuroscience,
https://doi.org/10.1007/978-3-030-54564-2_26

Besides pure entertainment, a wide variety of applications could be developed with virtual reality; examples include but not limited to:

- Scientific visualization
- Healthcare
- Training
- Education
- Media
- Telecommunication

Plenty of tools or frameworks related to designing virtual reality environments are available, yet still needed to be explored or tested with the up-to-date technologies. It is very promising to make use of these methods and let virtual reality better serve human beings.

1.1 Motivation

One in fifty-nine U.S. children has an autism spectrum disorder (ASD), which has increased by 150% from 2000 to 2014. Figure 1 shows an estimated prevalence of autism since 2000 by the CDC (CDCMMWR 2018). The public awareness of autism is also growing fast in the past decade. The public starts to pay attention to autistic children. It is devastating for their families since there is still no cure for autism. On the bright side, as a result of growing awareness, an increasing number of research projects were taken to assist the diagnosis of autism, as well as the therapeutic methods after diagnosis.

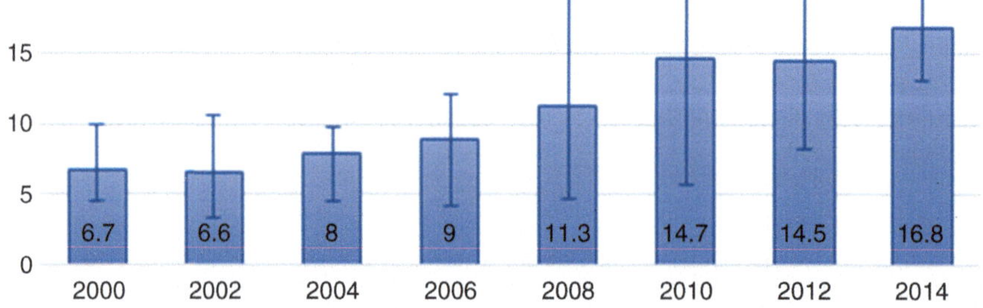

Fig. 1 Prevalence of ASD per 1000 children

Children on the ASD, even those who are diagnosed as high-functioning autism, suffer from impairment in social skills, emotion recognition, and expression and are at high risk of anxiety. Research indicates that many autistics are naturally visual learners. That is to say, they learn through the vision-based methods best among many other methods (Rao and Gagie 2016). Therefore, it inspires us to explore the possibility of using VR to create games that are dedicated to therapeutic or intervention purposes for autism. It is proven to be effective in a wide range of treatment sessions, such as exposure therapy for patients with addiction or social anxiety disorders.

VR provides a vital feature that suits perfectly for children on ASD—safely controlled environment. There are also training protocols utilizing VR equipment to train children with high-functioning autism social skills. Games designed for children on ASD often aim at their communication skills or activities of daily living. However, their lack of ability in emotional competence, including emotional recognition and expression, is hardly targeted.

It is very likely that the standard emotional response to the same VR scenarios between children with ASD and typical developed persons would be different. Thus, it is a reasonable approach to measure the differences in their emotional reactions, including physiological responses during VR environment immersion, in order to design VR-based affect training or therapy sessions optimum for children with ASD.

1.2 Objectives and Outline

In this chapter, an integrated multimodal model is proposed to enable autistic children to play games in a nonintrusive manner, while their progress and their reaction to different game scenarios are monitored and assessed. The goal is to design a gaming system that integrates cutting-edge gaming tools, psychophysiological analyzer, behavioral analyzer, and feedback mechanisms to learn and adapt to subjects' emotions dynamically. The gaming tools here refer to not only VR, but also other motion tracking devices such as Kinect. Some modules can be independent and optional; however, with more modules in place, the system should give more adequate data about user interactions.

Our research has been carried out in three phases out of four. Each phase described in the following paragraphs has achieved its objective, thus leading to a multimodal affective gaming application implemented in a healthcare theme. The state of the art of related works is covered in Sect. 2.

(a) Phase 1: Preliminary Affective Studies
During the first phase, we reviewed previous studies using traditional methodologies, including an audio-visual experiment, a music experiment, and a video experiment. Based on the accumulation of knowledge and research, we extended the existing methods and conducted a comparison between autistic children group and typically developing children group. We did not dig into the details

of this phase in this chapter. The results of this phase have been published elsewhere and provided evidence for choosing VR content to be used in later phases.

(b) Phase 2: VR Experiments
During the second phase, we chose several emotion-laden VR scenes to be shown in the experiment and recruited volunteers to interact with the VR context and recorded their physiological reaction data. These data were then analyzed offline and served as a benchmark for building the framework of the affective game. This phase is covered in Sect. 3.1.

(c) Phase 3: Game Framework Design
In the third phase, the interactive model for affective environment gaming design was established. The framework is generic and can be applied to many fields. This phase is covered in Sect. 3.2. A concept relevant game prototype for demonstration purposes was implemented.

(d) Phase 4: Game Implementation and Efficacy Test
The last phase is to design and develop a full game with adaptive emotional perceptions and adjustable difficulties. It will utilize the framework we proposed and will be dedicated to intervention for children with ASD. This is covered in the future work section.

Finally, in Sect. 4, we address the limitation of the current system and possible directions to improve it further. We also suggest potential applications to implement the proposed framework.

2 Literature Review

Psychology experiments have shown that many human reactions such as physiological signals, body movements, and facial expressions are related to emotions; thus, data can be collected and analyzed for emotion studies. Since the sympathetic and the parasympathetic nervous system are not easily controlled by awareness, they can become direct and robust measures to affect recognition.

This section provides a literature view of related works on affective computing, serious games, and most recent studies on using VR in autism intervention.

2.1 Affective Computing

Affective computing was introduced by Picard (2000) as a new interdisciplinary study area in the twenty-first century. There are plenty of affect recognition publications in the literature (Calvo and D'Mello 2010; Castellano et al. n.d.) The most commonly used methods are speech affect recognition(Nwe et al. 2003; Balti and Elmaghraby 2013), facial affect detection (Nwe et al. 2003; Milanova and Sirakov 2008; Li and Elmaghraby 2014), body gesture recognition (Kapur et al. 2005), and physiological monitoring (Kim et al. 2004; Li et al. 2015a). To utilize those methods and manage to form a more advanced system, researchers may integrate several of them together for a better computing performance.

Physiological responses are believed to be more reliable and more truthful because they can hardly be controlled by awareness. For example, one may hide his emotions by changing his facial expression, body gestures, as well as his vocal tones intentionally. Although recording physiological variables can be environment-dependent and very sensitive to surrounding conditions, studies found them reflecting cues of true emotions from subjects (Lang et al. 1998).

Adaptive environments in serious games are discussed in the current decade, as it is very helpful in training or therapy when the gaming environments are able to be adjusted in real-time according to users' responses.

Moghim et al. (2015) and Moghimi et al. (2020) have done similar work in an on-going research project that has planned to implement a physiological feed HCI system, which is closely related to the second phase of our research. They first conducted VR exposure experiments to test users' emotional reactions to various VR contents and analyzed the recorded physiological data with self-report questionnaires answered by users. A database was constructed to store the stimuli VR events and corresponding physiological reactions, as well as self-reported scaled emotional states. To analyze features being extracted and selected from the data, during the second phase, they compared and evaluated four different machine learning algorithms that are applicable in emotion classification. Our machine learning approach referred to their work. Based on the results of the machine learning algorithms, we stepped forward to bring up the adaptive interaction model and customizable user models for designing dynamic affective VR game systems. As their works are still on-going and continuously constructing the database, we will keep updated with their progress.

2.2 *Serious Games with VR*

Serious games can be defined as a gaming technology that is used for the purpose of education, training, and information, other than mere entertainment (Michael and Chen 2005).

For games with educational purposes, books in the cognitive and education field (Schutz and Pekrun 2007) point out that emotions are critical parts in the school context, including both teacher and student emotions. A review on affective educational games (Wilkinson 2013) summarized three key elements of cognition that these games would have impacts on: attention, memory storage and retrieval, and decision making.

Serious games are also used in safety training. Nasoz et al. (2010) reported their VR application with the focus on driving safety. The VR device they were using was not a small headset as what we are using today, but with a car-shaped machine that users can sit in to simulate driving. They managed to record similar physiological data such as skin conductance, heart rate, and temperature to recognize drivers' emotional states. By designing the VR context to elicit users' panic/fear, frustration/anger, and boredom/fatigue emotions, the application aimed to train drivers with possible incidents on the road to improve driving safety. This paper has introduced a driver's modeling using Bayesian Belief Network (BBN), which is referred to in the present research, where the customizable user modeling is derived.

There are a few VR games that are designed for information purposes. Lovreglio et al. (2018) proposed a VR game design for providing earthquake information based on a hospital case study. A VR intervention aimed to promote understanding and empathy for people with dementia (Wijma et al. 2017) was developed and reported a positive result in its pilot study. The presence of the visualization can improve the retainment rate of information, which also justifies the use of VR in intervention for subjects with emotional impairment.

2.3 VR Applications for ASD

VR is proven to be effective in a wide range of treatment sessions, such as virtual reality exposure therapy (VRET) for patients with addiction or social anxiety disorders. A more recent systematic review (Jerdan et al. 2018) compiled 82 studies from the past 5 years that used VR headset devices for interventions for both mental health and public health. The conditions addressed include anxiety, phobias, PTSD, addiction, depression, pain, burns, and more. It suggested that VRET has been used in a variety of interventions, and some of them show positive evidence. It also stated that VR is still in its infancy in mental health areas, despite the initial highlighted outcomes, yet more studies addressing depression and ASD with efficacy assessments are needed.

Experiments on using biofeedback or neurofeedback as a treatment for ASD have been conducted along many years, as stated in the review of the literature by Coben et al. (2009). In recent years, researchers started using virtual environments as a tool to assess different aspects of helping ASD. Topics include addressing communication skills (Zhang et al. 2019; Zhao et al. 2018; Kuriakose and Lahiri 2017), social attention (Ravindran et al. 2019; Kumazaki et al. 2018; Amaral et al. 2018), cognitive training (Nijman et al. 2019; Li and Yuan 2018), and more. Some of these studies are still in their early stages, but have shown the promising future of applying VR to intervention for patients with ASD. Making use of the interactive nature of VR games instead of the exposure-only VRET method will level up the efficacy and also make interventions more fun.

However, combining the emerging affect computing methods into gaming technologies such as VR as well as affective gaming is still a relatively new topic, especially for emotional competence training purposes. Designing dedicated games scenes that accommodate autistic children's emotional states will provide them a better and safer environment to train necessary social skills.

3 Methods

3.1 Experiments

We are mainly using physiological measurements in the current research. Electroencephalography (EEG) is also a popular method in monitoring brain activities in recent years. However, the installation of EEG requires direct connections of electrodes on the scalp. It is not feasible for any of the current VR headset hardware. Moreover, the EEG records suffer from extreme sensitivity to any non-brain electrical activity of the body and environment. Eyeblink, muscle movement, heartbeat, as well as surrounding electromagnetic fields are all possible to produce artifacts that impede the correct interpretation of brain activities. Therefore, we chose not to include EEG as a feature set until further improvement of VR hardware makes it ready to be adopted.

3.1.1 Physiological Parameters

The Autonomic Nervous System (ANS) consists of the sympathetic and the parasympathetic nervous systems that control and regulate most human organs' functions in competing and opposite ways. The ANS is believed to be associated with basic emotions (Ekman et al. 1983). In the proposed system, the ANS variables are detected during the game scenarios by sensors attached to subjects and recorded by monitoring devices. The corresponding emotional states are analyzed and used as feedback to interact with the game scenarios. In the current experiments, emotional states are generally divided into neutral, negative, and positive groups, which later can be extended to more specific groups, such as joy, fear, sadness, disgust, and anger.

Here are the physiological parameters recorded and analyzed in this research:

Electrodermal Activity

Electrodermal activity, also known as skin conductance response (SCR), or galvanic skin response (GSR), is commonly used to measure human ANS activity. It records continuous electricity changes on the skin and shows the activity of only sympathetic system. SCR is believed to be not under conscious control (Critchley 2002) and is highly related to emotional arousal, as mentioned in the literature, which is one of the three dimensions of the emotion mapping model. Thus, when exposed to certain stimuli, SCR is an important indicator of user's emotional reactions.

In this research, we are measuring different parameters under skin conductance, including skin conductance level (SCL), SCR peak average, i.e., the average of local maximum of SCR, SCR number per minute, i.e., nonspecific SCR frequency (NS. SCR freq.), which is measured as number of SCRs in 1 min.

Cardiovascular Activity

In the medical field, Heart rate variability (HRV) is normally measured by EKG or ECG, which requires several electrodes to be attached to a human body. As we aim to find portable devices for the whole adaptive system, we mainly rely on monitoring and recording blood volume pulse (BVP) to calculate heart rate and all its dependent HRV measurements. The ProComp BVP sensor we are using in this research is one single sensor that could be taped to a thumb.

HRV, calculated as cardiac autonomic control measures, can be analyzed in both time and frequency domains. In the time domain, RR interval is the time between successive heartbeats, and Fig. 2 shows an example of RR interval in a typical ECG image. The statistics of the RR intervals are reliable measures: mean value of RR intervals, standard deviation of RR intervals (SDNN(ms)), and root mean square of successive differences (RMSSD(ms)) (Medeiros 2010).

In the frequency domain, we look at high frequency (HF) components (0.15–0.40 Hz), low frequency (LF) components (0.04–0.15 Hz), and LF/HF ratio. See Fig. 3 as an example of these three components in the frequency domain. The very low frequency (VLF) (under 0.04 Hz) has not shown much physiological evidence related to emotional reactions, though some may claim that it reflects deficit energy states (Cornforth et al. 2015). Since it is still awaiting more experiments and proof, we chose not to include this measure in our analysis. HF component is generally believed to show the activity of the parasympathetic system, while the LF component is argued to be a general indicator of aggregate modulation of both the sympathetic and parasympathetic branch of ANS (Burr 2007). LF/HF ratio was initially believed to be measuring sympathovagal balance and was widely used as an indicator of stress in some earlier studies. However, in recent decades, studies are

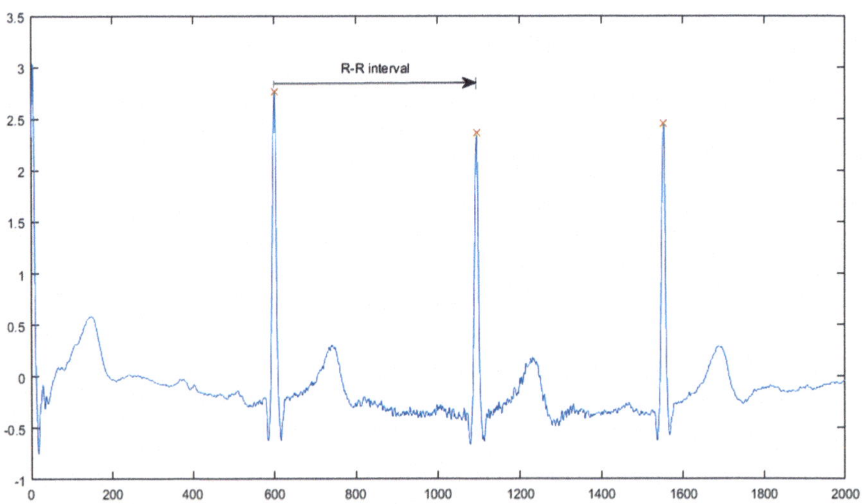

Fig. 2 RR interval example

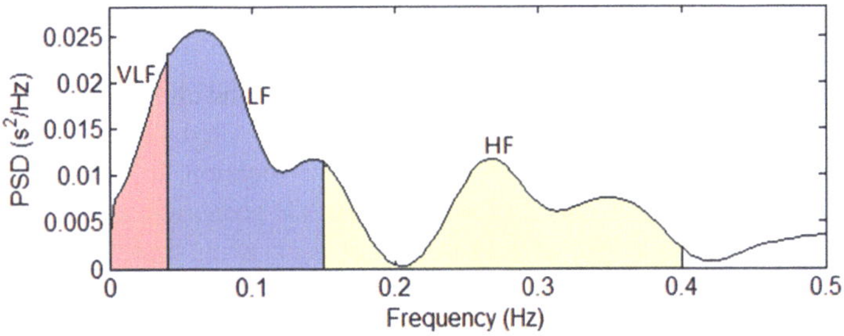

Fig. 3 HRV frequency domain example

arguing that as a univariate parameter, the LF/HF ratio lacks evidence to interpret physical and mental stress. Despite this controversy, we extracted LF/HF ratio as a feature, and let the feature selection process decide whether to keep it or not.

Fingertip Temperature (FTT)

Fingertip temperature is another popular parameter in physiological signals. Research projects (Shivakumar and Vijaya 2012, n.d.) have shown that if one person is comfortable, his/her blood vessels are dilated, and thus, the FTT should be relatively warmer. On the other hand, if this person is highly irritated or terrified, the vessels will be constricted so that the FTT will drop. By analyzing the variation of temperature, it should reflect a user's emotional reactions to stimulating events. In our experiments, the temperature sensor was taped to fingers and directly exposed to the environment taking place in a normal computer science laboratory, where a service center controlled the room temperature. The collected data are in Fahrenheit.

3.1.2 Experiments Using Traditional Media

In order to gain more data on emotional responses in physiological monitoring measures for biofeedback training, a series of experiments were conducted as preliminary studies before using virtual reality contents. These experiments are aimed to analyze emotions induced by traditional media, including: affective audio, music, visual expression pictures, and videos. Some of the experiments were conducted for other related projects and yielded quality publications. The experiments included a group of children with ASD and a control group. The results in this phase have provided evidence for choosing VR content to be used in later phases. Furthermore, they supported our assumption that VR contents are able to induce significantly stronger emotional reactivities in both groups.

3.1.3 Experiments Using Virtual Reality

Different emotionally laden VR scenarios are designed and chosen to induce different emotions. Two experiments were designed and conducted, one for training analysis, and the other one for extended testing and verification. In both experiments, users sit in front of a desk and explore different VR scenarios, which are chosen based on the experience of the above research projects using traditional media. The training experiment included three scenarios to induce three emotional states: neutral, negative, and positive. It served as a benchmark for more emotion labeling. The extended experiment took place later, using continuous non-labeled VR contents, which can be grouped into 16 scenarios with different events. It was conducted purposefully, cross-verifying the data in the training experiment. These events were then labeled to be in any one of the general emotions and then used in later game design. Physiological data were recorded by a monitoring device that has sensors attached to users' fingers during all experiment sessions, and features were extracted and analyzed for both experiments.

Instruments

The hardware to host VR is an Oculus Rift, and the software platform is Unity Engine by Unity Technologies (San Francisco, CA) with C# as developing language. In preliminary experiments that are used to determine the feasibility, SPSS was used in statistical analysis. Matlab 2015b 32-bit is used for data analysis, classifiers training, and emotion recognition.

Different physiological parameters are measured and recorded by a monitoring device that has sensors attached to subjects simultaneously with the VR scenes. The monitoring device used in experiments is ProComp Infiniti, and the software for data collection and storage is the BioGraph Infiniti produced by Thought Technology Ltd. (Montreal West, Quebec, Canada). Figure 4 shows different channels provided by ProComp hardware. We are using its B, E, F channels: blood volume pulse, skin conductance, and temperature, respectively.

VR Contents

(a) Training Experiment
Three VR scenarios were chosen and labeled as neutral, negative, and positive context:

- **ProComp Infiniti or ProComp+ with a BVP sensor:**

	Input B ***BVP***	Input C ***EMG***	Input D ***EMG***	Input E ***SC***	Input F ***Temp***	Input G ***Resp***	Input H ***Resp***

Fig. 4 ProComp device channels

- The neutral scene is simply exploring a neutral VR environment, wandering around on the terrain. This VR content is chosen to make users slightly more excited than baseline, but not enough to trigger tremendously distinct emotions.
- The negative scene makes use of affective audio and a series of events that cause fear, annoyance, disgust, and other uncomfortable experiences. Contents including spiders, flies, a dinosaur, darkness, and more were presented to users and expected to get dramatically negative emotions.
- The positive scene is a collection of happy scenes, including exciting toys and smiling people, with harmonious music. Positive emotions are expected to induce more arousal levels than those in the neutral scene.

  Figure 5 shows one example of the VR scenes.

(b) Testing/Verification Experiment

A clip of VR, around 13 min in length, with multiple continuous scenarios was presented in this experiment. The clip length varies each time slightly, depending on how fast the user triggers the next scene by exploring. The clip can be roughly categorized into 16 scenarios without emotional labels. It is expected to induce different emotions, which are still subject to the three general categories: neutral, negative, and positive.

Table 1 summarizes scenarios in short terms and shows an example of their time ranges recorded from one of the volunteers.

Fig. 5 VR scene examples

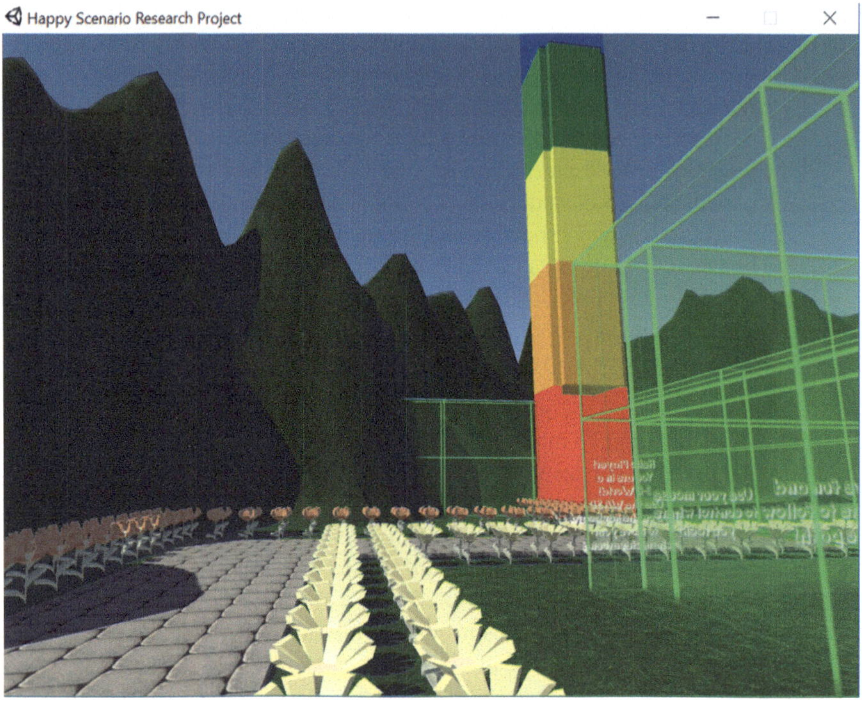

Fig. 5 (continued)

Table 1 Example of scenarios in the testing experiment time ranges

Scenes	Base	Instruction	Lights off	Particles
Time	0:01–0:59	1:56–2:36	2:53–3:12	3:13–3:58
Scenes	Grass dark	Butterfly/brightness	Birds/tower	Flowers/trees
Time	3:59–4:15	4:16–4:23	4:24–4:44	4:45–5:12
Scenes	Fog/crow/building	Thunderstorm	Car accident	Food
Time	5:18–5:33	5:38–6:01	6:02–6:16	6:45–6:58
Scenes	Space narrowing	Explosion/space	Net	Looking down from high
Time	7:28–8:18	8:22–10:25	10:27–11:15	11:17–11:56
Scenes	Drop from high			
Time	11:56–12:19			

Procedures

In the training experiment, one session requires exploration of all three scenarios and a 1-min baseline data recording. One scenario takes around 5 min for subjects to explore, which add up to 20 min in total, including configuration and mount time. Figure 6 shows a picture of an experiment session. Researchers follow these steps to conduct the experiment:

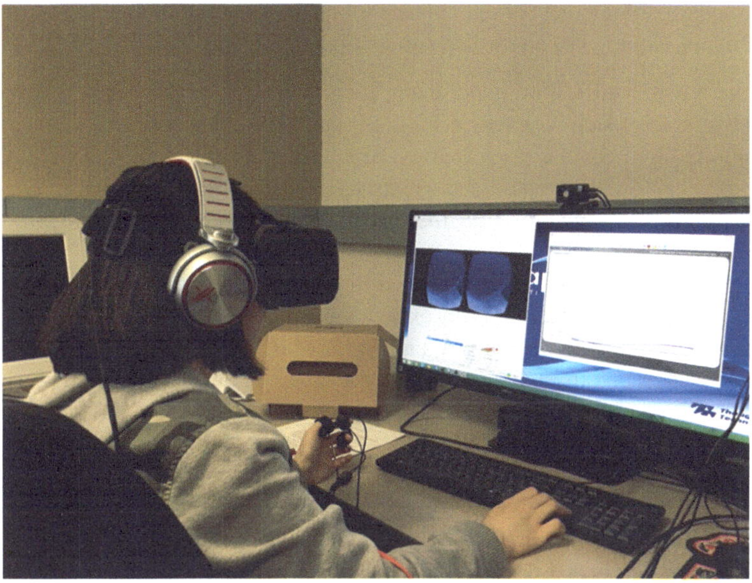

Fig. 6 An experiment session

(a) Equipment installation
A subject is instructed to adjust the Oculus Rift helmet, sitting in front of a desk and relax. Physiological sensors are attached to the subject's fingers. A simple VR room is presented on the helmet screen to help calibrating.

(b) Baseline testing
Before starting the VR exposure, a subject is asked to record at least 1 min baseline data in resting status. No VR scene is presented during this time. The subject is told to relax without verbal communications during this time.

(c) Scenarios exploration
After the baseline testing, VR scenes are presented on the helmet screen, which are 360° immersive, with stereo background music. The order of VR exposure is from neutral scenario, then negative scenario, and positive scenario by the end.
The neutral scenario serves as an introduction of VR, in order to eliminate the variation of first exposure. A subject would use his/her unoccupied hand to navigate through the scene and wander anywhere in the scene until time is up.
The other two scenarios mainly require subject to sit and explore by turning around his/her body. There is no input or oral communication needed from the subject. Researchers monitor the scenarios by streaming the screen of Oculus, instruct as needed, record both physiological data and screen, and mark event labels for synchronization purposes.

(d) Data synchronization and analysis
The recorded data are then analyzed along with the synchronized events and trigger from VR scenes. At this stage, all data are processed and analyzed

offline, in order to get better evidence for emotion recognition in later real-time affective games. The analysis result is separated in next section.

In the testing/verification experiment, it takes the same steps as in the training experiment except step 3, where in this case the VR content has been changed to a clip with multiple continuous scenarios. Similarly, the subject needs to sit and roll the chair to explore the content and trigger the next scenario. Due to different reaction times and style of exploration, the time to complete this experiment varies from 12 to 15 min.

The event markers are labeled when different scenarios are triggered. The synchronization is more difficult in this experiment since the duration of each scenario is flexible depending on the subject's reaction, while the training session has fixed length on each scenario except for slight differences in the negative one.

Preliminary Results

In order to testify the assumption of using physiological data to differentiate user emotional reactions toward VR contents and that VR contents are generally better stimuli than traditional media, an initial experiment using the same VR content with a small sample of participants was conducted, including a group of four children with ASD and a group of the same number of typically developed children. The data were analyzed with traditional statistical methods using software tools such as SPSS and Kubios for HRV analysis. The results served as an evidence showing prospective in emotional recognition with VR and the significant difference between ASD and the control group.

Figures 7, 8, 9, 10, 11, and 12 were previously reported in publication (Li et al. 2015a).

Figure 7 shows some samples of skin conductance level during the experiment session. The samples show common patterns as expected that SCL is low for neutral scenarios and high for negative scenarios. The major difference is that the autism group shows higher responses across three scenarios. Note that scales on y-axis are different in the two samples for the purpose of displaying.

Figure 8 shows samples of heart rate variability analyses for one subject in the control group. Dependent variables are extracted from the results to be analyzed within each group by different emotions.

Figures 9, 10, and 11 show significant statistic results for differentiating emotional states as reaction to VR scenarios. In Fig. 9, using nonspecific SCR signals, it is easy to distinguish negative emotions from neutral and positive emotions, while neutral and positive emotions are not significantly different from each other. Similarly, Fig. 10 demonstrates that both the LF component of HRV and LF/HF ratio are capable of distinguishing significant difference between negative and positive scenarios. On the other hand, Fig. 11 indicates that by using HRV in time domain, it is possible to differentiate neutral and positive emotions

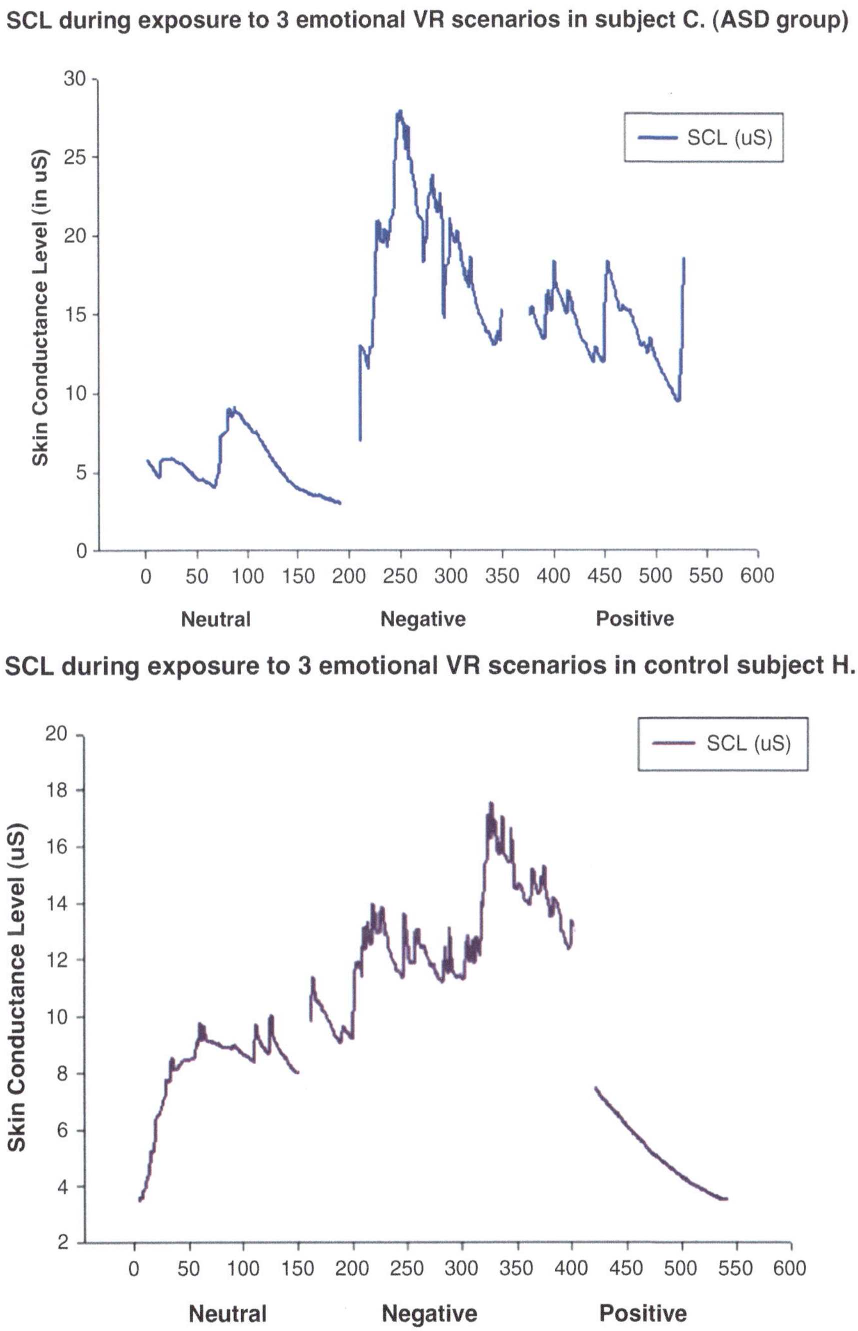

Fig. 7 (**a**) SCL result for ASD group. (**b**) SCL result for the control group

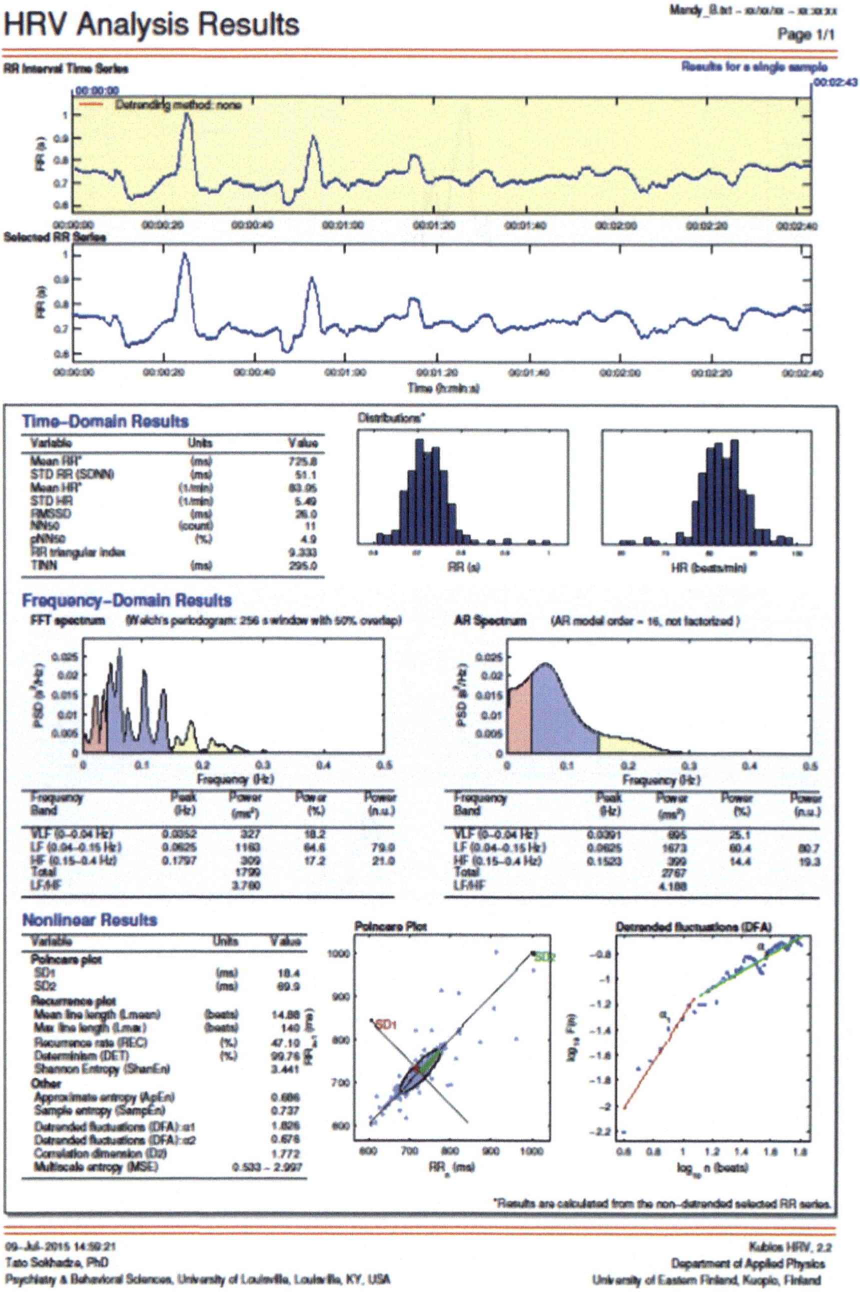

Fig. 8 (**a**) HRV result for one subject during neutral scene. (**b**) HRV result for one subject during negative scene

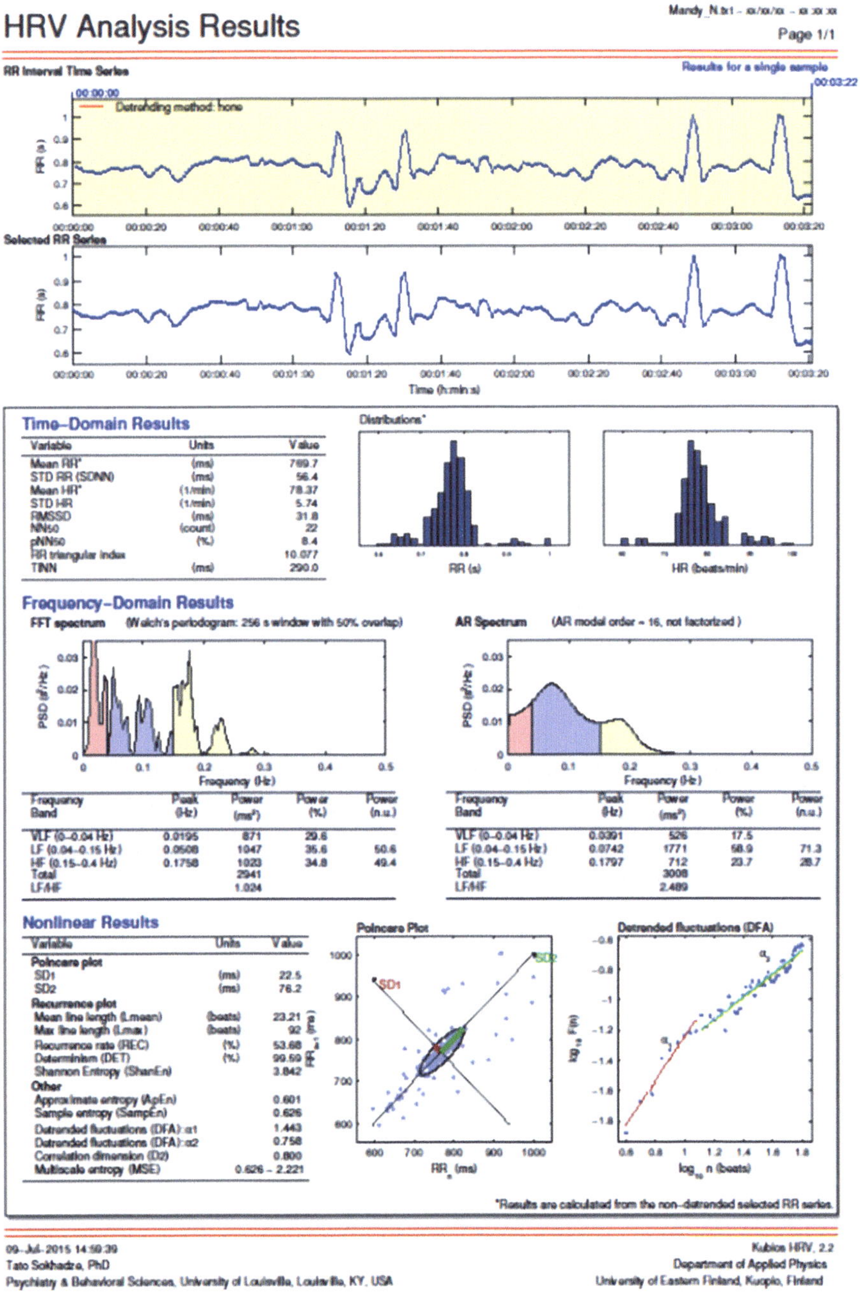

Fig. 8 (continued)

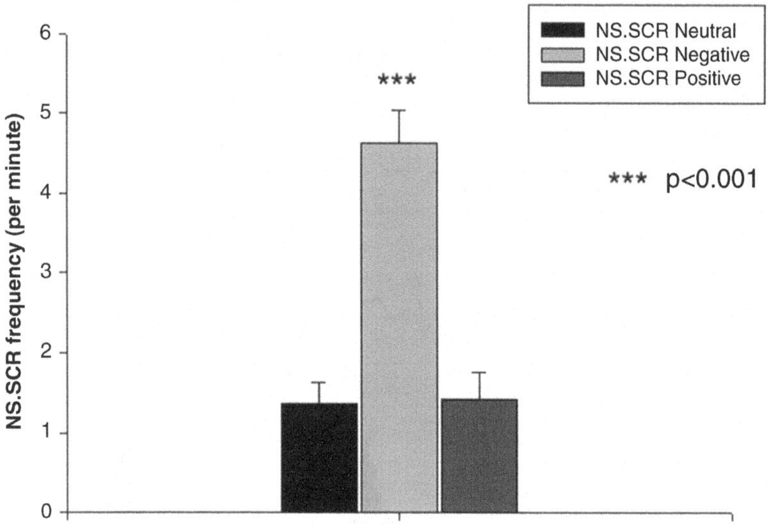

Fig. 9 Frequency of nonspecific SCR during exposure to emotional scenarios in VR

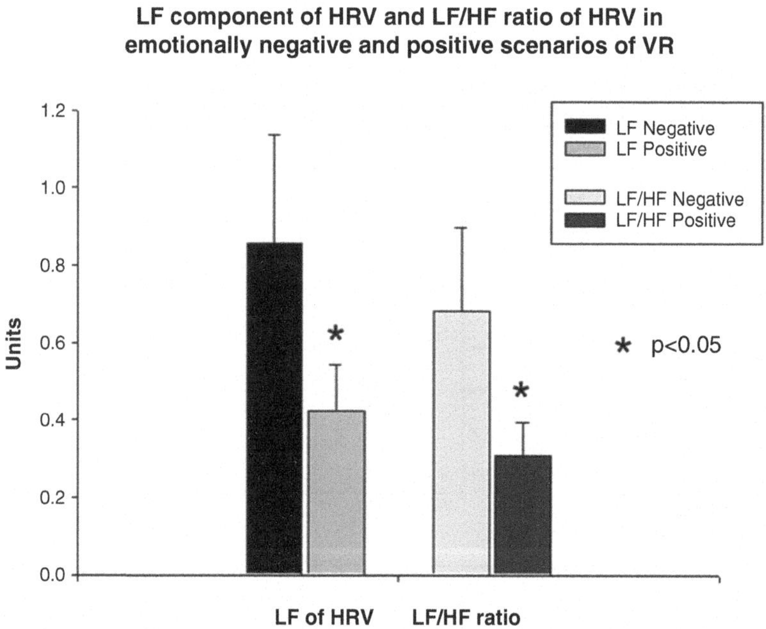

Fig. 10 HRV in frequency domain

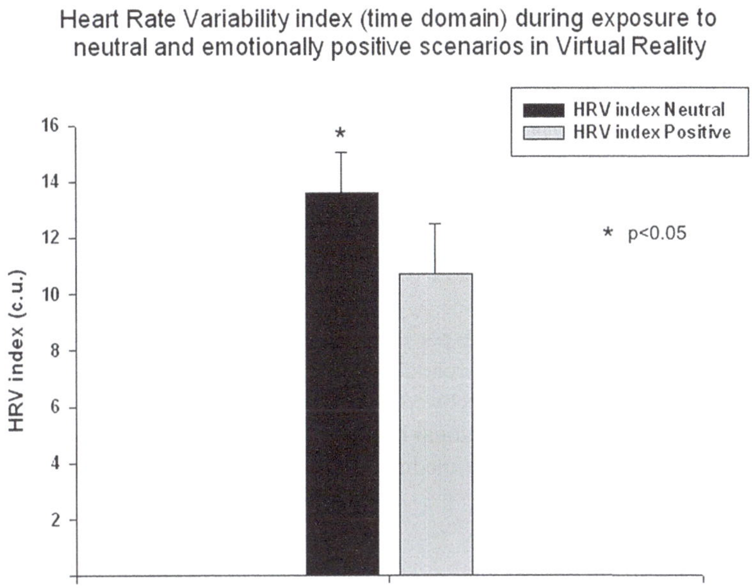

Fig. 11 HRV in time domain

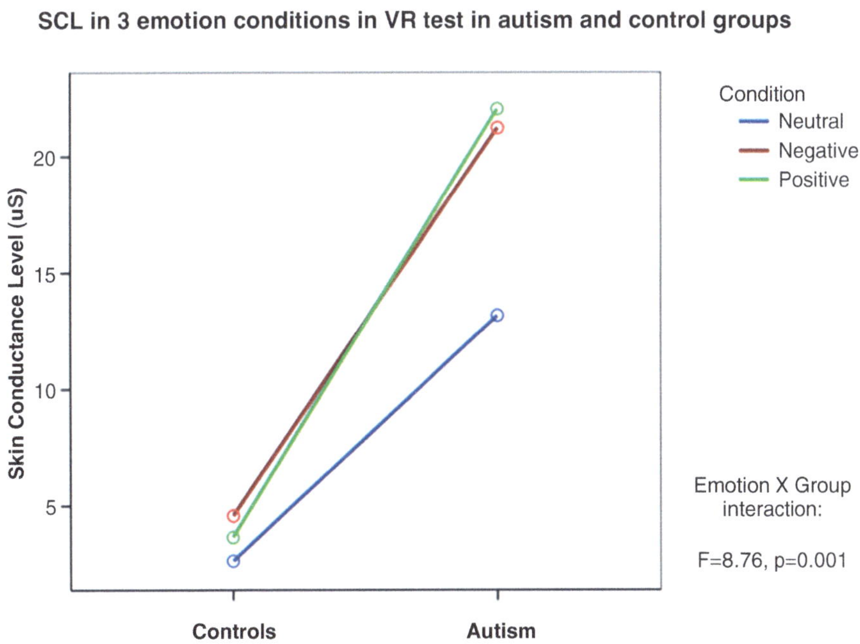

Fig. 12 Frequency of nonspecific SCR comparison between autism and control group

A comparison was made between children with ASD and normally developed individuals. Figure 12 compares SCL results between two groups. Generally, children with ASD have higher SCL responses for all three themes. It is also worth noticing that the trends for responses to positive and negative themes are different between the two groups. In fact, we are aware of the needs for gathering more data from those groups, like children with ASD, to build specialized models or databases for dedicated applications. However, due to the difficulties of getting IRB approval, we discontinued experiments on children with ASD after the preliminary experiment, only focusing on adults without emotional disorder for the rest of experiments.

With the results of the initial experiment on VR contents, it is feasible to use physiological reaction data to differentiate different emotions, as well as differentiate groups of people. What is more, by comparing the results with traditional media, VR contents trigger stronger emotional reactions in both experimental and control groups, and thus, could be a better medium to be used in treatment intervention in special groups of people.

Machine Learning Analysis Approach

This part is to demonstrate using the machine learning approach to train models and then categorize emotions. Due to the difficulties of finding enough sample subjects with ASD, we recruited a separate group to conduct both of the experiments. The participants are individuals between the ages from 9 to 66 without any reported mental disorders. The recorded data are used as training data and testing/verification data, respectively.

To achieve a better understanding and interpretation of user physiological reactions, we analyze the data by following machine learning steps as the flow chart shows in Fig. 13—a classic approach of machine learning. Twenty features were extracted from the raw data from the training experiment, shown in Table 2. Then they went through feature filter methods, including eliminating highly correlated features and a wrapper method. We trained models with three classic machine learning algorithms: K-nearest neighbor (KNN), Linear Discriminant Analysis (LDA), and Linear Support Vector Machine (SVM). Different sizes of feature subsets were used, and training results were compared in order to choose the best feature subset that yields the relatively best result while keeping the computing complexity relatively low. Figure 14 shows the comparison among classifiers over the feature subset size.
We then used the trained model to test the testing/verification of experiment data.

Figure 15 shows the emotion recognition results on each event. From event 2 to event 12, the dominating emotion state is neutral, which makes sense that most of the scenes are about scenery or weather: although there is bad weather or a car accident that may make fewer people feel uncomfortable. One surprising result is the event 12, in which the player is sitting in front of a dining table, where roasted chicken and ham are sizzling. Most volunteers had neutral feelings with this scenario, while a few

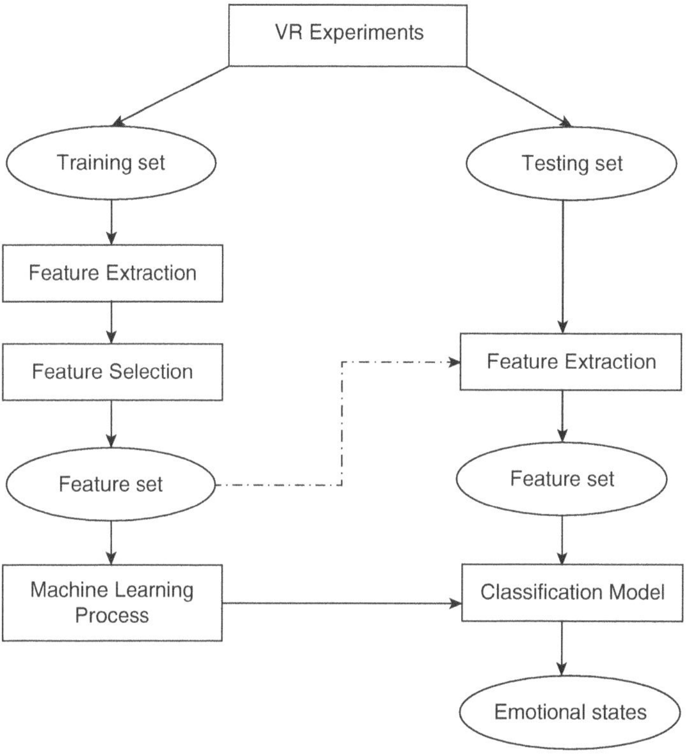

Fig. 13 Flow chart of classification modeling

Table 2 Summary of Extracted Features

Skin conductance					
f1	f2	f3	f4	f5	f6
Peak average	NS.SCR frequency	Average	Std. dev	Maximum	Minimum
Temperature					
f7	f8	f9	f10		
Average	Std. dev	Maximum	Minimum		
Heart rate					
f11	f12	f13	f14		
Average	Std. dev	Maximum	Minimum		
Heart rate variability					
f15	f16	f17	f18	f19	f20
RR average	SDNN	rmSSD	LF	HF	LF/HF ratio

felt negative, but no positive feelings were recognized in our current dataset. One reasonable guess is that they may feel bad when they are in front of food but it is not accessible to eat, or the food being presented is not attractive. The events 13–15 are in the space looking at stars and earth, and it is reasonable that most volunteers feel

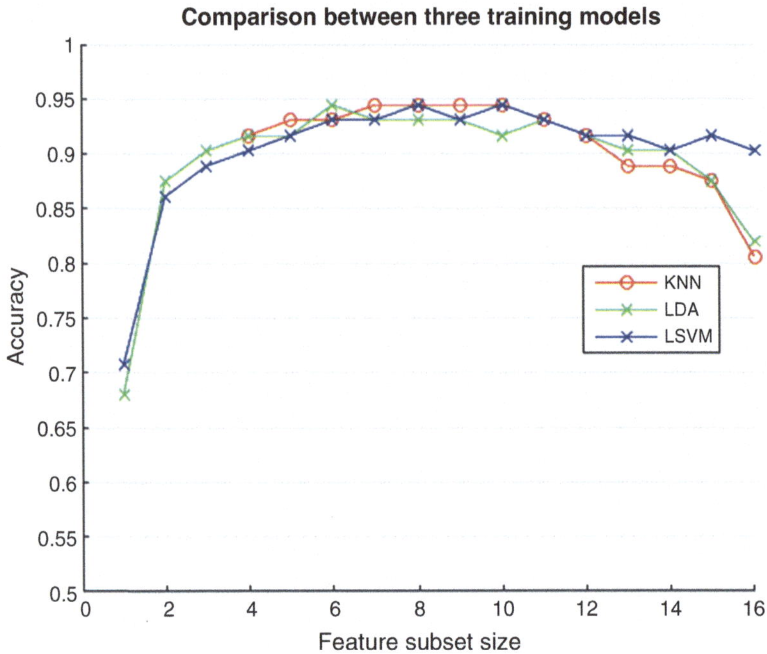

Fig. 14 Comparison among classifiers

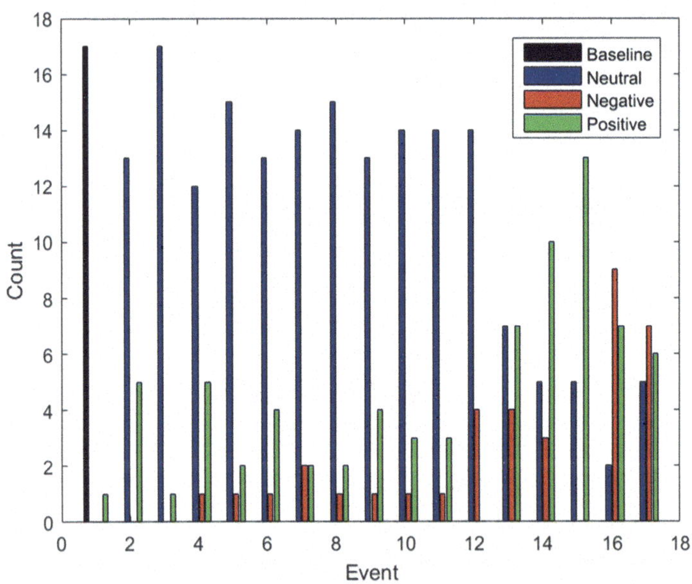

Fig. 15 Emotion recognition on each event

positive while some might be scared. Events 16 and 17 are more terrifying as the volunteers are sitting on the top of a building under construction with the risk of falling; however, a few may feel neutral because they know the danger is not real, and some may feel excited by nature.

3.2 *Game Design*

After the analysis of data, the next step is to construct a gaming model to serve as a guideline for adaptive game applications that interact with user emotions.

3.2.1 Game Interaction Modeling

Based on this concept, our current work presents a closed-loop affective computing system, shown in Fig. 16. This closed-loop affective game system consists of three essential modules: affect modeling module, affect recognition module, and affect control module. The dashed line indicates the optional offline path just to differentiate from the main loop.

Affect Recognition Module

This module is mainly applying the learned user model to recognize user emotional states. After a possible calibration process, a user could be categorized into user model presets or a dedicated model. The same set of features as the constructed model is extracted during game sessions, then sent for the recognition process. The results of recognition could be within a range of confidence. They can either be sent offline to improve the user model further if the offline mode is selected, or sent to generate feedback game scenario for real-time adaptivity.

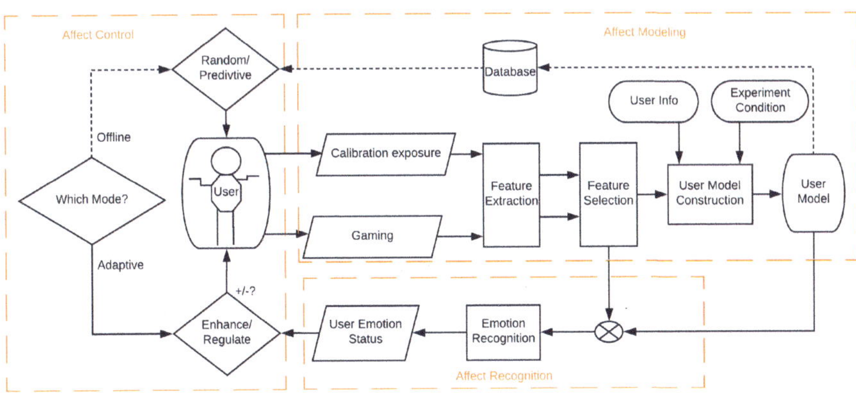

Fig. 16 A closed-loop affective game system

Affect Control Module

This module changes the environment according to user emotional states. The environment includes any individual or combined methods, such as visual exposure (picture, video, 3D or VR) and audio exposure (affective sounds or music) to induce users' emotions. The efficiency of each method is studied in both our preliminary and extended experiments, and the study results will be accumulated and used for possible future modifications.

This model should have the option to choose either a real-time adaptive or offline mode. Real-time adaptive mode means changing game scenarios immediately when emotion changes are detected. It also includes decision making on whether to enhance the current emotion intensity, or reduce emotion intensity for training purposes. Offline mode can either be randomly choosing the next scenario or running through a predictive path based on the offline dataset. Either way, the user data are collected and used to feed the user model construction.

If the desired emotional states are known beforehand, this framework could be applied to a training system; otherwise it is self-adaptive to be used as a testing system to keep track of changes in users' emotional states.

Every single individual may have different emotional reactions toward the same emotion stimuli. Therefore, in order to adapt the game scenarios to accommodate emotions, it is necessary to monitor user preference and construct user models to better understand the current emotional states and determine how to accordingly apply the adaptivity mechanism. It is nearly as crucial as designing an adaptivity mechanism from the game point of view to control the interactive environment based on the user models.

3.3 Customizable User Modeling

Besides the emotion recognition done by the last chapter, some users may follow similar patterns in their physiological reactions. We can cluster them into different groups to learn the pattern and to develop representation as the first step of constructing user models, especially when the dataset is not large enough.

As a simple example, the K-means clustering algorithm is selected in the present research to categorize user groups roughly. K-means is an unsupervised learning algorithm. It repeatedly calculates this distance and assigns the current data point to the nearest centroid, and then calculates new cluster centroids after one iteration until centroids do not move anymore. In this case, the data points are features extracted from physiological data, and the result cluster centroids are then stored to be representations of user groups. New users can get a few minutes of the same calibration VR exposure to place themselves in one group; then the adaptivity mechanism will be able to control game scenarios depending on the user model and the current user emotions.

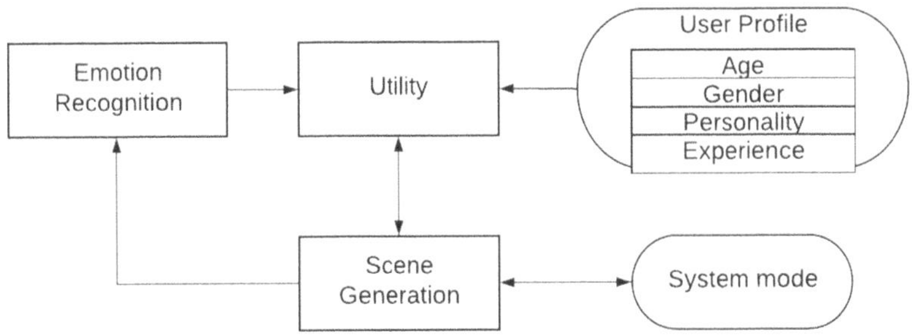

Fig. 17 BBN user model

To adjust user models more specific to individuals, a learning system can be designed to include more user profile information such as user preference, personality, previous experience of VR, as well as the current experiment condition. A Bayesian Belief Networks (BBN) (Pearl 2014) can be adopted in finalizing the model, for the reason that it is capable of handling uncertainty and extending to more variables when feasible.

In the present research case, an example of the BBN user model is shown in Fig. 17. The details of deriving formula and calculating the probabilities are beyond the scope of the chapter, and thus, are not discussed here.

A list of parameter nodes that can be chosen are:

- Emotion Recognition (e), including current recognized emotion and recognition accuracy;
- User Profile (p), as a combination of any known information like user age, user gender, user personality, and the user group assigned using clustering methods;
- System Mode (m), whether to enhance or reduce current emotion intensity;
- System Utility (U), to calculate the system utility and affect decision making for next scene generation;
- Scene Generation (s), including possible changing scenarios and their probability of inducing emotions.

3.4 Graph-Based Adaptivity Modeling

In order to adapt game scenarios depending on users' emotional reactions as a new type of game feedback, an adaptivity mechanism with thorough analysis and elaborate design is necessary. The Bayesian Belief Network described in the last section provides a solution for how to integrate known information to calculate the probability of generating each possible scene. Besides that, we need to look at the big picture of the gaming structure. This section is reproduced from Li et al. (2015b) with permission from IEEE.

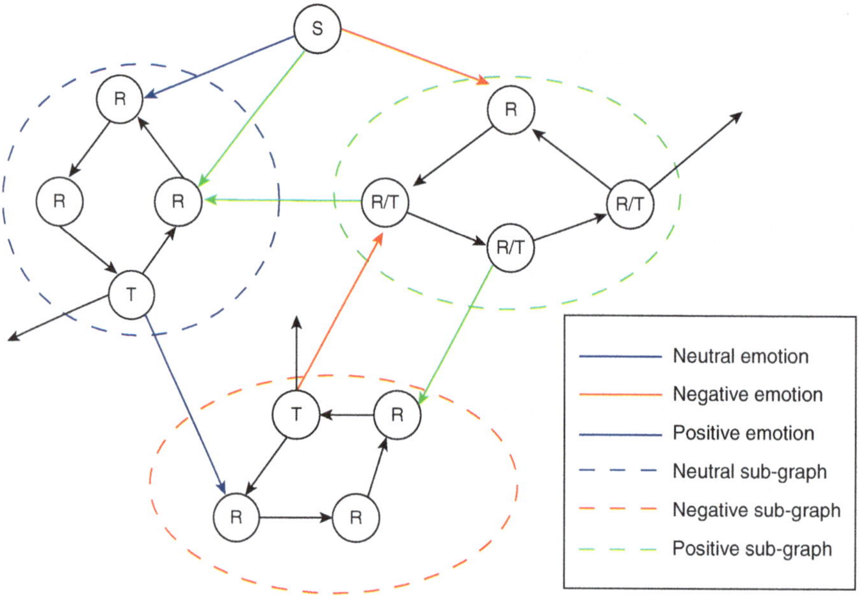

Fig. 18 Biofeedback structure

Table 3 Types of key scenario nodes

Notation	Representation	Action(s)
S	Source node	Decide which emotional subgraph to begin the game
R	Interactive node	Stimuli that either enhance or reduce current emotional intensity
T	Testing node	Decide whether to switch to another subgraph or stay in current subgraph

Based on the Dynamic Object-Oriented methodology (Arinbjarnar et al. 2009), our biofeedback structure is described as in Fig. 18. Key scenes are denoted as nodes in the figure, while emotional states indicated by biofeedback are coded by colored arrows. Subgraphs represent different types of scenario subgraphs that induce different emotions. Blue represents neutral subgraph or emotional state, while green represents positive, and red represents negative. Table 3 (Li et al., 2015) explains the different types of key scenarios and their actions.

The workflow of the Dynamic Biofeedback Structure is described below:

1. Let S be the source point of the game. Randomly assign one of the starting key scenarios from three subgraphs.
2. For each $r_{i,j} \in R_j$, where $i \geq 0$ represents nodes, and $j \geq 0$ represents subgraphs. Apply affective stimuli based on the current subgraph to enhance or reduce the emotional state, then move to a next node $r_{i+1,j}$. Keep track of enhancing time, node type, and affective scale.

3. When a $t_j \in T$ node occurs, test the current user emotional state by analyzing the psychophysiological signals collected during the time of past consecutive r_i, …, . r_{i+m} nodes, where m is a nonnegative integer. If the emotional state has changed during the period, decide whether to go to the corresponding subgraph or stay at the same subgraph by referring to the BBN model for next scene generation.
4. Continue looping previous steps, until either time is up or it reaches a desired emotional state, that is, the end node of the entire game.

The structure interacts with emotional states induced by key scenarios and initiates dynamic switch among key scenarios under constraints. A key scenario contains several VR attributes to stimulate a certain kind of emotion; stimuli are selected referring to the experiment outcomes of events voting.

Notably, in this presented design, there are a few reasons that the T nodes occur every fixed amount of time and analyze all factors during that range, and then make decisions based on the BBN model. First of all, for recorded experiments (usually with screen capture videos), it is relatively easier to sort out the time range when events begin to happen or finish, while real-time gaming will not have a clear timing for what is happening. Calculating emotions in every node in absolute real-time dramatically adds calculation complexity, which is not a good idea for the purpose of designing a portable gaming system. Meanwhile, a time window is needed anyway to calculate statistic features in order to recognize the current emotional state: too short a window will decrease the accuracy. The length of the window is adjustable and can be either fixed or relative. Moreover, making decisions based on every node leads to jumping around different subgraphs too soon to consummate a pleasant experience, since the rapidness may confuse users and get undesirable lousy looping.

In reality, any games, including serious games or normal entertainment games, can make use of the above user model and adaptivity mechanism, since it is one extra dimension of feedback. The storyline may or may not be subject to changes of emotions depending on the design needs. Subgraphs could be affecting the main storyline, changing the surroundings or leading to different side quests.

3.5 Game Application

In order to demonstrate this interactive game framework, a simple VR game application was developed, which consists of all essential modules described in previous chapters and sections. Figure 19 shows the essential parts of the game and its workflow.

A middleware was implemented with Matlab to form the affect recognition and affect control module. It connects ProComp sensors, reads real-time psychophysiological data, processes affect recognition, analyzes the user model, and then sends instructions to Unity, where the VR game scenarios are hosted and will be adaptively changed to coordinate with the affection status change.

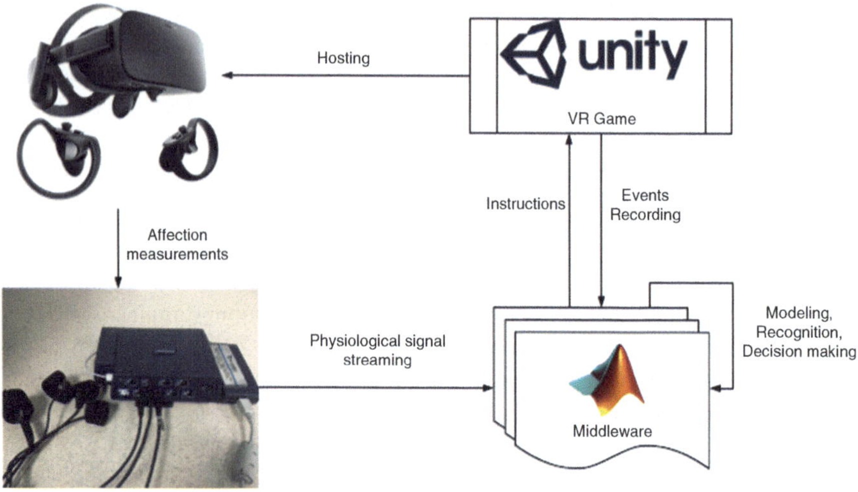

Fig. 19 VR game workflow

The demonstrative application changes the weather, tasks/messages, backgrounds, and music according to user emotion reactions while requiring physical movements to achieve health goals. For example, one of the game tasks requires a user to read out messages aloud. While the user is walking down the path, the environment, including the skybox, the background, the audio, and even the next message sign, will be changed according to the change of the user's emotional states. Figure 20 shows some of the changes in the game scene.

The game is able to show the basic concept of the introduced framework, but is still in need of further polish in regard to art design and intensity levels.

4 Discussion and Future Work

In this chapter, we introduced a novel approach for designing affective VR environments with feedback to user emotions based on the results of a series of VR exposure experiments. The interactive mechanism combines advantages of portable VR devices and reliable physiological sensors along with affective computing, to create immersive environments that interact with users' emotional states. The implementation of the system can either be applied to general emotional response testing games, or as a tool to assist intervention for children with ASD.

Based on the studies and analysis above, there are a few directions for future work to improve the model and optimize its applications:

First of all, as we stated and proved in the preliminary study, children with ASD have different emotional responses compared to typically developed individuals. In order to build a game dedicated to the ASD group, we will need more data collected

Fig. 20 (**a**) Negative scene. (**b**) Neutral scene. (**c**) Positive scene

from them to construct machine learning models, as well as customizable user models. Meanwhile, we need to scale up the group size to gain sufficient data for future statistical analysis.

Currently, three generic emotion categories are recognized and used to yield affective control of the game scenario. With the extension of the categories by including more detailed states, it is promising that game scenarios being designed in the future will introduce relatively more exhilarating experience and polished sensation to users.

In the laboratory experiments, we are using professional physiological sensors ProComp 5 series. As the purpose of the presented framework is to implement portable games so that they can be widely used in healthcare and other fields, looking for less expensive wireless sensors with acceptable accuracy as replacement is also a need. With the reduced size of the required feature set, a smartwatch or wristband may satisfy the needs. VR hardware is developing very fast in recent years. It is promising to see more portable VR systems on the market.

Last but not the least, as a multimodal model, behavioral analysis is also included as feature sets, but not implemented in the current project. In addition to using psychophysiological data, behavioral analysis can reveal user emotions as well. It can increase the accuracy of recognition and provide cross-verification. Examples of gaming tools that can be used are Kinect or Leap Motion. The most recent version of HTC Vive Pro has a built-in eye-tracking feature, which could be explored as a part of behavior analysis.

References

Amaral C, Mouga S, Simões M, Pereira HC, Bernardino I, Quental H, Playle R, McNamara R, Oliveira G, Castelo-Branco M (2018) A feasibility clinical trial to improve social attention in autistic spectrum disorder (ASD) using a brain computer interface. Front Neurosci 12:477. https://doi.org/10.3389/fnins.2018.00477

Arinbjarnar M, Barber H, Kudenko D (2009) A critical review of interactive drama systems. AISB 2009 symposium. AI & Games, Edinburgh, Citeseer

Balti H, Elmaghraby AS (2013) Speech emotion detection using time dependent self organizing maps. In: IEEE international symposium on signal processing and information technology, IEEE

Burr RL (2007) Interpretation of normalized spectral heart rate variability indices in sleep research: a critical review. Sleep 30(7):913–919. https://doi.org/10.1093/sleep/30.7.913

Calvo RA, D'Mello S (2010) Affect detection: an interdisciplinary review of models, methods, and their applications. IEEE Trans Affect Comput 1(1):18–37. https://doi.org/10.1109/t-affc.2010.1

Castellano G, Kessous L, Caridakis G (n.d.) Emotion recognition through multiple modalities: face, body gesture, speech. In: Affect and emotion in human-computer interaction. Springer, Berlin, pp 92–103. https://doi.org/10.1007/978-3-540-85099-1_8

CDCMMWR (2018) Correction and republication: prevalence and characteristics of autism spectrum disorder among children aged 8 years—autism and developmental disabilities monitoring network, 11 sites, United States, 2012. MMWR Morb Mortal Wkly Rep 67. https://doi.org/10.15585/mmwr.mm6745a7. https://www.cdc.gov/mmwr/volumes/67/wr/mm6745a7.htm. Accessed 21 Mar 2020

Coben R, Linden M, Myers TE (2009) Neurofeedback for autistic spectrum disorder: a review of the literature. Appl Psychophysiol Biofeedback 35(1):83–105. https://doi.org/10.1007/s10484-009-9117-y

Cornforth D, Jelinek H, Tarvainen M (2015) A comparison of nonlinear measures for the detection of cardiac autonomic neuropathy from heart rate variability. Entropy 17(3):1425–1440. https://doi.org/10.3390/e17031425

Critchley HD (2002) Review: electrodermal responses: what happens in the brain. Neuroscientist 8(2):132–142. https://doi.org/10.1177/107385840200800209

Ekman P, Levenson R, Friesen W (1983) Autonomic nervous system activity distinguishes among emotions. Science 221(4616):1208–1210. https://doi.org/10.1126/science.6612338

Jerdan SW, Grindle M, van Woerden HC, Kamel Boulos MN (2018) Head-mounted virtual reality and mental health: critical review of current research. JMIR Ser Games 6(3):e14. https://doi.org/10.2196/games.9226

Kapur A, Kapur A, Virji-Babul N, Tzanetakis G, Driessen PF (2005) Gesture-based affective computing on motion capture data. In: Affective computing and intelligent interaction. Springer, Berlin, pp 1–7. https://doi.org/10.1007/11573548_1

Kim KH, Bang SW, Kim SR (2004) Emotion recognition system using short-term monitoring of physiological signals. Med Biol Eng Comput 42(3):419–427. https://doi.org/10.1007/bf02344719

Kumazaki H, Warren Z, Swanson A, Yoshikawa Y, Matsumoto Y, Yoshimura Y, Shimaya J et al (2018) Brief report: evaluating the utility of varied technological agents to elicit social attention from children with autism spectrum disorders. J Autism Dev Disord 49(4):1700–1708. https://doi.org/10.1007/s10803-018-3841-1

Kuriakose S, Lahiri U (2017) Design of a physiology-sensitive VR-based social communication platform for children with autism. IEEE Trans Neural Syst Rehabil Eng 25(8):1180–1191. https://doi.org/10.1109/tnsre.2016.2613879

Lang PJ, Cuthbert BN, Bradley MM (1998) Measuring emotion in therapy: imagery, activation, and feeling. Behav Ther 29(4):655–674. https://doi.org/10.1016/S0005-7894(98)80024-5

Li Y, Elmaghraby AS (2014) A framework for using games for behavioral analysis of autistic children. In: 2014 computer games: AI, animation, mobile, multimedia, educational and serious games (CGAMES), IEEE

Li CC, Yuan SNV (2018) A case study on delivering virtual reality learning for children with autism spectrum disorder using virtual reality headsets. In: EDULEARN18 proceedings, IATED

Li Y, Elmaghraby AS, El-Baz A, Sokhadze EM (2015a) Using physiological signal analysis to design affective VR games. In: 2015 IEEE international symposium on signal processing and information technology (ISSPIT), IEEE, pp 57–62

Li Y, Elmaghraby AS, Sokhadze EM (2015b) Designing immersive affective environments with biofeedback. In: 2015 computer games: AI, animation, mobile, multimedia, educational and serious games (CGAMES), pp 73–77

Lovreglio R, Gonzalez V, Feng Z, Amor R, Spearpoint M, Thomas J, Trotter M, Sacks R (2018) Prototyping virtual reality serious games for building earthquake preparedness: the Auckland City Hospital Case Study. Adv Eng Inform 38:670–682. https://doi.org/10.1016/j.aei.2018.08.018

Medeiros JM (2010) Development of a heart rate variability analysis tool. Master's Thesis, University of Coimbra

Michael DR, Chen SL (2005) Serious games: games that educate, train, and inform. Muska & Lipman/Premier-Trade

Milanova M, Sirakov N (2008) Recognition of emotional states in natural human-computer interaction. In: 2008 IEEE international symposium on signal processing and information technology, IEEE

Moghim M, Stone R, Rotshtein P, Cooke N (2015) Adaptive virtual environments: a physiological feedback HCI system concept. In: 2015 7th computer science and electronic engineering conference (CEEC), IEEE

Moghimi M, Stone R, Rotshtein P (2020) Affective recognition in dynamic and interactive virtual environments. IEEE Trans Affect Comput 11(1):45–62. https://doi.org/10.1109/taffc.2017.2764896

Nasoz F, Lisetti CL, Vasilakos AV (2010) Affectively intelligent and adaptive car interfaces. Inform Sci 180(20):3817–3836. https://doi.org/10.1016/j.ins.2010.06.034

Nijman SA, Veling W, Greaves-Lord K, Vermeer RR, Vos M, Zandee CER, Zandstra DC, Geraets CNW, Pijnenborg GHM (2019) Dynamic interactive social cognition training in virtual reality (DiSCoVR) for social cognition and social functioning in people with a psychotic disorder: study protocol for a multicenter randomized controlled trial. BMC Psychiatry 19(1):272. https://doi.org/10.1186/s12888-019-2250-0

Nwe TL, Foo SW, Silva LCD (2003) Speech emotion recognition using hidden Markov models. Speech Commun 41(4):603–623. https://doi.org/10.1016/s0167-6393(03)00099-2

Pearl J (2014) Probabilistic reasoning in intelligent systems: networks of plausible inference. Elsevier

Picard RW (2000) Affective computing. MIT press

Rao SM, Gagie B (2016) Learning through seeing and doing: visual supports for children with autism. In: TEACHING exceptional children. https://journals.sagepub.com/doi/10.1177/004005990603800604. Accessed 21 Mar 2020

Ravindran V, Osgood M, Sazawal V, Solorzano R, Turnacioglu S (2019) Virtual reality support for joint attention using the floreo joint attention module: usability and feasibility pilot study. JMIR Pediatr Parent 2(2):e14429. https://doi.org/10.2196/14429

Schutz PA, Pekrun R (2007) Introduction to emotion in education. In: Emotion in education. Elsevier. pp 3–10. https://doi.org/10.1016/b978-012372545-5/50002-2

Shivakumar G, Vijaya PA (2012) Emotion recognition using finger tip temperature: firststep towards an automatic system. Int J Comput Electr Eng 4(3):252–255. https://doi.org/10.7763/ijcee.2012.v4.489

Shivakumar G, Vijaya PA (n.d.) Analysis of human emotions using galvanic skin response and finger tip temperature. In: Computer engineering, IGI Global. pp 792–802. https://doi.org/10.4018/978-1-61350-456-7.ch319

Wijma EM, Veerbeek MA, Prins M, Pot AM, Willemse BM (2017) A virtual reality intervention to improve the understanding and empathy for people with dementia in informal caregivers: results of a pilot study. Aging Ment Health 22(9):1121–1129. https://doi.org/10.1080/13607863.2017.1348470

Wilkinson P (2013) Affective educational games: utilizing emotions in game-based learning. In: 2013 5th international conference on games and virtual worlds for serious applications (VS-GAMES), IEEE

Zhang L, Weitlauf AS, Amat AZ, Swanson A, Warren ZE, Sarkar N (2019) Assessing social communication and collaboration in autism spectrum disorder using intelligent collaborative virtual environments. J Autism Dev Disord 50(1):199–211. https://doi.org/10.1007/s10803-019-04246-z

Zhao H, Swanson AR, Weitlauf AS, Warren ZE, Sarkar N (2018) Hand-in-hand: a communication-enhancement collaborative virtual reality system for promoting social interaction in children with autism spectrum disorders. IEEE Trans Hum Mach Syst 48(2):136–148. https://doi.org/10.1109/thms.2018.2791562

A Machine Learning Approach to Automatic Phobia Therapy with Virtual Reality

Oana Bălan, Alin Moldoveanu, and Marius Leordeanu

1 Phobias: Statistics, Effects, and Treatment

Phobia is a type of anxiety disorder manifested through an extreme, uncontrolled, and irrational fear that appears when the subject is exposed to certain stimuli—a naturalistic situation, the presence of people, animals, or objects. There are different types of phobias, such as *agoraphobia* – fear of crowds or open spaces, *social phobias*—fear of speaking in public, meeting people of higher authority, eating or using the telephone in front of others, and *specific phobias*—caused by various objects and situations (World Health Organization 2017). Social phobias affect all age categories, but the onset is usually in the adolescence (95% begin before the age of 20). In what concerns the sex categories, women are more affected than men (Olesen 2015). Also, anxiety disorders are more common in women—4.6% at the world level, compared to 2.6% in men. As for specific phobias, they occur at least once in a lifetime for 15–20% of the world's population (Olesen 2015). They have the following prevalence at the world level: acrophobia (fear of heights)—7.5%, arachnophobia (fear of spiders)—3.5%, aerophobia (fear of flying)—2.6%, astraphobia (fear of lightning and thunder)—2.1%, dentophobia (fear of dentist)—2.1% (Nation Wide Phobias Statistics 2019). Some phobias are connected: for example, acrophobia is related to fear of elevators and fear of flying (Muris et al. 1999). Specific phobias usually appear in childhood and prolong throughout the entire life (Olesen 2015). For acrophobia, various researchers supported hereditary and *nonassociative* factors in the development of this anxiety disorder, as the subjects were unable to account for a height-related experience triggering acrophobia. The subjects from the control group did not develop acrophobia, even if they have been exposed to heights (Menzies and Clarke 1993, 1995). Besides, Poulton et al. (1998) and Poulton and

O. Bălan (✉) · A. Moldoveanu · M. Leordeanu
Politehnica University of Bucharest, Bucharest, Romania
e-mail: oana.balan@cs.pub.ro

I. Opris et al. (eds.), *Modern Approaches to Augmentation of Brain Function*, Contemporary Clinical Neuroscience,
https://doi.org/10.1007/978-3-030-54564-2_27

Menzies (2002) showed that the lowest incidence of acrophobia was encountered for those who suffered heights-related injuries in childhood. Moreover, Menzies and Parker (2001) reported that the non-phobic subjects had the highest incidence of traumatic falls, without affecting their perception on heights. Many other studies supported the non-conditioning theory (Field et al. 2001; Graham and Gaffan 1997; Withers and Deane 1995) and claimed that phobias emerge as a result of other experiences that cannot be recalled or consciously brought into memory.

A phobia crisis causes both physical and emotional symptoms. Among its physical manifestations, we account for high heart rate, sweating, tremor, rapid breathing, or dizziness. On the other hand, the emotional symptoms could include anxiety attacks and difficulty in controlling one's emotional state despite intense efforts. The treatment for phobias comprises of medication (antianxiety and antidepressive drugs), in-vivo exposure in a controlled environment, Cognitive Behavioral Therapy (CBT), and virtual exposure. In 1958, Wolpe (1958) developed a technique called "systematic desensitization", based on deep mental and muscular relaxation. In 1977, Bandura (1977) proposed the "self-efficacy" theory that relies on one's confidence and personal judgment about the ability of overcoming the stressful stimuli. Another model of therapy was "reinforced practice", based on a continuous practice and improvement of the responses to certain therapeutic elements, such as the attitude toward stimuli, feedback to the therapist, and self-control (Leitenberg 1976). Ritter (1969) introduced the "contact desensitization" therapy, where the patient was assisted by the therapist who held his hand or arm during exposure. The desensitization method provided good results when the therapist behaved warmly or not with the patient (Morris and Magrath 1979) and even when the therapist was not present in the room—a tape recorder played the instructions for treatment (Baker et al. 1973). CBT is a strategy that encourages the subjects to change their attitude toward the aversive experience by replacing negative thoughts with positive ones. Only 23% of the people suffering from phobias seek treatment, especially medication and CBT. The study of Steinman and Teachman (2014) showed that CBT has the same rate of success for treating acrophobia as in-vivo exposure to heights.

2 Virtual Reality in Phobia Therapy

Virtual reality (VR) was used since the 1990s in phobia therapy. It benefits from some practical advantages such as a better control of the exposure, possibility to render situations that are not easily accessible, ability to provide stimuli of lower or higher magnitudes than in real-world settings (Choi et al. 2001), higher comfort for both the patient and the therapist, confidentiality, friendly environment, suitable especially for those who do not possess imaginative skills (Coelho et al. 2009). The idea of developing virtual worlds for training purposes dates back to the 1940s, when the American government invested in flight simulators in the context of the Second World War (Littman 1996). *Virtual Reality Exposure Therapy* systems (VRETs) emerged in 1996 when North and North (1996) observed that a flight simulating a virtual environment produced fear responses that were not associated with motion sickness.

According to Garcia-Palacios et al. (2002), 90% of respondents preferred VR exposure than in-vivo exposure for arachnophobia therapy. In Garcia-Palacios et al. (2001), over 80% of patients opted in favor of virtual exposure for acrophobia therapy. VRET systems offer similar results in the posttreatment assessments, comparable to those provided by CBT, with strong real-life impact and good stability of results in time (Opris et al. 2012).

In the virtual environment described in Hodges et al. (1995), there have been designed three situations to be used in acrophobia therapy: an elevator, a balcony, and a bridge. The participants were randomly divided into two groups: a treatment and a control group. The subjects from the treatment group used virtual therapy and they were free to spend as much time as they wanted in various sessions. The subjects from the control group received no treatment; they were only subjected to two evaluations after 7 weeks. The results of the study have shown that the VR-based treatment was as effective as traditional therapy (Hodges et al. 1995).

The quantitative meta-analysis performed in (Parsons and Rizzo 2008) highlighted that VRET has potential in treating anxiety and certain phobias, including acrophobia. The results of a VRET-based study are presented in Shiban et al. (2015), where the issue of Return of Fear (ROF) after successful treatment was thoroughly approached. All participants completed both a VR test and an in-vivo Behavioral Avoidance Test (BAT). The results of a meta-analysis demonstrated that VRET can produce significant behavior changes in real-life situations that support its applicability in treating specific phobias (Morina et al. 2015).

A comparison between an Augmented Reality (AR) and a VR system including acrophobic scenarios is presented in Carmen Juan and Perez (2010). There were no significant differences regarding the therapy results. In AR, the participants could see their hands, feet, and the scene is real, while in VR, all of these are simulated (Buna et al. 2017).

Nowadays, new and sophisticated VR devices have emerged. Their low price makes them affordable, so that they can be successfully used to build immersive VR environments for treating certain phobias, including acrophobia (Buna et al. 2017). The Climb (Robertson 2016) is a game that can be played on the Oculus device (Oculus Rift n.d.) with input from the Xbox gamepad. Richie's Plank Experience (n.d.) is a game for HTC Vive (n.d.) employing a customizable real plank replicated in the virtual environment. C2Phobia (n.d.) treats acrophobia. The player can exit on the balcony, take a transparent elevator, or move from one building to another using walkways with low walls, ropes, or without any protection. In the Stim Response Virtual Reality system (2BIOPAC n.d.), the events from VR and the physiological data are synchronized in real time and the scenes are adapted according to the player's biophysical output. A component of this system is VR-Acrophobia, which is fully modular and customizable, so that the therapist can create and recreate various scenes. The Virtual Reality Medical Center (VRMC) (Virtual Reality Medical Center n.d.) uses 3D computer simulation, biofeedback, and CBT to treat phobias and anxieties.

The above-mentioned systems demonstrate the advantages of using VR for treating phobias. However, they can only be used under medical surveillance, with guidance from a physician or psychologist.

3 Our Main Contributions

Our scientific contributions are presented in more detail in the next sections of the chapter. We also briefly enumerate them here:

1. We perform a comparison of several machine learning techniques (Support Vector Machine, Linear Discriminant Analysis, Random Forest, and k-Nearest Neighbors and 4 deep neural networks with different numbers of layers and neurons per layer), with and without feature selection, for classifying the six basic emotions (anger, joy, surprise, disgust, fear, and sadness). We classified the emotion of fear in two ways: first, a binary classification called the 2-level paradigm (0—no fear and 1—fear) and secondly, the 4-level paradigm (0—no fear, 1—low fear, 2—medium fear, 3—high fear).
2. We introduce the stages of development and evaluation of a virtual environment for treating acrophobia that relies on gradual exposure to stimuli, accompanied by physiological signals monitoring in a pilot experiment which involved the participation of 4 acrophobic subjects. Then, we present the design and development of a VR environment for acrophobia therapy in a naturalistic scenario—a mountain landscape (Fig. 1).
3. We introduced a novel approach toward using an intelligent virtual therapist that recognizes human emotions based on biophysical signals, provides encouragement, gives advice, changes his voice parameters, and adapts the scenario according to the subject's affective state.
4. We design a novel method for reducing in-game artifacts which consists in recognizing artifact patterns in the signals recorded during gameplay sessions, by

Fig. 1 User playing and the VR game—view from the cable car

aligning the biophysical data segments corresponding to the moments when the users performed head/hand/body movements with the artifacts recorded during a reference procedure.

5. Lastly, we introduce an approach for estimating respiration rate which consists in placing two HTC Vive trackers on the chest and on the back of the subjects and measure the distance between them. This distance varies during breathing—increases while inhaling and decreases during exhaling.

4 Emotion Models

Various emotion models have been issued throughout the years. The *discrete model*, proposed by Paul Ekman, consists of six basic emotions: sadness, happiness, disgust, anger, fear, and surprise (Ekman et al. 1969). The most well-known model for emotion classification is the *bipolar model* (Russell 1979). It considers two orthogonal dimensions, *arousal and valence*. Arousal ranges from "not excited" to "excited", while valence extends from "negative" to "positive". A third dimension, *dominance*, indicates how much the subject is in control over his emotions. Each emotion can be described as a combination of these three dimensions. For instance, fear is characterized by low valence, high arousal, and low dominance (Demaree et al. 2005). The *approach-withdrawal model* takes into account the motivating factor of emotions, reflecting the tendency to reach or reject a certain stimulus or situation (Davidson et al. 1990).

5 Biophysical Data

Biophysical data analysis is a more objective method of interpreting and assessing human emotions, compared to questionnaires or subjective ratings (Toth 2015). However, if used together, a wider perspective on the modality in which people decode the affective states can be obtained. According to Steimer (2002), fear causes a defensive behavior. The human body responds differently to fear, in an either active (high heart rate, increased sweat production, cortical activation) or passive modality (low pulse and respiration rate) (Kometer et al. 2010).

5.1 Galvanic Skin Response

Galvanic Skin Response (GSR) or *Electrodermal Activity* refers to a change in sweat glands activity or skin conductance, measured by electrodes applied on the skin. GSR has two components—a tonic (Skin Conductance Level—SCL) and a phasic one (Skin Conductance Response—SCR, a measure of arousal to stimuli, reflected in changes in the sympathetic nervous system's level of activation). Fear is mapped by an increase in the production of sweat and, consequently, in skin conductance (DiMeglio 2015). GSR has been intensively used in psychophysiological experiments, with high rates of success—in Healey (2009) and Fleureau et al. (2012) it has been the main classification factor for emotions, while in Westerink et al. (2009), the changes in GSR have been in line with the changes in arousal and also a comfortable type of measurement for the users, reliable in discriminating fear from other negative affective states (AlZoubi et al. 2012). The typical baseline values are around 0.03 and 0.05 microSiemens, while threatening stimuli produce a raise to around 2–3 microSiemens or extreme values of 8 microSiemens (Braithwaite et al. 2015). The subjects who watched a scary 2D video measured 8.05 microSiemens (1.57% ± 12.10 increase from baseline) while those who viewed a horror virtual reality video recorded on average 11.814 microSiemens (4.26% ± 6.31 increase from baseline) (Kometer et al. 2010).

5.2 *Blood Volume Pulse*

Blood volume pulse reflects the changes in the volume of the blood vessels and is recorded by a photoplethysmography (PPG) device, a noninvasive optical sensor that determines changes in the light absorption density of the skin (Agrafioti et al. 2012). PPG has been used in various experiments as a reliable estimator of emotional changes (Eum et al. 2011; Gouizi et al. 2011; Walter et al. 2013), being usually attached to the ear lobe or to the finger. Its values are converted into heart rate, measured in beats per minutes (bpm). High values of heart rate, over 90–100 bpm, indicate fear and anxiety (Wen et al. 2014; Rainville et al. 2006). The average heart rate was 80 bpm when the subjects watched a scary 2D video with a 6.97% ± 12.74 increase from baseline and 77.8 bpm with a 3.49% ±12.09 increase from baseline for those who watched a horror VR video (Kometer et al. 2010).

5.3 *Electroencephalography*

Electroencephalography (EEG) is a technique of recording and interpreting the electrical activity of the brain using electrodes placed on the scalp.

In the brain, fear is perceived first by the amygdala and then goes through the hypothalamus and midbrain (Quirk 2013). The right lobe mediates withdrawal, while the left side of the brain is involved in appetitive emotions and approach (Mauss and Robinson 2009). Phillips et al. (2003) pointed to a 2-way circuit for emotion regulation: a ventral one (including the amygdala, responsible for the identification of stimuli emotional significance) and a dorsal one (including the hippocampus, responsible for the regulation of affective states and behavior). In Petrantonakis and Hadjileontiadis (2009) and Chanel et al. (2011), it has been found that EEG was more reliable in fear classification than other biophysical features. EEG is commonly very susceptible to outside noise, especially body movements and artifacts introduced by the recording devices, but advanced filtering methods have emerged in order to remove them and obtain clearer signals. In a recent work (Cudlenco et al. 2020), it has been shown that EEG could also be used to predict the semantics of the visual input perceived by the human subject, even though the prediction is highly accurate only when combined with deep visual features directly extracted from the image.

The alpha waves (8–12 Hz) are neural oscillations that originate from the occipital lobe, being a reflection of the relaxation state of the individual, with high amplitudes when he has his eyes closed. Moreover, it has been demonstrated that the alpha waves are a marker of functional inhibition of the brain areas, involved in attentional processes (low alpha activity in the regions that are processing information and high alpha activity in the regions that are not involved in the current task) and anticipation of upcoming stimuli (Horschig et al. 2014).

When the subject performs mental processes, a phenomenon called *alpha blocking* occurs, which is reflected in a decrease of alpha amplitudes (Scott 1976). The cognitive states and the level of alpha waves are inversely related. The alpha waves have their origin in the occipital cortex and advance to the frontal lobes, the most evolved area of the brain, responsible for emotion, consciousness, and behavior. Usually there is a balance of alpha activation between the two hemispheres, but this balance impairs when emotional stimuli are provided. According to the approach/withdrawal model of *frontal alpha asymmetry* (Davidson 1993), left frontal activation corresponds to a positive approach to stimuli, while right frontal brain activation indicates negative affective responses (Bos 2006; Trainor and Schmidt 2003; Jones and Fox 1992; Canli et al. 1998). Both left and right activation correspond to low alpha levels.

The beta waves (13–30 Hz) are neural oscillations indicating wakefulness and consciousness, with average amplitudes around 20–200 μV. High levels of the beta waves indicate anxiety, alert, and fear (Arikan et al. 2006). In Kometer et al. (2010), for the beta band, horror virtual reality gameplay led to an increase of 33 μV from baseline.

The ratio of slow waves to fast waves (SW/FW) has a negative correlation with fear (Schutter and Van Honk 2005; Putman et al. 2010). There was a statistically significant reduction in the SW/FW ratio (delta/beta and theta/beta) in the left frontal lobe in an experiment where the EEG data has been recorded from a single electrode (Cheemalapati et al. 2016).

As suggested by Brouwer et al. (2015), body movements, mental states, subtle movements of sensors and wires are confounding factors that can affect the estimation of cognitive or affective states from neurophysiological signals. It is advisable to correctly detect and remove artifacts from the classification analysis and perform experiments where little movement of the body or recording devices is involved.

5.4 Biophysical Data and Virtual Reality

A Magnetic Resonance Imaging (MRI) experiment showed an increased level of emotional responses in the amygdala when VR stimuli have been presented to the subjects, compared to 2D videos (Dores et al. 2014). Coelho et al. (2008) found that movement in an acrophobia-simulated virtual environment conducted to a more realistic behavior of the subjects, similar to what has been observed in in-vivo exposure. In Costa et al. (2014), EEG and GSR data have been collected in real time while the users played an acrophobia-oriented game with a CAVE device. A VR system for treating stress-related disorders has been developed in Brouwer et al. (2011). Stress was induced by depicting a scenario simulating a bomb explosion, while associative stress was measured by immersing again the user in the scene after a period of time. Associative stress has been related to EEG mid-frontal alpha asymmetry and to an increase in heart rate variability.

6 Machine Learning Techniques for Emotion Classification

Emotion classification has been performed using various machine learning and feature selection algorithms in psychophysical experiments. In Koelstra et al. (2012), Fisher's linear discriminant was used for feature selection and the Naïve Bayes classifier for discriminating into low/high valence, arousal, and liking, with accuracies of 62%, 56%, and 55%. Atkinson and Campos (2016) used the minimum-Redundancy Maximum-Relevance (mRMR) method for feature selection and Support Vector Machines (SVM) for binary classification into low/high valence and arousal, with an accuracy of 73% for both. The study has been performed by extracting and processing the EEG features from the DEAP database. Yoon and Chung (2013) used the Pearson correlation coefficient (PCC) for feature extraction and a probabilistic classifier based on the Bayes theorem for resolving the binary classification problem of low/high valence and arousal discrimination, with an accuracy of 70% for both. Similarly, emotion recognition has been done based on the EEG data from the DEAP dataset. A similar approach is presented in Naser and Saha (2013), where the SVM algorithm conducted to an accuracy of 66%, 64%, and 70% for classifying valence, arousal, and liking into the low and high groups. By applying the SVM technique on the EEG features, a classification accuracy of 62 and 69% has been achieved during a music-induced affective states evaluation experiment where the users were required to rate their currently perceived emotion in terms of valence and arousal (Daly et al. 2015). Liu and Sourina (2013) conducted two experiments in which visual and audio stimuli have been used to evoke emotions. The SVM classifier, having as input Fractal Dimension Features, statistical and Higher Order Crossings extracted from the EEG signals provided the best accuracy of 53% for recognizing eight emotions—happy, surprised, satisfied, protected, angry, frightened, unconcerned, and sad. A comparative study of four machine learning methods showed that SVM offered the best accuracy—85%, followed by Regression Tree—83% for the classification of five types of emotions—anxiety, boredom, engagement, frustration, and anger into three categories—low, medium, and high (Liu et al. 2005). Soleymani et al. (2009) obtained an accuracy of 63% for differentiating three classes of emotions—calm, positive excited, and negative excited using a Bayesian classification method. A more complex SVM-based algorithm did not show improvements, compared to the Bayesian technique. In the case of binary classification into low/high valence, arousal, and liking, using EEG signals, the accuracy rates were 55%, 58%, and 49% with SVM. Having as input features the peripheral physiological responses, the classification accuracies recorded 58%, 54%, and 57% (Koelstra et al. 2010). Based on the MAHNOB dataset and using the SVM algorithm, Wiem and Lachiri (2017) reached a classification accuracy of 68% for valence and 64% for arousal when discriminating into the low/high groups and 56%, respectively, 54% for classifying into three groups. The most relevant features were the electrocardiogram and the respiration volume. In Alhagry et al. (2017), a deep learning method based on the Long-Short Term Memory (LSTM) networks was used for classifying low/high valence, arousal, and liking based on the EEG raw data from the DEAP dataset (Koelstra et al. 2012), with accuracies of around 85%. Jirayucharoensak et al. (2014) trained a deep neural network implemented with a stacked autoencoder based on the hierarchical feature learning approach. The input features were the power spectral densities of the EEG signals from the DEAP database, which were selected using Principal Component Analysis (PCA). The subjective ratings from 1 to 9 have been divided into three levels and mapped into "negative", "neutral", and "positive" for valence and into "passive", "neutral", and "active" for arousal. They were finally classified with an accuracy of 49% for valence and 46% for arousal.

7 Our Machine Learning Approach to Classifying the Six Basic Emotions

The evaluation of the users' emotional states is fundamental in VRET systems. In order to address the issue of emotion classification based on biophysical signals in terms of maximum accuracy and feature selection efficiency, we performed a comparison of several machine learning and deep learning techniques applied on the data from the DEAP database (Koelstra et al. 2012). The renowned DEAP database contains physiological recordings (GSR, PPG, skin temperature, breathing rate, electromyogram, data from 32 EEG channels) and subjective ratings from 32 subjects who watched 40 short videos eliciting various emotions. The participants were required to rate each video in terms of valence, arousal, and dominance on a scale from 1 to 9. By *combining the discrete model of emotions and the three-dimensional continuous model* (Ekman et al. 1969), we classified each of the six basic emotions into two groups—positive (the existence of emotion) and negative (lack of emotion), using four classical machine learning techniques (Support Vector Machine—SVM, Linear Discriminant Analysis—LDA, Random Forest—RF, and k-Nearest Neighbors—kNN) and four deep neural networks with different numbers of layers and neurons per layer, with and without feature selection. The feature selection algorithms were: Fisher score, Principal Component Analysis (PCA), and Sequential Forward Selection (SFS). Classification has been done based on the physiological data and subjective ratings of valence, arousal, and dominance from the DEAP database. From the EEG data, we extracted the Petrosian Fractal Dimension, Higuchi Fractal Dimension, and Approximate Entropy. The machine learning and deep learning algorithms have been trained and cross-validated, having as input the biophysical data and as output, two possible conditions: *0—negative or lack of emotion* and *1—positive or emotion*. The deep neural networks have been cross-validated using the *k*-fold and leave-one-subject-out methods, while for the machine learning techniques, the data have been divided into 70% training data and 30% test data. In the case of leave-one-subject-out, each classifier has been trained on the data of 31 subjects and tested on the data of the 32th user. Figure 2 presents the distribution of each of the six basic emotions—sadness, happiness, disgust, anger, fear, and surprise across the valence-arousal-dominance axis in the 3-dimensional continuous model of valence-arousal-dominance, as proposed by Russell and Mehrabian (1977).

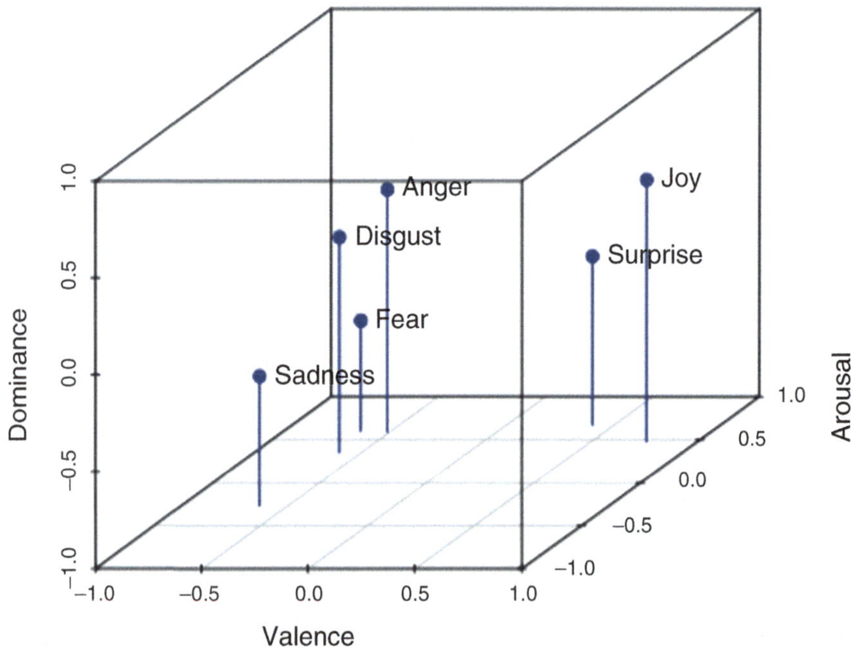

Fig. 2 The six basic emotions in the valence-arousal-dominance model

As the valence, arousal, and dominance dimensions have been rated on a scale from 1 to 9, we considered the following correspondences for each of the six basic emotions in terms of condition 1 (positive or the existence of emotion):

(a) Anger—low valence ([1; 5)), high arousal ([5; 9]), dominance in the interval [6;7],
(b) Joy—high valence ([5; 9]), high arousal ([5; 9]), dominance in the interval [6;7],
(c) Surprise—high valence ([5; 9]), high arousal ([5; 9]), dominance in the interval [4; 5],
(d) Disgust—low valence ([1; 5)), high arousal ([5; 9]), dominance in the interval [5; 6],
(e) Fear—low valence ([1; 5)), high arousal ([5; 9]), dominance in the interval [3; 4],
(f) Sadness—low valence ([1; 5)), low arousal ([1; 5)), dominance in the interval [3; 4].

The classification results showed that the highest F1 cross-validation scores were:

(a) Anger—Petrosian and Higuchi Fractal Dimension extracted from the EEG signals and peripheral biophysical data, using SVM (98.02%),
(b) Joy—kNN using Petrosian EEG values and peripheral data (87.9%),
(c) Surprise—kNN with raw EEG values and peripheral data (85.01%),
(d) Disgust—kNN with Petrosian EEG values and peripheral data (95%),
(e) Fear—kNN with raw EEG values and peripheral signals (90.75%),
(f) Sadness—SVM with Higuchi Fractal Dimensions extracted from the EEG signals and peripheral data (90.8%).

The *k*fold method provided higher F1 scores than Leave-One-Subject-Out. Without feature selection, kNN has provided the highest F1 scores in 13 cases, followed by Random Forest (seven times) and SVM (four times). With feature selection, kNN has provided the highest F1 scores in 12 of the tested cases, Random Forest in seven cases, SVM in five cases, and LDA only once. SFS has been selected two times and Fisher score 14 times. The most important classification features were:

(a) Anger—trapezius electromyography (EMG) and respiration rate,
(b) Joy—GSR and zygomaticus EMG,
(c) Surprise—GSR and FC1,
(d) Disgust—vertical and horizontal electrooculography (EOG),
(e) Fear—vertical and horizontal EOG, zygomaticus EMG, and activation of the frontal cortex (FC1, F4),
(f) Sadness—the left prefrontal cortex (FC1 and FP1).

The results obtained for classifying into two classes (the existence of emotion—positive condition and lack of emotion—negative condition) are higher than those obtained in the literature for classifying into low/high valence and arousal: 62%/56% (Koelstra et al. 2012), 73% (Atkinson and Campos 2016), 70% (Yoon and Chung 2013), 85% using the Long-Short Term Memory algorithm (all using the data from the DEAP database), 66%/64% (Daly et al. 2015), 62%/69% (Liu and Sourina 2013), 55%/58% (Soleymani et al. 2009), 68%/54% using the data from the MAHNOB database (Wiem and Lachiri 2017). A thorough description of the experiment, methods, and results, including a comparison between the cross-validation F1 scores achieved using the *k*fold and leave-one-subject-out techniques is provided in Bălan et al. (2020a).

8 Our Machine Learning Approach to Fear Level Classification

Using the same machine learning, deep learning, and feature selection algorithms applied on the biophysical recordings and subjective ratings from the DEAP database, we classified the emotion of *fear* in two ways: first, a binary classification called the *2-level paradigm (0—no fear and 1—fear)* and secondly, the *4-level paradigm (0—no fear, 1—low fear, 2—medium fear, 3—high fear)*. Considering the emotion dimensions from the 3-dimensional continuous model of emotions, fear was characterized by *low valence, high arousal, and low dominance*. The recordings have been assigned to either the 0—no fear or 1—fear group (in the case of the 2-level fear evaluation paradigm) or to the 0, 1, 2, or 3 classes (for the 4-level paradigm), considering the subjective ratings of valence, arousal, and dominance from the DEAP dataset. We applied the unsupervised K-means clustering algorithm on the data from DEAP and achieved a prediction accuracy of 87% for the 2-level evaluation modality. This means that 87% of the ratings proposed for the *fear* or *no fear* classes by the theory of low valence/high arousal/low dominance have been classified in the same cluster by the k-means technique. We used for training and cross-validation not only the raw EEG values, but also the peripheral signals. The EEG recordings have been decomposed into Power Spectral Densities of the alpha, beta, and theta frequencies, Petrosian Fractal Dimensions, Higuchi Fractal Dimension, and Approximate Entropy. *The highest F scores have been obtained by using the Random Forest Classifier—89.96%, having as input EEG Higuchi Fractal dimensions and peripheral data for the 2-level fear evaluation modality and 85.33% for the 4-level fear evaluation modality, both without feature selection. The most important classification features were the raw, alpha, and beta values in the left frontal hemisphere, GSR, and respiration rate*. We computed the difference in spectral power between the right and the left frontal hemispheres, for the alpha and theta frequency bands. We noticed that this difference increases with fear, being higher for the *medium fear* and *high fear* condition than for the *no fear* and *low fear* conditions. There was a higher level of alpha and theta activation in the left side of the brain. Moreover, the intensity of the left central and right frontal beta waves was directly associated with fear onset. In addition, a positive correlation between fear and the ratio of slow to fast waves has been observed, for both the 2-level and 4-level evaluation modalities. The purpose of this research approach was not just to classify the data into low/high valence/arousal/dominance, but also to combine these emotion dimensions and define a complex emotion such as fear. The results we obtained were similar to Alhagry et al. (2017), who achieved a classification accuracy of 85% by training and testing a Long-Short Term Memory network using raw EEG values. In the paper published in the Sensors journal in April 2019, we provide a full presentation of the research, results, and comparison with similar studies (Bălan et al. 1738).

9 Fear Level Classification in a VRET System for Acrophobia Therapy

In an experiment performed during June–August 2018, we trained and tested two classifiers: C1, which determines the patient's *current level of fear* and C2, which estimates *the next scenario of exposure* in a VR-based game for treating acrophobia. For this, we have collected biophysical data (EEG in the alpha, beta, and theta frequency ranges, GSR, and HR) from four subjects suffering from acrophobia, aged 22–50, in both in-vivo and virtual conditions. The subjects have been exposed to heights at the first, fourth, and sixth floors of a building, at 4 m, 2 m, and a few centimeters away from the balcony's railing. Besides, they played a VR-based game where they had to collect bronze, silver, and gold coins at the ground level and on terraces at the first, fourth, and sixth floor, as well as on the building's rooftop (Fig. 3).

During each trial of a session, the participants had to rate their perceived level of fear on a scale from 0 to 10 (the 11-choice scale), where 0 represents no fear at all and 10 stands for a high level of anxiety. The ratings on the 11-choice scale have been divided into the 2-choice and 4-choice scales. In the 2-choice scale, 0 means *relaxation* and 1 means *fear*. In the 4-choice scale, 0 stands for *relaxation*, 1—*low fear*, 2—*medium fear*, 3—*high fear* (Table 1).

Fig. 3 Our computer game for automatic acrophobia treatment

Table 1 Fear level classification scales

11-Choice-scale	4-Choice-scale	2-Choice-scale
0	0 (relaxation)	0 (relaxation)
1	1 (low fear)	
2		
3		
4	2 (medium fear)	1 (fear)
5		
6		
7		
8	3 (high fear)	
9		
10		

The data recorded during this preliminary experiment have been used for training classifier C1, on the 2-choice, 4-choice, and 11-choice scales. Classifier C1 received as input the EEG, GSR, and HR data and provided as output the perceived fear level. We have performed a comparative study of various classic machine learning and modern deep learning techniques as classification models: k Nearest Neighbors, Linear Discriminant Analysis, Random Forest, Support Vector Machine (with and without feature selection using the Sequential Forward Selection algorithm), and four types of deep neural networks—DNN_Model_1: 3 hidden layers, with 150 neurons on each hidden layer, DNN_Model_2: 3 hidden layers, with 300 neurons on each hidden layer, DNN_Model_3: 6 hidden layers, with 150 neurons on each hidden layer, DNN_Model_4: 6 hidden layers, with 300 neurons on each hidden layer. Our purpose was to automatically adapt the exposure scenarios according to the user's level of fear. For instance, if the patient is anxious and feeling as losing control of his emotional reactions, the level of exposure should be lowered. On the contrary, if he is in a relaxed state, the level of exposure should be increased. Classifier C2 automatically determines the next level of exposure, by taking into account the physiological data (EEG, GSR, and HR) and a parameter called target fear level (FL_t), computed using the current level of fear (FL_{cr}). FL_{cr} is determined using classifier C1. We have used the following formulas for the 2-choice and 4-choice scales:

2-Choice scale	4-Choice scale
If FL_{cr} = = 0 then FL_t = 1 If FL_{cr} = =1 then FL_t = 0	If FL_{cr} = = 0 or FL_{cr} = = 1 then $FL_t = FL_{cr} + 1$ If FL_{cr} = = 2 then $FL_t = FL_{cr}$ If FL_{cr}= = 3 then $FL_t = FL_{cr} - 1$

For classifier C2, we have used the same machine learning and deep learning algorithms as for C1 in our comparative study. For testing the accuracy of both classifiers, the acrophobic subjects have been required to play the VR game two times. Each session had a number of 10 trials. The game started at the ground floor where they had to collect bronze, silver, and gold coins, rate their perceived level of fear for ground truth acquisition and then, based on the FL_{cr} parameter estimated by C1 in real time and the computed Fl_t, classifier C2 determined the next level of the game where the players should be taken to. Classifier C1 has been cross-validated on the training dataset using the *k*fold method ($k = 10$) and tested on the test dataset obtained in the second experiment. On the other hand, classifier C2 has been only cross-validated on the training dataset acquired in the second experiment. As for now, we did not define a method for evaluating the test accuracy of C2. In the future, we will perform an experiment with a larger number of people and evaluate the therapeutic procedure. Thus, the users will play the VR game several times, across a certain number of days and then they will be exposed in real-world conditions to see whether their fear of heights has diminished. We consider that only by a final in-vivo exposure we can assess the efficiency of the VR therapy. We have used a *user-dependent* and a *user-independent* modality for assessing the classifiers' accuracy. In the case of the user-dependent modality, each classifier has been trained and tested on the same data—for each subject, on his own recordings. As for the user-independent modality, for each subject, each classifier has been trained on the data of the other three participants and tested on the recordings of the current subject. The highest cross-validation and test accuracies are presented in Table 2.

The results showed a very high cross-validation accuracy on the training set and good test accuracies, ranging from *42.5 to 89.5%*. For the 2-choice scale, the highest accuracy has been obtained by DNN_Model_4 (79.12%) for the player-independent

Table 2 Highest cross-validation and test accuracies

Method	**C 1**				
	2-Choice scale		**4-Choice scale**		**11-Choice scale**
	Cross-validation	**Test**	**Cross-validation**	**Test**	**Cross-validation**
Player-independent	kNN 99.5% RF 99.25%	DNN_ Model_4 79.12%	kNN 99% RF 99%	kNN 52.75%	kNN 98.25% RF 99%
Player-dependent	kNN 99.5% RF 99.75%	SVM 89.5%	kNN 99% RF 99.25%	SVM 42.5%	kNN 98.25%RF 99%
	C 2				
	2-Choice scale		**4-Choice scale**		**11-Choice scale**
	Cross-validation	**Test**	**Cross-validation**	**Test**	**Cross-validation**
Player-independent	RF 99.75%	–	RF 100%	–	RF 100%
Player-dependent	RF 99.75%	–	RF 99.75%	–	RF 100%

modality and SVM (89.5%) for the player-dependent modality. For the 4-choice scale, the highest accuracies were obtained using kNN (52.75%, player-independent modality) and SVM (42.5%, player-dependent modality).

The Radom Forest classifier adds the benefit of computing *feature importance*—how important is that feature for reducing impurity across the decision trees. For classifier C1, the most important features were *GSR, HR, and the EEG values in the beta frequency range, closely followed by the alpha and theta power spectral densities*. These findings are comparable to the results from other experiments (Arikan et al. 2006; Kometer et al. 2010). For C2, the most significant feature resulted to be $\mathbf{FL_t}$. It had a high importance index and also has been selected on all three fear estimation scales (2-choice, 4-choice, and 11-choice), for both the user-dependent and user-independent modalities.

Our results are comparable to those obtained by Liu et al. (2009), who reached a classification accuracy of 78% in a game where dynamic difficulty adjustment depended on simple "if" clauses and not on an automatic computation. Chanel et al. (2011) obtained an accuracy of 63% for classifying three classes of emotions in a study where 20 subjects played a Tetris game on three difficulty levels. In Hu et al. (2018), a convolutional deep neural network was used to classify fear ratings on a scale from 1 to 4. The EEG data of 60 subjects have been recorded while playing the Ritchie's Plank Experience VR game, with a classification accuracy of 88.77%. The system described in Šalkevicius et al. (2019) was used in the therapy of fear of public speaking. The GSR, blood volume pulse, and skin temperature of 30 subjects have been recorded and the current level of anxiety has been classified into four classes: low, mild, moderate, and high, using the SVM algorithm. The fusion of all three types of biophysical signals provided a classification accuracy of 86.3%.

10 Acrophobia Game in Naturalistic Landscape

During 2019, we refined the VRET system for acrophobia therapy, considering Jerald's statement: *We must create VR experiences with both emotion and logic*. For this, we adopted the Human-Centered Machine Learning approach that takes into account human interests in designing Machine Learning algorithms, making Machine Learning more useful and usable. According to this theory, humans and machines not only cooperate, but also adapt to each other—humans are able to alter the behavior of the machines and the machines modify human goals (Jerald 2016).

The VR game is rendered via the HTC Vive Head Mounted Display and contains a mountain environment with three scenes: a walk by foot—incorporating a path, a transparent platform across a canyon and a bridge, a ride by cable car (Fig. 4), and one by ski lift.

There are ten stops throughout each ride, where the user is asked to rate his fear level, valence, arousal, and dominance using the Self-Assessment Manikin, on a scale from 1 to 5 (Fig. 5).

Fig. 4 Our acrophobia computer game: view from the cable car

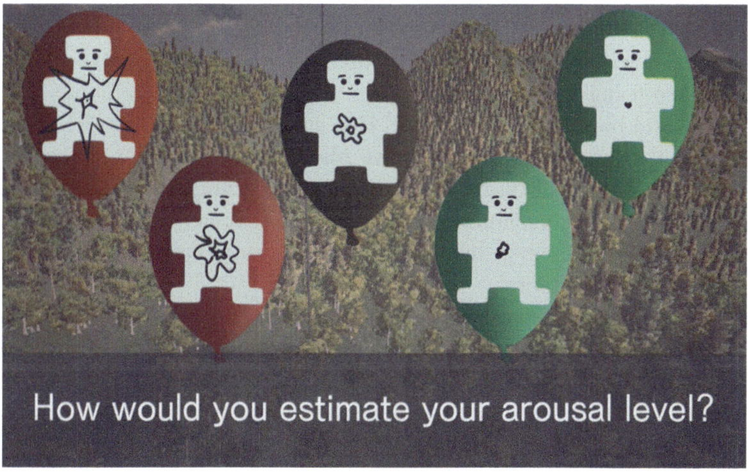

Fig. 5 Our acrophobia computer game: arousal rating

Also, the player has to give the answer to some short mathematical exercises. Logical thinking decouples cortical activation in the right brain hemisphere which is responsible with emotional processing. In this way, the subject begins to feel more detached from the anxiety-provoking experience, relaxes, and gains confidence. At this stage, in order to explore the environment and spend as much as time as possible immersed, the player is required to collect some small objects (stars, diamonds, and coins) that appear randomly and disappear as he fixes his gaze toward them. At any time, he can make use of some assistive elements—he can pause the game and listen to his favorite song, read an inspirational quote, and look at a nice picture. These elements are configured personally for each user apart and saved in his database profile.

We will perform a series of experiments to evaluate the efficacy of the VR environment in treating acrophobia. The participants will need to fill in the Heights Interpretation Questionnaire (Steinman and Teachman 2011), Visual Height Intolerance Severity Scale (Huppert et al. 2017), and Acrophobia Questionnaire (Cohen 1977). Then, they will be divided into three groups: Low, Medium, and High acrophobia. Also, they will initially pass through a mathematical test to evaluate their skills and divide them into Novice, Medium, and Expert. Based on their skill, in the game they will receive different numbers of simple, medium, and complicated exercises: Novice—3 simple exercises, Medium—1 simple and 2 medium, 1 medium and 2 complicated. The human-centered approach is ensured by having a virtual environment with a high level of realism in a real-world context (mountain site), with a scenario that is receptive to the player's needs—provides means of relaxation and exploration tasks, has a reward system, combines emotions with logical activity, applies the constructivist learning theory stating that knowledge and skills acquisition are gained by linking a new experience to a previous one, and the possibility of transferring the cognitive and emotional acquisition from the virtual to the real world (Bălan et al. 2019).

As the EEG recording device is rather cumbersome and difficult to be applied on the head when using the HTC Vive glasses at the same time, we chose to record solely peripheral biophysical data: GSR, HR, and Respiration Rate (RR). In order to determine the respiration rate, we have placed two HTC Vive trackers on the chest and on the back of the users and then, during breathing, measured the distance between these two trackers, normalized the values between $[-1; 1]$, applied several filters for smoothing the signal, and counted the number of peaks in the signal which represented the respiration rate.

We will perform a baseline recording when the user stands still in a relaxed position, for a time period of 3 min. As *artifacts identification* is an important step in obtaining clean physiological data, we have developed a method for artifacts reduction, which consists in recognizing artifact patterns in the signals recorded during gameplay sessions, by aligning the biophysical data segments corresponding to the moments when the users performed head/hand/body movements with the artifact signals recorded during a reference procedure. Physiological responses that are not correlated with the content of the game and the emotional responses generated in it make signal analysis very difficult. We define an artifact as *any misleading or confusing alteration in physiological data that appears as a result of external action such as head, hand, or body movements, being unrelated to the emotional effects that specific stimuli or the object under observation exert upon the user* (Balan et al. 2019). For validating this procedure, we have performed an experiment with five healthy subjects, aged 24–50. At first, we recorded a set of reference artifact measurements for each user, in order to acquire the physiological pattern (GSR, HR, and RR) for each artifact: *deep breath, head movement to the left, head movement to the right, head movement up, head movement down, click with the right hand on the HTC Vive controller, and right hand raise*. These are the common artifacts that can

Fig. 6 Acrophobia game: indication to move the head down during the VR game

occur during gameplay in VR. In the second phase of the experiment, each user played the VR game for acrophobia therapy and took the ride by cable car. During each of the ten stops, they were required to perform one of the tasks mentioned above (Fig. 6).

During the analysis step, we have aligned the reference artifacts to the biophysical data recorded during gameplay. Also, we have mapped the reference artifacts onto the gameplay data segments that start before and after the recorded timestamps, with one or two steps before and after and computed the Mean Absolute Error (MAE) and Mean Absolute Percentage Error (MAPE) for both GSR and HR. The results showed that the bias is lower on the perfectly aligned biophysical segments than on those located one or two steps before and after, for all seven types of tested artifacts. However, the results were not statistically significant in a Student t-test for independent means. Deep breath was the most relevant artifact introduced in the analysis, more prominent than the rest of the artifacts. During the VR experiment, different artifacts of breathing, head, hand, or body movements can be encountered. If the head movements do not produce significant artifacts and raising the right hand is not a frequent event during the VR game, breathing is a physiological artifact that must be taken into account to avoid overestimating the skin conductance responses during the experiments.

11 Intelligent Virtual Therapist for Acrophobia

We propose a novel and effective approach in which we replace the human therapist with a virtual one, called RoboTher. RoboTher has the appearance of a female avatar and a feminine voice as well. We choose to use a female voice because it is usually perceived as helping, not commanding (Borkowska and Pawlowski 2011), we are more familiar with a female voice because it is usually associated with the maternal presence (Lee and Kisilevsky 2014), the female voice is more pleasant, being processed in the same auditory area as music (Sokhi et al. 2005), and is perceived as offering more confidence than the male voice due to its higher pitch (Re et al. 2012).

RoboTher automatically identifies the users' emotional states (degree of relaxation or anxiety), compared to a physiological (GSR and HR) baseline recording, 3 min long, performed under resting conditions. RoboTher provides encouragement and is able to change its voice parameters—pitch, tempo, and volume according to the users' emotional states. It provides means of relaxation in the game, by inviting the player to take a break and listen to his favorite song, read his favorite quote, or look at a photo he likes. If these relaxation modalities are ineffective and the stress level is still high, the virtual therapist lowers the degree of exposure. Five subjects played the VR game (ride by cable car) under two conditions: control (one session, without RoboTher's assistance) and three sessions with assistance from RoboTher. We computed the difference, in percent, between the HR and GSR baseline values and the current ones during each game trial. A form of biofeedback has been provided to the subjects as bars of changeable colors (green, yellow, orange, red) for both GSR and HR that appeared in the left top corner of the visual field during the game sessions. If the percent was lower than 10%, the color was green, for a percent in the interval [10%; 40%], the color was yellow, [40%; 70%]—orange, higher than 70%—red (Fig. 7).

The Robot Interactions (RIs), which refer to the statements made by RoboTher and the changes in the voice parameters are presented in Table 3.

Fig. 7 Biofeedback automatically provided in the VR game

Table 3 Intelligent robot interactions based on HR and GSR signals

Robot interaction	Phrase	Voice parameters Pitch (%)	Tempo (%)	Volume (%)
RI1	"Good job! Keep going!"	+10	+10	+10
RI2	"Enjoy and relax for a while"	0	0	0
RI3	"Calm down and relax"	−10	−10	−10
RI4	"You are too tense. Take a deep breath and try to relax more"	−20	−20	−20

Table 4 RoboTher responses based on different Gameplay situations

	Condition		Procedure	
Situation no.	GSR color	HR color	Robot interaction	Change game level (only after the relaxation modalities are provided)
Situation1	Green	Green	RI1	No
Situation2	Green	Yellow	RI1	No
Situation3	Yellow	Green	RI1	No
Situation4	Yellow	Yellow	RI1	No
Situation5	Green	Orange	RI2	−1 level
Situation6	Orange	Green	RI2	−1 level
Situation7	Yellow	Orange	RI2	−1 level
Situation8	Orange	Yellow	RI2	−1 level
Situation9	Orange	Orange	RI3	−2 levels
Situation10	Orange	Red	RI3	−2 levels
Situation11	Red	Orange	RI3	−2 levels
Situation12	Green	Red	RI3	−2 levels
Situation13	Red	Green	RI3	−2 levels
Situation14	Yellow	Red	RI3	−2 levels
Situation15	Red	Yellow	RI3	−2 levels
Situation16	Red	Red	RI3	−3 levels

We identified 16 situations, corresponding to all possible combinations of HR and GSR colors. The Robot Interactions and the alteration in game levels (the automatic adaptation of scenario exposure) during gameplay are presented in Table 4.

The purpose of the therapy is to maintain the player within the green and yellow areas for both HR and GSR parameters throughout the entire game session.

For instance, in Situation5 to Situation8, one of GSRcolor or HRcolor is Green/Yellow and the other is Orange, which means that the subject tends to become anxious. RoboTher plays RI2 and then presents randomly either the user's favorite image, song, or quotation for 20 s. After these 20 s, the subject's emotional state is evaluated again. If it falls into Situation1–Situation4, RI1 appears and he may continue the game from there. If it falls into Situations 5–8, the player is taken to the previous level, so the level of exposure decreases with 1. If it falls into Situations 9–15, the level of exposure decreases with 2 and for Situation 16, decreases with 3.

The results showed that the subjects succeeded to finish the game quicker in the last game session with assistance from RoboTher. In the last session of the experiment, for all users, the most frequent situation was Situation1 (44%), followed by Situation2 (21%), Situation3 (17%), Situation6 (5%), Situation13 (9%), and Situation15 (4%). Both skin conductance parameters and heart rate decreased at the end of the 3 days of gameplay (from 1.68 to 0.9 uS for GSR and from 77.34 to 75.17 bpm for HR), compared to the control condition where the subjects initially played the game without support from RoboTher. These results were statistically significant in a paired-samples *t*-test (Bălan et al. 2020b).

12 Limitations of the Current Research

One limitation of the current research was the small number of subjects. The relatively small training and testing data size imposes a limit on the usage of modern deep neural networks, which usually need a much larger training set (at least in the order of tens of thousands) in order to generalize well and avoid overfitting. This might explain why in our experiments the more classical machine learning approaches, such as Support Vector Machines and Random Forests, achieved the best accuracy. The combination of HMD and EEG device was cumbersome to be worn on the head, causing serious discomfort to the volunteering patients for which we tested the system. Therefore, in the future we will explore VR-based smart therapy systems without the use of EEG. We also plan to extend the research with a larger number of participants (over 15), which is expected to improve the effectiveness of the deep learning approach and validate the effects of the VR therapy in real-world settings to see whether the level of acrophobia indeed decreased. Also, we will provide an introductory session that would familiarize the subjects with the perception of VR environments, apply questionnaires related to immersion and quality of presentation of the VR environment, give more attention to other GSR and HR features, such as inter-beat variability. Biofeedback can be provided not only as color bars, but also as changing elements in the environment—clouds and darker sky when the user is experiencing stress or clear, sunny weather when she or he is relaxed.

13 Limits of Current Phobia Therapy Systems

In a survey in which 19 psychologists from Romania replied, only two mentioned that they use VR to treat phobias: one is using the C2Care application (C2Phobia n.d.), and the other is using 3D mobile applications from Google Play. In the AcTiVity system (Activity System n.d.), exposure adjustment is not determined by the physiological recorded data. The biophysical data are solely recorded for analysis, as in the case of VRMC (Virtual Reality Medical Center n.d.), where physiological monitoring with visual feedback is used for acknowledging the patients when they are feeling stressed and not. In the system we envision, the next level of game exposure will be selected either by the psychologist, the user, or adjusted automatically, by an artificial intelligence algorithm, according to the patient's biophysical data. The psychologists appreciated that the most urgent phobias to treat are aerophobia (fear of flying by plane), fear of public speaking, claustrophobia, and agoraphobia. On a scale from 1 to 5, the most useful is to integrate an analysis tools addressed to the therapist, allowing him to analyze the patient's performance and evolution (4.5/5), followed by relaxation techniques (4.4/5), awareness techniques (4.25/5), games and rewards (4.1/5).

14 Proposed Future System

We aim to develop a system for phobia therapy that relies on gradual exposure in the Virtual Reality (VR), accompanied by physiological signals monitoring (pulse, electrodermal activity, and respiration rate) and real-time visual bio-feedback. The system can be used in the presence of the therapist or at home, for the patients who suffer from a mild phobia condition. During the therapy, the scenes from the virtual environment can be changed by the psychologist, the user himself or automatically, by a virtual therapist who adapts the scenario exposure based on the biophysical data recorded. Here, the human therapist is replaced by a virtual one with the shape of a game avatar, who offers support and encouragement to the patient. It directly interacts with the user and changes its voice parameters—pitch, tempo, and volume and facial expressions—according to the patient's emotional state. It identifies the current fear level and provides three modalities of relaxation—by determining the user to look at a favorite picture, listen to an enjoyable song, or read an inspirational quote. If the relaxation modalities fail to be effective, the virtual therapist automatically lowers the level of exposure according to a set of rules. The set of rules are part of an artificial intelligence future model, most likely trained using reinforcement learning and unsupervised learning techniques (Sutton and Barto 1998; Kallenberg et al. 2016; Erhan et al. 2010; Croitoru et al. 2019; Leordeanu et al. 2016) combined with either classical machine learning or deep neural networks, using biophysical user data and emotional ratings of valence/arousal/dominance (the emotion dimensions). A control panel allows introducing new patients, managing existing ones, recording sessions, replaying them, and generating statistics. An important aspect is the patients' gaze direction. We will record where the users are looking during the therapy and correlate it with the emotional state. Thus, our approach is at the confluence of psychology, artificial intelligence, and computer vision.

14.1 Methods and Instruments of Investigation

As *methods and instruments of investigation*, we will use game design, Virtual Reality integration, biophysical sensors that record electrodermal activity, heart rate and respiration rate (Shimmers Multisensory n.d.), artificial intelligence—classic machine learning and deep learning techniques (depending on the amount of training data available) for training and testing two classifiers: one that estimates the user's current emotional state (fear level) and one that determines the next exposure scenario according to the estimated fear level. Due to the strong limitations in supervised training signal and ground truth information (which usually comes from doctors), our intelligent system will learn and improve by itself during sessions, based on different self-supervised and reinforcement learning strategies, which we will explore. The combination of multiple sensors and actual user interaction with the VR system will enable the effectiveness of automatic self-training of the intelligent phobia therapist. The virtual therapist having the appearance of a female avatar will provide encouragement or invite the user to relax. The avatar will say expressions like "Well done! Keep going!" or "Calm down and try to relax". The virtual therapist's voice parameters—pitch, volume, tempo, and facial expression will change according to the user's emotional state. It selects relaxation modalities to provide the user during the game—a favorite song, image, or quote and then, based on the artificial intelligence models, estimates fear level and determines the next exposure scenario—whether the user will increase or decrease the level of exposure. During the game, the user is offered bio-feedback. The differences (in percent) between the current biophysical values and the baseline ones (recorded during a 3-min resting state) are presented as bars colored in green, yellow, orange, and red. Thus, the user can visualize his emotional state in a comprehensive way and struggle to relax in order to change the bars' color to green or yellow throughout the game session.

14.2 Potential Risks

The potential risks come from the very strong interdisciplinary nature of the project, combining psychology, medicine, engineering, and computer science. A potential risk resides in the uncontrolled reactions of the patient during exposure—motion sickness, anxiety, and inadaptation to VR. They will be minimized by pilot-testing the designed scenes using non-phobic persons or people who suffer from a mild phobic condition. The patients will be allowed to exit the virtual environment at any time. During the tests, a research assistant will monitor the users and ensure that they are feeling comfortable and safe. We will interact with psychologists in order to identify the potential risks that will be minimized by designing, implementing, and testing as many solutions as possible for each of the system's components. We will make sure that the virtual environment contains efficient gamification elements that engage the user in the therapy. The rendering quality will be enhanced by employing the most recent commercial technology. Perturbations and noises caused by the signal recording devices, disconnections, hardware, and physical limitations will be overcome by using advanced devices, monitoring the recording procedure, filtering, and post-processing the data. Another risk could be the slow learning rate of the intelligent system. As modern advanced machine learning algorithms significantly improve as training data size and diversity increases, it is possible that a long time will pass until sufficient data are captured for optimum performance. However, as the current system, trained on limited data, already shows good performance, we could realistically expect that future versions will improve considerably, once we have access to more data and develop more advanced self-supervised and reinforcement learning strategies for training a more powerful automated therapist.

14.3 High Future Gains in Phobia Therapy

The ability to bring the power of current computer systems and combine sophisticated technologies (such as virtual reality and artificial intelligence) with modern medicine, in order to treat various phobias and improve brain function, could have a tremendous positive impact in improving human life. Our proposed smart VR phobia therapist, with excellent initial results, even in the case of limited data, is a solid proof that automated technology will eventually improve current medicine and psychology practice. Doctors will be able to design more efficient treatments in combination with such smart robotic assistants. Time is on our side, as more powerful AI and VR algorithms and systems are created every day. As our understanding of artificial intelligence and the human brain will improve, along with our multidisciplinary experience and knowledge, the high gain of combining high tech with human sciences is almost certain. The only real question remaining is when should we expect current research to grow into a mature and well-trusted technology. Nevertheless, as history showed us time and again, research ideas with great potential to benefit human life, become a reality sooner rather than later.

Acknowledgements The work has been funded by the Operational Programme Human Capital of the Ministry of European Funds through the Financial Agreement 51675/09.07.2019, SMIS code 125125, UEFISCDI project 1/2018, and UPB CRC Research Grant 2017. This work has also been funded in part through UEFISCDI, from EEA Grants 2014–2021, project number EEA-RO-NO-2018-0496.

References

2BIOPAC (n.d.). https://www.biopac.com/application/virtual-reality/

Activity System (n.d.). https://www.unitylab.de/

Agrafioti F, Hatzinakos D, Anderson AK (2012) ECG pattern analysis for emotion detection. IEEE Trans Affect Comput 3(1):102–115

Alhagry S, Aly A, Reda A (2017) Emotion recognition based on EEG using LSTM recurrent neural network. Int J Adv Comput Sci Appl 8:355–358
AlZoubi O, D'Mello SK, Calvo RA (2012) Detecting naturalistic expressions of non-basic affect using physiological signals. IEEE Trans Affect Comput 3(3):298–310
Arikan K, Boutros NN, Bozhuyuk E, Poyraz BC, Savrun BM, Bayar R et al (2006) EEG correlates of startle reflex with reactivity to eye opening in psychiatric disorders: preliminary results. Clin EEG Neurosci 37:230–234
Atkinson J, Campos D (2016) Improving BCI-based emotion recognition by combining EEG feature selection and kernel classifiers. Expert Syst Appl 47:35–41
Baker BL, Cohen DC, Saunders JT (1973) Self-directed desensitization for acrophobia. Behav Res Ther 11(1):79–89
Bălan O, Moise G, Moldoveanu A, Leordeanu M, Moldoveanu F (1738) Fear level classification based on emotional dimensions and machine learning techniques. Sensors 2019:19
Bălan O, Moise G, Moldoveanu A, Leordeanu M, Moldoveanu F (2019) Challenges for ML-based emotion recognition systems in medicine. A human-centered approach. In: CHI'19 extended abstracts, may 4–9, 2019, Glasgow, Scotland, UK. ACM. ISBN: 978-1-4503-5971-9/19/05
Balan O, Moldoveanu A, Petrescu L, Moise G, Cristea S, Petrescu C, Moldoveanu F, Leordeanu M (2019) Sensors system methodology for artefacts identification in virtual reality games. ISAECT, Rome
Bălan O, Moise G, Petrescu L, Moldoveanu A, Leordeanu M, Moldoveanu F (2020a) Emotion classification based on biophysical signals and machine learning techniques. Symmetry 12:21
Bălan O, Cristea Ş, Moise G, Petrescu L, Moldoveanu A, Moldoveanu F, Leordeanu M (2020b) RoboTher—an assistive robot for acrophobia therapy in virtual reality. Submitted to ICRA
Bandura A (1977) Self-efficacy—toward a unifying theory of behavioral change. Psychol Rev 84(2):191–215
Borkowska B, Pawlowski B (2011) Female voice frequency in the context of dominance and attractiveness perception. Anim Behav 82(1):55–59
Bos DO (2006) EEG-based emotion recognition. The Influence of Visual and Auditory Stimuli University
Braithwaite JJ, Jones R, Rowe M, Watson DG (2015) A guide for analysing electrodermal activity (EDA) and skin conductance responses (SCRs) for psychological experiments. University of Birmingham, UK: Selective Attention and Awareness Laboratory
Brouwer AM, Neerincx MA, Kallen V, van der Leer L, ten Brinke M (2011) EEG alpha asymmetry, heart rate variability and cortisol in response to virtual reality induced stress. J Cyber Ther Rehabil 4:27–40
Brouwer A-M, Zander TO, van Erp JBF, Korteling JE, Bronkhorst AW (2015) Using neurophysiological signals that reflect cognitive or affective state: six recommendations to avoid common pitfalls. Front Neurosci 9:136. https://doi.org/10.3389/fnins.2015.00136
Buna P, Gorskia F, Grajewskia D, Wichniareka R, Zawadzkia P (2017) Low–cost devices used in virtual reality exposure therapy. Proc Comput Sci 104:445–451
C2Phobia (n.d.). https://www.c2.care/en/c2phobia-treating-phobias-in-virtual-reality/
Canli T, Desmond JE, Zhao Z, Glover G, Gabrieli JDE (1998) Hemispheric asymmetry for emotional stimuli detected with fMRI. Neuroreport 9(14):3233–3239
Carmen Juan M, Perez D (2010) Using augmented and virtual reality for the development of acrophobic scenarios. Comparison of the levels of presence and anxiety. Comput Graph 34:756–766
Chanel G, Rebetez C, Betrancourt M, Pun T (2011) Emotion assessment from physiological signals for adaptation of game difficulty. IEEE Trans Syst Man Cybern A Syst Hum 41(6):1052–1063
Cheemalapati S, Adithya PC, Valle MD, Gubanov M, Pyayt A (2016) Real time fear detection using wearable single channel electroencephalogram. Sensor Netw Data Commun 5:140. https://doi.org/10.4172/2090-4886.1000140
Choi YH, Jang DP, Ku JH, Shin MB, Kim SI (2001) Short-term treatment of acrophobia with virtual reality therapy (VRT): a case report. Cyberpsychol Behav 4(3):349–354

Coelho CM, Santos JA, Silva C, Wallis G, Tichon J, Hine TJ (2008) The role of self-motion in acrophobia treatment. Cyberpsychol Behav 11(6):723–725
Coelho CM, Waters AM, Hine TJ, Wallis G (2009) The use of virtual reality in acrophobia research and treatment. J Anxiety Disord 23:563–574
Cohen DC (1977) Comparison of self-report and behavioral procedures for assessing acrophobia. Behav Ther 8:17–23
Costa JP, Robb J, Nacke LE (2014) Physiological acrophobia evaluation through in vivo exposure in a VR CAVE. In: 2014 IEEE games media entertainment
Croitoru I, Bogolin SV, Leordeanu M (2019) Unsupervised learning of foreground object segmentation. Int J Comput Vision (IJCV) 127(9):1279–1302
Cudlenco N, Popescu N, Leordeanu M (2020) Reading into the mind's eye: boosting automatic visual recognition with EEG signals. Neurocomputing 386:281–292
Daly I, Malik A, Weaver J, Hwang F, Nasuto S, Williams D, Kirke A, Miranda E (2015) Identifying music-induced emotions from EEG for use in brain computer music interfacing. In: 6th affective computing and intelligent interaction
Davidson RJ (1993) Cerebral asymmetry and emotion: conceptual and methodological conundrums. Cognit Emot 7:115–138
Davidson RJ, Ekman P, Saron C, Senulis J, Friesen WV (1990) Approach/withdrawal and cerebral asymmetry: emotional expression and brain physiology, I. J Pers Soc Psychol 58:330–341
Demaree HA, Everhart DE, Youngstrom EA, Harrison DW (2005) Brain lateralization of emotional processing: Historical roots and a future incorporating "dominance". Behav Cogn Neurosci Rev 4:3–20
DiMeglio C (2015) Fear feedback loop: creative and dynamic fear experiences driven by user emotion, Master Thesis, Rochester Institute of Technology
Dores AR, Barbosa F, Monteiro L, Reis M, Coelho CM, Ribeiro E, Castro-Caldas A (2014) Amygdala activation in response to 2D and 3D emotion-inducing stimuli. PsychNol J 12(1–2):29–43
Ekman P, Sorenson ER, Friesen WV (1969) Pan-cultural elements in facial displays of emotions. Science 164:86–88
Erhan D, Bengio Y, Courville A, Manzagol PA, Vincent P, Bengio S (2010) Why does unsupervised pre-training help deep learning? J Mach Learn Res 11:625–660
Eum YJ, Jang EH, Park BJ, Choi SS, Kim SH, Sohn JH (2011) Emotion recognition by responses evoked by negative emotion. In Engineering and Industries (ICEI), 2011 international conference on IEEE. pp 1–4
Field AP, Argyris NG, Knowles KA (2001) Who's afraid of the big bad wolf: a prospective paradigm to test Rachman's indirect pathways in children. Behav Res Ther 39:1259–1276
Fleureau J, Philippe G, Huynh-Thu Q (2012) Physiological-based affect event detector for entertainment video applications. IEEE Trans Affect Comput 3(3):379–385
Garcia-Palacios HG, Hoffman S, Kwong See A, Botella C (2001) Redefining therapeutic success with virtual reality exposure therapy. Cyberpsychol Behav 4(3):341–348. http://online.liebertpub.com/doi/abs/10.1089/109493101300210231
Garcia-Palacios A, Hoffman H, Carlin A, Furness TA, Botella C (2002) Virtual reality in the treatment of spider phobia: a controlled study. Behav Res Ther 40:983–993
Gouizi K, Reguig F, Maaoui C (2011) Analysis physiological signals for emotion recognition. In: WOSSPA international workshop, IEEE. pp 147–150
Graham J, Gaffan EA (1997) Fear of water in children and adults: etiology and familial effects. Behav Res Ther 35(2):91–108
Healey J (2009) Affect detection in the real world: recording and processing physiological signals. In: Affective computing and intelligent interaction and workshops, 2009. ACII 2009. 3rd international conference on IEEE. pp 1–6
Hodges LF, Kooper R, Meyer TC, Rothbaum BO, Opdyke D, de Graaff JJ, Williford JS, North MM (1995) Virtual environments for treating the fear of heights. Computer 28(7):27–34

Horschig JM, Zumer JM, Bahramisharif A (2014) Hypothesis-driven methods to augment human cognition by optimizing cortical oscillations. Front Syst Neurosci 8:119. https://doi.org/10.3389/fnsys.2014.00119

HTC Vive (n.d.). https://www.vive.com/eu/

Hu F, Wang H, Chen J, Gong J (2018) Research on the characteristics of acrophobia in virtual altitude environment. In: Proceedings of the 2018 IEEE international conference on intelligence and safety for robotics, Shenyang, China, August, 24–27

Huppert D, Grill E, Brandt T (2017) A new questionnaire for estimating the severity of visual height intolerance and acrophobia by a metric interval scale. Front Neurol 8:211. https://doi.org/10.3389/fneur.2017.00211

Jerald J (2016) The VR book: human-centered design for virtual reality. ACM

Jirayucharoensak S, Pan-Ngum S, Israsena P (2014) EEG-based emotion recognition using deep learning network with principal component-based covariate shift adaptation. Sci World J 2014, article ID 627892, 10 pages

Jones NA, Fox NA (1992) Electroencephalogram asymmetry during emotionally evocative films and its relation to positive and negative affectivity. Brain Cogn 20(2):280–299

Kallenberg M, Petersen K, Nielsen M, Ng AY, Diao P, Igel C, Vachon CM, Holland K, Winkel RR, Karssemeijer N, Lillholm M (2016) Unsupervised deep learning applied to breast density segmentation and mammographic risk scoring. IEEE Trans Med Imaging 35(5):1322–1331

Koelstra S, Yazdani A, Soleymani M, Muhl C, Lee J-S, Nijholt A, Pun T, Ebrahimi T, Patras I (2010) Single trial classification of EEG and peripheral physiological signals for recognition of emotions induced by music videos. Brain Inform Ser Lect Notes Comput Sci 6334(9):89–100

Koelstra S, Muehl C, Soleymani M, Lee J-S, Yazdani A, Ebrahimi T, Pun T, Nijholt A, Patras I (2012) DEAP: a database for emotion analysis using physiological signals. IEEE Trans Affect Comput 3:18–31

Kometer H, Luedtke S, Stanuch K, Walczuk S, Wettstein J (2010) The effects virtual reality has on physiological responses as compared to two-dimensional video. University of Wisconsin School of Medicine and Public Health, Department of Physiology

Lee GY, Kisilevsky BS (2014) Fetuses respond to father's voice but prefer mother's voice after birth. Dev Psychobiol 56(1):1–11

Leitenberg H (1976) Behavioral approaches to treatment of neuroses. In: Leitenberg H (ed) Handbook of behavior modification and behavior therapy. Prentice-Hall, Englewood Cliffs NJ

Leordeanu M, Radu A, Baluja S, Sukthankar R (2016) Labeling the features not the samples: efficient video classification with minimal supervision. In: Thirtieth AAAI conference on artificial intelligence. AAAI

Littman MK (1996) Alternative meanings through the world of virtual reality. In: Vandergrift K (ed) Mosaics of meaning: enhancing the intellectual life of young adults through story. Scarecrow Press, Lanham, pp 425–455

Liu Y, Sourina O (2013) EEG databases for emotion recognition. In: 2013 international conference on cyberworlds (CW)

Liu C, Rani P, Sarkar N (2005) An empirical study of machine learning techniques for affect recognition in human-robot interaction. In: International conference on intelligent robots and systems, IEEE

Liu C, Agrawal P, Sarkar N, Chen S (2009) Dynamic difficulty adjustment in computer games through real-time anxiety-based affective feedback. Int J Hum Comput Interact 25(6):506–529

Mauss IB, Robinson MD (2009) Measures of emotion: a review. Cognit Emot 23(2):209–237

Menzies RG, Clarke JC (1993) The etiology of fear of heights and its relationship to severity and individual response patterns. Behav Res Ther 31(4):355–365

Menzies RG, Clarke JC (1995) The etiology of acrophobia and its relationship to severity and individual-response patterns. Behav Res Ther 33(7):795–803

Menzies RG, Parker L (2001) The origins of height fear: an evaluation of neoconditioning explanations. Behav Res Ther 39:185–199

Morina N, Ijntema H, Meyerbroker K, Emmelkamp PMG (2015) Can virtual reality exposure therapy gains be generalized to real-life? A meta-analysis of studies applying behavioral assessments. Behav Res Ther 74:18–24
Morris RJ, Magrath KH (1979) Contribution of therapist warmth to the contact desensitization-treatment of acrophobia. J Consult Clin Psychol 47(4):786–788
Muris P, Schmidt H, Merckelbach H (1999) The structure of specific phobia symptoms among children and adolescents. Behav Res Ther 37:863–868
Naser DS, Saha G (2013) Recognition of emotions induced by music videos using DT-CWPT. In: Medical informatics and telemedicine (ICMIT), 2013 Indian conference IEEE
Nation Wide Phobias Statistics (2019). https://blog.nationwide.com/common-phobias-statistics/
North MM, North SM (1996) Virtual psychotherapy. J Med Virtual Real 1:28–32
Oculus Rift (n.d.). https://www.oculus.com/rift/
Olesen J (2015) Phobia statistics and surprising facts about our biggest fears. http://www.fearof.net/phobia-statistics-and-surprising-facts-about-our-biggest-fears/
Opris D, Pintea S, Garcia-Palacios A, Botella C, Szamoskozi S, David D (2012) Virtual reality exposure therapy in anxiety disorders: a quantitative meta-analysis. Depress Anxiety 29:85–93
Parsons TD, Rizzo AA (2008) Affective outcomes of virtual reality exposure therapy for anxiety and specific phobias: a meta-analysis. J Behav Ther Exp Psychiatry 39(3):250–261
Petrantonakis PC, Hadjileontiadis LJ (2009) EEG-based emotion recognition using hybrid filtering and higher order crossings. In: Affective computing and intelligent interaction and workshops, 2009. ACII 2009. 3rd international conference on IEEE. pp 1–6
Phillips ML, Drevets WC, Rauch SL, Lane R (2003) Neurobiology of emotion perception II: implications for major psychiatric disorders. Biol Psychiatry 54:515–528
Poulton R, Menzies RG (2002) Non-associative fear acquisition: a review of the evidence from retrospective and longitudinal research. Behav Res Ther 40:127–149
Poulton R, Davies S, Menzies RG, Langley JD, Silva PA (1998) Evidence for a non-associative model of the acquisition of a fear of heights. Behav Res Ther 36(5):537–544
Putman P, Van PJ, Maimari I, Vander WS (2010) EEG Theta/Beta ratio in relation to fear-modulated response-inhibition, attentional control, and affective traits. Biol Psychol 83:73–78
Quirk GJ (2013) Fear. Neuroscience in the 21st century: from basic to clinical. pp 2009–2026
Rainville P, Bechara A, Naqvi N, Damasio AR (2006) Basic emotions are associated with distinct patterns of cardiorespiratory activity. Int J Psychophysiol 61(1):5–18
Re DE, O'Connor JJ, Bennett PJ, Feinberg DR (2012) Preferences for very low and very high voice pitch in humans. PLoS One 7(3):e31353
Ritchie's Plank Experience (n.d.). http://store.steampowered.com/app/517160/Richies_Plank_Experience/
Ritter B (1969) Treatment of acrophobia with contact desensitization. Behav Res Ther 7(1):41–45
Robertson A (2016) The climb turns virtual reality acrophobia into an extreme sport. https://www.theverge.com/2016/4/28/11526150/crytek-the-climb-vr-oculus-rift-review
Russell JA (1979) Affective space is bipolar. J Pers Soc Psychol 37(3):345–356
Russell JA, Mehrabian A (1977) Evidence for a three-factor theory of emotions. J Res Pers 11(3):273–294
Šalkevicius J, Damaševičius R, Maskeliunas R, Laukienė I (2019) Anxiety level recognition for virtual reality therapy system using physiological signals. Electronics 8:1039
Schutter DJ, Van Honk J (2005) Electrophysiological ratio markers for the balance between reward and punishment. Cogn Brain Res 24:685690
Scott D (1976) Understanding EEG: an introduction to electroencephalography. Duckworth. pp 18–32
Shiban Y, Schelhorn I, Pauli P, Mühlberger A (2015) Effect of combined multiple contexts and multiple stimuli exposure in spider phobia: a randomized clinical trial in virtual reality. Behav Res Ther 71:45–53
Shimmers Multisensory (n.d.). https://www.shimmersensing.com/products/shimmer3-wireless-gsr-sensor

Sokhi DS, Hunter MD, Wilkinson ID, Woodruff PW (2005) Male and female voices activate distinct regions in the male brain. NeuroImage 27(3):572–578
Soleymani M, Kierkels J, Chanel G, Pun T (2009) A Bayesian framework for video affective representation. In: Proceedings of the international conference on affective computing and intelligent interaction. pp 1–7
Steimer T (2002) The biology of fear and anxiety-related behaviors. Dialogues Clin Neurosci 4(3):231–249
Steinman SA, Teachman BA (2011) Cognitive processing and acrophobia: validating the heights interpretation questionnaire. J Anxiety Disord 25:896–902
Steinman SA, Teachman BA (2014) Reaching new heights: comparing interpretation bias modification to exposure therapy for extreme height fear. J Consult Clin Psychol 82(3):404–417. PMID: 24588406
Sutton RS, Barto AG (1998) Introduction to reinforcement learning, vol 2(4). MIT Press, Cambridge
Toth V (2015) Measurement of stress intensity using EEG, BSc Thesis
Trainor LJ, Schmidt LA (2003) Processing emotions induced by music. Cognitive neuroscience of music (Oxford). 317p
Virtual Reality Medical Center (n.d.). http://www.vrphobia.com/aboutus.htm
Walter S, Kim J, Hrabal D, Crawcour SC, Kessler H, Traue HC (2013) Trans-situational individual-specific biopsychological classification of emotions. IEEE Trans Syst Man Cybern Syst 43(4):988–995
Wen WH, Liu GY, Cheng NP, Wei J, Shangguan PC, Huang WJ (2014) Emotion recognition based on multi-variant correlation of physiological signals. IEEE Trans Affect Comput 5:126–140
Westerink J, Ouwerkerk M, de Vries GJ, de Waele S, van den Eerenbeemd J, van Boven M (2009) Emotion measurement platform for daily life situations. In: Affective computing and intelligent interaction and workshops, 2009. ACII 2009. 3rd international conference on IEEE. pp 1–6
Wiem MBH, Lachiri Z (2017) Emotion classification in arousal valence model using MAHNOB-HCI database. Int J Adv Comput Sci Appl Ijacsa 8(3)
Withers RD, Deane FP (1995) Origins of common fears: effects on severity, anxiety responses and memories of onset. Behav Res Ther 33(8):903–915
Wolpe J (1958) Psychotherapy by reciprocal inhibition. Stanford University Press, Palo Alto, CA
World Health Organization (2017) Depression and other common mental disorders: global health estimates. World Health Organization, Geneva. License: CC BY-NC-SA 3.0 IGO.PS
Yoon HJ, Chung SY (2013) EEG-based emotion estimation using Bayesian weighted-log-posterior function and perceptron convergence algorithm. Comput Biol Med 43(12):2230–2237

Part VII
Pharmacological Augmentation

Vision Augmentation by Pharmacological Enhancement of the Visual Experience

Elvire Vaucher

Abbreviations

5-HT	Serotonin
ACh	Acetylcholine
AChEI	Acetylcholinesterase inhibitor
BDNF	Brain-derived growth factor
BF	Basal forebrain
DPZ	Donepezil
E/I balance	Excitatory/inhibitory balance
LTD	Long-term depression
LTP	Long-term potentiation
mAChR	Muscarinic cholinergic receptors
nAChR	Nicotinic cholinergic receptors
NE	Noradrenaline (norepinephrine)
NGF	Nerve growth factor
SSRI	Serotonin reuptake transporter inhibitors
V1	Primary visual cortex
VEP	Visually evoked potential

The augmentation of vision would be useful for a wide range of activities, such as precision tasks and sports, and, naturally, for the alleviation of visual deficits. An estimated 246 million people worldwide have low vision resulting from ocular diseases, injuries, stroke, and concussion that negatively impact everyday activities required for independent living. Boosting of their residual vision is thus crucial to improve their daily living and social interactions.

Vision augmentation results from efficient attentional and learning capacities that enable perceptual learning, i.e., experience-driven acquisition of new

E. Vaucher (✉)
Laboratoire de Neurobiologie de la Cognition Visuelle, École d'optométrie, Université de Montréal, Montréal, QC, Canada
e-mail: elvire.vaucher@umontreal.ca

I. Opris et al. (eds.), *Modern Approaches to Augmentation of Brain Function*, Contemporary Clinical Neuroscience,
https://doi.org/10.1007/978-3-030-54564-2_28

information. A common approach to improve vision is visual training, in which repetition aids the encoding and storage of new visual information. The degree of vision improvement resulting from visual training depends on the capacity of neurons to adapt to new input, as well as to reorganize the structure and strength of neuronal outputs, i.e., brain plasticity. Visual training effects in adults are rapidly limited by poor brain plasticity. As cerebral plasticity is controlled by several central neuromodulator systems, which are also involved in attention and learning processes, potentiation of these systems during visual training has been suggested to reinforce training effects and to be beneficial for the augmentation of vision. In this chapter, we examine how neural plasticity and its enhancement by the pharmacological stimulation of central neuromodulator systems, such as the cholinergic and serotoninergic systems, might help to augment vision (Fig. 1) and perceptual learning in rodent models and humans.

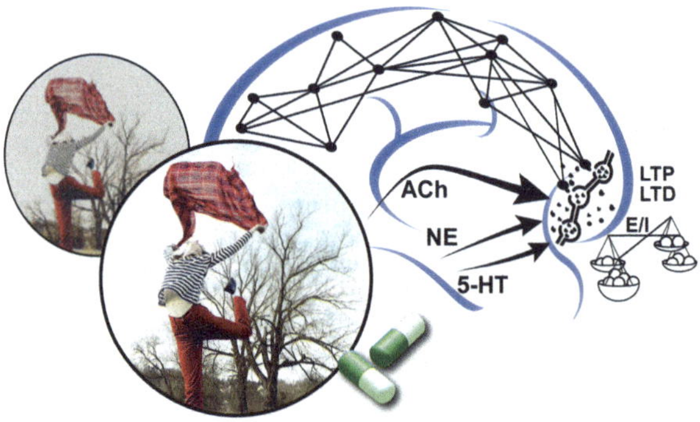

Fig. 1 Vision augmentation by the brain neuromodulator systems. Vision augmentation (large dancing woman image vs small one, left panels) results from efficient attentional, learning, and plastic capacities that improve visual cortex functioning (right edge of the symbolized brain). The degree of vision improvement results from long-term changes in synaptic strength (LTP, LTD), neuronal gain modulation and excitatory/inhibitory balance (*E/I*) changes, as well as to reorganization of the structure and strength of neuronal outputs and cortical circuits (symbolized neuronal network) up to the prefrontal cortex (left edge of the symbolized brain). Cerebral plasticity is controlled by several central neuromodulator systems, such as the cholinergic (ACh), the noradrenergic (NE), and serotoninergic (5-HT) systems which are also involved in attention and learning processes. Potentiation of these systems (represented by pills) during visual experience has been suggested to reinforce processing and to be beneficial for the augmentation of vision. Royalties free images from Pixabay: dancing woman from andreas160578; brain contours from OpenClipart. *5-HT* serotonin, *ACh* acetylcholine, *E/I* excitatory/inhibitory balance, *NE* noradrenalin (norepinephrine), *LTP* long-term potentiation, *LTD* long-term depression

1 Visual Plasticity: Neural Plasticity and Attentional Processes

The functional properties and organization of the primary visual cortex (V1) enable the early computing of visual inputs transmitted by the thalamus. Among the multiple competing inputs received by the V1, only those that elicit strong responses are further transmitted to higher associative visual and cognitive cortical areas, for the formation of conscious percepts and guiding of task-dependent visual behaviors (Glickfeld et al. 2013). Repetitive visual experiences and visual deficits modulate the efficiency and structure of neurons in the V1 (Nys et al. 2014; Keck et al. 2008; Gilbert and Li 2012) and cortical network. The neuronal plasticity in the V1 that sustains these changes involves complex, multifaceted processes at the synaptic and functional levels (Hubener and Bonhoeffer 2014). The success of visual capacity boosting thus relies on adequate triggering of these functional and plastic features.

1.1 Organization of the Visual Cortex

V1 neurons are organized in six layers through the depth of the cortex and compute the strength of excitatory thalamocortical inputs mediated by the excitatory pyramidal cells, relative to inhibitory local microcircuits and corticocortical influences. Sensory input entering the V1 in layer IV is transmitted by feedforward thalamic projections from the dorsal geniculate nucleus. Pyramidal neurons in the V1 (layers II/III, V, and VI) respond selectively to distinct features (e.g., contrast, orientation, spatial frequency) of a visual scene and monocular or binocular inputs via retinotopic mapping (Seabrook et al. 2017). These features of V1 neurons are universal in all mammals, but the visual pathways of primates and rodents differ in complexity. In particular, neurons with similar selective properties or ocular input are clustered into columns and blobs in the primate cortex, whereas the rodent V1 has a salt-and-pepper neuron distribution—although recent studies have demonstrated the segregation of feedforward and feedback afferences in the mouse visual cortex (Ji et al. 2015). Apart from the pyramidal cells that provide excitatory V1 output, various excitatory and inhibitory interneurons establish complex microcircuits of recurrent and horizontal (intralayer) connections that sustain the computation of visual inputs. In particular, GABAergic inhibitory interneurons play significant roles in the control of excitatory drive and the synchronization of neuronal responses within layers, columns, and areas of the cortex (Burkhalter 2016; Disney et al. 2006; Pfeffer et al. 2013). The correlation of neuronal discharge reinforces the strength of the output. Moreover, neuronal gain modulation underlies attentional control (Keitel et al. 2013) and network connectivity (Haider and McCormick 2009). Context-independent increases in thalamocortical pathway input during context-dependent intracortical suppression might facilitate the transmission of information related to novel stimuli. For example, optimization of the gain of supragranular pyramidal cells results in the detection of novel stimuli (Moran et al. 2013). The excitatory and inhibitory synaptic inputs tend to be equilibrated (*E*/*I* balance) in adults to optimally tune the neurons according to sensory experiences (Sun et al. 2010; Hensch et al. 1998). In addition to thalamic feedforward inputs, V1 neurons receive feedback connections from higher-level cortical areas that process complex visual information parameters. These corticocortical connections include top-down attentional control from the prefrontal cortex, which can strongly modulate the neuronal activity of the V1. Local and long-range recurrent circuits thus enable a primary level of sensory input integration, and enable or disable further transmission of a specific sensory input to higher associative areas (Priebe and McGee 2014).

1.2 Plasticity of the Visual Cortex

Cortical plasticity peaks during the critical postnatal period, when mammalian visual cortices acquire functional properties. Nevertheless, some plasticity mechanisms persist through adulthood, and reactivation of others might be triggered. The many mechanisms of efficient plasticity, enabling the long-term sustainment of neuronal responses to a particular stimulus, include variation in neuron selectivity (Froemke et al. 2007); the regulation of synapse quantity via synaptogenesis or synaptic pruning (Hofer et al. 2009); long-lasting increases in synaptic strength, reinforced or weakened by long-term potentiation or depression, respectively (Sale et al. 2011; Gagolewicz and Dringenberg 2011); and the formation of new neurites (Yamahachi et al. 2009; Gilbert and Li 2012). The mechanisms that control long-lasting synaptic transmission efficiency participate in learning and memory by consolidating the impact of a stimulus or by depressing it. The electrophysiological properties of neurons, such as the signal-to-noise ratio of responses to visual stimuli and correlated discharge of neurons, might also be adapted. These adaptive responses to visual stimulation alter the *E*/*I* balance. Disruption of the *E*/*I* balance has been suggested to reopen the critical period after maturation (Hensch 2004). During critical periods, a panoply of molecules that trigger plasticity, such as Lypd6 (Sadahiro et al. 2016), Arc (McCurry et al. 2010), NGF (Berardi et al. 1999; Lodovichi et al. 2000; Pizzorusso et al. 2002; Baho et al. 2019), and BDNF (Maya Vetencourt et al. 2008), is expressed from the extracellular matrix and cells. During cortical maturation, other molecules that reduce brain plasticity, such as Lynx1 (Morishita et al. 2010), Nogo-A (McGee et al. 2005), OTX2 (Beurdeley et al. 2012), and perineuronal nets (Carulli et al. 2010), are expressed. Researchers currently assume that plasticity could be reactivated in adults by altering the levels of these factors. Plasticity itself might be triggered or shut down by visual experience and neuromodulators, which has been termed "metaplasticity" (Bear 2003).

1.3 Neuromodulation of Visual Plasticity

Many plasticity processes in the V1 are controlled by neuromodulator systems, especially the cholinergic, serotoninergic, and noradrenergic systems (Gu 2002). Neuromodulators have permissive and terminating effects on neuroplasticity. Experimental approaches have been developed to potentiate the activity of these neuromodulators and enhance plasticity in adults. These approaches use electrical or optogenetic stimulation of selected neurons to release neuromodulators in the visual cortex. Pharmacological approaches, which are assets in the clinical setting, employ diverse strategies: the stimulation or antagonization of selective postsynaptic receptors, and augmentation of the extracellular concentrations of neuromodulators via reduction of their degradation or reuptake. Acetylcholine esterase inhibitors (AChEIs) and selective serotonin reuptake transporter inhibitors (SSRIs) are commonly used for such purposes.

1.3.1 Modulation of Visual Plasticity by Acetylcholine

Due to its roles in attention and plasticity, long-term V1 responsiveness, and the fine-tuning of cortical dynamics, the cholinergic system has the potential to promote neuroplasticity and the acquisition of new visual abilities. It is activated during sensory stimulation (Collier and Mitchell 1966; Jimenez-Capdeville et al. 1997; Laplante et al. 2005), according to stimulus novelty and relevance (Hasselmo and Sarter 2011), and immediately after unexpected reward or punishment (Hangya et al. 2015; Chubykin et al. 2013).

The mechanisms of stimulus reinforcement by the neurotransmitter acetylcholine (ACh) include the adaptation of neuronal receptive fields in the corresponding sensory area (Groleau et al. 2014), increases in neuronal gain (Gritton et al. 2016), and modulation of synaptic strength (Kang and Vaucher 2009; Gagolewicz and Dringenberg 2009; Stewart and Dringenberg 2016). ACh influences V1 neuronal function in terms of the intensity of activity (Kirkwood et al. 1999; Kimura et al. 1999; Brocher et al. 1992; Thiel and Fink 2008; Gil et al. 1997; Soma et al. 2013a; Pinto et al. 2013), preferred responses (Roberts et al. 2005), receptive-field properties (Thiel and Fink 2008; Herrero et al. 2008), (de)synchronization of networks (gamma oscillations) (Rodriguez 2004), and behavioral performance in visual learning and memory tasks (Dotigny et al. 2008; Bentley et al. 2004; Thiel and Fink 2008). Some of these mechanisms are linked directly to visual attention and the arousal state. Recent studies have demonstrated the cholinergic dependency of visual attention mechanisms in the V1 (Pinto et al. 2013; Lindner et al. 2017; Herrero et al. 2008, 2017). Moreover, cholinergic activation increases visual response (signal) correlations without affecting response variability between trials (noise) (Minces et al. 2017; van Kempen et al. 2017), thereby enabling the encoding of information. Response gain modulation by ACh has frequently been observed (Soma et al. 2013a; Disney et al. 2007; Bhattacharyya et al. 2013; Aggelopoulos et al. 2011) and follows the gain control model, at least in terms of the contrast-response function. In the V1, context-dependent (i.e., increased maximal response) and -independent (i.e., increased baseline response) gain control (80% and 20%, respectively) due to cholinergic effects without laminar bias has been observed (Soma et al. 2013b). These processes could be related to optimization of the gain of supragranular pyramidal cells controlled by ACh, which could result in the detection of novel stimuli and, hence, perceptual learning (Moran et al. 2013). The large gain that results from the amplification of excited neurons' responses is similar to attention processes (Eldar et al. 2013; Servan-Schreiber et al. 1990), and hence facilitates learning.

These effects are mediated by complex interaction of ACh with nicotinic receptors (nAChRs) and muscarinic receptors (mAChRs) at different levels of the thalamic and corticocortical fibers, excitatory cells, and GABAergic interneurons (Gil et al. 1997; Groleau et al. 2015; Roberts et al. 2005; Coppola et al. 2016; Obermayer et al. 2017). Both cholinergic receptor types can influence visual processes, with variable intensity and kinetics. mAChRs have a weaker, but more persistent, influence due to their metabotropic action, and nAChRs may act as short, but strong, inducers of juvenile-like cortical plasticity. mAChRs have important effects on sensory learning and the retrieval of information acquired from experience (Leaderbrand et al. 2016; Groleau et al. 2015). They also play roles in the establishment of the visual field and visual acuity (Groleau et al. 2014); especially, the deletion of type 2 mAChRs leads to significant modification of the apparent visual field (Groleau et al. 2014), and the deletion of type 1/3 mAChRs leads to decreased visual stimulus detection and long-term plasticity changes. Similar results have been obtained by administration of a high dose of scopolamine, an mAChR antagonist (Robinson et al. 2004), and by deletion of nicotinic subunit α7 (Origlia et al. 2012). nAChRs also participate in visual cortex plasticity by strongly modulating the *E*/*I* balance and by exerting control via plasticity brakes, such as Lynx1 and Lypd6, expressed by GABAergic cells (Morishita et al. 2010; Sadahiro et al. 2016).

Cholinergic potentiation by systemic drugs may affect other cerebral structures and visual pathways, such as the dorsal geniculate nucleus of the thalamus, the prefrontal cortex, and even the retina (Faiq et al. 2019). These structures, in turn, may contribute to the modulation of neuronal activity in the V1 and perceptual function in higher cognitive areas. Cholinergic activation also shapes cortical dynamics by increasing neural efficiency in the whole cortex, thereby reducing activation in regions involved in attention (Furey et al. 2008b; Ricciardi et al. 2013).

1.3.2 Modulation of Visual Plasticity by Monoamines

Monoamines, i.e., noradrenaline, serotonin (5-HT), and dopamine, are potent substances that modulate cortical circuits and behavior related to attention, decision making, mental flexibility, and emotional state regulation. Their roles in visual plasticity (for an extensive review, see Gu (2002)) have been studied thoroughly using the monocular deprivation model, in which ocular dominance is shifted toward the non-deprived eye. All of these substances can also modulate the *E/I* balance and long-lasting NMDA receptor-mediated activity, thereby enabling the occurrence of high plasticity periods. As dopamine virtually does not innervate the V1, we do not describe its effects here, although its role in the regulation of the prefrontal cortex certainly has a retrograde influence on V1 functioning and plasticity.

Noradrenaline is involved in many steps of sensory processing and behavioral control (Treviño et al. 2019). It plays roles in attention, arousal, vigilance, and hunger/satiety. It exerts excitatory drive mediated by α1 or β2 receptors, and inhibitory influence mediated by α2 or β1 receptors (Warren and Dykes 1996; Gu 2002). Noradrenergic input enhances the signal-to-noise ratio of target cells in the cortex (Salgado et al. 2012; Treviño et al. 2019) and facilitates activity-dependent synaptic modifications in the visual cortex (Greuel et al. 1988). Endogenous noradrenaline enhances visual detectability, depending on stimulus spatial properties, mainly via β-adrenergic receptors (Mizuyama et al. 2016). Noradrenergic activation reduces the gain in visual response without affecting contrast sensitivity (Treviño et al. 2019). It enhances ocular dominance plasticity in the kitten visual cortex after monocular deprivation (the ocular dominance shift toward the non-deprived eye is reversed by the withdrawal of ocular deprivation and noradrenergic system activation) (Shirokawa et al. 1989). The α2-receptor agonist clonidine, which decreases noradrenaline release from presynaptic terminals, has also been found to reduce ocular dominance plasticity (Nelson et al. 1985). Use-dependent receptive-field modifications in the visual cortex have been induced by the pairing of visual stimuli with iontophoretic application of noradrenaline (Greuel et al. 1988). These results indicate that noradrenaline plays a permissive role in activity-dependent modifications of neuronal connections in the visual cortex, and that β1 receptors most likely mediate these effects.

5-HT has been implicated in the regulation of memory, aggression, anxiety, depression, sleep, the neuroendocrine system, sex behavior, pain perception, body temperature, and feeding behavior. These effects are mediated through G-protein-coupled 5-HT receptors (except the ionotropic 5-HT3), distributed into seven families (5-HT1–7); among them, genes for 5-HT1B and 5-HT2A have been identified in the V1 (Shimegi et al. 2016). These receptors can enhance or depress neuronal activity upon the iontophoretic application of 5-HT to cortical neurons (Shimegi et al. 2016). The activation of 5-HT1B seems to ameliorate contrast-related image segmentation by enhancing high-contrast and depressing low-contrast stimuli, whereas 5-HT2A activation tends to suppress neurons with strong responses and facilitate those with weak responses. In addition to gain control activity, the serotonergic system in the V1 is characterized by a dependency on visual activity; all processes, including receptor expression, 5-HT release, and visual gain control, are regulated by visually evoked activity. The role of 5-HT in ocular dominance plasticity has also been investigated in the kitten and rat visual cortices. Electrophysiological recordings in 5-HT-depleted cortical regions and during 5-HT antagonist administration have shown no blockage of visual input activity, but inhibition of plasticity, as ocular dominance was not shifted after monocular deprivation (Gu and Singer 1995). On the other hand, chronic administration of fluoxetine, a 5-HT reuptake inhibitor, reinstates ocular dominance plasticity in adult rats and promotes the recovery of visual function in adult amblyopic animals (Maya Vetencourt et al. 2008, 2011). These effects were accompanied by reduced intracortical inhibition and increased expression of brain-derived neurotrophic factor in the visual cortex. Moreover, fluoxetine promotes adaptation-induced orientation plasticity by favoring the adaptation of neurons to a nonoptimal orientation (induction of shifts of tuning curves toward the adapting orientation) (Bachatene et al. 2013).

1.3.3 Modulation of Visual Plasticity by GABA and Glutamate

Intrinsic cortical interneurons containing GABA or glutamate are strongly involved in plasticity processes, via direct participation in *E*/*I* balance or promotion of NMDA receptor-dependent long-term potentiation and depression of synaptic strength.

Recent studies have demonstrated that the GABAergic inhibitory system is key for cortical plasticity (Hensch 2005; Sale et al. 2010). The onset of the critical period is accelerated by GABAA inhibitory receptor activation (Fagiolini and Hensch 2000; Iwai et al. 2003). Conversely, plasticity can be restored after the critical period by reducing the inhibitory drive via the injection of GABAA receptor antagonists (Harauzov et al. 2010). GABA potentiation by diazepam inhibits the ocular dominance plasticity induced by fluoxetine (Maya Vetencourt et al. 2008). Numerous studies have shown that changes in the *E*/*I* balance are potent inducers of plastic changes. In addition, GABAergic neurons express plasticity brakes, such as the Lynx1 protein (Takesian and Hensch 2013), which reduce nAChR efficiency (Morishita et al. 2010). The inhibitory neurons are strongly modulated by neuromodulator systems, such as those of nAChRs (Arroyo et al. 2012; Christophe et al. 2002), mAChRs (Salgado et al. 2007), and fluoxetine (Maya Vetencourt et al. 2008; Sale et al. 2010), but most neuromodulators can modulate the *E*/*I* balance and facilitate cortical plasticity in adults, promoting perceptual learning.

The glutamatergic system is also strongly involved in plasticity mechanisms, particularly the synaptic plasticity related to the NMDA receptor (Gu et al. 1989; Bear et al. 1990; Quinlan et al. 1999; Fong et al. 2019). Long-term NMDA receptor-dependent modifications of postsynaptic glutamatergic neurons are related to memory formation. The opening of the NMDA receptor launches a second messenger cascade and guides the expression of synaptic glutamate receptors (Zhong et al. 2006; Regehr and Tank 1990), but also activates autoregulated kinases that confer persistent improved neuronal responses to stimuli. Ocular dominance plasticity depends on LTP and LTD changes (Di Cristo et al. 2001; Dan and Poo 2006), especially relative to NMDA receptors distributed in V1 layer IV (Fong et al. 2019). ACh, 5-HT, and noradrenaline may also support cortical plasticity through an NMDA receptor-gated mechanism (Kang and Vaucher 2009; Origlia et al. 2006; Criscuolo et al. 2015; Bear and Singer 1986; Kirkwood et al. 1999; Moreau et al. 2013).

2 Vision Augmentation by Pharmacological Enhancement

Rodent studies have provided very robust and promising results showing the influences of neuromodulator enhancement on learning and perceptual learning. Evidence for such effects in humans is also accumulating, although outcomes have been more puzzling. Different strategies can be used to facilitate the transfer of rodent outcomes to human studies. Nevertheless, the administration of drugs can induce cortical plasticity and perceptual enhancement in humans, even in healthy young adults. As neuromodulators essentially act on neural systems to increase or decrease their efficiency, they should be potentiated when visual circuits are activated, i.e., during a visual experience or training. Different training paradigms have been used in humans, ranging from attentional tasks (Gratton et al. 2017) to perceptual learning tasks (Rokem and Silver 2010; Chung et al. 2017) and 3D training in virtual reality chambers, such as the 3D multiple object tracking task (Faubert and Sidebottom 2012; Parsons et al. 2016), binocular function and ocular dominance evaluations (Levi et al. 2015; Chadnova et al. 2017; Spiegel et al. 2017), and video game practice (Vedamurthy et al. 2015; Föcker et al. 2018; West et al. 2018; Diarra et al. 2019; Guo et al. 2016; Gao et al. 2018); the latter includes attention, motivation, and reward mechanisms. Administration of these molecules during visual rehabilitation training could theoretically aid functional restoration in patients with visual input or cognitive impairments (Whelan et al. 2000).

2.1 Vision Augmentation by Cholinergic Drugs

In animal models, chronic or acute cholinergic enhancement by drug administration or electrical stimulation of cholinergic neurons can induce long-lasting potentiation of the behavioral visual response and consolidation of information on repeated stimulation (Dotigny et al. 2008; Kang and Vaucher 2009; Kang et al. 2014, 2015; Goard and Dan 2009). The use of cholinergic enhancement in a visual learning paradigm has a similar effect, confirming the involvement of cholinergic neurons in experience-dependent plasticity. The use of donepezil (DPZ) augments discrimination abilities after visual deficit in rats (Chamoun et al. 2016, 2017b; Soma et al. 2013c) (for an extensive review of our laboratory work, see Vaucher et al. (2019)). The elegant work of Bear's group on visual response plasticity (Cooke and Bear 2010; Cooke et al. 2015) showed that daily presentation of an oriented drifting grating enhanced the visual evoked potential (VEP) elicited by this specific orientation, and that this plasticity in the V1 influenced modification of the behavioral response. Cholinergic projection from the basal forebrain to the V1 has proven to be essential for visually acquired behavioral reinforcement, without influencing the persistence of previously acquired stimulus responses (Chubykin et al. 2013).

In the human setting, acute administration of 5 mg DPZ improves the speed of learning when consistently paired with a perceptual-cognitive task, without interacting with basic visual processing (Chamoun et al. 2017a), and can modulate experience-driven plasticity (Sheynin et al. 2019). Importantly, although an acute dose of DPZ can potentiate the speed of perceptual learning (Chamoun et al. 2017a) and reduce the ocular dominance shift induced by short-term monocular deprivation (Sheynin et al. 2019), it does not play a role in covert attention shifts (Vaucher et al. 2019). Thus, DPZ-based enhancement of perceptual learning may result in a balance between attention and plasticity mechanisms, which shifts to one particular side depending on the task. Silver et al.'s extensive research on the effects of DPZ on visuospatial tasks has revealed effects on the spatial precision of visual cortical representations and visual perception. They have reported significant effects of a single 5 mg dose of DPZ on endogenous spatial attention and visual perception (Rokem et al. 2010; Rokem and Silver 2013), behavioral measures of surround suppression (Kosovicheva et al. 2012), and the spatial extent of facilitatory target/flanker interactions on visual perception (Gratton et al. 2017). However, no effect on perceptual learning in patients with amblyopia (Chung et al. 2017), or on spatial memory (Harewood Smith et al. 2017) has been observed. Cholinergic processes, though still puzzling, may support the role of the cholinergic system in circuit refinement and encoding (Dannenberg et al. 2016; Hasselmo and Sarter 2011; Minces et al. 2017), as well as increased efficiency due to reduced functional connectivity (Furey et al. 2000; Ricciardi et al. 2013). The effects of 5 mg DPZ on visual cognitive mechanisms in humans remain poorly understood. Several studies have demonstrated that cholinergic modulation depends on task difficulty (Bentley et al. 2004) and the level of attentional processes required to perform a task (Boucart et al. 2015), as ACh is released more abundantly in the presence of high attentional demand (Himmelheber et al. 2000). Moreover, ACh exertion might be selective to some types of attentional or learning process. As such, DPZ has proven to be effective during specific attentional processes (e.g., selective and voluntary attention) (Rokem et al. 2010; Bentley et al. 2004; Furey et al. 2008a), but ineffective during other tasks (e.g., involuntary attention) in human studies. Other clinical studies have demonstrated that cholinergic system enhancement improves perception by modulating extravisual areas (Furey et al. 2000, 2008a; Handjaras et al. 2013; Ricciardi et al. 2013).

2.2 Vision Augmentation by Monaminergic Drugs

The enhancement of monoaminergic transmission with catecholamine reuptake blockers has been found to boost sensory-evoked responses in the somatosensory cortex in anesthetized rats (Drouin et al. 2007). Methylphenidate, a noradrenaline (and dopamine) reuptake inhibitor, improves reaction times for correct signal detection in operant conditioning, indicating more efficient behavioral performance, correlated strongly with faster VEP latencies (Navarra et al. 2017). Clonidine, an α2-adrenergic receptor agonist, impairs temporal attention during multiple target detection in a rapid serial visual presentation task, through a decrease in tonic alertness (Brown et al. 2016). In awake mice, visual responses have been found to be enhanced substantially during running compared to stationary phases (Niell and Stryker 2010), which might be related to noradrenergic transmission (Polack et al. 2013). In humans, norepinephrine–dopamine reuptake inhibition and caffeine were found to prevent fatigue-related decrements in eye movement velocity. Pursuit eye movements and visual attention were, however, unaffected (Connell et al. 2017). Similarly, noradrenaline increases the consistency of electroencephalographic VEPs; detection sensitivity, discrimination accuracy, and subjective visibility changed in accordance with noradrenaline levels, whereas decision bias (criterion) was not affected, suggesting that noradrenaline is a key factor causally linking visual awareness to external world events (Gelbard-Sagiv et al. 2018). However, norepinephrine manipulations with the α2-adrenergic receptor agonist guanfacine did not improve target detection performance in a task requiring sustained attention to discriminate a target with flanking distractors, and did not alter the spatial profiles of perceptual interactions between targets and distractors across all target–flanker distances (Gratton et al. 2017).

Relative to the effect of 5-HT on vision enhancement, systemic administration of an SSRI was recently observed to slow the reaction times and reduce the perceptual performance of macaques performing a color discrimination task (Costa et al. 2016; Normann et al. 2007). In humans, the SSRI sertraline enhanced the effect of a visual stimulation protocol designed to induce long-term potentiation in the V1. However, the SSRI fluoxetine did not increase the rate or magnitude of visual perceptual learning compared with placebo in observers with normal vision (Lagas et al. 2016) or with amblyopia (Huttunen et al. 2018). Administration of the SSRI citalopram combined with 2 weeks of amblyopic eye patching did not alter visual acuity or a range of secondary outcome measures in adults with amblyopia (Lagas et al. 2019).

2.3 Vision Augmentation by Glutamatergic and GABAergic Drugs

All drugs that enhance LTP or LTD mechanisms, including 5-HT, cholinergic drugs, and noradrenaline, are assumed to enhance visual perception. Ketamine, which antagonizes NMDA receptors in excitatory neurons in the cerebral cortex, was recently shown to alter VEPs in humans (Sumner et al. 2019). It also impaired performance in a texture discrimination task compared with the placebo condition, but had a much lesser negative effect on performance in a control fixation task (Meuwese et al. 2013). Glutamate levels in the early visual cortex correlate with the strength of perceptual suppression (Robertson et al. 2016) and the overall duration of dominant percepts during rivalry (van Loon et al. 2013). In addition, transcranial magnetic stimulation, which increases cortical excitability, transiently improves contrast sensitivity in adult humans with amblyopia, likely acting on the *E*/*I* balance (Thompson et al. 2008; Spiegel et al. 2013). On the other hand, drugs that increase GABAergic transmission, such as benzodiazepines, abolish visual plasticity (Maya Vetencourt et al. 2008; Sale et al. 2010) and increase perceptual suppression during rivalry relative to placebo (Mentch et al. 2019).

2.4 *Vision Augmentation in Rodents Vs. Humans*

Translation of laboratory findings to human studies primarily requires determination of a pharmacological regimen in humans that favors a drug concentration efficient enough to bind to neuronal receptors and induce the expression of plasticity-enhancing molecules. For example, AChEIs mainly stimulate long-lasting action of ACh, which might not be as efficient as phasic actions. Other pharmacological approaches, such as the activation of postsynaptic receptors (Sarter and Lustig 2019), could thus be used. The lack of selective drugs that can cross the blood–brain barrier and/or specifically target the visual cortex, however, limits the use of such approaches in human studies. In addition, the biological effects of neuromodulators may be more likely to induce plastic events in rodents than in humans. Apart from the dose limitation for humans, the neuronal organization of the V1 (i.e., with shorter and more direct connections) renders plastic mechanisms less straightforward in the human than in the rodent brain. Moreover, rodents have substantially more interconnections between low and high visual areas, whereas primates have more linear connections throughout their visual streams. Even if the acute administration of cholinergic or serotoninergic drugs does not affect particular cognitive tests, however, it has been demonstrated to deeply change network connectivity (Klaassens et al. 2017).

3 Conclusion

In this chapter, we proposed that neuromodulators involved in attention, plasticity, and learning might participate in and promote perceptual learning and vision. Much remains to be uncovered regarding whether these neuromodulators have the potential to improve brain function and speed vision rehabilitation in the clinical setting. Multiple factors, such as subjects' age and cognitive functioning levels, as well as the relative/actual difficulty of tasks, might influence the cholinergic response to current clinical drugs, such as AChEIs (Bentley et al. 2011), in humans. However, experimental and clinical investigations are warranted to better define the use of neuromodulators in the improvement of visual function and vision rehabilitation.

Acknowledgments We thank the following organizations for their financial support: the Canadian Institutes of Health Research, the Natural Sciences and Engineering Research Council of Canada, the Fondation INLB (Institute Nazareth et Louis Braille), the CRIR (Centre for Interdisciplinary Research in Rehabilitation), the FRQS Vision Research Network, the FRQS Bio-imaging network, and the École d'optométrie, Université de Montréal.

References

Aggelopoulos NC, Liebe S, Logothetis NK, Rainer G (2011) Cholinergic control of visual categorization in macaques. Front Behav Neurosci 5:73. https://doi.org/10.3389/Fnbeh.2011.00073

Arroyo S, Bennett C, Aziz D, Brown SP, Hestrin S (2012) Prolonged disynaptic inhibition in the cortex mediated by slow, non-Alpha7 nicotinic excitation of a specific subset of cortical interneurons. J Neurosci 32(11):3859–3864. https://doi.org/10.1523/Jneurosci.0115-12.2012

Bachatene L, Bharmauria V, Cattan S, Molotchnikoff S (2013) Fluoxetine and serotonin facilitate attractive-adaptation-induced orientation plasticity in adult cat visual cortex. Eur J Neurosci 38(1):2065–2077. https://doi.org/10.1111/Ejn.12206

Baho E, Chattopadhyaya B, Lavertu-Jolin M, Mazziotti R, Awad PN, Chehrazi P, Groleau M, Jahannault-Talignani C, Vaucher E, Ango F, Pizzorusso T, Baroncelli L, Di Cristo G (2019) P75 neurotrophin receptor activation regulates the timing of the maturation of cortical parvalbumin interneuron connectivity and promotes juvenile-like plasticity in adult visual cortex. J Neurosci 39(23):4489–4510. https://doi.org/10.1523/Jneurosci.2881-18.2019

Bear MF (2003) Bidirectional synaptic plasticity: from theory to reality. Philos Trans R Soc Lond Ser B Biol Sci 358(1432):649–655. https://doi.org/10.1098/Rstb.2002.1255

Bear MF, Singer W (1986) Modulation of visual cortical plasticity by acetylcholine and noradrenaline. Nature 320(6058):172–176

Bear MF, Kleinschmidt A, Gu QA, Singer W (1990) Disruption of experience-dependent synaptic modifications in striate cortex by infusion of An NMDA receptor antagonist. J Neurosci 10(3):909–925

Bentley P, Husain M, Dolan RJ (2004) Effects of cholinergic enhancement on visual stimulation, spatial attention, and spatial working memory. Neuron 41(6):969–982

Bentley P, Driver J, Dolan RJ (2011) Cholinergic modulation of cognition: insights from human pharmacological functional neuroimaging. Prog Neurobiol 94(4):360–388. https://doi.org/10.1016/J.Pneurobio.2011.06.002

Berardi N, Lodovichi C, Caleo M, Pizzorusso T, Maffei L (1999) Role of neurotrophins in neural plasticity: what we learn from the visual cortex. Restor Neurol Neurosci 15(2–3):125–136

Beurdeley M, Spatazza J, Lee HH, Sugiyama S, Bernard C, Di Nardo AA, Hensch TK, Prochiantz A (2012) Otx2 binding to perineuronal nets persistently regulates plasticity in the mature visual cortex. J Neurosci 32(27):9429–9437. https://doi.org/10.1523/Jneurosci.0394-12.2012

Bhattacharyya A, Veit J, Kretz R, Bondar I, Rainer G (2013) Basal forebrain activation controls contrast sensitivity in primary visual cortex. BMC Neurosci 14:55. https://doi.org/10.1186/1471-2202-14-55

Boucart M, Michael GA, Bubicco G, Ponchel A, Waucquier N, Deplanque D, Deguil J, Bordet R (2015) Cholinergic modulation of stimulus-driven attentional capture. Behav Brain Res 283c:47–52. https://doi.org/10.1016/J.Bbr.2015.01.024

Brocher S, Artola A, Singer W (1992) Agonists of cholinergic and noradrenergic receptors facilitate synergistically the induction of long-term potentiation in slices of rat visual cortex. Brain Res 573(1):27–36

Brown SB, Slagter HA, Van Noorden MS, Giltay EJ, Van Der Wee NJ, Nieuwenhuis S (2016) Effects of clonidine and scopolamine on multiple target detection in rapid serial visual presentation. Psychopharmacology 233(2):341–350. https://doi.org/10.1007/S00213-015-4111-Y

Burkhalter A (2016) The network for intracortical communication in mouse visual cortex. In: Kennedy H, Van Essen DC, Christen Y (eds) Micro-, meso- and macro-connectomics of the brain. Springer, Cham, pp 31–43. https://doi.org/10.1007/978-3-319-27777-6_4

Carulli D, Pizzorusso T, Kwok JCF, Putignano E, Poli A, Forostyak S, Andrews MR, Deepa SS, Glant T, Fawcett JW (2010) Animals lacking link protein have attenuated perineuronal nets and persistent plasticity. Brain 133(Pt 8):2331–2347. https://doi.org/10.1093/Brain/Awq145

Chadnova E, Reynaud A, Clavagnier S, Hess RF (2017) Short-term monocular occlusion produces changes in ocular dominance by a reciprocal modulation of interocular inhibition. Sci Rep 7:41747. https://doi.org/10.1038/Srep41747

Chamoun M, Groleau M, Bhat M, Vaucher E (2016) Dose-dependent effect of donepezil administration on long-term enhancement of visually evoked potentials and cholinergic receptor overexpression in rat visual cortex. J Physiol Paris 110(1–2):65–74. https://doi.org/10.1016/J.Jphysparis.2016.11.010

Chamoun M, Huppe-Gourgues F, Legault I, Rosa-Neto P, Dumbrava D, Faubert J, Vaucher E (2017a) Cholinergic potentiation improves perceptual-cognitive training of healthy young adults in three dimensional multiple object tracking. Front Hum Neurosci 11:128. https://doi.org/10.3389/Fnhum.2017.00128

Chamoun M, Sergeeva EG, Henrich-Noack P, Jia S, Grigartzik L, Ma J, You Q, Huppe-Gourgues F, Sabel BA, Vaucher E (2017b) Cholinergic potentiation of restoration of visual function after optic nerve damage in rats. Neural Plast 2017:6928489. https://doi.org/10.1155/2017/6928489

Christophe E, Roebuck A, Staiger JF, Lavery DJ, Charpak S, Audinat E (2002) Two types of nicotinic receptors mediate an excitation of neocortical layer I interneurons. J Neurophysiol 88(3):1318–1327

Chubykin AA, Roach EB, Bear MF, Shuler MG (2013) A cholinergic mechanism for reward timing within primary visual cortex. Neuron 77(4):723–735. https://doi.org/10.1016/J.Neuron.2012.12.039

Chung STL, Li RW, Silver MA, Levi DM (2017) Donepezil does not enhance perceptual learning in adults with amblyopia: a pilot study. Front Neurosci 11:448. https://doi.org/10.3389/Fnins.2017.00448

Collier B, Mitchell JF (1966) The central release of acetylcholine during stimulation of the visual pathway. J Physiol 184(1):239–254

Connell CJ, Thompson B, Turuwhenua J, Srzich A, Gant N (2017) Fatigue-related impairments in oculomotor control are prevented by norepinephrine-dopamine reuptake inhibition. Sci Rep 7:42726. https://doi.org/10.1038/Srep42726

Cooke SF, Bear MF (2010) Visual experience induces long-term potentiation in the primary visual cortex. J Neurosci 30(48):16304–16313. https://doi.org/10.1523/Jneurosci.4333-10.2010

Cooke SF, Komorowski RW, Kaplan ES, Gavornik JP, Bear MF (2015) Visual recognition memory, manifested as long-term habituation, requires synaptic plasticity in V1. Nat Neurosci 18(2):262–271. https://doi.org/10.1038/Nn.3920

Coppola JJ, Ward NJ, Jadi MP, Disney AA (2016) Modulatory compartments in cortex and local regulation of cholinergic tone. J Physiol Paris 110(1–2):3–9. https://doi.org/10.1016/J.Jphysparis.2016.08.001

Costa VD, Kakalios LC, Averbeck BB (2016) Blocking serotonin but not dopamine reuptake alters neural processing during perceptual decision making. Behav Neurosci 130(5):461–468. https://doi.org/10.1037/Bne0000162

Criscuolo C, Accorroni A, Domenici L, Origlia N (2015) Impaired synaptic plasticity in the visual cortex of mice lacking Alpha7-nicotinic receptor subunit. Neuroscience 294:166–171. https://doi.org/10.1016/J.Neuroscience.2015.03.022

Dan Y, Poo MM (2006) Spike timing-dependent plasticity: from synapse to perception. Physiol Rev 86(3):1033–1048. https://doi.org/10.1152/Physrev.00030.2005

Dannenberg H, Hinman JR, Hasselmo ME (2016) Potential roles of cholinergic modulation in the neural coding of location and movement speed. J Physiol Paris 110(1–2):52–64. https://doi.org/10.1016/J.Jphysparis.2016.09.002

Di Cristo G, Berardi N, Cancedda L, Pizzorusso T, Putignano E, Ratto GM, Maffei L (2001) Requirement of Erk activation for visual cortical plasticity. Science 292(5525):2337–2340

Diarra M, Zendel BR, Benady-Chorney J, Blanchette CA, Lepore F, Peretz I, Belleville S, West GL (2019) Playing Super Mario increases oculomotor inhibition and frontal eye field grey matter in older adults. Exp Brain Res 237(3):723–733. https://doi.org/10.1007/S00221-018-5453-6

Disney AA, Domakonda KV, Aoki C (2006) Differential expression of muscarinic acetylcholine receptors across excitatory and inhibitory cells in visual cortical areas V1 and V2 of the macaque monkey. J Comp Neurol 499(1):49–63. https://doi.org/10.1002/Cne.21096

Disney AA, Aoki C, Hawken MJ (2007) Gain modulation by nicotine in macaque V1. Neuron 56(4):701–713. https://doi.org/10.1016/J.Neuron.2007.09.034

Dotigny F, Ben Amor AY, Burke M, Vaucher E (2008) Neuromodulatory role of acetylcholine in visually-induced cortical activation: behavioral and neuroanatomical correlates. Neuroscience 154(4):1607–1618. https://doi.org/10.1016/J.Neuroscience.2008.04.030
Drouin C, Wang D, Waterhouse BD (2007) Neurophysiological actions of methylphenidate in the primary somatosensory cortex. Synapse 61(12):985–990. https://doi.org/10.1002/Syn.20454
Eldar E, Cohen JD, Niv Y (2013) The effects of neural gain on attention and learning. Nat Neurosci 16(8):1146–1153. https://doi.org/10.1038/Nn.3428
Fagiolini M, Hensch TK (2000) Inhibitory threshold for critical-period activation in primary visual cortex. Nature 404(6774):183–186
Faiq MA, Wollstein G, Schuman JS, Chan KC (2019) Cholinergic nervous system and glaucoma: from basic science to clinical applications. Prog Retin Eye Res 72:100767. https://doi.org/10.1016/J.Preteyeres.2019.06.003
Faubert J, Sidebottom L (2012) Perceptual-cognitive training of athletes. J Clin Sport Psychol 6:85–102
Föcker J, Cole D, Beer AL, Bavelier D (2018) Neural bases of enhanced attentional control: lessons from action video game players. Brain Behav 8(7):E01019. https://doi.org/10.1002/Brb3.1019
Fong MF, Finnie PS, Kim T, Thomazeau A, Kaplan ES, Cooke SF, Bear MF (2019) Distinct laminar requirements for NMDA receptors in experience-dependent visual cortical plasticity. Cereb Cortex 30(2):2555–2572. https://doi.org/10.1093/Cercor/Bhz260
Froemke RC, Merzenich MM, Schreiner CE (2007) A synaptic memory trace for cortical receptive field plasticity. Nature 450(7168):425–429. https://doi.org/10.1038/Nature06289
Furey ML, Pietrini P, Haxby JV (2000) Cholinergic enhancement and increased selectivity of perceptual processing during working memory. Science 290(5500):2315–2319. https://doi.org/10.1126/Science.290.5500.2315
Furey ML, Pietrini P, Haxby JV, Drevets WC (2008a) Selective effects of cholinergic modulation on task performance during selective attention. Neuropsychopharmacology 33(4):913–923. https://doi.org/10.1038/Sj.Npp.1301461
Furey ML, Ricciardi E, Schapiro MB, Rapoport SI, Pietrini P (2008b) Cholinergic enhancement eliminates modulation of neural activity by task difficulty in the prefrontal cortex during working memory. J Cogn Neurosci 20(7):1342–1353. https://doi.org/10.1162/Jocn.2008.20092
Gagolewicz PJ, Dringenberg HC (2009) Selective potentiation of crossed vs. uncrossed inputs from lateral geniculate nucleus to visual cortex by the basal forebrain: potential facilitation of rodent binocularity. Neurosci Lett 463(2):130–134. https://doi.org/10.1016/J.Neulet.2009.07.052
Gagolewicz PJ, Dringenberg HC (2011) Nr2b-subunit dependent facilitation of long-term potentiation in primary visual cortex following visual discrimination training of adult rats. Eur J Neurosci 34(8):1222–1229. https://doi.org/10.1111/J.1460-9568.2011.07842.X
Gao T, Guo CX, Babu RJ, Black JM, Bobier WR, Chakraborty A, Dai S, Hess RF, Jenkins M, Jiang Y, Kearns LS, Kowal L, Lam CSY, Pang PCK, Parag V, Pieri R, Raveendren RN, South J, Staffieri SE, Wadham A, Walker N, Thompson B, Team BS (2018) Effectiveness of a binocular video game vs placebo video game for improving visual functions in older children, teenagers, and adults with amblyopia: a randomized clinical trial. JAMA Ophthalmol 136(2):172–181. https://doi.org/10.1001/Jamaophthalmol.2017.6090
Gelbard-Sagiv H, Magidov E, Sharon H, Hendler T, Nir Y (2018) Noradrenaline modulates visual perception and late visually evoked activity. Curr Biol 28(14):2239–2249.E2236. https://doi.org/10.1016/J.Cub.2018.05.051
Gil Z, Connors BW, Amitai Y (1997) Differential regulation of neocortical synapses by neuromodulators and activity. Neuron 19(3):679–686
Gilbert CD, Li W (2012) Adult visual cortical plasticity. Neuron 75(2):250–264. https://doi.org/10.1016/J.Neuron.2012.06.030
Glickfeld LL, Histed MH, Maunsell JH (2013) Mouse primary visual cortex is used to detect both orientation and contrast changes. J Neurosci 33(50):19416–19422. https://doi.org/10.1523/Jneurosci.3560-13.2013
Goard M, Dan Y (2009) Basal forebrain activation enhances cortical coding of natural scenes. Nat Neurosci 12(11):1444–1449. https://doi.org/10.1038/Nn.2402

Gratton C, Yousef S, Aarts E, Wallace DL, D'esposito M, Silver MA (2017) Cholinergic, but not dopaminergic or noradrenergic, enhancement sharpens visual spatial perception in humans. J Neurosci 37(16):4405–4415. https://doi.org/10.1523/Jneurosci.2405-16.2017
Greuel JM, Luhmann HJ, Singer W (1988) Pharmacological induction of use-dependent receptive field modifications in the visual cortex. Science 242(4875):74–77
Gritton HJ, Howe WM, Mallory CS, Hetrick VL, Berke JD, Sarter M (2016) Cortical cholinergic signaling controls the detection of cues. Proc Natl Acad Sci U S A 113(8):E1089–E1097. https://doi.org/10.1073/Pnas.1516134113
Groleau M, Nguyen HN, Vanni MP, Huppe-Gourgues F, Casanova C, Vaucher E (2014) Impaired functional organization in the visual cortex of muscarinic receptor knock-out mice. Neuroimage 98:233–242. https://doi.org/10.1016/J.Neuroimage.2014.05.016
Groleau M, Kang JI, Huppe-Gourgues F, Vaucher E (2015) Distribution and effects of the muscarinic receptor subtypes in the primary visual cortex. Front Synaptic Neurosci 7:10. https://doi.org/10.3389/Fnsyn.2015.00010
Gu Q (2002) Neuromodulatory transmitter systems in the cortex and their role in cortical plasticity. Neuroscience 111(4):815–835
Gu Q, Singer W (1995) Involvement of serotonin in developmental plasticity of kitten visual cortex. Eur J Neurosci 7(6):1146–1153. https://doi.org/10.1111/J.1460-9568.1995.Tb01104.X
Gu QA, Bear MF, Singer W (1989) Blockade of NMDA-receptors prevents scularity changes in kitten visual cortex after reversed monocular deprivation. Brain Res Dev Brain Res 47(2):281–288. https://doi.org/10.1016/0165-3806(89)90183-1
Guo CX, Babu RJ, Black JM, Bobier WR, Lam CS, Dai S, Gao TY, Hess RF, Jenkins M, Jiang Y, Kowal L, Parag V, South J, Staffieri SE, Walker N, Wadham A, Thompson B, Team BS (2016) Binocular treatment of amblyopia using videogames (BRAVO): study protocol for a randomised controlled trial. Trials 17(1):504. https://doi.org/10.1186/S13063-016-1635-3
Haider B, Mccormick DA (2009) Rapid neocortical dynamics: cellular and network mechanisms. Neuron 62(2):171–189. https://doi.org/10.1016/J.Neuron.2009.04.008
Handjaras G, Ricciardi E, Szczepanik J, Pietrini P, Furey ML (2013) Cholinergic enhancement differentially modulates neural response to encoding during face identity and face location working memory tasks. Exp Biol Med (Maywood) 238(9):999–1008. https://doi.org/10.1177/1535370213497326
Hangya B, Ranade SP, Lorenc M, Kepecs A (2015) Central cholinergic neurons are rapidly recruited by reinforcement feedback. Cell 162(5):1155–1168. https://doi.org/10.1016/J.Cell.2015.07.057
Harauzov A, Spolidoro M, Dicristo G, De Pasquale R, Cancedda L, Pizzorusso T, Viegi A, Berardi N, Maffei L (2010) Reducing intracortical inhibition in the adult visual cortex promotes ocular dominance plasticity. J Neurosci 30(1):361–371. https://doi.org/10.1523/Jneurosci.2233-09.2010
Harewood Smith AN, Challa JA, Silver MA (2017) Neither cholinergic nor dopaminergic enhancement improve spatial working memory precision in humans. Front Neural Circuits 11:94. https://doi.org/10.3389/Fncir.2017.00094
Hasselmo ME, Sarter M (2011) Modes and models of forebrain cholinergic neuromodulation of cognition. Neuropsychopharmacology 36(1):52–73. https://doi.org/10.1038/Npp.2010.104
Hensch TK (2004) Critical period regulation. Annu Rev Neurosci 27:549–579. https://doi.org/10.1146/Annurev.Neuro.27.070203.144327
Hensch TK (2005) Critical period plasticity in local cortical circuits. Nat Rev Neurosci 6(11):877–888. https://doi.org/10.1038/Nrn1787
Hensch TK, Fagiolini M, Mataga N, Stryker MP, Baekkeskov S, Kash SF (1998) Local GABA circuit control of experience-dependent plasticity in developing visual cortex. Science 282(5393):1504–1508
Herrero JL, Roberts MJ, Delicato LS, Gieselmann MA, Dayan P, Thiele A (2008) Acetylcholine contributes through muscarinic receptors to attentional modulation in V1. Nature 454(7208):1110–1114. https://doi.org/10.1038/Nature07141

Herrero JL, Gieselmann MA, Thiele A (2017) Muscarinic and nicotinic contribution to contrast sensitivity of macaque area V1 neurons. Front Neural Circuits 11:106. https://doi.org/10.3389/Fncir.2017.00106

Himmelheber AM, Sarter M, Bruno JP (2000) Increases in cortical acetylcholine release during sustained attention performance in rats. Brain Res Cogn Brain Res 9(3):313–325

Hofer SB, Mrsic-Flogel TD, Bonhoeffer T, Hubener M (2009) Experience leaves a lasting structural trace in cortical circuits. Nature 457(7227):313–317. https://doi.org/10.1038/Nature07487

Hubener M, Bonhoeffer T (2014) Neuronal plasticity: beyond the critical period. Cell 159(4):727–737. https://doi.org/10.1016/J.Cell.2014.10.035

Huttunen HJ, Palva JM, Lindberg L, Palva S, Saarela V, Karvonen E, Latvala ML, Liinamaa J, Booms S, Castrén E, Uusitalo H (2018) Fluoxetine does not enhance the effect of perceptual learning on visual function in adults with amblyopia. Sci Rep 8(1):12830. https://doi.org/10.1038/S41598-018-31169-Z

Iwai Y, Fagiolini M, Obata K, Hensch TK (2003) Rapid critical period induction by tonic inhibition in visual cortex. J Neurosci 23(17):6695–6702. https://doi.org/10.1523/JNEUROSCI.23-17-06695.2003

Ji W, Gamanut R, Bista P, D'souza RD, Wang Q, Burkhalter A (2015) Modularity in the organization of mouse primary visual cortex. Neuron 87(3):632–643. https://doi.org/10.1016/J.Neuron.2015.07.004

Jimenez-Capdeville ME, Dykes RW, Myasnikov AA (1997) Differential control of cortical activity by the basal forebrain in rats: a role for both cholinergic and inhibitory influences. J Comp Neurol 381(1):53–67

Kang JI, Vaucher E (2009) Cholinergic pairing with visual activation results in long-term enhancement of visual evoked potentials. PLoS One 4(6):E5995. https://doi.org/10.1371/Journal.Pone.0005995

Kang JI, Groleau M, Dotigny F, Giguere H, Vaucher E (2014) Visual training paired with electrical stimulation of the basal forebrain improves orientation-selective visual acuity in the rat. Brain Struct Funct 219(4):1493–1507. https://doi.org/10.1007/S00429-013-0582-Y

Kang JI, Huppe-Gourgues F, Vaucher E (2015) Pharmacological mechanisms of cortical enhancement induced by the repetitive pairing of visual/cholinergic stimulation. PLoS One 10(10):E0141663. https://doi.org/10.1371/Journal.Pone.0141663

Keck T, Mrsic-Flogel TD, Vaz Afonso M, Eysel U, Bonhoeffer T, Hubener M (2008) Massive restructuring of neuronal circuits during functional reorganization of adult visual cortex. Nat Neurosci 11(10):1162–1167. https://doi.org/10.1038/Nn.2181

Keitel C, Andersen SK, Quigley C, Muller MM (2013) Independent effects of attentional gain control and competitive interactions on visual stimulus processing. Cereb Cortex 23(4):940–946. https://doi.org/10.1093/Cercor/Bhs084

Kimura F, Fukuda M, Tsumoto T (1999) Acetylcholine suppresses the spread of excitation in the visual cortex revealed by optical recording: possible differential effect depending on the source of input. Eur J Neurosci 11(10):3597–3609. https://doi.org/10.1046/J.1460-9568.1999.00779.X

Kirkwood A, Rozas C, Kirkwood J, Perez F, Bear MF (1999) Modulation of long-term synaptic depression in visual cortex by acetylcholine and norepinephrine. J Neurosci 19(5):1599–1609

Klaassens BL, Rombouts SA, Winkler AM, Van Gorsel HC, Van Der Grond J, Van Gerven JM (2017) Time related effects on functional brain connectivity after serotonergic and cholinergic neuromodulation. Hum Brain Mapp 38(1):308–325. https://doi.org/10.1002/Hbm.23362

Kosovicheva AA, Sheremata SL, Rokem A, Landau AN, Silver MA (2012) Cholinergic enhancement reduces orientation-specific surround suppression but not visual crowding. Front Behav Neurosci 6:61. https://doi.org/10.3389/Fnbeh.2012.00061

Lagas AK, Black JM, Byblow WD, Fleming MK, Goodman LK, Kydd RR, Russell BR, Stinear CM, Thompson B (2016) Fluoxetine does not enhance visual perceptual learning and triazolam specifically impairs learning transfer. Front Hum Neurosci 10:532. https://doi.org/10.3389/Fnhum.2016.00532

Lagas AK, Black JM, Russell BR, Kydd RR, Thompson B (2019) The effect of combined patching and citalopram on visual acuity in adults with amblyopia: a randomized, crossover, placebo-controlled trial. Neural Plast 2019:5857243. https://doi.org/10.1155/2019/5857243

Laplante F, Morin Y, Quirion R, Vaucher E (2005) Acetylcholine release is elicited in the visual cortex, but not in the prefrontal cortex, by patterned visual stimulation: a dual microdialysis study with functional correlates in the rat brain. Neuroscience 132(2):501–510. https://doi.org/10.1016/J.Neuroscience.2004.11.059

Leaderbrand K, Chen HJ, Corcoran KA, Guedea AL, Jovasevic V, Wess J, Radulovic J (2016) Muscarinic acetylcholine receptors act in synergy to facilitate learning and memory. Learn Mem 23(11):631–638. https://doi.org/10.1101/Lm.043133.116

Levi DM, Knill DC, Bavelier D (2015) Stereopsis and amblyopia: a mini-review. Vis Res 114:17–30. https://doi.org/10.1016/J.Visres.2015.01.002

Lindner M, Bell T, Iqbal S, Mullins PG, Christakou A (2017) In vivo functional neurochemistry of human cortical cholinergic function during visuospatial attention. PLoS One 12(2):E0171338. https://doi.org/10.1371/Journal.Pone.0171338

Lodovichi C, Berardi N, Pizzorusso T, Maffei L (2000) Effects of neurotrophins on cortical plasticity: same or different? J Neurosci 20(6):2155–2165

Maya Vetencourt JF, Sale A, Viegi A, Baroncelli L, De Pasquale R, O'leary OF, Castren E, Maffei L (2008) The antidepressant fluoxetine restores plasticity in the adult visual cortex. Science 320(5874):385–388. https://doi.org/10.1126/Science.1150516

Maya Vetencourt JF, Tiraboschi E, Spolidoro M, Castren E, Maffei L (2011) Serotonin triggers a transient epigenetic mechanism that reinstates adult visual cortex plasticity in rats. Eur J Neurosci 33(1):49–57. https://doi.org/10.1111/J.1460-9568.2010.07488.X

Mccurry CL, Shepherd JD, Tropea D, Wang KH, Bear MF, Sur M (2010) Loss of arc renders the visual cortex impervious to the effects of sensory experience or deprivation. Nat Neurosci 13(4):450–457. https://doi.org/10.1038/Nn.2508

Mcgee AW, Yang Y, Fischer QS, Daw NW, Strittmatter SM (2005) Experience-driven plasticity of visual cortex limited by myelin and Nogo receptor. Science 309(5744):2222–2226. https://doi.org/10.1126/Science.1114362

Mentch J, Spiegel A, Ricciardi C, Robertson CE (2019) Gabaergic inhibition gates perceptual awareness during binocular rivalry. J Neurosci 39(42):8398–8407. https://doi.org/10.1523/Jneurosci.0836-19.2019

Meuwese JD, Van Loon AM, Scholte HS, Lirk PB, Vulink NC, Hollmann MW, Lamme VA (2013) NMDA receptor antagonist ketamine impairs feature integration in visual perception. PLoS One 8(11):E79326. https://doi.org/10.1371/Journal.Pone.0079326

Minces V, Pinto L, Dan Y, Chiba AA (2017) Cholinergic shaping of neural correlations. Proc Natl Acad Sci U S A 114(22):5725–5730. https://doi.org/10.1073/Pnas.1621493114

Mizuyama R, Soma S, Suemastu N, Shimegi S (2016) Noradrenaline improves behavioral contrast sensitivity via the β-adrenergic receptor. PLoS One 11(12):E0168455. https://doi.org/10.1371/Journal.Pone.0168455

Moran RJ, Campo P, Symmonds M, Stephan KE, Dolan RJ, Friston KJ (2013) Free energy, precision and learning: the role of cholinergic neuromodulation. J Neurosci 33(19):8227–8236. https://doi.org/10.1523/Jneurosci.4255-12.2013

Moreau AW, Amar M, Callebert J, Fossier P (2013) Serotonergic modulation of LTP at excitatory and inhibitory synapses in the developing rat visual cortex. Neuroscience 238:148–158. https://doi.org/10.1016/J.Neuroscience.2013.02.013

Morishita H, Miwa JM, Heintz N, Hensch TK (2010) Lynx1, a cholinergic brake, limits plasticity in adult visual cortex. Science 330(6008):1238–1240. https://doi.org/10.1126/Science.1195320

Navarra RL, Clark BD, Gargiulo AT, Waterhouse BD (2017) Methylphenidate enhances early-stage sensory processing and rodent performance of a visual signal detection task. Neuropsychopharmacology 42(6):1326–1337. https://doi.org/10.1038/Npp.2016.267

Nelson SB, Schwartz MA, Daniels JD (1985) Clonidine and cortical plasticity: possible evidence for noradrenergic involvement. Brain Res 355(1):39–50. https://doi.org/10.1016/0165-3806(85)90005-7

Niell CM, Stryker MP (2010) Modulation of visual responses by behavioral state in mouse visual cortex. Neuron 65(4):472–479. https://doi.org/10.1016/J.Neuron.2010.01.033

Normann C, Schmitz D, Fürmaier A, Döing C, Bach M (2007) Long-term plasticity of visually evoked potentials in humans is altered in major depression. Biol Psychiatry 62(5):373–380. https://doi.org/10.1016/J.Biopsych.2006.10.006

Nys J, Aerts J, Ytebrouck E, Vreysen S, Laeremans A, Arckens L (2014) The cross-modal aspect of mouse visual cortex plasticity induced by monocular enucleation is age dependent. J Comp Neurol 522(4):950–970. https://doi.org/10.1002/Cne.23455

Obermayer J, Verhoog MB, Luchicchi A, Mansvelder HD (2017) Cholinergic modulation of cortical microcircuits is layer-specific: evidence from rodent, monkey and human brain. Front Neural Circuits 11:100. https://doi.org/10.3389/Fncir.2017.00100

Origlia N, Kuczewski N, Aztiria E, Gautam D, Wess J, Domenici L (2006) Muscarinic acetylcholine receptor knockout mice show distinct synaptic plasticity impairments in the visual cortex. J Physiol 577(3):829–840. https://doi.org/10.1113/Jphysiol.2006.117119

Origlia N, Valenzano DR, Moretti M, Gotti C, Domenici L (2012) Visual acuity is reduced in alpha 7 nicotinic receptor knockout mice. Invest Ophthalmol Vis Sci 53(3):1211–1218. https://doi.org/10.1167/Iovs.11-8007

Parsons B, Magill T, Boucher A, Zhang M, Zogbo K, Berube S, Scheffer O, Beauregard M, Faubert J (2016) Enhancing cognitive function using perceptual-cognitive training. Clin EEG Neurosci 47(1):37–47. https://doi.org/10.1177/1550059414563746

Pfeffer CK, Xue M, He M, Huang ZJ, Scanziani M (2013) Inhibition of inhibition in visual cortex: the logic of connections between molecularly distinct interneurons. Nat Neurosci 16(8):1068–1076. https://doi.org/10.1038/Nn.3446

Pinto L, Goard MJ, Estandian D, Xu M, Kwan AC, Lee SH, Harrison TC, Feng G, Dan Y (2013) Fast modulation of visual perception by basal forebrain cholinergic neurons. Nat Neurosci 16(12):1857–1863. https://doi.org/10.1038/Nn.3552

Pizzorusso T, Medini P, Berardi N, Chierzi S, Fawcett JW, Maffei L (2002) Reactivation of ocular dominance plasticity in the adult visual cortex. Science 298(5596):1248–1251. https://doi.org/10.1126/Science.1072699

Polack PO, Friedman J, Golshani P (2013) Cellular mechanisms of brain state-dependent gain modulation in visual cortex. Nat Neurosci 16(9):1331–1339. https://doi.org/10.1038/Nn.3464

Priebe NJ, Mcgee AW (2014) Mouse vision as a gateway for understanding how experience shapes neural circuits. Front Neural Circuits 8:123. https://doi.org/10.3389/Fncir.2014.00123

Quinlan EM, Olstein DH, Bear MF (1999) Bidirectional, experience-dependent regulation of N-methyl-D-aspartate receptor subunit composition in the rat visual cortex during postnatal development. Proc Natl Acad Sci U S A 96(22):12876–12880

Regehr WG, Tank DW (1990) Postsynaptic NMDA receptor-mediated calcium accumulation in hippocampal CA1 pyramidal cell dendrites. Nature 345(6278):807–810. https://doi.org/10.1038/345807a0

Ricciardi E, Handjaras G, Bernardi G, Pietrini P, Furey ML (2013) Cholinergic enhancement reduces functional connectivity and bold variability in visual extrastriate cortex during selective attention. Neuropharmacology 64:305–313. https://doi.org/10.1016/J.Neuropharm.2012.07.003

Roberts MJ, Zinke W, Guo K, Robertson R, Mcdonald JS, Thiele A (2005) Acetylcholine dynamically controls spatial integration in marmoset primary visual cortex. J Neurophysiol 93(4):2062–2072. https://doi.org/10.1152/Jn.00911.2004

Robertson CE, Ratai EM, Kanwisher N (2016) Reduced GABAergic action in the autistic brain. Curr Biol 26(1):80–85. https://doi.org/10.1016/J.Cub.2015.11.019

Robinson L, Harbaran D, Riedel G (2004) Visual acuity in the water maze: sensitivity to muscarinic receptor blockade in rats and mice. Behav Brain Res 151(1–2):277–286. https://doi.org/10.1016/J.Bbr.2003.09.001

Rodriguez R (2004) Short- and long-term effects of cholinergic modulation on gamma oscillations and response synchronization in the visual cortex. J Neurosci 24(46):10369–10378. https://doi.org/10.1523/Jneurosci.1839-04.2004

Rokem A, Silver MA (2010) Cholinergic enhancement augments magnitude and specificity of visual perceptual learning in healthy humans. Curr Biol 20(19):1723–1728. https://doi.org/10.1016/J.Cub.2010.08.027
Rokem A, Silver MA (2013) The benefits of cholinergic enhancement during perceptual learning are long-lasting. Front Comput Neurosci 7:66. https://doi.org/10.3389/Fncom.2013.00066
Rokem A, Landau AN, Garg D, Prinzmetal WM, Silver MLA (2010) Cholinergic enhancement increases the effects of voluntary attention but does not affect involuntary attention. Neuropsychopharmacology 35(13):2538–2544. https://doi.org/10.1038/Npp.2010.118
Sadahiro M, Sajo M, Morishita H (2016) Nicotinic regulation of experience-dependent plasticity in visual cortex. J Physiol Paris 110(1–2):29–36. https://doi.org/10.1016/J.Jphysparis.2016.11.003
Sale A, Berardi N, Spolidoro M, Baroncelli L, Maffei L (2010) Gabaergic inhibition in visual cortical plasticity. Front Cell Neurosci 4:10. https://doi.org/10.3389/Fncel.2010.00010
Sale A, De Pasquale R, Bonaccorsi J, Pietra G, Olivieri D, Berardi N, Maffei L (2011) Visual perceptual learning induces long-term potentiation in the visual cortex. Neuroscience 172:219–225. https://doi.org/10.1016/J.Neuroscience.2010.10.078
Salgado H, Bellay T, Nichols JA, Bose M, Martinolich L, Perrotti L, Atzori M (2007) Muscarinic M2 and M1 receptors reduce GABA release by Ca^{2+} channel modulation through activation of PI3K/Ca^{2+}-independent and PLC/Ca^{2+}-dependent PKC. J Neurophysiol 98(2):952–965
Salgado H, Köhr G, Treviño M (2012) Noradrenergic 'tone' determines dichotomous control of cortical spike-timing-dependent plasticity. Sci Rep 2:417. https://doi.org/10.1038/Srep00417
Sarter M, Lustig C (2019) Forebrain cholinergic signaling: wired and phasic, not tonic, and causing behavior. Preprints: 2010040010
Seabrook TA, Burbridge TJ, Crair MC, Huberman AD (2017) Architecture, function, and assembly of the mouse visual system. Annu Rev Neurosci 40:499–538. https://doi.org/10.1146/Annurev-Neuro-071714-033842
Servan-Schreiber D, Printz H, Cohen JD (1990) A network model of catecholamine effects: gain, signal-to-noise ratio, and behavior. Science 249(4971):892–895
Sheynin Y, Chamoun M, Baldwin AS, Rosa-Neto P, Hess RF, Vaucher E (2019) Cholinergic potentiation alters perceptual eye dominance plasticity induced by a few hours of monocular patching in adults. Front Neurosci 13:22. https://doi.org/10.3389/Fnins.2019.00022
Shimegi S, Kimura A, Sato A, Aoyama C, Mizuyama R, Tsunoda K, Ueda F, Araki S, Goya R, Sato H (2016) Cholinergic and serotonergic modulation of visual information processing in monkey V1. J Physiol Paris 110(1–2):44–51. https://doi.org/10.1016/J.Jphysparis.2016.09.001
Shirokawa T, Kasamatsu T, Kuppermann BD, Ramachandran VS (1989) Noradrenergic control of ocular dominance plasticity in the visual cortex of dark-reared cats. Brain Res Dev Brain Res 47(2):303–308. https://doi.org/10.1016/0165-3806(89)90187-9
Soma S, Shimegi S, Suematsu N, Sato H (2013a) Cholinergic modulation of response gain in the rat primary visual cortex. Sci Rep 3:1138. https://doi.org/10.1038/Srep01138
Soma S, Shimegi S, Suematsu N, Tamura H, Sato H (2013b) Modulation-specific and laminar-dependent effects of acetylcholine on visual responses in the rat primary visual cortex. PLoS One 8(7):E68430. https://doi.org/10.1371/Journal.Pone.0068430
Soma S, Suematsu N, Shimegi S (2013c) Cholinesterase inhibitor, donepezil, improves visual contrast detectability in freely behaving rats. Behav Brain Res 256:362–367. https://doi.org/10.1016/J.Bbr.2013.08.022
Spiegel DP, Li J, Hess RF, Byblow WD, Deng D, Yu M, Thompson B (2013) Transcranial direct current stimulation enhances recovery of stereopsis in adults with amblyopia. Neurotherapeutics 10(4):831–839. https://doi.org/10.1007/S13311-013-0200-Y
Spiegel DP, Baldwin AS, Hess RF (2017) Ocular dominance plasticity: inhibitory interactions and contrast equivalence. Sci Rep 7:39913. https://doi.org/10.1038/Srep39913
Stewart MR, Dringenberg HC (2016) Potential role of synaptic activity to inhibit LTD induction in rat visual cortex. Neural Plast 2016:1401935. https://doi.org/10.1155/2016/1401935
Sumner RL, Mcmillan R, Spriggs MJ, Campbell D, Malpas G, Maxwell E, Deng C, Hay J, Ponton R, Kirk IJ, Sundram F, Muthukumaraswamy SD (2019) Ketamine enhances visual sensory evoked

potential long-term potentiation in patients with major depressive disorder. Biol Psychiatry Cogn Neurosci Neuroimaging 5(1):45–55. https://doi.org/10.1016/J.Bpsc.2019.07.002

Sun YJ, Wu GK, Liu BH, Li P, Zhou M, Xiao Z, Tao HW, Zhang LI (2010) Fine-tuning of pre-balanced excitation and inhibition during auditory cortical development. Nature 465(7300):927–931. https://doi.org/10.1038/Nature09079

Takesian AE, Hensch TK (2013) Balancing plasticity/stability across brain development. Prog Brain Res 207:3–34. https://doi.org/10.1016/B978-0-444-63327-9.00001-1

Thiel CM, Fink GR (2008) Effects of the cholinergic agonist nicotine on reorienting of visual spatial attention and top-down attentional control. Neuroscience 152(2):381–390. https://doi.org/10.1016/J.Neuroscience.2007.10.061

Thompson B, Mansouri B, Koski L, Hess RF (2008) Brain plasticity in the adult: modulation of function in amblyopia with RTMS. Curr Biol 18(14):1067–1071. https://doi.org/10.1016/J.Cub.2008.06.052

Treviño M, Medina-Coss Y, León R, Lezama E (2019) Adrenergic modulation of visually-guided behavior. Front Synaptic Neurosci 11:9. https://doi.org/10.3389/Fnsyn.2019.00009

Van Kempen J, Panzeri S, Thiele A (2017) Cholinergic control of information coding. Trends Neurosci 40(9):522–524. https://doi.org/10.1016/J.Tins.2017.06.006

Van Loon AM, Knapen T, Scholte HS, St John-Saaltink E, Donner TH, Lamme VA (2013) GABA shapes the dynamics of bistable perception. Curr Biol 23(9):823–827. https://doi.org/10.1016/J.Cub.2013.03.067

Vaucher E, Laliberte G, Higgins MC, Maheux M, Jolicoeur P, Chamoun M (2019) Cholinergic potentiation of visual perception and vision restoration in rodents and humans. Restor Neurol Neurosci 37(6):553–569. https://doi.org/10.3233/Rnn-190947

Vedamurthy I, Nahum M, Huang SJ, Zheng F, Bayliss J, Bavelier D, Levi DM (2015) A dichoptic custom-made action video game as a treatment for adult amblyopia. Vis Res 114:173–187. https://doi.org/10.1016/J.Visres.2015.04.008

Warren RA, Dykes RW (1996) Transient and long-lasting effects of iontophoretically administered norepinephrine on somatosensory cortical neurons in halothane-anesthetized cats. Can J Physiol Pharmacol 74(1):38–57

West GL, Konishi K, Diarra M, Benady-Chorney J, Drisdelle BL, Dahmani L, Sodums DJ, Lepore F, Jolicoeur P, Bohbot VD (2018) Impact of video games on plasticity of the hippocampus. Mol Psychiatry 23(7):1566–1574. https://doi.org/10.1038/Mp.2017.155

Whelan FJ, Walker MS, Schultz SK (2000) Donepezil in the treatment of cognitive dysfunction associated with traumatic brain injury. Ann Clin Psychiatry 12(3):131–135

Yamahachi H, Marik SA, Mcmanus JN, Denk W, Gilbert CD (2009) Rapid axonal sprouting and pruning accompany functional reorganization in primary visual cortex. Neuron 64(5):719–729. https://doi.org/10.1016/J.Neuron.2009.11.026

Zhong WX, Dong ZF, Tian M, Cao J, Xu L, Luo JH (2006) N-methyl-D-aspartate receptor-dependent long-term potentiation in CA1 region affects synaptic expression of glutamate receptor subunits and associated proteins in the whole hippocampus. Neuroscience 141(3):1399–1413. https://doi.org/10.1016/J.Neuroscience.2006.04.070

Cognitive-Enhancing Substances and the Developing Brain: Risks and Benefits

Kimberly R. Urban and Wen-Jun Gao

1 Introduction

Humans have long sought ways to improve not only physical but cognitive prowess, from ancient civilizations consuming hallucinogenic plants as a means to alter consciousness so they could attempt to commune with their gods, to the popularity of caffeine-containing substances (coffee, soda, energy drinks), to the more recent interest in pharmaceutical nootropics such as psychostimulants, ampakines, and wakefulness-promoting drugs. However, along with the interest in cognitive enhancement come multiple ethical considerations. While one side of the argument states that self-improvement is merely an inherent human desire and cognitive enhancement is simply the next step in this age-old search, the other side argues that use of these substances may impart an unfair advantage to those wealthy enough to afford them, while others without means and access will be unable to achieve academic and career (Banjo et al. 2010; Butcher 2003; Cakic 2009; Franke et al. 2011; Goodman 2010; Schelle et al. 2015). Should cognitive-enhancing substances in academic and business settings be viewed the same way as steroids are in athletics? Do they constitute cheating, or are they merely helping individuals achieve their personal peak?

With the increasingly competitive nature of the workplace and education in the twenty-first century, interest in cognitive enhancement is rising, especially among the adolescent population (see Table 1). According to a 2013 study by the Center on

K. R. Urban
Division of Stress Neurobiology, Department of General Anesthesia, Children's Hospital of Philadelphia, Philadelphia, PA, USA

W.-J. Gao (✉)
Department of Neurobiology and Anatomy, Drexel University College of Medicine, Philadelphia, PA, USA
e-mail: wg38@drexel.edu

I. Opris et al. (eds.), *Modern Approaches to Augmentation of Brain Function*, Contemporary Clinical Neuroscience,
https://doi.org/10.1007/978-3-030-54564-2_29

Table 1 Misuse of prescription medications by age in 2017–2018 (percentages of United States population)

Drug	Middle school (2018) (%) (NIDA 2018)	10th graders (%) (NIDA 2018)	12th graders (%) (NIDA 2018)	Adults 18–25 (%) (SAMHSA 2018)	Adults over 25 (%) (SAMHSA 2018)
Adderall®	1.8	4.1	4.6	14.7	5.5
Ritalin®	0.5	0.8	0.9	14.7	5.5
Other Rx drugs	–	–	15.5	41.1	44.5
Amphetamine	5.9	8.6	8.6	–	–

[a]Statistics on abuse or misuse of common prescription cognitive-enhancing drugs in adolescents and adults in the United States in 2017–2018. Detailed statistics on newer nootropics such as ampakines and cholinesterase inhibitors are not available, highlighting the need for continued research. https://www.cdc.gov/nchs/fastats/drug-use-therapeutic.htm

Young Adult Health and Development at the University of Maryland, 61% of college students will be offered prescription stimulant drugs and over 30% will use these substances as "study drugs" at least once during their college years (Garnier-Dykstra et al. 2012). Nationwide, on the other hand, illicit drug use has decreased in 2018 among middle and high schoolers, including amphetamine-type drugs (The National Institute of Drug Addiction (NIDA) 2018). This survey covers the use of only illegal drugs, however, and may not reflect true black market use of prescription stimulants and related drugs. Furthermore, much of our current research into the safety and efficacy of cognitive-enhancing pharmaceuticals and supplements has been conducted in adults, leaving large gaps in the knowledge of how these substances may affect brain maturation in adolescents, who are often more likely to abuse them (Agay et al. 2010; Camp-Bruno and Herting 1994; Elliott et al. 1997; Kirschner et al. 2003; Linssen et al. 2012; Studer et al. 2010; Tomasi et al. 2010). In addition to ethical arguments, the prevalence of the unregulated use of stimulants and related cognitive-enhancing drugs raises multiple public health concerns. In this chapter, we will examine some of the most commonly used cognitive-enhancing drugs, and discuss their potential risks and benefits in the adolescent population (see Table 2 for a summary).

2 Cognitive Executive Function and the Prefrontal Cortex

The brain has an incredible capacity to form mental representations of the world around us, to hold information briefly while we process it and determine a reaction. For example, if someone tells you their phone number, you must briefly hold it in your mind so you can write it down. Or when you are driving to work, you must create a mental image of the route, and constantly update as you go, making turns where needed. This process of creating mental images and holding pertinent data briefly in the mind is dubbed "working memory" (Arnsten and Jin 2014). The

Table 2 Mechanisms of actions of selected purported nootropic drugs and their risks

Drug	Neurotransmitter(s) affected	Action	Risk(s)
Methylphenidate (FDA 2011; Novartis 2019a, b; Urban et al. 2012a, b, 2017)	Dopamine, norepinephrine	Reuptake inhibitor, increases in available neurotransmitter	Cardiac distress, insomnia, anorexia, impaired executive cognition, altered PFC development
Amphetamine (Advokat 2007; Calipari and Ferris 2013)	Dopamine, norepinephrine, serotonin	Reuptake inhibitor, increases in available neurotransmitter	Addiction, cardiac distress, insomnia, anorexia, impaired executive cognition, altered PFC development
Modafinil (Baranski et al. 2004; Battleday and Brem 2015; Esposito et al. 2013; FDA 2007)	Orexins, dopamine	Weak increases in neurotransmitter	Potential for similar actions as stimulants
Donepizil/galantamine (E.C. Ltd 2010; FDA 2014)	Acetylcholine	Acetylcholinesterase inhibitor, increases in neurotransmitter	Cholinergic crisis, sometimes known by the mnemonic "sludge syndrome"
Memantine (Aracava et al. 2005; Danysz and Parsons 2003; Ota and Godwin 2006; Rogawski and Wenk 2003)	Glutamate	NMDA receptor blocker, decreases in glutamate function	Reduced neuronal plasticity, learning and memory impairments
Ampakines (Lynch and Gall 2006)	Glutamate	AMPA receptor agonist, increases in glutamate	Glutamate excitotoxicity, overactivation of receptors, neuronal damage

[a]Detailed information on the most commonly sought prescription candidate drugs for cognitive enhancement, their actions on neurotransmitters, and risks to cognition and brain development. https://www.cdc.gov/nchs/fastats/drug-use-therapeutic.htm

prefrontal cortex, located in the anterior-most portion of the frontal lobe, behind the forehead, is the center of working memory. Working memory cannot only utilize information from an event that has just occurred, but it can reactivate stored memories and allow for the application of learned information to the current situation. This process is critical for behavioral inhibition and decision-making, one of the executive functions. Other executive functions include direction of attention—one must be able to shift focus from one stimulus to another as the salience (importance) of those stimuli changes (Arnsten and Li 2005; Funahashi and Kubota 1994). For example, a student in a classroom is taking notes while the teacher lectures. Suddenly, the fire alarm goes off, a much louder, more salient stimulus. The student must shift attention from the teacher's voice to the fire alarm and determine an appropriate response: get up and follow the teacher safely out of the building. Humans have a uniquely large prefrontal cortex, with an increased number of neurons, dendritic branches and connections between neurons compared to other mammals like rodents (Elston 2000, 2003); thus, human working memory may be more precise while other mammals may need to rely more on generalized strategies.

3 Development of the Prefrontal Cortex and a Potential "Vulnerable Period"

Despite how it may feel as we grow and age, the brain does not develop and mature in a purely linear fashion; rather, it matures in bursts, with certain regions finishing before other cortical regions (Kolb et al. 2012). Once one learns how to walk, or pick up an object, it would be very inconvenient if these neural pathways were susceptible to change. You wouldn't want to throw across the room the coffee cup you were trying to drink from! On the other hand, you need the ability to form new memories, and to modify or extinguish existing memories, throughout life. You also need the ability to respond to situations individually; there is no one-size-fits-all approach. Therefore, neural pathways involved in these "executive functions" of judgment, decision-making, behavioral inhibition and the like remain plastic throughout life. It is the prefrontal cortex, located in the anterior-most portion of the frontal lobe, which controls executive functions and initiates memory formation. Excitatory pyramidal neurons in the prefrontal cortex have a unique composition of glutamate *N*-methyl-D-aspartate (NMDA) receptors that impart a high level of plasticity on them throughout adulthood, allowing for working memory and executive functions (Wang et al. 2008). This high level of plasticity, although necessary for proper control of behavior and attention, may increase the vulnerability of prefrontal cortex to perturbations such as stress and drugs of abuse (Monaco et al. 2015). In fact, many stress-related psychiatric disorders such as depression and anxiety syndromes have symptoms of impairments in executive function. Furthermore, medications for attention-deficit/hyperactivity disorder that improve attention and cognitive focus have been shown to have an inverted-U function on prefrontal cortex (Urban et al. 2014, 2017; Urban and Gao 2014): small doses improve executive function while increasingly higher doses begin to impair these same cognitive processes.

Furthermore, not only is prefrontal cortical function highly tuned with an optimal range, and similar behavioral impairments on both sides of this range, but the prefrontal cortex itself does not finish developing until much later than other cortical regions (Fig. 1) (Kolb et al. 2012; Monaco et al. 2015; Lewis 1997). There is a large surge of prefrontal cortical maturation shortly after birth, then a plateau throughout childhood, and a final surge during puberty that lasts until early adulthood. In humans, prefrontal cortex finishes maturing during the early part of the third decade of life (Lewis 1997). Pathways involved in motivation and reward mature first, while those involved in decision-making and behavioral inhibition fin-

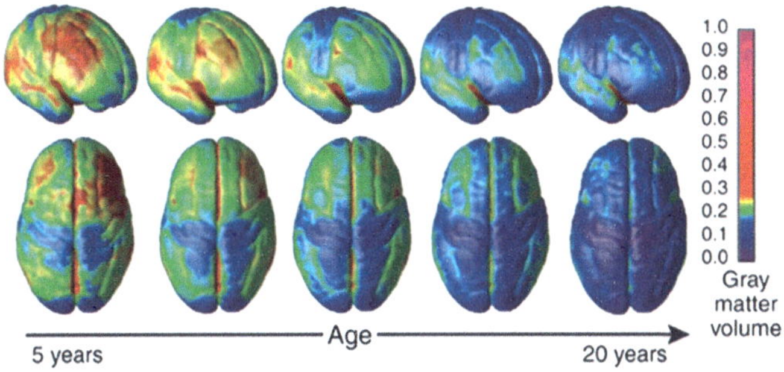

Fig. 1 Grey matter development in human cortex. Anatomical MRI scans taken from 13 individuals age 4–21 were repeated at 2-year intervals to examine the development of the grey matter in their cortices. During maturation, the cortex grey matter thins due to the pruning of excessive synaptic connectivity. This process allows for the fine-tuning of neural circuits; with only neurons that are highly connected and active in functional circuits remaining, signals traveling along these circuits can be strengthened, while less active connections can be removed. The prefrontal cortex is the final cortical region to undergo this maturation process. In humans, this occurs during the third decade of life. [*Reproduced from Tau and Peterson 2009 (Copyright 2010) Neuropsychopharmacology and Gogtay* et al *(2004) (Copyright 2004) National Academy of Sciences, USA* (Permission obtained from the Copyright Clearance Center's *RightsLink®* service)]

ish last. During this period of maturation, the levels of dopamine and norepinephrine rise, and are critically important for proper development (Xing et al. 2016); thus, chemicals or stressors that alter these neurotransmitters may have more damaging impacts during adolescence, when prefrontal cortex is still maturing (Arnsten 2015; Arnsten et al. 2015; Connor et al. 2015). In the next sections, we will discuss the main classes of drugs being used for cognitive enhancement, their potential risks, and their efficacy.

4 Psychostimulants

One of the first, and possibly most well-known, class of cognitive-enhancing drugs is the psychostimulants: methylphenidate (Ritalin®, Concerta®) and amphetamine (Adderall®, Vyvanse®). These medications work by inhibiting reuptake transporters for norepinephrine and dopamine, effectively raising the available concentration of the neurotransmitters in synapses. Amphetamine raises levels of dopamine, norepinephrine, and serotonin, while methylphenidate acts only on dopamine and norepinephrine, which gives it a lower potential for addiction and abuse. Both medications are approved for the treatment of attention-deficit/hyperactivity disorder; amphetamine was approved first, and methylphenidate emerged in the 1960s for use as an ADHD treatment, after being developed to treat narcolepsy, depression, and barbiturate-induced comas (Fry 1998; Lange et al. 2010; Mitler 1994). A large body of research details the efficacy of both psychostimulants for the treatment of ADHD symptoms, which are thought to arise from hypoactivity of the dopamine and norepinephrine systems in the prefrontal cortex. There is little doubt that low, therapeutic doses of these psychostimulants (5–30 mg twice daily for up to 60 mg per day for immediate-release formulas (Kidd 2000) or 10–40 mg once daily for extended-release formulas (Anderson and Keating 2006; Kowalik et al. 2006)) provide relief of locomotor hyperactivity, inattention, and lack of focus in affected individuals (Cepeda et al. 2000; de Sonneville et al. 1994; Kramer et al. 2001; Malone and Swanson 1993; Scheres et al. 2003; Tannock et al. 2000; Trommer et al. 1991). However, not all individuals with ADHD will display every symptom, and not every patient will note improvement of all of their symptoms with psychostimulant treatment; thus, doctors balance overall improvement with experience of side effects (Swanson et al. 2011). These side effects include lack of appetite and weight loss, insomnia, increased blood pressure, tachycardia, sweating, anxiety, and even tics (Novartis 2019b; Urban and Gao 2015). At higher doses, both drugs have been known to produce stereotypic (repetitive, non-goal-oriented) behaviors (Novartis 2019a; Shire 2013).

Interest in the psychostimulants as nootropics emerged from the plethora of studies performed on healthy adult animals and adult humans that showed that methylphenidate improved aspects of cognition in these individuals (Agay et al. 2010; Linssen et al. 2012; Urban and Gao 2015; Arakawa 1994; Askenasy et al. 2007; Navarra et al. 2017). However, it is important to note that these studies examined effects in adults, whereas psychostimulants are often abused by adolescents, who obtain them from classmates or friends with prescriptions, in order to attempt to boost their scholastic performance (Franke et al. 2011; Goodman 2010; Schelle et al. 2015; Liakoni et al. 2015; Partridge et al. 2011). Due to the delayed maturation of the prefrontal cortex, and the critical involvement of dopamine and norepinephrine in this development, abuse during adolescence raises the risk of lasting damage (Fig. 2). Indeed, multiple studies have shown differential effects of psychostimulants in adolescents versus adults. Methylphenidate exposure during adolescence increases anxiety, even in adulthood (Bolanos et al. 2003; Vendruscolo et al. 2008). It was also shown to reduce social play behavior, impair pattern learning, and lead to enhanced reactivity to stimulants during adulthood (Griggs et al. 2010; Izquierdo et al. 2016; Rowan et al. 2015; Vanderschuren et al. 2008). Even a low therapeutic dose that improves cognition in the adult can potentially impair the adolescent brain. Methylphenidate during early adolescence reduced the activity of excitatory neurons in the prefrontal cortex while increasing the activity of inhibitory neurons, and impairing plasticity (Urban et al. 2012a, b, 2017). Overall, the current literature suggests that the adolescent brain may be more sensitive to the effects of psychostimulants than the adult brain, and that abuse during this time period may lead to persistent changes in prefrontal cortex-controlled behaviors.

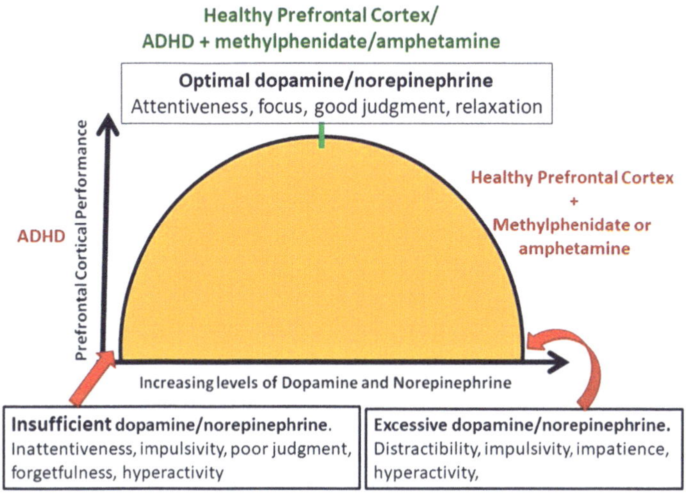

Fig. 2 How psychostimulants affect the adolescent brain. Prefrontal cortical function as a result of levels of dopamine and norepinephrine can be represented with an "inverted-U curve." When dopamine and norepinephrine levels are insufficient, as is thought to be the case with ADHD, the prefrontal cortical activity is low, and symptoms of impulsivity, hyperactivity, poor judgment and inhibition, and hyperactivity are evident as a result of executive function failure. As levels of dopamine and norepinephrine rise, as in the healthy prefrontal cortex, or an ADHD patient being treated with psychostimulants, executive functions improve along with prefrontal cortical activity, leading to proper behavioral control, attentiveness, and ability to focus. However, if levels of dopamine and norepinephrine become excessive beyond the optimal point, as can happen if a healthy individual ingests psychostimulants, we begin to see overactivation of prefrontal cortex and the emergence of symptoms akin to those seen with insufficient activation: impulsivity, hyperactivity, impatience, distractibility

5 Modafinil and Wakefulness-Promoting Drugs

Another class of medications commonly considered to be cognitively enhancing is the wakefulness-promoting drugs, the most common of which is Modafinil (Provigil®, Alertec®). Modafinil was approved in 1998 for the treatment of narcolepsy, shift work sleep disorder, and sleepiness associated with sleep apnea (FDA 2007). It is also used in the military to help troops maintain vigilance on long, intense missions (Estrada et al. 2012; Gore et al. 2010; Murillo-Rodriguez et al. 2018; Parker and Parker 2017). The complete mechanism of action of modafinil remains unknown; however, it does act as a weak dopamine reuptake inhibitor, and it is thought that this mechanism indirectly increases the release of arousing orexins in the lateral hypothalamus (Gerrard and Malcolm 2007; Ishizuka et al. 2012). This use in the military, along with studies showing that modafinil could improve mood and wakefulness in healthy adults, led to an interest in its usefulness as a nootropic. Modafinil has been shown to improve recall and memory of multiple domains including patterns, letters, and digits, and it has also been shown to improve "mental manipulation" of numbers (performing basic mathematical functions such as addition, subtraction, multiplication, and division in the mind) (Baranski et al. 2004; Muller et al. 2004; Randall et al. 2005a; Turner et al. 2003). However, a recent meta-analysis calls into question the efficacy of modafinil in improving actual cognitive function (Battleday and Brem 2015). They showed that studies that relied on simple psychometric measures, such as alertness (attention towards an isolated stimulus), selective attention (ability to isolate stimulus during distractions), and sustained attention (ability to maintain attention to stimulus over time), generally reported no effect, or very limited improvement, of modafinil treatment (Baranski et al. 2004; Muller et al. 2004; Randall et al. 2003, 2004, 2005a, b; Turner et al. 2003; Gilleen et al. 2014; Liepert et al. 2004; Theunissen et al. 2009; Winder-Rhodes et al. 2010).

However, when more complex measures were examined, such as executive functions, modafinil appears to have more beneficial effects (Battleday and Brem 2015). Modafinil improves working memory, inhibitory control, planning, and decision-making (Esposito et al. 2013; Muller et al. 2004, 2013; Turner et al. 2003; Winder-Rhodes et al. 2010; Rycroft et al. 2007). Thus, it can be generally stated that modafinil's utility as a nootropic is limited to its complex effects on higher-order executive functions, and that it may actually impair certain simpler aspects of attention (Battleday and Brem 2015).

Along with its vague mechanism of action, modafinil has a profile of side effects and potential for addiction, along with risk to development that is quite similar to the psychostimulants methylphenidate and amphetamine, due to its actions on the dopamine reuptake transporter. These include insomnia, nausea, diarrhea, nervousness, increased heart rate, and sweating, along with skin rashes (FDA 2007). There is very little information regarding potential developmental effects of modafinil, with most studies having been performed on adults; however, due to its similarities to the psychostimulants, it may be likely that adolescent abuse of modafinil can result in similar impairments in prefrontal cortical function as methylphenidate (Urban et al. 2012a, b, 2017; Urban and Gao 2014).

6 Glutamate-Modulating Drugs

In addition to psychostimulants and wakefulness-promoting drugs, medications used to treat Alzheimer's are sometimes considered for cognitive enhancement in healthy individuals. Donepezil (Aricept ©) and galantamine (Razadyne ©) are both anticholinesterases, leading to an increase in acetylcholine at synapses (E.C. Ltd 2010; FDA 2014). In 2002, a study be Yesavage et al. reported that healthy pilots performed complex memory retention tasks in a flight simulator better after receiving donepezil, and this led to reports of varying accuracy stating the drug's utility as a general nootropic (Wade et al. 2014; Yesavage et al. 2002). Enhancement of cholinergic transmission has been shown to improve complex visual processing and spatial perception, linked to the hippocampus and medial septum (Chamoun et al. 2017; Gratton et al. 2017; Sors et al. 2016). However, due to their activity on acetylcholine, drugs like donepezil and galantamine can result in cholinergic crisis, a syndrome of excessive sweating, salivation, severe nausea and vomiting, and respiratory distress (Adeyinka and Kondamudi 2018; Ohbe et al. 2018). Other compounds commonly responsible for cholinergic crisis include nerve agents like sarin and some snake venoms. In addition, these drugs do not seem effective for mild cognitive impairment, and the side effect profile outweighs any benefits (Tricco et al. 2012, 2013). While it is unlikely that any particular age-related risks exist with these medications, it is also unlikely that they will be attractive options to youth seeking cognitive enhancement, due to the incidence of nausea, vomiting, and other gastrointestinal symptoms.

Another Alzheimer's medication, memantine (Axura®, Ebixa®, Namenda®) may have more insidious risks, especially to developing young brains. Memantine is an inhibitor of glutamate *N*-methyl-D-aspartate (NMDA) receptors, reducing excitotoxicity that is thought to be a component of Alzheimer's (Rogawski and Wenk 2003). Memantine also acts as an antagonist at the serotonin 5-HT3 receptor and at nicotinic acetylcholine receptors, similar to donepezil (Aracava et al. 2005; Danysz and Parsons 2003; Rogawski and Wenk 2003; Rammes et al. 2001; Seeman et al. 2008). However, it is the NMDA receptor antagonism that is particularly concerning regarding development and cognitive function. Glutamate is the primary excitatory neurotransmitter in the brain, and NMDA receptors, in particular, are critical for synaptic modulation associated with learning. NMDA receptors are particularly important for the modulation of long-term potentiation and depression, and remain in particularly high concentration in the prefrontal cortex throughout adulthood, allowing for continuing plasticity and learning throughout life (Wang et al. 2008; Cull-Candy et al. 2001; Paoletti and Neyton 2007). Hypofunction of NMDA receptors has been associated with schizophrenia, depression, and age-related dementia (Billard and Rouaud 2007; Brigman et al. 2010; Murai et al. 2007; Nabeshima et al. 2006). Thus, giving memantine to a healthy individual, particularly a young individual whose brain is still developing, may result in not only impairments in certain aspects of cognitive performance reliant on learning and memory, but also raise the risk for development of schizophrenia or early-onset dementia.

How then could this drug even be considered for nootropic use? The answer seems to lie in its effects on *N*-acetyl-aspartate (NAA), a neuronal marker thought to aid in energy production from glutamate (Rao et al. 2001). Patients with Alzheimer's show a steady decline in levels of NAA, and healthy individuals with high intelligence scores show a higher level of NAA than those with lower intelligence scores (Jung et al. 1999). However, the potential benefit of richer synaptic innervation and reduction of the risk of glutamate excitotoxicity, which is not an issue in the healthy developing brain, are likely not outweighed by the risks of depleting such a critical receptor as the NMDA receptor and very little work has been done systematically evaluating the efficacy of memantine in healthy individuals, especially young ones.

Another class of glutamate-modulating drugs that have been considered for nootropic use is the ampakines. Ampakines are positive allosteric modulators of the α-amino-3-hydroxy-5-methyl-4-isoxazolepropionic acid (AMPA) receptor (Suppiramaniam et al. 2001). AMPA receptors, like NMDA receptors, are critically involved in neuronal plasticity underlying learning and memory (Sheng and Lee 2001). Ampakines, like piracetam (Nootropil®), have been investigated for the treatment of depression, Parkinson's disease, schizophrenia, and ADHD, and have been examined in limited studies in healthy adults (Lynch and Gall 2006; Fond et al. 2015). However, these medications are largely in development, with few widely available, and little is known about their potential risks. One major concern with increasing glutamate transmission is the risk of excitotoxicity, neuronal death that occurs from excessive calcium influx brought on by over-transmission of glutamate (Manev et al. 1989; Watkins and Jane 2006). It is not known how quickly or at what dosage ampakines might lead to excess glutamate transmission in the healthy brain, but far more research is needed before these medications can be fully evaluated for cognitive-enhancing capabilities.

In conclusion, many classes of drugs are considered for potential cognitive enhancement, but of these, the psychostimulants are not only the most widely available, but also the best studied and most likely to offer mild cognitive enhancement in adults. Despite the public acceptance of these drugs, interest in cognitive enhancement, and the continued development of new classes of drugs with potential benefits, there is still much work to be done in order to evaluate them.

7 The Future of Cognitive Enhancers

Increased competitiveness in school and the workplace has led to increased interest in cognitive-enhancing substances, particularly in high-stress, high stakes fields such as high school and college competitions, medical professionals, military services, and lawyers (Estrada et al. 2012; Baker and Forbes-Ewan 2017). This has led to ethical concerns, such as whether taking such substances constitutes cheating in educational settings, and whether or not access to cognitive enhancers gives an unfair edge to those wealthy enough to afford them (Banjo et al. 2010; Butcher 2003; Cakic 2009; Franke et al. 2011; Goodman 2010). Steroids have been ruled cheating in sports, as they give athletes an unnatural advantage in muscle building, stamina, and performance, and cognitive-enhancing substances purportedly increase an individual's capacity for attention, focus, and executive functions beyond that individual's normal ceiling, so should they be considered "steroids for the brain" (Goodman 2010)? How would we go about testing for the presence of cognitive-enhancing substances? Modafinil is not regularly included in drug screenings, but due to its half-life of 10–15 h, it would theoretically be detectable in urine for up to 30 h after its ingestion (Pro 2018; Moeller et al. 2008). Methylphenidate and amphetamine would show up in the amphetamines screening of drug tests, and may be detectable in urine for 1–3 days after use (Moeller et al. 2008; Buddy 2019). There is currently no urine or blood test standardized for ampakines or other glutamatergic drugs. In addition, the last few years have seen an increase in the number and variety of over-the-counter nootropic supplements available for consumers, and these contain largely plant extracts and vitamins that would not show up on standard drug tests. Therefore, we are tasked with the need to develop reliable, noninvasive panels to test for a wide variety of cognitive-enhancing substances if we wish to label them illegal and ensure they are not being used in academic settings, although this would raise its own set of ethical concerns.

In addition to ethical concerns, the availability of over-the-counter nootropic supplements raises the question of whether any of these may be effective. A quick Google search results in dozens of brands being offered. These supplements promise increases in focus, mood, memory, mental clarity, and creativity, but most contain large doses of B vitamins, amino acids, and as much caffeine as a cup of coffee. It is likely that the caffeine is providing the biggest boost to cognitive function, but it is temporary, and unlikely to provide lasting changes. Furthermore, these substances are not subject to FDA regulations, and therefore may be dangerously variable in their actual concentrations of ingredients (Brodwin 2017; Guallar et al. 2013). Since there is no governmental regulation on over-the-counter supplements, there is currently no reliable, controlled and peer-reviewed scientific research into their safety or efficacy, and this may be particularly concerning for adolescents whose brains are still developing.

It is, therefore, the responsibility of scientists to push for continued research into the age-specific effects of drugs considered for cognitive enhancement, so that we can determine not only their safety and efficacy, but their potential ramifications for brain development and maturation. In the last few years, studies have increased, and many have shown potential harmful effects of stimulants on adolescent brains (Schmitz et al. 2012, 2016, 2017; Yang et al. 2011). However, there is currently no research into harmful effects of modafinil or glutamatergic drugs on the juvenile brain, which is concerning considering the interest in these drugs as cognitive enhancers, and their availability.

In addition, it is the consumer's responsibility to carefully examine ingredients lists on over-the-counter nootropics and be wary of any grand claims these companies make. Many list scientific studies purportedly supporting their claims; however, these studies may examine a single ingredient at higher doses than is in the final product, or may actually be unrelated, and many media outlets have begun reporting on these discrepancies (Watling 2019; Dreyfuss 2017).

What is the struggling student or stressed worker to do? While one may wish that a pill could make them genius and ease their stress, there is simply no miracle for enhancing the brain. While one can hone certain aspects of cognitive ability through practice and hard work, each of us is born with a limit to our capabilities, and getting past that limit is currently not possible. Currently, the best research suggests that regular physical exercise, healthy social relationships, and the use of puzzles (number, crossword, word find, etc.) are the best ways to maintain cognitive function and prevent age-related declines (Alkadhi 2018; Brooker et al. 2019; Herting and Chu 2017; Kelly et al. 2017). Thus, there is no magic pill for cognition, but exercising the brain along with exercising the body may provide the most reliable and long-lasting benefits to our cognitive power. We must continue to be skeptical of supplements claiming outrageous results and wary of prescription medications that may alter brain development and maturation processes, and push for continued controlled peer-reviewed, large-scale scientific studies to put these substances to the test.

Acknowledgments This study is supported by the NIH R01MH085666 to W. J. Gao.

Financial Disclosures and Conflict of Interests: The authors declare no financial and any other conflict of interest.

Author Contribution: Both KRU and WJG wrote the paper.

References

Adeyinka A, Kondamudi NP (2018) Cholinergic crisis. In: StatPearls. StatPearls Publishing LLC, Treasure Island, FL

Advokat C (2007) Update on amphetamine neurotoxicity and its relevance to the treatment of ADHD. J Atten Disord 11(1):8–16

Agay N et al (2010) Non-specific effects of methylphenidate (Ritalin) on cognitive ability and decision-making of ADHD and healthy adults. Psychopharmacology 210(4):511–519

Alkadhi KA (2018) Exercise as a positive modulator of brain function. Mol Neurobiol 55(4):3112–3130

Anderson VR, Keating GM (2006) Methylphenidate controlled-delivery capsules (EquasymXL, Metadate CD): a review of its use in the treatment of children and adolescents with attention-deficit hyperactivity disorder. Paediatr Drugs 8(5):319–333

Aracava Y et al (2005) Memantine blocks alpha7* nicotinic acetylcholine receptors more potently than n-methyl-D-aspartate receptors in rat hippocampal neurons. J Pharmacol Exp Ther 312(3):1195–1205

Arakawa O (1994) Effects of methamphetamine and methylphenidate on single and paired rat open-field behaviors. Physiol Behav 55(3):441–446

Arnsten AF (2015) Stress weakens prefrontal networks: molecular insults to higher cognition. Nat Neurosci 18(10):1376–1385

Arnsten AF, Jin LE (2014) Molecular influences on working memory circuits in dorsolateral prefrontal cortex. Prog Mol Biol Transl Sci 122:211–231

Arnsten AF, Li BM (2005) Neurobiology of executive functions: catecholamine influences on prefrontal cortical functions. Biol Psychiatry 57(11):1377–1384

Arnsten AF et al (2015) The effects of stress exposure on prefrontal cortex: translating basic research into successful treatments for post-traumatic stress disorder. Neurobiol Stress 1:89–99

Askenasy EPT, Taber KH, Yang PB, Dafny N (2007) Methylphenidate (Ritalin): behavioral studies in the rat. Int J Neurosci 117:757–794

Baker B, Forbes-Ewan C (2017) Military effectiveness of five dietary supplements purported to aid cognitive and physical performance. J Mil Veterans Health 25(2):35–47

Banjo OC, Nadler R, Reiner PB (2010) Physician attitudes towards pharmacological cognitive enhancement: safety concerns are paramount. PLoS One 5(12):e14322

Baranski JV et al (2004) Effects of modafinil on cognitive and meta-cognitive performance. Hum Psychopharmacol 19(5):323–332

Battleday RM, Brem AK (2015) Modafinil for cognitive neuroenhancement in healthy non-sleep-deprived subjects: a systematic review. Eur Neuropsychopharmacol 25(11):1865–1881

Billard JM, Rouaud E (2007) Deficit of NMDA receptor activation in CA1 hippocampal area of aged rats is rescued by D-cycloserine. Eur J Neurosci 25(8):2260–2268

Bolanos CA et al (2003) Methylphenidate treatment during pre- and periadolescence alters behavioral responses to emotional stimuli at adulthood. Biol Psychiatry 54(12):1317–1329

Brigman JL, Wright T, Talani G, Prasad-Mulcare S, Jinde S, Seabold GK, Mathur P, Davis MI, Bock R, Gustin RM, Colbran RJ, Alvarez VA, Nakazawa K, Delpire E, Lovinger DM, Holmes A (2010) Loss of Glu-N2B-containing NMDA receptors in CA1 hippocampus and cortex impairs long-term depression, reduces dendritic spine density, and disrupts learning. J Neurosci 30:4590–4600

Brodwin E (2017) The $37 billion supplement industry is barely regulated—and it's allowing dangerous products to slip through the cracks. Business Insider
Brooker H et al (2019) The relationship between the frequency of number puzzle use and baseline cognitive function in a large online sample of adults aged 50 and over. Int J Geriatr Psychiatry
Buddy T (2019) How long does methylphenidate stay in your system? Very well mind 2019 [cited 27 Feb 2019]. https://www.verywellmind.com/how-long-does-methylphenidate-stay-in-your-system-80285
Butcher J (2003) Cognitive enhancement raises ethical concerns. Academics urge pre-emptive debate on neurotechnologies. Lancet 362(9378):132–133
Cakic V (2009) Smart drugs for cognitive enhancement: ethical and pragmatic considerations in the era of cosmetic neurology. J Med Ethics 35(10):611–615
Calipari ES, Ferris MJ (2013) Amphetamine mechanisms and actions at the dopamine terminal revisited. J Neurosci 33(21):8923–8925
Camp-Bruno JA, Herting RL (1994) Cognitive effects of milacemide and methylphenidate in healthy young adults. Psychopharmacology 115(1–2):46–52
Cepeda NJ, Cepeda ML, Kramer AF (2000) Task switching and attention deficit hyperactivity disorder. J Abnorm Child Psychol 28(3):213–226
Chamoun M et al (2017) Cholinergic potentiation improves perceptual-cognitive training of healthy young adults in three dimensional multiple object tracking. Front Hum Neurosci 11:128
Connor DF et al (2015) An update on posttraumatic stress disorder in children and adolescents. Clin Pediatr (Phila) 54(6):517–528
Cull-Candy S, Brickley S, Farrant M (2001) NMDA receptor subunits: diversity, development and disease. Curr Opin Neurobiol 11(3):327–335
Danysz W, Parsons CG (2003) The NMDA receptor antagonist memantine as a symptomatological and neuroprotective treatment for Alzheimer's disease: preclinical evidence. Int J Geriatr Psychiatry 18(Suppl 1):S23–S32
de Sonneville LM, Njiokiktjien C, Bos H (1994) Methylphenidate and information processing. Part 1: differentiation between responders and nonresponders; part 2: efficacy in responders. J Clin Exp Neuropsychol 16(6):877–897
Dreyfuss E (2017) Don't fall for the 'memory' pills targeting baby boomers. Wired
E.C. Ltd (2010) Aricept (donepezil hydrochloride) package insert. In: E.C. Ltd (ed). Woodcliff Lake, NJ
Elliott R et al (1997) Effects of methylphenidate on spatial working memory and planning in healthy young adults. Psychopharmacology 131(2):196–206
Elston GN (2000) Pyramidal cells of the frontal lobe: all the more spinous to think with. J Neurosci 20(18):RC95
Elston GN (2003) Cortex, cognition and the cell: new insights into the pyramidal neuron and prefrontal function. Cereb Cortex 13(11):1124–1138
Esposito R et al (2013) Acute effects of modafinil on brain resting state networks in young healthy subjects. PLoS One 8(7):e69224
Estrada A et al (2012) Modafinil as a replacement for dextroamphetamine for sustaining alertness in military helicopter pilots. Aviat Space Environ Med 83(6):556–564
FDA (2007) Provigil (R) (modafinil) Tablets [C-IV]. In: F.D. Administration (ed). Federal Drug Administration
FDA (2011) Ritalin-SR medication guide, rev. 2007 May 31
FDA (2014) Galantamine hydrobromide (trademark). In: U.S.F.a.D. Administration (ed). Washington, DC
Fond G et al (2015) Innovative mechanisms of action for pharmaceutical cognitive enhancement: a systematic review. Psychiatry Res 229(1–2):12–20
Franke AG et al (2011) Non-medical use of prescription stimulants and illicit use of stimulants for cognitive enhancement in pupils and students in Germany. Pharmacopsychiatry 44(2):60–66
Fry JM (1998) Treatment modalities for narcolepsy. Neurology 50(2 Suppl 1):S43–S48
Funahashi S, Kubota K (1994) Working memory and prefrontal cortex. Neurosci Res 21(1):1–11

Garnier-Dykstra LM et al (2012) Nonmedical use of prescription stimulants during college: four-year trends in exposure opportunity, use, motives, and sources. J Am Coll Heal 60(3):226–234
Gerrard P, Malcolm R (2007) Mechanisms of modafinil: a review of current research. Neuropsychiatr Dis Treat 3(3):349–364
Gilleen J et al (2014) Modafinil combined with cognitive training is associated with improved learning in healthy volunteers—a randomised controlled trial. Eur Neuropsychopharmacol 24(4):529–539
Goodman R (2010) Cognitive enhancement, cheating, and accomplishment. Kennedy Inst Ethics J 20(2):145–160
Gore RK, Webb TS, Hermes ED (2010) Fatigue and stimulant use in military fighter aircrew during combat operations. Aviat Space Environ Med 81(8):719–727
Gratton C et al (2017) Cholinergic, but not dopaminergic or noradrenergic, enhancement sharpens visual spatial perception in humans. J Neurosci 37(16):4405–4415
Griggs R et al (2010) Intermittent methylphenidate during adolescent development produces locomotor hyperactivity and an enhanced response to cocaine compared to continuous treatment in rats. Pharmacol Biochem Behav 96(2):166–174
Guallar E et al (2013) Enough is enough: stop wasting money on vitamin and mineral supplements. Ann Intern Med 159(12):850–851
Herting MM, Chu X (2017) Exercise, cognition, and the adolescent brain. Birth Defects Res 109(20):1672–1679
Ishizuka T, Murotani T, Yamatodani A (2012) Action of modafinil through histaminergic and orexinergic neurons. Vitam Horm 89:259–278
Izquierdo A et al (2016) Sex differences, learning flexibility, and striatal dopamine D1 and D2 following adolescent drug exposure in rats. Behav Brain Res 308:104–114
Jung RE et al (1999) Biochemical markers of intelligence: a proton MR spectroscopy study of normal human brain. Proc Biol Sci 266(1426):1375–1379
Kelly ME et al (2017) The impact of social activities, social networks, social support and social relationships on the cognitive functioning of healthy older adults: a systematic review. Syst Rev 6(1):259
Kidd P (2000) Attention deficit/hyperactivity disorder (ADHD) in children: rationale for its integrative management. Altern Med Rev 5(5):26
Kirschner J et al (2003) Methylphenidate enhances both intracortical inhibition and facilitation in healthy adults. Pharmacopsychiatry 36(2):79–82
Kolb B et al (2012) Experience and the developing prefrontal cortex. Proc Natl Acad Sci U S A 109(Suppl 2):17186–17193
Kowalik S, Minami H, Silva R (2006) Dexmethylphenidate extended-release capsules for the treatment of attention deficit hyperactivity disorder. Expert Opin Pharmacother 7(18):2547–2557
Kramer AF, Cepeda NJ, Cepeda ML (2001) Methylphenidate effects on task-switching performance in attention-deficit/hyperactivity disorder. J Am Acad Child Adolesc Psychiatry 40(11):1277–1284
Lange KW et al (2010) The history of attention deficit hyperactivity disorder. Atten Defic Hyperact Disord 2(4):241–255
Lewis DA (1997) Development of the prefrontal cortex during adolescence: insights into vulnerable neural circuits in schizophrenia. Neuropsychopharmacology 16(6):385–398
Liakoni E et al (2015) The use of prescription drugs, recreational drugs, and "soft enhancers" for cognitive enhancement among Swiss secondary school students. PLoS One 10(10):e0141289
Liepert J, Allstadt-Schmitz J, Weiller C (2004) Motor excitability and motor behaviour after modafinil ingestion—a double-blind placebo-controlled cross-over trial. J Neural Transm (Vienna) 111(6):703–711
Linssen AM et al (2012) Methylphenidate produces selective enhancement of declarative memory consolidation in healthy volunteers. Psychopharmacology 221(4):611–619
Lynch G, Gall CM (2006) Ampakines and the threefold path to cognitive enhancement. Trends Neurosci 29(10):554–562

Malone MA, Swanson JM (1993) Effects of methylphenidate on impulsive responding in children with attention-deficit hyperactivity disorder. J Child Neurol 8(2):157–163

Manev H et al (1989) Delayed increase of Ca^{2+} influx elicited by glutamate: role in neuronal death. Mol Pharmacol 36(1):106–112

Mitler MM (1994) Evaluation of treatment with stimulants in narcolepsy. Sleep 17(8 Suppl):S103–S106

Moeller KE, Lee KC, Kissack JC (2008) Urine drug screening: practical guide for clinicians. Mayo Clin Proc 83(1):66–76

Monaco SA, Gulchina Y, Gao WJ (2015) NR2B subunit in the prefrontal cortex: a double-edged sword for working memory function and psychiatric disorders. Neurosci Biobehav Rev 56:127–138

Muller U et al (2004) Effects of modafinil on working memory processes in humans. Psychopharmacology 177(1–2):161–169

Muller U et al (2013) Effects of modafinil on non-verbal cognition, task enjoyment and creative thinking in healthy volunteers. Neuropharmacology 64:490–495

Murai R et al (2007) Hypofunctional glutamatergic neurotransmission in the prefrontal cortex is involved in the emotional deficit induced by repeated treatment with phencyclidine in mice: implications for abnormalities of glutamate release and NMDA-CaMKII signaling. Behav Brain Res 180(2):152–160

Murillo-Rodriguez E et al (2018) An overview of the clinical uses, pharmacology, and safety of modafinil. ACS Chem Neurosci 9(2):151–158

Nabeshima T et al (2006) Animal model of schizophrenia: dysfunction of NMDA receptor-signaling in mice following withdrawal from repeated administration of phencyclidine. Ann N Y Acad Sci 1086:160–168

Navarra RL et al (2017) Methylphenidate enhances early-stage sensory processing and rodent performance of a visual signal detection task. Neuropsychopharmacology 42(6):1326–1337

NIDA (2018) Monitoring the future study: trends in prevalence of various drugs. In: NIDA (ed). NIDA, Washington, DC

Novartis (2019a) Ritalin(c) LA. Highlights of prescribing information. Novartis

Novartis (2019b) Ritalin LA (R) (methylphenidate hydrochloride) extended-release capsules for oral use, CII. In: F.D. Administration (ed). Novartis

Ohbe H et al (2018) Cholinergic crisis caused by cholinesterase inhibitors: a retrospective nation-wide database study. J Med Toxicol 14(3):237–241

Ota KS, Godwin T (2006) Memantine: the next trend in academic performance enhancement? J Am Osteopath Assoc 106(6):358–359

Paoletti P, Neyton J (2007) NMDA receptor subunits: function and pharmacology. Curr Opin Pharmacol 7(1):39–47

Parker RS, Parker P (2017) The impact of sleep deprivation in military surgical teams: a systematic review. J R Army Med Corps 163(3):158–163

Partridge BJ et al (2011) Smart drugs "as common as coffee": media hype about neuroenhancement. PLoS One 6(11):e28416

Pro A (2018) Does modafinil show up on drug tests. 25 Aug 2018. [cited 27 Feb 2019]. https://modafinil.org/modafinil-drug-test/

Rammes G et al (2001) The N-methyl-D-aspartate receptor channel blockers memantine, MRZ 2/579 and other amino-alkyl-cyclohexanes antagonise 5-HT(3) receptor currents in cultured HEK-293 and N1E-115 cell systems in a non-competitive manner. Neurosci Lett 306(1–2):81–84

Randall DC et al (2003) Modafinil affects mood, but not cognitive function, in healthy young volunteers. Hum Psychopharmacol 18(3):163–173

Randall DC et al (2004) The cognitive-enhancing properties of modafinil are limited in non-sleep-deprived middle-aged volunteers. Pharmacol Biochem Behav 77(3):547–555

Randall DC et al (2005a) Does modafinil enhance cognitive performance in young volunteers who are not sleep-deprived? J Clin Psychopharmacol 25(2):175–179

Randall DC, Shneerson JM, File SE (2005b) Cognitive effects of modafinil in student volunteers may depend on IQ. Pharmacol Biochem Behav 82(1):133–139
Rao VL et al (2001) Neuroprotection by memantine, a non-competitive NMDA receptor antagonist after traumatic brain injury in rats. Brain Res 911(1):96–100
Rogawski MA, Wenk GL (2003) The neuropharmacological basis for the use of memantine in the treatment of Alzheimer's disease. CNS Drug Rev 9(3):275–308
Rowan JD et al (2015) Adolescent exposure to methylphenidate impairs serial pattern learning in the serial multiple choice (SMC) task in adult rats. Neurotoxicol Teratol 51:21–26
Rycroft N et al (2007) Non-cholinergic modulation of antisaccade performance: a modafinil-nicotine comparison. Psychopharmacology 195(2):245–253
SAMHSA (2018) National survey on drug use and health: 2017. SAMHSA, Rockville, MD
Schelle KJ et al (2015) A survey of substance use for cognitive enhancement by university students in the Netherlands. Front Syst Neurosci 9:10
Scheres A et al (2003) The effect of methylphenidate on three forms of response inhibition in boys with AD/HD. J Abnorm Child Psychol 31(1):105–120
Schmitz F et al (2012) Methylphenidate induces lipid and protein damage in prefrontal cortex, but not in cerebellum, striatum and hippocampus of juvenile rats. Metab Brain Dis 27(4):605–612
Schmitz F et al (2016) Chronic treatment with a clinically relevant dose of methylphenidate increases glutamate levels in cerebrospinal fluid and impairs glutamatergic homeostasis in prefrontal cortex of juvenile rats. Mol Neurobiol 53(4):2384–2396
Schmitz F et al (2017) Methylphenidate causes behavioral impairments and neuron and astrocyte loss in the hippocampus of juvenile rats. Mol Neurobiol 54(6):4201–4216
Seeman P, Caruso C, Lasaga M (2008) Memantine agonist action at dopamine D2High receptors. Synapse 62(2):149–153
Sheng ML, Lee SH (2001) AMPA receptor trafficking and the control of synaptic transmission. Cell 105:825–828
Shire (2013) Adderal XR prescribing information. In: U.S.F.a.D. Administration (ed). Shire
Sors A et al (2016) The synergistic enhancing-memory effect of donepezil and S 38093 (a histamine H3 antagonist) is mediated by increased neural activity in the septo-hippocampal circuitry in middle-aged mice. Front Pharmacol 7:492
Studer P et al (2010) ERP effects of methylphenidate and working memory load in healthy adults during a serial visual working memory task. Neurosci Lett 482(2):172–176
Suppiramaniam V et al (2001) Member of the Ampakine class of memory enhancers prolongs the single channel open time of reconstituted AMPA receptors. Synapse 40(2):154–158
Swanson JB, Baler RD, Volkow ND (2011) Understanding the effects of stimulant medications on cognition in individuals with attention-deficit hyperactivity disorder: a decade of progress. Neuropsychopharmacology 36:207–226
Tannock R, Martinussen R, Frijters J (2000) Naming speed performance and stimulant effects indicate effortful, semantic processing deficits in attention-deficit/hyperactivity disorder. J Abnorm Child Psychol 28(3):237–252
The National Institute of Drug Addiction (NIDA) (2018) Monitoring the future survey: high school and youth trends. [cited 5 Mar 2019], https://www.drugabuse.gov/publications/drugfacts/monitoring-future-survey-high-school-youth-trends
Theunissen EL et al (2009) Comparing the stimulant effects of the H1-antagonist fexofenadine with 2 psychostimulants, modafinil and methylphenidate. J Clin Psychopharmacol 29(5):439–443
Tomasi D et al (2010) Methylphenidate enhances brain activation and deactivation responses to visual attention and working memory tasks in healthy controls. NeuroImage 54(4):3101–3110
Tricco AC et al (2012) Use of cognitive enhancers for mild cognitive impairment: protocol for a systematic review and network meta-analysis. Syst Rev 1:25
Tricco AC et al (2013) Efficacy and safety of cognitive enhancers for patients with mild cognitive impairment: a systematic review and meta-analysis. CMAJ 185(16):1393–1401
Trommer BL, Hoeppner JA, Zecker SG (1991) The go-no go test in attention deficit disorder is sensitive to methylphenidate. J Child Neurol 6(Suppl):S128–S131

Turner DC et al (2003) Cognitive enhancing effects of modafinil in healthy volunteers. Psychopharmacology 165(3):260–269
Urban KR, Gao WJ (2014) Performance enhancement at the cost of potential brain plasticity: neural ramifications of nootropic drugs in the healthy developing brain. Front Syst Neurosci 8:38
Urban KR, Gao WJ (2015) Evolution of the study of methylphenidate and its actions on the adult versus juvenile brain. J Atten Disord 19(7):603–619
Urban KR, Gao WJ, Waterhouse BW (2012a) Distinct age-dependent effects of methylphenidate on developing and adult prefrontal neurons. Biol Psychiatry 72:880–888
Urban KR, Li YC, Gao WJ (2012b) Treatment with a clinically-relevant dose of methylphenidate alters NMDA receptor composition and synaptic plasticity in the juvenile rat prefrontal cortex. Neurobiol Learn Mem 101:65–74
Urban KR, Layfield DM, Griffin AL (2014) Transient inactivation of the medial prefrontal cortex impairs performance on a working memory-dependent conditional discrimination task. Behav Neurosci 128(6):639–643
Urban KR et al (2017) A clinically-relevant dose of methylphenidate enhances synaptic inhibition in the juvenile rat prefrontal cortex. J Reward Defic Syndr Addict Sci 2(3):69–77
Vanderschuren LJ et al (2008) Methylphenidate disrupts social play behavior in adolescent rats. Neuropsychopharmacology 33(12):2946–2956
Vendruscolo LF et al (2008) Chronic methylphenidate treatment during adolescence increases anxiety-related behaviors and ethanol drinking in adult spontaneously hypertensive rats. Behav Pharmacol 19(1):21–27
Wade L, Forlini C, Racine E (2014) Generating genius: how an Alzheimer's drug became considered a 'cognitive enhancer' for healthy individuals. BMC Med Ethics 15:37
Wang H et al (2008) A specialized NMDA receptor function in layer 5 recurrent microcircuitry of the adult rat prefrontal cortex. Proc Natl Acad Sci U S A 105(43):16791–16796
Watkins JC, Jane DE (2006) The glutamate story. Br J Pharmacol 147(Suppl 1):S100–S108
Watling E (2019) Nootropics: do 'smart drugs' really work? Newsweek
Winder-Rhodes SE et al (2010) Effects of modafinil and prazosin on cognitive and physiological functions in healthy volunteers. J Psychopharmacol 24(11):1649–1657
Xing B, Li YC, Gao WJ (2016) Norepinephrine versus dopamine and their interaction in modulating synaptic function in the prefrontal cortex. Brain Res 1641(Pt B):217–233
Yang HJ et al (2011) Abnormal behaviors and microstructural changes in white matter of juvenile mice repeatedly exposed to amphetamine. Schizophr Res Treat 2011:542896
Yesavage JA et al (2002) Donepezil and flight simulator performance: effects on retention of complex skills. Neurology 59(1):123–125

Pharmacological Approaches in the Augmentation and Recovery of Brain Function

Dafin F. Mureșanu, Codruța Bîrle, Livia Livinț Popa, Olivia Verișezan-Roșu, and Ștefan Strilciuc

1 Introduction

In the active yet continually ageing society we live in, the need for recovery, preservation, and improvement of our brains poses a significant challenge. Today, more than ever, pharmacological intervention for cognitive enhancement raises heated discussion in the fields of neuroscience, law, ethics, and others on topics such as medical safety, equality, fairness, and civil liberties (Schelle et al. 2014). Cognition is a highly dynamic brain function that undergoes constant modulation and changes in the morpho-functional architecture of neuronal circuits and brain areas. Recent advancements in molecular biology, electrophysiology, and imaging tools such as fluorescent microscopy, functional magnetic resonance imaging (fMRI), positron emission tomography (PET), and magnetoencephalography (MEG), have provided valuable insight into the functioning of the brain, particularly by enhancing understanding of the structure–function relationship of resting-state networks and the dynamic cross talk between other networks, and by highlighting disruptions in functional connectivity that occur with various diseases.

D. F. Mureșanu (✉) · C. Bîrle · L. L. Popa · O. Verișezan-Roșu · Ș. Strilciuc
RoNeuro Institute for Neurological Research and Diagnostic, Cluj-Napoca, Romania

Department of Neurosciences, Iuliu Hațieganu University of Medicine and Pharmacy, Cluj-Napoca, Romania
e-mail: dafinm@ssnn.ro

I. Opris et al. (eds.), *Modern Approaches to Augmentation of Brain Function*, Contemporary Clinical Neuroscience,
https://doi.org/10.1007/978-3-030-54564-2_30

1.1 Substrates of Cognitive Function

Cognition is a highly complex brain function that implies coordination and switch between multiple subprocesses of memory, attention, and executive function. Attention implies selection of relevant stimuli, divided attention, attention switching (when more tasks or relevant external stimuli are simultaneous), and sustained attention over an extended period. Memory implies encoding, storing, and retrieval of information. Its two main subdomains are working and long-term memory. The latter may also be divided in episodic, semantic, autobiographical, procedural, and prospective memory. Executive function is a multicomponent construct that implies planning, organization, coordination, implementation, and evaluation (Glisky 2007).

Cognitive function also involves perception, metacognition, and mind wandering. Despite perception initially being thought to occur separately and before cognition, studies have more recently highlighted a concomitant alteration of perception along with cognitive impairment (Glisky 2007; Goldstone et al. 2015), suggesting an interactive activation between them. Metacognition is related to "awareness, understanding, control and manipulation of the individual cognitive processes" (Osman and Hannafin 1992). The term was first defined by John Flavell in 1979, who described four classes of metacognitive phenomena: metacognitive knowledge, metacognitive experiences, task or goals, strategies or activities (Flavell 1979). An essential factor for both consciousness and flexibility of cognition, mind wandering implies shifting attention from external or internal tasks to unrelated thoughts and feelings (Smallwood and Schooler 2015).

Additionally, it has a vital role in autobiographical planning and creative problem-solving (Marron et al. 2018; Mooneyham and Schooler 2013). Even if mind wandering and metacognition appear to be anticorrelated, recent studies have shown that there is also a facilitating interaction between them (Allen et al. 2013; Fox and Christoff 2014). Mind wandering is involved in monitoring own thoughts, thus being helpful in problem-solving (Kudesia et al. 2015; Marron et al. 2018).

Three intercorrelated levels of organization have been described for brain function: cellular and molecular, circuitry, and the dynamic network, all of which have been implicated both in maintaining endogenous homeostasis and in pathophysiological processes. Alterations that directly affect the cellular and molecular level also affect the circuitry and dynamic network levels.

Functional connectomics is the dynamic balance between anticorrelated networks that function by activation (synchronization) and deactivation (desynchronization) in response to different tasks (Fox et al. 2005). Synchronization is responsible for the integration of information, while desynchronization is responsible for segregation of information. The integration–segregation balance provides both stability and flexibility within brain circuits (Tognoli and Kelso 2014).

This dynamic balance was intensively studied between resting-state networks (RSN—characterized by spontaneous activity that increases in the absence of a task) and task-positive networks (TPN). The default mode network (DMN) uses direct structural connection from other RSNs, involving several brain regions:

precuneus, posterior cingulate (retrosplenial cortex), ventromedial prefrontal cortex, inferior parietal lobes (angular gyrus), and lateral temporal cortex (Kabbara et al. 2017; Sporns 2013).

Examples of TPN networks include the salience network (SN), the central executive network (CEN), the dorsal and ventral attention networks, and the motor networks (MOT). The SN is activated by external stimuli and contains the dorsal anterior cingulate and the fronto-insular cortexes. The right anterior insula (AI) is believed to be a key node, acting as a switch between the DMN and the CEN (He et al. 2014; Jilka et al. 2014). The anterior cingulate cortex has been associated with cognitive functions, especially attention, executive function, and metacognition (Kerns et al. 2004; Silton et al. 2010). The CEN is activated by cognitive tasks and is comprised by the dorsolateral prefrontal cortex and the posterior parietal cortex. In cognitively impaired patients, structural and functional alterations of the SN such as increased and decreased intraconnectivity are present. These alterations may lead to disconnection between the insula, the DMN, and the CEN (He et al. 2014). Attention may be driven by task-relevant stimuli (either endogenous or goal-driven attention) or by unexpected events. Therefore, a continuous switch between two types of networks localized in the dorsal frontoparietal regions is needed: the dorsal attention network (DAN, which can be activated by both types of stimuli) and the ventral attention network (VAN, which is activated just by unexpected targets) (Corbetta and Shulman 2002; Daitch et al. 2013). The MOT includes the bilateral primary network cortex, a part of the bilateral primary somatosensory cortex and the supplementary motor area.

However, the segregation between RSN and TPN networks is relative. Recent studies have highlighted that parts or RSN and TPN may be activated together and facilitate each other's activity. For example, although the DMN is considered to function in opposition to the SN, parts of DMN are activated along with task-activated regions during cognitive activities: the ventral posterior cingulate cortex during attentionally demanding tasks (Leech et al. 2011), the inferior parietal cortex (Piccoli et al. 2015), the medial prefrontal cortex and posterior cingulate cortex (Koshino et al. 2014) during preparation and retrieval phases of working memory, and the inferior parietal cortex during error awareness, a metacognitive ability (Allen et al. 2013; Thurm et al. 2013). Moreover, mind wandering is not only associated with an increased activity of the DMN, but also with decreased connectivity in various regions within this network, as a probable consequence of coactivation during new tasks (de Pasquale et al. 2012).

The DMN, along with the VAN and the MOT, achieve the integration of internal, otherwise segregated cognitive processes, with sensory and motor information through several hubs characterized by rich club organization (de Pasquale et al. 2012, 2013, 2016; Scalf et al. 2014). Hubs actively change the degree of connectivity, switching their network roles from high-centrality to a low-centrality and vice versa to achieve this dynamic (de Pasquale et al. 2017; Kabbara et al. 2017). Through this "dynamic core", the RSN presents a dynamic behavior, continuously coupling and decoupling in order to optimize the global flow of interaction.

Spontaneous brain activity is characterized by both oscillatory patterns and by arrhythmic firing with scale-free activity. It appears that during task-driven activities (i.e. motor, cognitive tasks), the brain may operate in a rhythmic mode by coupling and decoupling higher with lower frequencies in the alertness state, in which attention must be maximized (Chacko et al. 2018). A specific task determines simultaneous phase modulation and synchronization of task-relevant brain regions, translating into enhanced connectivity (Daitch et al. 2013).

Brain network strength is modulated by synaptic communication through LTP and LTD and by resting membrane potential. These are determined by the expression of genes closely linked to neurotransmitters and ion channel activity. Recent studies have proved the association between functional networks (including resting-state networks) and gene networks, with direct implications on neuro-psychiatric pathologies (Chatterjee et al. 2017; Richiardi et al. 2015; Waldron 2015). Additionally, neurodegenerative diseases such as Alzheimer's Disease (AD) appear to be correlated with the accumulation of functional alterations rather than particular, independent mutations in the genetic regulatory network (Pérez-Palma et al. 2014; Pita-Juárez et al. 2018). The notion of gene networks is derived from the observation of dynamical regulatory interactions between genes (Fu et al. 2014; Nido et al. 2015; Wildenhain and Crampin 2006). Similar to brain networks, gene networks are characterized by oscillatory and non-oscillatory activity. Synchronization of neurons by temporal coordination during gamma oscillations is modulated by the interplay between inhibitory and excitatory neurotransmitters. GABA levels have been proved to modulate RSN. High GABA concentrations in the posterior cingulate cortex and the precuneus are associated with DMN deactivation, which is essential for the good performance of task-related activities (Hu et al. 2013).

Moreover, GABA is involved in the modulation of the basal ganglia network, cortico-striatal connectivity, and thalamo-striatal connectivity, playing an essential role in the executive function (Dharmadhikari et al. 2015; Haag et al. 2015). However, the function of inhibitory fast-spiking interneurons which are responsible for the release of GABA (Gulyás et al. 2010) is dependent on high energy expenditure, rendering neurons vulnerable in the elderly and individual with various neurological conditions (e.g. neurodegenerative diseases, stroke, vascular dementia) (Kann 2016). Overexpression of glutamate signaling in AD contributes to the imbalance of neurotransmitters, with subsequent alteration of synchronization in different frequency bands (López et al. 2014), and large-scale networks (Deleglise et al. 2018; Kapogiannis et al. 2013).

Moreover, recent research has identified cortical areas that are important for cognitive function, in addition to ones that had previously been identified in stroke studies. For example, the angular gyrus (AG) has a vital role in semantic comprehension (Price et al. 2016) and is also an essential hub of the DMN, showing decreased connectivity in patients with mild cognitive impairment (Lee et al. 2016). The occipital cortex is involved in visual memory, and the temporoparietal junction represents an important hub for attention networks and is also involved in spatial memory in relation to visuospatial attention (Committeri et al. 2015).

These breakthroughs prove that our understanding of the substrates and processes involved in cognition is still constantly shifting. We therefore suggest that all pharmacological approaches for cognitive recovery, preservation, or enhancement should be mindful of the brain's natural response, rather than reliant on isolated mechanisms of action. Given that we are far from exhaustively unraveling the intricate ways of our minds, the epistemological foundation of pharmacological neuromodulation should be mirroring, rather than interceding endogenous processes.

1.2 Pharmacological Modulation of the Endogenous Defense Activity

The concept of endogenous neuromodulation refers to the brain's capacity to balance anticorrelated processes, such as pro-survival versus pro-death signaling mechanisms at the cellular and molecular level, LTP versus LTD at the local circuit level, and synchronization versus desynchronization at the dynamic network level (Fig. 1). Every level in turn is comprised by several sublevels, each of which may be

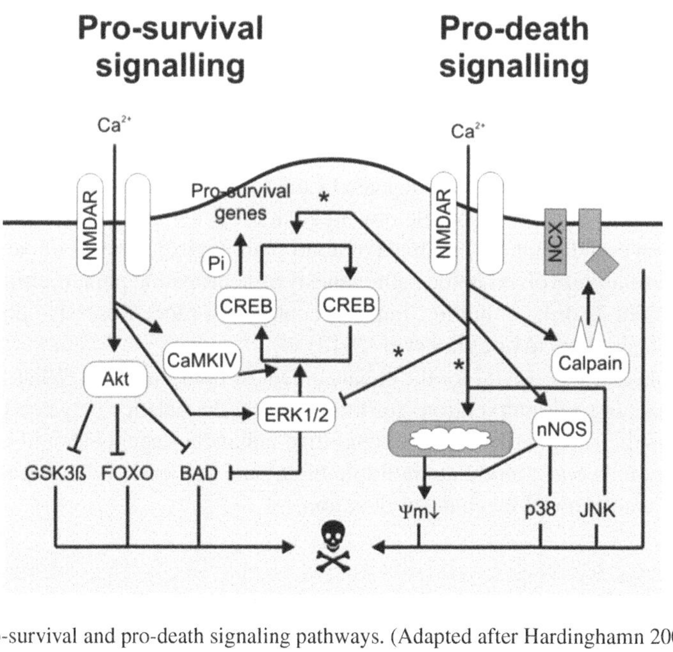

Fig. 1 Pro-survival and pro-death signaling pathways. (Adapted after Hardinghamn 2006)

characterized by a multitude of such anticorrelated processes (Muresanu et al. 2012). At the cellular and molecular level, endogenous neuromodulation involves the optimization of biological processes that may lead to cell death or promote neurodegeneration. Additionally, at the circuitries and dynamic network levels, it implies rebalancing functional connectivity in resting-state and task-positive networks.

The endogenous defense activity (EDA) is a continuous mechanism of the brain which counteracts pathophysiological events by modulating four fundamental biological processes that share common features and act together under genetic control: neurotrophicity, neuroprotection, neuroplasticity, and neurogenesis (Muresanu 2007). Neurotrophicity is a continuous process that allows cells to regulate gene expression in order to maintain their phenotype. Through neuroprotection, neuronal, glial, and endothelial cell components in the CNS are shielded against noxious stimuli or harmful factors. Neurogenesis is the process by which new nervous tissue cells (e.g. neurons, astrocytes, oligodendrocytes) are created from stem cells. In a strict sense, neurogenesis refers only to new neuron creation. Neuroplasticity is the brain's ability to adopt functional or structural responses that promote adaptive behavior, learning, and memory. These complex processes that imply molecular, structural, and functional integration are carried out in neurons, glia, and synapses. As synaptic connections are continually being removed and recreated, the harmony between opposing actions is the foundation of synaptic plasticity. The activity-dependence of synaptic plasticity is a central point of general neuroplasticity, and memory and learning theories based on experience-induced changes in synaptic structure and function.

The genetic programs underlying EDA can switch from neuroprotective to neuroplastic patterns using the same molecules for up and downregulation. In this process, hundreds of molecules (e.g. transcription factor, kinase network molecules, neurotransmitter receptors, growth factors/receptors, growth-associated cytoskeletal molecules, synapse-related molecules, adhesion molecules) are regulated by immediate early genes and late activated genes with the help of neurotrophic factors (Fig. 2). Regulation disturbances in each of the four components of EDA may trigger pathological conditions. While downregulation generates a deficit of recovery, upregulation could generate hundreds of neuropathological patterns of pathological plasticity, usually involved in the pathogenesis of neuropathic pain, multiple sclerosis, movement disorders, tinnitus, impulse control disorders, obsessive-compulsive disorders, and others (Muresanu et al. 2012).

Neuroplasticity is crucial for the enhancement of brain function. When considering pharmacological interventions for this purpose, the balance between EDA processes must be safeguarded. Hence, cognitive enhancing agents should ideally be able to simultaneously modulate multiple biological processes, eliciting a pharmacologically multimodal mechanism of action.

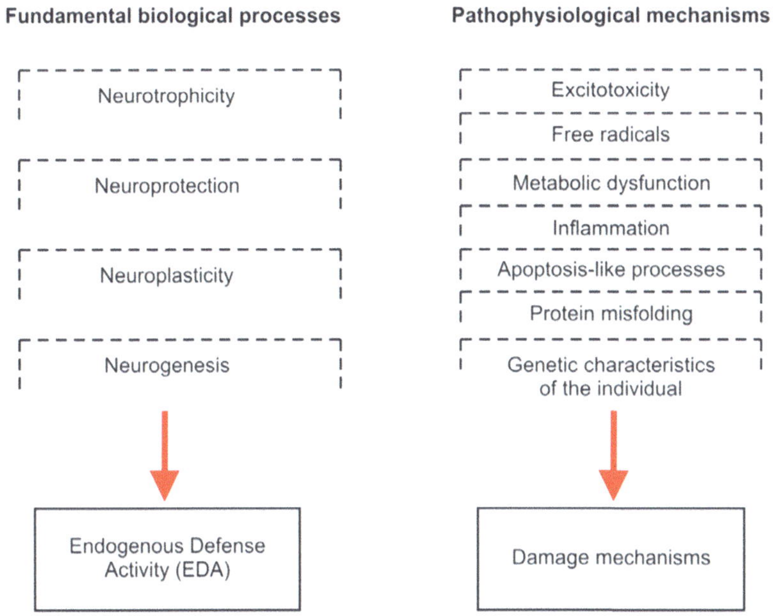

Fig. 2 Pathophysiological processes (Muresanu et al. 2012)

2 Pharmacological Agents for Brain Function Enhancement and Recovery

2.1 Brain Function Enhancement

2.1.1 Ampakines

AMPA receptors are glutamatergic-type receptors that mediate fast synaptic transmission, being involved in LTP, a form of synaptic plasticity, by diminishing deactivation and desensitization. Attempts to modulate these receptors through Ampakines have showed some promising results regarding cognitive enhancement. In rats, they improved short-term and long-term memory. In primates and humans may promote visual, olfactory, and spatial memory. Also, there are studies which revealed that ampakine increases the expression of BDNF and cell proliferation, having a neuroprotective effect. Until now, they were evaluated for treatment of PD and Alzheimer's disease (AD), schizophrenia, depression, autism, ADHD or Rhett syndrome. Boyle et al. tested CX717, a positive allosteric AMPA receptor modulator, on healthy adults in order to see its influence on arousal, recovery sleep, and cognitive function during an extended wakefulness period. The only dose that made a difference was 1000 mg of CX717, demonstrating an enhancement of attention-based task performance, and counteracting the effect of sleep-deprivation (Boyle et al. 2012; Urban and Gao 2014). The same substance was tested in monkeys, revealing a dose-dependent improvement of task-performance, a relief of performance impairment caused by sleep deprivation, and increased activity in medial temporal lobe, cortex, and dorsal striatum (Porrino et al. 2005). Nevertheless, after decades of intensive clinical trial research, treatment with ampakines did not yield credible brain-enhancing results in humans, but remain valid prospects for other indications such as respiratory depression in humans treated with opioids, prevention and rescue propofol-induced severe apnea, or as respiratory stimulants in postoperative patients (Knafo and Esteban 2015).

2.1.2 Amphetamines

Considering amphetamine's capacity to enhance memory, attention, or executive functions, and it being an effective long-term therapy for ADHD, many young people have misused the psychostimulant in order to perform better in their educational activities. Contrary to student assumptions and beliefs, the intake of stimulants has been proved to be adversely correlated with academic performances. Ilieva at al. analyzed the effects of amphetamine on cognition in healthy adults. No significant improvement of objective cognitive functions (episodic and working memory, creativity, intelligence) was found in their study. Nonetheless, participants mentioned a subjective, perceived cognitive enhancement, similar to a placebo effect (Ilieva et al. 2013).

In 2018, Weyandt et al. performed a double-blind, placebo-controlled study, to determine if Adderall could improve cognitive and emotional functions in healthy college subjects. The results were mixed, with minimal improvement of attention performance and worsening of working memory, all together with significant autonomic and emotional effects, and subjective positive experiences (Weyandt et al. 2018). The risks of psychostimulant abuse, particularly in healthy adolescents and young adults is an important factor to take into account, as the effects of amphetamine on developing brains are not yet fully understood. Extrapolating information from clinical trials, the safety profile of amphetamines does not warrant their use as cognitive enhancers in healthy subjects (Berman et al. 2009).

2.1.3 Citicoline

Due to its presumed role in neuroplasticity and cognitive recovery in cerebral injuries, several studies have investigated the potential cognitive-enhancing properties of citicoline in healthy persons. McGlade et al. conducted a study on 60 adult women, who received 250 mg citicoline, 500 mg citicoline, or placebo for 28 days. Their results suggest that this nutraceutical may improve attentional performance in healthy subjects (McGlade et al. 2012). Same effects, along with reduced impulsivity and enhanced psychomotor speed, were also obtained in 2019 in a study performed on 75 healthy adolescent males with similar dosage and duration of treatment

(McGlade et al. 2019). Further research is warranted to establish a clear recommendation for supplementation with citicoline for cognitive enhancement.

2.1.4 Methylphenidate

Methylphenidate (MPH) is the most common treatment for ADHD, wielding its effects by blocking the reuptake of dopamine and norepinephrine, and increasing the availability of these neurotransmitters in the synaptic zone (McDonald et al. 2016). Multiple studies on healthy adult animals or humans have assessed its capacity to enhance cognitive functions. While lower doses of MPH result in a minimal increase of distinct neurotransmitters in the prefrontal cortex with enhanced performance on an attention task, executive control, working memory, and no effect on motor activity, high doses of the drug intensify locomotor activity and diminish attention and performance of cognitive tasks (Urban and Gao 2014).

Nowadays, methylphenidate is abusively consumed by teenagers and young people in order to boost attention, memory, and energy for tests and exams. Overuse at a young age is a potentially perilous habit since the prefrontal cortex is fully developed only around the third decade of life. Administering a drug like MPH might break the physiological maturation and generate persistent behavioral consequences (Urban and Gao 2014).

In 2017, Franke et al. performed a clinical trial in which they compared methylphenidate, modafinil, and caffeine with placebo for cognitive enhancement in healthy subjects that played chess. All substances increased average reflection time per game, but only the first two influenced chess performance, resulting in significantly better scores compared to placebo (Franke et al. 2017). As with amphetamines, important questions must be raised regarding the safety profile of this intervention.

2.1.5 Modafinil

Modafinil is a nootropic drug with recognized wake-promoting effects. Its mechanisms of action are not fully understood, but it seems to interact with various central neurochemical projections. Among them, the most important one is the dopaminergic pathway. Modafinil inhibits dopamine reuptake resulting in increased dopamine levels in several brain areas, including the prefrontal cortex and nucleus accumbens, and also downregulates D1 and D2 receptors. These may result in enhanced executive functions. Also, Modafinil interferes with noradrenaline, GABA, glutamate, serotonin, orexin/hypocretin circuits and raises hypothalamic histamine levels. It is usually prescribed for excessive day sleepiness as a result of sleep disorders like obstructive sleep apnea or narcolepsy. However, it was proved to be useful in psychiatric and neurologic pathologies—ADHD, major depressive disorder, bipolar disorder, schizophrenia, Lewy Body dementia, or multiple sclerosis, for adjusting sleep and for augmenting working memory performance (Sousa and Dinis-Oliveira 2020; Urban and Gao 2014).

Such as the case of methylphenidate, Modafinil is used for "academic doping" by students who aspire to better concentration, improved memory, and longer stamina for learning. Recent studies on healthy subjects showed that the drug might increase alertness and improve pattern recognition memory, mental digit manipulation, and other higher executive functions, but is sensible to baseline performance (Sousa and Dinis-Oliveira 2020; Urban and Gao 2014). No specific dose of Modafinil was established to have cognitive effects. If some studies obtained an improvement of attentional set-shifting at 200 mg, others used 400 mg to reduce impulsivity and 1-mg for better attentional visual flexibility (Nikiforuk et al. 2017). Nevertheless, modafinil maintains the same risk for young people. In addition to the risks of addiction, its misuse may impact cortical plasticity and damage cognition, emotions, and behavior (Urban and Gao 2014).

2.2 *Brain Function Recovery*

2.2.1 Amphetamines

Current approaches in stroke recovery highlight the brain's ability to undergo dynamic plastic changes in order to regain functioning. Aside from physical rehabilitation, the concept of pharmacological intervention may represent a significant opportunity for neurorecovery. Experimental studies on the pharmacological rehabilitation of stroke have suggested that amphetamines may improve recovery by increasing noradrenergic transmission, blocking the reuptake of dopamine, and facilitating brain plasticity. As cerebral ischemia produces a deficit of catecholamines, administrating noradrenergic agonists may result in structural and physiological brain changes, like enhancing neural sprouting, synaptogenesis, and facilitating LTP (Gladstone et al. 2006; Hart et al. 2018; Sprigg et al. 2007).

In several studies, amphetamine was administered in different doses combined with physical therapy in patients with ischemic stroke. In an older study from 1988, 8 patients showed improvement in motor scores as compared to controls (Crisostomo et al. 1988). Daily doses of 10 mg amphetamine for 5 weeks did not influence the recovery of motor function and ADL in stroke patients (Sonde et al. 2001). In another study performed on 71 patients and 10 mg dextroamphetamine twice/week, for 5 weeks, significant results were obtained for upper limb motor recovery in patients with moderate hemiparesis. For severe cases and cortical stroke, there was no difference in recovery between treatment groups (Gladstone et al. 2006). The same results for severely disabled stroke cases were confirmed by the study of Martinsson et al., in which all patients received dexamphetamine in order to increase alertness. The difference between groups consisted in the amount of physical therapy. The outcomes at 3 and 12 months showed improvement in both groups, but without statistical significance (Martinsson et al. 2003). However, the sympathomimetic effects of amphetamine on blood pressure and heart rate might represent an important influencing factor for recovery, because increased BP and HR may aggravate the outcome (Sprigg et al. 2007).

Psychostimulants were also tested for their potential to enhance cognitive functions in patients with traumatic brain injury by increasing norepinephrine and dopamine, neurotransmitters implicated in neuroplasticity that regulate prefrontal cortical circuit activity (Hart et al. 2018; Hylin et al. 2017; Marshall et al. 2019). For example, 10 mg of dextroamphetamine or placebo were administered for 3 weeks in 32 patients with traumatic brain injury (TBI). There was no evidence that the drug would accelerate recovery, but it improved the speed of information processing (Hart et al. 2018).

2.2.2 Antidepressants

Several studies on stroke patients have tested whether antidepressants may have a role in neurorehabilitation. Despite the exact mechanism of action not being fully understood, studies on animal models have suggested possible processes that could augment recovery after stroke. Among these, the most notable were inducing arteriolar dilatation by modulating NO-muscarinic signaling, enhancing neurogenesis and angiogenesis in the adjacent area, or increasing the secretion of neurotrophic factors like BDNF (Kumar and Kitago 2019; Savadi Oskouie et al. 2017). Serotonin-reuptake inhibitors in particular are recognized to elicit a neuroprotective effect through their anti-inflammatory action. By heightening the amount of serotonin, a valuable cerebral monoamine, these drugs may influence both short-term and long-term facilitation processes involved in motor and cognitive rehabilitation (Chollet et al. 2011).

Savadi et al. evaluated the effectiveness of 3-month administration of daily 20 mg of Citalopram in non-depressive patients with ischemic stroke, along with standard care. The results were in favor of Citalopram group, with a significant reduction of NIHSS score at 3 months, mainly in scores of motor and language components (Savadi Oskouie et al. 2017). A similar study was TALOS, a double-blind, placebo-controlled study of Citalopram in nondepressed ischemic stroke patients, with daily administration for 6 months. In comparison with the previous study, citalopram did not improve functional recovery as assessed by the modified Rankin Scale at 6 months (Kraglund et al. 2018). Regarding the restoration of cognitive functions, Jorge et al. evaluated the effects of escitalopram on 129 patients. The active drug group showed higher scores in global cognitive function, verbal and visual memory, which resulted in ameliorated daily living activities. Possibly, the subjacent mechanism may be represented by increased neurogenesis in the dentate gyrus, respectively, by remodeling hippocampal circuitries (Jorge et al. 2010).

In the FLAME trial, the investigators have analyzed the effect of Fluoxetine administration combined with physiotherapy in patients with hemiplegia or hemiparesis poststroke, quantified through Fugl-Meyer motor scale. The increase of FMMS at day 90 (the primary outcome) was significantly higher in the fluoxetine group for both superior and inferior limb, with better results in NIHSS and mRS, too (Chollet et al. 2011). In 2018, Asadollahi et al. conducted a comparative study of oral citalopram, fluoxetine, and placebo, in patients with poststroke motor deficit. Both antidepressants had improved scores as compared to placebo, with a significant increase in FMMS score at 90 days, but without statistical significance (Asadollahi et al. 2018).

In 2019, Mead et al. reported a systematic review on the ability of Fluoxetine to reduce disability in stroke patients. Despite Fluoxetine being associated with better neurological outcome, high-quality clinical trials did not register improved functional outcomes, thus failing to support its use either for the prevention of poststroke depression or functional neurorecovery (Mead et al. 2019).

2.2.3 Cannabidiol

Cannabidiol (CBD) is a non-psychotomimetic substance present in *Cannabis sativa* plant (Mechoulam and Gaoni 1965; Pertwee et al. 2005). In the last 15 years, many researches have shown potential therapeutic effects of CBD in stroke (Alvarez et al. 2008a; Castillo et al. 2010; Hayakawa et al. 2008; Lafuente et al. 2011; Mishima et al. 2005; Pazos et al. 2013), neurodegenerative disorders—Parkinson's disease (Garcia-Arencibia et al. 2007) and Alzheimer's disease (Martín-Moreno et al. 2011), multiple sclerosis and epilepsy (Leo et al. 2016), but also in neuropsychiatric conditions (Campos et al. 2012, 2016; Fernández-Ruiz et al. 2013).

In ischemic cerebrovascular disease, CBD limits the neuronal loss (Alvarez et al. 2008a; Castillo et al. 2010; Lafuente et al. 2011; Pazos et al. 2013). In models of stroke-induced by middle cerebral artery occlusion (MCAo) in mice (Hayakawa et al. 2008, 2009; Mishima et al. 2005), cannabidiol decreased infarct volume, reduced microglial activation, improved neurological functioning and increased survival rates (Mori et al. 2017).

In mice that suffered a bilateral common carotid arterial occlusion (Mori et al. 2017; Schiavon et al. 2014), CBD prevents hippocampal cell loss and white matter degeneration and promotes neurogenesis and reduces astrogliosis, with protective effects against memory impairment and emotional alteration (anxiety-like behavior and despair-like behavior).

The pharmacological mechanisms of CBD reach multiple targets in the brain ischemic injury cascade. Cannabidiol influences the endocannabinoid system and stimulates 5-hydroxytryptamine 1A (5-HT1A) receptors, adenosine receptors, and peroxisome proliferator-activated receptors (PPARs) in neuronal nuclei (Fernández-Ruiz et al. 2013). By reducing glutamate excitotoxicity, oxidative stress, and inflammation, it exerts a triple effect (antioxidant, anti-inflammatory, and neuroprotective) against ischemic injury (Pazos et al. 2013). Overall, it contributes to global functional recovery in several weeks after brain ischemia (Balkaya et al. 2013; Kronenberg et al. 2014; Mori et al. 2017).

2.2.4 Cerebrolysin

Neurotrophic factors are polypeptides, part of the growth factor family, that regulate the proliferation, survival, migration, and differentiation of cells in the nervous system. Their gene expression is enhanced after brain injury, but in most cases, this endogenous overexpression is not sufficient to meet the demands of neuroprotection and neurorepair (Nieto-Sampedro et al. 1982; Truettner et al. 1999). Cerebrolysin is a pharmacological agent that contains active fragments of various neurotrophic factors, obtained using a standardized biological method of controlled breakdown of highly purified lipid-free brain proteins (Muresanu et al. 2020). Active neurotrophic factor fragments (peptides) and amino acids quickly cross the blood–brain barrier (BBB) and bind to specific receptors on different membranes of the nervous system. Each fragment specifically initiates an intracellular signaling pathway via the phosphorylation of the involved protein kinases, which ultimately leads to the activation of transcription factors and the production of proteins involved in processes such as the maintenance of cellular neurotrophicity, neuroprotection, neuroplasticity, and neurogenesis. Two essential roles are exerted in neuromodulation (changes in neuronal and synaptic plasticity) and metabolic regulation (against lactic acidosis and increase in resilience against hypoxic conditions (Muresanu et al. 2020).

Cerebrolysin has a pharmacologically multimodal mechanism of action, influencing EDA in the post-lesional brain via pleiotropic therapeutic effects (Fig. 3) by simultaneously modulating several components of the pathological cascade in stroke, TBI, and neurodegenerative diseases (Muresanu et al. 2012; Riley et al. 2006; Wronski et al. 2000). The agent inhibits APP phosphorylation and its transport from the cell body to the synapse, thus decreasing the production of β42 amyloidal, which is involved in the molecular mechanisms underlying dementia (Masliah and Diez-Tejedor 2012).

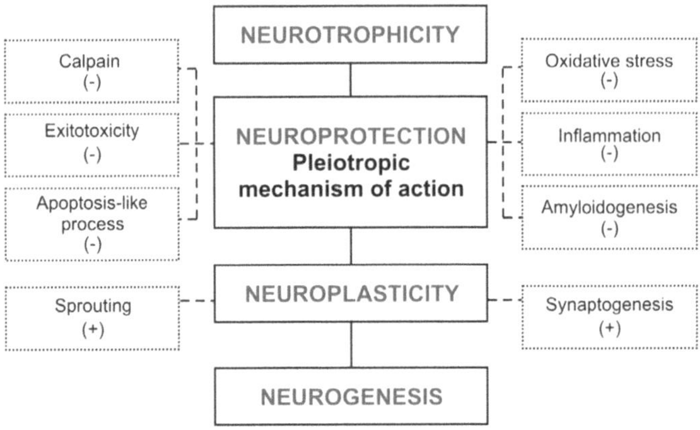

Fig. 3 The modulating, pleiotropic, neuroprotective, and multimodal mechanism of action of Cerebrolysin (Muresanu et al. 2012)

Multiple clinical studies have confirmed the safety and efficacy of Cerebrolysin treatment in acute, subacute, and chronic therapeutic strategies in neurorecovery. Clinical trial data suggest that Cerebrolysin is an effective adjunctive treatment for ischemic stroke, improving cognitive performance and clinical outcome, while not showing serious adverse effects and being safe when used in combination with recombinant tissue-type plasminogen activators or cholinesterase inhibitors, such as donepezil or rivastigmine (Bornstein and Poon 2012). Furthermore, Cerebrolysin has a rapid onset of action and offers pharmacological support for rehabilitation (Labiche and Grotta 2004).

In CARS (Cerebrolysin and Recovery After Stroke), a prospective, double-blind, placebo-controlled study, patients with acute ischemic stroke received either Cerebrolysin 30 mL/day or placebo, for 21 days, starting in the first 24 to 72 h after the stroke onset and their assessment was based on 12 different outcome scales. The scores of the Action Research Arm Test, the primary study endpoint on day 90, proved that Cerebrolysin had a beneficial effect on early motor recovery. The multivariate analysis of all scales showed a medium-to-moderate superiority on the global status of the patients, in favor of the active treatment (Muresanu et al. 2016).

The neuroprotective and the neurorestorative effects of Cerebrolysin, given as add-on to standardized rehabilitation therapy, were highlighted by Chang W.H. et al. in a prospective, multicenter, randomized, double-blind, placebo-controlled, parallel-group study published in 2016. Participants with subacute stroke and moderate-to-severe motor impairment were included in this phase IV trial. They received Cerebrolysin 30 mL/day or placebo, in addition to standardized rehabilitation therapy, for 21 days (days 8–28 after stroke). Assessments were performed at baseline, immediately after treatment and at 2 and 3 months after stroke onset, respectively. The motor system neuroplasticity was monitored by DTI and with resting state functional magnetic resonance imaging (rsfMRI) at the same intervals. Significant improvement in motor function was seen in both groups, but in the stroke patients with severe motor impairment, the Cerebrolysin group exhibited significantly more improvement in motor function compared with the placebo group. Further, plastic changes of the corticospinal tract (restricted increments of corticospinal diffusivity) and recovery of the sensorimotor connectivity were observed in this category of patients (Chang et al. 2016).

A meta-analysis of the CARS trials regarding the safety and efficacy of Cerebrolysin in motor function recovery after stroke confirmed that Cerebrolysin had a beneficial effect on motor function and neurological status in early rehabilitation patients after acute ischemic stroke. Safety aspects were comparable to placebo, showing a favorable benefit/risk ratio (Guekht et al. 2017).

Another meta-analysis published in 2018 included nine studies assessing the effects of Cerebrolysin in acute ischemic stroke, comprising total 1879 patients. Only randomized, double-blind, placebo-controlled, completed studies were considered for the final analysis. Results showed strong evidence for the positive effect of Cerebrolysin on early neurological function. Neurological deficits, as measured by the NIHSS (National Institute of Health Status Scale) scores at 30 and 21 days, respectively, were improved, as well as functional outcome at 90 days, as measured by modified Rankin Scale (Bornstein et al. 2018).

The clinical efficacy of Cerebrolysin in dementia has also been evaluated in several studies in patients with AD and vascular dementia. The assessment of Cerebrolysin's therapeutic efficacy was performed in the domains of cognition, activities of daily life, global assessment, and behavior (Alvarez et al. 2006; Muresanu et al. 2008; Ruether et al. 2002).

Highly promising results regarding the efficacy and safety of Cerebrolysin in vascular dementia were published in a Cochrane review in 2013. Six randomized clinical trials involving a total of 597 subjects were included in this meta-analysis (Chen et al. 2013b). The conclusion was that the safety profile of Cerebrolysin is excellent, and Cerebrolysin treatment vastly improves global cognitive function and executive function in patients with mild and moderate vascular dementia.

The beneficial clinical effects of Cerebrolysin in vascular dementia can be explained by its extremely well-documented mechanisms of action, such as (1) reducing amyloid plaque formation; (2) stimulating the proliferation, differentiation, and migration of neural progenitor cells from the subventricular zone (Darsalia et al. 2005); and (3) promoting the growth and sprouting and supporting the formation of synaptic contacts (Zhang et al. 2013). These actions protect neurons from apoptosis and degeneration by the inhibition of calpain (Masliah and Diez-Tejedor 2012).

Several cohort and clinical studies on patients with moderate and severe TBI showed that the administration of Cerebrolysin was associated with lower mortality and a favorable outcome (Khalili et al. 2017; Muresanu et al. 2015; Wong et al. 2005). In a double-blind, placebo-controlled study on mild TBI, results demonstrated improved cognitive abilities, improved drawing function, and long-term memory in patients treated with Cerebrolysin compared to placebo (Chen et al. 2013a).

The CAPTAIN trial series employed a sophisticated ensemble of functional and neuropsychological scales in patients with moderate-to-severe TBI, at different points in time (10, 30, 90 days) (Muresanu et al. 2020; Poon et al. 2020). Particularly, the CAPTAIN II trial conducted by Mureșanu D.F. and his collaborators had as primary objective assessment of Cerebrolysin effects on general and neurocognitive outcomes after moderate-to-severe TBI. This research reiterated the results from a previous study performed on a sample of Asian patients (Muresanu et al. 2020; Poon et al. 2020): the treatment with Cerebrolysin improves global outcome at 90 days as compared to placebo in people with this condition. A statistically significant improvement on six individual outcome scales (measuring cognitive speed, attention, and depression) was found 90 days after baseline (Muresanu et al. 2020).

The hypothesis that Cerebrolysin may enhance brain function recovery after TBI was launched in 2003, when an exploratory study exhibited positive effects of Cerebrolysin on cognition, clinical outcome, and electroencephalogram slowing, in patients with postacute TBI (Alvarez et al. 2003). Later studies used neurophysiological methods, among which quantitative EEG, in order to demonstrate the efficacy of Cerebrolysin and other nootropic drugs in the TBI therapy (Alvarez et al.

2008b; Iznak et al. 2010). A systematic review finished in 2016 showed potential neuroprotective properties of statins, *N*-acetyl cysteine, nitric oxide synthase inhibitor (VAS203), Enzogenol, and Cerebrolysin in people with TBI history (Gruenbaum et al. 2016).

Two years later, El Sayed and his team published a meta-analysis of Cerebrolysin, citicoline, and piracetam, concerning the effectiveness of Cerebrolysin, citicoline, and piracetam on Glasgow outcome score (GOS), cognitive performance, and survival in TBI patients. Studies included patients in all age groups and various severity degree of trauma. Cerebrolysin was superior to the other two neuroprotective drugs in threefold cognitive improvement and favorable GOS score (El Sayed et al. 2016).

The most recent meta-analysis performed exclusively on Cerebrolysin was mainly based on cohort studies and some randomized clinical trial regarding the effects of Cerebrolysin on functional outcome in patients with moderate and severe TBI. The conclusion indicated that this agent's administration significantly increased GOS score and decreased mRS score (Ghaffarpasand et al. 2018).

These new-generation researches, designed around multidimensional outcome measures (e.g. CAPTAIN trial series), allowed superior assessment of patients' functional and cognitive status and have shown a beneficial effect of Cerebrolysin in TBI (Muresanu et al. 2020).

2.2.5 Citicoline

Citicoline (CDP-choline or Cytidine-5′-diphosphocholine) is an acetylcholine precursor, a key neurotransmitter in the normal functioning of the central and peripheral nervous systems (motor and cognitive functioning, arousal, memory). Due to its high bioavailability and to its potential implications in neural repair, it may be used in different neurological disorders with chronic disabilities—stroke, TBI, and AD (Arenth et al. 2011). Choline, a quaternary ammonium salt, is an essential component of membrane phospholipids, essential for cellular structure and cellular signaling (Tayebati et al. 2015). Its structural and functional role suggests the possible involvement in neural stabilization after CNS injury (disruption of cell membrane integrity, neurochemical signaling, cellular metabolism) (Lakhan et al. 2009).

The advantage of CDP-choline is that it efficiently crosses the BBB into the CNS, after oral, intravenous, or intramuscular administration. When taken orally, it is hydrolyzed and absorbed as choline and cytidine (Arenth et al. 2011). It has similar adverse effects to other cholinomimetic drugs—most commonly mild gastrointestinal symptoms (Cho and Kim 2009).

In acute ischemic stroke, citicoline exerts different therapeutic actions at several stages of physiopathological cascade (Alvarez-Sabín and Román 2013). By stimulating the synthesis of phosphatidylcholine and sphingomyelin (Secades and Lorenzo 2006) and by suppressing the release of free fatty acids, it may stabilize cell membranes. By maintaining membrane integrity, citicoline inhibits glutamate release during ischemia and limits the infarcted area (Hurtado et al. 2005). Citicoline has been shown to decrease the level of caspase activation products, which have a damaging effect in human stroke (Krupinski et al. 2002). It facilitates the synthesis of nucleic acids and proteins and diminishes free radical production (Hurtado et al. 2007). In that way, it protects the injured tissue against early and delayed mechanisms responsible for ischemic stroke. Citicoline contributes to neuroplasticity by enhancing neural synaptic spines (Hurtado et al. 2007) and, subsequently, improving motor, behavioral, and mnesic performances (Garcia-Cobos et al. 2010).

A meta-analysis of ten controlled clinical trials using citicoline included 2279 participants with either ischemic or hemorrhagic stroke, patients who received citicoline had a significant reduction in the frequency of death or disability at follow-up in comparison with placebo and no significant difference in the safety profile (Alvarez-Sabín and Román 2013; Saver 2008). Beside the neuroprotective effects, citicoline has a neuroregenerative potential. In an experimental stroke models, administration of citicoline decreased neuronal apoptosis, endorsed internal brain repair, and synaptic outgrowth in treated animals (Hurtado et al. 2007).

Another role of citicoline is to stimulate proliferation of endothelial progenitor cells (EPCs) and their differentiation into mature endothelial cells. By this process of vasculogenesis, it contributes to capillary and vascular recovery of the ischemic zone. EPCs favor growth factor release and neurogenesis. The increase of EPCs in peripheral blood after acute ischemic stroke is accompanied by reduction of infarcted area and an excellent functional outcome (Sobrino et al. 2007). CDP-choline influences specific brain plasticity markers: it increases synaptophysin levels. It decreases glial fibrillary acidic protein (GFAP) levels in the peri-infarct area of the ischemic stroke, as well as the expression of low-density lipoprotein receptor-related protein (LRP) (Gutierrez-Fernandez et al. 2012).

2.2.6 Curcumin

Among dietary polyphenols (flavonoids, phenolic acids, phenolic alcohols, and lignans), Curcumin, the primary extract of turmeric, Curcuma longa (Zingiberaceae), was intensively studied for its beneficial health effects. Its multi-target therapeutic potential (represented by antioxidant, anti-inflammatory, immunomodulatory, neurotrophic, antiapoptotic, and antiproliferative activities) is shaded by its low oral bioavailability that restricts its utilization in humans. Future solutions for enhancing bioavailability are the nanotechnologies and a targeted drug delivery system (Mandal et al. 2020). Curcumin was reported to exert cytoprotective effects by modulating multiple signaling pathways in several neurological conditions. It modulates the expression of transcriptional factors, antioxidants, inflammatory cytokines, growth factors, and anti-apoptotic proteins (Mandal et al. 2020).

Curcumin is a selective activator of Nrf2/Keap1/ARE, and, by activating the heme oxygenase-1, mitigates the oxidative stress-mediated neuronal injury (Abrahams et al. 2019) and limits the apoptosis and the reactive astrogliosis, protects the astrocytes from oxidative damage and mitochondrial dysfunction (Daverey and Agrawal 2016). In stroke, curcumin exerts pleiotropic effects through neuroprotective, anti-inflammatory, antilipemic, and antiplatelet mechanisms (Mandal et al.

2020). The neuroprotection is due to the inhibition of lipid peroxidation and NO synthase activation, together with the increased synthesis of GSH, which is a key intracellular antioxidant (Strimpakos and Sharma 2008). In rats with focal cerebral ischemia, administration of curcumin reduced the infarct size, decreased the water volume in the brain, prevented the BBB damage, improved neurological performances, and reduced mortality (Jiang et al. 2007).

The inflammatory pathway is inhibited by curcumin through two significant interventions: the blockage in generating proinflammatory cytokines (IL-β, IL-8, TNF-α, and reactive GFAP), and the suppression of transcriptional factors (NF-kB and AP-1) (Strimpakos and Sharma 2008). Curcumin has a protective role against the axon degeneration via inhibition of NO release (mediated by microglial MyD88/p38 MAPK signaling) and suppression of JNK phosphorylation (Mandal et al. 2020; Tegenge et al. 2014). The essential neuroprotective role of curcumin in AD is to prevent β-amyloid-induced oxidative damage and neuroinflammation. It blocks the generation hyperphosphorylated *tau* protein and amyloid. Other mechanisms meant to shield the neurons of degeneration are: copper binding, cholesterol-lowering, modulation of microglial activity, acetylcholinesterase inhibition, upregulation of insulin-signaling pathway, and antioxidant effect (Tang and Taghibiglou 2017). The neuroprotective potential of curcumin in PD is derived from its antioxidant property. In a 6-hydroxydopamine (6-OHDA) rat model of PD, the antioxidant effect is seen in the increase of dopamine levels in the nigrostriatal tract and the reduction of Fe^{3+} level via chelation (Zbarsky et al. 2005).

The role of curcumin in multiple sclerosis is a complex one, involving anti-inflammatory, antioxidant, antidifferentiation, and anti-proliferative mechanisms that drive to neuroprotection via transcriptional factors, growth factors and receptors, proteins, enzymes, protein kinases, and inflammatory cytokines (Mandal et al. 2020). The critical elements in the pathogenesis of MS are the circulatory augmentation of Th17 cells and the generation of IL-17. In a Lewis rats model of experimental autoimmune encephalomyelitis (EAE) (Xie et al. 2009), after curcumin therapy, it was noticed a decrease in inflammatory cells in the brain, myelin basic protein, cytokines (IL-21, IL-6, IL-17), transforming growth factor, RAR-related orphan receptor γ, signal transducer, and activator of transcription protein 3 (STAT3) expression (Mandal et al. 2020).

The antidepressant effect of curcumin could be resumed in following actions: neurogenesis in the hippocampus and frontal cortex, restoration of hypothalamic-pituitary-adrenal axis dysfunction (increasing BDNF and phosphorylated cAMP response element-binding protein), modulation of neurotransmitters (NA, DA, and 5HT release), inhibition of MAO-A and MAO-B, anti-inflammatory action (against IL-1β, COX-2, MCP-1, TNF-α), antioxidant effects (enhancing the action of antioxidant enzymes SOD and GPX), reduction of NO and deactivation of nitration action (modulation of NOS), regulation of transcriptional factors, growth factors, and protein kinases (NF-kB, TGF, Nrf2, STAT) (Mandal et al. 2020).

2.2.7 Epigallocatechin-3-Gallate (EGCG)

Epigallatocathechin-3-gallate (EGCG), the primary polyphenol of green tea leaves (Islam 2012), seems to be a powerful antioxidant, anti-inflammatory, antibacterial, and antiviral agent. It is capable of modulating pro-survival/differentiation signal transduction pathways (e.g. ERK, PI3K/AKT, and DYRK1A) (Shankar et al. 2007) and changing the lipids' metabolism (Legeay et al. 2015).

Over the past few years, this compound has garnered significant scientific interest as a therapeutic option for several neurological disorders (Granja et al. 2017; Gueroux et al. 2017; Trovo et al. 2020).

The neuroprotective properties of the EGCG are related to its antioxidant, anti-inflammatory, and iron-chelating effects (Lee et al. 2009). It may cross the BBB, but the exact mechanism of the passage of this hydrophilic compound through the BBB is not known (Giunta et al. 2010).

In the literature, EGCG was described as being more efficient in radical scavenging than vitamins C and E. This phenol may be used as a therapeutic agent in AD and PD, due to its iron-chelating ability (Granja et al. 2017; Weinreb et al. 2004).

Its compelling neuroprotective effects are based on cellular modulation of different pathways.

Several researches confirm that EGCG has neuroprotective properties in humans, increasing global cerebral activity and stimulating cognition after oral administration (Scholey et al. 2012).

EGCG is a pleiotropic compound with numerous cellular targets, including the dual-specificity tyrosine-phosphorylation regulated kinase 1A (DYRK1A). DYRK1A controls dendritic development and spine formation. It was postulated that its deregulation and malfunctioning are involved in neurodevelopmental and degenerative diseases. This kinase is selectively inhibited by EGCG (Trovo et al. 2020). EGCG therapy is currently under testing in more than 90 clinical trials, including Fragile X and Down syndrome (DS) (Trovo et al. 2020). Associated to cognitive training, it significantly improves visual recognition memory, inhibitory control, and adaptive behavior in DS patients. Its action occurs through the inhibition of DYRK1A (de la Torre et al. 2016).

A recent study published in 2020 indicated that DYRK1A deregulation might contribute to the neurodevelopmental alterations caused by CDKL5 deficiency. Trovò L et al. demonstrated that EGCG treatment could restore defects in dendritic and synaptic development of *Cdkl5*-KO hippocampal neurons, with proper spine formation and function (Trovo et al. 2020).

2.2.8 Levodopa

Levodopa (L-DOPA) is a dopamine precursor generally indicated in the therapy of Parkinson's disease (PD). Once transformed in active dopamine, it wields several effects on the cerebral functions such as motor activity, cognition, or emotions. Therefore, it has been hypothesized that it may have a role in stroke recovery. In a mice model, L-DOPA proved to improve the neurological functions as compared to placebo. A possible mechanism of action for this might be via modulation of neurotransmission in the peri-infarct area, by downregulating a protein named synaptogyrin in the affected hemisphere, inhibiting the reuptake of dopamine and hence potentiating its local action (Häggman Henrikson et al. 2020). Additionally, another preclinical study with the same animal model found reactive astrocytes in the peri-infarct area expressed numerous D1 and D2 receptors that after L-DOPA treatment (Viale et al. 2017).

In human clinical trials, the administration of levodopa was evaluated for efficacy on recovery of both motor and cognitive functions. In 2001, a study on 53 stroke patients which received 100 mg/day levodopa demonstrated that the drug enhanced the motor recovery. Conversely, a trial with a larger sample size (593 patients) with the same dosage of levodopa coupled with rehabilitation therapy showed no difference between groups regarding the primary outcome: independent walking (Lin David et al. 2018). The same results were obtained by Sonde et al. in a four-arm design clinical trial. No significant difference motor function or activities of daily living was detected between treatment groups: 20 mg of D-amphetamine, 100 mg of levodopa, 10 mg of D-amphetamine + 50 mg of levodopa, and placebo (Sonde and Lokk 2007).

Several studies tested various administration schemes of 100 mg of Levodopa for cognitive rehabilitation in patients with stroke. The results show promising improvement of manual dexterity and procedural motor learning and encoding, without any effects on long-term memory or selective attention (Fond et al. 2015; Viale et al. 2017).

2.2.9 Methylphenidate

Alterations of motor control network (MCN) are frequent after traumatic brain injury, resulting from cortical contusions, hemorrhagic lesions, diffuse axonal injuries, or hypoxic-ischemic injuries. In previous fMRI studies, Dorer at al. detected less interhemispheric interactions and lack of anticorrelation between motor control network regions. Recent studies have established that the pathological processes might progress up to 2 years post-TBI. Due to the similar cognitive symptoms between TBI and ADHD, methylphenidate, the specific treatment of ADHD, was evaluated in order to find out if it could enhance recovery in TBI patients. The results from fMRI, DTI scans, and sustained attention task showed faster reaction times, increased activation of left inferior frontal cortex and greater functional connectivity in MCN in favor of the methylphenidate group (Dorer et al. 2018).

Another mechanism that might influence the effect of MPH on cognitive impairment after TBI was studied by Jenkins et al., who measured dopamine transporters levels using SPECT. Cognitive enhancement, measured by improvement of processing speed and apathy, was seen only in patients with low caudate dopamine transporter, with no change in the normal dopamine transporter group. These results suggest the importance of stratifying the factors that influence the outcome after a traumatic brain injury in order to personalize pharmacological therapies (Jenkins et al. 2019).

In 2016, the placebo-controlled study of McDonald et al. compared the efficacy of two cognitive recovery interventions (Memory and Attention Adaptation Training and Attention Builders Training), coupled with pharmacological rehabilitation with methylphenidate in patients with cognitive dysfunction after traumatic brain injury. Their results revealed that a combined approach could be superior in improving memory, attention, and executive functioning (McDonald et al. 2016). Likewise, in order to resume all the possible benefits of MPH in traumatic brain injury, Huang et al. analyzed ten double-blind, randomized clinical trials in a meta-analysis from 2016. Vigilance and attention were significantly improved, whereas, for memory and procession speed, the impact was minor (Huang et al. 2016).

2.2.10 NeuroAiD (MLC601, MLC901)

Systematic reviews of Traditional Chinese Medicine (TCM) treatments for stroke recovery have revealed conflicting results (Han et al. 2017). A systematic review performed by Gonzales-Fraile et al. in 2016 that compared the effect of MLC601 (marketed in China as Danqi Piantang Jiaonang) in stroke rehabilitation as compared to placebo showed that it has limited therapeutic benefit (Gonzalez-Fraile et al. 2016).

First trials of NeuroAiD, performed in China on poststroke patients demonstrated its safety and efficacy in independence and motor recovery (Chen et al. 2009; Siow 2008). In terms of safety and tolerability, studies showed no effect of Neuroaid on biochemistry, hematology, and hemostasis in healthy subjects and stroke patients (Gan et al. 2008).

Neuroaid associated to other medications—anticoagulant (Warfarin), antiplatelet (Clopidogrel, Aspirin, Dypiridamole), antihypertensive, lipid-lowering, antidiabetic, and antidepressant medications—for 2–3 months, demonstrated a good tolerability (Siow 2008).

The anti-inflammatory, antioxidant, and antiglutamate effects of TCM were highlighted in pharmacological studies (Chen et al. 2009; Rausch et al. 2006). It was proven that TCM produces widening of blood vessels, suppression of platelet aggregation, protects against ischemic reperfusion injury, and boosts the tolerance of ischemic tissue to hypoxia (Bei et al. 2007; Chen et al. 2009).

These properties could be used to adapt therapeutic strategies for reducing the long-term disability of stroke or other brain injuries (Heurteaux et al. 2013). The neurorestorative and neuroprotective properties of MLC601 in humans was confirmed in the Chinese Medicine NeuroAiD Efficacy on Stroke recovery—Extension (CHIMES-E) study (Venketasubramanian et al. 2015). This research that included 880 subjects with acute ischemic stroke showed that 3 months therapy with MLC601 contributed to an increase in functional independence at 6 months and beyond (Venketasubramanian et al. 2015, 2016).

MLC901 (NeuroAiD II™) is a second-generation, herbal-only version of MLC601 that also proved its utility in poststroke recovery (Navarro et al. 2012). It was demonstrated that it could promote hippocampal neurogenesis in normal C57BL/6 mice (Lorivel et al. 2015) and may inhibit tau hyperphosphorylation, raising the possibility of usage against cognitive impairment in neurodegenerative diseases, especially in Alzheimer's disease AD and frontotemporal dementia (Lee et al. 2017).

In 2017, Theadom et al. provided Class I/II evidence that, for patients with mild-to-moderate TBI, 6 months of MLC901 improved cognitive functioning, especially complex attention and executive functioning, in the most affected post-TBI. The results suggested a cumulative therapeutic effect, with a rate of improvement following cessation of treatment (Theadom et al. 2018).

In 2018, Suwanwela and his collaborators proved that combining MLC601 with rehabilitation is more efficient on functional recovery and independence compared to rehabilitation only in people having suffered an acute ischemic stroke. After 3 months of combined therapy, the beneficial effect increased over time, peaking at 1 year (Suwanwela et al. 2018).

The pathophysiological cascade following brain ischemia involves excitotoxicity, acidosis, depolarization in the peri-infarct area, oxidative stress, inflammation, and apoptosis (Zuany-Amorim et al. 2002). MLC901 exerts beneficial effects at different levels in this cascade, and it could be considered a 'multitarget' therapeutic option. MLC901 determines the activation of K ATP channels and a wide hyperpolarization, which prevents the massive release of excitotoxic glutamate and the glutamate-triggered calcium influx (Cekanaviciute and Buckwalter 2016; Rothwell et al. 1997). On the other hand, in vivo experiments in the model of global ischemia have shown that MLC901 treatment stimulated the serine/threonine kinase Akt (protein kinase B) pathway in the vulnerable brain regions, favorizing the cell survival after cerebral ischemia (Chen et al. 2009).

Another mechanism of action consists of the decrease of apoptotic pathways—by reducing the level of the Bax protein (a proapoptotic molecule) and of TUNEL labeling (a marker of DNA degradation) (Schnell et al. 1999). The antioxidant effect is reflected in the drastic decrease, during MLC901 treatment, of malondialdehyde production induced by ischemia (Schnell et al. 1999).

After brain ischemia, a reorganization of the cortical maps and an expansion in many dendrites and synapses is taking place. Some studies emphasized that MLC901 was able to promote basal neurogenesis. It was proven that MLC901 administrated in drinking water for 6 weeks doubled the number of newborn cells, which differentiate into mature neurons in 3 weeks (Rothwell et al. 1997). In rats with global ischemia, MLC901 stimulated neurogenesis in the subgranular zone of the dentate gyrus (Schnell et al. 1999). After reperfusion, this treatment increased the number of BrdU-positive neuronal precursors. MLC901 had a positive effect on the number of neural progenitors derived from human embryonic stem cells and induced neurogenic processes in cortical neurons (Rothwell et al. 1997). All these studies conclude that MLC901 contains key molecules that have enriched the microenvironment to develop a neurogenic niche, by amplification and differentiation of neural progenitors (Widmann et al. 2018).

2.2.11 Piracetam

Another representative of the nootropic drug class is Piracetam (2-oxo-1-pyrrolidine acetamide), formerly used as a medicine in AD and other dementias (Wilms et al. 2019). Nowadays, it is prescribed in European countries for cognitive impairment and dementia, while in the USA, it is considered a dietary supplement. The daily dose may vary between 800 and 2400 mg, up to 4800 mg, with adjustment based on renal function. Adverse effects of piracetam include anxiety, agitation, insomnia, depression, drowsiness, and weight gain (Cohen et al. 2019).

Its mechanism of action is to facilitate the neurotransmission in the brain (in cholinergic, dopaminergic, noradrenergic systems) and to maintain the optimal functioning of the neural receptors. Secondly, it protects the neurons from toxins and helps in re-establishing impaired neurotransmission (Giurgea 1982; Lanni et al. 2008).

This compound is a precursor for the whole racetams group, to which also belong substances such as oxiracetam, aniracetam, levetiracetam, or pramiracetam.

Piracetam has beneficial effects on neurotransmission, neuroplasticity, and microcirculation (Fessel 2019; Winblad 2005).

In AD model of cultured cells, it was noticed that piracetam restored the mitochondrial dysfunction, shifted the balance of mitochondrial fission toward fusion, more favorable for ATP production, decreased mitochondrial permeability, and increased neurite outgrowth (Stockburger et al. 2016).

In another research conducted by the same team on mice, simultaneously with the increased neurite length after administration of piracetam, an increased expression of the synaptic marker, GAP43, was observed (Stockburger et al. 2016).

A study published in 2019 by Uniyala et al. was aimed to explore the effect of piracetam, risperidone, and their combinations in experimentally induced PTSD-like phenotype in rats (Uniyal et al. 2019). Neurotransmitters (dopamine, serotonin, and their metabolites), BDNF, proinflammatory cytokines (TNF-α, IL-6), caspase-3, and markers for oxidative stress were assessed in the cortex and hippocampus, while corticosterone and nitrite levels were measured in plasma. Both medicines were found to be effective in restoring the physiological pathways in hippocampus and cortex, and in suppressing the aversive memory and the symptom cluster (anxiety, avoidance-like behavior, social withdrawal) of PTSD. Hence, piracetam and risperidone reversed the extinction deficit, behavioral alterations, impaired oxidative stress markers, neurotransmitter levels, plasma corticosterone, and nitrite levels. The anti-inflammatory and anti-apoptotic effects attenuated the progression of the disease. It was proven that the high dose of combined treatment has a potentiating effect and elevates the therapeutic efficacy (Uniyal et al. 2019).

2.2.12 Resveratrol

Resveratrol is a polyphenolic compound found in more than 70 different species of plants (*Polygonum cuspidatum*, eucalyptus, and *Picea excelsa*), and fruits (mulberries, raspberries, pines, peanuts, blueberries, and grapes). It has been purported to possess multiple therapeutic potential—anti-inflammatory, antiviral, antimicrobial, antitumor, estrogenic, cardioprotective, neuroprotective, and immunomodulatory (Perrone et al. 2017). Resveratrol has been shown to improve structural and functional recovery after acute CNS injury, including stroke (Simao et al. 2013), traumatic brain injury (Girbovan et al. 2012), spinal cord injury (Ates et al. 2006), and subarachnoid hemorrhage (Shao et al. 2014).

The neuroprotective effects in AD were described in detail in different researches. In vivo studies pointed out that a diet reach in resveratrol reduced oligomerization of Aβ peptide, amyloid deposition, the number of activated microglia, and the serum level of TNF-α (Ma et al. 2014). Concurrently, the presence of resveratrol in food or drink determined an increase in microvascular density and a decrease of vacuolar abnormalities (Oomen et al. 2009), an enhancement in mitochondrial function, with an improved aerobic capacity and sensorimotor function (Lagouge et al. 2006), a better spatial orientation and memory performance (Oomen et al. 2009).

Oral administration of resveratrol led to a reduction in malondialdehyde (MDA) levels, with GSH and AchE activity recovered (Kumar et al. 2007), and proved some positive effects on the expression of BDNF (Rahvar et al. 2011). If administrated intraperitoneally, resveratrol modulated cholinergic neurotransmission and improved cognition (Schmatz et al. 2009).

In vitro studies on hippocampal cells demonstrated a reduction in Aβ25–35-induced cell death and a decrease in the phosphorylation of PKC-δ (43), downregulation of ERK activation, decreasing in IL-1β expression, and downregulation of MCP-1 in the hippocampus (Lee et al. 2010). Resveratrol administered on cortical neurons produced an increase in SIRT1 activity and prevented cognitive decline (Kim et al. 2007), inhibited the elevation of intracellular calcium, and ROS formation (Ban et al. 2008). In microglial cells, it inhibited PGE2 and production of free radicals, and reduced LPS-mediated expression of mPGES-1 and COX-1 (Candelario-Jalil et al. 2007).

In a recent research published by Zhang et al. in 2019, resveratrol treatment significantly improved the spatial learning and memory in a rat model of vascular dementia (VD) (Zhang et al. 2019).

The pathogenesis of VD is based on oxidative stress, mitochondrial dysfunction, and apoptosis (Bennett et al. 2009; Wang et al. 2009). The intrinsic mechanism of resveratrol associated with neuroprotection may be related to the inhibition of the apoptosis pathway and oxidative stress injury (Zhang et al. 2019).

2.2.13 Sex Hormones

It is well known that estrogen receptors are extensively scattered over the brain, including regions like the amygdala and hippocampus, recognized for their contribution to learning and memory (Hsu et al. 2018). The administration of estrogen may have bidirectional, opposing effects on cognitive functions. While a high dose of estrogen improves the capacity of solving "hippocampus-sensitive tasks" like working memory, spatially driven or specific tasks, the same quantity impairs the ability to solve "striatum sensitive tasks" such as egocentric navigation rules or stimulus-response strategies. These anticorrelated effects of estrogen support the idea that there are multiple cerebral memory systems, with different structures optimized for specific cognitive functions (Korol and Pisani 2015; Luine 2016).

Cognitive impairment is frequently associated with schizophrenia, and its mechanism seems to correlate with sex hormone levels. The hormonal–inflammation interactions appear to modulate the substrates of cognitive deficits. Reports described a significant increase in markers of inflammation in the brains of schizophrenia patients (Weickert et al. 2016). Starting from the anti-inflammatory properties of estrogen, several studies on postmenopausal women with cognitive impairment associated with schizophrenia tested the role of Raloxifene, a selective estrogen receptor modulator, on cognition. Significant results were found in memory and executive domains. Nevertheless, the administration of dehydroepiandrosterone, a precursor for androids and estrogen, did not affect cognitive performance, when administered to patients with schizophrenia. However, in another study with women with AD, the administration of Raloxifene did not show any difference in terms of cognitive functions (Hsu et al. 2018).

Preclinical trials have tested the influence of progesterone over functional performance in rats with cerebral ischemia, hypothesizing that it may regulate cerebral plasticity and promote the survival of newly hippocampal neurons by favoring this microenvironment. Additionally, it was reported that progesterone modulates the synthesis of BDNF and VEGF after ischemia, affecting neurogenesis and vascular remodeling, and that it changes microglial phenotype, from a pro-inflammatory one to an anti-inflammatory role. Rats treated with progesterone after severe global cerebral ischemia showed significantly better results in spatial learning and memory tests and a significant number of new, mature neurons in the dentate gyrus. Due to its pleiotropic neuroprotective effects, progesterone may represent a promising therapeutic alternative for patients with ischemic stroke (Jiang et al. 2016; Montes et al. 2019).

3 Conclusion

This chapter has touched upon several pharmacological interventions in healthy and ill individuals, targeted at the augmentation, preservation, or recovery of brain function. While it is not an exhaustive collection of representatives from large drug classes such as antidepressants, nootropics, or psychostimulants, this exposition illustrates the current variety of approaches in tackling enhancement of motor, cognitive, and emotional domains in scientific, clinical, and informal professional or recreational settings alike.

While the discussion regarding off-label and nonprescription use in teenagers and young adults revolves around safety and its importance in the developing brain, neurorecovery is confronted with the limited efficacy of interventions across the extremely complex pathophysiology of various diseases among heterogenous populations. This has challenged researchers to design superior methods to study the multifaceted interaction between pharmacology, the dimensions of the brain's post-lesional response, and the biological reserve of individuals.

We conclude that to elicit sustainable and continued effects, agents used for brain function augmentation must work as the brain does, modulating multiple anticorrelated processes. While many interventions presented in this chapter loosely fit the definition of pharmacologically multimodal mechanisms of action, the benefit–risk balance is skewed in favor of non-synthetic agents with superior safety profiles, a key factor in determining their recommendation for brain enhancement in human subjects. As research progresses, the paradigm of "good-dirty" interventions might emerge as the new horizon in this field.

References

Abrahams S, Haylett WL, Johnson G, Carr JA, Bardien S (2019) Antioxidant effects of curcumin in models of neurodegeneration, aging, oxidative and nitrosative stress: a review. Neuroscience 406:1–21. https://doi.org/10.1016/j.neuroscience.2019.02.020

Allen M, Smallwood J, Christensen J, Gramm D, Rasmussen B, Jensen CG, Roepstorff A, Lutz A (2013) The balanced mind: the variability of task-unrelated thoughts predicts error monitoring. Front Hum Neurosci 7:743. https://doi.org/10.3389/fnhum.2013.00743

Alvarez XA, Sampedro C, Perez P, Laredo M, Couceiro V, Hernandez A, Figueroa J, Varela M, Arias D, Corzo L, Zas R, Lombardi V, Fernandez-Novoa L, Pichel V, Cacabelos R, Windisch M, Aleixandre M, Moessler H (2003) Positive effects of Cerebrolysin on electroencephalogram slowing, cognition and clinical outcome in patients with postacute traumatic brain injury: an exploratory study. Int Clin Psychopharmacol 18(5):271–278. https://doi.org/10.1097/00004850-200309000-00003

Alvarez XA, Cacabelos R, Laredo M, Couceiro V, Sampedro C, Varela M, Corzo L, Fernandez-Novoa L, Vargas M, Aleixandre M, Linares C, Granizo E, Muresanu D, Moessler H (2006) A 24-week, double-blind, placebo-controlled study of three dosages of Cerebrolysin in patients with mild to moderate Alzheimer's disease. Eur J Neurol 13(1):43–54. https://doi.org/10.1111/j.1468-1331.2006.01222.x

Alvarez FJ, Lafuente H, Rey-Santano MC, Mielgo VE, Gastiasoro E, Rueda M, Pertwee RG, Castillo AI, Romero J, Martinez-Orgado J (2008a) Neuroprotective effects of the nonpsychoactive cannabinoid cannabidiol in hypoxic-ischemic newborn piglets. Pediatr Res 64(6):653–658. https://doi.org/10.1203/PDR.0b013e318186e5dd

Alvarez X, Sampedro C, Figueroa J, Tellado I, González A, García-Fantini M, Cacabelos R, Muresanu D, Moessler H (2008b) Reductions in qEEG slowing over 1 year and after treatment with Cerebrolysin in patients with moderate-severe traumatic brain injury. J Neural Transm 115:683–692. https://doi.org/10.1007/s00702-008-0024-9

Alvarez-Sabín J, Román GC (2013) The role of citicoline in neuroprotection and neurorepair in ischemic stroke. Brain Sci 3(3):1395–1414. https://doi.org/10.3390/brainsci3031395

Arenth PM, Russell KC, Ricker JH, Zafonte RD (2011) CDP-Choline as a biological supplement during neurorecovery: a focused review. PM&R 3(6S):S123–S131. https://doi.org/10.1016/j.pmrj.2011.03.012

Asadollahi M, Ramezani M, Khanmoradi Z, Karimialavijeh E (2018) The efficacy comparison of citalopram, fluoxetine, and placebo on motor recovery after ischemic stroke: a double-blind placebo-controlled randomized controlled trial. Clin Rehabil 32(8):1069–1075. https://doi.org/10.1177/0269215518777791

Ates O, Cayli S, Altinoz E, Gurses I, Yucel N, Kocak A, Yologlu S, Turkoz Y (2006) Effects of resveratrol and methylprednisolone on biochemical, neurobehavioral and histopathological recovery after experimental spinal cord injury. Acta Pharmacol Sin 27(10):1317–1325. https://doi.org/10.1111/j.1745-7254.2006.00416.x

Balkaya M, Krober J, Gertz K, Peruzzaro S, Endres M (2013) Characterization of long-term functional outcome in a murine model of mild brain ischemia. J Neurosci Methods 213(2):179–187. https://doi.org/10.1016/j.jneumeth.2012.12.021

Ban JY, Cho SO, Choi S-H, Ju HS, Kim JY, Bae K, Song K-S, Seong YH (2008) Neuroprotective effect of Smilacis chinae rhizome on NMDA-induced neurotoxicity in vitro and focal cerebral ischemia in vivo. J Pharmacol Sci 106(1):68–77. https://doi.org/10.1254/jphs.FP0071206

Bei W, Peng W, Zang L, Xie Z, Hu D, Xu A (2007) Neuroprotective effects of a standardized extract of Diospyros kaki leaves on MCAO transient focal cerebral ischemic rats and cultured neurons injured by glutamate or hypoxia. Planta Med 73(7):636–643. https://doi.org/10.1055/s-2007-981532

Bennett S, Grant MM, Aldred S (2009) Oxidative stress in vascular dementia and Alzheimer's disease: a common pathology. J Alzheimers Dis 17(2):245–257. https://doi.org/10.3233/JAD-2009-1041

Berman SM, Kuczenski R, McCracken JT, London ED (2009) Potential adverse effects of amphetamine treatment on brain and behavior: a review. Mol Psychiatry 14(2):123–142. https://doi.org/10.1038/mp.2008.90

Bornstein N, Poon WS (2012) Accelerated recovery from acute brain injuries: clinical efficacy of neurotrophic treatment in stroke and traumatic brain injuries. Drugs Today 48(Suppl A):43–61. https://doi.org/10.1358/dot.2012.48(Suppl.A).1739723

Bornstein NM, Guekht A, Vester J, Heiss W-D, Gusev E, Hömberg V, Rahlfs VW, Bajenaru O, Popescu BO, Muresanu D (2018) Safety and efficacy of Cerebrolysin in early post-stroke recovery: a meta-analysis of nine randomized clinical trials. Neurol Sci 39(4):629–640. https://doi.org/10.1007/s10072-017-3214-0

Boyle J, Stanley N, James LM, Wright N, Johnsen S, Arbon EL, Dijk D-J (2012) Acute sleep deprivation: the effects of the AMPAKINE compound CX717 on human cognitive performance, alertness and recovery sleep. J Psychopharmacol 26(8):1047–1057. https://doi.org/10.1177/0269881111405353

Campos AC, Ferreira FR, Guimaraes FS (2012) Cannabidiol blocks long-lasting behavioral consequences of predator threat stress: possible involvement of 5HT1A receptors. J Psychiatr Res 46(11):1501–1510. https://doi.org/10.1016/j.jpsychires.2012.08.012

Campos AC, Fogaca MV, Sonego AB, Guimaraes FS (2016) Cannabidiol, neuroprotection and neuropsychiatric disorders. Pharmacol Res 112:119–127. https://doi.org/10.1016/j.phrs.2016.01.033

Candelario-Jalil E, de Oliveira A, Graef S, Hüll M, Muñoz E, Fiebich B, Bhatia H (2007) Resveratrol potently reduces prostaglandin E2 production and free radical formation in lipopolysaccharide-activated primary rat microglia. J Neuroinflammation 4:25. https://doi.org/10.1186/1742-2094-4-25

Castillo A, Tolon MR, Fernandez-Ruiz J, Romero J, Martinez-Orgado J (2010) The neuroprotective effect of cannabidiol in an in vitro model of newborn hypoxic-ischemic brain damage in mice is mediated by CB(2) and adenosine receptors. Neurobiol Dis 37(2):434–440. https://doi.org/10.1016/j.nbd.2009.10.023

Cekanaviciute E, Buckwalter MS (2016) Astrocytes: integrative regulators of neuroinflammation in stroke and other neurological diseases. Neurotherapeutics 13(4):685–701. https://doi.org/10.1007/s13311-016-0477-8

Chacko RV, Kim B, Jung SW, Daitch AL, Roland JL, Metcalf NV, Corbetta M, Shulman GL, Leuthardt EC (2018) Distinct phase-amplitude couplings distinguish cognitive processes in human attention. NeuroImage 175:111–121. https://doi.org/10.1016/j.neuroimage.2018.03.003

Chang WH, Park C-H, Kim D, Shin Y, Ko M-H, Lee A, Jang S, Kim Y-H (2016) Cerebrolysin combined with rehabilitation promotes motor recovery in patients with severe motor impairment after stroke. BMC Neurol 16:31. https://doi.org/10.1186/s12883-016-0553-z

Chatterjee P, Roy D, Bhattacharyya M, Bandyopadhyay S (2017) Biological networks in Parkinson's disease: an insight into the epigenetic mechanisms associated with this disease. BMC Genomics 18(1):721–721. https://doi.org/10.1186/s12864-017-4098-3

Chen C, Venketasubramanian N, Gan RN, Lambert C, Picard D, Chan BPL, Chan E, Bousser MG, Xuemin S (2009) Danqi Piantang Jiaonang (DJ), a traditional Chinese medicine, in poststroke recovery. Stroke 40(3):859–863. https://doi.org/10.1161/STROKEAHA.108.531616

Chen C-C, Wei S-T, Tsaia S-C, Chen X-X, Cho D-Y (2013a) Cerebrolysin enhances cognitive recovery of mild traumatic brain injury patients: double-blind, placebo-controlled, randomized study. Br J Neurosurg 27(6):803–807. https://doi.org/10.3109/02688697.2013.793287

Chen N, Yang M, Guo J, Zhou M, Zhu C, He L (2013b) Cerebrolysin for vascular dementia. Cochrane Database Syst Rev 1. https://doi.org/10.1002/14651858.CD008900.pub2

Cho H-J, Kim YJ (2009) Efficacy and safety of oral citicoline in acute ischemic stroke: drug surveillance study in 4,191 cases. Methods Find Exp Clin Pharmacol 31(3):171–176. https://doi.org/10.1358/mf.2009.31.3.1364241

Chollet F, Tardy J, Albucher J-F, Thalamas C, Berard E, Lamy C, Bejot Y, Deltour S, Jaillard A, Niclot P, Guillon B, Moulin T, Marque P, Pariente J, Arnaud C, Loubinoux I (2011) Fluoxetine for motor recovery after acute ischaemic stroke (FLAME): a randomised placebo-controlled trial. Lancet Neurol 10(2):123–130. https://doi.org/10.1016/S1474-4422(10)70314-8

Cohen PA, Zakharevich I, Gerona R (2019) Presence of piracetam in cognitive enhancement dietary supplements. JAMA Intern Med 180(3):458–459. https://doi.org/10.1001/jamainternmed.2019.5507

Committeri G, Piccardi L, Galati G, Guariglia C (2015) Where did you "left" Piazza del Popolo? At your "right" temporo-parietal junction. Cortex 73:106–111. https://doi.org/10.1016/j.cortex.2015.08.009

Corbetta M, Shulman GL (2002) Control of goal-directed and stimulus-driven attention in the brain. Nat Rev Neurosci 3(3):201–215. https://doi.org/10.1038/nrn755

Crisostomo EA, Duncan PW, Propst M, Dawson DV, Davis JN (1988) Evidence that amphetamine with physical therapy promotes recovery of motor function in stroke patients. Ann Neurol 23(1):94–97. https://doi.org/10.1002/ana.410230117

Daitch AL, Sharma M, Roland JL, Astafiev SV, Bundy DT, Gaona CM, Snyder AZ, Shulman GL, Leuthardt EC, Corbetta M (2013) Frequency-specific mechanism links human brain networks for spatial attention. Proc Natl Acad Sci U S A 110(48):19585–19590. https://doi.org/10.1073/pnas.1307947110

Darsalia V, Heldmann U, Lindvall O, Kokaia Z (2005) Stroke-induced neurogenesis in aged brain. Stroke 36(8):1790–1795. https://doi.org/10.1161/01.STR.0000173151.36031.be

Daverey A, Agrawal SK (2016) Curcumin alleviates oxidative stress and mitochondrial dysfunction in astrocytes. Neuroscience 333:92–103. https://doi.org/10.1016/j.neuroscience.2016.07.012

de la Torre R, de Sola S, Hernandez G, Farre M, Pujol J, Rodriguez J, Espadaler JM, Langohr K, Cuenca-Royo A, Principe A, Xicota L, Janel N, Catuara-Solarz S, Sanchez-Benavides G, Blehaut H, Duenas-Espin I, Del Hoyo L, Benejam B, Blanco-Hinojo L et al (2016) Safety and efficacy of cognitive training plus epigallocatechin-3-gallate in young adults with Down's syndrome (TESDAD): a double-blind, randomised, placebo-controlled, phase 2 trial. Lancet Neurol 15(8):801–810. https://doi.org/10.1016/S1474-4422(16)30034-5

de Pasquale F, Della Penna S, Snyder AZ, Marzetti L, Pizzella V, Romani GL, Corbetta M (2012) A cortical core for dynamic integration of functional networks in the resting human brain. Neuron 74(4):753–764. https://doi.org/10.1016/j.neuron.2012.03.031

de Pasquale F, Sabatini U, Della Penna S, Sestieri C, Caravasso CF, Formisano R, Peran P (2013) The connectivity of functional cores reveals different degrees of segregation and integration in the brain at rest. NeuroImage 69:51–61. https://doi.org/10.1016/j.neuroimage.2012.11.051

de Pasquale F, Della Penna S, Sporns O, Romani GL, Corbetta M (2016) A dynamic core network and global efficiency in the resting human brain. Cereb Cortex 26(10):4015–4033. https://doi.org/10.1093/cercor/bhv185

de Pasquale F, Corbetta M, Betti V, Della Penna S (2017) Cortical cores in network dynamics. NeuroImage 180(Pt B):370–382. https://doi.org/10.1016/j.neuroimage.2017.09.063

Deleglise B, Lassus B, Soubeyre V, Doulazmi M, Brugg B, Vanhoutte P, Peyrin J-M (2018) Dysregulated neurotransmission induces trans-synaptic degeneration in reconstructed neuronal networks. Sci Rep 8(1):11596. https://doi.org/10.1038/s41598-018-29918-1

Dharmadhikari S, Ma R, Yeh C-L, Stock A-K, Snyder S, Zauber SE, Dydak U, Beste C (2015) Striatal and thalamic GABA level concentrations play differential roles for the modulation of response selection processes by proprioceptive information. NeuroImage 120:36–42. https://doi.org/10.1016/j.neuroimage.2015.06.066

Dorer CL, Manktelow AE, Allanson J, Sahakian BJ, Pickard JD, Bateman A, Menon DK, Stamatakis EA (2018) Methylphenidate-mediated motor control network enhancement in patients with traumatic brain injury. Brain Injury 32(8):1040–1049. https://doi.org/10.1080/02699052.2018.1469166

El Sayed I, Zaki A, Fayed A, Shehata G, Abdelmonem S (2016) A meta-analysis of the effect of different neuroprotective drugs in management of patients with traumatic brain injury. Neurosurg Rev 41(2):427–438. https://doi.org/10.1007/s10143-016-0775-y

Fernández-Ruiz J, Sagredo O, Pazos MR, García C, Pertwee R, Mechoulam R, Martínez-Orgado J (2013) Cannabidiol for neurodegenerative disorders: important new clinical applications for this phytocannabinoid? Br J Clin Pharmacol 75(2):323–333. https://doi.org/10.1111/j.1365-2125.2012.04341.x

Fessel J (2019) Prevention of Alzheimer's disease by treating mild cognitive impairment with combinations chosen from eight available drugs. Alzheimers Dement 5:780–788. https://doi.org/10.1016/j.trci.2019.09.019

Flavell JH (1979) Metacognition and cognitive monitoring: a new area of cognitive–developmental inquiry. Am Psychol 34(10):906–911. https://doi.org/10.1037/0003-066X.34.10.906

Fond G, Micoulaud-Franchi J-A, Brunel L, Macgregor A, Miot S, Lopez R, Richieri R, Abbar M, Lancon C, Repantis D (2015) Innovative mechanisms of action for pharmaceutical cognitive enhancement: a systematic review. Psychiatry Res 229(1–2):12–20. https://doi.org/10.1016/j.psychres.2015.07.006

Fox KCR, Christoff K (2014) Metacognitive facilitation of spontaneous thought processes: when metacognition helps the wandering mind find its way. In: The cognitive neuroscience of metacognition. Springer, Berlin, pp 293–319. https://doi.org/10.1007/978-3-642-45190-4_13

Fox MD, Snyder AZ, Vincent JL, Corbetta M, Van Essen DC, Raichle ME (2005) The human brain is intrinsically organized into dynamic, anticorrelated functional networks. Proc Natl Acad Sci U S A 102(27):9673–9678. https://doi.org/10.1073/pnas.0504136102

Franke AG, Gränsmark P, Agricola A, Schühle K, Rommel T, Sebastian A, Balló HE, Gorbulev S, Gerdes C, Frank B, Ruckes C, Tüscher O, Lieb K (2017) Methylphenidate, modafinil, and caffeine for cognitive enhancement in chess: a double-blind, randomised controlled trial. Eur Neuropsychopharmacol 27(3):248–260. https://doi.org/10.1016/j.euroneuro.2017.01.006

Fu D, Tan P, Kuznetsov A, Molkov YI (2014) Chaos and robustness in a single family of genetic oscillatory networks. PLoS One 9(3):e90666. https://doi.org/10.1371/journal.pone.0090666

Gan R, Lambert C, Lianting J, Chan E, Venketasubramanian N, Chen C, Chan B, Samama M, Bousser M (2008) Danqi Piantan Jiaonang does not modify hemostasis, hematology, and biochemistry in normal subjects and stroke patients. Cerebrovasc Dis 25:450–456. https://doi.org/10.1159/000126919

Garcia-Arencibia M, Gonzalez S, de Lago E, Ramos JA, Mechoulam R, Fernandez-Ruiz J (2007) Evaluation of the neuroprotective effect of cannabinoids in a rat model of Parkinson's disease: importance of antioxidant and cannabinoid receptor-independent properties. Brain Res 1134(1):162–170. https://doi.org/10.1016/j.brainres.2006.11.063

Garcia-Cobos R, Frank-Garcia A, Gutierrez-Fernandez M, Diez-Tejedor E (2010) Citicoline, use in cognitive decline: vascular and degenerative. J Neurol Sci 299(1–2):188–192. https://doi.org/10.1016/j.jns.2010.08.027

Ghaffarpasand F, Torabi S, Rasti A, Niakan MH, Aghabaklou S, Pakzad F, Beheshtian MS, Tabrizi R (2018) Effects of cerebrolysin on functional outcome of patients with traumatic brain injury: a systematic review and meta-analysis. Neuropsychiatr Dis Treat 15:127–135. https://doi.org/10.2147/NDT.S186865

Girbovan C, Morin L, Plamondon H (2012) Repeated resveratrol administration confers lasting protection against neuronal damage but induces dose-related alterations of behavioral impairments after global ischemia. Behav Pharmacol 23(1):1–13. https://doi.org/10.1097/FBP.0b013e32834eafa3

Giunta B, Hou H, Zhu Y, Salemi J, Ruscin A, Shytle RD, Tan J (2010) Fish oil enhances anti-amyloidogenic properties of green tea EGCG in Tg2576 mice. Neurosci Lett 471(3):134–138. https://doi.org/10.1016/j.neulet.2010.01.026

Giurgea CE (1982) The nootropic concept and its prospective implications. Drug Dev Res 2(5):441–446. https://doi.org/10.1002/ddr.430020505

Gladstone DJ, Danells CJ, Armesto A, McIlroy WE, Staines WR, Graham SJ, Herrmann N, Szalai JP, Black SE (2006) Physiotherapy coupled with dextroamphetamine for rehabilitation after hemiparetic stroke: a randomized, double-blind, placebo-controlled trial. Stroke 37(1):179–185. https://doi.org/10.1161/01.STR.0000195169.42447.78

Glisky EL (2007) Changes in cognitive function in human aging. In: Riddle DR (ed) Brain aging: models, methods, and mechanisms. CRC Press/Taylor & Francis, New York. http://www.ncbi.nlm.nih.gov/books/NBK3885/

Goldstone RL, de Leeuw JR, Landy DH (2015) Fitting perception in and to cognition. Cognition 135:24–29. https://doi.org/10.1016/j.cognition.2014.11.027

Gonzalez-Fraile E, Martin-Carrasco M, Ballesteros J (2016) Efficacy of MLC601 on functional recovery after stroke: a systematic review and meta-analysis of randomized controlled trials. Brain Injury 30(3):267–270. https://doi.org/10.3109/02699052.2015.1118764

Granja A, Frias I, Neves AR, Pinheiro M, Reis S (2017) Therapeutic potential of epigallocatechin gallate nanodelivery systems. BioMed Res Int 2017:5813793. https://doi.org/10.1155/2017/5813793

Gruenbaum SE, Zlotnik A, Gruenbaum BF, Hersey D, Bilotta F (2016) Pharmacologic neuroprotection for functional outcomes after traumatic brain injury: a systematic review of the clinical literature. CNS Drugs 30(9):791–806. https://doi.org/10.1007/s40263-016-0355-2

Guekht A, Vester J, Heiss W-D, Gusev E, Hoemberg V, Rahlfs VW, Bajenaru O, Popescu BO, Doppler E, Winter S, Moessler H, Muresanu D (2017) Safety and efficacy of Cerebrolysin in motor function recovery after stroke: a meta-analysis of the CARS trials. Neurol Sci 38(10):1761–1769. https://doi.org/10.1007/s10072-017-3037-z

Gueroux M, Fleau C, Slozeck M, Laguerre M, Pianet I (2017) Epigallocatechin 3-gallate as an inhibitor of Tau phosphorylation and aggregation: a molecular and structural insight. J Prevent Alzheimers Dis 4(4):218–225. https://doi.org/10.14283/jpad.2017.35

Gulyás AI, Szabó GG, Ulbert I, Holderith N, Monyer H, Erdélyi F, Szabó G, Freund TF, Hájos N (2010) Parvalbumin-containing fast-spiking basket cells generate the field potential oscillations induced by cholinergic receptor activation in the hippocampus. J Neurosci 30(45):15134–15145. https://doi.org/10.1523/JNEUROSCI.4104-10.2010

Gutierrez-Fernandez M, Rodriguez-Frutos B, Fuentes B, Vallejo-Cremades MT, Alvarez-Grech J, Exposito-Alcaide M, Diez-Tejedor E (2012) CDP-choline treatment induces brain plasticity markers expression in experimental animal stroke. Neurochem Int 60(3):310–317. https://doi.org/10.1016/j.neuint.2011.12.015

Haag L, Quetscher C, Dharmadhikari S, Dydak U, Schmidt-Wilcke T, Beste C (2015) Interrelation of resting state functional connectivity, striatal GABA levels, and cognitive control processes. Hum Brain Mapp 36(11):4383–4393. https://doi.org/10.1002/hbm.22920

Häggman Henrikson J, Pombo Antunes AR, Wieloch T, Ruscher K (2020) Enhanced functional recovery by levodopa is associated with decreased levels of synaptogyrin following stroke in aged mice. Brain Res Bull 155:61–66. https://doi.org/10.1016/j.brainresbull.2019.11.019

Han S-Y, Hong Z-Y, Xie Y-H, Zhao Y, Xu X (2017) Therapeutic effect of Chinese herbal medicines for post stroke recovery: a traditional and network meta-analysis. Medicine 96(49):e8830. https://doi.org/10.1097/MD.0000000000008830

Hart T, Whyte J, Watanabe T, Chervoneva I (2018) Effects of dextroamphetamine in subacute traumatic brain injury: a randomized, placebo-controlled pilot study. J Neurosci Res 96(4):702–710. https://doi.org/10.1002/jnr.24102

Hayakawa K, Mishima K, Irie K, Hazekawa M, Mishima S, Fujioka M, Orito K, Egashira N, Katsurabayashi S, Takasaki K, Iwasaki K, Fujiwara M (2008) Cannabidiol prevents a post-ischemic injury progressively induced by cerebral ischemia via a high-mobility group box1-inhibiting mechanism. Neuropharmacology 55(8):1280–1286. https://doi.org/10.1016/j.neuropharm.2008.06.040

Hayakawa K, Irie K, Sano K, Watanabe T, Higuchi S, Enoki M, Nakano T, Harada K, Ishikane S, Ikeda T, Fujioka M, Orito K, Iwasaki K, Mishima K, Fujiwara M (2009) Therapeutic time window of cannabidiol treatment on delayed ischemic damage via high-mobility group box1-inhibiting mechanism. Biol Pharm Bull 32(9):1538–1544. https://doi.org/10.1248/bpb.32.1538

He X, Qin W, Liu Y, Zhang X, Duan Y, Song J, Li K, Jiang T, Yu C (2014) Abnormal salience network in normal aging and in amnestic mild cognitive impairment and Alzheimer's disease. Hum Brain Mapp 35(7):3446–3464. https://doi.org/10.1002/hbm.22414

Heurteaux C, Widmann C, Moha ou Maati H, Quintard H, Gandin C, Borsotto M, Veyssiere J, Onteniente B, Lazdunski M (2013) NeuroAiD: properties for neuroprotection and neurorepair. Cerebrovasc Dis 35(Suppl 1):1–7. https://doi.org/10.1159/000346228

Hsu W-Y, Lane H-Y, Lin C-H (2018) Medications used for cognitive enhancement in patients with schizophrenia, bipolar disorder, Alzheimer's disease, and Parkinson's disease. Front Psychiatry 9:91. https://doi.org/10.3389/fpsyt.2018.00091

Hu Y, Chen X, Gu H, Yang Y (2013) Resting-state glutamate and GABA concentrations predict task-induced deactivation in the default mode network. J Neurosci 33(47):18566–18573. https://doi.org/10.1523/JNEUROSCI.1973-13.2013

Huang C, Huang C-C, Sun C-K, Lin G-H, Hou W-H (2016) Methylphenidate on cognitive improvement in patients with traumatic brain injury: a meta-analysis. Curr Neuropharmacol 14:272. https://doi.org/10.2174/1570159X13666150514233033

Hurtado O, Moro MA, Cardenas A, Sanchez V, Fernandez-Tome P, Leza JC, Lorenzo P, Secades JJ, Lozano R, Davalos A, Castillo J, Lizasoain I (2005) Neuroprotection afforded by prior citicoline administration in experimental brain ischemia: effects on glutamate transport. Neurobiol Dis 18(2):336–345. https://doi.org/10.1016/j.nbd.2004.10.006

Hurtado O, Cardenas A, Pradillo JM, Morales JR, Ortego F, Sobrino T, Castillo J, Moro MA, Lizasoain I (2007) A chronic treatment with CDP-choline improves functional recovery and increases neuronal plasticity after experimental stroke. Neurobiol Dis 26(1):105–111. https://doi.org/10.1016/j.nbd.2006.12.005

Hylin MJ, Brenneman MM, Corwin JV (2017) Noradrenergic antagonists mitigate amphetamine-induced recovery. Behav Brain Res 334:61–71. https://doi.org/10.1016/j.bbr.2017.07.035

Ilieva I, Boland J, Farah MJ (2013) Objective and subjective cognitive enhancing effects of mixed amphetamine salts in healthy people. Neuropharmacology 64:496–505. https://doi.org/10.1016/j.neuropharm.2012.07.021

Islam M (2012) Cardiovascular effects of green tea catechins: progress and promise. Recent Pat Cardiovasc Drug Discov 7:88–99. https://doi.org/10.2174/157489012801227292

Iznak EV, Iznak AF, Pankratova EA, Zavadenko NN, Guzilova LS, Guzilova II (2010) [Electrophysiological correlates of efficacy of nootropic drugs in the treatment of consequences of traumatic brain injury in adolescents]. Zhurnal nevrologii i psikhiatrii imeni S.S. Korsakova 110(5 Pt 1):27–32.

Jenkins P, Simoni S, Bourke N, Fleminger J, Scott G, Towey D, Svensson W, Khan S, Patel M, Greenwood R, Friedland D, Hampshire A, Cole J, Sharp D (2019) Stratifying drug treatment of cognitive impairments after traumatic brain injury using neuroimaging. Brain 142(8):2367–2379. https://doi.org/10.1093/brain/awz149

Jiang J, Wang W, Sun YJ, Hu M, Li F, Zhu DY (2007) Neuroprotective effect of curcumin on focal cerebral ischemic rats by preventing blood-brain barrier damage. Eur J Pharmacol 561(1–3):54–62. https://doi.org/10.1016/j.ejphar.2006.12.028

Jiang C, Zuo F, Wang Y, Lu H, Yang Q, Wang J (2016) Progesterone changes VEGF and BDNF expression and promotes neurogenesis after ischemic stroke. Mol Neurobiol 54:571–581. https://doi.org/10.1007/s12035-015-9651-y

Jilka SR, Scott G, Ham T, Pickering A, Bonnelle V, Braga RM, Leech R, Sharp DJ (2014) Damage to the salience network and interactions with the default mode network. J Neurosci 34(33):10798–10807. https://doi.org/10.1523/JNEUROSCI.0518-14.2014

Jorge RE, Acion L, Moser D, Adams HPJ, Robinson RG (2010) Escitalopram and enhancement of cognitive recovery following stroke. Arch Gen Psychiatry 67(2):187–196. https://doi.org/10.1001/archgenpsychiatry.2009.185

Kabbara A, El Falou W, Khalil M, Wendling F, Hassan M (2017) The dynamic functional core network of the human brain at rest. Sci Rep 7(1):2936. https://doi.org/10.1038/s41598-017-03420-6

Kann O (2016) The interneuron energy hypothesis: implications for brain disease. Neurobiol Dis 90:75–85. https://doi.org/10.1016/j.nbd.2015.08.005

Kapogiannis D, Reiter DA, Willette AA, Mattson MP (2013) Posteromedial cortex glutamate and GABA predict intrinsic functional connectivity of the default mode network. NeuroImage 64:112–119. https://doi.org/10.1016/j.neuroimage.2012.09.029

Kerns JG, Cohen JD, MacDonald AW, Cho RY, Stenger VA, Carter CS (2004) Anterior cingulate conflict monitoring and adjustments in control. Science 303(5660):1023–1026. https://doi.org/10.1126/science.1089910

Khalili H, Niakan A, Ghaffarpasand F (2017) Effects of cerebrolysin on functional recovery in patients with severe disability after traumatic brain injury: a historical cohort study. Clin Neurol Neurosurg 152:34–38. https://doi.org/10.1016/j.clineuro.2016.11.011

Kim D, Nguyen MD, Dobbin MM, Fischer A, Sananbenesi F, Rodgers JT, Delalle I, Baur JA, Sui G, Armour SM, Puigserver P, Sinclair DA, Tsai L-H (2007) SIRT1 deacetylase protects against neurodegeneration in models for Alzheimer's disease and amyotrophic lateral sclerosis. EMBO J 26(13):3169–3179. https://doi.org/10.1038/sj.emboj.7601758

Knafo S, Esteban JA (2015) Chapter 3—molecular mechanisms of drug-induced cognitive enhancement. In: Knafo S, Venero C (eds) Cognitive enhancement. Academic Press, pp 43–59. https://doi.org/10.1016/B978-0-12-417042-1.00003-6

Korol DL, Pisani SL (2015) Estrogens and cognition: friends or foes?: an evaluation of the opposing effects of estrogens on learning and memory. Horm Behav 74:105–115. https://doi.org/10.1016/j.yhbeh.2015.06.017

Koshino H, Minamoto T, Yaoi K, Osaka M, Osaka N (2014) Coactivation of the default mode network regions and working memory Network regions during task preparation. Sci Rep 4:5954. https://doi.org/10.1038/srep05954

Kraglund KL, Mortensen JK, Damsbo AG, Modrau B, Simonsen SA, Iversen HK, Madsen M, Grove EL, Johnsen SP, Andersen G (2018) Neuroregeneration and vascular protection by citalopram in acute ischemic stroke (TALOS). Stroke 49(11):2568–2576. https://doi.org/10.1161/STROKEAHA.117.020067

Kronenberg G, Gertz K, Heinz A, Endres M (2014) Of mice and men: modelling post-stroke depression experimentally. Br J Pharmacol 171(20):4673–4689. https://doi.org/10.1111/bph.12775

Krupinski J, Ferrer I, Barrachina M, Secades JJ, Mercadal J, Lozano R (2002) CDP-choline reduces pro-caspase and cleaved caspase-3 expression, nuclear DNA fragmentation, and specific PARP-cleaved products of caspase activation following middle cerebral artery occlusion in the rat. Neuropharmacology 42(6):846–854. https://doi.org/10.1016/s0028-3908(02)00032-1

Kudesia RS, Baer M, Elfenbein HA (2015) A wandering mind does not stray far from home: the value of metacognition in distant search. PLoS One 10(5):e0126865. https://doi.org/10.1371/journal.pone.0126865

Kumar A, Kitago T (2019) Pharmacological enhancement of stroke recovery. Curr Neurol Neurosci Rep 19(7):43. https://doi.org/10.1007/s11910-019-0959-2

Kumar A, Naidu PS, Seghal N, Padi SSV (2007) Neuroprotective effects of resveratrol against intracerebroventricular colchicine-induced cognitive impairment and oxidative stress in rats. Pharmacology 79(1):17–26. https://doi.org/10.1159/000097511

Labiche LA, Grotta JC (2004) Clinical trials for cytoprotection in stroke. NeuroRx 1(1):46–70. https://doi.org/10.1602/neurorx.1.1.46

Lafuente H, Alvarez FJ, Pazos MR, Alvarez A, Rey-Santano MC, Mielgo V, Murgia-Esteve X, Hilario E, Martinez-Orgado J (2011) Cannabidiol reduces brain damage and improves functional recovery after acute hypoxia-ischemia in newborn pigs. Pediatr Res 70(3):272–277. https://doi.org/10.1203/PDR.0b013e3182276b11

Lagouge M, Argmann C, Gerhart-Hines Z, Meziane H, Lerin C, Daussin F, Messadeq N, Milne J, Lambert P, Elliott P, Geny B, Laakso M, Puigserver P, Auwerx J (2006) Resveratrol improves mitochondrial function and protects against metabolic disease by activating SIRT1 and PGC-1alpha. Cell 127(6):1109–1122. https://doi.org/10.1016/j.cell.2006.11.013

Lakhan SE, Kirchgessner A, Hofer M (2009) Inflammatory mechanisms in ischemic stroke: therapeutic approaches. J Transl Med 7:97–97. https://doi.org/10.1186/1479-5876-7-97

Lanni C, Lenzken SC, Pascale A, Del Vecchio I, Racchi M, Pistoia F, Govoni S (2008) Cognition enhancers between treating and doping the mind. Pharmacol Res 57(3):196–213. https://doi.org/10.1016/j.phrs.2008.02.004

Lee JW, Lee YK, Ban JO, Ha TY, Yun YP, Han SB, Oh KW, Hong JT (2009) Green tea (-)-epigallocatechin-3-gallate inhibits beta-amyloid-induced cognitive dysfunction through modification of secretase activity via inhibition of ERK and. J Nutr 139(10):1987–1993. https://doi.org/10.3945/jn.109.109785

Lee EO, Park HJ, Kang JL, Kim H-S, Chong YH (2010) Resveratrol reduces glutamate-mediated monocyte chemotactic protein-1 expression via inhibition of extracellular signal-regulated kinase 1/2 pathway in rat hippocampal slice cultures. J Neurochem 112(6):1477–1487. https://doi.org/10.1111/j.1471-4159.2009.06564.x

Lee E-S, Yoo K, Lee Y-B, Chung J, Lim J-E, Yoon B, Jeong Y (2016) Default mode network functional connectivity in early and late mild cognitive impairment: results from the Alzheimer's disease neuroimaging initiative. Alzheimer Dis Assoc Disord 30(4):289–296. https://doi.org/10.1097/WAD.0000000000000143

Lee WT, Hsian CCL, Lim Y-A (2017) The effects of MLC901 on tau phosphorylation. Neuroreport 28(16):1043–1048. https://doi.org/10.1097/WNR.0000000000000884

Leech R, Kamourieh S, Beckmann CF, Sharp DJ (2011) Fractionating the default mode network: distinct contributions of the ventral and dorsal posterior cingulate cortex to cognitive control. J Neurosci 31(9):3217–3224. https://doi.org/10.1523/JNEUROSCI.5626-10.2011

Legeay S, Rodier M, Fillon L, Faure S, Clere N (2015) Epigallocatechin gallate: a review of its beneficial properties to prevent metabolic syndrome. Nutrients 7(7):5443–5468. https://doi.org/10.3390/nu7075230

Leo A, Russo E, Elia M (2016) Cannabidiol and epilepsy: rationale and therapeutic potential. Pharmacol Res 107:85–92. https://doi.org/10.1016/j.phrs.2016.03.005

Lin David J, Finklestein Seth P, Cramer Steven C (2018) New directions in treatments targeting stroke recovery. Stroke 49(12):3107–3114. https://doi.org/10.1161/STROKEAHA.118.021359

López ME, Garcés P, Cuesta P, Castellanos NP, Aurtenetxe S, Bajo R, Marcos A, Montenegro M, Yubero R, del Pozo F, Sancho M, Maestú F (2014) Synchronization during an internally directed cognitive state in healthy aging and mild cognitive impairment: a MEG study. Age 36(3):9643–9643. https://doi.org/10.1007/s11357-014-9643-2

Lorivel T, Gandin C, Veyssière J, Lazdunski M, Heurteaux C (2015) Positive effects of the traditional Chinese medicine MLC901 in cognitive tasks. J Neurosci Res 93(11):1648–1663. https://doi.org/10.1002/jnr.23591

Luine V (2016) Estradiol: mediator of memories, spine density and cognitive resilience to stress in female rodents. J Steroid Biochem Mol Biol 160:189–195. https://doi.org/10.1016/j.jsbmb.2015.07.022

Ma T, Tan M-S, Yu J-T, Tan L (2014) Resveratrol as a therapeutic agent for Alzheimer's disease. BioMed Res Int 2014:350516. https://doi.org/10.1155/2014/350516

Mandal M, Jaiswal P, Mishra A (2020) Role of curcumin and its nanoformulations in neurotherapeutics: a comprehensive review. J Biochem Mol Toxicol 34(6):e22478. https://doi.org/10.1002/jbt.22478

Marron TR, Lerner Y, Berant E, Kinreich S, Shapira-Lichter I, Hendler T, Faust M (2018) Chain free association, creativity, and the default mode network. Neuropsychologia 118(Pt A):40–58. https://doi.org/10.1016/j.neuropsychologia.2018.03.018

Marshall CA, Brodnik ZD, Mortensen OV, Reith MEA, Shumsky JS, Waterhouse BD, Espana RA, Kortagere S (2019) Selective activation of Dopamine D3 receptors and norepinephrine transporter blockade enhances sustained attention. Neuropharmacology 148:178–188. https://doi.org/10.1016/j.neuropharm.2019.01.003

Martín-Moreno AM, Reigada D, Ramírez BG, Mechoulam R, Innamorato N, Cuadrado A, de Ceballos ML (2011) Cannabidiol and other cannabinoids reduce microglial activation in vitro and in vivo: relevance to Alzheimer's disease. Mol Pharmacol 79(6):964–973. https://doi.org/10.1124/mol.111.071290

Martinsson L, Eksborg S, Wahlgren NG (2003) Intensive early physiotherapy combined with dexamphetamine treatment in severe stroke: a randomized, controlled pilot study. Cerebrovasc Dis 16(4):338–345. https://doi.org/10.1159/000072555

Masliah E, Diez-Tejedor E (2012) The pharmacology of neurotrophic treatment with Cerebrolysin: brain protection and repair to counteract pathologies of acute and chronic neurological disorders. Drugs Today 48(Suppl A):3–24. https://doi.org/10.1358/dot.2012.48(Suppl.A).1739716

McDonald B, Flashman L, Arciniegas D, Ferguson R, Xing L, Harezlak J, Sprehn G, Maerlender A, Kruck C, Gillock K, Frey K, Wall R, Saykin A, Mcallister T (2016) Methylphenidate and memory and attention adaptation training for persistent cognitive symptoms after traumatic brain injury: a randomized, placebo-controlled trial. Neuropsychopharmacology 42(9):1766–1775. https://doi.org/10.1038/npp.2016.261

McGlade E, Locatelli A, Hardy J, Kamiya T, Morita M, Morishita K, Sugimura Y, Yurgelun-Todd D (2012) Improved attentional performance following citicoline administration in healthy adult women. Food Nutr Sci 3(6):769–773. https://doi.org/10.4236/fns.2012.36103

McGlade E, Agoston AM, DiMuzio J, Kizaki M, Nakazaki E, Kamiya T, Yurgelun-Todd D (2019) The effect of citicoline supplementation on motor speed and attention in adolescent males. J Attent Disord 23(2):121–134. https://doi.org/10.1177/1087054715593633

Mead GE, Legg L, Tilney R, Hsieh CF, Wu S, Lundstrom E, Rudberg AS, Kutlubaev M, Dennis MS, Soleimani B, Barugh A, Hackett ML, Hankey GJ (2019) Fluoxetine for stroke recovery: meta-analysis of randomized controlled trials. Int J Stroke 15(4):365–376. https://doi.org/10.1177/1747493019879655

Mechoulam R, Gaoni Y (1965) A total synthesis of dl-delta-1-tetrahydrocannabinol, the active constituent of hashish. J Am Chem Soc 87:3273–3275. https://doi.org/10.1021/ja01092a065

Mishima K, Hayakawa K, Abe K, Ikeda T, Egashira N, Iwasaki K, Fujiwara M (2005) Cannabidiol prevents cerebral infarction via a serotonergic 5-hydroxytryptamine1A receptor-dependent mechanism. Stroke 36(5):1077–1082. https://doi.org/10.1161/01.STR.0000163083.59201.34

Montes P, Vigueras-Villaseñor R, Rojas-Castaneda J, Monfil T, Cervantes M, Morali G (2019) Progesterone treatment in rats after severe global cerebral ischemia promotes hippocampal dentate gyrus neurogenesis and functional recovery. Neurol Res 41:1–8. https://doi.org/10.1080/01616412.2019.1576356

Mooneyham BW, Schooler JW (2013) The costs and benefits of mind-wandering: a review. Can J Exp Psychol 67(1):11–18. https://doi.org/10.1037/a0031569

Mori MA, Meyer E, Soares LM, Milani H, Guimarães FS, de Oliveira RMW (2017) Cannabidiol reduces neuroinflammation and promotes neuroplasticity and functional recovery after brain ischemia. Prog Neuropsychopharmacol Biol Psychiatry 75:94–105. https://doi.org/10.1016/j.pnpbp.2016.11.005

Muresanu DF (2007) Neuroprotection and neuroplasticity—a holistic approach and future perspectives. J Neurol Sci 257(1):38–43. https://doi.org/10.1016/j.jns.2007.01.041

Muresanu DF, Alvarez XA, Moessler H, Buia M, Stan A, Pintea D, Moldovan F, Popescu BO (2008) A pilot study to evaluate the effects of Cerebrolysin on cognition and qEEG in vascular dementia: cognitive improvement correlates with qEEG acceleration. J Neurol Sci 267(1–2):112–119. https://doi.org/10.1016/j.jns.2007.10.016

Muresanu DF, Buzoianu A, Florian SI, von Wild T (2012) Towards a roadmap in brain protection and recovery. J Cell Mol Med 16(12):2861–2871. https://doi.org/10.1111/j.1582-4934.2012.01605.x

Muresanu DF, Ciurea AV, Gorgan RM, Gheorghita E, Florian SI, Stan H, Blaga A, Ianovici N, Iencean SM, Turliuc D, Davidescu HB, Mihalache C, Brehar FM, Mihaescu AS, Mardare DC, Anghelescu A, Chiparus C, Lapadat M, Pruna V et al (2015) A retrospective, multi-center cohort study evaluating the severity-related effects of cerebrolysin treatment on clinical outcomes in traumatic brain injury. CNS Neurol Disord Drug Targets 14(5):587–599. https://doi.org/10.2174/1871527314666150430162531

Muresanu DF, Heiss W-D, Hoemberg V, Bajenaru O, Popescu CD, Vester JC, Rahlfs VW, Doppler E, Meier D, Moessler H, Guekht A (2016) Cerebrolysin and recovery after stroke (CARS): a randomized, placebo-controlled, double-blind, multicenter trial. Stroke 47(1):151–159. https://doi.org/10.1161/STROKEAHA.115.009416

Muresanu DF, Florian S, Homberg V, Matula C, von Steinbuchel N, Vos PE, von Wild K, Birle C, Muresanu I, Slavoaca D, Rosu OV, Strilciuc S, Vester J (2020) Efficacy and safety of cerebrolysin in neurorecovery after moderate-severe traumatic brain injury: results from the CAPTAIN II trial. Neurol Sci 41(5):1171–1181. https://doi.org/10.1007/s10072-019-04181-y

Navarro JC, Molina MC, Baroque Ii AC, Lokin JK (2012) The use of NeuroAiD (MLC601) in post-ischemic stroke patients. Rehabil Res Pract 2012:506387. https://doi.org/10.1155/2012/506387

Nido G, Ryan M, Benuskova L, Williams J (2015) Dynamical properties of gene regulatory networks involved in long-term potentiation. Front Mol Neurosci 8:42. https://doi.org/10.3389/fnmol.2015.00042

Nieto-Sampedro M, Lewis E, Cotman C, Manthorpe M, Skaper S, Barbin G, Longo F, Varon S (1982) Brain injury causes a time-dependent increase in neuronotrophic activity at the lesion site. Science 217(4562):860–861. https://doi.org/10.1126/science.7100931

Nikiforuk A, Kalaba P, Ilic M, Korz V, Dragačević V, Wackerlig J, Langer T, Höger H, Golebiowska J, Popik P, Lubec G (2017) A novel dopamine transporter inhibitor CE-123 improves cognitive flexibility and maintains impulsivity in healthy male rats. Front Behav Neurosci 11:222. https://doi.org/10.3389/fnbeh.2017.00222

Oomen C, Farkas E, Roman V, Van Der Beek E, Luiten P, Meerlo P (2009) Resveratrol preserves cerebrovascular density and cognitive function in aging mice. Front Aging Neurosci 1:4. https://doi.org/10.3389/neuro.24.004.2009

Osman ME, Hannafin MJ (1992) Metacognition research and theory: analysis and implications for instructional design. Educ Technol Res Dev 40(2):83–99. https://doi.org/10.1007/BF02297053

Pazos MR, Mohammed N, Lafuente H, Santos M, Martinez-Pinilla E, Moreno E, Valdizan E, Romero J, Pazos A, Franco R, Hillard CJ, Alvarez FJ, Martinez-Orgado J (2013) Mechanisms of cannabidiol neuroprotection in hypoxic-ischemic newborn pigs: role of 5HT(1A) and CB2 receptors. Neuropharmacology 71:282–291. https://doi.org/10.1016/j.neuropharm.2013.03.027

Pérez-Palma E, Bustos BI, Villamán CF, Alarcón MA, Avila ME, Ugarte GD, Reyes AE, Opazo C, De Ferrari GV, Alzheimer's Disease Neuroimaging Initiative, & NIA-LOAD/NCRAD Family Study Group (2014) Overrepresentation of glutamate signaling in Alzheimer's disease: network-based pathway enrichment using meta-analysis of genome-wide association studies. PLoS One 9(4):e95413. https://doi.org/10.1371/journal.pone.0095413

Perrone D, Fuggetta MP, Ardito F, Cottarelli A, De Filippis A, Ravagnan G, De Maria S, Lo Muzio L (2017) Resveratrol (3,5,4′-trihydroxystilbene) and its properties in oral diseases. Exp Ther Med 14(1):3–9. https://doi.org/10.3892/etm.2017.4472

Pertwee RG, Thomas A, Stevenson LA, Maor Y, Mechoulam R (2005) Evidence that (−)-7-hydroxy-4′-dimethylheptyl-cannabidiol activates a non-CB1, non-CB2, non-TRPV1 target in the mouse vas deferens. Future Direct Cannabinoid Ther 48(8):1139–1146. https://doi.org/10.1016/j.neuropharm.2005.01.010

Piccoli T, Valente G, Linden DEJ, Re M, Esposito F, Sack AT, Di Salle F (2015) The default mode network and the working memory network are not anti-correlated during all phases of a working memory task. PLoS One 10(4):e0123354. https://doi.org/10.1371/journal.pone.0123354

Pita-Juárez Y, Altschuler G, Kariotis S, Wei W, Koler K, Green C, Tanzi RE, Hide W (2018) The pathway coexpression network: revealing pathway relationships. PLoS Comput Biol 14(3):e1006042. https://doi.org/10.1371/journal.pcbi.1006042

Poon W, Matula C, Vos PE, Muresanu DF, von Steinbuchel N, von Wild K, Homberg V, Wang E, Lee TMC, Strilciuc S, Vester JC (2020) Safety and efficacy of Cerebrolysin in acute brain injury and neurorecovery: CAPTAIN I-a randomized, placebo-controlled, double-blind, Asian-Pacific trial. Neurol Sci 41(2):281–293. https://doi.org/10.1007/s10072-019-04053-5

Porrino L, Daunais J, Rogers G, Hampson R, Deadwyler S (2005) Facilitation of task performance and removal of the effects of sleep deprivation by an ampakine (CX717) in nonhuman primates. PLoS Biol 3(9):e299

Price AR, Peelle JE, Bonner MF, Grossman M, Hamilton RH (2016) Causal evidence for a mechanism of semantic integration in the angular gyrus as revealed by high-definition transcranial direct current stimulation. J Neurosci 36(13):3829–3838. https://doi.org/10.1523/JNEUROSCI.3120-15.2016

Rahvar M, Nikseresht M, Shafiee SM, Naghibalhossaini F, Rasti M, Panjehshahin MR, Owji AA (2011) Effect of oral resveratrol on the BDNF gene expression in the hippocampus of the rat brain. Neurochem Res 36(5):761–765. https://doi.org/10.1007/s11064-010-0396-8

Rausch W-D, Liu S, Gille G, Radad K (2006) Neuroprotective effects of ginsenosides. Acta Neurobiol Exp 66(4):369–375

Richiardi J, Altmann A, Milazzo A-C, Chang C, Chakravarty MM, Banaschewski T, Barker GJ, Bokde ALW, Bromberg U, Büchel C, Conrod P, Fauth-Bühler M, Flor H, Frouin V, Gallinat J, Garavan H, Gowland P, Heinz A, Lemaître H et al (2015) Brain Networks. Correlated gene expression supports synchronous activity in brain networks. Science 348(6240):1241–1244. https://doi.org/10.1126/science.1255905

Riley C, Hutter-Paier B, Windisch M, Doppler E, Moessler H, Wronski R (2006) A peptide preparation protects cells in organotypic brain slices against cell death after glutamate intoxication. J Neural Transm 113(1):103–110. https://doi.org/10.1007/s00702-005-0302-8

Rothwell NJ, Loddick SA, Stroemer P (1997) Interleukins and cerebral ischaemia. Int Rev Neurobiol 40:281–298. https://doi.org/10.1016/s0074-7742(08)60724-2

Ruether E, Alvarez X, Rainer M, Moessler H (2002) Sustained improvement of cognition and global function in patients with moderately severe Alzheimer's disease: a double-blind, placebo-controlled study with the neurotrophic agent Cerebrolysin. J Neural Transm Suppl 62:265–275. https://doi.org/10.1007/978-3-7091-6139-5_24

Savadi Oskouie D, Sharifipour E, Sadeghi Bazargani H, Hashemilar M, Nikanfar M, Ghazanfari Amlashi S, Abbaszade Z, Sadeghihokmabadi E, Rikhtegar R, Golzari SEJ (2017) Efficacy of citalopram on acute ischemic stroke outcome: a randomized clinical trial. Neurorehabil Neural Repair 31(7):638–647. https://doi.org/10.1177/1545968317704902

Saver JL (2008) Citicoline: update on a promising and widely available agent for neuroprotection and neurorepair. Rev Neurol Dis 5(4):167–177

Scalf PE, Ahn J, Beck DM, Lleras A (2014) Trial history effects in the ventral attentional network. J Cogn Neurosci 26(12):2789–2797. https://doi.org/10.1162/jocn_a_00678

Schelle KJ, Faulmüller N, Caviola L, Hewstone M (2014) Attitudes toward pharmacological cognitive enhancement—a review. Front Syst Neurosci 8:53–53. https://doi.org/10.3389/fnsys.2014.00053

Schiavon AP, Soares LM, Bonato JM, Milani H, Guimaraes FS, Weffort de Oliveira RM (2014) Protective effects of cannabidiol against hippocampal cell death and cognitive impairment induced by bilateral common carotid artery occlusion in mice. Neurotox Res 26(4):307–316. https://doi.org/10.1007/s12640-014-9457-0

Schmatz R, Mazzanti CM, Spanevello R, Stefanello N, Gutierres J, Correa M, da Rosa MM, Rubin MA, Chitolina Schetinger MR, Morsch VM (2009) Resveratrol prevents memory deficits and the increase in acetylcholinesterase activity in streptozotocin-induced diabetic rats. Eur J Pharmacol 610(1–3):42–48. https://doi.org/10.1016/j.ejphar.2009.03.032

Schnell L, Fearn S, Schwab ME, Perry VH, Anthony DC (1999) Cytokine-induced acute inflammation in the brain and spinal cord. J Neuropathol Exp Neurol 58(3):245–254. https://doi.org/10.1097/00005072-199903000-00004

Scholey A, Downey LA, Ciorciari J, Pipingas A, Nolidin K, Finn M, Wines M, Catchlove S, Terrens A, Barlow E, Gordon L, Stough C (2012) Acute neurocognitive effects of epigallocatechin gallate (EGCG). Appetite 58(2):767–770. https://doi.org/10.1016/j.appet.2011.11.016

Secades JJ, Lorenzo JL (2006) Citicoline: pharmacological and clinical review, 2006 update. Methods Find Exp Clin Pharmacol 28(Suppl B):1–56

Shankar S, Suthakar G, Srivastava RK (2007) Epigallocatechin-3-gallate inhibits cell cycle and induces apoptosis in pancreatic cancer. Front Biosci 12:5039–5051. https://doi.org/10.2741/2446

Shao A-W, Wu H-J, Chen S, Ammar A, Zhang J-M, Hong Y (2014) Resveratrol attenuates early brain injury after subarachnoid hemorrhage through inhibition of NF-κB-dependent inflammatory/MMP-9 pathway. CNS Neurosci Ther 20(2):182–185. https://doi.org/10.1111/cns.12194

Silton RL, Heller W, Towers DN, Engels AS, Spielberg JM, Edgar JC, Sass SM, Stewart JL, Sutton BP, Banich MT, Miller GA (2010) The time course of activity in dorsolateral prefrontal cortex and anterior cingulate cortex during top-down attentional control. NeuroImage 50(3):1292–1302. https://doi.org/10.1016/j.neuroimage.2009.12.061

Simao F, Matte A, Breier AC, Kreutz F, Trindade VMT, Netto CA, Salbego CG (2013) Resveratrol prevents global cerebral ischemia-induced decrease in lipid content. Neurol Res 35(1):59–64. https://doi.org/10.1179/1743132812Y.0000000116

Siow CHC (2008) Neuroaid in stroke recovery. Eur Neurol 60(5):264–266. https://doi.org/10.1159/000155220

Smallwood J, Schooler JW (2015) The science of mind wandering: empirically navigating the stream of consciousness. Annu Rev Psychol 66:487–518. https://doi.org/10.1146/annurev-psych-010814-015331

Sobrino T, Hurtado O, Moro MA, Rodriguez-Yanez M, Castellanos M, Brea D, Moldes O, Blanco M, Arenillas JF, Leira R, Davalos A, Lizasoain I, Castillo J (2007) The increase of circulating endothelial progenitor cells after acute ischemic stroke is associated with good outcome. Stroke 38(10):2759–2764. https://doi.org/10.1161/STROKEAHA.107.484386

Sonde L, Lokk J (2007) Effects of amphetamine and/or L-dopa and physiotherapy after stroke—a blinded randomized study. Acta Neurol Scand 115(1):55–59. https://doi.org/10.1111/j.1600-0404.2006.00728.x

Sonde L, Nordstrom M, Nilsson CG, Lokk J, Viitanen M (2001) A double-blind placebo-controlled study of the effects of amphetamine and physiotherapy after stroke. Cerebrovasc Dis 12(3):253–257. https://doi.org/10.1159/000047712

Sousa A, Dinis-Oliveira R (2020) Pharmacokinetic and pharmacodynamic of the cognitive enhancer modafinil: relevant clinical and forensic aspects. Subst Abuse 41(2):155–173. https://doi.org/10.1080/08897077.2019.1700584

Sporns O (2013) Structure and function of complex brain networks. Dialogues Clin Neurosci 15(3):247–262

Sprigg N, Willmot MR, Gray LJ, Sunderland A, Pomeroy V, Walker M, Bath PMW (2007) Amphetamine increases blood pressure and heart rate but has no effect on motor recovery or cerebral haemodynamics in ischaemic stroke: a randomized controlled trial (ISRCTN 36285333). J Hum Hypert 21(8):616–624. https://doi.org/10.1038/sj.jhh.1002205

Stockburger C, Miano D, Pallas T, Friedland K, Müller WE (2016) Enhanced neuroplasticity by the metabolic enhancer piracetam associated with improved mitochondrial dynamics and altered permeability transition pore function. Neural Plast 2016:8075903. https://doi.org/10.1155/2016/8075903

Strimpakos AS, Sharma RA (2008) Curcumin: preventive and therapeutic properties in laboratory studies and clinical trials. Antioxid Redox Signal 10(3):511–545. https://doi.org/10.1089/ars.2007.1769

Suwanwela NC, Chen CLH, Lee CF, Young SH, Tay SS, Umapathi T, Lao AY, Gan HH, Baroque Ii AC, Navarro JC, Chang HM, Advincula JM, Muengtaweepongsa S, Chan BPL, Chua CL, Wijekoon N, de Silva HA, Hiyadan JHB, Wong KSL et al (2018) Effect of combined treatment with MLC601 (NeuroAiDTM) and rehabilitation on post-stroke recovery: the CHIMES and CHIMES-E studies. Cerebrovasc Dis 46(1–2):82–88. https://doi.org/10.1159/000492625

Tang M, Taghibiglou C (2017) The mechanisms of action of curcumin in Alzheimer's disease. J Alzheimers Dis 58(4):1003–1016. https://doi.org/10.3233/JAD-170188

Tayebati SK, Marucci G, Santinelli C, Buccioni M, Amenta F (2015) Choline-containing phospholipids: structure-activity relationships versus therapeutic applications. Curr Med Chem 22(38):4328–4340. https://doi.org/10.2174/0929867322666151029104152

Tegenge MA, Rajbhandari L, Shrestha S, Mithal A, Hosmane S, Venkatesan A (2014) Curcumin protects axons from degeneration in the setting of local neuroinflammation. Exp Neurol 253:102–110. https://doi.org/10.1016/j.expneurol.2013.12.016

Theadom A, Barker-Collo S, Jones KM, Parmar P, Bhattacharjee R, Feigin VL (2018) MLC901 (NeuroAiD II^{TM}) for cognition after traumatic brain injury: a pilot randomized clinical trial. Eur J Neurol 25(8):1055–1e82. https://doi.org/10.1111/ene.13653

Thurm F, Antonenko D, Schlee W, Kolassa S, Elbert T, Kolassa I-T (2013) Effects of aging and mild cognitive impairment on electrophysiological correlates of performance monitoring. J Alzheimers Dis 35(3):575–587. https://doi.org/10.3233/JAD-121348

Tognoli E, Kelso JAS (2014) The metastable brain. Neuron 81(1):35–48. https://doi.org/10.1016/j.neuron.2013.12.022

Trovo L, Fuchs C, De Rosa R, Barbiero I, Tramarin M, Ciani E, Rusconi L, Kilstrup-Nielsen C (2020) The green tea polyphenol epigallocatechin-3-gallate (EGCG) restores. Neurobiol Dis 138:104791. https://doi.org/10.1016/j.nbd.2020.104791

Truettner J, Schmidt-Kastner R, Busto R, Alonso O, Loor J, Dietrich W, Ginsberg M (1999) Expression of brain-derived neurotrophic factor, nerve growth factor, and heat shock protein HSP70 following fluid percussion brain injury in rats. J Neurotrauma 16(6):471–486. https://doi.org/10.1089/neu.1999.16.471

Uniyal A, Singh R, Akhtar A, Bansal Y, Kuhad A, Sah SP (2019) Co-treatment of piracetam with risperidone rescued extinction deficits in experimental paradigms of post-traumatic stress disorder by restoring the physiological alterations in cortex and hippocampus. Pharmacol Biochem Behav 185:172763. https://doi.org/10.1016/j.pbb.2019.172763

Urban K, Gao W-J (2014) Performance enhancement at the cost of potential brain plasticity: neural ramifications of nootropic drugs in the healthy developing brain. Front Syst Neurosci 8:38. https://doi.org/10.3389/fnsys.2014.00038

Venketasubramanian N, Young SH, Tay SS, Umapathi T, Lao AY, Gan HH, Baroque AC II, Navarro JC, Chang HM, Advincula JM, Muengtaweepongsa S, Chan BPL, Chua CL, Wijekoon N, de Silva HA, Hiyadan JHB, Suwanwela NC, Wong KSL, Poungvarin N et al (2015) CHInese Medicine NeuroAiD Efficacy on stroke recovery—extension study (CHIMES-E): a multicenter study of long-term efficacy. Cerebrovasc Dis 39(5–6):309–318. https://doi.org/10.1159/000382082

Venketasubramanian N, Lee C, Young S, Tay SS, Umapathi T, Lao A, Gan H, Baroque AI, Navarro J, Chang H, Advincula J, Muengtaweepongsa S, Chan B, Chua C, Wijekoon N, de Silva H, Hiyadan J, Suwanwela N, Wong K, Chen C (2016) Prognostic factors and pattern of long-term recovery with MLC601 (NeuroAiD™) in the Chinese Medicine NeuroAiD efficacy on stroke recovery—extension study. Cerebrovasc Dis 43:36–42. https://doi.org/10.1159/000452285

Viale L, Catoira N, Girolamo G, Gonzalez C (2017) Pharmacotherapy and motor recovery after stroke. Exp Rev Neurother 18:1–18. https://doi.org/10.1080/14737175.2018.1400910

Waldron D (2015) Gene expression and functional brain networks. Nat Rev Genet 16(8):439–439. https://doi.org/10.1038/nrg3986

Wang J, Zhang H-Y, Tang X-C (2009) Cholinergic deficiency involved in vascular dementia: possible mechanism and strategy of treatment. Acta Pharmacol Sin 30(7):879–888. https://doi.org/10.1038/aps.2009.82

Weickert TW, Allen KM, Weickert CS (2016) Potential role of oestrogen modulation in the treatment of neurocognitive deficits in schizophrenia. CNS Drugs 30(2):125–133. https://doi.org/10.1007/s40263-016-0312-0

Weinreb O, Mandel S, Amit T, Youdim MBH (2004) Neurological mechanisms of green tea polyphenols in Alzheimer's and Parkinson's diseases. J Nutr Biochem 15(9):506–516. https://doi.org/10.1016/j.jnutbio.2004.05.002

Weyandt LL, White TL, Gudmundsdottir BG, Nitenson AZ, Rathkey ES, De Leon KA, Bjorn SA (2018) Neurocognitive, autonomic, and mood effects of Adderall: a pilot study of healthy college students. Pharmacy 6(3):58. https://doi.org/10.3390/pharmacy6030058

Widmann C, Gandin C, Petit-Paitel A, Lazdunski M, Heurteaux C (2018) The Traditional Chinese Medicine MLC901 inhibits inflammation processes after focal cerebral ischemia. Sci Rep 8(1):18062. https://doi.org/10.1038/s41598-018-36138-0

Wildenhain J, Crampin EJ (2006) Reconstructing gene regulatory networks: from random to scale-free connectivity. Syst Biol 153(4):247–256. https://doi.org/10.1049/ip-syb:20050092

Wilms W, Wozniak-Karczewska M, Corvini PF-X, Chrzanowski L (2019) Nootropic drugs: methylphenidate, modafinil and piracetam—population use trends, occurrence in the environment, ecotoxicity and removal methods—a review. Chemosphere 233:771–785. https://doi.org/10.1016/j.chemosphere.2019.06.016

Winblad B (2005) Piracetam: a review of pharmacological properties and clinical uses. CNS Drug Rev 11(2):169–182. https://doi.org/10.1111/j.1527-3458.2005.tb00268.x

Wong GKC, Zhu X, Poon W (2005) Beneficial effect of Cerebrolysin on moderate and severe head injury patients: result of a cohort study. Acta Neurochir Suppl 95:59–60. https://doi.org/10.1007/3-211-32318-x_13

Wronski R, Tompa P, Hutter-Paier B, Crailsheim K, Friedrich P, Windisch M (2000) Inhibitory effect of a brain derived peptide preparation on the Ca++-dependent protease, calpain. J Neural Transm 107(2):145–157. https://doi.org/10.1007/s007020050013

Xie L, Li X-K, Funeshima-Fuji N, Kimura H, Matsumoto Y, Isaka Y, Takahara S (2009) Amelioration of experimental autoimmune encephalomyelitis by curcumin treatment through inhibition of IL-17 production. Int Immunopharmacol 9(5):575–581. https://doi.org/10.1016/j.intimp.2009.01.025

Zbarsky V, Datla KP, Parkar S, Rai DK, Aruoma OI, Dexter DT (2005) Neuroprotective properties of the natural phenolic antioxidants curcumin and naringenin but not quercetin and fisetin

in a 6-OHDA model of Parkinson's disease. Free Radical Res 39(10):1119–1125. https://doi.org/10.1080/10715760500233113
Zhang L, Chopp M, Meier DH, Winter S, Wang L, Szalad A, Lu M, Wei M, Cui Y, Zhang ZG (2013) Sonic hedgehog signaling pathway mediates cerebrolysin-improved neurological function after stroke. Stroke 44(7):1965–1972. https://doi.org/10.1161/STROKEAHA.111.000831
Zhang Y, Li Y, Wang Y, Wang G, Mao L, Zhang D, Wang J (2019) Effects of resveratrol on learning and memory in rats with vascular dementia. Mol Med Rep 20(5):4587–4593. https://doi.org/10.3892/mmr.2019.10723
Zuany-Amorim C, Hastewell J, Walker C (2002) Toll-like receptors as potential therapeutic targets for multiple diseases. Nat Rev Drug Discov 1(10):797–807. https://doi.org/10.1038/nrd914

Correction to: Neuromodulation for Gait Disorders

Stephano J. Chang, Ioan Opris, James D. Guest, and Brian R. Noga

Correction to: Chapter 23 in: I. Opris et al. (eds.), *Modern Approaches to Augmentation of Brain Function*, Contemporary Clinical Neuroscience, https://doi.org/10.1007/978-3-030-54564-2_23

Multiple corrections provided by the author were not updated in this chapter, which are mentioned hereby. The below errors have been corrected in this chapter.

I. In Figure 3 caption, reference citations "Hong et al. 2009" and "MacKinnon et al. 2018" were deleted. These references were removed as the original text referred to stimulation sites labeled "c" and "d" in part A of the figure and correspondingly as "3" and "4" in part B of the figure (not reference numbers). So, the figure legend should read:

"Fig. 3 Electrical mapping of the MLR and surrounding region in the cat. (Adapted from Opris et al. 2019, with permission). (a) Stimulation of the CnF (site c) and Subcuneiform (site d) produces the largest locomotor responses and associated pressor responses (from arterial line), shown in (b) (sites 3 and 4, respectively). (c) Detailed plot of blood pressure and EMG responses to stimulation of the CnF (b: site 3). *R* right, *L* left, *Pbst* posterior biceps/semitendinosus, *Smab* semimembranosus/anterior biceps, *Sart* sartorius, *Quad* quadriceps"

II. Lines 171–174 were revised as: "Although this close intermingling of inhibitory and excitatory populations in the medulla may hinder attempts to evoke locomotor activity electrically (Capelli et al. 2017), stimulation of medullary cells either electrically or pharmacologically in the cat can produce locomotion (Noga et al. 1988)."

The updated online version of this chapter can be found at
https://doi.org/10.1007/978-3-030-54564-2_23

I. Opris et al. (eds.), *Modern Approaches to Augmentation of Brain Function*,
Contemporary Clinical Neuroscience,
https://doi.org/10.1007/978-3-030-54564-2_31

III. The following reference was added:

Noga BR, Kettler J, Jordan LM (1988) Locomotion produced in mesencephalic cats by injections of putative transmitter substances and antagonists into the medial reticular formation and the pontomedullary locomotor strip. J Neurosci 8(6):2074–2086

IV. Figure 4 legend (just after line 521) was mislabeled as Fig. 5.

V. Figure 5 legend (immediately following) was mislabeled as Fig. 4.

VI. The sentence was revised as: "MLR DBS improves partially weight-supported gait…"

VII. Ln. 646. The following sentence was inserted before the sentence beginning with "In intact rodents…":

"In decerebrate cats, electrical or pharmacological activation of RST neurons in this area can initiate locomotion (Noga et al. 1988), which can be blocked by spinal intrathecal injection of glutamatergic antagonists (Douglas et al. 1993)."

VIII. The following reference was added:

Douglas JR, Noga BR, Dai X, Jordan LM (1993) The effects of intrathecal administration of excitatory amino acid agonists and antagonists on the initiation of locomotion in the adult cat. J Neurosci 13(3):990–1000

Conclusion

With the development of neuroscience and modern technologies, the prospect of augmenting brain functions has become realistic, with benefits for healthy humans and patients suffering from neurological conditions.

I. Opris et al. (eds.), *Modern Approaches to Augmentation of Brain Function*, Contemporary Clinical Neuroscience,
https://doi.org/10.1007/978-3-030-54564-2

Index

I. Opris et al. (eds.), *Modern Approaches to Augmentation of Brain Function*, Contemporary Clinical Neuroscience,
https://doi.org/10.1007/978-3-030-54564-2

E

N

O

P

Q

R

S

The manufacturer's authorised representative in the EU is Springer Nature Customer Service Centre GmbH, Europaplatz 3, 69115 Heidelberg, Germany. If you have any concerns regarding our products, please contact ProductSafety@springernature.com

Printed and bound by CPI Group (UK) Ltd, Croydon, CR0 4YY

30/09/2026

02225506-0019